T0140548

Studies in Computational Intelligence

Volume 609

Series editor

Janusz Kacprzyk, Polish Academy of Sciences, Warsaw, Poland
e-mail: kacprzyk@ibspan.waw.pl

About this Series

The series "Studies in Computational Intelligence" (SCI) publishes new developments and advances in the various areas of computational intelligence—quickly and with a high quality. The intent is to cover the theory, applications, and design methods of computational intelligence, as embedded in the fields of engineering, computer science, physics and life sciences, as well as the methodologies behind them. The series contains monographs, lecture notes and edited volumes in computational intelligence spanning the areas of neural networks, connectionist systems, genetic algorithms, evolutionary computation, artificial intelligence, cellular automata, self-organizing systems, soft computing, contributors and the readership are the short publication timeframe and the worldwide distribution, which enable both wide and rapid dissemination of research output.

More information about this series at http://www.springer.com/series/7092

George A. Anastassiou

Intelligent Comparisons:
Analytic Inequalities

George A. Anastassiou
Department of Mathematical Sciences
University of Memphis
Memphis, TN
USA

ISSN 1860-949X   ISSN 1860-9503   (electronic)
Studies in Computational Intelligence
ISBN 978-3-319-37060-6   ISBN 978-3-319-21121-3   (eBook)
DOI 10.1007/978-3-319-21121-3

Springer Cham Heidelberg New York Dordrecht London
© Springer International Publishing Switzerland 2016
Softcover reprint of the hardcover 1st edition 2016

Printed on acid-free paper

Springer International Publishing AG Switzerland is part of Springer Science+Business Media
(www.springer.com)

To my wife Koula and my daughters
Angela and Peggy

Preface

In this monograph we present the recent work of the past 5 years of the author on inequalities in real, functional and fractional analysis. It is the natural outgrowth of his related publications. Chapters are self-contained and can be read independently and several advanced courses can be taught out of this book. An extensive list of references is given per chapter.

The inequalities covered are diverse. A list of these follows:

Fractional Polya type integral inequality.

Univariate fractional Polya type integral inequalities.

Multivariate generalized fractional Polya type integral inequalities.

Balanced Canavati type fractional Opial inequalities.

Fractional representation formulae under initial conditions and fractional Ostrowski type inequalities.

Basic fractional integral inequalities.

Harmonic multivariate Ostrowski and Grüss type inequalities for several functions.

Fractional Ostrowski and Grüss type inequalities involving several functions.

Further interpretation of some fractional Ostrowski and Grüss type inequalities.

Multivariate fractional representation formula and Ostrowski type inequality.

Multivariate weighted fractional representation formulae and Ostrowski type inequalities.

Multivariate Lyapunov inequalities.

Ostrowski inequalities for semigroups.

Ostrowski inequalities for cosine and sine operator functions.

Hilbert-Pachpatte type inequalities for semigroups, cosine and sine operator functions.

Ostrowski and Landau inequalities for Banach space valued functions.

Multidimensional Ostrowski inequalities for Banach space valued functions.

Fractional representation formulae and right fractional inequalities.

Canavati fractional Ostrowski type inequalities.

Most general fractional representation formula for functions and implications.

Rational inequalities for integral operators under convexity.

Fractional integral inequalities involving convexity.

Vectorial inequalities for integral operators involving ratios of functions and convexity.

Vectorial splitting rational L_p inequalities for integral operators.

Separating rational L_p inequalities for integral operators.

Vectorial Hardy type fractional inequalities.

Vectorial fractional integral inequalities with convexity.

This book's results are expected to find applications in many areas of pure and applied mathematics, especially in ordinary and partial differential equations and fractional differential equations. As such this monograph is suitable for researchers, graduate students, and seminars of the above subjects, also to be in all science and engineering libraries.

The preparation of this book took place during 2014–2015 in Memphis, Tennessee, USA.

I would like to thank Prof. Alina Lupas, of University of Oradea, Romania, for checking and reading the manuscript.

Memphis, USA   George A. Anastassiou
April 2015

Contents

About the Author

George A. Anastassiou was born in Athens, Greece in 1952. He received his B.Sc. degree in Mathematics from Athens University, Greece in 1975. He received his Diploma in Operations Research from Southampton University, UK in 1976. He also received his MA in Mathematics from University of Rochester, USA in 1981. He was awarded his Ph.D. in Mathematics from University of Rochester, USA in 1984. During 1984–1986 he served as a visiting assistant professor at the University of Rhode Island, USA. Since 1986 till now in 2015, he is a faculty member at the University of Memphis, USA. He is currently a full Professor of Mathematics since 1994. His research area is "Computational Analysis" in a very broad sense. He has published over 400 research articles in international mathematical journals and over 27 monographs, proceedings and textbooks in well-known publishing houses. Several awards have been awarded to George Anastassiou. In 2007 he received the Honorary Doctoral Degree from University of Oradea, Romania. He is associate editor in over 60 international mathematical journals and editor in-chief in three journals, most notably in the well-known "Journal of Computational Analysis and Applications".

Chapter 1
Fractional Polya Integral Inequality

Here we establish a fractional Polya type integral inequality with the help of generalised right and left fractional derivatives. The amazing fact here is that we do not
need any boundary conditions as the classical Polya integral inequality requires. It
follows [3].

1.1 Introduction

We mention the following famous Polya's integral inequality, see [8], [9, p. 62], [10]
and [11, p. 83].

Theorem 1.1 *Let* $f(x)$ *be differentiable and not identically a constant on* $[a, b]$
with $f(a) = f(b) = 0$. *Then the exists at least one point* $\xi \in [a, b]$ *such that*

$$\left| f'(\xi) \right| > \frac{4}{(b-a)^2} \int_a^b f(x)\, dx. \tag{1.1}$$

In [12], Feng Qi presents the following very interesting Polya type integral
inequality (1.2), which generalizes (1.1).

Theorem 1.2 *Let* $f(x)$ *be differentiable and not identically constant on* $[a, b]$ *with*
$f(a) = f(b) = 0$ *and* $M = \sup\limits_{x \in [a,b]} \left| f'(x) \right|$. *Then*

$$\left| \int_a^b f(x)\, dx \right| \leq \frac{(b-a)^2}{4} M, \tag{1.2}$$

where $\frac{(b-a)^2}{4}$ *in (1.2) is the best constant.*

© Springer International Publishing Switzerland 2016
G.A. Anastassiou, *Intelligent Comparisons: Analytic Inequalities*,
Studies in Computational Intelligence 609,
DOI 10.1007/978-3-319-21121-3_1

In this short note we present a fractional Polya type integral inequality, similar to (1.2), without the boundary conditions $f(a) = f(b) = 0$.

For the last we need the following fractional calculus background. Here $[\cdot]$ denotes the integral part of the number.

Let $\alpha > 0$, $m = [\alpha]$, $\beta = \alpha - m$, $0 < \beta < 1$, $f \in C([a, b])$, $[a, b] \subset \mathbb{R}$, $x \in [a, b]$. The gamma function Γ is given by $\Gamma(\alpha) = \int_0^\infty e^{-t} t^{\alpha-1} dt$. We define the left Riemann-Liouville integral

$$\left(J_\alpha^{a+} f\right)(x) = \frac{1}{\Gamma(\alpha)} \int_a^x (x - t)^{\alpha-1} f(t) \, dt, \tag{1.3}$$

$a \le x \le b$. We define the subspace $C_{a+}^\alpha([a.b])$ of $C^m([a, b])$:

$$C_{a+}^\alpha([a, b]) = \left\{ f \in C^m([a, b]) : J_{1-\beta}^{a+} f^{(m)} \in C^1([a, b]) \right\}. \tag{1.4}$$

For $f \in C_{a+}^\alpha([a, b])$, we define the left generalized α-fractional derivative of f over $[a, b]$ as

$$D_{a+}^\alpha f := \left(J_{1-\beta}^{a+} f^{(m)}\right)', \tag{1.5}$$

see [1], p. 24. Canavati first in [4] introduced the above over $[0, 1]$.

Notice that $D_{a+}^\alpha f \in C([a, b])$.

We need the following left fractional Taylor's formula, see [1], pp. 8–10, and in [4] the same over $[0, 1]$ that appeared first.

Theorem 1.3 Let $f \in C_{a+}^\alpha([a, b])$.

(i) If $\alpha \ge 1$, then

$$f(x) = f(a) + f'(a)(x - a) + f''(a)\frac{(x - a)^2}{2} + \cdots + f^{(m-1)}(a)\frac{(x - a)^{m-1}}{(m - 1)!} \tag{1.6}$$

$$+ \frac{1}{\Gamma(\alpha)} \int_a^x (x - t)^{\alpha-1} \left(D_{a+}^\alpha f\right)(t) \, dt, \quad all \ x \in [a, b].$$

(ii) If $0 < \alpha < 1$, we have

$$f(x) = \frac{1}{\Gamma(\alpha)} \int_a^x (x - t)^{\alpha-1} \left(D_{a+}^\alpha f\right)(t) \, dt, \quad all \ x \in [a, b]. \tag{1.7}$$

We will use (1.7).

Notice that

$$\int_a^x (x - t)^{\alpha-1} \left(D_{a+}^\alpha f\right)(t) \, dt = \int_a^x \left(D_{a+}^\alpha f\right)(t) \, d\left(\frac{(x - t)^\alpha}{-\alpha}\right)$$

$$= \left(D_{a+}^{\alpha} f\right)(\xi_x) \frac{(x-a)^{\alpha}}{\alpha}, \quad \text{where} \quad \xi_x \in [a, x], \tag{1.8}$$

by first integral mean value theorem. Hence, when $0 < \alpha < 1$, we get

$$f(x) = \left(D_{a+}^{\alpha} f\right)(\xi_x) \frac{(x-a)^{\alpha}}{\Gamma(\alpha+1)}, \quad \text{all} \quad x \in [a, b]. \tag{1.9}$$

Furthermore we need:

Let again $\alpha > 0$, $m = [\alpha]$, $\beta = \alpha - m$, $f \in C([a, b])$, call the right Riemann-Liouville fractional integral operator by

$$\left(J_{b-}^{\alpha} f\right)(x) := \frac{1}{\Gamma(\alpha)} \int_x^b (t-x)^{\alpha-1} f(t)\, dt, \tag{1.10}$$

$x \in [a, b]$, see also [2, 5–7, 13]. Define the subspace of functions

$$C_{b-}^{\alpha}([a, b]) := \left\{ f \in C^m([a, b]) : J_{b-}^{1-\beta} f^{(m)} \in C^1([a, b]) \right\}. \tag{1.11}$$

Define the right generalized α-fractional derivative of f over $[a, b]$ as

$$D_{b-}^{\alpha} f = (-1)^{m-1} \left(J_{b-}^{1-\beta} f^{(m)} \right)', \tag{1.12}$$

see [2]. We set $D_{b-}^0 f = f$. Notice that $D_{b-}^{\alpha} f \in C([a, b])$.

From [2], we need the following right Taylor fractional formula.

Theorem 1.4 *Let* $f \in C_{b-}^{\alpha}([a, b])$, $\alpha > 0$, $m := [\alpha]$. *Then*

(i) If $\alpha \geq 1$, *we get*

$$f(x) = \sum_{k=0}^{m-1} \frac{f^{(k)}(b)}{k!} (x-b)^k + \left(J_{b-}^{\alpha} D_{b-}^{\alpha} f\right)(x), \quad \text{all} \quad x \in [a, b]. \tag{1.13}$$

(ii) If $0 < \alpha < 1$, *we get*

$$f(x) = J_{b-}^{\alpha} D_{b-}^{\alpha} f(x) = \frac{1}{\Gamma(\alpha)} \int_x^b (t-x)^{\alpha-1} \left(D_{b-}^{\alpha} f\right)(t)\, dt, \quad \text{all} \quad x \in [a, b]. \tag{1.14}$$

We will use (1.14).

Notice that

$$\int_x^b (t-x)^{\alpha-1} \left(D_{b-}^\alpha f\right)(t)\, dt = \int_x^b \left(D_{b-}^\alpha f\right)(t)\, d\left(\frac{(t-x)^\alpha}{\alpha}\right)$$

$$= \left(D_{b-}^\alpha f\right)(\eta_x)\, \frac{(b-x)^\alpha}{\alpha} \text{where} \quad \eta_x \in [x, b], \tag{1.15}$$

by first integral mean value theorem. Hence, when $0 < \alpha < 1$, we obtain

$$f(x) = \left(D_{b-}^\alpha f\right)(\eta_x)\, \frac{(b-x)^\alpha}{\Gamma(\alpha+1)}, \quad \text{all} \;\; x \in [a, b]. \tag{1.16}$$

1.2 Main Result

We present the following fractional Polya type integral inequality without any boundary conditions.

Theorem 1.5 *Let* $0 < \alpha < 1$, $f \in C([a, b])$. *Assume* $f \in C_{a+}^\alpha \left(\left[a, \frac{a+b}{2}\right]\right)$ *and* $f \in C_{b-}^\alpha \left(\left[\frac{a+b}{2}, b\right]\right)$. *Set*

$$M(f) = \max\left\{\left\|D_{a+}^\alpha f\right\|_{\infty, \left[a, \frac{a+b}{2}\right]}, \left\|D_{b-}^\alpha f\right\|_{\infty, \left[\frac{a+b}{2}, b\right]}\right\}. \tag{1.17}$$

Then

$$\left|\int_a^b f(x)\, dx\right| \le \int_a^b |f(x)|\, dx \le M(f)\, \frac{(b-a)^{\alpha+1}}{\Gamma(\alpha+2)\, 2^\alpha}. \tag{1.18}$$

Inequality (1.18) is sharp, namely it is attained by

$$f_*(x) = \begin{cases} (x-a)^\alpha, & x \in \left[a, \frac{a+b}{2}\right] \\ (b-x)^\alpha, & x \in \left[\frac{a+b}{2}, b\right] \end{cases}, 0 < \alpha < 1. \tag{1.19}$$

Clearly here non zero constant functions f *are excluded.*

Proof By (1.9) we get

$$|f(x)| \le \left\|D_{a+}^\alpha f\right\|_{\infty, \left[a, \frac{a+b}{2}\right]} \frac{(x-a)^\alpha}{\Gamma(\alpha+1)}, \quad \text{for any} \;\; x \in \left[a, \frac{a+b}{2}\right]. \tag{1.20}$$

By (1.16) we derive

$$|f(x)| \le \left\|D_{b-}^\alpha f\right\|_{\infty, \left[\frac{a+b}{2}, b\right]} \frac{(b-x)^\alpha}{\Gamma(\alpha+1)}, \quad \text{for any} \;\; x \in \left[\frac{a+b}{2}, b\right]. \tag{1.21}$$

Hence we get

$$\int_a^b |f(x)|\,dx = \int_a^{\frac{a+b}{2}} |f(x)|\,dx + \int_{\frac{a+b}{2}}^b |f(x)|\,dx$$

(by (1.20), (1.21))

$$\leq \frac{\left\|D_{a+}^\alpha f\right\|_{\infty,\left[a,\frac{a+b}{2}\right]}}{\Gamma(\alpha+1)} \int_a^{\frac{a+b}{2}} (x-a)^\alpha\,dx + \frac{\left\|D_{b-}^\alpha f\right\|_{\infty,\left[\frac{a+b}{2},b\right]}}{\Gamma(\alpha+1)} \int_{\frac{a+b}{2}}^b (b-x)^\alpha\,dx \tag{1.22}$$

$$= \frac{\left\|D_{a+}^\alpha f\right\|_{\infty,\left[a,\frac{a+b}{2}\right]}}{(\Gamma(\alpha+1))(\alpha+1)} \left(\frac{b-a}{2}\right)^{\alpha+1} + \frac{\left\|D_{b-}^\alpha f\right\|_{\infty,\left[\frac{a+b}{2},b\right]}}{(\Gamma(\alpha+1))(\alpha+1)} \left(\frac{b-a}{2}\right)^{\alpha+1}$$

$$= \frac{\left(\left\|D_{a+}^\alpha f\right\|_{\infty,\left[a,\frac{a+b}{2}\right]} + \left\|D_{b-}^\alpha f\right\|_{\infty,\left[\frac{a+b}{2},b\right]}\right)}{\Gamma(\alpha+2)} \left(\frac{b-a}{2}\right)^{\alpha+1}. \tag{1.23}$$

So we have proved that

$$\int_a^b |f(x)|\,dx \leq \max\left\{\left\|D_{a+}^\alpha f\right\|_{\infty,\left[a,\frac{a+b}{2}\right]}, \left\|D_{b-}^\alpha f\right\|_{\infty,\left[\frac{a+b}{2},b\right]}\right\} \frac{(b-a)^{\alpha+1}}{\Gamma(\alpha+2)\,2^\alpha}, \tag{1.24}$$

proving (1.18).

Notice that

$$f_*\left(\left(\frac{a+b}{2}\right)_-\right) = f_*\left(\left(\frac{a+b}{2}\right)_+\right) = \left(\frac{b-a}{2}\right)^\alpha,$$

so that $f_* \in C([a,b])$.

Here $m = 0$. We see that

$$\left(J_{1-\beta}^{\alpha+} (\cdot - a)^\alpha\right)(x) = \left(J_{1-\alpha}^{a+} (\cdot - a)^\alpha\right)(x) =$$

$$\frac{1}{\Gamma(1-\alpha)} \int_a^x (x-t)^{-\alpha} (t-a)^\alpha\,dt =$$

$$\frac{1}{\Gamma(1-\alpha)} \int_a^x (x-t)^{(1-\alpha)-1} (t-a)^{(\alpha+1)-1}\,dt =$$

(by [14], p. 256)

$$\frac{1}{\Gamma(1-\alpha)} \frac{\Gamma(1-\alpha)\,\Gamma(\alpha+1)}{\Gamma(2)} (x-a) = \Gamma(\alpha+1)(x-a).$$

Hence
$$D_{a+}^{\alpha} (x - a)^{\alpha} = \Gamma (\alpha + 1), \quad \text{for all} \quad x \in \left[a, \frac{a+b}{2} \right]. \tag{1.25}$$

Therefore
$$\left\| D_{a+}^{\alpha} (x - a)^{\alpha} \right\|_{\infty, \left[a, \frac{a+b}{2} \right]} = \Gamma (a + 1). \tag{1.26}$$

Furthermore we have
$$\left(J_{b-}^{1-\alpha} (b - \cdot)^{\alpha} \right) (x) = \frac{1}{\Gamma (1 - \alpha)} \int_{x}^{b} (t - x)^{-\alpha} (b - t)^{\alpha} \, dt =$$

$$\frac{1}{\Gamma (1 - \alpha)} \int_{x}^{b} (b - t)^{(\alpha+1)-1} (t - x)^{(1-\alpha)-1} \, dt =$$

(by [14], p. 256)
$$\frac{1}{\Gamma (1 - \alpha)} \frac{\Gamma (\alpha + 1) \Gamma (1 - \alpha)}{\Gamma (2)} (b - x) = \Gamma (\alpha + 1) (b - x).$$

Therefore
$$D_{b-}^{\alpha} (b - x)^{\alpha} = \Gamma (\alpha + 1), \quad \text{for all} \quad x \in \left[\frac{a+b}{2}, b \right], \tag{1.27}$$

and
$$\left\| D_{b-}^{\alpha} (b - x)^{\alpha} \right\|_{\infty, \left[\frac{a+b}{2}, b \right]} = \Gamma (a + 1). \tag{1.28}$$

Consequently we find that
$$M (f_*) = \Gamma (\alpha + 1). \tag{1.29}$$

Applying f_* into (1.18) we obtain:
$$\text{R.H.S. (1.18) for } f_* = \Gamma (\alpha + 1) \frac{(b - a)^{\alpha+1}}{\Gamma (\alpha + 2) \, 2^{\alpha}} = \frac{(b - a)^{\alpha+1}}{(\alpha + 1) \, 2^{\alpha}}, \tag{1.30}$$

while we get the same result from
$$\text{L.H.S. (1.18) for } f_* = \left| \int_{a}^{b} f_* (x) \, dx \right| =$$

$$\int_{a}^{\frac{a+b}{2}} (x - a)^{\alpha} \, dx + \int_{\frac{a+b}{2}}^{b} (b - x)^{\alpha} \, dx = \frac{(b - a)^{\alpha+1}}{(\alpha + 1) \, 2^{\alpha}}, \tag{1.31}$$

proving sharpness of (1.18). ■

We make

Remark 1.6 When $\alpha \geq 1$, thus $m = [\alpha] \geq 1$, and by assuming that $f^{(k)}(a) = f^{(k)}(b) = 0$, $k = 0, 1, \ldots, m - 1$, we can prove the same statements as in Theorem 1.5. If we set there $\alpha = 1$ we derive exactly Theorem 1.2. So we generalize Theorem 1.2. Again here $f^{(m)}$ cannot be a constant different than zero, equivalently, f cannot be a non-trivial polynomial of degree m.

References

1. G.A. Anastassiou, *Fractional Differentiation Inequalities* (Springer, New York, 2009)
2. G.A. Anastassiou, On right fractional calculus. Chaos, Solitons Fractals **42**, 365–376 (2009)
3. G.A. Anastassiou, Fractional Polya type integral inequality. J. Comput. Anal. Appl. **17**(4), 736–742 (2014)
4. J.A. Canavati, The Riemann-Liouville integral. Nieuw Archief Voor Wiskunde **5**(1), 53–75 (1987)
5. A.M.A. El-Sayed, M. Gaber, On the finite Caputo and finite Riesz derivatives. Electr. J. Theor. Phys. **3**(12), 81–95 (2006)
6. G.S. Frederico, D.F.M. Torres, Fractional optimal control in the sense of caputo and the fractional Noether's theorem. Int. Math. Forum **3**(10), 479–493 (2008)
7. R. Gorenflo, F. Mainardi, Essentials of Fractional Calculus, 2000, Maphysto Center, http://www.maphysto.dk/oldpages/events/LevyCAC2000/MainardiNotes/fm2k0a.ps
8. G. Polya, Ein mittelwertsatz für Funktionen mehrerer Veränderlichen. Tohoku Math. J. **19**, 1–3 (1921)
9. G. Polya, G. Szegö, *Aufgaben und Lehrsä tze aus der Analysis*, vol. I (Springer, Berlin, 1925). (German)
10. G. Polya, G. Szegö, *Problems and Theorems in Analysis*, vol. I (Springer, Berlin, 1972)
11. G. Polya, G. Szegö, *Problems and Theorems in Analysis*, vol. I, Chinese edn. (1984)
12. F. Qi, *Polya Type Integral Inequalities: Origin, Variants, Proofs, Refinements, Generalizations, Equivalences and Applications*, article no. 20, 16th vol. 2013, RGMIA, Res. Rep. Coll., http://rgmia.org/v16.php
13. S.G. Samko, A.A. Kilbas, O.I. Marichev, *Fractional Integrals and Derivatives, Theory and Applications*, (Gordon and Breach, Amsterdam, 1993) [English translation from the Russian, Integrals and Derivatives of Fractional Order and Some of Their Applications (Nauka i Tekhnika, Minsk, 1987)]
14. E.T. Whittaker, G.N. Watson, *A Course in Modern Analysis* (Cambridge University Press, Cambridge, 1927)

Chapter 2
Univariate Fractional Polya Integral Inequalities

Here we establish a series of various fractional Polya type integral inequalities with the help of generalised right and left fractional derivatives. We give an application to complex valued functions defined on the unit circle. It follows [5].

2.1 Introduction

We mention the following famous Polya's integral inequality, see [13], [14, p. 62], [15] and [16, p. 83].

Theorem 2.1 *Let* $f(x)$ *be differentiable and not identically a constant on* $[a, b]$ *with* $f(a) = f(b) = 0$. *Then the exists at least one point* $\xi \in [a, b]$ *such that*

$$\left| f'(\xi) \right| > \frac{4}{(b-a)^2} \int_a^b f(x) \, dx. \tag{2.1}$$

In [17], Feng Qi presents the following very interesting Polya type integral inequality (2.2), which generalizes (2.1).

Theorem 2.2 *Let* $f(x)$ *be differentiable and not identically constant on* $[a, b]$ *with* $f(a) = f(b) = 0$ *and* $M = \sup_{x \in [a,b]} \left| f'(x) \right|$. *Then*

$$\left| \int_a^b f(x) \, dx \right| \leq \frac{(b-a)^2}{4} M, \tag{2.2}$$

where $\frac{(b-a)^2}{4}$ *in (2.2) is the best constant.*

© Springer International Publishing Switzerland 2016
G.A. Anastassiou, *Intelligent Comparisons: Analytic Inequalities*,
Studies in Computational Intelligence 609,
DOI 10.1007/978-3-319-21121-3_2

The above motivate the current chapter.

In this chapter we present univariate fractional Polya type integral inequalities in various cases, similar to (2.2).

For this purpose we need the following fractional calculus background.

Let $\alpha > 0$, $m = [\alpha]$, $\beta = \alpha - m$, $0 < \beta < 1$, $f \in C([a, b])$, $[a, b] \subset \mathbb{R}$, $x \in [a, b]$. The gamma function Γ is given by $\Gamma(\alpha) = \int_0^\infty e^{-t} t^{\alpha-1} dt$. We define the left Riemann-Liouville integral

$$\left(J_a^{a+} f\right)(x) = \frac{1}{\Gamma(\alpha)} \int_a^x (x - t)^{\alpha-1} f(t) \, dt, \tag{2.3}$$

$a \le x \le b$. We define the subspace $C_{a+}^\alpha([a.b])$ of $C^m([a, b])$:

$$C_{a+}^\alpha([a, b]) = \left\{ f \in C^m([a, b]) : J_{1-\beta}^{a+} f^{(m)} \in C^1([a, b]) \right\}. \tag{2.4}$$

For $f \in C_{a+}^\alpha([a, b])$, we define the left generalized α-fractional derivative of f over $[a, b]$ as

$$D_{a+}^\alpha f := \left(J_{1-\beta}^{a+} f^{(m)} \right)', \tag{2.5}$$

see [1], p. 24. Canavati first in [6], introduced the above over $[0, 1]$.

Notice that $D_{a+}^\alpha f \in C([a, b])$.

We need the following left fractional Taylor's formula, see [1], pp. 8–10, and in [6] the same formula over $[0, 1]$ that appeared first.

Theorem 2.3 Let $f \in C_{a+}^\alpha([a, b])$.

(i) If $\alpha \ge 1$, then

$$f(x) = f(a) + f'(a)(x - a) + f''(a) \frac{(x - a)^2}{2} + \cdots + f^{(m-1)}(a) \frac{(x - a)^{m-1}}{(m - 1)!}$$
$$+ \frac{1}{\Gamma(\alpha)} \int_a^x (x - t)^{\alpha-1} \left(D_{a+}^\alpha f \right)(t) \, dt, \quad \text{all } x \in [a, b]. \tag{2.6}$$

(ii) If $0 < \alpha < 1$, we have

$$f(x) = \frac{1}{\Gamma(\alpha)} \int_a^x (x - t)^{\alpha-1} \left(D_{a+}^\alpha f \right)(t) \, dt, \quad \text{all } x \in [a, b]. \tag{2.7}$$

Let again $\alpha > 0$, $m = [\alpha]$, $\beta = \alpha - m$, $f \in C([a, b])$, call the right Riemann-Liouville fractional integral operator by

$$\left(J_{b-}^\alpha f\right)(x) := \frac{1}{\Gamma(\alpha)} \int_x^b (t - x)^{\alpha-1} f(t) \, dt, \tag{2.8}$$

$x \in [a, b]$, see also [2, 9–11, 18]. Define the subspace of functions

$$C_{b-}^{\alpha} ([a, b]) := \left\{ f \in C^m ([a, b]) : J_{b-}^{1-\beta} f^{(m)} \in C^1 ([a, b]) \right\}. \qquad (2.9)$$

Define the right generalized α-fractional derivative of f over $[a, b]$ as

$$D_{b-}^{\alpha} f = (-1)^{m-1} \left(J_{b-}^{1-\beta} f^{(m)} \right)', \qquad (2.10)$$

see [2]. We set $D_{b-}^0 f = f$. Notice that $D_{b-}^{\alpha} f \in C ([a, b])$.

We need the following right Taylor fractional formula from [2].

Theorem 2.4 *Let $f \in C_{b-}^{\alpha} ([a, b])$, $\alpha > 0$, $m := [\alpha]$. Then*

(i) If $\alpha \geq 1$, we get

$$f (x) = \sum_{k=0}^{m-1} \frac{f^{(k)} (b)}{k!} (x - b)^k + \left(J_{b-}^{\alpha} D_{b-}^{\alpha} f \right) (x), \quad all\, x \in [a, b]. \quad (2.11)$$

(ii) If $0 < \alpha < 1$, we get

$$f (x) = J_{b-}^{\alpha} D_{b-}^{\alpha} f (x) = \frac{1}{\Gamma (\alpha)} \int_x^b (t - x)^{\alpha-1} \left(D_{b-}^{\alpha} f \right) (t) \, dt, \quad all\, x \in [a, b]. \qquad (2.12)$$

Definition 2.5 ([3]) *Let $f \in C ([a, b])$, $x \in [a, b]$, $\alpha > 0$, $m := [a]$. Assume that $f \in C_{b-}^{\alpha} \left(\left[\frac{a+b}{2}, b \right] \right)$ and $f \in C_{a+}^{\alpha} \left(\left[a, \frac{a+b}{2} \right] \right)$. We define the balanced Canavati type fractional derivative by*

$$D^{\alpha} f (x) := \begin{cases} D_{b-}^{\alpha} f (x), & \text{for } \frac{a+b}{2} \leq x \leq b, \\ D_{a+}^{\alpha} f (x), & \text{for } a \leq x < \frac{a+b}{2}. \end{cases} \qquad (2.13)$$

In [4] we proved the following fractional Polya type integral inequality without any boundary conditions.

Theorem 2.6 *Let $0 < \alpha < 1$, $f \in C ([a, b])$. Assume $f \in C_{a+}^{\alpha} \left(\left[a, \frac{a+b}{2} \right] \right)$ and $f \in C_{b-}^{\alpha} \left(\left[\frac{a+b}{2}, b \right] \right)$. Set*

$$M_1 (f) = \max \left\{ \left\| D_{a+}^{\alpha} f \right\|_{\infty, \left[a, \frac{a+b}{2} \right]}, \left\| D_{b-}^{\alpha} f \right\|_{\infty, \left[\frac{a+b}{2}, b \right]} \right\}. \qquad (2.14)$$

Then

$$\left| \int_a^b f (x) \, dx \right| \leq \int_a^b |f (x)| \, dx \leq \qquad (2.15)$$

$$\frac{\left(\left\|D_{a+}^{\alpha}f\right\|_{\infty,\left[a,\frac{a+b}{2}\right]}+\left\|D_{b-}^{\alpha}f\right\|_{\infty,\left[\frac{a+b}{2},b\right]}\right)}{\Gamma(\alpha+2)}\left(\frac{b-a}{2}\right)^{\alpha+1}\leq M_1(f)\frac{(b-a)^{\alpha+1}}{\Gamma(\alpha+2)\,2^{\alpha}}.$$

$$(2.16)$$

Inequalities (2.15), (2.16) are sharp, namely they are attained by

$$f_*(x)=\begin{cases}(x-a)^{\alpha},&x\in\left[a,\frac{a+b}{2}\right]\\(b-x)^{\alpha},&x\in\left[\frac{a+b}{2},b\right]\end{cases},\quad 0<\alpha<1.\qquad(2.17)$$

Clearly here non zero constant functions f are excluded.

The last result also motivates this chapter.

Remark 2.7 (see [4]) When $\alpha\geq 1$, thus $m=\lceil\alpha\rceil\geq 1$, and by assuming that $f^{(k)}(a)=f^{(k)}(b)=0$, $k=0,1,\ldots,m-1$, we can prove the same statements (2.15–2.17) as in Theorem 2.6. If we set there $\alpha=1$ we derive exactly Theorem 2.2 . So we have generalized Theorem 2.2. Again here $f^{(m)}$ cannot be a constant different than zero, equivalently, f cannot be a non-trivial polynomial of degree m.

We continue here with other interesting univariate fractional Polya type inequalities.

2.2 Main Results

We present our first main result.

Theorem 2.8 *Let $\alpha\geq 1$, $m=\lceil\alpha\rceil$, $f\in C([a,b])$. Assume $f\in C_{a+}^{\alpha}\left(\left[a,\frac{a+b}{2}\right]\right)$ and $f\in C_{b-}^{\alpha}\left(\left[\frac{a+b}{2},b\right]\right)$, such that $f^{(k)}(a)=f^{(k)}(b)=0$, $k=0,1,\ldots,m-1$. Set*

$$M_2(f)=\max\left\{\left\|D_{a+}^{\alpha}f\right\|_{L_1\left(\left[a,\frac{a+b}{2}\right]\right)},\left\|D_{b-}^{\alpha}f\right\|_{L_1\left(\left[\frac{a+b}{2},b\right]\right)}\right\}.\qquad(2.18)$$

Then

$$\left|\int_a^b f(x)\,dx\right|\leq\frac{1}{\Gamma(\alpha+1)}\left(\frac{b-a}{2}\right)^{\alpha}\left\|D^{\alpha}f\right\|_{L_1([a,b])}\leq\frac{M_2(f)}{\Gamma(\alpha+1)\,2^{\alpha-1}}(b-a)^{\alpha}.$$

$$(2.19)$$

Here f cannot be a non-trivial polynomial of degree m.

Proof By assumption and Theorem 2.3 we have

$$f(x)=\frac{1}{\Gamma(\alpha)}\int_a^x(x-t)^{\alpha-1}\left(D_{a+}^{\alpha}f\right)(t)\,dt,\quad\text{all }x\in\left[a,\frac{a+b}{2}\right],\qquad(2.20)$$

also it holds, by assumption and Theorem 2.4, that

$$f(x) = \frac{1}{\Gamma(\alpha)} \int_x^b (t-x)^{\alpha-1} \left(D_{b-}^{\alpha} f\right)(t)\, dt, \quad \text{all} \quad x \in \left[\frac{a+b}{2}, b\right]. \quad (2.21)$$

By (2.20) we get

$$|f(x)| \le \frac{1}{\Gamma(\alpha)} \int_a^x (x-t)^{\alpha-1} \left|\left(D_{a+}^{\alpha} f\right)(t)\right| dt$$

$$\le \frac{(x-a)^{\alpha-1}}{\Gamma(\alpha)} \int_a^{\frac{a+b}{2}} \left|\left(D_{a+}^{\alpha} f\right)(t)\right| dt, \quad \text{all } x \in \left[a, \frac{a+b}{2}\right]. \quad (2.22)$$

By (2.21) we derive

$$|f(x)| \le \frac{1}{\Gamma(\alpha)} \int_x^b (t-x)^{\alpha-1} \left|\left(D_{b-}^{\alpha} f\right)(t)\right| dt$$

$$\le \frac{(b-x)^{\alpha-1}}{\Gamma(\alpha)} \int_{\frac{a+b}{2}}^b \left|\left(D_{b-}^{\alpha} f\right)(t)\right| dt, \quad \text{all } x \in \left[\frac{a+b}{2}, b\right]. \quad (2.23)$$

Consequently we have

$$\int_a^{\frac{a+b}{2}} |f(x)|\, dx \le \frac{1}{\Gamma(\alpha)} \left(\int_a^{\frac{a+b}{2}} (x-a)^{\alpha-1}\, dx \right) \left\| D_{a+}^{\alpha} f \right\|_{L_1\left(\left[a, \frac{a+b}{2}\right]\right)} \quad (2.24)$$

$$= \frac{1}{\Gamma(\alpha+1)} \left(\frac{b-a}{2}\right)^{\alpha} \left\| D_{a+}^{\alpha} f \right\|_{L_1\left(\left[a, \frac{a+b}{2}\right]\right)},$$

and

$$\int_{\frac{a+b}{2}}^b |f(x)|\, dx \le \frac{1}{\Gamma(\alpha)} \left(\int_{\frac{a+b}{2}}^b (b-x)^{\alpha-1}\, dx \right) \left\| D_{b-}^{\alpha} f \right\|_{L_1\left(\left[\frac{a+b}{2}, b\right]\right)} \quad (2.25)$$

$$= \frac{1}{\Gamma(\alpha+1)} \left(\frac{b-a}{2}\right)^{\alpha} \left\| D_{b-}^{\alpha} f \right\|_{L_1\left(\left[\frac{a+b}{2}, b\right]\right)}.$$

Therefore we obtain by adding (2.24) and (2.25) that

$$\int_a^b |f(x)|\, dx \le \frac{1}{\Gamma(\alpha+1)} \left(\frac{b-a}{2}\right)^{\alpha} \left[\left\| D_{a+}^{\alpha} f \right\|_{L_1\left(\left[a, \frac{a+b}{2}\right]\right)} + \left\| D_{b-}^{\alpha} f \right\|_{L_1\left(\left[\frac{a+b}{2}, b\right]\right)} \right]$$

$$= \frac{1}{\Gamma(\alpha+1)} \left(\frac{b-a}{2}\right)^{\alpha} \left\| D^{\alpha} f \right\|_{L_1([a,b])} \le$$

$$\max\left\{\left\|D_{a+}^{\alpha}f\right\|_{L_1\left(\left[a,\frac{a+b}{2}\right]\right)}, \left\|D_{b-}^{\alpha}f\right\|_{L_1\left(\left[\frac{a+b}{2},b\right]\right)}\right\}\frac{(b-a)^{\alpha}}{\Gamma(\alpha+1)\,2^{\alpha-1}}, \qquad (2.26)$$

proving the claim. ∎

We continue with

Theorem 2.9 *Let $p, q > 1 : \frac{1}{p} + \frac{1}{q} = 1$, $\alpha > \frac{1}{q}$, $m = [\alpha]$, $f \in C([a, b])$. Assume $f \in C_{a+}^{\alpha}\left(\left[a, \frac{a+b}{2}\right]\right)$ and $f \in C_{b-}^{\alpha}\left(\left[\frac{a+b}{2}, b\right]\right)$, such that $f^{(k)}(a) = f^{(k)}(b) = 0$, $k = 0, 1, \ldots, m-1$. When $\frac{1}{q} < \alpha < 1$, the last boundary conditions are void.*
   Set

$$M_3(f) = \max\left\{\left\|D_{a+}^{\alpha}f\right\|_{L_q\left(\left[a,\frac{a+b}{2}\right]\right)}, \left\|D_{b-}^{\alpha}f\right\|_{L_q\left(\left[\frac{a+b}{2},b\right]\right)}\right\}. \qquad (2.27)$$

Then

$$\int_a^b |f(x)|\,dx \le \frac{1}{\Gamma(\alpha)\,(p(\alpha-1)+1)^{\frac{1}{p}}\left(\alpha+\frac{1}{p}\right)}\left(\frac{b-a}{2}\right)^{\alpha+\frac{1}{p}}. \qquad (2.28)$$

$$\left[\left\|D_{a+}^{\alpha}f\right\|_{L_q\left(\left[a,\frac{a+b}{2}\right]\right)} + \left\|D_{b-}^{\alpha}f\right\|_{L_q\left(\left[\frac{a+b}{2},b\right]\right)}\right] \le$$

$$\frac{M_3(f)}{\Gamma(\alpha)\,(p(\alpha-1)+1)^{\frac{1}{p}}\left(\alpha+\frac{1}{p}\right)2^{\alpha-\frac{1}{q}}}\,(b-a)^{\alpha+\frac{1}{p}}. \qquad (2.29)$$

Again here f cannot be a non-trivial polynomial of degree m.

Proof By Theorem 2.3 we have

$$|f(x)| \le \frac{1}{\Gamma(\alpha)}\int_a^x (x-t)^{\alpha-1}\left|(D_{a+}^{\alpha}f)(t)\right|dt$$

$$\le \frac{1}{\Gamma(\alpha)}\left(\int_a^x (x-t)^{p(\alpha-1)}\,dt\right)^{\frac{1}{p}}\left(\int_a^x \left|(D_{a+}^{\alpha}f)(t)\right|^q\,dt\right)^{\frac{1}{q}}$$

$$\le \frac{1}{\Gamma(\alpha)}\frac{(x-a)^{\frac{(p(\alpha-1)+1)}{p}}}{(p(\alpha-1)+1)^{\frac{1}{p}}}\left\|D_{a+}^{\alpha}f\right\|_{L_q\left(\left[a,\frac{a+b}{2}\right]\right)}, \quad \text{for all } x \in \left[a, \frac{a+b}{2}\right]. \quad (2.30)$$

That is

$$|f(x)| \leq \frac{1}{\Gamma(\alpha)} \frac{(x-a)^{\alpha-1+\frac{1}{p}}}{(p(\alpha-1)+1)^{\frac{1}{p}}} \|D_{a+}^{\alpha}f\|_{L_q([a,\frac{a+b}{2}])}, \quad \text{for all } x \in \left[a, \frac{a+b}{2}\right]$$
(2.31)

Similarly from Theorem 2.4 we get

$$|f(x)| \leq \frac{1}{\Gamma(\alpha)} \int_x^b (t-x)^{\alpha-1} \left|(D_{b-}^{\alpha}f)(t)\right| dt$$

$$\leq \frac{1}{\Gamma(\alpha)} \left(\int_x^b (t-x)^{p(\alpha-1)} dt\right)^{\frac{1}{p}} \left(\int_x^b \left|(D_{b-}^{\alpha}f)(t)\right|^q dt\right)^{\frac{1}{q}}$$
(2.32)

$$\leq \frac{1}{\Gamma(\alpha)} \frac{(b-x)^{(\alpha-1)+\frac{1}{p}}}{(p(\alpha-1)+1)^{\frac{1}{p}}} \|D_{b-}^{\alpha}f\|_{L_q([\frac{a+b}{2},b])}, \quad \text{for all } x \in \left[\frac{a+b}{2}, b\right].$$
(2.33)

That is

$$|f(x)| \leq \frac{1}{\Gamma(\alpha)} \frac{(b-x)^{\alpha-1+\frac{1}{p}}}{(p(\alpha-1)+1)^{\frac{1}{p}}} \|D_{b-}^{\alpha}f\|_{L_q([\frac{a+b}{2},b])}, \quad \text{for all } x \in \left[\frac{a+b}{2}, b\right].$$
(2.34)

Consequently we obtain by (2.31) that

$$\int_a^{\frac{a+b}{2}} |f(x)| dx \leq$$

$$\frac{1}{\Gamma(\alpha)(p(\alpha-1)+1)^{\frac{1}{p}}} \left(\int_a^{\frac{a+b}{2}} (x-a)^{\alpha-1+\frac{1}{p}} dx\right) \|D_{a+}^{\alpha}f\|_{L_q([a,\frac{a+b}{2}])} =$$

$$\frac{1}{\Gamma(\alpha)(p(\alpha-1)+1)^{\frac{1}{p}} \left(\alpha+\frac{1}{p}\right)} \left(\frac{b-a}{2}\right)^{\alpha+\frac{1}{p}} \|D_{a+}^{\alpha}f\|_{L_q([a,\frac{a+b}{2}])}.$$
(2.35)

Similarly it holds by (2.34) that

$$\int_{\frac{a+b}{2}}^b |f(x)| dx \leq$$

$$\frac{1}{\Gamma(\alpha)(p(\alpha-1)+1)^{\frac{1}{p}}} \left(\int_{\frac{a+b}{2}}^b (b-x)^{\alpha-1+\frac{1}{p}} dx\right) \|D_{b-}^{\alpha}f\|_{L_q([\frac{a+b}{2},b])} =$$

$$\frac{1}{\Gamma\left(\alpha\right)\left(p\left(\alpha-1\right)+1\right)^{\frac{1}{p}}\left(\alpha+\frac{1}{p}\right)}\left(\frac{b-a}{2}\right)^{\alpha+\frac{1}{p}}\left\|D_{b-}^{\alpha}f\right\|_{L_q\left(\left[\frac{a+b}{2},b\right]\right)}\cdot \qquad (2.36)$$

Adding (2.35) and (2.36) we have

$$\int_a^b |f(x)|\,dx \le \frac{1}{\Gamma\left(\alpha\right)\left(p\left(\alpha-1\right)+1\right)^{\frac{1}{p}}\left(\alpha+\frac{1}{p}\right)}\left(\frac{b-a}{2}\right)^{\alpha+\frac{1}{p}} \qquad (2.37)$$

$$\cdot\left[\left\|D_{a+}^{\alpha}f\right\|_{L_q\left(\left[a,\frac{a+b}{2}\right]\right)}+\left\|D_{b-}^{\alpha}f\right\|_{L_q\left(\left[\frac{a+b}{2},b\right]\right)}\right] \le$$

$$\frac{\max\left\{\left\|D_{a+}^{\alpha}f\right\|_{L_q\left(\left[a,\frac{a+b}{2}\right]\right)},\left\|D_{b-}^{\alpha}f\right\|_{L_q\left(\left[\frac{a+b}{2},b\right]\right)}\right\}}{\Gamma\left(\alpha\right)\left(p\left(\alpha-1\right)+1\right)^{\frac{1}{p}}\left(\alpha+\frac{1}{p}\right)2^{\alpha-\frac{1}{q}}}\left(b-a\right)^{\alpha+\frac{1}{p}}, \qquad (2.38)$$

proving the claim. ∎

Combining Theorem 2.6, Remark 2.7, Theorems 2.8 and 2.9, we obtain

Theorem 2.10 *Let any* $p,q>1:\frac{1}{p}+\frac{1}{q}=1$, $\alpha\ge 1$, $m=\lceil\alpha\rceil$, $f\in C\left([a,b]\right)$. *Assume* $f\in C_{a+}^{\alpha}\left(\left[a,\frac{a+b}{2}\right]\right)$ *and* $f\in C_{b-}^{\alpha}\left(\left[\frac{a+b}{2},b\right]\right)$, *such that* $f^{(k)}(a)=f^{(k)}(b)=0$, $k=0,1,\dots,m-1$. *Then*

$$\int_a^b |f(x)|\,dx \le$$

$$\min\left\{\frac{\left(\left\|D_{a+}^{\alpha}f\right\|_{\infty,\left[a,\frac{a+b}{2}\right]}+\left\|D_{b-}^{\alpha}f\right\|_{\infty,\left[\frac{a+b}{2},b\right]}\right)}{\Gamma\left(\alpha+2\right)}\left(\frac{b-a}{2}\right)^{\alpha+1},\right.$$

$$\frac{1}{\Gamma\left(\alpha+1\right)}\left(\frac{b-a}{2}\right)^{\alpha}\left\|D^{\alpha}f\right\|_{L_1([a,b])}, \qquad (2.39)$$

$$\left.\frac{\left[\left\|D_{a+}^{\alpha}f\right\|_{L_q\left(\left[a,\frac{a+b}{2}\right]\right)}+\left\|D_{b-}^{\alpha}f\right\|_{L_q\left(\left[\frac{a+b}{2},b\right]\right)}\right]}{\Gamma\left(\alpha\right)\left(p\left(\alpha-1\right)+1\right)^{\frac{1}{p}}\left(\alpha+\frac{1}{p}\right)}\left(\frac{b-a}{2}\right)^{\alpha+\frac{1}{p}}\right\} \le$$

$$\min\left\{M_1\left(f\right)\frac{(b-a)^{\alpha+1}}{\Gamma\left(\alpha+2\right)2^{\alpha}},\frac{M_2\left(f\right)}{\Gamma\left(\alpha+1\right)2^{\alpha-1}}\left(b-a\right)^{\alpha},\right.$$

$$\frac{M_3\left(f\right)}{\Gamma\left(\alpha\right)\left(p\left(\alpha-1\right)+1\right)^{\frac{1}{p}}\left(\alpha+\frac{1}{p}\right)2^{\alpha-\frac{1}{q}}}\left(b-a\right)^{\alpha+\frac{1}{p}}\right\}, \tag{2.40}$$

where $M_1\left(f\right)$ as in (2.14), $M_2\left(f\right)$ as in (2.18) and $M_3\left(f\right)$ as in (2.27).
Here f cannot be a non-trivial polynomial of degree m.

Corollary 2.11 *Here all as in Theorem 2.10. Then*

$$\left|\frac{1}{b-a}\int_a^b f\left(x\right)dx\right|\leq\frac{1}{b-a}\int_a^b|f\left(x\right)|dx\leq$$

$$\min\left\{\frac{\left(\left\|D_{a+}^\alpha f\right\|_{\infty,\left[a,\frac{a+b}{2}\right]}+\left\|D_{b-}^\alpha f\right\|_{\infty,\left[\frac{a+b}{2},b\right]}\right)}{\Gamma\left(\alpha+2\right)2^{\alpha+1}}\left(b-a\right)^\alpha,\right. \tag{2.41}$$

$$\frac{1}{2^\alpha\Gamma\left(\alpha+1\right)}\left(b-a\right)^{\alpha-1}\left\|D^\alpha f\right\|_{L_1\left(\left[a,b\right]\right)},$$

$$\frac{\left[\left\|D_{a+}^\alpha f\right\|_{L_q\left(\left[a,\frac{a+b}{2}\right]\right)}+\left\|D_{b-}^\alpha f\right\|_{L_q\left(\left[\frac{a+b}{2},b\right]\right)}\right]}{\Gamma\left(\alpha\right)\left(p\left(\alpha-1\right)+1\right)^{\frac{1}{p}}\left(\alpha+\frac{1}{p}\right)2^{\alpha+\frac{1}{p}}}\left(b-a\right)^{\alpha+\frac{1}{p}-1}\right\}\leq$$

$$\min\left\{M_1\left(f\right)\frac{\left(b-a\right)^\alpha}{\Gamma\left(\alpha+2\right)2^\alpha},\frac{M_2\left(f\right)}{\Gamma\left(\alpha+1\right)2^{\alpha-1}}\left(b-a\right)^{\alpha-1},\right.$$

$$\left.\frac{M_3\left(f\right)}{\Gamma\left(\alpha\right)\left(p\left(\alpha-1\right)+1\right)^{\frac{1}{p}}\left(\alpha+\frac{1}{p}\right)2^{\alpha-\frac{1}{q}}}\left(b-a\right)^{\alpha+\frac{1}{p}-1}\right\}. \tag{2.42}$$

In 1938, Ostrowski [12] proved the following important inequality.

Theorem 2.12 *Let* $f:[a,b]\to\mathbb{R}$ *be continuous on* $[a,b]$ *and differentiable on* (a,b) *whose derivative* $f':(a,b)\to\mathbb{R}$ *is bounded on* (a,b), *i.e.,* $\left\|f'\right\|_\infty:=\sup_{t\in(a,b)}\left|f'\left(t\right)\right|<+\infty.$ *Then*

$$\left|\frac{1}{b-a}\int_a^b f\left(t\right)dt-f\left(x\right)\right|\leq\left[\frac{1}{4}+\frac{\left(x-\frac{a+b}{2}\right)^2}{\left(b-a\right)^2}\right]\cdot\left(b-a\right)\left\|f'\right\|_\infty, \tag{2.43}$$

for any $x\in[a,b]$. *The constant* $\frac{1}{4}$ *is the best possible.*

In (2.43) for $x = \frac{a+b}{2}$ we get

$$\left| \frac{1}{b-a} \int_a^b f(t)\, dt - f\left(\frac{a+b}{2}\right) \right| \le \left(\frac{b-a}{4}\right) \|f'\|_\infty. \tag{2.44}$$

We have proved the following

Theorem 2.13 *Let $f \in C^1([a, b])$, with $f\left(\frac{a+b}{2}\right) = 0$. Then*

$$\left| \int_a^b f(t)\, dt \right| \le \frac{(b-a)^2}{4} \|f'\|_\infty, \tag{2.45}$$

where the constant $\frac{1}{4}$ is the best possible.

So we proved once again (2.2) with only one initial condition.

2.3 Application

Inequalities for complex valued functions defined on the unit circle were studied extensively by Dragomir, see [7, 8].

We give here our version for these functions involved in a Polya type inequality, by applying a result of this chapter.

Let $t \in [a, b] \subseteq [0, 2\pi)$, the unit circle arc $A = \{z \in \mathbb{C} : z = e^{it},\ t \in [a, b]\}$, and $f : A \to \mathbb{C}$ be a continuous function. Clearly here there exist functions $u, v : A \to \mathbb{R}$ continuous, the real and the complex part of f, respectively, such that

$$f\left(e^{it}\right) = u\left(e^{it}\right) + iv\left(e^{it}\right). \tag{2.46}$$

So that f is continuous, iff u, v are continuous.

Call $g(t) = f\left(e^{it}\right)$, $l_1(t) = u\left(e^{it}\right)$, $l_2(t) = v\left(e^{it}\right)$, $t \in [a, b]$; so that $g : [a, b] \to \mathbb{C}$ and $l_1, l_2 : [a, b] \to \mathbb{R}$ are continuous functions in t.

If g has a derivative with respect to t, then l_1, l_2 have also derivatives with respect to t. In that case

$$f_t\left(e^{it}\right) = u_t\left(e^{it}\right) + iv_t\left(e^{it}\right), \tag{2.47}$$

(i.e. $g'(t) = l_1'(t) + il_2'(t)$), which means

$$f_t(\cos t + i \sin t) = u_t(\cos t + i \sin t) + iv_t(\cos t + i \sin t). \tag{2.48}$$

Let us call $x = \cos t$, $y = \sin t$. Then

$$u_t\left(e^{it}\right) = u_t(\cos t + i \sin t) = u_t(x + iy) = u_t(x, y) =$$

$$\frac{\partial u}{\partial x}\frac{\partial x}{\partial t} + \frac{\partial u}{\partial y}\frac{\partial y}{\partial t} = \frac{\partial u\left(e^{it}\right)}{\partial x}\left(-\sin t\right) + \frac{\partial u\left(e^{it}\right)}{\partial y}\cos t. \tag{2.49}$$

Similarly we find that

$$v_t\left(e^{it}\right) = \frac{\partial v\left(e^{it}\right)}{\partial x}\left(-\sin t\right) + \frac{\partial v\left(e^{it}\right)}{\partial y}\cos t. \tag{2.50}$$

Since g is continuous on $[a, b]$, then $\int_a^b f\left(e^{it}\right) dt$ exists. Furthermore it holds

$$\int_a^b f\left(e^{it}\right) dt = \int_a^b u\left(e^{it}\right) dt + i \int_a^b v\left(e^{it}\right) dt. \tag{2.51}$$

We have here that

$$\left| \int_a^b f\left(e^{it}\right) dt \right| \leq \int_a^b \left| f\left(e^{it}\right) \right| dt \leq \tag{2.52}$$

$$\int_a^b \left| u\left(e^{it}\right) \right| dt + \int_a^b \left| v\left(e^{it}\right) \right| dt = \int_a^b |l_1(t)| \, dt + \int_a^b |l_2(t)| \, dt. \tag{2.53}$$

We give the following application of Theorem 2.10.

Theorem 2.14 *Let $f \in C(A, \mathbb{C})$, $[a, b] \subseteq [0, 2\pi)$; any $p, q > 1 : \frac{1}{p} + \frac{1}{q} = 1$, $\alpha \geq 1$, $m = [\alpha]$. Assume $l_1, l_2 \in C_{a+}^{\alpha}\left([a, \frac{a+b}{2}]\right)$ and $l_1, l_2 \in C_{b-}^{\alpha}\left([\frac{a+b}{2}, b]\right)$, such that $l_1^{(k)}(a) = l_2^{(k)}(a) = l_1^{(k)}(b) = l_2^{(k)}(b) = 0$, $k = 0, 1, \ldots, m-1$. Then*

$$\left| \int_a^b f\left(e^{it}\right) dt \right| \leq \int_a^b \left| f\left(e^{it}\right) \right| dt \leq$$

$$\min \left\{ \frac{\left(\left\| D_{a+}^{\alpha} l_1 \right\|_{\infty, \left[a, \frac{a+b}{2}\right]} + \left\| D_{b-}^{\alpha} l_1 \right\|_{\infty, \left[\frac{a+b}{2}, b\right]} \right)}{\Gamma(\alpha + 2)} \left(\frac{b-a}{2} \right)^{\alpha+1}, \tag{2.54} \right.$$

$$\frac{1}{\Gamma(\alpha+1)} \left(\frac{b-a}{2} \right)^{\alpha} \left\| D^{\alpha} l_1 \right\|_{L_1([a,b])},$$

$$\left. \frac{\left[\left\| D_{a+}^{\alpha} l_1 \right\|_{L_q\left([a, \frac{a+b}{2}]\right)} + \left\| D_{b-}^{\alpha} l_1 \right\|_{L_q\left([\frac{a+b}{2}, b]\right)} \right]}{\Gamma(\alpha)(p(\alpha-1)+1)^{\frac{1}{p}}\left(\alpha + \frac{1}{p}\right)} \left(\frac{b-a}{2} \right)^{\alpha+\frac{1}{p}} \right\} +$$

$$\min\left\{\frac{\left(\left\|D_{a+}^{\alpha}l_2\right\|_{\infty,\left[a,\frac{a+b}{2}\right]}+\left\|D_{b-}^{\alpha}l_2\right\|_{\infty,\left[\frac{a+b}{2},b\right]}\right)}{\Gamma\left(\alpha+2\right)}\left(\frac{b-a}{2}\right)^{\alpha+1},\right.$$

$$\frac{1}{\Gamma\left(\alpha+1\right)}\left(\frac{b-a}{2}\right)^{\alpha}\left\|D^{\alpha}l_2\right\|_{L_1([a,b])},$$

$$\left.\frac{\left[\left\|D_{a+}^{\alpha}l_2\right\|_{L_q\left(\left[a,\frac{a+b}{2}\right]\right)}+\left\|D_{b-}^{\alpha}l_2\right\|_{L_q\left(\left[\frac{a+b}{2},b\right]\right)}\right]}{\Gamma\left(\alpha\right)\left(p\left(\alpha-1\right)+1\right)^{\frac{1}{p}}\left(\alpha+\frac{1}{p}\right)}\left(\frac{b-a}{2}\right)^{\alpha+\frac{1}{p}}\right\}\leq$$

$$\min\left\{M_1\left(l_1\right)\frac{\left(b-a\right)^{\alpha+1}}{\Gamma\left(\alpha+2\right)2^{\alpha}},\frac{M_2\left(l_1\right)}{\Gamma\left(\alpha+1\right)2^{\alpha-1}}\left(b-a\right)^{\alpha},\right.$$

$$\left.\frac{M_3\left(l_1\right)}{\Gamma\left(\alpha\right)\left(p\left(\alpha-1\right)+1\right)^{\frac{1}{p}}\left(\alpha+\frac{1}{p}\right)2^{\alpha-\frac{1}{q}}}\left(b-a\right)^{\alpha+\frac{1}{p}}\right\}+ \qquad (2.55)$$

$$\min\left\{M_1\left(l_2\right)\frac{\left(b-a\right)^{\alpha+1}}{\Gamma\left(\alpha+2\right)2^{\alpha}},\frac{M_2\left(l_2\right)}{\Gamma\left(\alpha+1\right)2^{\alpha-1}}\left(b-a\right)^{\alpha},\right.$$

$$\left.\frac{M_3\left(l_2\right)}{\Gamma\left(\alpha\right)\left(p\left(\alpha-1\right)+1\right)^{\frac{1}{p}}\left(\alpha+\frac{1}{p}\right)2^{\alpha-\frac{1}{q}}}\left(b-a\right)^{\alpha+\frac{1}{p}}\right\},$$

where $M_1\left(l_i\right)$ as in (2.14), $M_2\left(l_i\right)$ as in (2.18) and $M_3\left(l_i\right)$ as in (2.27), $i=1,2$. Here l_1, l_2 cannot be non-trivial polynomials of degree m.

References

1. G.A. Anastassiou, *Fractional Differentiation Inequalities* (Research Monograph, Springer, New York, 2009)
2. G.A. Anastassiou, On right fractional calculus. Chaos, Solitons Fractals **42**, 365–376 (2009)
3. G.A. Anastassiou, Balanced canavati type fractional opial inequalities. J. Appl. Funct. Anal. **9**(3–4), 230–238 (2014)
4. G.A. Anastassiou, Fractional Polya type integral inequality. J. Comput. Anal. Appl. **17**(4), 736–742 (2014)
5. G.A. Anastassiou, Univariate fractional Polya type integral inequalities. Mat. Vesnik **66**(4), 387–396 (2014)
6. J.A. Canavati, The Riemann-Liouville integral. Nieuw Archief Voor Wiskunde **5**(1), 53–75 (1987)

7. S.S. Dragomir, *Ostrowski's Type Inequalities for Complex Functions Defined on unit Circle with Applications for Unitary Operators in Hilbert Spaces*, article no. 6, 16th vol. 2013, RGMIA, Res. Rep. Coll., http://rgmia.org/v16.php

8. S.S. Dragomir, *Generalized Trapezoidal Type Inequalities for Complex Functions Defined on Unit Circle with Applications for Unitary Operators in Hilbert Spaces*, article no. 9, 16th vol. 2013, RGMIA, Res. Rep. Coll., http://rgmia.org/v16.php

9. A.M.A. El-Sayed, M. Gaber, On the finite Caputo and finite Riesz derivatives. Electron. J. Theor. Phys. **3**(12), 81–95 (2006)

10. G.S. Frederico, D.F.M. Torres, Fractional optimal control in the sense of caputo and the fractional Noether's theorem. Int. Math. Forum **3**(10), 479–493 (2008)

11. R. Gorenflo, F. Mainardi, *Essentials of Fractional Calculus*, 2000, Maphysto Center, http://www.maphysto.dk/oldpages/events/LevyCAC2000/MainardiNotes/fm2k0a.ps

12. A. Ostrowski, Über die Absolutabweichung einer differentiabaren Funcktion von ihrem Integralmittelwert. Comment. Math. Helv. **10**, 226–227 (1938)

13. G. Polya, Ein mittelwertsatz für Funktionen mehrerer Veränderlichen. Tohoku Math. J. **19**, 1–3 (1921)

14. G. Polya, G. Szegö, *Aufgaben und Lehrsätze aus der Analysis*, vol. I (Springer, Berlin, 1925). (German)

15. G. Polya, G. Szegö, *Problems and Theorems in Analysis*, vol. I, Classics in Mathematics (Springer, Berlin, 1972)

16. G. Polya, G. Szegö, *Problems and Theorems in Analysis*, vol. I, Chinese edn. (1984)

17. F. Qi, Polya type integral inequalities: origin, variants, proofs, refinements, generalizations, equivalences, and applications, article no. 20, 16th, vol. 2013, RGMIA, Res. Rep. Coll., http://rgmia.org/v16.php

18. S.G. Samko, A.A. Kilbas, O.I. Marichev, *Fractional Integrals and Derivatives, Theory and Applications* (Gordon and Breach, Amsterdam, 1993) [English translation from the Russian, Integrals and Derivatives of Fractional Order and Some of Their Applications (Nauka i Tekhnika, Minsk, 1987)]

Chapter 3
About Multivariate General Fractional Polya Integral Inequalities

Here we present a set of multivariate general fractional Polya type integral inequalities on the ball and shell. We treat both the radial and non-radial cases in all possibilities. We give also estimates for the related averages. It follows [4].

3.1 Introduction

We mention the following famous Polya's integral inequality, see [10], [11, p. 62], [12] and [13, p. 83].

Theorem 3.1 *Let* $f(x)$ *be differentiable and not identically a constant on* $[a, b]$ *with* $f(a) = f(b) = 0$. *Then the exists at least one point* $\xi \in [a, b]$ *such that*

$$\left| f'(\xi) \right| > \frac{4}{(b-a)^2} \int_a^b f(x)\, dx. \tag{3.1}$$

In [14], Qi presents the following very interesting Polya type integral inequality (3.2), which generalizes (3.1).

Theorem 3.2 *Let* $f(x)$ *be differentiable and not identically constant on* $[a, b]$ *with* $f(a) = f(b) = 0$ *and* $M = \sup_{x \in [a,b]} \left| f'(x) \right|$. *Then*

$$\left| \int_a^b f(x)\, dx \right| \le \frac{(b-a)^2}{4} M, \tag{3.2}$$

where $\frac{(b-a)^2}{4}$ *in (3.2) is the best constant.*

© Springer International Publishing Switzerland 2016
G.A. Anastassiou, *Intelligent Comparisons: Analytic Inequalities*,
Studies in Computational Intelligence 609,
DOI 10.1007/978-3-319-21121-3_3

The above motivate the current chapter.

In this chapter we present multivariate fractional Polya type integral inequalities in various cases, similar to (3.2).

For the last we need the following fractional calculus background.

Let $\alpha > 0$, $m = [\alpha]$ ($[\cdot]$ is the integral part), $\beta = \alpha - m$, $0 < \beta < 1$, $f \in C([a, b])$, $[a, b] \subset \mathbb{R}$, $x \in [a, b]$. The gamma function Γ is given by $\Gamma(\alpha) = \int_0^\infty e^{-t} t^{\alpha-1} dt$. We define the left Riemann-Liouville integral

$$\left(J_\alpha^{a+} f \right) (x) = \frac{1}{\Gamma(\alpha)} \int_a^x (x - t)^{\alpha-1} f(t)\, dt, \tag{3.3}$$

$a \le x \le b$. We define the subspace $C_{a+}^\alpha ([a.b])$ of $C^m ([a, b])$:

$$C_{a+}^\alpha ([a, b]) = \left\{ f \in C^m ([a, b]) : J_{1-\beta}^{a+} f^{(m)} \in C^1 ([a, b]) \right\}. \tag{3.4}$$

For $f \in C_{a+}^\alpha ([a, b])$, we define the left generalized α-fractional derivative of f over $[a, b]$ as

$$D_{a+}^\alpha f := \left(J_{1-\beta}^{a+} f^{(m)} \right)', \tag{3.5}$$

see [1], p. 24. Canavati first in [6], introduced the above over $[0, 1]$.

Notice that $D_{a+}^\alpha f \in C([a, b])$.

We need the following left fractional Taylor's formula, see [1], pp. 8–10, and in [6] the same over $[0, 1]$ that appeared first.

Theorem 3.3 *Let* $f \in C_{a+}^\alpha ([a, b])$.

(i) If $\alpha \ge 1$, *then*

$$f(x) = f(a) + f'(a)(x - a) + f''(a) \frac{(x - a)^2}{2} + \ldots + f^{(m-1)}(a) \frac{(x - a)^{m-1}}{(m - 1)!} \tag{3.6}$$

$$+ \frac{1}{\Gamma(\alpha)} \int_a^x (x - t)^{\alpha-1} \left(D_{a+}^\alpha f \right) (t)\, dt, \ \ all\ x \in [a, b].$$

(ii) If $0 < \alpha < 1$, *we have*

$$f(x) = \frac{1}{\Gamma(\alpha)} \int_a^x (x - t)^{\alpha-1} \left(D_{a+}^\alpha f \right) (t)\, dt, \ \ all\ x \in [a, b]. \tag{3.7}$$

Furthermore we need:

Let again $\alpha > 0$, $m = [\alpha]$, $\beta = \alpha - m$, $f \in C([a, b])$, call the right Riemann-Liouville fractional integral operator by

$$\left(J_{b-}^\alpha f \right) (x) := \frac{1}{\Gamma(\alpha)} \int_x^b (t - x)^{\alpha-1} f(t)\, dt, \tag{3.8}$$

$x \in [a, b]$, see also [2, 7–9, 16]. Define the subspace of functions

$$C_{b-}^{\alpha} \left([a, b] \right) := \left\{ f \in C^m \left([a, b] \right) : J_{b-}^{1-\beta} f^{(m)} \in C^1 \left([a, b] \right) \right\}. \quad (3.9)$$

Define the right generalized α-fractional derivative of f over $[a, b]$ as

$$D_{b-}^{\alpha} f = (-1)^{m-1} \left(J_{b-}^{1-\beta} f^{(m)} \right)', \quad (3.10)$$

see [2]. We set $D_{b-}^0 f = f$. Notice that $D_{b-}^{\alpha} f \in C \left([a, b] \right)$.

From [2], we need the following right Taylor fractional formula.

Theorem 3.4 Let $f \in C_{b-}^{\alpha} \left([a, b] \right)$, $\alpha > 0$, $m := [\alpha]$. Then

(i) If $\alpha \geq 1$, we get

$$f(x) = \sum_{k=0}^{m-1} \frac{f^{(k)}(b)}{k!} (x - b)^k + \left(J_{b-}^{\alpha} D_{b-}^{\alpha} f \right)(x), \ all \ x \in [a, b]. \quad (3.11)$$

(ii) If $0 < \alpha < 1$, we get

$$f(x) = J_{b-}^{\alpha} D_{b-}^{\alpha} f(x) = \frac{1}{\Gamma(\alpha)} \int_x^b (t - x)^{\alpha-1} \left(D_{b-}^{\alpha} f \right)(t) \, dt, all \ x \in [a, b]. \quad (3.12)$$

We need from [3]

Definition 3.5 Let $f \in C \left([a, b] \right)$, $x \in [a, b]$, $\alpha > 0$, $m := [\alpha]$. Assume that $f \in C_{b-}^{\alpha} \left(\left[\frac{a+b}{2}, b \right] \right)$ and $f \in C_{a+}^{\alpha} \left(\left[a, \frac{a+b}{2} \right] \right)$. We define the balanced Canavati type fractional derivative by

$$D^{\alpha} f(x) := \begin{cases} D_{b-}^{\alpha} f(x), & for \ \frac{a+b}{2} \leq x \leq b, \\ D_{a+}^{\alpha} f(x), & for \ a \leq x < \frac{a+b}{2}. \end{cases} \quad (3.13)$$

In [5] we proved the following fractional Polya type integral inequality without any boundary conditions.

Theorem 3.6 Let $0 < \alpha < 1$, $f \in C \left([a, b] \right)$. Assume $f \in C_{a+}^{\alpha} \left(\left[a, \frac{a+b}{2} \right] \right)$ and $f \in C_{b-}^{\alpha} \left(\left[\frac{a+b}{2}, b \right] \right)$. Set

$$M_1(f) = \max \left\{ \left\| D_{a+}^{\alpha} f \right\|_{\infty, \left[a, \frac{a+b}{2} \right]}, \left\| D_{b-}^{\alpha} f \right\|_{\infty, \left[\frac{a+b}{2}, b \right]} \right\}. \quad (3.14)$$

Then

$$\left| \int_a^b f(x) \, dx \right| \leq \int_a^b |f(x)| \, dx \leq \quad (3.15)$$

$$\frac{\left(\|D_{a+}^{\alpha}f\|_{\infty,\left[a,\frac{a+b}{2}\right]}+\|D_{b-}^{\alpha}f\|_{\infty,\left[\frac{a+b}{2},b\right]}\right)}{\Gamma(\alpha+2)}\left(\frac{b-a}{2}\right)^{\alpha+1}\leq M_1(f)\frac{(b-a)^{\alpha+1}}{\Gamma(\alpha+2)2^{\alpha}}.$$

$$\tag{3.16}$$

Inequalities (3.15), (3.16) are sharp, namely they are attained by

$$f_*(x)=\begin{cases}(x-a)^{\alpha},\, x\in\left[a,\frac{a+b}{2}\right]\\ (b-x)^{\alpha},\, x\in\left[\frac{a+b}{2},b\right]\end{cases},\, 0<\alpha<1.\tag{3.17}$$

Clearly here non zero constant functions f are excluded.

The last result also motivates this chapter.

Remark 3.7 (see [5]) When $\alpha\geq 1$, thus $m=[\alpha]\geq 1$, and by assuming that $f^{(k)}(a)=f^{(k)}(b)=0$, $k=0,1,...,m-1$, we can prove the same statements (3.15)–(3.17) as in Theorem 3.6. If we set there $\alpha=1$ we derive exactly Theorem 3.2. So we have generalized Theorem 3.2. Again here $f^{(m)}$ cannot be a constant different than zero, equivalently, f cannot be a non-trivial polynomial of degree m.

We present Polya type integral inequalities on the ball and shell.

3.2 Main Results

We need

Remark 3.8 We define the ball $B(0,R)=\{x\in\mathbb{R}^N:|x|<R\}\subseteq\mathbb{R}^N$, $N\geq 2$, $R>0$, and the sphere

$$S^{N-1}:=\{x\in\mathbb{R}^N:|x|=1\},$$

where $|\cdot|$ is the Euclidean norm. Let $d\omega$ be the element of surface measure on S^{N-1} and let

$$\omega_N=\int_{S^{N-1}}d\omega=\frac{2\pi^{\frac{N}{2}}}{\Gamma\left(\frac{N}{2}\right)}.$$

For $x\in\mathbb{R}^N-\{0\}$ we can write uniquely $x=r\omega$, where $r=|x|>0$ and $\omega=\frac{x}{r}\in S^{N-1}$, $|\omega|=1$. Note that $\int_{B(0,R)}dy=\frac{\omega_N R^N}{N}$ is the Lebesgue measure of the ball.

Following [15, pp. 149–150, exercise 6] and [17, pp. 87–88, Theorem 5.2.2] we can write $F:\overline{B(0,R)}\to\mathbb{R}$ a Lebesgue integrable function that

$$\int_{B(0,R)}F(x)\,dx=\int_{S^{N-1}}\left(\int_0^R F(r\omega)\,r^{N-1}dr\right)d\omega;\tag{3.18}$$

we use this formula a lot.

Initially the function $f : \overline{B(0, R)} \to \mathbb{R}$ is radial; that is, there exists a function g such that $f(x) = g(r)$, where $r = |x|, r \in [0, R], \forall\, x \in \overline{B(0, R)}, \alpha > 0, m = [\alpha]$. Here we assume that $g \in C([0, R])$ with $g \in C_{0+}^\alpha\left([0, \frac{R}{2}]\right)$ and $g \in C_{R-}^\alpha\left([\frac{R}{2}, R]\right)$, such that $g^{(k)}(0) = g^{(k)}(R) = 0, k = 0, 1, ..., m - 1$. In case of $0 < \alpha < 1$ then the last boundary conditions are void.

By assumption here and Theorem 3.3 we have

$$g(s) = \frac{1}{\Gamma(\alpha)} \int_0^s (s - t)^{\alpha-1} \left(D_{0+}^\alpha g\right)(t)\, dt, \tag{3.19}$$

all $s \in \left[0, \frac{R}{2}\right]$,

also it holds, by assumption and Theorem 3.4, that

$$g(s) = \frac{1}{\Gamma(\alpha)} \int_s^R (t - s)^{\alpha-1} \left(D_{R-}^\alpha g\right)(t)\, dt, \tag{3.20}$$

all $s \in \left[\frac{R}{2}, R\right]$.

By (3.19) we get

$$|g(s)| \leq \frac{1}{\Gamma(\alpha)} \int_0^s (s - t)^{\alpha-1} \left|\left(D_{0+}^\alpha g\right)(t)\right| dt$$

$$\leq \left\|D_{0+}^\alpha g\right\|_{\infty,\left[0,\frac{R}{2}\right]} \frac{1}{\Gamma(\alpha)} \int_0^s (s - t)^{\alpha-1}\, dt = \frac{\left\|D_{0+}^\alpha g\right\|_{\infty,\left[0,\frac{R}{2}\right]}}{\Gamma(\alpha + 1)} s^\alpha, \tag{3.21}$$

for any $s \in \left[0, \frac{R}{2}\right]$.

That is

$$|g(s)| \leq \frac{\left\|D_{0+}^\alpha g\right\|_{\infty,\left[0,\frac{R}{2}\right]}}{\Gamma(\alpha + 1)} s^\alpha, \tag{3.22}$$

for any $s \in \left[0, \frac{R}{2}\right]$.

Similarly we obtain

$$|g(s)| \leq \frac{1}{\Gamma(\alpha)} \int_s^R (t - s)^{\alpha-1} \left|\left(D_{R-}^\alpha g\right)(t)\right| dt$$

$$\leq \frac{\left\|D_{R-}^\alpha g\right\|_{\infty,\left[\frac{R}{2},R\right]}}{\Gamma(\alpha)} \int_s^R (t - s)^{\alpha-1}\, dt = \frac{\left\|D_{R-}^\alpha g\right\|_{\infty,\left[\frac{R}{2},R\right]}}{\Gamma(\alpha + 1)} (R - s)^\alpha, \tag{3.23}$$

for any $s \in \left[\frac{R}{2}, R\right]$.

I.e. it holds

$$|g(s)| \leq \frac{\left\| D_{R-}^{\alpha} g \right\|_{\infty, \left[\frac{R}{2}, R \right]}}{\Gamma(\alpha + 1)} (R - s)^{\alpha}, \tag{3.24}$$

for any $s \in \left[\frac{R}{2}, R \right]$.

Next we observe that

$$\left| \int_{B(0,R)} f(y) \, dy \right| \leq \int_{B(0,R)} |f(y)| \, dy \overset{(3.18)}{=}$$

$$\int_{S^{N-1}} \left(\int_0^R |g(s)| s^{N-1} ds \right) d\omega = \left(\int_0^R |g(s)| s^{N-1} ds \right) \int_{S^{N-1}} d\omega =$$

$$\left(\int_0^R |g(s)| s^{N-1} ds \right) \frac{2\pi^{\frac{N}{2}}}{\Gamma\left(\frac{N}{2} \right)} = \tag{3.25}$$

$$\frac{2\pi^{\frac{N}{2}}}{\Gamma\left(\frac{N}{2} \right)} \left\{ \int_0^{\frac{R}{2}} |g(s)| s^{N-1} ds + \int_{\frac{R}{2}}^R |g(s)| s^{N-1} ds \right\} \overset{\text{(by (3.22) and (3.24))}}{\leq}$$

$$\frac{2\pi^{\frac{N}{2}}}{\Gamma\left(\frac{N}{2} \right) \Gamma(\alpha + 1)} \left\{ \left\| D_{0+}^{\alpha} g \right\|_{\infty, \left[0, \frac{R}{2} \right]} \int_0^{\frac{R}{2}} s^{\alpha + N - 1} ds + \right.$$

$$\left. \left\| D_{R-}^{\alpha} g \right\|_{\infty, \left[\frac{R}{2}, R \right]} \int_{\frac{R}{2}}^R (R - s)^{\alpha} \left(\left(s - \frac{R}{2} \right) + \frac{R}{2} \right)^{N-1} ds \right\} = \tag{3.26}$$

$$\frac{2\pi^{\frac{N}{2}}}{\Gamma\left(\frac{N}{2} \right) \Gamma(\alpha + 1)} \left[\frac{\left\| D_{0+}^{\alpha} g \right\|_{\infty, \left[0, \frac{R}{2} \right]}}{(\alpha + N)} \left(\frac{R}{2} \right)^{\alpha + N} + \left\| D_{R-}^{\alpha} g \right\|_{\infty, \left[\frac{R}{2}, R \right]} \cdot \right.$$

$$\left. \left[\sum_{k=0}^{N-1} \binom{N-1}{k} \left(\frac{R}{2} \right)^k \int_{\frac{R}{2}}^R (R - s)^{(\alpha+1)-1} \left(s - \frac{R}{2} \right)^{N-k-1} ds \right] \right\} =$$

$$\frac{2\pi^{\frac{N}{2}}}{\Gamma\left(\frac{N}{2} \right) \Gamma(\alpha + 1)} \left\{ \frac{\left\| D_{0+}^{\alpha} g \right\|_{\infty, \left[0, \frac{R}{2} \right]}}{(\alpha + N)} \left(\frac{R}{2} \right)^{\alpha + N} + \right. \tag{3.27}$$

$$\left. \left\| D_{R-}^{\alpha} g \right\|_{\infty, \left[\frac{R}{2}, R \right]} \left[\sum_{k=0}^{N-1} \binom{N-1}{k} \left(\frac{R}{2} \right)^k \frac{\Gamma(\alpha + 1) \Gamma(N - k)}{\Gamma(\alpha + N + 1 - k)} \left(\frac{R}{2} \right)^{\alpha + N - k} \right] \right\}$$

$$= \frac{\pi^{\frac{N}{2}} R^{\alpha+N}}{2^{\alpha+N-1}\Gamma\left(\frac{N}{2}\right)} \left\{ \frac{\left\| D_{0+}^{\alpha} g \right\|_{\infty,\left[0,\frac{R}{2}\right]}}{(\alpha+N)\,\Gamma\,(\alpha+1)} + \right.$$

$$\left. \left\| D_{R-}^{\alpha} g \right\|_{\infty,\left[\frac{R}{2},R\right]} (N-1)! \left[\sum_{k=0}^{N-1} \frac{1}{k!\,\Gamma\,(\alpha+N+1-k)} \right] \right\}. \tag{3.28}$$

We have proved that

$$\left| \int_{B(0,R)} f\,(y)\,dy \right| \le \int_{B(0,R)} |f\,(y)|\,dy \le$$

$$\frac{\pi^{\frac{N}{2}} R^{\alpha+N}}{2^{\alpha+N-1}\Gamma\left(\frac{N}{2}\right)} \left\{ \frac{\left\| D_{0+}^{\alpha} g \right\|_{\infty,\left[0,\frac{R}{2}\right]}}{(\alpha+N)\,\Gamma\,(\alpha+1)} + \right. \tag{3.29}$$

$$\left. (N-1)! \left\| D_{R-}^{\alpha} g \right\|_{\infty,\left[\frac{R}{2},R\right]} \left[\sum_{k=0}^{N-1} \frac{1}{k!\,\Gamma\,(\alpha+N+1-k)} \right] \right\}.$$

Consider now

$$g_*\,(s) = \begin{cases} s^{\alpha}, & s \in \left[0,\frac{R}{2}\right], \\ (R-s)^{\alpha}, & s \in \left[\frac{R}{2},R\right], \end{cases} \alpha > 0. \tag{3.30}$$

We have as in [5] that

$$D_{0+}^{\alpha} s^{\alpha} = \Gamma\,(\alpha+1)\,,\, all\, s \in \left[0,\frac{R}{2}\right], \tag{3.31}$$

and

$$\left\| D_{0+}^{\alpha} s^{\alpha} \right\|_{\infty,\left[0,\frac{R}{2}\right]} = \Gamma\,(\alpha+1)\,.$$

Similarly we get

$$D_{R-}^{\alpha}\,(R-s)^{\alpha} = \Gamma\,(\alpha+1)\,,\, all\, s \in \left[\frac{R}{2},R\right], \tag{3.32}$$

and

$$\left\| D_{R-}^{\alpha}\,(R-s)^{\alpha} \right\|_{\infty,\left[\frac{R}{2},R\right]} = \Gamma\,(\alpha+1)\,. \tag{3.33}$$

That is

$$\left\| D_{0+}^{\alpha} g_* \right\|_{\infty,\left[0,\frac{R}{2}\right]} = \left\| D_{R-}^{\alpha} g_* \right\|_{\infty,\left[\frac{R}{2},R\right]} = \Gamma\,(\alpha+1)\,. \tag{3.34}$$

Consequently we find that

$$R.H.S.\,(3.29) = \frac{\pi^{\frac{N}{2}} R^{\alpha+N}}{2^{\alpha+N-1}\Gamma\left(\frac{N}{2}\right)} \left\{ \frac{1}{(\alpha+N)} + \right.$$

$$(N-1)!\,\Gamma\,(\alpha+1)\left[\sum_{k=0}^{N-1} \frac{1}{k!\,\Gamma\,(\alpha+N+1-k)}\right]\bigg\}. \tag{3.35}$$

Let $f_* : \overline{B\,(0,R)} \to \mathbb{R}$ be radial such that $f_*\,(x) = g_*\,(s)$, $s = |x|$, $s \in [0,R]$, $\forall$ $x \in \overline{B\,(0,R)}$.

Then we have

$$L.H.S.\,(3.29) = \int_{B(0,R)} f_*\,(y)\,dy \overset{(3.18)}{=}$$

$$\left(\int_0^R g_*\,(s)\,s^{N-1}ds\right)\frac{2\pi^{\frac{N}{2}}}{\Gamma\left(\frac{N}{2}\right)} =$$

$$\frac{2\pi^{\frac{N}{2}}}{\Gamma\left(\frac{N}{2}\right)}\left\{\int_0^{\frac{R}{2}} s^{\alpha+N-1}ds + \int_{\frac{R}{2}}^R (R-s)^\alpha\,s^{N-1}ds\right\} = \tag{3.36}$$

$$\frac{2\pi^{\frac{N}{2}}}{\Gamma\left(\frac{N}{2}\right)}\left\{\left(\frac{R}{2}\right)^{\alpha+N}\frac{1}{(\alpha+N)} + \int_{\frac{R}{2}}^R (R-s)^\alpha\left(\left(s-\frac{R}{2}\right)+\frac{R}{2}\right)^{N-1}ds\right\} =$$

$$\frac{2\pi^{\frac{N}{2}}}{\Gamma\left(\frac{N}{2}\right)}\left\{\frac{R^{\alpha+N}}{2^{\alpha+N}\,(\alpha+N)} + \right.$$

$$\sum_{k=0}^{N-1}\binom{N-1}{k}\left(\frac{R}{2}\right)^k\int_{\frac{R}{2}}^R (R-s)^{(\alpha+1)-1}\left(s-\frac{R}{2}\right)^{N-k-1}ds\bigg\} = \tag{3.37}$$

$$\frac{2\pi^{\frac{N}{2}}}{\Gamma\left(\frac{N}{2}\right)}\left\{\frac{R^{\alpha+N}}{2^{\alpha+N}\,(\alpha+N)} + \right.$$

$$\sum_{k=0}^{N-1}\binom{N-1}{k}\left(\frac{R}{2}\right)^k \frac{\Gamma\,(\alpha+1)\,\Gamma\,(N-k)}{\Gamma\,(\alpha+N+1-k)}\left(\frac{R}{2}\right)^{\alpha+N-k}\bigg\} =$$

$$\frac{\pi^{\frac{N}{2}} R^{\alpha+N}}{\Gamma\left(\frac{N}{2}\right)2^{\alpha+N-1}}\left\{\frac{1}{(\alpha+N)} + \right.$$

$$(N-1)!\Gamma(\alpha+1)\left[\sum_{k=0}^{N-1}\frac{1}{k!\Gamma(\alpha+N+1-k)}\right]\right\}\overset{(3.35)}{=}R.H.S.\,(3.29),\quad(3.38)$$

proving (3.29) sharp, infact it is attained.

We have proved the following main result.

Theorem 3.9 *Let* $f:\overline{B(0,R)}\to\mathbb{R}$ *be radial; that is, there exists a function* g *such that* $f(x)=g(s)$, $s=|x|$, $s\in[0,R]$, $\forall\,x\in\overline{B(0,R)}$, $\alpha>0$. *Assume that* $g\in C([0,R])$, *with* $g\in C_{0+}^{\alpha}\left([0,\frac{R}{2}]\right)$ *and* $g\in C_{R-}^{\alpha}\left([\frac{R}{2},R]\right)$, *such that* $g^{(k)}(0)=g^{(k)}(R)=0$, $k=0,1,\dots,m-1$, $m=[\alpha]$. *When* $0<\alpha<1$ *the last boundary conditions are void. Then*

$$\left|\int_{B(0,R)}f(y)\,dy\right|\le\int_{B(0,R)}|f(y)|\,dy\le$$

$$\frac{\pi^{\frac{N}{2}}R^{\alpha+N}}{2^{\alpha+N-1}\Gamma\left(\frac{N}{2}\right)}\left\{\frac{\left\|D_{0+}^{\alpha}g\right\|_{\infty,\left[0,\frac{R}{2}\right]}}{(\alpha+N)\,\Gamma(\alpha+1)}+\right.\qquad(3.39)$$

$$\left.(N-1)!\left\|D_{R-}^{\alpha}g\right\|_{\infty,\left[\frac{R}{2},R\right]}\left[\sum_{k=0}^{N-1}\frac{1}{k!\Gamma(\alpha+N+1-k)}\right]\right\}.$$

Inequalities (3.39) *are sharp, namely they are attained by a radial function* f_* *such that* $f_*(x)=g_*(s)$, *all* $s\in[0,R]$, *where*

$$g_*(s)=\begin{cases}s^{\alpha},\;s\in\left[0,\frac{R}{2}\right],\\(R-s)^{\alpha}\,,\;s\in\left[\frac{R}{2},R\right].\end{cases}\qquad(3.40)$$

We continue with

Remark 3.10 (Continuation of Remark 3.8) Here we assume that $\alpha\ge1$. By (3.19) we get

$$|g(s)|\le\frac{s^{\alpha-1}}{\Gamma(\alpha)}\left\|D_{0+}^{\alpha}g\right\|_{L_1\left([0,\frac{R}{2}]\right)},\qquad(3.41)$$

all $s\in\left[0,\frac{R}{2}\right]$.

Also, by (3.20), we obtain

$$|g(s)|\le\frac{(R-s)^{\alpha-1}}{\Gamma(\alpha)}\left\|D_{R-}^{\alpha}g\right\|_{L_1\left([\frac{R}{2},R]\right)},\qquad(3.42)$$

all $s\in\left[\frac{R}{2},R\right]$.

Hence as in (3.25) we get

$$\int_{B(0,R)} |f(y)|\,dy \le \frac{2\pi^{\frac{N}{2}}}{\Gamma\left(\frac{N}{2}\right)} \left(\int_0^R |g(s)|\,s^{N-1}\,ds\right) = \qquad (3.43)$$

$$\frac{2\pi^{\frac{N}{2}}}{\Gamma\left(\frac{N}{2}\right)} \left\{\int_0^{\frac{R}{2}} |g(s)|\,s^{N-1}\,ds + \int_{\frac{R}{2}}^R |g(s)|\,s^{N-1}\,ds\right\} \overset{\text{(by (3.41), (3.42))}}{\le}$$

$$\frac{2\pi^{\frac{N}{2}}}{\Gamma\left(\frac{N}{2}\right)\Gamma(\alpha)} \left\{\left(\int_0^{\frac{R}{2}} s^{N+\alpha-2}\,ds\right) \left\|D_{0+}^{\alpha}g\right\|_{L_1\left(\left[0,\frac{R}{2}\right]\right)} + \qquad (3.44)$$

$$\left(\int_{\frac{R}{2}}^R (R-s)^{\alpha-1}\,s^{N-1}\,ds\right) \left\|D_{R-}^{\alpha}g\right\|_{L_1\left(\left[\frac{R}{2},R\right]\right)}\right\} =$$

(acting the same as before, see (3.26)–(3.28))

$$\frac{\pi^{\frac{N}{2}}R^{\alpha+N-1}}{2^{\alpha+N-2}\Gamma\left(\frac{N}{2}\right)} \left\{ \frac{\left\|D_{0+}^{\alpha}g\right\|_{L_1\left(\left[0,\frac{R}{2}\right]\right)}}{(\alpha+N-1)\,\Gamma(\alpha)} + \right.$$

$$\left. (N-1)!\left\|D_{R-}^{\alpha}g\right\|_{L_1\left(\left[\frac{R}{2},R\right]\right)} \left[\sum_{k=0}^{N-1} \frac{1}{k!\,\Gamma(\alpha+N-k)}\right]\right\} \overset{(3.13)}{=} \qquad (3.45)$$

$$\frac{\pi^{\frac{N}{2}}R^{\alpha+N-1}}{2^{\alpha+N-2}\Gamma\left(\frac{N}{2}\right)} \left\{ \frac{\left\|D^{\alpha}g\right\|_{L_1\left(\left[0,\frac{R}{2}\right]\right)}}{(\alpha+N-1)\,\Gamma(\alpha)} + \right.$$

$$\left. (N-1)!\left\|D^{\alpha}g\right\|_{L_1\left(\left[\frac{R}{2},R\right]\right)} \left[\sum_{k=0}^{N-1} \frac{1}{k!\,\Gamma(\alpha+N-k)}\right]\right\}. \qquad (3.46)$$

We have proved

Theorem 3.11 *Here all terms and assumptions as in Theorem 3.9, however with* $\alpha \ge 1$. *Then*

$$\int_{B(0,R)} |f(y)|\,dy \le \frac{\pi^{\frac{N}{2}}R^{\alpha+N-1}}{2^{\alpha+N-2}\Gamma\left(\frac{N}{2}\right)} \left\{ \frac{\left\|D_{0+}^{\alpha}g\right\|_{L_1\left(\left[0,\frac{R}{2}\right]\right)}}{(\alpha+N-1)\,\Gamma(\alpha)} + \right.$$

$$(N-1)! \left\| D_{R-}^{\alpha} g \right\|_{L_1\left(\left[\frac{R}{2},R\right]\right)} \left[\sum_{k=0}^{N-1} \frac{1}{k! \Gamma(\alpha+N-k)} \right] \right\} . \tag{3.47}$$

We continue with

Remark 3.12 (Also a continuation of Remark 3.8) Let here $p, q > 1 : \frac{1}{p} + \frac{1}{q} = 1$, with $\alpha > \frac{1}{q}$. By (3.19) we have

$$|g(s)| \le \frac{1}{\Gamma(\alpha)} \int_0^s (s-t)^{\alpha-1} \left| \left(D_{0+}^{\alpha} g \right)(t) \right| dt \le$$

$$\frac{1}{\Gamma(\alpha)} \left(\int_0^s (s-t)^{p(\alpha-1)} dt \right)^{\frac{1}{p}} \left(\int_0^s \left| \left(D_{0+}^{\alpha} g \right)(t) \right|^q dt \right)^{\frac{1}{q}} =$$

$$\frac{1}{\Gamma(\alpha)} \frac{s^{\alpha-1+\frac{1}{p}}}{(p(\alpha-1)+1)^{\frac{1}{p}}} \left\| D_{0+}^{\alpha} g \right\|_{L_q\left(\left[0,\frac{R}{2}\right]\right)}, \tag{3.48}$$

all $s \in \left[0, \frac{R}{2}\right]$.

Similarly by (3.20) we obtain

$$|g(s)| \le \frac{1}{\Gamma(\alpha)} \int_s^R (t-s)^{\alpha-1} \left| \left(D_{R-}^{\alpha} g \right)(t) \right| dt \le$$

$$\frac{1}{\Gamma(\alpha)} \left(\int_s^R (t-s)^{p(\alpha-1)} dt \right)^{\frac{1}{p}} \left(\int_s^R \left| \left(D_{R-}^{\alpha} g \right)(t) \right|^q dt \right)^{\frac{1}{q}} =$$

$$\frac{1}{\Gamma(\alpha)} \frac{(R-s)^{\alpha-1+\frac{1}{p}}}{(p(\alpha-1)+1)^{\frac{1}{p}}} \left\| D_{R-}^{\alpha} g \right\|_{L_q\left(\left[\frac{R}{2},R\right]\right)}, \tag{3.49}$$

all $s \in \left[\frac{R}{2}, R\right]$.

Hence it holds

$$\int_{B(0,R)} |f(y)| dy \overset{(3.25)}{=}$$

$$\frac{2\pi^{\frac{N}{2}}}{\Gamma\left(\frac{N}{2}\right)} \left\{ \int_0^{\frac{R}{2}} |g(s)| s^{N-1} ds + \int_{\frac{R}{2}}^R |g(s)| s^{N-1} ds \right\} \overset{\text{(by (3.48), (3.49))}}{\le}$$

$$\frac{2\pi^{\frac{N}{2}}}{\Gamma(\alpha)\,\Gamma\left(\frac{N}{2}\right)(p(\alpha-1)+1)^{\frac{1}{p}}}\left\{\left(\int_0^{\frac{R}{2}}s^{\alpha+N-2+\frac{1}{p}}\,ds\right)\left\|D_{0+}^{\alpha}g\right\|_{L_q\left(\left[0,\frac{R}{2}\right]\right)}+\right.$$

(3.50)

$$\left.\left(\int_{\frac{R}{2}}^{R}(R-s)^{\alpha-1+\frac{1}{p}}s^{N-1}\,ds\right)\left\|D_{R-}^{\alpha}g\right\|_{L_q\left(\left[\frac{R}{2},R\right]\right)}\right\}=$$

$$\frac{2\pi^{\frac{N}{2}}}{\Gamma(\alpha)\,\Gamma\left(\frac{N}{2}\right)(p(\alpha-1)+1)^{\frac{1}{p}}}\left\{\frac{\left(\frac{R}{2}\right)^{\left(\alpha+N-\frac{1}{q}\right)}}{\left(\alpha+N-\frac{1}{q}\right)}\left\|D_{0+}^{\alpha}g\right\|_{L_q\left(\left[0,\frac{R}{2}\right]\right)}+\right.$$

$$\left[\sum_{k=0}^{N-1}\binom{N-1}{k}\left(\frac{R}{2}\right)^{k}\left(\int_{\frac{R}{2}}^{R}(R-s)^{\left(\alpha+\frac{1}{p}-1\right)}\left(s-\frac{R}{2}\right)^{N-k-1}\,ds\right)\right]$$

$$\left.\left\|D_{R-}^{\alpha}g\right\|_{L_q\left(\left[\frac{R}{2},R\right]\right)}\right\}=$$

$$\frac{2\pi^{\frac{N}{2}}}{\Gamma(\alpha)\,\Gamma\left(\frac{N}{2}\right)(p(\alpha-1)+1)^{\frac{1}{p}}}\left\{\frac{R^{\left(\alpha+N-\frac{1}{q}\right)}}{\left(\alpha+N-\frac{1}{q}\right)2^{\left(\alpha+N-\frac{1}{q}\right)}}\left\|D_{0+}^{\alpha}g\right\|_{L_q\left(\left[0,\frac{R}{2}\right]\right)}+\right.$$

(3.51)

$$\left[\sum_{k=0}^{N-1}\frac{(N-1)!}{k!\,(N-k-1)!}\left(\frac{R}{2}\right)^{k}\frac{\Gamma\left(\alpha+\frac{1}{p}\right)\Gamma(N-k)}{\Gamma\left(\alpha+\frac{1}{p}+N-k\right)}\left(\frac{R}{2}\right)^{\alpha+\frac{1}{p}+N-k-1}\right]$$

$$\left.\left\|D_{R-}^{\alpha}g\right\|_{L_q\left(\left[\frac{R}{2},R\right]\right)}\right\}=$$

$$\frac{2\pi^{\frac{N}{2}}}{\Gamma(\alpha)\,\Gamma\left(\frac{N}{2}\right)(p(\alpha-1)+1)^{\frac{1}{p}}}\left\{\frac{R^{\left(\alpha+N-\frac{1}{q}\right)}}{\left(\alpha+N-\frac{1}{q}\right)2^{\left(\alpha+N-\frac{1}{q}\right)}}\left\|D_{0+}^{\alpha}g\right\|_{L_q\left(\left[0,\frac{R}{2}\right]\right)}+\right.$$

(3.52)

$$(N-1)!\,\Gamma\left(\alpha+\frac{1}{p}\right)\left(\frac{R^{\alpha+N-\frac{1}{q}}}{2^{\alpha+N-\frac{1}{q}}}\right).$$

$$\left[\sum_{k=0}^{N-1}\frac{1}{k!\,\Gamma\left(\alpha+\frac{1}{p}+N-k\right)}\right]\left\|D_{R-}^{\alpha}g\right\|_{L_q\left(\left[\frac{R}{2},R\right]\right)}\right\}=$$

$$\frac{\pi^{\frac{N}{2}}R^{\alpha+N-\frac{1}{q}}}{\Gamma\left(\alpha\right)\Gamma\left(\frac{N}{2}\right)\left(p\left(\alpha-1\right)+1\right)^{\frac{1}{p}}2^{\alpha+N-\frac{1}{q}-1}}\left\{\frac{\left\|D_{0+}^{\alpha}g\right\|_{L_q\left(\left[0,\frac{R}{2}\right]\right)}}{\left(\alpha+N-\frac{1}{q}\right)}+\right.$$

$$\left.(N-1)!\,\Gamma\left(\alpha+\frac{1}{p}\right)\left[\sum_{k=0}^{N-1}\frac{1}{k!\,\Gamma\left(\alpha+\frac{1}{p}+N-k\right)}\right]\left\|D_{R-}^{\alpha}g\right\|_{L_q\left(\left[\frac{R}{2},R\right]\right)}\right\}.$$

$$(3.53)$$

We have proved the following

Theorem 3.13 *Let $p,q>1:\frac{1}{p}+\frac{1}{q}=1$, $\alpha>\frac{1}{q}$. All other terms and assumptions as in Theorem 3.9. Then*

$$\int_{B(0,R)}|f\left(y\right)|\,dy\le$$

$$\frac{\pi^{\frac{N}{2}}R^{\alpha+N-\frac{1}{q}}}{\Gamma\left(\alpha\right)\Gamma\left(\frac{N}{2}\right)\left(p\left(\alpha-1\right)+1\right)^{\frac{1}{p}}2^{\alpha+N-\frac{1}{q}-1}}\left\{\frac{\left\|D_{0+}^{\alpha}g\right\|_{L_q\left(\left[0,\frac{R}{2}\right]\right)}}{\left(\alpha+N-\frac{1}{q}\right)}+\right.$$

$$\left.(N-1)!\,\Gamma\left(\alpha+\frac{1}{p}\right)\left[\sum_{k=0}^{N-1}\frac{1}{k!\,\Gamma\left(\alpha+\frac{1}{p}+N-k\right)}\right]\left\|D_{R-}^{\alpha}g\right\|_{L_q\left(\left[\frac{R}{2},R\right]\right)}\right\}.$$

$$(3.54)$$

Combining Theorems 3.9, 3.11, 3.13 we derive

Theorem 3.14 *Let any $p,q>1:\frac{1}{p}+\frac{1}{q}=1$ and $\alpha\ge1$. And let $f:\overline{B\left(0,R\right)}\to\mathbb{R}$ be radial; that is, there exists a function g such that $f\left(x\right)=g\left(s\right)$, $s=|x|$, $s\in[0,R]$, $\forall\,x\in\overline{B\left(0,R\right)}$. Assume that $g\in C\left([0,R]\right)$, with $g\in C_{0+}^{\alpha}\left(\left[0,\frac{R}{2}\right]\right)$ and $g\in C_{R-}^{\alpha}\left(\left[\frac{R}{2},R\right]\right)$, such that $g^{(k)}\left(0\right)=g^{(k)}\left(R\right)=0$, $k=0,1,...,m-1$, $m=[\alpha]$. When $0<\alpha<1$ the last boundary conditions are void. Then*

$$\left|\int_{B(0,R)}f\left(y\right)dy\right|\le\int_{B(0,R)}|f\left(y\right)|\,dy\le$$

$$\min\left\{\frac{\pi^{\frac{N}{2}}R^{\alpha+N}}{2^{\alpha+N-1}\Gamma\left(\frac{N}{2}\right)}\left\{\frac{\left\|D_{0+}^{\alpha}g\right\|_{\infty,\left[0,\frac{R}{2}\right]}}{(\alpha+N)\,\Gamma\left(\alpha+1\right)}+\right.\right.$$

$$(N-1)! \left\| D_{R-}^{\alpha} g \right\|_{\infty, \left[\frac{R}{2}, R\right]} \left[\sum_{k=0}^{N-1} \frac{1}{k! \Gamma (\alpha + N + 1 - k)} \right] \right\},$$

$$\frac{\pi^{\frac{N}{2}} R^{\alpha + N - 1}}{2^{\alpha + N - 2} \Gamma \left(\frac{N}{2}\right)} \left\{ \frac{\left\| D_{0+}^{\alpha} g \right\|_{L_1 \left(\left[0, \frac{R}{2}\right]\right)}}{(\alpha + N - 1) \Gamma (\alpha)} + \right.$$

$$(N-1)! \left\| D_{R-}^{\alpha} g \right\|_{L_1 \left(\left[\frac{R}{2}, R\right]\right)} \left[\sum_{k=0}^{N-1} \frac{1}{k! \Gamma (\alpha + N - k)} \right] \right\},$$

$$\frac{\pi^{\frac{N}{2}} R^{\alpha + N - \frac{1}{q}}}{\Gamma (\alpha) \Gamma \left(\frac{N}{2}\right) (p (\alpha - 1) + 1)^{\frac{1}{p}} 2^{\alpha + N - \frac{1}{q} - 1}} \left\{ \frac{\left\| D_{0+}^{\alpha} g \right\|_{L_q \left(\left[0, \frac{R}{2}\right]\right)}}{\left(\alpha + N - \frac{1}{q}\right)} + \right.$$

$$(N-1)! \Gamma \left(\alpha + \frac{1}{p}\right) \left[\sum_{k=0}^{N-1} \frac{1}{k! \Gamma \left(\alpha + \frac{1}{p} + N - k\right)} \right] \left\| D_{R-}^{\alpha} g \right\|_{L_q \left(\left[\frac{R}{2}, R\right]\right)} \right\} \right\}.$$

$$(3.55)$$

Note 3.15 *It holds*

$$Vol (B (0, R)) = \frac{2 \pi^{\frac{N}{2}} R^N}{\Gamma \left(\frac{N}{2}\right) N}. \tag{3.56}$$

The corresponding estimate on the average follows

Corollary 3.16 *Let all terms and assumptions as in Theorem 3.14. Then*

$$\left| \frac{1}{Vol (B (0, R))} \int_{B(0,R)} f (y) \, dy \right| \leq \frac{1}{Vol (B (0, R))} \int_{B(0,R)} |f (y)| \, dy \leq$$

$$\min \left\{ \frac{N R^{\alpha}}{2^{\alpha + N}} \left\{ \frac{\left\| D_{0+}^{\alpha} g \right\|_{\infty, \left[0, \frac{R}{2}\right]}}{(\alpha + N) \Gamma (\alpha + 1)} + \right. \right.$$

$$(N-1)! \left\| D_{R-}^{\alpha} g \right\|_{\infty, \left[\frac{R}{2}, R\right]} \left[\sum_{k=0}^{N-1} \frac{1}{k! \Gamma (\alpha + N + 1 - k)} \right] \right\},$$

$$\frac{N R^{\alpha-1}}{2^{\alpha+N-1}} \left\{ \frac{\left\| D_{0+}^{\alpha} g \right\|_{L_1\left(\left[0,\frac{R}{2}\right]\right)}}{(\alpha+N-1)\,\Gamma(\alpha)} + \right.$$

$$\left. (N-1)! \left\| D_{R-}^{\alpha} g \right\|_{L_1\left(\left[\frac{R}{2},R\right]\right)} \left[\sum_{k=0}^{N-1} \frac{1}{k!\,\Gamma(\alpha+N-k)} \right] \right\},$$

$$\frac{N R^{\alpha-\frac{1}{q}}}{\Gamma(\alpha)\,(p(\alpha-1)+1)^{\frac{1}{p}}\,2^{\alpha+N-\frac{1}{q}}} \left\{ \frac{\left\| D_{0+}^{\alpha} g \right\|_{L_q\left(\left[0,\frac{R}{2}\right]\right)}}{\left(\alpha+N-\frac{1}{q}\right)} + \right.$$

$$\left. (N-1)!\,\Gamma\left(\alpha+\frac{1}{p}\right) \left[\sum_{k=0}^{N-1} \frac{1}{k!\,\Gamma\left(\alpha+\frac{1}{p}+N-k\right)} \right] \left\| D_{R-}^{\alpha} g \right\|_{L_q\left(\left[\frac{R}{2},R\right]\right)} \right\} \right\}.$$

$$(3.57)$$

We continue with Polya type inequalities on the ball for non-radial functions.

Theorem 3.17 *Let $f \in C\left(\overline{B(0,R)}\right)$ that is not necessarily radial, $0 < \alpha < 2$. Assume for any $\omega \in S^{N-1}$, that $f(\cdot\omega) \in C_{0+}^{\alpha}\left(\left[0,\frac{R}{2}\right]\right)$ and $f(\cdot\omega) \in C_{R-}^{\alpha}\left(\left[\frac{R}{2},R\right]\right)$, such that $f(0) = f(R\omega) = 0$. When $0 < \alpha < 1$ the last boundary conditions are void. We further assume that*

$$\left\| \frac{\partial_{0+}^{\alpha} f(r\omega)}{\partial r^{\alpha}} \right\|_{\infty,\left(r\in\left[0,\frac{R}{2}\right]\right)}, \quad \left\| \frac{\partial_{R-}^{\alpha} f(r\omega)}{\partial r^{\alpha}} \right\|_{\infty,\left(r\in\left[\frac{R}{2},R\right]\right)} \leq K, \qquad (3.58)$$

for every $\omega \in S^{N-1}$, where $K > 0$.
   Then
(i)

$$\int_{B(0,R)} |f(y)|\,dy \leq \frac{K\pi^{\frac{N}{2}} R^{\alpha+N}}{2^{\alpha+N-1}\Gamma\left(\frac{N}{2}\right)}. \qquad (3.59)$$

$$\left\{ \frac{1}{(\alpha+N)\,\Gamma(\alpha+1)} + (N-1)! \left[\sum_{k=0}^{N-1} \frac{1}{k!\,\Gamma(\alpha+N+1-k)} \right] \right\},$$

and
(ii)

$$\left| \frac{1}{Vol(B(0,R))} \int_{B(0,R)} f(y)\,dy \right| \leq \frac{1}{Vol(B(0,R))} \int_{B(0,R)} |f(y)|\,dy \leq \qquad (3.60)$$

$$\frac{KNR^\alpha}{2^{\alpha+N}}\left\{\frac{1}{(\alpha+N)\,\Gamma\,(\alpha+1)}+(N-1)!\left[\sum_{k=0}^{N-1}\frac{1}{k!\Gamma\,(\alpha+N+1-k)}\right]\right\}.$$

Proof In Remark 3.8, see (3.25)–(3.28), we proved that

$$\int_0^R |g\,(s)|\,s^{N-1}ds \le \left(\frac{R}{2}\right)^{\alpha+N}.$$

$$\left\{\frac{\left\|D_{0+}^\alpha g\right\|_{\infty,\left[0,\frac{R}{2}\right]}}{(\alpha+N)\,\Gamma\,(\alpha+1)}+\left\|D_{R-}^\alpha g\right\|_{\infty,\left[\frac{R}{2},R\right]}(N-1)!\left[\sum_{k=0}^{N-1}\frac{1}{k!\Gamma\,(\alpha+N+1-k)}\right]\right\}.$$

$$\tag{3.61}$$

In the above (3.61) we plug in $g\,(\cdot) = f\,(\cdot\omega)$, for $\omega \in S^{N-1}$ fixed, and we get

$$\int_0^R |f\,(s\omega)|\,s^{N-1}ds \stackrel{(3.58)}{\le} K\left(\frac{R}{2}\right)^{\alpha+N}.$$

$$\left\{\frac{1}{(\alpha+N)\,\Gamma\,(\alpha+1)}+(N-1)!\left[\sum_{k=0}^{N-1}\frac{1}{k!\Gamma\,(\alpha+N+1-k)}\right]\right\} =: \lambda_1. \quad (3.62)$$

Consequently we obtain

$$\int_{B(0,R)} |f\,(y)|\,dy = \int_{S^{N-1}}\left(\int_0^R |f\,(s\omega)|\,s^{N-1}ds\right)d\omega \le$$

$$\lambda_1\int_{S^{N-1}} d\omega = \lambda_1\frac{2\pi^{\frac{N}{2}}}{\Gamma\left(\frac{N}{2}\right)}, \quad (3.63)$$

proving the claims. ■

We continue with

Theorem 3.18 *Let* $f \in C\left(\overline{B\,(0,R)}\right)$ *that is not necessarily radial,* $1 \le \alpha < 2$. *Assume for any* $\omega \in S^{N-1}$, *that* $f\,(\cdot\omega) \in C_{0+}^\alpha\left(\left[0,\frac{R}{2}\right]\right)$ *and* $f\,(\cdot\omega) \in C_{R-}^\alpha\left(\left[\frac{R}{2},R\right]\right)$, *such that* $f\,(0) = f\,(R\omega) = 0$. *We further assume*

$$\left\|\frac{\partial_{0+}^\alpha f\,(\cdot\omega)}{\partial r^\alpha}\right\|_{L_1\left(\left[0,\frac{R}{2}\right]\right)}, \quad \left\|\frac{\partial_{R-}^\alpha f\,(\cdot\omega)}{\partial r^\alpha}\right\|_{L_1\left(\left[\frac{R}{2},R\right]\right)} \le M, \quad (3.64)$$

for every $\omega \in S^{N-1}$, *where* $M > 0$.

Then

(i)

$$\int_{B(0,R)} |f(y)|\,dy \le \frac{M\pi^{\frac{N}{2}} R^{\alpha+N-1}}{2^{\alpha+N-2}\Gamma\left(\frac{N}{2}\right)}.\tag{3.65}$$

$$\left\{\frac{1}{(\alpha+N-1)\,\Gamma(\alpha)} + (N-1)!\left[\sum_{k=0}^{N-1}\frac{1}{k!\,\Gamma(\alpha+N-k)}\right]\right\},$$

and

(ii)

$$\frac{1}{Vol\,(B(0,R))}\int_{B(0,R)} |f(y)|\,dy \le \tag{3.66}$$

$$\frac{MNR^{\alpha-1}}{2^{\alpha+N-1}}\left\{\frac{1}{(\alpha+N-1)\,\Gamma(\alpha)} + (N-1)!\left[\sum_{k=0}^{N-1}\frac{1}{k!\,\Gamma(\alpha+N-k)}\right]\right\}.$$

Proof In Remark 3.10, see (3.43)–(3.45), we proved that

$$\int_0^R |g(s)|\,s^{N-1}ds \le \left(\frac{R}{2}\right)^{\alpha+N-1}.$$

$$\left\{\frac{\left\|D_{0+}^{\alpha}g\right\|_{L_1\left(\left[0,\frac{R}{2}\right]\right)}}{(\alpha+N-1)\,\Gamma(\alpha)} + \left\|D_{R-}^{\alpha}g\right\|_{L_1\left(\left[\frac{R}{2},R\right]\right)}(N-1)!\left[\sum_{k=0}^{N-1}\frac{1}{k!\,\Gamma(\alpha+N-k)}\right]\right\}.$$

$$\tag{3.67}$$

In the above (3.67) we plug in $g(\cdot) = f(\cdot\omega)$, for $\omega \in S^{N-1}$ fixed, and we derive

$$\int_0^R |f(s\omega)|\,s^{N-1}ds \overset{(3.64)}{\le} M\left(\frac{R}{2}\right)^{\alpha+N-1}.$$

$$\left\{\frac{1}{(\alpha+N-1)\,\Gamma(\alpha)} + (N-1)!\left[\sum_{k=0}^{N-1}\frac{1}{k!\,\Gamma(\alpha+N-k)}\right]\right\} =: \lambda_2.\tag{3.68}$$

Hence

$$\int_{B(0,R)} |f(y)|\,dy = \int_{S^{N-1}}\left(\int_0^R |f(s\omega)|\,s^{N-1}ds\right)d\omega \le$$

$$\lambda_2\int_{S^{N-1}} d\omega = \lambda_2\frac{2\pi^{\frac{N}{2}}}{\Gamma\left(\frac{N}{2}\right)},\tag{3.69}$$

proving the claims. ∎

We further have

Theorem 3.19 *Let $p, q > 1 : \frac{1}{p} + \frac{1}{q} = 1$, and $\frac{1}{q} < \alpha < 2$. Let $f \in C\left(\overline{B\left(0, R\right)}\right)$ that is not necessarily radial. Assume for any $\omega \in S^{N-1}$, that $f\left(\cdot\omega\right) \in C_{0+}^{\alpha}\left(\left[0, \frac{R}{2}\right]\right)$ and $f\left(\cdot\omega\right) \in C_{R-}^{\alpha}\left(\left[\frac{R}{2}, R\right]\right)$, such that $f\left(0\right) = f\left(R\omega\right) = 0$. When $\frac{1}{q} < \alpha < 1$ the last boundary condition is void. We further assume*

$$\left\|\frac{\partial_{0+}^{\alpha} f\left(\cdot\omega\right)}{\partial r^{\alpha}}\right\|_{L_q\left(\left[0, \frac{R}{2}\right]\right)}, \quad \left\|\frac{\partial_{R-}^{\alpha} f\left(\cdot\omega\right)}{\partial r^{\alpha}}\right\|_{L_q\left(\left[\frac{R}{2}, R\right]\right)} \leq \Phi, \tag{3.70}$$

for every $\omega \in S^{N-1}$, where $\Phi > 0$.
   Then
(i)

$$\int_{B(0,R)} |f\left(y\right)| dy \leq \frac{\Phi \pi^{\frac{N}{2}} R^{\alpha+N-\frac{1}{q}}}{\Gamma\left(\alpha\right)\Gamma\left(\frac{N}{2}\right)\left(p\left(\alpha-1\right)+1\right)^{\frac{1}{p}} 2^{\alpha+N-\frac{1}{q}-1}}. \tag{3.71}$$

$$\left\{\frac{1}{\left(\alpha+N-\frac{1}{q}\right)} + (N-1)!\Gamma\left(\alpha+\frac{1}{p}\right)\left[\sum_{k=0}^{N-1} \frac{1}{k!\Gamma\left(\alpha+\frac{1}{p}+N-k\right)}\right]\right\},$$

and
(ii)

$$\frac{1}{Vol\left(B\left(0, R\right)\right)} \int_{B(0,R)} |f\left(y\right)| dy \leq \frac{\Phi N R^{\alpha-\frac{1}{q}}}{\Gamma\left(\alpha\right)\left(p\left(\alpha-1\right)+1\right)^{\frac{1}{p}} 2^{\alpha+N-\frac{1}{q}}}. \tag{3.72}$$

$$\left\{\frac{1}{\left(\alpha+N-\frac{1}{q}\right)} + (N-1)!\Gamma\left(\alpha+\frac{1}{p}\right)\left[\sum_{k=0}^{N-1} \frac{1}{k!\Gamma\left(\alpha+\frac{1}{p}+N-k\right)}\right]\right\}.$$

Proof In Remark 3.12, see (3.50)–(3.53), we proved that

$$\int_0^R |g\left(s\right)| s^{N-1} ds \leq$$

$$\left(\frac{R}{2}\right)^{\alpha+N-\frac{1}{q}} \frac{1}{\Gamma\left(\alpha\right)\left(p\left(\alpha-1\right)+1\right)^{\frac{1}{p}}} \cdot \left\{\frac{\left\|D_{0+}^{\alpha} g\right\|_{L_q\left(\left[0, \frac{R}{2}\right]\right)}}{\left(\alpha+N-\frac{1}{q}\right)} + \right. \tag{3.73}$$

$$(N-1)!\Gamma\left(\alpha+\frac{1}{p}\right)\left[\sum_{k=0}^{N-1}\frac{1}{k!\Gamma\left(\alpha+\frac{1}{p}+N-k\right)}\right]\left\|D_{R-}^{\alpha}g\right\|_{L_q\left(\left[\frac{R}{2},R\right]\right)}\right\}.$$

In the above (3.73) we plug in $g(\cdot) = f(\cdot\omega)$, for $\omega \in S^{N-1}$ fixed, and we find

$$\int_0^R |f(s\omega)| s^{N-1}ds \overset{(3.70)}{\leq} \Phi\left(\frac{R}{2}\right)^{\alpha+N-\frac{1}{q}} \frac{1}{\Gamma(\alpha)(p(\alpha-1)+1)^{\frac{1}{p}}}$$

$$\left\{\frac{1}{\left(\alpha+N-\frac{1}{q}\right)}+(N-1)!\Gamma\left(\alpha+\frac{1}{p}\right)\left[\sum_{k=0}^{N-1}\frac{1}{k!\Gamma\left(\alpha+\frac{1}{p}+N-k\right)}\right]\right\} =: \lambda_3.$$

(3.74)

Thus

$$\int_{B(0,R)} |f(y)|\,dy = \int_{S^{N-1}}\left(\int_0^R |f(s\omega)| s^{N-1}ds\right)d\omega \leq$$

$$\lambda_3 \int_{S^{N-1}} d\omega = \lambda_3 \frac{2\pi^{\frac{N}{2}}}{\Gamma\left(\frac{N}{2}\right)},$$

(3.75)

proving the claims. ∎

We make

Remark 3.20 Let the spherical shell $A := B(0, R_2) - \overline{B(0, R_1)}$, $0 < R_1 < R_2$, $A \subseteq \mathbb{R}^N$, $N \geq 2$, $x \in \overline{A}$. Consider first that $f : \overline{A} \to \mathbb{R}$ is radial; that is, there exists g such that $f(x) = g(r)$, $r = |x|$, $r \in [R_1, R_2]$, $\forall x \in \overline{A}$. Here x can be written uniquely as $x = r\omega$, where $r = |x| > 0$ and $\omega = \frac{x}{r} \in S^{N-1}$, $|\omega| = 1$, see ([1], pp. 149–150 and [15], p. 421), furthermore for general $F : \overline{A} \to \mathbb{R}$ Lebesgue integrable function we have that

$$\int_A F(x)\,dx = \int_{S^{N-1}}\left(\int_{R_1}^{R_2} F(r\omega) r^{N-1}dr\right)d\omega.$$

(3.76)

Let $d\omega$ be the element of surface measure on S^{N-1}, then

$$\omega_N := \int_{S^{N-1}} d\omega = \frac{2\pi^{\frac{N}{2}}}{\Gamma\left(\frac{N}{2}\right)}.$$

(3.77)

Here

$$Vol(A) = \frac{\omega_N\left(R_2^N - R_1^N\right)}{N} = \frac{2\pi^{\frac{N}{2}}\left(R_2^N - R_1^N\right)}{N\Gamma\left(\frac{N}{2}\right)}.$$

(3.78)

We assume that $g \in C([R_1, R_2])$, and $\alpha > 0$, $m = [\alpha]$, such that $g \in C^\alpha_{R_1+}\left(\left[R_1, \frac{R_1+R_2}{2}\right]\right)$ and $g \in C^\alpha_{R_2-}\left(\left[\frac{R_1+R_2}{2}, R_2\right]\right)$, with $g^{(k)}(R_1) = g^{(k)}(R_2) = 0$, $k = 0, 1, ..., m - 1$. When $0 < \alpha < 1$ the last boundary conditions are void.

By assumption here and Theorem 3.3 we have

$$g(s) = \frac{1}{\Gamma(\alpha)} \int_{R_1}^{s} (s - t)^{\alpha-1} \left(D^\alpha_{R_1+}g\right)(t) \, dt, \tag{3.79}$$

all $s \in \left[R_1, \frac{R_1+R_2}{2}\right]$,

also it holds, by assumption and Theorem 3.4, that

$$g(s) = \frac{1}{\Gamma(\alpha)} \int_{s}^{R_2} (t - s)^{\alpha-1} \left(D^\alpha_{R_2-}g\right)(t) \, dt, \tag{3.80}$$

all $s \in \left[\frac{R_1+R_2}{2}, R_2\right]$.

By (3.79) we get

$$|g(s)| \le \frac{1}{\Gamma(\alpha)} \int_{R_1}^{s} (s - t)^{\alpha-1} \left|\left(D^\alpha_{R_1+}g\right)(t)\right| \, dt \tag{3.81}$$

$$\le \left\|D^\alpha_{R_1+}g\right\|_{\infty,\left[R_1, \frac{R_1+R_2}{2}\right]} \frac{(s - R_1)^\alpha}{\Gamma(\alpha + 1)}, \tag{3.82}$$

for any $s \in \left[R_1, \frac{R_1+R_2}{2}\right]$.

Similarly we obtain by (3.80) that

$$|g(s)| \le \frac{1}{\Gamma(\alpha)} \int_{s}^{R_2} (t - s)^{\alpha-1} \left|\left(D^\alpha_{R_2-}g\right)(t)\right| \, dt \tag{3.83}$$

$$\le \left\|D^\alpha_{R_2-}g\right\|_{\infty,\left[\frac{R_1+R_2}{2}, R_2\right]} \frac{(R_2 - s)^\alpha}{\Gamma(\alpha + 1)}, \tag{3.84}$$

for any $s \in \left[\frac{R_1+R_2}{2}, R_2\right]$.

Next we observe that

$$\left|\int_A f(y) \, dy\right| \le \int_A |f(y)| \, dy \overset{(3.76)}{=} \tag{3.85}$$

$$\int_{S^{N-1}} \left(\int_{R_1}^{R_2} |g(s)| s^{N-1} ds\right) d\omega = \left(\int_{R_1}^{R_2} |g(s)| s^{N-1} ds\right) \frac{2\pi^{\frac{N}{2}}}{\Gamma\left(\frac{N}{2}\right)} = \tag{3.86}$$

$$\frac{2\pi^{\frac{N}{2}}}{\Gamma\left(\frac{N}{2}\right)}\left\{\int_{R_1}^{\frac{R_1+R_2}{2}}|g(s)|\,s^{N-1}ds+\int_{\frac{R_1+R_2}{2}}^{R_2}|g(s)|\,s^{N-1}ds\right\}\overset{\text{(by (3.82) and (3.84))}}{\leq}$$

$$\frac{2\pi^{\frac{N}{2}}}{\Gamma\left(\frac{N}{2}\right)\Gamma(\alpha+1)}\left\{\left\|D_{R_1+}^\alpha g\right\|_{\infty,\left[R_1,\frac{R_1+R_2}{2}\right]}\int_{R_1}^{\frac{R_1+R_2}{2}}(s-R_1)^\alpha s^{N-1}ds\right.$$

$$\left.+\left\|D_{R_2-}^\alpha g\right\|_{\infty,\left[\frac{R_1+R_2}{2},R_2\right]}\int_{\frac{R_1+R_2}{2}}^{R_2}(R_2-s)^\alpha s^{N-1}ds\right\}= \qquad (3.87)$$

$$\frac{\pi^{\frac{N}{2}}(N-1)!}{\Gamma\left(\frac{N}{2}\right)2^{\alpha+N-1}}\left\{\left\|D_{R_1+}^\alpha g\right\|_{\infty,\left[R_1,\frac{R_1+R_2}{2}\right]}\right.$$

$$\left(\sum_{k=0}^{N-1}(-1)^{N+k-1}\frac{(R_1+R_2)^k(R_2-R_1)^{N-k+\alpha}}{k!\Gamma(N-k+\alpha+1)}\right)+$$

$$\left.\left\|D_{R_2-}^\alpha g\right\|_{\infty,\left[\frac{R_1+R_2}{2},R_2\right]}\left[\sum_{k=0}^{N-1}\frac{(R_1+R_2)^k(R_2-R_1)^{\alpha+N-k}}{k!\Gamma(\alpha+1+N-k)}\right]\right\}. \qquad (3.88)$$

We have proved that

$$\left|\int_A f(y)\,dy\right|\leq\int_A|f(y)|\,dy\leq\frac{\pi^{\frac{N}{2}}(N-1)!}{\Gamma\left(\frac{N}{2}\right)2^{\alpha+N-1}}\cdot$$

$$\left\{\left\|D_{R_1+}^\alpha g\right\|_{\infty,\left[R_1,\frac{R_1+R_2}{2}\right]}\left(\sum_{k=0}^{N-1}\frac{(-1)^{N+k-1}(R_1+R_2)^k(R_2-R_1)^{N-k+\alpha}}{k!\Gamma(N-k+\alpha+1)}\right)+\right.$$

$$\left.\left\|D_{R_2-}^\alpha g\right\|_{\infty,\left[\frac{R_1+R_2}{2},R_2\right]}\left[\sum_{k=0}^{N-1}\frac{(R_1+R_2)^k(R_2-R_1)^{\alpha+N-k}}{k!\Gamma(\alpha+1+N-k)}\right]\right\}. \qquad (3.89)$$

Consider now $f_*:\overline{A}\to\mathbb{R}$ be radial such that $f_*(x)=g_*(s)$, $s=|x|$, $s\in[R_1,R_2]$, $\forall\,x\in\overline{A}$, where

$$g_*(s)=\begin{cases}(s-R_1)^\alpha\,,\ s\in\left[R_1,\frac{R_1+R_2}{2}\right],\\[2mm](R_2-s)^\alpha\,,\ s\in\left[\frac{R_1+R_2}{2},R_2\right],\ \alpha>0.\end{cases} \qquad (3.90)$$

We have, as in [5], that

$$\left\| D_{R_1+}^{\alpha} g_* \right\|_{\infty, \left[R_1, \frac{R_1+R_2}{2} \right]} = \Gamma\left(\alpha + 1 \right), \text{ and } \left\| D_{R_2-}^{\alpha} g_* \right\|_{\infty, \left[\frac{R_1+R_2}{2}, R_2 \right]} = \Gamma\left(\alpha + 1 \right).$$

$$(3.91)$$

Hence

$$R.H.S. \ (3.89) \ (applied \ on \ g_*) = \frac{\Gamma\left(\alpha + 1 \right) \pi^{\frac{N}{2}} \left(N - 1 \right)!}{\Gamma\left(\frac{N}{2} \right) 2^{\alpha + N - 1}} \cdot$$

$$\left\{ \sum_{k=0}^{N-1} \left(1 + (-1)^{N+k-1} \right) \frac{(R_1 + R_2)^k (R_2 - R_1)^{\alpha + N - k}}{k! \Gamma\left(\alpha + 1 + N - k \right)} \right\}. \qquad (3.92)$$

Furthermore we find

$$L.H.S. \ (3.89) \ (applied \ on \ f_*) = \int_A f_* \left(y \right) dy =$$

$$\left(\int_{R_1}^{R_2} g_* \left(s \right) s^{N-1} ds \right) \frac{2\pi^{\frac{N}{2}}}{\Gamma\left(\frac{N}{2} \right)} =$$

$$\frac{2\pi^{\frac{N}{2}}}{\Gamma\left(\frac{N}{2} \right)} \left\{ \int_{R_1}^{\frac{R_1+R_2}{2}} (s - R_1)^{\alpha} s^{N-1} ds + \int_{\frac{R_1+R_2}{2}}^{R_2} (R_2 - s)^{\alpha} s^{N-1} ds \right\} = \qquad (3.93)$$

$$\frac{\pi^{\frac{N}{2}} \left(N - 1 \right)! \Gamma\left(\alpha + 1 \right)}{\Gamma\left(\frac{N}{2} \right) 2^{N+\alpha-1}} \left\{ \left(\sum_{k=0}^{N-1} \frac{(-1)^{N+k-1} (R_1 + R_2)^k (R_2 - R_1)^{N+\alpha-k}}{k! \Gamma\left(N + \alpha + 1 - k \right)} \right) + \right.$$

$$(3.94)$$

$$\left. \left(\sum_{k=0}^{N-1} \frac{(R_1 + R_2)^k (R_2 - R_1)^{\alpha + N - k}}{k! \Gamma\left(\alpha + 1 + N - k \right)} \right) \right\} =$$

$$\frac{\pi^{\frac{N}{2}} \left(N - 1 \right)! \Gamma\left(\alpha + 1 \right)}{\Gamma\left(\frac{N}{2} \right) 2^{N+\alpha-1}} \left\{ \sum_{k=0}^{N-1} \left((-1)^{N+k-1} + 1 \right) \frac{(R_1 + R_2)^k (R_2 - R_1)^{N+\alpha-k}}{k! \Gamma\left(N + \alpha + 1 - k \right)} \right\}.$$

$$(3.95)$$

So that we find

$$R.H.S. \ (3.89) \ (applied \ on \ g_*) = L.H.S. \ (3.89) \ (applied \ on \ f_*), \qquad (3.96)$$

proving sharpness of (3.89).

We have proved the following

Theorem 3.21 Let $f : \overline{A} \to \mathbb{R}$ be radial; that is, there exists a function g such that $f(x) = g(s)$, $s = |x|$, $s \in [R_1, R_2]$, $\forall x \in \overline{A}$, $\alpha > 0$, $m = [\alpha]$. We assume that $g \in C\left([R_1, R_2] \right)$, such that $g \in C_{R_1+}^{\alpha} \left(\left[R_1, \frac{R_1+R_2}{2} \right] \right)$ and $g \in C_{R_2-}^{\alpha} \left(\left[\frac{R_1+R_2}{2}, R_2 \right] \right)$, with $g^{(k)}(R_1) = g^{(k)}(R_2) = 0$, $k = 0, 1, ..., m - 1$. When $0 < \alpha < 1$ the last boundary conditions are void. Then

$$\left| \int_A f(y) \, dy \right| \leq \int_A |f(y)| \, dy \leq \frac{\pi^{\frac{N}{2}}(N-1)!}{\Gamma\left(\frac{N}{2}\right) 2^{\alpha+N-1}} \cdot$$

$$\left\{ \left\| D_{R_1+}^{\alpha} g \right\|_{\infty, \left[R_1, \frac{R_1+R_2}{2}\right]} \left(\sum_{k=0}^{N-1} \frac{(-1)^{N+k-1}(R_1+R_2)^k(R_2-R_1)^{N-k+\alpha}}{k! \, \Gamma(N-k+\alpha+1)} \right) + \right.$$

$$\left. \left\| D_{R_2-}^{\alpha} g \right\|_{\infty, \left[\frac{R_1+R_2}{2}, R_2\right]} \left(\sum_{k=0}^{N-1} \frac{(R_1+R_2)^k(R_2-R_1)^{\alpha+N-k}}{k! \, \Gamma(\alpha+1+N-k)} \right) \right\}.$$

(3.97)

Inequalities (3.97) are sharp, namely they are attained by the radial function f_* : $\overline{A} \to \mathbb{R}$ such that $f_*(x) = g_*(s)$, $s = |x|$, $s \in [R_1, R_2]$, $\forall x \in \overline{A}$, where

$$g_*(s) = \begin{cases} (s-R_1)^{\alpha}, & s \in \left[R_1, \frac{R_1+R_2}{2}\right], \\ (R_2-s)^{\alpha}, & \in \left[\frac{R_1+R_2}{2}, R_2\right]. \end{cases}$$

(3.98)

We continue with

Remark 3.22 Here $\alpha \geq 1$. By (3.81) we get

$$|g(s)| \leq \frac{(s-R_1)^{\alpha-1}}{\Gamma(\alpha)} \left\| D_{R_1+}^{\alpha} g \right\|_{L_1\left(\left[R_1, \frac{R_1+R_2}{2}\right]\right)},$$

(3.99)

for any $s \in \left[R_1, \frac{R_1+R_2}{2}\right]$.

And by (3.83) we derive

$$|g(s)| \leq \frac{(R_2-s)^{\alpha-1}}{\Gamma(\alpha)} \left\| D_{R_2-}^{\alpha} g \right\|_{L_1\left(\left[\frac{R_1+R_2}{2}, R_2\right]\right)},$$

(3.100)

for any $s \in \left[\frac{R_1+R_2}{2}, R_2\right]$.

Hence

$$\int_A |f(y)| \, dy \stackrel{(3.86)}{=}$$

$$\frac{2\pi^{\frac{N}{2}}}{\Gamma\left(\frac{N}{2}\right)} \left\{ \int_{R_1}^{\frac{R_1+R_2}{2}} |g(s)| s^{N-1} ds + \int_{\frac{R_1+R_2}{2}}^{R_2} |g(s)| s^{N-1} ds \right\} \stackrel{\text{(by (3.99) and (3.100))}}{\leq}$$

(3.101)

$$\frac{2\pi^{\frac{N}{2}}}{\Gamma\left(\frac{N}{2}\right)\Gamma\left(\alpha\right)}\left\{\left\|D_{R_1+}^{\alpha}g\right\|_{L_1\left(\left[R_1,\frac{R_1+R_2}{2}\right]\right)}\left(\int_{R_1}^{\frac{R_1+R_2}{2}}(s-R_1)^{\alpha-1}s^{N-1}ds\right)+\right.$$

(3.102)

$$\left.\left\|D_{R_2-}^{\alpha}g\right\|_{L_1\left(\left[\frac{R_1+R_2}{2},R_2\right]\right)}\left(\int_{\frac{R_1+R_2}{2}}^{R_2}(R_2-s)^{\alpha-1}s^{N-1}ds\right)\right\}=$$

$$\frac{\pi^{\frac{N}{2}}(N-1)!}{\Gamma\left(\frac{N}{2}\right)2^{\alpha+N-2}}\left\{\left\|D_{R_1+}^{\alpha}g\right\|_{L_1\left(\left[R_1,\frac{R_1+R_2}{2}\right]\right)}\right.$$

$$\left(\sum_{k=0}^{N-1}(-1)^{N+k-1}\frac{(R_1+R_2)^k(R_2-R_1)^{N+\alpha-k-1}}{k!\Gamma(N+\alpha-k)}\right)+$$

(3.103)

$$\left.\left\|D_{R_2-}^{\alpha}g\right\|_{L_1\left(\left[\frac{R_1+R_2}{2},R_2\right]\right)}\left(\sum_{k=0}^{N-1}\frac{(R_1+R_2)^k(R_2-R_1)^{N+\alpha-k-1}}{k!\Gamma(N+\alpha-k)}\right)\right\}.$$

We have proved that

Theorem 3.23 *All terms and assumptions here as in Theorem 3.21, but with* $\alpha\geq 1$. *Then*

$$\int_A|f(y)|dy\leq\frac{\pi^{\frac{N}{2}}(N-1)!}{\Gamma\left(\frac{N}{2}\right)2^{\alpha+N-2}}.$$

$$\left\{\left\|D_{R_1+}^{\alpha}g\right\|_{L_1\left(\left[R_1,\frac{R_1+R_2}{2}\right]\right)}\left(\sum_{k=0}^{N-1}(-1)^{N+k-1}\frac{(R_1+R_2)^k(R_2-R_1)^{N+\alpha-k-1}}{k!\Gamma(N+\alpha-k)}\right)+\right.$$

$$\left.\left\|D_{R_2-}^{\alpha}g\right\|_{L_1\left(\left[\frac{R_1+R_2}{2},R_2\right]\right)}\left(\sum_{k=0}^{N-1}\frac{(R_1+R_2)^k(R_2-R_1)^{N+\alpha-k-1}}{k!\Gamma(N+\alpha-k)}\right)\right\}.$$ (3.104)

We continue with

Remark 3.24 Let $p,q>1:\frac{1}{p}+\frac{1}{q}=1$. Let $\alpha>\frac{1}{q}$. By (3.81) we get

$$|g(s)|\leq\frac{(s-R_1)^{\alpha-1+\frac{1}{p}}}{\Gamma(\alpha)(p(\alpha-1)+1)^{\frac{1}{p}}}\left\|D_{R_1+}^{\alpha}g\right\|_{L_q\left(\left[R_1,\frac{R_1+R_2}{2}\right]\right)},$$ (3.105)

for any $s\in\left[R_1,\frac{R_1+R_2}{2}\right]$.

Similarly by (3.83) we derive

$$|g(s)| \leq \frac{(R_2 - s)^{\alpha - 1 + \frac{1}{p}}}{\Gamma(\alpha)(p(\alpha - 1) + 1)^{\frac{1}{p}}} \left\| D_{R_2-}^{\alpha} g \right\|_{L_q\left(\left[\frac{R_1+R_2}{2}, R_2\right]\right)}, \tag{3.106}$$

for any $s \in \left[\frac{R_1+R_2}{2}, R_2\right]$. Hence

$$\int_A |f(y)| \, dy =$$

$$\frac{2\pi^{\frac{N}{2}}}{\Gamma\left(\frac{N}{2}\right)} \left\{ \int_{R_1}^{\frac{R_1+R_2}{2}} |g(s)| s^{N-1} ds + \int_{\frac{R_1+R_2}{2}}^{R_2} |g(s)| s^{N-1} ds \right\} \leq$$

$$\frac{2\pi^{\frac{N}{2}}}{\Gamma\left(\frac{N}{2}\right) \Gamma(\alpha)(p(\alpha-1)+1)^{\frac{1}{p}}} \cdot$$

$$\left\{ \left\| D_{R_1+}^{\alpha} g \right\|_{L_q\left(\left[R_1, \frac{R_1+R_2}{2}\right]\right)} \left(\int_{R_1}^{\frac{R_1+R_2}{2}} (s - R_1)^{\alpha - 1 + \frac{1}{p}} s^{N-1} ds \right) + \right.$$

$$\left. \left\| D_{R_2-}^{\alpha} g \right\|_{L_q\left(\left[\frac{R_1+R_2}{2}, R_2\right]\right)} \left(\int_{\frac{R_1+R_2}{2}}^{R_2} (R_2 - s)^{\alpha - 1 + \frac{1}{p}} s^{N-1} ds \right) \right\} = \tag{3.107}$$

$$\frac{2\pi^{\frac{N}{2}}}{\Gamma\left(\frac{N}{2}\right) \Gamma(\alpha)(p(\alpha-1)+1)^{\frac{1}{p}}} \left\{ \left\| D_{R_1+}^{\alpha} g \right\|_{L_q\left(\left[R_1, \frac{R_1+R_2}{2}\right]\right)} \left(\frac{(N-1)!\Gamma\left(\alpha+\frac{1}{p}\right)}{2^{\alpha+N-\frac{1}{q}}} \right) \cdot \right.$$

$$\left(\sum_{k=0}^{N-1} \frac{(-1)^{N+k-1} (R_1 + R_2)^k (R_2 - R_1)^{N-k+\alpha-\frac{1}{q}}}{k! \Gamma\left(N + \alpha + \frac{1}{p} - k\right)} \right) + \tag{3.108}$$

$$\left\| D_{R_2-}^{\alpha} g \right\|_{L_q\left(\left[\frac{R_1+R_2}{2}, R_2\right]\right)} \left(\frac{(N-1)!\Gamma\left(\alpha+\frac{1}{p}\right)}{2^{\alpha+N-\frac{1}{q}}} \right) \cdot$$

$$\left. \left(\sum_{k=0}^{N-1} \frac{(R_1 + R_2)^k (R_2 - R_1)^{\alpha+N-k-\frac{1}{q}}}{k! \Gamma\left(\alpha + \frac{1}{p} + N - k\right)} \right) \right\} =$$

$$\frac{\pi^{\frac{N}{2}} (N-1)! \Gamma\left(\alpha + \frac{1}{p}\right)}{\Gamma\left(\frac{N}{2}\right) \Gamma(\alpha)(p(\alpha-1)+1)^{\frac{1}{p}} 2^{\alpha+N-\frac{1}{q}-1}} \cdot$$

$$\left\{ \left\| D^{\alpha}_{R_1+} g \right\|_{L_q\left(\left[R_1, \frac{R_1+R_2}{2} \right] \right)} \left(\sum_{k=0}^{N-1} \frac{(-1)^{N+k-1} (R_1 + R_2)^k (R_2 - R_1)^{N+\alpha-k-\frac{1}{q}}}{k! \Gamma \left(N + \alpha + \frac{1}{p} - k \right)} \right) + \right.$$

$$\left. \left\| D^{\alpha}_{R_2-} g \right\|_{L_q\left(\left[\frac{R_1+R_2}{2}, R_2 \right] \right)} \left(\sum_{k=0}^{N-1} \frac{(R_1 + R_2)^k (R_2 - R_1)^{N+\alpha-k-\frac{1}{q}}}{k! \Gamma \left(\alpha + N + \frac{1}{p} - k \right)} \right) \right\}. \qquad (3.109)$$

We have proved

Theorem 3.25 Let $p, q > 1 : \frac{1}{p} + \frac{1}{q} = 1$, $\alpha > \frac{1}{q}$. All terms and assumptions as in Theorem 3.21. Then

$$\int_A |f(y)| \, dy \leq \frac{\pi^{\frac{N}{2}} (N-1)! \Gamma \left(\alpha + \frac{1}{p} \right)}{\Gamma \left(\frac{N}{2} \right) \Gamma (\alpha) (p(\alpha-1)+1)^{\frac{1}{p}} 2^{\alpha+N-\frac{1}{q}-1}} \cdot$$

$$\left\{ \left\| D^{\alpha}_{R_1+} g \right\|_{L_q\left(\left[R_1, \frac{R_1+R_2}{2} \right] \right)} \left(\sum_{k=0}^{N-1} \frac{(-1)^{N+k-1} (R_1 + R_2)^k (R_2 - R_1)^{N+\alpha-k-\frac{1}{q}}}{k! \Gamma \left(N + \alpha + \frac{1}{p} - k \right)} \right) + \right.$$

$$\left. \left\| D^{\alpha}_{R_2-} g \right\|_{L_q\left(\left[\frac{R_1+R_2}{2}, R_2 \right] \right)} \left(\sum_{k=0}^{N-1} \frac{(R_1 + R_2)^k (R_2 - R_1)^{N+\alpha-k-\frac{1}{q}}}{k! \Gamma \left(\alpha + N + \frac{1}{p} - k \right)} \right) \right\}. \qquad (3.110)$$

Combining Theorems 3.21, 3.23, 3.25 we derive

Theorem 3.26 Let any $p, q > 1 : \frac{1}{p} + \frac{1}{q} = 1$. And let $f : \overline{A} \to \mathbb{R}$ be radial; that is, there exists a function g such that $f(x) = g(s)$, $s = |x|$, $s \in [R_1, R_2]$, $\forall x \in \overline{A}$; $\alpha \geq 1$, $m = [\alpha]$. We assume that $g \in C([R_1, R_2])$, such that $g \in C^{\alpha}_{R_1+} \left(\left[R_1, \frac{R_1+R_2}{2} \right] \right)$ and $g \in C^{\alpha}_{R_2-} \left(\left[\frac{R_1+R_2}{2}, R_2 \right] \right)$, with $g^{(k)}(R_1) = g^{(k)}(R_2) = 0$, $k = 0, 1, ..., m-1$. Then

$$\left| \int_A f(y) \, dy \right| \leq \int_A |f(y)| \, dy \leq \min \left\{ \frac{\pi^{\frac{N}{2}} (N-1)!}{\Gamma \left(\frac{N}{2} \right) 2^{\alpha+N-1}} \cdot \right.$$

$$\left\{ \left\| D^{\alpha}_{R_1+} g \right\|_{\infty, \left[R_1, \frac{R_1+R_2}{2} \right]} \left(\sum_{k=0}^{N-1} \frac{(-1)^{N+k-1} (R_1 + R_2)^k (R_2 - R_1)^{N-k+\alpha}}{k! \Gamma (N-k+\alpha+1)} \right) + \right.$$

$$\left. \left. \left\| D^{\alpha}_{R_2-} g \right\|_{\infty, \left[\frac{R_1+R_2}{2}, R_2 \right]} \left(\sum_{k=0}^{N-1} \frac{(R_1 + R_2)^k (R_2 - R_1)^{\alpha+N-k}}{k! \Gamma (\alpha+1+N-k)} \right) \right\} \right\},$$

$$\frac{\pi^{\frac{N}{2}} (N-1)!}{\Gamma\left(\frac{N}{2}\right) 2^{\alpha+N-2}} \left\{ \left\| D_{R_1+}^{\alpha} g \right\|_{L_1\left(\left[R_1, \frac{R_1+R_2}{2}\right]\right)} \cdot \right.$$

$$\left(\sum_{k=0}^{N-1} (-1)^{N+k-1} \frac{(R_1+R_2)^k (R_2-R_1)^{N+\alpha-k-1}}{k! \Gamma(N+\alpha-k)} \right) +$$

$$\left. \left\| D_{R_2-}^{\alpha} g \right\|_{L_1\left(\left[\frac{R_1+R_2}{2}, R_2\right]\right)} \left(\sum_{k=0}^{N-1} \frac{(R_1+R_2)^k (R_2-R_1)^{N+\alpha-k-1}}{k! \Gamma(N+\alpha-k)} \right) \right\},$$

$$\frac{\pi^{\frac{N}{2}} (N-1)! \Gamma\left(\alpha+\frac{1}{p}\right)}{\Gamma\left(\frac{N}{2}\right) \Gamma(\alpha) (p(\alpha-1)+1)^{\frac{1}{p}} 2^{\alpha+N-\frac{1}{q}-1}} \cdot$$

$$\left\{ \left\| D_{R_1+}^{\alpha} g \right\|_{L_q\left(\left[R_1, \frac{R_1+R_2}{2}\right]\right)} \left(\sum_{k=0}^{N-1} \frac{(-1)^{N+k-1} (R_1+R_2)^k (R_2-R_1)^{N+\alpha-k-\frac{1}{q}}}{k! \Gamma\left(N+\alpha+\frac{1}{p}-k\right)} \right) \right.$$

$$\left. \left. + \left\| D_{R_2-}^{\alpha} g \right\|_{L_q\left(\left[\frac{R_1+R_2}{2}, R_2\right]\right)} \left(\sum_{k=0}^{N-1} \frac{(R_1+R_2)^k (R_2-R_1)^{N+\alpha-k-\frac{1}{q}}}{k! \Gamma\left(\alpha+N+\frac{1}{p}-k\right)} \right) \right\} \right\}.$$

$$(3.111)$$

The corresponding estimate on the average follows

Corollary 3.27 *Let all terms and assumptions as in Theorem 3.26. Then*

$$\left| \frac{1}{Vol(A)} \int_A f(y)\,dy \right| \le \frac{1}{Vol(A)} \int_A |f(y)|\,dy \le \left(\frac{N!}{2^{\alpha+N} (R_2^N - R_1^N)} \right) \cdot$$

$$\min \left\{ \left\{ \left\| D_{R_1+}^{\alpha} g \right\|_{\infty, \left[R_1, \frac{R_1+R_2}{2}\right]} \left(\sum_{k=0}^{N-1} \frac{(-1)^{N+k-1} (R_1+R_2)^k (R_2-R_1)^{N-k+\alpha}}{k! \Gamma(N-k+\alpha+1)} \right) \right. \right.$$

$$\left. + \left\| D_{R_2-}^{\alpha} g \right\|_{\infty, \left[\frac{R_1+R_2}{2}, R_2\right]} \left(\sum_{k=0}^{N-1} \frac{(R_1+R_2)^k (R_2-R_1)^{\alpha+N-k}}{k! \Gamma(\alpha+1+N-k)} \right) \right\},$$

$$2 \left\{ \left\| D_{R_1+}^{\alpha} g \right\|_{L_1\left(\left[R_1, \frac{R_1+R_2}{2}\right]\right)} \left(\sum_{k=0}^{N-1} (-1)^{N+k-1} \frac{(R_1+R_2)^k (R_2-R_1)^{N+\alpha-k-1}}{k! \Gamma(N+\alpha-k)} \right) \right.$$

$$\left. + \left\| D_{R_2-}^{\alpha} g \right\|_{L_1\left(\left[\frac{R_1+R_2}{2}, R_2\right]\right)} \left(\sum_{k=0}^{N-1} \frac{(R_1+R_2)^k (R_2-R_1)^{N+\alpha-k-1}}{k! \Gamma(N+\alpha-k)} \right) \right\},$$

$$\frac{\Gamma\left(\alpha+\frac{1}{p}\right)2^{\frac{1}{q}}}{\Gamma(\alpha)(p(\alpha-1)+1)^{\frac{1}{p}}}\cdot$$

$$\left\{\left\|D^{\alpha}_{R_1+}g\right\|_{L_q\left(\left[R_1,\frac{R_1+R_2}{2}\right]\right)}\left(\sum_{k=0}^{N-1}\frac{(-1)^{N+k-1}(R_1+R_2)^k(R_2-R_1)^{N+\alpha-k-\frac{1}{q}}}{k!\Gamma\left(N+\alpha+\frac{1}{p}-k\right)}\right)+\right.$$

$$\left.\left\|D^{\alpha}_{R_2-}g\right\|_{L_q\left(\left[\frac{R_1+R_2}{2},R_2\right]\right)}\left(\sum_{k=0}^{N-1}\frac{(R_1+R_2)^k(R_2-R_1)^{N+\alpha-k-\frac{1}{q}}}{k!\Gamma\left(\alpha+N+\frac{1}{p}-k\right)}\right)\right\}\right\}. \quad (3.112)$$

We need

Definition 3.28 (see [1], p. 287) Let $\alpha>0$, $m=[\alpha]$, $\beta:=\alpha-m$, $f\in C^m\left(\overline{A}\right)$, and A is a spherical shell. Assume that there exists $\frac{\partial^{\alpha}_{R_1+}f(x)}{\partial r^{\alpha}}\in C\left(\overline{A}\right)$, given by

$$\frac{\partial^{\alpha}_{R_1+}f(x)}{\partial r^{\alpha}}:=\frac{1}{\Gamma(1-\beta)}\frac{\partial}{\partial r}\left(\int_{R_1}^r(r-t)^{-\beta}\frac{\partial^m f(t\omega)}{\partial r^m}dt\right), \quad (3.113)$$

where $x\in\overline{A}$; that is, $x=r\omega$, $r\in[R_1,R_2]$, and $\omega\in S^{N-1}$.

We call $\frac{\partial^{\alpha}_{R_1+}f}{\partial r^{\alpha}}$ the left radial generalised fractional derivative of f of order α.

We also need to introduce

Definition 3.29 Let $\alpha>0$, $m=[\alpha]$, $\beta:=\alpha-m$, $f\in C^m\left(\overline{A}\right)$, and A is a spherical shell. Assume that there exists $\frac{\partial^{\alpha}_{R_2-}f(x)}{\partial r^{\alpha}}\in C\left(\overline{A}\right)$, given by

$$\frac{\partial^{\alpha}_{R_2-}f(x)}{\partial r^{\alpha}}:=(-1)^{m-1}\frac{1}{\Gamma(1-\beta)}\frac{\partial}{\partial r}\left(\int_r^{R_2}(t-r)^{-\beta}\frac{\partial^m f(t\omega)}{\partial r^m}dt\right), \quad (3.114)$$

where $x\in\overline{A}$; that is, $x=r\omega$, $r\in[R_1,R_2]$, and $\omega\in S^{N-1}$.

We call $\frac{\partial^{\alpha}_{R_2-}f}{\partial r^{\alpha}}$ the right radial generalised fractional derivative of f of order α.

We present

Theorem 3.30 *Let the sperical shells* $A:=B(0,R_2)-\overline{B(0,R_1)}$, $0<R_1<R_2$, $A\subseteq\mathbb{R}^N$, $N\geq 2$; $A_1:=B\left(0,\frac{R_1+R_2}{2}\right)-\overline{B(0,R_1)}$, $A_2:=B(0,R_2)-\overline{B\left(0,\frac{R_1+R_2}{2}\right)}$. *Let* $f\in C\left(\overline{A}\right)$, *not necessarily radial,* $\alpha>0$, $m=[\alpha]$. *Assume that* $\frac{\partial^{\alpha}_{R_1+}f}{\partial r^{\alpha}}\in C\left(\overline{A_1}\right)$, $\frac{\partial^{\alpha}_{R_2-}f}{\partial r^{\alpha}}\in C\left(\overline{A_2}\right)$. *For each* $\omega\in S^{N-1}$, *we assume further that* $f(\cdot\omega)\in C^{\alpha}_{R_1+}\left(\left[R_1,\frac{R_1+R_2}{2}\right]\right)$ *and* $f(\cdot\omega)\in C^{\alpha}_{R_2-}\left(\left[\frac{R_1+R_2}{2},R_2\right]\right)$, *with*

$\frac{\partial^k f(R_1\omega)}{\partial r^k} = \frac{\partial^k f(R_2\omega)}{\partial r^k} = 0$, $k = 0, 1, ..., m - 1$. When $0 < \alpha < 1$ *the last boundary conditions are void. Then*

(i)

$$\left| \int_A f(y)\, dy \right| \le \int_A |f(y)|\, dy \le \frac{\pi^{\frac{N}{2}}(N-1)!}{\Gamma\left(\frac{N}{2}\right) 2^{\alpha+N-1}} \cdot$$

$$\left\{ \left\| \frac{\partial^\alpha_{R_1+} f}{\partial r^\alpha} \right\|_{\infty, \overline{A_1}} \left(\sum_{k=0}^{N-1} \frac{(-1)^{N+k-1}(R_1+R_2)^k(R_2-R_1)^{N+\alpha-k}}{k!\Gamma(N+\alpha+1-k)} \right) + \right.$$

$$\left. \left\| \frac{\partial^\alpha_{R_2-} f}{\partial r^\alpha} \right\|_{\infty, \overline{A_2}} \left(\sum_{k=0}^{N-1} \frac{(R_1+R_2)^k(R_2-R_1)^{N+\alpha-k}}{k!\Gamma(N+\alpha+1-k)} \right) \right\}, \tag{3.115}$$

and

(ii)

$$\left| \frac{1}{Vol(A)} \int_A f(y)\, dy \right| \le \frac{1}{Vol(A)} \int_A |f(y)|\, dy \le \left(\frac{N!}{2^{\alpha+N}\left(R_2^N - R_1^N\right)} \right) \cdot \tag{3.116}$$

$$\left\{ \left\| \frac{\partial^\alpha_{R_1+} f}{\partial r^\alpha} \right\|_{\infty, \overline{A_1}} \left(\sum_{k=0}^{N-1} \frac{(-1)^{N+k-1}(R_1+R_2)^k(R_2-R_1)^{N+\alpha-k}}{k!\Gamma(N+\alpha+1-k)} \right) + \right.$$

$$\left. \left\| \frac{\partial^\alpha_{R_2-} f}{\partial r^\alpha} \right\|_{\infty, \overline{A_2}} \left(\sum_{k=0}^{N-1} \frac{(R_1+R_2)^k(R_2-R_1)^{N+\alpha-k}}{k!\Gamma(N+\alpha+1-k)} \right) \right\}.$$

Proof By (3.86)–(3.88) we get

$$\int_{R_1}^{R_2} |g(s)| s^{N-1} ds \le \left(\frac{\Gamma\left(\frac{N}{2}\right)}{2\pi^{\frac{N}{2}}} \right) \left(\frac{\pi^{\frac{N}{2}}(N-1)!}{\Gamma\left(\frac{N}{2}\right) 2^{\alpha+N-1}} \right). \tag{3.117}$$

$$\left\{ \left\| D^\alpha_{R_1+} g \right\|_{\infty, \left[R_1, \frac{R_1+R_2}{2}\right]} \left(\sum_{k=0}^{N-1} \frac{(-1)^{N+k-1}(R_1+R_2)^k(R_2-R_1)^{N+\alpha-k}}{k!\Gamma(N+\alpha+1-k)} \right) + \right.$$

$$\left. \left\| D^\alpha_{R_2-} g \right\|_{\infty, \left[\frac{R_1+R_2}{2}, R_2\right]} \left[\sum_{k=0}^{N-1} \frac{(R_1+R_2)^k(R_2-R_1)^{N+\alpha-k}}{k!\Gamma(N+\alpha+1-k)} \right] \right\}.$$

For fixed $\omega \in S^{N-1}$, $f(\cdot\omega)$ sets like a radial function on $\overline{A}$. Thus plugging $f(\cdot\omega)$ into (3.117), we get

$$\int_{R_1}^{R_2} |f(s\omega)| \, s^{N-1} ds \le \left(\frac{\Gamma\left(\frac{N}{2}\right)}{2\pi^{\frac{N}{2}}} \right) \left(\frac{\pi^{\frac{N}{2}} (N-1)!}{\Gamma\left(\frac{N}{2}\right) 2^{\alpha+N-1}} \right).$$ (3.118)

$$\left\{ \left\| \frac{\partial_{R_1+}^{\alpha} f}{\partial r^{\alpha}} \right\|_{\infty, \overline{A_1}} \left(\sum_{k=0}^{N-1} \frac{(-1)^{N+k-1} (R_1+R_2)^k (R_2-R_1)^{N+\alpha-k}}{k! \Gamma(N+\alpha+1-k)} \right) + \right.$$

$$\left. \left\| \frac{\partial_{R_2-}^{\alpha} f}{\partial r^{\alpha}} \right\|_{\infty, \overline{A_2}} \left(\sum_{k=0}^{N-1} \frac{(R_1+R_2)^k (R_2-R_1)^{N+\alpha-k}}{k! \Gamma(N+\alpha+1-k)} \right) \right\} =: \gamma_1.$$

Therefore by (3.76) and (3.118) we derive

$$\int_A |f(y)| \, dy = \int_{S^{N-1}} \left(\int_{R_1}^{R_2} |f(s\omega)| \, s^{N-1} ds \right) d\omega \le$$

$$\gamma_1 \int_{S^{N-1}} d\omega = \gamma_1 \frac{2\pi^{\frac{N}{2}}}{\Gamma\left(\frac{N}{2}\right)} = \left(\frac{\pi^{\frac{N}{2}} (N-1)!}{\Gamma\left(\frac{N}{2}\right) 2^{\alpha+N-1}} \right).$$ (3.119)

$$\left\{ \left\| \frac{\partial_{R_1+}^{\alpha} f}{\partial r^{\alpha}} \right\|_{\infty, \overline{A_1}} \left(\sum_{k=0}^{N-1} \frac{(-1)^{N+k-1} (R_1+R_2)^k (R_2-R_1)^{N+\alpha-k}}{k! \Gamma(N+\alpha+1-k)} \right) + \right.$$

$$\left. \left\| \frac{\partial_{R_2-}^{\alpha} f}{\partial r^{\alpha}} \right\|_{\infty, \overline{A_2}} \left(\sum_{k=0}^{N-1} \frac{(R_1+R_2)^k (R_2-R_1)^{N+\alpha-k}}{k! \Gamma(N+\alpha+1-k)} \right) \right\},$$

proving the claims of the theorem. ■

We give also

Theorem 3.31 *Let* $f \in C\left(\overline{A}\right)$, *not necessarily radial,* $\alpha \ge 1$, $m = [\alpha]$. *For each* $\omega \in S^{N-1}$, *we assume that* $f(\cdot\omega) \in C_{R_1+}^{\alpha}\left(\left[R_1, \frac{R_1+R_2}{2}\right]\right)$ *and* $f(\cdot\omega) \in C_{R_2-}^{\alpha}\left(\left[\frac{R_1+R_2}{2}, R_2\right]\right)$, *with* $\frac{\partial^k f(R_1\omega)}{\partial r^k} = \frac{\partial^k f(R_2\omega)}{\partial r^k} = 0$, $k = 0, 1, ..., m-1$. *We further assume*

$$\left\| \frac{\partial_{R_1+}^{\alpha} f(\cdot\omega)}{\partial r^{\alpha}} \right\|_{L_1\left(\left[R_1, \frac{R_1+R_2}{2}\right]\right)}, \left\| \frac{\partial_{R_2-}^{\alpha} f(\cdot\omega)}{\partial r^{\alpha}} \right\|_{L_1\left(\left[\frac{R_1+R_2}{2}, R_2\right]\right)} \le \Psi_1,$$ (3.120)

for every $\omega \in S^{N-1}$, *where* $\Psi_1 > 0$.
    Then
(i)

$$\int_A |f(y)| \, dy \le \frac{\Psi_1 \pi^{\frac{N}{2}} (N-1)!}{\Gamma\left(\frac{N}{2}\right) 2^{\alpha+N-2}}.$$ (3.121)

$$\left\{ \left(\sum_{k=0}^{N-1} \frac{(-1)^{N+k-1} (R_1 + R_2)^k (R_2 - R_1)^{N+\alpha-k-1}}{k!\,\Gamma\,(N+\alpha-k)} \right) + \right.$$

$$\left. \left(\sum_{k=0}^{N-1} \frac{(R_1 + R_2)^k (R_2 - R_1)^{N+\alpha-k-1}}{k!\,\Gamma\,(N+\alpha-k)} \right) \right\},$$

and
(ii)

$$\frac{1}{Vol\,(A)} \int_A |f\,(y)|\,dy \le \frac{\Psi_1 N!}{2^{\alpha+N-1} \left(R_2^N - R_1^N\right)}. \qquad (3.122)$$

$$\left\{ \left(\sum_{k=0}^{N-1} \frac{(-1)^{N+k-1} (R_1 + R_2)^k (R_2 - R_1)^{N+\alpha-k-1}}{k!\,\Gamma\,(N+\alpha-k)} \right) + \right.$$

$$\left. \left(\sum_{k=0}^{N-1} \frac{(R_1 + R_2)^k (R_2 - R_1)^{N+\alpha-k-1}}{k!\,\Gamma\,(N+\alpha-k)} \right) \right\}.$$

Proof Similar to Theorem 3.30, using (3.101)–(3.103). ∎

We finish with

Theorem 3.32 *Let* $f \in C\left(\overline{A}\right)$, *not necessarily radial,* $\alpha > \frac{1}{q}$, *where* $p, q > 1 : \frac{1}{p} + \frac{1}{q} = 1$, $m = [\alpha]$. *For each* $\omega \in S^{N-1}$, *we assume that* $f\,(\cdot\omega) \in C_{R_1+}^{\alpha}\left(\left[R_1, \frac{R_1+R_2}{2}\right]\right)$ *and* $f\,(\cdot\omega) \in C_{R_2-}^{\alpha}\left(\left[\frac{R_1+R_2}{2}, R_2\right]\right)$, *with* $\frac{\partial^k f(R_1\omega)}{\partial r^k} = \frac{\partial^k f(R_2\omega)}{\partial r^k} = 0$, $k = 0, 1, ..., m-1$. *When* $\frac{1}{q} < \alpha < 1$ *the last boundary conditions is void. We further assume*

$$\left\| \frac{\partial_{R_1+}^{\alpha} f\,(\cdot\omega)}{\partial r^{\alpha}} \right\|_{L_q\left(\left[R_1, \frac{R_1+R_2}{2}\right]\right)}, \left\| \frac{\partial_{R_2-}^{\alpha} f\,(\cdot\omega)}{\partial r^{\alpha}} \right\|_{L_q\left(\left[\frac{R_1+R_2}{2}, R_2\right]\right)} \le \Psi_2, \qquad (3.123)$$

for every $\omega \in S^{N-1}$, *where* $\Psi_2 > 0$.
   Then
(i)

$$\int_A |f\,(y)|\,dy \le \frac{\Psi_2 \pi^{\frac{N}{2}} (N-1)!\,\Gamma\left(\alpha + \frac{1}{p}\right)}{\Gamma\left(\frac{N}{2}\right) \Gamma\,(\alpha)\,(p\,(\alpha-1)+1)^{\frac{1}{p}}\,2^{\alpha+N-\frac{1}{q}-1}}. \qquad (3.124)$$

$$\left\{ \left(\sum_{k=0}^{N-1} \frac{(-1)^{N+k-1} (R_1 + R_2)^k (R_2 - R_1)^{N+\alpha-k-\frac{1}{q}}}{k!\,\Gamma\left(N+\alpha+\frac{1}{p}-k\right)} \right) + \right.$$

$$\left(\sum_{k=0}^{N-1} \frac{(R_1 + R_2)^k (R_2 - R_1)^{N+\alpha-k-\frac{1}{q}}}{k!\Gamma\left(\alpha + N + \frac{1}{p} - k\right)}\right)\right\},$$

and
(ii)

$$\frac{1}{Vol(A)} \int_A |f(y)| \, dy \leq \frac{N!\Gamma\left(\alpha + \frac{1}{p}\right)\Psi_2}{2^{\alpha+N-\frac{1}{q}}\left(R_2^N - R_1^N\right)\Gamma(\alpha)(p(\alpha-1)+1)^{\frac{1}{p}}} \cdot \quad (3.125)$$

$$\left\{\left(\sum_{k=0}^{N-1} \frac{(-1)^{N+k-1}(R_1 + R_2)^k (R_2 - R_1)^{N+\alpha-k-\frac{1}{q}}}{k!\Gamma\left(N + \alpha + \frac{1}{p} - k\right)}\right) + \right.$$

$$\left.\left(\sum_{k=0}^{N-1} \frac{(R_1 + R_2)^k (R_2 - R_1)^{N+\alpha-k-\frac{1}{q}}}{k!\Gamma\left(\alpha + N + \frac{1}{p} - k\right)}\right)\right\}.$$

Proof Similar to Theorem 3.30, using (3.107)–(3.109). ■

References

1. G.A. Anastassiou, *Fractional Differentiation Inequalities* (Springer, New York, 2009)
2. G.A. Anastassiou, On right fractional calculus. Chaos, Solitons Fractals **42**, 365–376 (2009)
3. G.A. Anastassiou, *Balanced Canavati type fractional Opial inequalities*, J. Appl. Funct. Anal. **9**(3–4), 230–238 (2014)
4. G.A. Anastassiou, Multivariate generalised fractional Polya type integral inequalities. Stud. Math. Babes Bolyai **58**(3), 297–323 (2013)
5. G.A. Anastassiou, Fractional Polya type integral inequality. J. Comput. Anal. Appl. **17**(4), 736–742 (2014)
6. J.A. Canavati, The Riemann-Liouville integral. Nieuw Archief Voor Wiskunde **5**(1), 53–75 (1987)
7. A.M.A. El-Sayed, M. Gaber, On the finite Caputo and finite Riesz derivatives. Electron. J. Theor. Phys. **3**(12), 81–95 (2006)
8. G.S. Frederico, D.F.M. Torres, Fractional optimal control in the sense of Caputo and the fractional Noether's theorem. Int. Math. Forum **3**(10), 479–493 (2008)
9. R. Gorenflo, F. Mainardi, Essentials of fractional calculus (Maphysto Center, 2000), http://www.maphysto.dk/oldpages/events/LevyCAC2000/MainardiNotes/fm2k0a.ps
10. G. Polya, Ein mittelwertsatz für Funktionen mehrerer Veränderlichen. Tohoku Math. J. **19**, 1–3 (1921)
11. G. Polya, G. Szegö, *Aufgaben und Lehrs ätze aus der Analysis* (Springer, Berlin, 1925). (German)
12. G. Polya, G. Szegö, *Problems and Theorems in Analysis*, vol. I (Springer, Berlin, 1972)
13. G. Polya, G. Szegö, *Problems and Theorems in Analysis*, vol. I, Chinese Edition (Chinese Academy of Sciences, Beijing, 1984)

14. F. Qi, Polya type integral inequalities: origin, variants, proofs, refinements, generalizations, equivalences, and applications. RGMIA Res. Rep. Coll. **16**(2013), article no. 20. http://rgmia.org/v16.php
15. W. Rudin, *Real and Complex Analysis* (McGraw Hill, London, 1970)
16. S.G. Samko, A.A. Kilbas, O.I. Marichev, *Fractional Integrals and Derivatives, Theory and Applications*, (Gordon and Breach, Amsterdam, 1993) [English translation from the Russian, Integrals and Derivatives of Fractional Order and Some of Their Applications (Nauka i Tekhnika, Minsk, 1987)]
17. D. Stroock, *A Concise Introduction to the Theory of Integration*, 3rd edn. (Birkhaüser, Boston, 1999)

Chapter 4
Balanced Canavati Fractional Opial Inequalities

Here we present L_p, $p > 1$, fractional Opial type inequalities subject to high order boundary conditions. They involve the right and left Canavati type generalised fractional derivatives. These derivatives are mixed together into the balanced Canavati type generalised fractional derivative. This balanced fractional derivative is introduced and activated here for the first time. It follows [4].

4.1 Introduction

This chapter is inspired by the famous theorem of Opial [11], 1960, which follows

Theorem 4.1 *Let $x(t) \in C^1([0, h])$ be such that $x(0) = x(h) = 0$, and $x(t) > 0$ in $(0, h)$. Then*

$$\int_0^h |x(t) x'(t)| \, dt \leq \frac{h}{4} \int_0^h (x'(t))^2 \, dt. \qquad (4.1)$$

In (4.1), the constant $\frac{h}{4}$ is the best possible. Inequality (4.1) holds as equality for the optimal function

$$x(t) = \begin{cases} ct, & 0 \leq t \leq \frac{h}{2}, \\ c(h-t), & \frac{h}{2} \leq t \leq h, \end{cases}$$

where $c > 0$ is an arbitrary constant.

To prove easier Theorem 4.1, Beesack [5] proved the following well-known Opial type inequality which is used very commonly.

This is another inspiration to our chapter.

© Springer International Publishing Switzerland 2016
G.A. Anastassiou, *Intelligent Comparisons: Analytic Inequalities*,
Studies in Computational Intelligence 609,
DOI 10.1007/978-3-319-21121-3_4

Theorem 4.2 *Let $x(t)$ be absolutely continuous in $[0, a]$, and $x(0) = 0$. Then*

$$\int_0^a \left| x(t) x'(t) \right| dt \le \frac{a}{2} \int_0^a \left(x'(t) \right)^2 dt. \tag{4.2}$$

Inequality (4.2) is sharp, it is attained by $x(t) = ct$, $c > 0$ is an arbitrary constant.

Opial type inequalities are used a lot in proving uniqueness of solutions to differential equations, also to give upper bounds to their solutions.

By themselves have made a great subject of intensive research and there exists a great literature about them.

Typical and great sources on them are the monographs [1, 2].

We define here the balanced Canavati type fractional derivative and we prove related Opial type inequalities subject to boundary conditions.

These have smaller constants than in other Opial inequalities when using traditional fractional derivatives.

4.2 Background

Let $\nu > 0$, $n := [\nu]$ (integral part of ν), and $\alpha := \nu - n$ ($0 < \alpha < 1$). The gamma function Γ is given by $\Gamma(\nu) = \int_0^\infty e^{-t} t^{\nu-1} dt$. Here $[a, b] \subseteq \mathbb{R}$, $x, x_0 \in [a, b]$ such that $x \ge x_0$, where x_0 is fixed. Let $f \in C([a, b])$ and define the left Riemann-Liouville integral

$$\left(J_\nu^{x_0} f \right)(x) := \frac{1}{\Gamma(\nu)} \int_{x_0}^x (x - t)^{\nu-1} f(t) dt, \tag{4.3}$$

$x_0 \le x \le b$. We define the subspace $C_{x_0}^\nu([a.b])$ of $C^n([a, b])$:

$$C_{x_0}^\nu([a, b]) := \left\{ f \in C^n([a, b]) : J_{1-\alpha}^{x_0} f^{(n)} \in C^1([x_0, b]) \right\}. \tag{4.4}$$

For $f \in C_{x_0}^\nu([a, b])$, we define the left generalized ν-fractional derivative of f over $[x_0, b]$ as

$$D_{x_0}^\nu f := \left(J_{1-\alpha}^{x_0} f^{(n)} \right)', \tag{4.5}$$

see [2], p. 24, and Canavati derivative in [6].

Notice that $D_{x_0}^\nu f \in C([x_0, b])$.

We need the following generalization of Taylor's formula at the fractional level, see [2], pp. 8–10, and [6].

Theorem 4.3 *Let* $f \in C_{x_0}^{\nu}([a, b])$, $x_0 \in [a, b]$ *fixed.*

(i) If $\nu \geq 1$ *then*

$$f(x) = f(x_0) + f'(x_0)(x - x_0) + f''(x_0)\frac{(x - x_0)^2}{2} + \ldots + f^{(n-1)}(x_0)\frac{(x - x_0)^{n-1}}{(n-1)!}$$
$$+ \left(J_{\nu}^{x_0} D_{x_0}^{\nu} f\right)(x), \text{ all } x \in [a, b] : x \geq x_0. \tag{4.6}$$

(ii) If $0 < \nu < 1$ *we get*

$$f(x) = \left(J_{\nu}^{x_0} D_{x_0}^{\nu} f\right)(x), \text{ all } x \in [a, b] : x \geq x_0 \tag{4.7}$$

We will use (4.6) and (4.7).

Furthermore we need:

Let $\alpha > 0$, $m = [\alpha]$, $\beta = \alpha - m$, $0 < \beta < 1$, $f \in C([a, b])$, call the right Riemann-Liouville fractional integral operator by

$$\left(J_{b-}^{\alpha} f\right)(x) := \frac{1}{\Gamma(\alpha)} \int_x^b (J - x)^{\alpha-1} f(J) dJ, \tag{4.8}$$

$x \in [a, b]$, see also [3, 7–9, 12]. Define the subspace of functions

$$C_{b-}^{\alpha}([a, b]) := \left\{ f \in C^m([a, b]) : J_{b-}^{1-\beta} f^{(m)} \in C^1([a, b]) \right\}. \tag{4.9}$$

Define the right generalized α-fractional derivative of f over $[a, b]$ as

$$D_{b-}^{\alpha} f := (-1)^{m-1} \left(J_{b-}^{1-\beta} f^{(m)}\right)', \tag{4.10}$$

see [3]. We set $D_{b-}^0 f = f$. Notice that $D_{b-}^{\alpha} f \in C([a, b])$.

From [3], we need the following Taylor fractional formula.

Theorem 4.4 *Let* $f \in C_{b-}^{\alpha}([a, b])$, $\alpha > 0$, $m := [\alpha]$. *Then*

(1) If $\alpha \geq 1$, *we get*

$$f(x) = \sum_{k=0}^{m-1} \frac{f^{(k)}(b_-)}{k!}(x - b)^k + \left(J_{b-}^{\alpha} D_{b-}^{\alpha} f\right)(x), \quad \forall x \in [a, b]. \tag{4.11}$$

(2) If $0 < \alpha < 1$, *we get*

$$f(x) = J_{b-}^{\alpha} D_{b-}^{\alpha} f(x), \quad \forall x \in [a, b]. \tag{4.12}$$

We will use (4.11) and (4.12).

We introduce a new concept

Definition 4.5 Let $f \in C([a,b])$, $x \in [a,b]$, $\alpha > 0$, $m := [\alpha]$. Assume that $f \in C_{b-}^{\alpha}\left(\left[\frac{a+b}{2}, b\right]\right)$ and $f \in C_a^{\alpha}\left(\left[a, \frac{a+b}{2}\right]\right)$. We define the balanced Canavati type fractional derivative by

$$D^{\alpha} f(x) := \begin{cases} D_{b-}^{\alpha} f(x), & \text{for } \frac{a+b}{2} \leq x \leq b, \\ D_a^{\alpha} f(x), & \text{for } a \leq x < \frac{a+b}{2}. \end{cases} \tag{4.13}$$

4.3 Main Result

We give our main result

Theorem 4.6 Let $f \in C([a,b])$, $\alpha > 0$, $m := [\alpha]$. Assume that $f \in C_{b-}^{\alpha}\left(\left[\frac{a+b}{2}, b\right]\right)$ and $f \in C_a^{\alpha}\left(\left[a, \frac{a+b}{2}\right]\right)$. Assume further that

$$f^{(k)}(a) = f^{(k)}(b) = 0, \quad k = 0, 1, \dots, m-1; \tag{4.14}$$

$$p, q > 1 : \frac{1}{p} + \frac{1}{q} = 1, \quad \text{and} \quad \alpha > \frac{1}{q}.$$

(i) Case of $1 < q \leq 2$. Then

$$\int_a^b |f(\omega)| \left| D^{\alpha} f(\omega) \right| d\omega \leq \tag{4.15}$$

$$\frac{2^{-\left(\alpha + \frac{1}{p}\right)} (b-a)^{\left(\frac{p(\alpha-1)+2}{p}\right)}}{\Gamma(\alpha) \left[(p(\alpha-1)+1)(p(\alpha-1)+2) \right]^{\frac{1}{p}}} \left(\int_a^b |D^{\alpha} f(\omega)|^q \, d\omega \right)^{\frac{2}{q}}.$$

(ii) Case of $q > 2$. Then

$$\int_a^b |f(\omega)| \left| D^{\alpha} f(\omega) \right| d\omega \leq \tag{4.16}$$

$$\frac{2^{-\left(\alpha + \frac{1}{q}\right)} (b-a)^{\left(\frac{p(\alpha-1)+2}{p}\right)}}{\Gamma(\alpha) \left[(p(\alpha-1)+1)(p(\alpha-1)+2) \right]^{\frac{1}{p}}} \left(\int_a^b |D^{\alpha} f(\omega)|^q \, d\omega \right)^{\frac{2}{q}}.$$

(iii) When $p = q = 2$, $\alpha > \frac{1}{2}$, then

$$\int_a^b |f(\omega)| \left| D^{\alpha} f(\omega) \right| d\omega \leq \tag{4.17}$$

$$\frac{2^{-\left(\alpha+\frac{1}{2}\right)} (b-a)^\alpha}{\Gamma(\alpha) \left[\sqrt{2\alpha (2\alpha - 1)}\right]} \left(\int_a^b |D^\alpha f(\omega)|^2 d\omega\right).$$

Remark 4.7 Let us say that $\alpha = 1$, then by (4.17) we obtain

$$\int_a^b |f(\omega)| |f'(\omega)| d\omega \leq \frac{(b-a)}{4} \left(\int_a^b (f'(\omega))^2 d\omega\right), \tag{4.18}$$

that is reproving and recovering Opial's inequality (4.1), see [11], see also Olech's result [10].

Proof of Theorem 4.6 Let $x \in \left[a, \frac{a+b}{2}\right]$, we have by assumption $f^{(k)}(a) = 0$, $k = 0, 1, ..., m - 1$ and Theorem 4.3 that

$$f(x) = \frac{1}{\Gamma(\alpha)} \int_a^x (x - \tau)^{\alpha-1} D_a^\alpha f(\tau) d\tau. \tag{4.19}$$

Let $x \in \left[\frac{a+b}{2}, b\right]$, we have by assumption $f^{(k)}(b) = 0$, $k = 0, 1, ..., m - 1$ and Theorem 4.4 that

$$f(x) = \frac{1}{\Gamma(\alpha)} \int_x^b (\tau - x)^{\alpha-1} D_{b-}^\alpha f(\tau) d\tau. \tag{4.20}$$

Using Hölder's inequality on (4.19) we get

$$|f(x)| \leq \frac{1}{\Gamma(\alpha)} \int_a^x (x - \tau)^{\alpha-1} |D_a^\alpha f(\tau)| d\tau \leq$$

$$\frac{1}{\Gamma(\alpha)} \left(\int_a^x \left((x - \tau)^{\alpha-1}\right)^p d\tau\right)^{\frac{1}{p}} \left(\int_a^x |D_a^\alpha f(\tau)|^q d\tau\right)^{\frac{1}{q}} =$$

$$\frac{1}{\Gamma(\alpha)} \frac{(x - a)^{\frac{p(\alpha-1)+1}{p}}}{(p(\alpha-1)+1)^{\frac{1}{p}}} \left(\int_a^x |D_a^\alpha f(\tau)|^q d\tau\right)^{\frac{1}{q}}. \tag{4.21}$$

Set

$$z(x) := \int_a^x |D_a^\alpha f(\tau)|^q d\tau, \quad (z(a) = 0).$$

Then

$$z'(x) = |D_a^\alpha f(x)|^q,$$

and

$$\left|D_a^\alpha f(x)\right| = \left(z'(x)\right)^{\frac{1}{q}}, \text{ all } a \le x \le \frac{a+b}{2}.$$

Therefore by (4.21) we have

$$|f(\omega)|\left|D_a^\alpha f(\omega)\right| \le \frac{1}{\Gamma(\alpha)} \frac{(\omega - a)^{\frac{p(\alpha-1)+1}{p}}}{(p(\alpha-1)+1)^{\frac{1}{p}}} \left(z(\omega) z'(\omega)\right)^{\frac{1}{q}}, \tag{4.22}$$

all $a \le \omega \le x \le \frac{a+b}{2}$.

Next working similarly with (4.20) we obtain

$$|f(x)| \le \frac{1}{\Gamma(\alpha)} \int_x^b (\tau - x)^{\alpha-1} \left|D_{b-}^\alpha f(\tau)\right| d\tau \le$$

$$\frac{1}{\Gamma(\alpha)} \left(\int_x^b \left((\tau - x)^{\alpha-1}\right)^p d\tau\right)^{\frac{1}{p}} \left(\int_x^b \left|D_{b-}^\alpha f(\tau)\right|^q d\tau\right)^{\frac{1}{q}} =$$

$$\frac{1}{\Gamma(\alpha)} \frac{(b - x)^{\frac{p(\alpha-1)+1}{p}}}{(p(\alpha-1)+1)^{\frac{1}{p}}} \left(\int_x^b \left|D_{b-}^\alpha f(\tau)\right|^q d\tau\right)^{\frac{1}{q}}. \tag{4.23}$$

Set

$$\lambda(x) := \int_x^b \left|D_{b-}^\alpha f(\tau)\right|^q d\tau = - \int_b^x \left|D_{b-}^\alpha f(\tau)\right|^q d\tau, \ (\lambda(b) = 0).$$

Then

$$\lambda'(x) = - \left|D_{b-}^\alpha f(x)\right|^q$$

and

$$\left|D_{b-}^\alpha f(x)\right| = \left(-\lambda'(x)\right)^{\frac{1}{q}}, \text{ all } \frac{a+b}{2} \le x \le b.$$

Therefore by (4.23) we have

$$|f(\omega)|\left|D_{b-}^\alpha f(\omega)\right| \le \frac{1}{\Gamma(\alpha)} \frac{(b - \omega)^{\frac{p(\alpha-1)+1}{p}}}{(p(\alpha-1)+1)^{\frac{1}{p}}} \left(-\lambda(\omega) \lambda'(\omega)\right)^{\frac{1}{q}}, \tag{4.24}$$

all $\frac{a+b}{2} \le x \le \omega \le b$.

Next we integrate (4.22) over $[a, x]$ to obtain

$$\int_a^x |f(\omega)| \, |D_a^\alpha f(\omega)| \, d\omega \leq$$

$$\frac{1}{\Gamma(\alpha)(p(\alpha-1)+1)^{\frac{1}{p}}} \int_a^x (\omega-a)^{\frac{p(\alpha-1)+1}{p}} (z(\omega) z'(\omega))^{\frac{1}{q}} \, d\omega \leq$$

$$\frac{1}{\Gamma(\alpha)(p(\alpha-1)+1)^{\frac{1}{p}}} \left(\int_a^x (\omega-a)^{p(\alpha-1)+1} \, d\omega \right)^{\frac{1}{p}} \left(\int_a^x z(\omega) z'(\omega) \, d\omega \right)^{\frac{1}{q}} =$$

$$\frac{1}{\Gamma(\alpha)(p(\alpha-1)+1)^{\frac{1}{p}}} \frac{(x-a)^{\frac{p(\alpha-1)+2}{p}}}{(p(\alpha-1)+2)^{\frac{1}{p}}} \frac{z(x)^{\frac{2}{q}}}{2^{\frac{1}{q}}} =$$

$$\frac{2^{-\frac{1}{q}} (x-a)^{\frac{p(\alpha-1)+2}{p}}}{\Gamma(\alpha)[(p(\alpha-1)+1)(p(\alpha-1)+2)]^{\frac{1}{p}}} \left(\int_a^x |D_a^\alpha f(\omega)|^q \, d\omega \right)^{\frac{2}{q}}. \quad (4.25)$$

So we have proved

$$\int_a^x |f(\omega)| \, |D_a^\alpha f(\omega)| \, d\omega \leq$$

$$\frac{2^{-\frac{1}{q}} (x-a)^{\frac{p(\alpha-1)+2}{p}}}{\Gamma(\alpha)[(p(\alpha-1)+1)(p(\alpha-1)+2)]^{\frac{1}{p}}} \left(\int_a^x |D_a^\alpha f(\omega)|^q \, d\omega \right)^{\frac{2}{q}}, \quad (4.26)$$

for all $a \leq x \leq \frac{a+b}{2}$.

By (4.26) we get

$$\int_a^{\frac{a+b}{2}} |f(\omega)| \, |D_a^\alpha f(\omega)| \, d\omega \leq$$

$$\frac{(b-a)^{\frac{(p(\alpha-1)+2)}{p}} 2^{-\left[\frac{p(\alpha-1)+2}{p}+\frac{1}{q}\right]}}{\Gamma(\alpha)[(p(\alpha-1)+1)(p(\alpha-1)+2)]^{\frac{1}{p}}} \left(\int_a^{\frac{a+b}{2}} |D_a^\alpha f(\omega)|^q \, d\omega \right)^{\frac{2}{q}}. \quad (4.27)$$

Similarly we integrate (4.24) over $[x, b]$ to obtain

$$\int_x^b |f(\omega)| \, |D_{b-}^\alpha f(\omega)| \, d\omega \leq$$

$$\frac{1}{\Gamma(\alpha)(p(\alpha-1)+1)^{\frac{1}{p}}} \int_x^b (b-\omega)^{\frac{p(\alpha-1)+1}{p}} (-\lambda(\omega) \lambda'(\omega))^{\frac{1}{q}} \, d\omega \leq$$

$$\frac{1}{\Gamma(\alpha)(p(\alpha-1)+1)^{\frac{1}{p}}}\left(\int_x^b (b-\omega)^{p(\alpha-1)+1}d\omega\right)^{\frac{1}{p}}\left(\int_x^b -\lambda(\omega)\lambda'(\omega)d\omega\right)^{\frac{1}{q}}=$$

$$\frac{1}{\Gamma(\alpha)(p(\alpha-1)+1)^{\frac{1}{p}}}\frac{(b-x)^{\frac{p(\alpha-1)+2}{p}}}{(p(\alpha-1)+2)^{\frac{1}{p}}}\frac{(\lambda(x))^{\frac{2}{q}}}{2^{\frac{1}{q}}}. \tag{4.28}$$

We have proved that

$$\int_x^b |f(\omega)|\left|D_{b-}^\alpha f(\omega)\right|d\omega \le$$

$$\frac{2^{-\frac{1}{q}}(b-x)^{\frac{p(\alpha-1)+2}{p}}}{\Gamma(\alpha)[(p(\alpha-1)+1)(p(\alpha-1)+2)]^{\frac{1}{p}}}\left(\int_x^b \left|D_{b-}^\alpha f(\omega)\right|^q d\omega\right)^{\frac{2}{q}}, \tag{4.29}$$

for all $\frac{a+b}{2}\le x \le b$.

By (4.29) we get

$$\int_{\frac{a+b}{2}}^b |f(\omega)|\left|D_{b-}^\alpha f(\omega)\right|d\omega \le$$

$$\frac{(b-a)^{\frac{(p(\alpha-1)+2)}{p}}2^{-\left[\frac{p(\alpha-1)+2}{p}+\frac{1}{q}\right]}}{\Gamma(\alpha)[(p(\alpha-1)+1)(p(\alpha-1)+2)]^{\frac{1}{p}}}\left(\int_{\frac{a+b}{2}}^b \left|D_{b-}^\alpha f(\omega)\right|^q d\omega\right)^{\frac{2}{q}}. \tag{4.30}$$

Adding (4.27) and (4.30) we get

$$\int_a^b |f(\omega)|\left|D^\alpha f(\omega)\right|d\omega \le \frac{2^{-\left(\alpha+\frac{1}{p}\right)}(b-a)^{\left(\frac{p(\alpha-1)+2}{p}\right)}}{\Gamma(\alpha)[(p(\alpha-1)+1)(p(\alpha-1)+2)]^{\frac{1}{p}}}\cdot$$

$$\left[\left(\int_a^{\frac{a+b}{2}} \left|D_a^\alpha f(\omega)\right|^q d\omega\right)^{\frac{2}{q}}+\left(\int_{\frac{a+b}{2}}^b \left|D_{b-}^\alpha f(\omega)\right|^q d\omega\right)^{\frac{2}{q}}\right] =: (*.) \tag{4.31}$$

Assume $1 < q \le 2$, then $\frac{2}{q} \ge 1$.

Therefore we get

$$(*) \le \frac{2^{-\left(\alpha+\frac{1}{p}\right)}(b-a)^{\left(\frac{p(\alpha-1)+2}{p}\right)}}{\Gamma(\alpha)[(p(\alpha-1)+1)(p(\alpha-1)+2)]^{\frac{1}{p}}}\cdot$$

$$\left[\int_a^{\frac{a+b}{2}} \left| D_a^\alpha f(\omega) \right|^q d\omega + \int_{\frac{a+b}{2}}^b \left| D_{b-}^\alpha f(\omega) \right|^q d\omega \right]^{\frac{2}{q}} = \tag{4.32}$$

$$\frac{2^{-\left(\alpha+\frac{1}{p}\right)} (b-a)^{\left(\frac{p(\alpha-1)+2}{p}\right)}}{\Gamma(\alpha) \left[(p(\alpha-1)+1)(p(\alpha-1)+2) \right]^{\frac{1}{p}}} \left(\int_a^b \left| D^\alpha f(\omega) \right|^q d\omega \right)^{\frac{2}{q}}. \tag{4.33}$$

So for $1 < q \le 2$ we have proved (4.15).

Assume now $q > 2$, then $0 < \frac{2}{q} < 1$.

Therefore we get

$$(*) \le \frac{2^{-\left(\alpha+\frac{1}{p}\right)} (b-a)^{\left(\frac{p(\alpha-1)+2}{p}\right)} 2^{1-\frac{2}{q}}}{\Gamma(\alpha) \left[(p(\alpha-1)+1)(p(\alpha-1)+2) \right]^{\frac{1}{p}}} \cdot$$

$$\left[\int_a^{\frac{a+b}{2}} \left| D_a^\alpha f(\omega) \right|^q d\omega + \int_{\frac{a+b}{2}}^b \left| D_{b-}^\alpha f(\omega) \right|^q d\omega \right]^{\frac{2}{q}} =$$

$$\frac{2^{-\left(\alpha+\frac{1}{q}\right)} (b-a)^{\left(\frac{p(\alpha-1)+2}{p}\right)}}{\Gamma(\alpha) \left[(p(\alpha-1)+1)(p(\alpha-1)+2) \right]^{\frac{1}{p}}} \left(\int_a^b \left| D^\alpha f(\omega) \right|^q d\omega \right)^{\frac{2}{q}}. \tag{4.34}$$

So when $q > 2$ we have established (4.16).

(iii) The case of $p = q = 2$, see (4.17), is obvious, it derives from (4.15) immediately. ∎

References

1. R.P. Agarwal, P.Y.H. Pang, *Opial Inequalities with Applications in Differential and Difference Equations* (Kluwer, Dordrecht, 1995)
2. G.A. Anastassiou, *Fractional Differentiation Inequalities*, Research Monograph (Springer, New York, 2009)
3. G.A. Anastassiou, On right fractional calculus. Chaos, Solitons Fractals **42**, 365–376 (2009)
4. G.A. Anastassiou, Balanced Canavati type fractional opial inequalities. J. Appl. Func. Anal. **9**(3-4), 230-238 (2014)
5. P.R. Beesack, On an integral inequality of Z. Opial. Trans. Am. Math. Soc. **104**, 470–475 (1962)
6. J.A. Canavati, The Riemann-Liouville Integral. Nieuw Archief Voor Wiskunde **5**(1), 53–75 (1987)
7. A.M.A. El-Sayed, M. Gaber, On the finite Caputo and finite Riesz derivatives. Electron. J. Theor. Phys. **3**(12), 81–95 (2006)
8. G.S. Frederico, D.F.M. Torres, Fractional optimal control in the sense of Caputo and the fractional Noether's theorem. Int. Math. Forum **3**(10), 479–493 (2008)
9. R. Gorenflo, F. Mainardi, Essentials of fractional calculus (Maphysto Center, 2000), http://www.maphysto.dk/oldpages/events/LevyCAC2000/MainardiNotes/fm2k0a.ps

10. C. Olech, A simple proof of a certain result of Z. Opial. Ann. Polon. Math. **8**, 61–63 (1960)
11. Z. Opial, Sur une inegalite. Ann. Polon. Math. **8**, 29–32 (1960)
12. S.G. Samko, A.A. Kilbas, O.I. Marichev, *Fractional Integrals and Derivatives, Theory and Applications*, (Gordon and Breach, Amsterdam, 1993) [English translation from the Russian, Integrals and Derivatives of Fractional Order and Some of Their Applications (Nauka i Tekhnika, Minsk, 1987)]

Chapter 5
Fractional Representation Formulae Using Initial Conditions and Fractional Ostrowski Inequalities

Here we present very general fractional representation formulae for a function in terms of the fractional Riemann-Liouville integrals of different orders of the function and its ordinary derivatives under initial conditions. Based on these we derive general fractional Ostrowski type inequalities with respect to all basic norms. It follows [2].

5.1 Introduction

Let $f : [a, b] \to \mathbb{R}$ be differentiable on $[a, b]$, and $f' : [a, b] \to \mathbb{R}$ be integrable on $[a, b]$, then the following Montgomery identity holds [3]:

$$f(x) = \frac{1}{b-a} \int_a^b f(t)\, dt + \int_a^b P_1(x, t) f'(t)\, dt, \qquad (5.1)$$

where $P_1(x, t)$ is the Peano kernel

$$P_1(x, t) = \begin{cases} \frac{t-a}{b-a}, & a \leq t \leq x, \\ \frac{t-b}{b-a}, & x < t \leq b, \end{cases} \qquad (5.2)$$

The Riemann-Liouville integral operator of order $\alpha > 0$ with anchor point $a \in \mathbb{R}$ is defined by

$$J_a^\alpha f(x) := \frac{1}{\Gamma(\alpha)} \int_a^x (x - t)^{\alpha - 1} f(t)\, dt, \qquad (5.3)$$

$$J_a^0 f(x) := f(x), x \in [a, b]. \qquad (5.4)$$

Properties of the above operator can be found in [4].

When $\alpha = 1$, J_a^1 reduces to the classical integral.

© Springer International Publishing Switzerland 2016
G.A. Anastassiou, *Intelligent Comparisons: Analytic Inequalities*,
Studies in Computational Intelligence 609,
DOI 10.1007/978-3-319-21121-3_5

In [1] we proved the following fractional representation formula of Montgomery identity type.

Theorem 5.1 *Let* $f : [a, b] \to \mathbb{R}$ *be differentiable on* $[a, b]$, *and* $f' : [a, b] \to \mathbb{R}$ *be integrable on* $[a, b]$, $\alpha \geq 1$, $x \in [a, b)$. *Then*

$$f(x) =$$

$$(b - x)^{1-\alpha} \, \Gamma(\alpha) \left\{ \frac{J_a^\alpha f(b)}{b - a} - J_a^{\alpha-1} (P_1(x, b) f(b)) + J_a^\alpha (P_1(x, b) f'(b)) \right\}.$$

$$(5.5)$$

When $\alpha = 1$ *the last* (5.5) *reduces to classic Montgomery identity* (1).

In this chapter we find higher order fractional representation for $f(x)$, similar to basic (5.5), and from there we derive interesting fractional Ostrowski type inequalities.

5.2 Main Results

Next we give higher order fractional representation of f subject to initial conditions.

Theorem 5.2 *Let* $\alpha > 2$, $x \in [a, b)$ *fixed,* $f : [a, b] \to \mathbb{R}$ *twice differentiable, with* $f'' : [a, b] \to \mathbb{R}$ *integrable on* $[a, b]$. *Assume* $f'(x) = 0$. *Then*

$$f(x) = \frac{(b - x)^{2-\alpha}}{\alpha - 1} \left[-(b - a)^{\alpha-2} f(a) + \Gamma(\alpha) \left\{ \frac{2}{b - a} J_a^{\alpha-1} f(b) - \right. \right. \quad (5.6)$$

$$\left. \left. J_a^{\alpha-2} (P_1(x, b) f(b)) + J_a^\alpha (P_1(x, b) f''(b)) \right\} \right].$$

Proof Let here $\alpha > 2$ and there exists $f'' : [a, b] \to \mathbb{R}$ that is integrable on $[a, b]$. We have

$$\Gamma(\alpha) J_a^\alpha (P_1(x, b) f''(b)) = \int_a^b (b - t)^{\alpha-1} P_1(x, t) f''(t) \, dt = \quad (5.7)$$

$$\int_a^x \left(\frac{t - a}{b - a} \right) (b - t)^{\alpha-1} f''(t) \, dt + \int_x^b \left(\frac{t - b}{b - a} \right) (b - t)^{\alpha-1} f''(t) \, dt =$$

$$\int_a^x \left(\frac{t - a}{b - a} \right) (b - t)^{\alpha-1} f''(t) \, dt + \int_a^b \left(\frac{t - b}{b - a} \right) (b - t)^{\alpha-1} f''(t) \, dt \quad (5.8)$$

$$-\int_a^x \left(\frac{t-b}{b-a}\right) (b-t)^{\alpha-1} f''(t)\, dt =$$

$$\int_a^x (b-t)^{\alpha-1} f''(t)\, dt - \frac{1}{(b-a)} \int_a^b (b-t)^{\alpha} f''(t)\, dt.$$

That is

$$\Gamma(\alpha)\, J_a^{\alpha}\left(P_1(x,b)\, f''(b)\right) =$$

$$\int_a^x (b-t)^{\alpha-1} f''(t)\, dt - \frac{1}{(b-a)} \int_a^b (b-t)^{\alpha} f''(t)\, dt =: (\xi_1). \qquad (5.9)$$

Next we use integration by parts, plus the assumption $f'(x) = 0$. We have

$$\int_a^x (b-t)^{\alpha-1} f''(t)\, dt = \int_a^x (b-t)^{\alpha-1}\, df'(t) =$$

$$-(b-a)^{\alpha-1} f'(a) - \int_a^x f'(t)\, d(b-t)^{\alpha-1} = -(b-a)^{\alpha-1} f'(a) \qquad (5.10)$$

$$+(\alpha-1)\int_a^x (b-t)^{\alpha-2}\, df(t) = -(b-a)^{\alpha-1} f'(a)$$

$$+(\alpha-1)\left[(b-x)^{\alpha-2} f(x) - (b-a)^{\alpha-2} f(a) - \int_a^x f(t)\, d(b-t)^{\alpha-2}\right] =$$

$$-(b-a)^{\alpha-1} f'(a) + (\alpha-1)(b-x)^{\alpha-2} f(x) - (\alpha-1)(b-a)^{\alpha-2} f(a) + \qquad (5.11)$$

$$(\alpha-1)(\alpha-2)\int_a^x (b-t)^{\alpha-3} f(t)\, dt.$$

That is

$$\int_a^x (b-t)^{\alpha-1} f''(t)\, dt = -(b-a)^{\alpha-1} f'(a) + (\alpha-1)(b-x)^{\alpha-2} f(x) -$$

$$(\alpha-1)(b-a)^{\alpha-2} f(a) + (\alpha-1)(\alpha-2)\int_a^x (b-t)^{\alpha-3} f(t)\, dt =: (\lambda_1).$$

$$(5.12)$$

Next we observe

$$-\frac{1}{(b-a)}\int_a^b (b-t)^{\alpha} f''(t)\, dt = -\frac{1}{(b-a)}\left[\int_a^b (b-t)^{\alpha}\, df'(t)\right] =$$

$$-\frac{1}{(b-a)}\left[-(b-a)^\alpha \, f'(a) + \alpha \int_a^b (b-t)^{\alpha-1} \, df(t)\right] = -\frac{1}{(b-a)} \cdot$$

$$\left[-(b-a)^\alpha \, f'(a) - \alpha \, (b-a)^{\alpha-1} \, f(a) + \alpha \, (\alpha-1) \int_a^b (b-t)^{\alpha-2} \, f(t) \, dt\right] =$$

(5.13)

$$(b-a)^{\alpha-1} \, f'(a) + \alpha \, (b-a)^{\alpha-2} \, f(a) - \frac{\alpha \, (\alpha-1)}{(b-a)} \int_a^b (b-t)^{\alpha-2} \, f(t) \, dt.$$

That is

$$-\frac{1}{(b-a)} \int_a^b (b-t)^\alpha \, f''(t) \, dt = (b-a)^{\alpha-1} \, f'(a) +$$

$$\alpha \, (b-a)^{\alpha-2} \, f(a) - \frac{\alpha \, (\alpha-1)}{(b-a)} \int_a^b (b-t)^{\alpha-2} \, f(t) \, dt =: (\lambda_2) \,.$$

(5.14)

We have that

$$(\xi_1) = (\lambda_1) + (\lambda_2) \,.$$

Thus

$$\Gamma(\alpha) \, J_a^\alpha \left(P_1(x,b) \, f''(b)\right) = (\alpha-1)(b-x)^{\alpha-2} \, f(x) + (b-a)^{\alpha-2} \, f(a) +$$

(5.15)

$$(\alpha-1)(\alpha-2) \int_a^x (b-t)^{\alpha-3} \, f(t) \, dt - \frac{\alpha \, (\alpha-1)}{b-a} \int_a^b (b-t)^{\alpha-2} \, f(t) \, dt.$$

Notice that

$$-\alpha \, (\alpha-1) = -(\alpha-1)(\alpha-2) - 2 \, (\alpha-1) \,.$$

(5.16)

We split

$$-\frac{\alpha \, (\alpha-1)}{b-a} \int_a^b (b-t)^{\alpha-2} \, f(t) \, dt = -\frac{(\alpha-1)(\alpha-2)}{b-a} \int_a^b (b-t)^{\alpha-2} \, f(t) \, dt$$

$$-\frac{2 \, (\alpha-1)}{b-a} \int_a^b (b-t)^{\alpha-2} \, f(t) \, dt.$$

(5.17)

But we see that

$$-\frac{(\alpha-1)(\alpha-2)}{b-a} \int_a^b (b-t)^{\alpha-2} \, f(t) \, dt =$$

(5.18)

$$-\frac{(\alpha-1)(\alpha-2)}{b-a}\left[\int_a^x (b-t)^{\alpha-2} f(t)\,dt + \int_x^b (b-t)^{\alpha-2} f(t)\,dt\right] =$$

$$-\frac{(\alpha-1)(\alpha-2)}{b-a}\left[\int_a^x (b-t)(b-t)^{\alpha-3} f(t)\,dt + \int_x^b (b-t)(b-t)^{\alpha-3} f(t)\,dt\right]$$
(5.19)

$$= -(\alpha-1)(\alpha-2)\cdot$$

$$\left[\int_a^x \left(1-\left(\frac{t-a}{b-a}\right)\right)(b-t)^{\alpha-3} f(t)\,dt - \int_x^b \left(\frac{t-b}{b-a}\right)(b-t)^{\alpha-3} f(t)\,dt\right] =$$

$$-(\alpha-1)(\alpha-2)\left[\int_a^x (b-t)^{\alpha-3} f(t)\,dt - \left[\int_a^x \left(\frac{t-a}{b-a}\right)(b-t)^{\alpha-3} f(t)\,dt\right.\right.$$
(5.20)

$$\left.\left.+ \int_x^b \left(\frac{t-b}{b-a}\right)(b-t)^{\alpha-3} f(t)\,dt\right]\right] =$$

$$-(\alpha-1)(\alpha-2)\int_a^x (b-t)^{\alpha-3} f(t)\,dt +$$

$$(\alpha-1)(\alpha-2)\int_a^b P_1(x,t)(b-t)^{\alpha-3} f(t)\,dt.$$
(5.21)

Therefore

$$-\frac{\alpha(\alpha-1)}{b-a}\int_a^b (b-t)^{\alpha-2} f(t)\,dt = -\frac{2(\alpha-1)}{b-a}\int_a^b (b-t)^{\alpha-2} f(t)\,dt +$$

$$(\alpha-1)(\alpha-2)\int_a^b P_1(x,t)(b-t)^{\alpha-3} f(t)\,dt$$
(5.22)

$$-(\alpha-1)(\alpha-2)\int_a^x (b-t)^{\alpha-3} f(t)\,dt.$$

Hence it holds

$$\Gamma(\alpha)J_a^\alpha\left(P_1(x,b)f''(b)\right) = (\alpha-1)(b-x)^{\alpha-2} f(x) + (b-a)^{\alpha-2} f(a) -$$

$$\frac{2(\alpha-1)}{b-a}\int_a^b (b-t)^{\alpha-2} f(t)\,dt + (\alpha-1)(\alpha-2)\int_a^b P_1(x,t)(b-t)^{\alpha-3} f(t)\,dt =$$
(5.23)

$$(\alpha - 1)(b - x)^{\alpha-2} f(x) + (b - a)^{\alpha-2} f(a) - \frac{2(\alpha - 1)\Gamma(\alpha - 1)}{b - a} J_a^{\alpha-1} f(b) +$$

$$\text{(5.24)}$$

$$(\alpha - 1)(\alpha - 2)\Gamma(\alpha - 2) J_a^{\alpha-2} (P_1(x, b) f(b)) =$$

$$(\alpha - 1)(b - x)^{\alpha-2} f(x) + (b - a)^{\alpha-2} f(a) -$$

$$\frac{2\Gamma(\alpha)}{b - a} J_a^{\alpha-1} f(b) + \Gamma(\alpha) J_a^{\alpha-2} (P_1(x, b) f(b)). \qquad \text{(5.25)}$$

We have proved that

$$(\alpha - 1)(b - x)^{\alpha-2} f(x) = -(b - a)^{\alpha-2} f(a) + \frac{2\Gamma(\alpha)}{b - a} J_a^{\alpha-1} f(b) -$$

$$\Gamma(\alpha) J_a^{\alpha-2} (P_1(x, b) f(b)) + \Gamma(\alpha) J_a^{\alpha} (P_1(x, b) f''(b)) = \qquad \text{(5.26)}$$

$$-(b - a)^{\alpha-2} f(a) +$$

$$\Gamma(\alpha) \left\{ \frac{2}{b - a} J_a^{\alpha-1} f(b) - J_a^{\alpha-2} (P_1(x, b) f(b)) + J_a^{\alpha} (P_1(x, b) f''(b)) \right\}.$$

$$\text{(5.27)}$$

We have produced (5.6). ∎

We continue with

Theorem 5.3 *Let $\alpha > 3$, $x \in [a, b)$ fixed, $f : [a, b] \to \mathbb{R}$ three times differentiable, with $f''' : [a, b] \to \mathbb{R}$ integrable on $[a, b]$. Assume $f'(x) = f''(x) = 0$. Then*

$$f(x) = \frac{(b - x)^{3-\alpha}}{(\alpha - 1)(\alpha - 2)} \left\{ -2(\alpha - 1)(b - a)^{\alpha-3} f(a) - (b - a)^{\alpha-2} f'(a) + \right.$$

$$\left. \Gamma(\alpha) \left\{ \frac{3}{(b - a)} J_a^{\alpha-2} f(b) - J_a^{\alpha-3} (P_1(x, b) f(b)) + J_a^{\alpha} (P_1(x, b) f'''(b)) \right\} \right\}.$$

$$\text{(5.28)}$$

Proof Let here $\alpha > 3$ and there exists $f''' : [a, b] \to \mathbb{R}$ that is integrable on $[a, b]$. We have as before that

$$\Gamma(\alpha) J_a^{\alpha} (P_1(x, b) f'''(b)) =$$

$$\int_a^x (b - t)^{\alpha-1} f'''(t)\, dt - \frac{1}{(b - a)} \int_a^b (b - t)^{\alpha} f'''(t)\, dt =: (\xi_2). \qquad \text{(5.29)}$$

By assumption we have $f'(x) = f''(x) = 0$. We use repeatedly integration by parts next

$$\int_a^x (b-t)^{\alpha-1} f'''(t)\, dt = \int_a^x (b-t)^{\alpha-1} df''(t) =$$

$$-(b-a)^{\alpha-1} f''(a) + (\alpha-1) \int_a^x (b-t)^{\alpha-2} f''(t)\, dt =$$

$$-(b-a)^{\alpha-1} f''(a) + (\alpha-1) \int_a^x (b-t)^{\alpha-2} df'(t) =$$

$$-(b-a)^{\alpha-1} f''(a) +$$

$$(\alpha-1) \left[-(b-a)^{\alpha-2} f'(a) + (\alpha-2) \int_a^x (b-t)^{\alpha-3} f'(t)\, dt \right] = \quad (5.30)$$

$$-(b-a)^{\alpha-1} f''(a) - (\alpha-1)(b-a)^{\alpha-2} f'(a) +$$

$$(\alpha-1)(\alpha-2) \int_a^x (b-t)^{\alpha-3} df(t) =$$

$$-(b-a)^{\alpha-1} f''(a) - (\alpha-1)(b-a)^{\alpha-2} f'(a) + (\alpha-1)(\alpha-2) \left[(b-x)^{\alpha-3} f(x) \right.$$

$$\left. -(b-a)^{\alpha-3} f(a) + (\alpha-3) \int_a^x (b-t)^{\alpha-4} f(t)\, dt \right] =$$

$$-(b-a)^{\alpha-1} f''(a) - (\alpha-1)(b-a)^{\alpha-2} f'(a) + (\alpha-1)(\alpha-2)(b-x)^{\alpha-3} f(x) -$$

$$(\alpha-1)(\alpha-2)(b-a)^{\alpha-3} f(a) + (\alpha-1)(\alpha-2)(\alpha-3) \int_a^x (b-t)^{\alpha-4} f(t)\, dt.$$

$$(5.31)$$

That is

$$\int_a^x (b-t)^{\alpha-1} f'''(t)\, dt = -(b-a)^{\alpha-1} f''(a) - (\alpha-1)(b-a)^{\alpha-2} f'(a) +$$

$$(\alpha-1)(\alpha-2)(b-x)^{\alpha-3} f(x) - (\alpha-1)(\alpha-2)(b-a)^{\alpha-3} f(a) + \quad (5.32)$$

$$(\alpha-1)(\alpha-2)(\alpha-3) \int_a^x (b-t)^{\alpha-4} f(t)\, dt =: (\omega_1).$$

Similarly we find

$$-\frac{1}{(b-a)} \int_a^b (b-t)^\alpha f'''(t) \, dt = -\frac{1}{(b-a)} \int_a^b (b-t)^\alpha df''(t) = \quad (5.33)$$

$$-\frac{1}{(b-a)} \left[-(b-a)^\alpha f''(a) + \alpha \int_a^b (b-t)^{\alpha-1} f''(t) \, dt \right] =$$

$$(b-a)^{\alpha-1} f''(a) - \frac{\alpha}{(b-a)} \int_a^b (b-t)^{\alpha-1} df'(t) =$$

$$(b-a)^{\alpha-1} f''(a) - \frac{\alpha}{(b-a)} \left[-(b-a)^{\alpha-1} f'(a) + (\alpha-1) \int_a^b (b-t)^{\alpha-2} f'(t) \, dt \right]$$
$$(5.34)$$

$$= (b-a)^{\alpha-1} f''(a) + \alpha (b-a)^{\alpha-2} f'(a) - \frac{\alpha(\alpha-1)}{(b-a)} \int_a^b (b-t)^{\alpha-2} df(t) =$$

$$(b-a)^{\alpha-1} f''(a) + \alpha (b-a)^{\alpha-2} f'(a) - \quad (5.35)$$

$$\frac{\alpha(\alpha-1)}{(b-a)} \left[-(b-a)^{\alpha-2} f(a) + (\alpha-2) \int_a^b (b-t)^{\alpha-3} f(t) \, dt \right] =$$

$$(b-a)^{\alpha-1} f''(a) + \alpha (b-a)^{\alpha-2} f'(a) +$$

$$\alpha(\alpha-1)(b-a)^{\alpha-3} f(a) - \frac{\alpha(\alpha-1)(\alpha-2)}{(b-a)} \int_a^b (b-t)^{\alpha-3} f(t) \, dt. \quad (5.36)$$

That is we found

$$-\frac{1}{(b-a)} \int_a^b (b-t)^\alpha f'''(t) \, dt = (b-a)^{\alpha-1} f''(a) + \alpha (b-a)^{\alpha-2} f'(a) +$$

$$\alpha(\alpha-1)(b-a)^{\alpha-3} f(a) - \frac{\alpha(\alpha-1)(\alpha-2)}{(b-a)} \int_a^b (b-t)^{\alpha-3} f(t) \, dt =: (\omega_2) \,.$$
$$(5.37)$$

Notice that

$$(\xi_2) = (\omega_1) + (\omega_2) \,.$$

We have

$$\Gamma(\alpha) J_a^\alpha \left(P_1(x, b) f'''(b) \right) = (b-a)^{\alpha-2} f'(a) + (\alpha-1)(\alpha-2)(b-x)^{\alpha-3} f(x) +$$

$$2\left(\alpha-1\right)\left(b-a\right)^{\alpha-3} f\left(a\right) + \left(a-1\right)\left(\alpha-2\right)\left(\alpha-3\right)\int_{a}^{x}\left(b-t\right)^{\alpha-4} f\left(t\right) dt -$$

$$\frac{\alpha\left(a-1\right)\left(\alpha-2\right)}{\left(b-a\right)}\int_{a}^{b}\left(b-t\right)^{\alpha-3} f\left(t\right) dt. \tag{5.38}$$

We notice that

$$-\alpha\left(\alpha-1\right)\left(\alpha-2\right) = -3\left(\alpha-1\right)\left(\alpha-2\right) - \left(\alpha-1\right)\left(\alpha-2\right)\left(\alpha-3\right). \tag{5.39}$$

Hence

$$-\frac{\alpha\left(\alpha-1\right)\left(\alpha-2\right)}{\left(b-a\right)}\int_{a}^{b}\left(b-t\right)^{\alpha-3} f\left(t\right) dt =$$

$$-\frac{3\left(\alpha-1\right)\left(\alpha-2\right)}{\left(b-a\right)}\int_{a}^{b}\left(b-t\right)^{\alpha-3} f\left(t\right) dt - \tag{5.40}$$

$$\frac{\left(\alpha-1\right)\left(\alpha-2\right)\left(\alpha-3\right)}{\left(b-a\right)}\int_{a}^{b}\left(b-t\right)^{\alpha-3} f\left(t\right) dt.$$

But we see that

$$-\frac{\left(\alpha-1\right)\left(\alpha-2\right)\left(\alpha-3\right)}{\left(b-a\right)}\int_{a}^{b}\left(b-t\right)^{\alpha-3} f\left(t\right) dt =$$

$$-\frac{\left(\alpha-1\right)\left(\alpha-2\right)\left(\alpha-3\right)}{\left(b-a\right)}\left[\int_{a}^{x}\left(b-t\right)^{\alpha-3} f\left(t\right) dt + \int_{x}^{b}\left(b-t\right)^{\alpha-3} f\left(t\right) dt\right] =$$

$$-\frac{\left(\alpha-1\right)\left(\alpha-2\right)\left(\alpha-3\right)}{\left(b-a\right)}. \tag{5.41}$$

$$\left[\int_{a}^{x}\left(b-t\right)\left(b-t\right)^{\alpha-4} f\left(t\right) dt + \int_{x}^{b}\left(b-t\right)\left(b-t\right)^{\alpha-4} f\left(t\right) dt\right] = \tag{5.42}$$

$$-\frac{\left(\alpha-1\right)\left(\alpha-2\right)\left(\alpha-3\right)}{\left(b-a\right)}.$$

$$\left[\int_{a}^{x}\left(\left(b-a\right)-\left(t-a\right)\right)\left(b-t\right)^{\alpha-4} f\left(t\right) dt - \int_{x}^{b}\left(t-b\right)\left(b-t\right)^{\alpha-4} f\left(t\right) dt\right] =$$

$$-\left(\alpha-1\right)\left(\alpha-2\right)\left(\alpha-3\right)\left[\int_{a}^{x}\left(1-\left(\frac{t-a}{b-a}\right)\right)\left(b-t\right)^{\alpha-4} f\left(t\right) dt -\right.$$

$$\left. \int_x^b \left(\frac{t-b}{b-a} \right) (b-t)^{\alpha-4} f(t)\, dt \right] = \tag{5.43}$$

$$- (\alpha-1)(\alpha-2)(\alpha-3) \left[\int_a^x (b-t)^{\alpha-4} f(t)\, dt - \int_a^b P_1(x,t)(b-t)^{\alpha-4} f(t)\, dt \right].$$

We derived that

$$- \frac{(\alpha-1)(\alpha-2)(\alpha-3)}{(b-a)} \int_a^b (b-t)^{\alpha-3} f(t)\, dt = \tag{5.44}$$

$$- (\alpha-1)(\alpha-2)(\alpha-3) \int_a^x (b-t)^{\alpha-4} f(t)\, dt +$$

$$(\alpha-1)(\alpha-2)(\alpha-3) \int_a^b P_1(x,t)(b-t)^{\alpha-4} f(t)\, dt.$$

Therefore we obtain

$$- \frac{\alpha(\alpha-1)(\alpha-2)}{(b-a)} \int_a^b (b-t)^{\alpha-3} f(t)\, dt =$$

$$- \frac{3(\alpha-1)(\alpha-2)}{(b-a)} \int_a^b (b-t)^{\alpha-3} f(t)\, dt - \tag{5.45}$$

$$(\alpha-1)(\alpha-2)(\alpha-3) \int_a^x (b-t)^{\alpha-4} f(t)\, dt +$$

$$(\alpha-1)(\alpha-2)(\alpha-3) \int_a^b P_1(x,t)(b-t)^{\alpha-4} f(t)\, dt.$$

Combining (5.38) and (5.45) we find

$$\Gamma(\alpha) J_a^\alpha \left(P_1(x,b) f'''(b) \right) = (b-a)^{\alpha-2} f'(a) + (\alpha-1)(\alpha-2)(b-x)^{\alpha-3} f(x) +$$

$$2(\alpha-1)(b-a)^{\alpha-3} f(a) - \frac{3(\alpha-1)(\alpha-2)}{(b-a)} \int_a^b (b-t)^{\alpha-3} f(t)\, dt + \tag{5.46}$$

$$(\alpha-1)(\alpha-2)(\alpha-3) \int_a^b P_1(x,t)(b-t)^{\alpha-4} f(t)\, dt =$$

$$(b-a)^{\alpha-2} f'(a) + (\alpha-1)(\alpha-2)(b-x)^{\alpha-3} f(x) + 2(\alpha-1)(b-a)^{\alpha-3} f(a) -$$

$$\frac{3\,(\alpha-1)\,(\alpha-2)\,\Gamma\,(\alpha-2)}{b-a}J_a^{\alpha-2}f\,(b)+ \tag{5.47}$$

$$(\alpha-1)\,(\alpha-2)\,(\alpha-3)\,\Gamma\,(\alpha-3)\,J_a^{\alpha-3}\,(P_1\,(x,b)\,f\,(b))=$$

$$(b-a)^{\alpha-2}\,f'\,(a)+(\alpha-1)\,(\alpha-2)\,(b-x)^{\alpha-3}\,f\,(x)+2\,(\alpha-1)\,(b-a)^{\alpha-3}\,f\,(a)-$$

$$\frac{3\Gamma\,(\alpha)}{(b-a)}J_a^{\alpha-2}f\,(b)+\Gamma\,(\alpha)\,J_a^{\alpha-3}\,(P_1\,(x,b)\,f\,(b)) . \tag{5.48}$$

Consequently we get

$$(\alpha-1)\,(\alpha-2)\,(b-x)^{\alpha-3}\,f\,(x)=-(b-a)^{\alpha-2}\,f'\,(a)-2\,(\alpha-1)\,(b-a)^{\alpha-3}\,f\,(a)$$

$$+\frac{3\Gamma\,(\alpha)}{(b-a)}J_a^{\alpha-2}f\,(b)-\Gamma\,(\alpha)\,J_a^{\alpha-3}\,(P_1\,(x,b)\,f\,(b))+\Gamma\,(\alpha)\,J_a^{\alpha}\,(P_1\,(x,b)\,f'''\,(b))=$$
$$\tag{5.49}$$
$$-(b-a)^{\alpha-2}\,f'\,(a)-2\,(\alpha-1)\,(b-a)^{\alpha-3}\,f\,(a)+$$

$$\Gamma\,(\alpha)\left\{\frac{3}{(b-a)}J_a^{\alpha-2}f\,(b)-J_a^{\alpha-3}\,(P_1\,(x,b)\,f\,(b))+J_a^{\alpha}\,(P_1\,(x,b)\,f'''\,(b))\right\},$$
$$\tag{5.50}$$

proving the claim. ∎

We continue with

Theorem 5.4 *Let* $\alpha>4$, $x\in[a,b)$ *fixed,* $f:[a,b]\rightarrow\mathbb{R}$ *four times differentiable, with* $f^{(4)}:[a,b]\rightarrow\mathbb{R}$ *integrable on* $[a,b]$. *Assume* $f'\,(x)=f''\,(x)=f'''\,(x)=0$. *Then*

$$f\,(x)=\frac{(b-x)^{4-\alpha}}{(\alpha-1)\,(\alpha-2)\,(\alpha-3)}\left\{-3\,(\alpha-1)\,(\alpha-2)\,(b-a)^{\alpha-4}\,f\,(\alpha)-\right.$$

$$2\,(\alpha-1)\,(b-a)^{\alpha-3}\,f'\,(a)-(b-a)^{\alpha-2}\,f^{(2)}\,(a)+$$

$$\left.\Gamma\,(\alpha)\left\{\frac{4J_a^{\alpha-3}\,(f\,(b))}{(b-a)}-J_a^{\alpha-4}\,(P_1\,(x,b)\,f\,(b))+J_a^{\alpha}\,\left(P_1\,(x,b)\,f^{(4)}\,(b)\right)\right\}\right\}.$$
$$\tag{5.51}$$

Proof Let here $\alpha>4$ and there exists $f^{(4)}:[a,b]\rightarrow\mathbb{R}$ that is integrable on $[a,b]$. We have as before that

$$\Gamma\left(\alpha\right) J_a^\alpha \left(P_1\left(x, b\right) f^{(4)}\left(b\right)\right) = \int_a^x \left(b - t\right)^{\alpha - 1} f^{(4)}\left(t\right) dt -$$

$$\frac{1}{\left(b - a\right)} \int_a^b \left(b - t\right)^\alpha f^{(4)}\left(t\right) dt =: \left(\xi_3\right).$$

By assumption we have $f'\left(x\right) = f''\left(x\right) = f'''\left(x\right) = 0$. We use repeatedly integration by parts next

$$\int_a^x \left(b - t\right)^{\alpha - 1} f^{(4)}\left(t\right) dt = \int_a^x \left(b - t\right)^{\alpha - 1} df^{(3)}\left(t\right) =$$

$$-\left(b - a\right)^{\alpha - 1} f^{(3)}\left(a\right) + \left(\alpha - 1\right) \int_a^x \left(b - t\right)^{\alpha - 2} df^{(2)}\left(t\right) = -\left(b - a\right)^{\alpha - 1} f^{(3)}\left(a\right) +$$
$$\tag{5.52}$$

$$\left(\alpha - 1\right) \left[-\left(b - a\right)^{\alpha - 2} f^{(2)}\left(a\right) + \left(\alpha - 2\right) \int_a^x \left(b - t\right)^{\alpha - 3} df'\left(t\right)\right] =$$

$$-\left(b - a\right)^{\alpha - 1} f^{(3)}\left(a\right) - \left(\alpha - 1\right) \left(b - a\right)^{\alpha - 2} f^{(2)}\left(a\right) +$$

$$\left(\alpha - 1\right) \left(\alpha - 2\right) \int_a^x \left(b - t\right)^{\alpha - 3} df'\left(t\right) =$$

$$-\left(b - a\right)^{\alpha - 1} f^{(3)}\left(a\right) - \left(\alpha - 1\right) \left(b - a\right)^{a - 2} f^{(2)}\left(a\right) +$$

$$\left(\alpha - 1\right) \left(\alpha - 2\right) \left[-\left(b - a\right)^{\alpha - 3} f'\left(a\right) + \left(\alpha - 3\right) \int_a^x \left(b - t\right)^{\alpha - 4} df\left(t\right)\right] =$$
$$\tag{5.53}$$
$$-\left(b - a\right)^{\alpha - 1} f^{(3)}\left(a\right) - \left(\alpha - 1\right) \left(b - a\right)^{\alpha - 2} f^{(2)}\left(a\right) - \left(\alpha - 1\right) \left(\alpha - 2\right) \left(b - a\right)^{\alpha - 3} f'\left(a\right)$$

$$+\left(\alpha - 1\right) \left(\alpha - 2\right) \left(\alpha - 3\right) \int_a^x \left(b - t\right)^{\alpha - 4} df\left(t\right) =$$

$$-\left(b - a\right)^{\alpha - 1} f^{(3)}\left(a\right) - \left(\alpha - 1\right) \left(b - a\right)^{\alpha - 2} f^{(2)}\left(a\right) - \left(\alpha - 1\right) \left(\alpha - 2\right) \left(b - a\right)^{\alpha - 3} f'\left(a\right)$$

$$+\left(\alpha - 1\right) \left(\alpha - 2\right) \left(\alpha - 3\right) \left(b - x\right)^{\alpha - 4} f\left(x\right) -$$

$$\left(\alpha - 1\right) \left(\alpha - 2\right) \left(\alpha - 3\right) \left(b - a\right)^{\alpha - 4} f\left(a\right) +$$

$$\left(\alpha - 1\right) \left(\alpha - 2\right) \left(\alpha - 3\right) \left(\alpha - 4\right) \int_a^x \left(b - t\right)^{\alpha - 5} f\left(t\right) dt. \tag{5.54}$$

We find that

$$\int_a^x (b-t)^{\alpha-1} f^{(4)}(t)\, dt = -(b-a)^{\alpha-1} f^{(3)}(a) - (\alpha-1)(b-a)^{\alpha-2} f^{(2)}(a) -$$

$$(\alpha-1)(\alpha-2)(b-a)^{\alpha-3} f'(a) + (\alpha-1)(\alpha-2)(\alpha-3)(b-x)^{\alpha-4} f(x) -$$

$$(\alpha-1)(\alpha-2)(\alpha-3)(b-a)^{\alpha-4} f(a) +$$

$$(\alpha-1)(\alpha-2)(\alpha-3)(\alpha-4) \int_a^x (b-t)^{\alpha-5} f(t)\, dt =: (\theta_1). \qquad (5.55)$$

Next we observe that

$$-\frac{1}{(b-a)} \int_a^b (b-t)^{\alpha} f^{(4)}(t)\, dt = -\frac{1}{(b-a)} \int_a^b (b-t)^{\alpha}\, df^{(3)}(t) =$$

$$-\frac{1}{(b-a)} \left[-(b-a)^{\alpha} f^{(3)}(a) + \alpha \int_a^b (b-t)^{\alpha-1}\, df^{(2)}(t) \right] = \qquad (5.56)$$

$$(b-a)^{\alpha-1} f^{(3)}(a) - \frac{\alpha}{b-a} \int_a^b (b-t)^{\alpha-1}\, df^{(2)}(t) = (b-a)^{\alpha-1} f^{(3)}(a) -$$

$$\frac{\alpha}{b-a} \left[-(b-a)^{\alpha-1} f^{(2)}(a) + (\alpha-1) \int_a^b (b-t)^{\alpha-2}\, df'(t) \right] =$$

$$(b-a)^{\alpha-1} f^{(3)}(a) + \alpha (b-a)^{\alpha-2} f^{(2)}(a) - \frac{\alpha(\alpha-1)}{(b-a)} \int_a^b (b-t)^{\alpha-2}\, df'(t) =$$

$$(b-a)^{\alpha-1} f^{(3)}(a) + \alpha (b-a)^{\alpha-2} f^{(2)}(a) -$$

$$\frac{\alpha(\alpha-1)}{(b-a)} \left[-(b-a)^{\alpha-2} f'(a) + (\alpha-2) \int_a^b (b-t)^{\alpha-3}\, df(t) \right] = \qquad (5.57)$$

$$(b-a)^{\alpha-1} f^{(3)}(a) + \alpha (b-a)^{\alpha-2} f^{(2)}(a) + \alpha(\alpha-1)(b-a)^{\alpha-3} f'(a) -$$

$$\frac{\alpha(\alpha-1)(\alpha-2)}{(b-a)} \int_a^b (b-t)^{\alpha-3}\, df(t) =$$

$$(b-a)^{a-1} f^{(3)}(a) + \alpha (b-a)^{\alpha-2} f^{(2)}(a) + \alpha(\alpha-1)(b-a)^{\alpha-3} f'(a) -$$

$$\qquad\qquad\qquad\qquad\qquad\qquad\qquad\qquad\qquad (5.58)$$

$$\frac{\alpha(\alpha-1)(\alpha-2)}{(b-a)} \left[-(b-a)^{\alpha-3} f(a) + (\alpha-3) \int_a^b (b-t)^{\alpha-4} f(t)\, dt \right] =$$

$$(b-a)^{\alpha-1} f^{(3)}(a) + \alpha (b-a)^{\alpha-2} f^{(2)}(a) + \alpha (\alpha-1)(b-a)^{\alpha-3} f'(a) +$$

$$\alpha (\alpha-1)(\alpha-2)(b-a)^{\alpha-4} f(a) - \frac{\alpha(\alpha-1)(\alpha-2)(\alpha-3)}{(b-a)} \int_a^b (b-t)^{\alpha-4} f(t)\, dt.$$

That is

$$-\frac{1}{(b-a)} \int_a^b (b-t)^{\alpha} f^{(4)}(t)\, dt = (b-a)^{\alpha-1} f^{(3)}(a) + \alpha (b-a)^{\alpha-2} f^{(2)}(a) +$$

$$\alpha (\alpha-1)(b-a)^{\alpha-3} f'(a) + \alpha (\alpha-1)(\alpha-2)(b-a)^{\alpha-4} f(a) -$$

$$\frac{\alpha(\alpha-1)(\alpha-2)(\alpha-3)}{(b-a)} \int_a^b (b-t)^{\alpha-4} f(t)\, dt =: (\theta_2). \tag{5.59}$$

Notice that

$$(\xi_3) = (\theta_1) + (\theta_2). \tag{5.60}$$

We find that

$$\Gamma(\alpha) J_a^{\alpha}\left(P_1(x,b) f^{(4)}(b)\right) = (b-a)^{\alpha-2} f^{(2)}(a) + 2(\alpha-1)(b-a)^{\alpha-3} f'(a) +$$

$$3(\alpha-1)(\alpha-2)(b-a)^{\alpha-4} f(a) + (\alpha-1)(\alpha-2)(\alpha-3)(b-x)^{\alpha-4} f(x) + \tag{5.61}$$

$$(\alpha-1)(\alpha-2)(\alpha-3)(\alpha-4) \int_a^x (b-t)^{\alpha-5} f(t)\, dt$$

$$-\frac{\alpha(\alpha-1)(\alpha-2)(\alpha-3)}{(b-a)} \int_a^b (b-t)^{\alpha-4} f(t)\, dt.$$

We have

$$-\alpha(\alpha-1)(\alpha-2)(\alpha-3) =$$

$$-4(\alpha-1)(\alpha-2)(\alpha-3) - (\alpha-1)(\alpha-2)(\alpha-3)(\alpha-4). \tag{5.62}$$

and

$$\frac{-\alpha(\alpha-1)(\alpha-2)(\alpha-3)}{(b-a)} \int_a^b (b-t)^{\alpha-4} f(t)\, dt = \tag{5.63}$$

$$-\frac{4(\alpha-1)(\alpha-2)(\alpha-3)}{(b-a)} \int_a^b (b-t)^{\alpha-4} f(t)\, dt$$

$$-\frac{(\alpha-1)(\alpha-2)(\alpha-3)(\alpha-4)}{(b-a)} \int_a^b (b-t)^{\alpha-4} f(t)\, dt.$$

But we see that

$$-\frac{(\alpha-1)(\alpha-2)(\alpha-3)(\alpha-4)}{(b-a)}\int_a^b (b-t)^{\alpha-4} f(t)\,dt =$$

$$-\frac{(\alpha-1)(\alpha-2)(\alpha-3)(\alpha-4)}{(b-a)}\left[\int_a^x ((b-a)-(t-a))(b-t)^{\alpha-5} f(t)\,dt\right.$$

$$\left.-\int_x^b (t-b)(b-t)^{\alpha-5} f(t)\,dt\right] =$$

$$-(\alpha-1)(\alpha-2)(\alpha-3)(\alpha-4)\left[\int_a^x (b-t)^{\alpha-5} f(t)\,dt\right. \tag{5.64}$$

$$\left.-\int_a^b P_1(x,t)(b-t)^{\alpha-5} f(t)\,dt\right].$$

Therefore it holds

$$-\frac{\alpha(\alpha-1)(\alpha-2)(\alpha-3)}{(b-a)}\int_a^b (b-t)^{\alpha-4} f(t)\,dt =$$

$$-\frac{4(\alpha-1)(\alpha-2)(\alpha-3)}{(b-a)}\int_a^b (b-t)^{\alpha-4} f(t)\,dt$$

$$-(\alpha-1)(\alpha-2)(\alpha-3)(\alpha-4)\int_a^x (b-t)^{\alpha-5} f(t)\,dt$$

$$+(\alpha-1)(\alpha-2)(\alpha-3)(\alpha-4)\int_a^b P_1(x,t)(b-t)^{\alpha-5} f(t)\,dt. \tag{5.65}$$

Consequently we get

$$\Gamma(\alpha) J_a^\alpha\left(P_1(x,b) f^{(4)}(b)\right) = (b-a)^{\alpha-2} f^{(2)}(a) + 2(\alpha-1)(b-a)^{\alpha-3} f'(a) +$$

$$3(\alpha-1)(\alpha-2)(b-a)^{\alpha-4} f(a) + (\alpha-1)(\alpha-2)(\alpha-3)(b-x)^{\alpha-4} f(x) -$$

$$\frac{4(\alpha-1)(\alpha-2)(\alpha-3)}{(b-a)}\int_a^b (b-t)^{\alpha-4} f(t)\,dt+ \tag{5.66}$$

$$(\alpha-1)(\alpha-2)(\alpha-3)(\alpha-4)\int_a^b P_1(x,t)(b-t)^{\alpha-5} f(t)\,dt =$$

$$(b-a)^{\alpha-2} f^{(2)}(a) + 2(\alpha-1)(b-a)^{\alpha-3} f'(a) + 3(\alpha-1)(\alpha-2)(b-a)^{\alpha-4} f(a)$$

$$+ (\alpha-1)(\alpha-2)(\alpha-3)(b-x)^{\alpha-4} f(x) -$$

$$\frac{4(\alpha-1)(\alpha-2)(\alpha-3)\Gamma(\alpha-3)}{(b-a)} J_a^{\alpha-3}(f(b)) +$$

$$(\alpha-1)(\alpha-2)(\alpha-3)(\alpha-4)\Gamma(\alpha-4) J_a^{\alpha-4}(P_1(x,b) f(b)) = \qquad (5.67)$$

$$(b-a)^{\alpha-2} f^{(2)}(a) + 2(\alpha-1)(b-a)^{\alpha-3} f'(a) + 3(\alpha-1)(\alpha-2)(b-a)^{\alpha-4} f(a)$$

$$+ (\alpha-1)(\alpha-2)(\alpha-3)(b-x)^{\alpha-4} f(x) -$$

$$\frac{4\Gamma(\alpha)}{(b-a)} J_a^{\alpha-3}(f(b)) + \Gamma(\alpha) J_a^{\alpha-4}(P_1(x,b) f(b)). \qquad (5.68)$$

That is

$$\Gamma(\alpha) J_a^\alpha \left(P_1(x,b) f^{(4)}(b) \right) = (b-a)^{\alpha-2} f^{(2)}(a) + 2(\alpha-1)(b-a)^{\alpha-3} f'(a) +$$

$$3(\alpha-1)(\alpha-2)(b-a)^{\alpha-4} f(a) + (\alpha-1)(\alpha-2)(\alpha-3)(b-x)^{\alpha-4} f(x) +$$

$$\Gamma(\alpha) \left\{ -\frac{4 J_a^{\alpha-3}(f(b))}{(b-a)} + J_a^{\alpha-4}(P_1(x,b) f(b)) \right\}, \qquad (5.69)$$

proving the claim. ∎

We continue with

Theorem 5.5 Let $\alpha > 5$, $x \in [a,b)$ fixed, $f : [a,b] \to \mathbb{R}$ five times differentiable, with $f^{(5)} : [a,b] \to \mathbb{R}$ integrable on $[a,b]$. Assume $f^{(j)}(x) = 0$, $j = 1,2,3,4$. Then

$$f(x) = \frac{(b-x)^{5-\alpha}}{\displaystyle\prod_{j=1}^{4}(\alpha-j)} \left\{ -4 \prod_{j=1}^{3}(\alpha-j)(b-a)^{\alpha-5} f(a) - \right.$$

$$3 \prod_{j=1}^{2}(\alpha-j)(b-a)^{\alpha-4} f'(a) - 2(\alpha-1)(b-a)^{\alpha-3} f^{(2)}(a) - (b-a)^{\alpha-2} f^{(3)}(a) +$$

$$\Gamma(\alpha) \left\{ \frac{5}{(b-a)} \left(J_a^{\alpha-4}(f(b)) \right) - J_a^{\alpha-5}(P_1(x,b) f(b)) + J_a^\alpha \left(P_1(x,b) f^{(5)}(b) \right) \right\} \right\}.$$

$$(5.70)$$

Proof Let here $\alpha > 5$ and there exists $f^{(5)} : [a, b] \to \mathbb{R}$ that is integrable on $[a, b]$. We have as before that

$$\Gamma(\alpha) J_a^\alpha \left(P_1(x, b) f^{(5)}(b) \right) =$$

$$\int_a^x (b-t)^{\alpha-1} f^{(5)}(t)\, dt - \frac{1}{(b-a)} \int_a^b (b-t)^\alpha f^{(5)}(t)\, dt =: (\xi_4) . \quad (5.71)$$

By assumption we have $f^{(j)}(x) = 0$, $j = 1, 2, 3, 4$. We use repeatedly integration by parts next

$$\int_a^x (b-t)^{\alpha-1} f^{(5)}(t)\, dt = \int_a^x (b-t)^{\alpha-1} df^{(4)}(t) =$$

$$-(b-a)^{\alpha-1} f^{(4)}(a) + (\alpha-1) \int_a^x (b-t)^{\alpha-2} df^{(3)}(t) = -(b-a)^{\alpha-1} f^{(4)}(a) +$$

$$(5.72)$$

$$(\alpha-1) \left\{ -(b-a)^{\alpha-2} f^{(3)}(a) + (\alpha-2) \int_a^x (b-t)^{\alpha-3} df^{(2)}(t) \right\} =$$

$$- (b-a)^{\alpha-1} f^{(4)}(a) - (\alpha-1)(b-a)^{\alpha-2} f^{(3)}(a) +$$

$$(\alpha-1)(\alpha-2) \int_a^x (b-t)^{\alpha-3} df^{(2)}(t) =$$

$$- (b-a)^{\alpha-1} f^{(4)}(a) - (\alpha-1)(b-a)^{\alpha-2} f^{(3)}(a) +$$

$$(\alpha-1)(\alpha-2) \left\{ -(b-a)^{\alpha-3} f^{(2)}(a) + (\alpha-3) \int_a^x (b-t)^{\alpha-4} df'(t) \right\} =$$

$$(5.73)$$

$$- (b-a)^{\alpha-1} f^{(4)}(a) - (\alpha-1)(b-a)^{\alpha-2} f^{(3)}(a) -$$

$$(\alpha-1)(\alpha-2)(b-a)^{\alpha-3} f^{(2)}(a) + (\alpha-1)(\alpha-2)(\alpha-3) \int_a^x (b-t)^{\alpha-4} df'(t)$$

$$= - (b-a)^{\alpha-1} f^{(4)}(a) - (\alpha-1)(b-a)^{\alpha-2} f^{(3)}(a)$$

$$- (\alpha-1)(\alpha-2)(b-a)^{\alpha-3} f^{(2)}(a) +$$

$$(\alpha-1)(\alpha-2)(\alpha-3) \left\{ -(b-a)^{\alpha-4} f'(a) + (\alpha-4) \int_a^x (b-t)^{\alpha-5} df(t) \right\} =$$

$$- (b-a)^{\alpha-1} f^{(4)}(a) - (\alpha-1)(b-a)^{\alpha-2} f^{(3)}(a) -$$

$$(\alpha - 1)(\alpha - 2)(b - a)^{\alpha - 3} f^{(2)}(a) - (\alpha - 1)(\alpha - 2)(\alpha - 3)(b - a)^{\alpha - 4} f'(a)$$
$$(5.74)$$

$$+ (\alpha - 1)(\alpha - 2)(\alpha - 3)(\alpha - 4) \int_a^x (b - t)^{\alpha - 5} df(t) =$$

$$- (b - a)^{\alpha - 1} f^{(4)}(a) - (\alpha - 1)(b - a)^{\alpha - 2} f^{(3)}(a) -$$

$$(\alpha - 1)(\alpha - 2)(b - a)^{\alpha - 3} f^{(2)}(a) -$$

$$(\alpha - 1)(\alpha - 2)(\alpha - 3)(b - a)^{\alpha - 4} f'(a) + (\alpha - 1)(\alpha - 2)(\alpha - 3)(\alpha - 4) \cdot$$

$$\left\{ (b - x)^{\alpha - 5} f(x) - (b - a)^{\alpha - 5} f(a) + (\alpha - 5) \int_a^x (b - t)^{\alpha - 6} f(t) dt \right\}.$$

That is

$$\int_a^x (b - t)^{\alpha - 1} f^{(5)}(t) dt = - (b - a)^{\alpha - 1} f^{(4)}(a) - (\alpha - 1)(b - a)^{\alpha - 2} f^{(3)}(a) -$$

$$(\alpha - 1)(\alpha - 2)(b - a)^{\alpha - 3} f^{(2)}(a) - (\alpha - 1)(\alpha - 2)(\alpha - 3)(b - a)^{\alpha - 4} f'(a) +$$
$$(5.75)$$

$$(\alpha - 1)(\alpha - 2)(\alpha - 3)(\alpha - 4)(b - x)^{\alpha - 5} f(x) -$$

$$(\alpha - 1)(\alpha - 2)(\alpha - 3)(\alpha - 4)(b - a)^{\alpha - 5} f(a) +$$

$$(\alpha - 1)(\alpha - 2)(\alpha - 3)(\alpha - 4)(\alpha - 5) \int_a^x (b - t)^{\alpha - 6} f(t) dt =: (\eta_1).$$

Next we observe that

$$-\frac{1}{(b - a)} \int_a^b (b - t)^{\alpha} f^{(5)}(t) dt = -\frac{1}{(b - a)} \int_a^b (b - t)^{\alpha} df^{(4)}(t) =$$

$$-\frac{1}{(b - a)} \left\{ - (b - a)^{\alpha} f^{(4)}(a) + \alpha \int_a^b (b - t)^{\alpha - 1} df^{(3)}(t) \right\} =$$

$$(b - a)^{\alpha - 1} f^{(4)}(a) - \frac{\alpha}{(b - a)} \int_a^b (b - t)^{\alpha - 1} df^{(3)}(t) = \qquad (5.76)$$

$$(b - a)^{\alpha - 1} f^{(4)}(a) -$$

$$\frac{\alpha}{(b - a)} \left\{ - (b - a)^{\alpha - 1} f^{(3)}(a) + (\alpha - 1) \int_a^b (b - t)^{\alpha - 2} df^{(2)}(t) \right\} =$$

$$(b-a)^{\alpha-1} f^{(4)}(a) + \alpha (b-a)^{\alpha-2} f^{(3)}(a) - \frac{\alpha(\alpha-1)}{(b-a)} \int_a^b (b-t)^{\alpha-2} df^{(2)}(t) =$$

$$(b-a)^{\alpha-1} f^{(4)}(a) + \alpha (b-a)^{\alpha-2} f^{(3)}(a) -$$

$$\frac{\alpha(\alpha-1)}{(b-a)} \left\{ -(b-a)^{\alpha-2} f^{(2)}(a) + (\alpha-2) \int_a^b (b-t)^{\alpha-3} df'(t) \right\} =$$

$$(b-a)^{\alpha-1} f^{(4)}(a) + \alpha (b-a)^{\alpha-2} f^{(3)}(a) + \alpha(\alpha-1)(b-a)^{\alpha-3} f^{(2)}(a) -$$

$$\frac{\alpha(\alpha-1)(\alpha-2)}{(b-a)} \int_a^b (b-t)^{\alpha-3} df'(t) =$$

$$(b-a)^{\alpha-1} f^{(4)}(a) + \alpha (b-a)^{\alpha-2} f^{(3)}(a) + \alpha(\alpha-1)(b-a)^{\alpha-3} f^{(2)}(a) -$$

$$\frac{\alpha(\alpha-1)(\alpha-2)}{(b-a)} \left\{ -(b-a)^{\alpha-3} f'(a) + (\alpha-3) \int_a^b (b-t)^{\alpha-4} df(t) \right\} =$$

$$(b-a)^{\alpha-1} f^{(4)}(a) + \alpha (b-a)^{\alpha-2} f^{(3)}(a) + \alpha(\alpha-1)(b-a)^{\alpha-3} f^{(2)}(a) + \tag{5.77}$$

$$\alpha(\alpha-1)(\alpha-2)(b-a)^{\alpha-4} f'(a) - \frac{\alpha(\alpha-1)(\alpha-2)(\alpha-3)}{(b-a)} \int_a^b (b-t)^{\alpha-4} df(t)$$

$$= (b-a)^{\alpha-1} f^{(4)}(a) + \alpha (b-a)^{\alpha-2} f^{(3)}(a) + \alpha(\alpha-1)(b-a)^{\alpha-3} f^{(2)}(a) +$$

$$\alpha(\alpha-1)(\alpha-2)(b-a)^{\alpha-4} f'(a) - \frac{\alpha(\alpha-1)(\alpha-2)(\alpha-3)}{(b-a)}.$$

$$\left\{ -(b-a)^{\alpha-4} f(a) + (\alpha-4) \int_a^b (b-t)^{\alpha-5} f(t) dt \right\}. \tag{5.78}$$

We proved that

$$-\frac{1}{(b-a)} \int_a^b (b-t)^{\alpha} f^{(5)}(t) dt = (b-a)^{\alpha-1} f^{(4)}(a) + \alpha (b-a)^{\alpha-2} f^{(3)}(a) +$$

$$\alpha(\alpha-1)(b-a)^{\alpha-3} f^{(2)}(a) + \alpha(\alpha-1)(\alpha-2)(b-a)^{\alpha-4} f'(a) +$$

$$\alpha(\alpha-1)(\alpha-2)(\alpha-3)(b-a)^{\alpha-5} f(a) - \tag{5.79}$$

$$\frac{\alpha\,(\alpha-1)\,(\alpha-2)\,(\alpha-3)\,(\alpha-4)}{(b-a)}\int_a^b (b-t)^{\alpha-5}\,f(t)\,dt =: (\eta_2)\,.$$

We have

$$(\xi_4) = (\eta_1) + (\eta_2)\,.$$

Therefore it holds

$$\Gamma(\alpha)\,J_a^\alpha\left(P_1(x,b)\,f^{(5)}(b)\right) = (b-a)^{\alpha-2}\,f^{(3)}(a) + \qquad (5.80)$$

$$2\,(\alpha-1)\,(b-a)^{\alpha-3}\,f^{(2)}(a) + 3\,(\alpha-1)\,(\alpha-2)\,(b-a)^{\alpha-4}\,f'(a)$$

$$+\,(\alpha-1)\,(\alpha-2)\,(\alpha-3)\,(\alpha-4)\,(b-x)^{\alpha-5}\,f(x) +$$

$$4\,(\alpha-1)\,(\alpha-2)\,(\alpha-3)\,(b-a)^{\alpha-5}\,f(a) +$$

$$(\alpha-1)\,(\alpha-2)\,(\alpha-3)\,(\alpha-4)\,(\alpha-5)\int_a^x (b-t)^{\alpha-6}\,f(t)\,dt$$

$$-\frac{\alpha\,(\alpha-1)\,(\alpha-2)\,(\alpha-3)\,(\alpha-4)}{(b-a)}\int_a^b (b-t)^{\alpha-5}\,f(t)\,dt\,.$$

We see that

$$-\frac{\alpha\,(\alpha-1)\,(\alpha-2)\,(\alpha-3)\,(\alpha-4)}{(b-a)}\int_a^b (b-t)^{\alpha-5}\,f(t)\,dt =$$

$$-\frac{5\,(\alpha-1)\,(\alpha-2)\,(\alpha-3)\,(\alpha-4)}{(b-a)}\int_a^b (b-t)^{\alpha-5}\,f(t)\,dt \qquad (5.81)$$

$$-\frac{(\alpha-1)\,(\alpha-2)\,(\alpha-3)\,(\alpha-4)\,(\alpha-5)}{(b-a)}\int_a^b (b-t)^{\alpha-5}\,f(t)\,dt\,.$$

We have

$$-\frac{\prod_{j=1}^5 (\alpha-j)}{(b-a)}\int_a^b (b-t)^{\alpha-5}\,f(t)\,dt = -\frac{\prod_{j=1}^5 (\alpha-j)}{(b-a)}\cdot$$

$$\left[\int_a^x ((b-a)-(t-a))\,(b-t)^{\alpha-6}\,f(t)\,dt - \int_x^b (t-b)\,(b-t)^{\alpha-6}\,f(t)\,dt\right] =$$

$$-\prod_{j=1}^{5} (\alpha - j) \left[\int_a^x (b-t)^{\alpha-6} f(t) \, dt - \int_a^b P_1(x,t)(b-t)^{\alpha-6} f(t) \, dt \right].$$

(5.82)

Therefore it holds

$$-\frac{\alpha(\alpha-1)(\alpha-2)(\alpha-3)(\alpha-4)}{(b-a)} \int_a^b (b-t)^{\alpha-5} f(t) \, dt =$$

$$-\frac{5(\alpha-1)(\alpha-2)(\alpha-3)(\alpha-4)}{(b-a)} \int_a^b (b-t)^{\alpha-5} f(t) \, dt -$$

$$\prod_{j=1}^{5} (\alpha-j) \int_a^x (b-t)^{\alpha-6} f(t) \, dt + \prod_{j=1}^{5} (\alpha-j) \int_a^b P_1(x,t)(b-t)^{\alpha-6} f(t) \, dt.$$

(5.83)

Consequently we get

$$\Gamma(\alpha) J_a^\alpha \left(P_1(x,b) f^{(5)}(b) \right) = (b-a)^{\alpha-2} f^{(3)}(a) +$$

$$2(\alpha-1)(b-a)^{\alpha-3} f^{(2)}(a) + 3(\alpha-1)(\alpha-2)(b-a)^{\alpha-4} f'(a)$$

$$+ (\alpha-1)(\alpha-2)(\alpha-3)(\alpha-4)(b-x)^{\alpha-5} f(x) + \qquad (5.84)$$

$$4(\alpha-1)(\alpha-2)(\alpha-3)(b-a)^{\alpha-5} f(a) -$$

$$\frac{5(\alpha-1)(\alpha-2)(\alpha-3)(\alpha-4)}{(b-a)} \int_a^b (b-t)^{\alpha-5} f(t) \, dt$$

$$+ \prod_{j=1}^{5} (\alpha-j) \int_a^b P_1(x,t)(b-t)^{\alpha-6} f(t) \, dt.$$

So that

$$\Gamma(\alpha) J_a^\alpha \left(P_1(x,b) f^{(5)}(b) \right) = (b-a)^{\alpha-2} f^{(3)}(a) +$$

$$2(\alpha-1)(b-a)^{\alpha-3} f^{(2)}(a) + 3(\alpha-1)(\alpha-2)(b-a)^{\alpha-4} f'(a)$$

$$+ (\alpha-1)(\alpha-2)(\alpha-3)(\alpha-4)(b-x)^{\alpha-5} f(x) + \qquad (5.85)$$

$$4(\alpha-1)(\alpha-2)(\alpha-3)(b-a)^{\alpha-5} f(a) -$$

$$\frac{5(\alpha-1)(\alpha-2)(\alpha-3)(\alpha-4)\Gamma(\alpha-4)}{(b-a)} \left(J_a^{\alpha-4}(f(b)) \right)$$

$$+\prod_{j=1}^{5} (\alpha - j) \, \Gamma (\alpha - 5) \, J_a^{\alpha-5} \left(P_1 (x, b) \, f (b) \right).$$

And finally we derive

$$\prod_{j=1}^{4} (\alpha - j) \, (b - x)^{\alpha-5} \, f (x) = -4 \, (\alpha - 1) \, (\alpha - 2) \, (\alpha - 3) \, (b - a)^{\alpha-5} \, f (a)$$

(5.86)

$$-3 \, (\alpha - 1) \, (\alpha - 2) \, (b - a)^{\alpha-4} \, f' (a)$$

$$-2 \, (\alpha - 1) \, (b - a)^{\alpha-3} \, f^{(2)} (a) - (b - a)^{\alpha-2} \, f^{(3)} (a) +$$

$$\Gamma (\alpha) \left\{ \frac{5}{(b - a)} \left(J_a^{\alpha-4} (f (b)) \right) - J_a^{\alpha-5} (P_1 (x, b) \, f (b)) + J_a^{\alpha} \left(P_1 (x, b) \, f^{(5)} (b) \right) \right\},$$

proving the claim.   ∎

In general holds the following fractional representation formula

Theorem 5.6 *Let $\alpha > n$, $n \in \mathbb{N}$, $x \in [a, b)$ fixed, $f : [a, b] \to \mathbb{R}$ n-times differentiable, with $f^{(n)} : [a, b] \to \mathbb{R}$ integrable on $[a, b]$. Assume $f^{(j)} (x) = 0$, $j = 1, ..., n - 1$. Then*

$$f (x) = \frac{(b - x)^{n-\alpha}}{\prod\limits_{j=1}^{n-1} (\alpha - j)} \left\{ -(n - 1) \prod_{j=1}^{n-2} (\alpha - j) \, (b - a)^{\alpha-n} \, f (a) - \right.$$

(5.87)

$$(n - 2) \prod_{j=1}^{n-3} (\alpha - j) \, (b - a)^{\alpha-n+1} \, f' (a) - (n - 3) \prod_{j=1}^{n-4} (\alpha - j) \, (b - a)^{\alpha-n+2} \, f^{(2)} (a)$$

$$- (n - 4) \prod_{j=1}^{n-5} (\alpha - j) \, (b - a)^{\alpha-n+3} \, f^{(3)} (a) - \ldots$$

$$- (b - a)^{\alpha-2} \, f^{(n-2)} (a) + \Gamma (\alpha) \left\{ \frac{n}{b - a} \left(J_a^{\alpha-n+1} (f (b)) \right) - J_a^{\alpha-n} (P_1 (x, b) \, f (b)) \right.$$

$$\left. \left. + J_a^{\alpha} \left(P_1 (x, b) \, f^{(n)} (b) \right) \right\} \right\}.$$

Above we assume that $\prod\limits_{j=1}^{0} (\alpha - j) = 1$, *and* $\prod\limits_{j=1}^{k} (\alpha - j) = 0$ *if* $k \in \{-1, -2, ...\}$.

Also set $f^{(-1)} (a) := 0$.

Proof Based on Theorems 5.1–5.5. ∎

Theorems 5.1–5.5 are special cases of Theorem 5.6.
We give applications of Theorem 5.6 for $n = 6, 7$.

Theorem 5.7 *Let* $\alpha > 6$, $x \in [a, b)$ *fixed*, $f : [a, b] \to \mathbb{R}$ *six times differentiable*, *with* $f^{(6)} : [a, b] \to \mathbb{R}$ *integrable on* $[a, b]$. *Assume* $f^{(j)} (x) = 0$, $j = 1, ..., 5$. *Then*

$$f(x) = \frac{(b-x)^{6-\alpha}}{\prod\limits_{j=1}^{5} (\alpha - j)} \left\{ -5 \prod\limits_{j=1}^{4} (\alpha - j)(b-a)^{\alpha-6} f(a) - \right.$$

$$4 \prod\limits_{j=1}^{3} (\alpha - j)(b-a)^{\alpha-5} f'(a) - 3 \prod\limits_{j=1}^{2} (\alpha - j)(b-a)^{\alpha-4} f^{(2)}(a) -$$

$$2(\alpha - 1)(b-a)^{\alpha-3} f^{(3)}(a) - (b-a)^{\alpha-2} f^{(4)}(a) + \qquad (5.88)$$

$$\left. \Gamma(\alpha) \left\{ \frac{6}{b-a} \left(J_a^{\alpha-5}(f(b)) \right) - J_a^{\alpha-6}(P_1(x,b) f(b)) + J_a^{\alpha}\left(P_1(x,b) f^{(6)}(b) \right) \right\} \right\}.$$

Theorem 5.8 *Let* $\alpha > 7$, $x \in [a, b)$ *fixed*, $f : [a, b] \to \mathbb{R}$ *seven times differentiable*, *with* $f^{(7)} : [a, b] \to \mathbb{R}$ *integrable on* $[a, b]$. *Assume* $f^{(j)} (x) = 0$, $j = 1, ..., 6$. *Then*

$$f(x) = \frac{(b-x)^{7-\alpha}}{\prod\limits_{j=1}^{6} (\alpha - j)} \left\{ -6 \prod\limits_{j=1}^{5} (\alpha - j)(b-a)^{\alpha-7} f(a) - \qquad (5.89) \right.$$

$$5 \prod\limits_{j=1}^{4} (\alpha - j)(b-a)^{\alpha-6} f'(a) - 4 \prod\limits_{j=1}^{3} (\alpha - j)(b-a)^{\alpha-5} f''(a)$$

$$-3 \prod\limits_{j=1}^{2} (\alpha - j)(b-a)^{\alpha-4} f^{(3)}(a) - 2(\alpha - 1)(b-a)^{\alpha-3} f^{(4)}(a)$$

$$- (b-a)^{\alpha-2} f^{(5)}(a) + \Gamma(\alpha) \left\{ \frac{7}{b-a} \left(J_a^{\alpha-6}(f(b)) \right) - J_a^{\alpha-7}(P_1(x,b) f(b)) \right.$$

$$+J_a^\alpha \left(P_1\left(x,b\right) f^{(7)}\left(b\right)\right)\Big\}\Bigg].$$

We make

Remark 5.9 We rewrite (5.87) as follows:

$$E_n\left(f,\alpha,x\right) := f\left(x\right) + \frac{(b-x)^{n-\alpha}}{\prod\limits_{j=1}^{n-1}(\alpha-j)}\left\{(n-1)\prod_{j=1}^{n-2}(\alpha-j)(b-a)^{\alpha-n}\,f\left(a\right)+\right.$$

$$(5.90)$$

$$(n-2)\prod_{j=1}^{n-3}(\alpha-j)(b-a)^{\alpha-n+1}\,f'\left(a\right) + (n-3)\prod_{j=1}^{n-4}(\alpha-j)(b-a)^{\alpha-n+2}\,f^{(2)}\left(a\right)$$

$$+\left(n-4\right)\prod_{j=1}^{n-5}(\alpha-j)(b-a)^{\alpha-n+3}\,f^{(3)}\left(a\right) + \ldots + (b-a)^{\alpha-2}\,f^{(n-2)}\left(a\right)+$$

$$+\Gamma\left(\alpha\right)\left\{-\frac{n}{b-a}\left(J_a^{\alpha-n+1}\left(f\left(b\right)\right)\right) + J_a^{\alpha-n}\left(P_1\left(x,b\right) f\left(b\right)\right)\right\}\right\} =$$

$$\frac{(b-x)^{n-\alpha}\,\Gamma\left(\alpha\right)}{\prod\limits_{j=1}^{n-1}(\alpha-j)}\,J_a^\alpha\left(P_1\left(x,b\right) f^{(n)}\left(b\right)\right) =$$

$$\frac{(b-x)^{n-\alpha}}{\prod\limits_{j=1}^{n-1}(\alpha-j)}\int_a^b (b-t)^{\alpha-1}\,P_1\left(x,t\right) f^{(n)}\left(t\right) dt. \qquad (5.91)$$

We upper bound $E_n\left(f,\alpha,x\right)$, that is we upper bound the right hand side of (5.91).

Consequently we produce fractional Ostrowski type inequalities motivated by [1] done there for $n=1$.

Theorem 5.10 *Let* $\alpha > n$, $n \in \mathbb{N}$, $x \in [a,b)$ *fixed*, $f : [a,b] \to \mathbb{R}$ *n-times differentiable, with* $f^{(n)} : [a,b] \to \mathbb{R}$ *integrable on* $[a,b]$. *Assume* $f^{(j)}\left(x\right) = 0$, $j = 1, \ldots, n-1$, *and* $\left\| f^{(n)} \right\|_\infty < \infty$. *Then*

$$|E_n\left(f,\alpha,x\right)| \le$$

$$\frac{\|f^{(n)}\|_\infty}{\prod_{j=1}^{n-1}(\alpha-j)}\left[\frac{(b-x)^{n-\alpha}(b-a)^\alpha}{\alpha(\alpha+1)}-\frac{(b-x)^n}{\alpha}+\frac{2(b-x)^{n+1}}{(b-a)(\alpha+1)}\right], \quad (5.92)$$

where $E_n(f,\alpha,x)$ as in (5.90).

Proof We have that

$$|E_n(f,\alpha,x)| \le \frac{(b-x)^{n-\alpha}}{\prod_{j=1}^{n-1}(\alpha-j)}\int_a^b (b-t)^{\alpha-1}|P_1(x,t)|\left|f^{(n)}(t)\right|dt \le$$

$$\frac{(b-x)^{n-\alpha}\|f^{(n)}\|_\infty}{(b-a)\left(\prod_{j=1}^{n-1}(\alpha-j)\right)}\left[\int_a^x (b-t)^{\alpha-1}(t-a)\,dt+\int_x^b (b-t)^\alpha\,dt\right]= \quad (5.93)$$

$$\frac{(b-x)^{n-\alpha}\|f^{(n)}\|_\infty}{(b-a)\left(\prod_{j=1}^{n-1}(\alpha-j)\right)}\left[\int_a^x (b-t)^{\alpha-1}((b-a)-(b-t))\,dt+\frac{(b-x)^{\alpha+1}}{\alpha+1}\right]=$$

$$\frac{\|f^{(n)}\|_\infty (b-x)^{n-\alpha}}{(b-a)\left(\prod_{j=1}^{n-1}(\alpha-j)\right)}\left[(b-a)\int_a^x (b-t)^{\alpha-1}\,dt-\int_a^x (b-t)^\alpha\,dt+\frac{(b-x)^{\alpha+1}}{\alpha+1}\right]$$

$$=\frac{\|f^{(n)}\|_\infty (b-x)^{n-\alpha}}{(b-a)\left(\prod_{j=1}^{n-1}(\alpha-j)\right)}\cdot$$

$$\left[(b-a)\left(\frac{(b-a)^\alpha}{\alpha}-\frac{(b-x)^\alpha}{\alpha}\right)-\frac{(b-a)^{\alpha+1}}{\alpha+1}+\frac{2(b-x)^{\alpha+1}}{\alpha+1}\right]= \quad (5.94)$$

$$\frac{(b-x)^{n-\alpha}\|f^{(n)}\|_\infty}{\prod_{j=1}^{n-1}(\alpha-j)}\left[\frac{(b-a)^\alpha}{\alpha(\alpha+1)}-\frac{(b-x)^\alpha}{\alpha}+\frac{2(b-x)^{\alpha+1}}{(b-a)(\alpha+1)}\right]= \quad (5.95)$$

$$\frac{\left\| f^{(n)} \right\|_\infty}{\prod\limits_{j=1}^{n-1} (\alpha - j)} \left[\frac{(b-x)^{n-\alpha} (b-a)^\alpha}{\alpha (\alpha + 1)} - \frac{(b-x)^n}{\alpha} + \frac{2 (b-x)^{n+1}}{(b-a) (\alpha + 1)} \right].$$

■

Theorem 5.11 *Let all as in Theorem 5.6. Then*

$$|E_n (f, \alpha, x)| \leq \left(\frac{(b-x)^{n-\alpha} (b-a)^{\alpha-2}}{2 \prod\limits_{j=1}^{n-1} (\alpha - j)} \right) (b - a + |a + b - 2x|) \left\| f^{(n)} \right\|_{L_1([a,b])}.$$

(5.96)

Proof We have that

$$|E_n (f, \alpha, x)| \leq \frac{(b-x)^{n-\alpha}}{\prod\limits_{j=1}^{n-1} (\alpha - j)} \int_a^b (b-t)^{\alpha-1} |P_1 (x, t)| \left| f^{(n)} (t) \right| dt \leq \quad (5.97)$$

$$\left(\frac{(b-x)^{n-\alpha}}{\prod\limits_{j=1}^{n-1} (\alpha - j)} \right) (b-a)^{\alpha-2} \max \{x - a, b - x\} \left\| f^{(n)} \right\|_{L_1([a,b])} = \quad (5.98)$$

$$\left(\frac{(b-x)^{n-\alpha}}{\prod\limits_{j=1}^{n-1} (\alpha - j)} \right) (b-a)^{\alpha-2} \left(\frac{b - a + |a + b - 2x|}{2} \right) \left\| f^{(n)} \right\|_{L_1([a,b])}.$$

■

Theorem 5.12 *Let $p, q, r > 1$ such that $\frac{1}{p} + \frac{1}{q} + \frac{1}{r} = 1$. Let all as in Theorem 5.6, but now $f^{(n)} \in L_r ([a, b])$. Then*

$$|E_n (f, \alpha, x)| \leq$$

$$
\left(\frac{(b-x)^{n-\alpha}}{\prod\limits_{j=1}^{n-1}(\alpha-j)} \right) \frac{(b-a)^{\alpha-2+\frac{1}{p}}}{(p(\alpha-1)+1)^{\frac{1}{p}}} \left\{ \frac{(b-x)^{q+1}+(x-a)^{q+1}}{(q+1)} \right\}^{\frac{1}{q}} \left\| f^{(n)} \right\|_{L_r([a,b])}.
$$

$$(5.99)$$

Proof We have

$$
|E_n(f,\alpha,x)| \overset{(5.97)}{\leq} \left(\frac{(b-x)^{n-\alpha}}{\prod\limits_{j=1}^{n-1}(\alpha-j)} \right) \left(\int_a^b (b-t)^{(\alpha-1)p}\,dt \right)^{\frac{1}{p}} \cdot
$$

$$
\left(\int_a^b |P_1(x,t)|^q\,dt \right)^{\frac{1}{q}} \left\| f^{(n)} \right\|_{L_r([a,b])} = \tag{5.100}
$$

$$
\left(\frac{(b-x)^{n-\alpha}}{\prod\limits_{j=1}^{n-1}(\alpha-j)} \right) \frac{(b-a)^{(\alpha-2)+\frac{1}{p}}}{(p(\alpha-1)+1)^{\frac{1}{p}}} \cdot
$$

$$
\left(\int_a^x (t-a)^q\,dt + \int_x^b (b-t)^q\,dt \right)^{\frac{1}{q}} \left\| f^{(n)} \right\|_{L_r([a,b])} =
$$

$$
\left(\frac{(b-x)^{n-\alpha}}{\prod\limits_{j=1}^{n-1}(\alpha-j)} \right) \frac{(b-a)^{(\alpha-2)+\frac{1}{p}}}{(p(\alpha-1)+1)^{\frac{1}{p}}} \left(\frac{(x-a)^{q+1}}{(q+1)} + \frac{(b-x)^{q+1}}{(q+1)} \right)^{\frac{1}{q}} \left\| f^{(n)} \right\|_{L_r([a,b])} =
$$

$$
\left(\frac{(b-x)^{n-\alpha}}{\prod\limits_{j=1}^{n-1}(\alpha-j)} \right) \frac{(b-a)^{\alpha-2+\frac{1}{p}}}{(p(\alpha-1)+1)^{\frac{1}{p}}} \left\{ \frac{(b-x)^{q+1}+(x-a)^{q+1}}{(q+1)} \right\}^{\frac{1}{q}} \left\| f^{(n)} \right\|_{L_r([a,b])},
$$

$$(5.101)$$

proving the claim. ∎

References

1. G. Anastassiou, M. Hooshmandasl, A. Ghasemi, F. Moftakharzadeh, Montgomery identities for fractional integrals and related fractional inequalities. J. Inequal. Pure Appl. Math. **10**(4), Article 97, 6 (2009)
2. G.A. Anastassiou, in *Fractional Representation Formulae Under Initial Conditions and Fractional Ostrowski Type Inequalities*, Demonstratio Mathematica, 2014
3. D.S. Mitrinovic, J.E. Pecaric, A.M. Fink, *Inequalities for Functions and Their Integrals and Derivatives* (Kluwer Academic Publishers, Dordrecht, 1994)
4. S. Miller, B. Ross, *An Introduction to the Fractional Calculus and Fractional Differential Equations* (Wiley, USA, 1993)

Chapter 6
Basic Fractional Integral Inequalities

Here we present basic L_p fractional integral inequalities for left and right Riemann-Liouville, generalized Riemann-Liouville, Hadamard, Erdelyi-Kober and multivariate Riemann-Liouville fractional integrals. Then we derive basic L_p fractional inequalities regarding the left Riemann-Liouville, the left and right Caputo and the left and right Canavati type fractional derivatives. It follows [7].

6.1 Introduction

We start with some facts about fractional integrals needed in the sequel, for more details see, for instance [1, 12].

Let $a < b$, $a, b \in \mathbb{R}$. By $C^N([a, b])$, we denote the space of all functions on $[a, b]$ which have continuous derivatives up to order N, and $AC([a, b])$ is the space of all absolutely continuous functions on $[a, b]$. By $AC^N([a, b])$, we denote the space of all functions g with $g^{(N-1)} \in AC([a, b])$. For any $\alpha \in \mathbb{R}$, we denote by $[\alpha]$ the integral part of α (the integer k satisfying $k \leq \alpha < k + 1$), and $\lceil \alpha \rceil$ is the ceiling of α ($\min\{n \in \mathbb{N}, n \geq \alpha\}$). By $L_1(a, b)$, we denote the space of all functions integrable on the interval (a, b), and by $L_\infty(a, b)$ the set of all functions measurable and essentially bounded on (a, b). Clearly, $L_\infty(a, b) \subset L_1(a, b)$.

We start with the definition of the Riemann-Liouville fractional integrals, see [15]. Let $[a, b]$, $(-\infty < a < b < \infty)$ be a finite interval on the real axis $\mathbb{R}$. The Riemann-Liouville fractional integrals $I_{a+}^\alpha f$ and $I_{b-}^\alpha f$ of order $\alpha > 0$ are defined by

$$\left(I_{a+}^\alpha f\right)(x) = \frac{1}{\Gamma(\alpha)} \int_a^x f(t)(x - t)^{\alpha - 1} dt, \quad (x > a), \tag{6.1}$$

$$\left(I_{b-}^\alpha f\right)(x) = \frac{1}{\Gamma(\alpha)} \int_x^b f(t)(t - x)^{\alpha - 1} dt, \quad (x < b), \tag{6.2}$$

© Springer International Publishing Switzerland 2016
G.A. Anastassiou, *Intelligent Comparisons: Analytic Inequalities*,
Studies in Computational Intelligence 609,
DOI 10.1007/978-3-319-21121-3_6

respectively. Here $\Gamma(\alpha)$ is the Gamma function. These integrals are called the left-sided and the right-sided fractional integrals. We mention some properties of the operators $I^{\alpha}_{a+}f$ and $I^{\alpha}_{b-}f$ of order $\alpha > 0$, see also [17]. The first result yields that the fractional integral operators $I^{\alpha}_{a+}f$ and $I^{\alpha}_{b-}f$ are bounded in $L_p(a, b)$, $1 \le p \le \infty$, that is

$$\left\| I^{\alpha}_{a+}f \right\|_p \le K \left\| f \right\|_p, \quad \left\| I^{\alpha}_{b-}f \right\|_p \le K \left\| f \right\|_p, \tag{6.3}$$

where

$$K = \frac{(b-a)^{\alpha}}{\alpha \Gamma(\alpha)}. \tag{6.4}$$

Inequality (6.3), that is the result involving the left-sided fractional integral, was proved by H. G. Hardy in one of his first papers, see [13].

In this chapter we prove basic Hardy type fractional integral inequalities and we are motivated by [5, 6, 13, 14].

6.2 Main Results

We present our first result.

Theorem 6.1 *Let* $p, q > 1$ *such that* $\frac{1}{p} + \frac{1}{q} = 1$; $\alpha_i > 0$, $i = 1, \ldots, m$. *Let* $f_i :$ $(a, b) \to \mathbb{R}$, *be Lebesgue measurable functions so that* $\| f_i \|_q$ *is finite,* $i = 1, \ldots, m$. *Then*

$$\left\| \prod_{i=1}^{m} \left(I^{\alpha_i}_{a+} f_i \right) \right\|_p \le \frac{(b-a)^{\sum\limits_{i=1}^{m} \alpha_i + m\left(\frac{1}{p}-1\right)+\frac{1}{p}}}{\left[\left(p \sum\limits_{i=1}^{m} \alpha_i + m(1-p) + 1 \right)^{\frac{1}{p}} \left(\prod\limits_{i=1}^{m} \Gamma(\alpha_i) (p(\alpha_i - 1) + 1)^{\frac{1}{p}} \right) \right]}$$

$$\cdot \left(\prod_{i=1}^{m} \| f_i \|_q \right). \tag{6.5}$$

Proof By (6.1) we have

$$\left(I^{\alpha_i}_{a+} f_i \right)(x) = \frac{1}{\Gamma(\alpha_i)} \int_a^x (x-t)^{\alpha_i - 1} f_i(t) \, dt, \tag{6.6}$$

$x > a$, $i = 1, \ldots, m$.

We have that

$$\left|\left(I_{a+}^{\alpha_i} f_i\right)(x)\right| \le \frac{1}{\Gamma(\alpha_i)} \int_a^x (x-t)^{\alpha_i-1} |f_i(t)|\, dt, \tag{6.7}$$

$x > a, i = 1, \dots, m$.

By Hölder's inequality we get

$$\left|\left(I_{a+}^{\alpha_i} f_i\right)(x)\right| \le \frac{1}{\Gamma(\alpha_i)} \left(\int_a^x (x-t)^{p(\alpha_i-1)}\, dt\right)^{\frac{1}{p}} \left(\int_a^x |f_i(t)|^q\, dt\right)^{\frac{1}{q}}$$

$$\le \frac{1}{\Gamma(\alpha_i)} \frac{(x-a)^{(\alpha_i-1)+\frac{1}{p}}}{(p(\alpha_i-1)+1)^{\frac{1}{p}}} \left(\int_a^b |f_i(t)|^q\, dt\right)^{\frac{1}{q}}, \tag{6.8}$$

$x > a, i = 1, \dots, m$.

Therefore

$$\prod_{i=1}^m \left|\left(I_{a+}^{\alpha_i} f_i\right)(x)\right|^p \le \frac{1}{\left(\prod_{i=1}^m \Gamma(\alpha_i)\right)^p} \frac{(x-a)^{p\sum_{i=1}^m \alpha_i+m(1-p)}}{\prod_{i=1}^m (p(\alpha_i-1)+1)} \left(\prod_{i=1}^m \int_a^b |f_i(t)|^q\, dt\right)^{\frac{p}{q}}, \tag{6.9}$$

$x \in (a, b)$.

Consequently we get

$$\int_a^b \left(\prod_{i=1}^m \left|\left(I_{a+}^{\alpha_i} f_i\right)(x)\right|^p\right) dx \le \left(\frac{1}{\prod_{i=1}^m \left(\Gamma(\alpha_i)^p (p(\alpha_i-1)+1)\right)}\right)$$

$$\cdot \left(\int_a^b (x-a)^{p\sum_{i=1}^m \alpha_i+m(1-p)}\, dx\right) \left(\prod_{i=1}^m \int_a^b |f_i(t)|^q\, dt\right)^{\frac{p}{q}} \tag{6.10}$$

$$= \frac{(b-a)^{p\sum_{i=1}^m \alpha_i+m(1-p)+1} \left(\prod_{i=1}^m \int_a^b |f_i(t)|^q\, dt\right)^{\frac{p}{q}}}{\left[\left(p\sum_{i=1}^m \alpha_i+m(1-p)+1\right)\left(\prod_{i=1}^m \left(\Gamma(\alpha_i)^p (p(\alpha_i-1)+1)\right)\right)\right]}, \tag{6.11}$$

proving the claim. ∎

We give also the following general variant in

Theorem 6.2 *Let $p, q > 1$ such that $\frac{1}{p} + \frac{1}{q} = 1, r > 0; \alpha_i > 0, i = 1, \ldots, m$. Let $f_i : (a, b) \to \mathbb{R}$, be Lebesgue measurable functions so that $\| f_i \|_q$ is finite, $i = 1, \ldots, m$. Then*

$$\left\| \prod_{i=1}^{m} \left(I_{a+}^{\alpha_i} f_i \right) \right\|_r \leq \frac{(b-a)^{\sum\limits_{i=1}^{m} \alpha_i - m + \frac{m}{p} + \frac{1}{r}}}{\left[\left(r \left(\sum\limits_{i=1}^{m} \alpha_i - m + \frac{m}{p} \right) + 1 \right)^{\frac{1}{r}} \left(\prod\limits_{i=1}^{m} \Gamma(\alpha_i) \left(p(\alpha_i - 1) + 1 \right)^{\frac{1}{p}} \right) \right]}$$

$$\cdot \left(\prod_{i=1}^{m} \| f_i \|_q \right). \tag{6.12}$$

Proof Using $r > 0$ and (6.8) we get

$$\left| \left(I_{a+}^{\alpha_i} f_i \right)(x) \right|^r \leq \frac{1}{\Gamma(\alpha_i)^r} \frac{(x-a)^{r\left((\alpha_i - 1) + \frac{1}{p} \right)}}{(p(\alpha_i - 1) + 1)^{\frac{r}{p}}} \left(\int_a^b |f_i(t)|^q \, dt \right)^{\frac{r}{q}}, \tag{6.13}$$

and

$$\prod_{i=1}^{m} \left| \left(I_{a+}^{\alpha_i} f_i \right)(x) \right|^r \leq$$

$$\frac{1}{\prod\limits_{i=1}^{m} \Gamma(\alpha_i)^r} \frac{(x-a)^{r\left(\sum\limits_{i=1}^{m} \alpha_i - m + \frac{m}{p} \right)}}{\left(\prod\limits_{i=1}^{m} (p(\alpha_i - 1) + 1) \right)^{\frac{r}{p}}} \left(\prod_{i=1}^{m} \left(\int_a^b |f_i(t)|^q \, dt \right)^{\frac{1}{q}} \right)^r. \tag{6.14}$$

Consequently

$$\int_a^b \left(\prod_{i=1}^{m} \left| \left(I_{a+}^{\alpha_i} f_i \right)(x) \right|^r \right) dx \leq \frac{\left(\int_a^b (x-a)^{r\left(\sum\limits_{i=1}^{m} \alpha_i - m + \frac{m}{p} \right)} dx \right)}{\left(\prod\limits_{i=1}^{m} \Gamma(\alpha_i)^r \right) \left(\prod\limits_{i=1}^{m} (p(\alpha_i - 1) + 1) \right)^{\frac{r}{p}}}$$

$$\cdot \left(\prod_{i=1}^{m} \left(\int_a^b |f_i(t)|^q \, dt \right)^{\frac{1}{q}} \right)^r \tag{6.15}$$

$$= \frac{(b-a)^{r\left(\sum\limits_{i=1}^{m} \alpha_i - m + \frac{m}{p}\right)+1}}{\left(r\left(\sum\limits_{i=1}^{m} \alpha_i - m + \frac{m}{p}\right)+1\right)\left(\prod\limits_{i=1}^{m} \Gamma(\alpha_i)(p(\alpha_i-1)+1)^{\frac{1}{p}}\right)^r},$$ (6.16)

$$\cdot \left(\prod_{i=1}^{m}\left(\int_a^b |f_i(t)|^q \, dt\right)^{\frac{1}{q}}\right)^r.$$

The claim is proved. ■

We continue with

Theorem 6.3 *Let* $p, q > 1$ *such that* $\frac{1}{p} + \frac{1}{q} = 1$; $\alpha_i > 0, i = 1, \ldots, m$. *Let* $f_i :$ $(a, b) \to \mathbb{R}$, *be Lebesgue measurable functions so that* $\|f_i\|_q$ *is finite,* $i = 1, \ldots, m$. *Then*

$$\left\|\prod_{i=1}^{m}\left(I_{b-}^{\alpha_i} f_i\right)\right\|_p \leq \frac{(b-a)^{\sum\limits_{i=1}^{m} \alpha_i + m\left(\frac{1}{p}-1\right)+\frac{1}{p}}}{\left[\left(p\sum\limits_{i=1}^{m} \alpha_i + m(1-p)+1\right)^{\frac{1}{p}}\left(\prod\limits_{i=1}^{m} \Gamma(\alpha_i)(p(\alpha_i-1)+1)^{\frac{1}{p}}\right)\right]}$$

$$\cdot \left(\prod_{i=1}^{m} \|f_i\|_q\right).$$ (6.17)

Proof By (6.2) we have

$$\left(I_{b-}^{\alpha_i} f_i\right)(x) = \frac{1}{\Gamma(\alpha_i)} \int_x^b (t-x)^{\alpha_i-1} f_i(t) \, dt,$$ (6.18)

$x < b, i = 1, \ldots, m$.

We have that

$$\left|\left(I_{b-}^{\alpha_i} f_i\right)(x)\right| \leq \frac{1}{\Gamma(\alpha_i)} \int_x^b (t-x)^{\alpha_i-1} |f_i(t)| \, dt,$$ (6.19)

$x < b, i = 1, \ldots, m$.

By Hölder's inequality we get

$$\left|\left(I_{b-}^{\alpha_i} f_i\right)(x)\right| \leq \frac{1}{\Gamma(\alpha_i)}\left(\int_x^b (t-x)^{p(\alpha_i-1)} \, dt\right)^{\frac{1}{p}}\left(\int_x^b |f_i(t)|^q \, dt\right)^{\frac{1}{q}}$$ (6.20)

$$\leq \frac{1}{\Gamma(\alpha_i)} \frac{(b-x)^{\alpha_i - 1 + \frac{1}{p}}}{(p(\alpha_i - 1) + 1)^{\frac{1}{p}}} \left(\int_a^b |f_i(t)|^q \, dt \right)^{\frac{1}{q}}, \tag{6.21}$$

$x < b, i = 1, \ldots, m.$

Therefore

$$\prod_{i=1}^m |(I_{b-}^{\alpha_i} f_i)(x)|^p \leq \frac{1}{\left(\prod\limits_{i=1}^m \Gamma(\alpha_i) \right)^p} \frac{(b-x)^{p \sum\limits_{i=1}^m \alpha_i + m(1-p)}}{\prod\limits_{i=1}^m (p(\alpha_i - 1) + 1)} \left(\prod_{i=1}^m \int_a^b |f_i(t)|^q \, dt \right)^{\frac{p}{q}}, \tag{6.22}$$

$x \in (a, b).$

Consequently we get

$$\int_a^b \left(\prod_{i=1}^m |(I_{b-}^{\alpha_i} f_i)(x)|^p \right) dx \leq \left(\frac{1}{\left(\prod\limits_{i=1}^m \Gamma(\alpha_i) \right)^p \left(\prod\limits_{i=1}^m (p(\alpha_i - 1) + 1) \right)} \right)$$

$$\cdot \left(\int_a^b (b-x)^{p \sum\limits_{i=1}^m \alpha_i + m(1-p)} \, dx \right) \left(\prod_{i=1}^m \int_a^b |f_i(t)|^q \, dt \right)^{\frac{p}{q}} \tag{6.23}$$

$$= \frac{(b-a)^{p \sum\limits_{i=1}^m \alpha_i + m(1-p) + 1} \left(\prod\limits_{i=1}^m \int_a^b |f_i(t)|^q \, dt \right)^{\frac{p}{q}}}{\left[\left(p \sum\limits_{i=1}^m \alpha_i + m(1-p) + 1 \right) \left(\prod\limits_{i=1}^m \left(\Gamma(\alpha_i)^p (p(\alpha_i - 1) + 1) \right) \right) \right]}, \tag{6.24}$$

proving the claim. ∎

It follows

Theorem 6.4 *Let* $p, q > 1$ *such that* $\frac{1}{p} + \frac{1}{q} = 1, r > 0; \alpha_i > 0, i = 1, \ldots, m.$ *Let* $f_i : (a, b) \to \mathbb{R}$, *be Lebesgue measurable functions so that* $\|f_i\|_q$ *is finite,* $i = 1, \ldots, m.$ *Then*

$$\left\| \prod_{i=1}^m (I_{b-}^{\alpha_i} f_i) \right\|_r \leq \frac{(b-a)^{\sum\limits_{i=1}^m \alpha_i - m + \frac{m}{p} + \frac{1}{r}}}{\left[\left(r \left(\sum\limits_{i=1}^m \alpha_i - m + \frac{m}{p} \right) + 1 \right)^{\frac{1}{r}} \left(\prod\limits_{i=1}^m \Gamma(\alpha_i)(p(\alpha_i - 1) + 1)^{\frac{1}{p}} \right) \right]}$$

$$\cdot \left(\prod_{i=1}^m \|f_i\|_q \right). \tag{6.25}$$

Proof Using $r > 0$ and (6.21) we get

$$\left|\left(I_{b-}^{\alpha_i} f_i\right)(x)\right|^r \leq \frac{1}{\Gamma(\alpha_i)^r} \frac{(b-x)^{r\left((\alpha_i-1)+\frac{1}{p}\right)}}{(p(\alpha_i-1)+1)^{\frac{r}{p}}} \left(\int_a^b |f_i(t)|^q \, dt\right)^{\frac{r}{q}}, \tag{6.26}$$

and

$$\prod_{i=1}^m \left|\left(I_{b-}^{\alpha_i} f_i\right)(x)\right|^r \leq$$

$$\frac{1}{\prod\limits_{i=1}^m \Gamma(\alpha_i)^r} \frac{(b-x)^{r\left(\sum\limits_{i=1}^m \alpha_i - m + \frac{m}{p}\right)}}{\left(\prod\limits_{i=1}^m (p(\alpha_i-1)+1)\right)^{\frac{r}{p}}} \left(\prod_{i=1}^m \left(\int_a^b |f_i(t)|^q \, dt\right)^{\frac{1}{q}}\right)^r. \tag{6.27}$$

Consequently it holds

$$\int_a^b \left(\prod_{i=1}^m \left|\left(I_{b-}^{\alpha_i} f_i\right)(x)\right|^r\right) dx \leq \frac{\left(\int_a^b (b-x)^{r\left(\sum\limits_{i=1}^m \alpha_i - m + \frac{m}{p}\right)} dx\right)}{\left(\prod\limits_{i=1}^m \Gamma(\alpha_i)^r\right) \left(\prod\limits_{i=1}^m (p(\alpha_i-1)+1)\right)^{\frac{r}{p}}}$$

$$\cdot \left(\prod_{i=1}^m \left(\int_a^b |f_i(t)|^q \, dt\right)^{\frac{1}{q}}\right)^r \tag{6.28}$$

$$= \frac{(b-a)^{r\left(\sum\limits_{i=1}^m \alpha_i - m + \frac{m}{p}\right)+1}}{\left(r\left(\sum\limits_{i=1}^m \alpha_i - m + \frac{m}{p}\right)+1\right) \left(\prod\limits_{i=1}^m \Gamma(\alpha_i)(p(\alpha_i-1)+1)^{\frac{1}{p}}\right)^r}, \tag{6.29}$$

$$\cdot \left(\prod_{i=1}^m \left(\int_a^b |f_i(t)|^q \, dt\right)^{\frac{1}{q}}\right)^r.$$

The claim is proved. ∎

We need

Definition 6.5 ([16, p. 99]) The fractional integrals of a function f with respect to given function g are defined as follows:

Let $a, b \in \mathbb{R}$, $a < b$, $\alpha > 0$. Here g is an increasing function on $[a, b]$ and $g \in C^1([a, b])$. The left- and right-sided fractional integrals of a function f with respect to another function g in $[a, b]$ are given by

$$\left(I_{a+;g}^{\alpha} f\right)(x) = \frac{1}{\Gamma(\alpha)} \int_a^x \frac{g'(t) f(t) \, dt}{(g(x) - g(t))^{1-\alpha}}, \quad x > a, \tag{6.30}$$

$$\left(I_{b-;g}^{\alpha} f\right)(x) = \frac{1}{\Gamma(\alpha)} \int_x^b \frac{g'(t) f(t) \, dt}{(g(t) - g(x))^{1-\alpha}}, \quad x < b, \tag{6.31}$$

respectively.

We present

Theorem 6.6 *Let $p, q > 1$ such that $\frac{1}{p} + \frac{1}{q} = 1$; $\alpha_i > 0$, $i = 1, \ldots, m$. Here $a, b \in \mathbb{R}$ and strictly increasing g with $I_{a+;g}^{\alpha_i}$ as in Definition 6.5, see (6.30). Let $f_i : (a, b) \to \mathbb{R}$, be Lebesgue measurable functions so that $\|f_i\|_{L_q(g)}$ is finite, $i = 1, \ldots, m$. Then*

$$\left\| \prod_{i=1}^m \left(I_{a+;g}^{\alpha_i} f_i\right) \right\|_{L_p(g)} \leq$$

$$\frac{(g(b) - g(a))^{\sum_{i=1}^m \alpha_i + m\left(\frac{1}{p} - 1\right) + \frac{1}{p}}}{\left[\left(p \sum_{i=1}^m \alpha_i + m(1-p) + 1\right)^{\frac{1}{p}} \left(\prod_{i=1}^m \Gamma(\alpha_i)(p(\alpha_i - 1) + 1)^{\frac{1}{p}}\right)\right]} \left(\prod_{i=1}^m \|f_i\|_{L_q(g)}\right). \tag{6.32}$$

Proof By (6.30) we have

$$\left(I_{a+;g}^{\alpha_i} f_i\right)(x) = \frac{1}{\Gamma(\alpha_i)} \int_a^x \frac{g'(t) f_i(t)}{(g(x) - g(t))^{1-\alpha_i}} \, dt, \tag{6.33}$$

$x > a$, $i = 1, \ldots, m$.

We have that

$$\left|\left(I_{a+;g}^{\alpha_i} f_i\right)(x)\right| \leq \frac{1}{\Gamma(\alpha_i)} \int_a^x (g(x) - g(t))^{\alpha_i - 1} g'(t) |f_i(t)| \, dt$$

$$= \frac{1}{\Gamma(\alpha_i)} \int_a^x (g(x) - g(t))^{\alpha_i - 1} |f_i(t)| \, dg(t), \tag{6.34}$$

$x > a$, $i = 1, \ldots, m$.

By Hölder's inequality we get

$$\left|\left(I_{a+;g}^{\alpha_i} f_i\right)(x)\right| \leq$$

$$\frac{1}{\Gamma(\alpha_i)} \left(\int_a^x (g(x) - g(t))^{p(\alpha_i - 1)} \, dg(t)\right)^{\frac{1}{p}} \left(\int_a^x |f_i(t)|^q \, dg(t)\right)^{\frac{1}{q}}$$

$$\leq \frac{1}{\Gamma(\alpha_i)} \frac{(g(x) - g(a))^{\alpha_i - 1 + \frac{1}{p}}}{(p(\alpha_i - 1) + 1)^{\frac{1}{p}}} \left(\int_a^b |f_i(t)|^q \, dg(t)\right)^{\frac{1}{q}} \tag{6.35}$$

$$= \frac{1}{\Gamma(\alpha_i)} \frac{(g(x) - g(a))^{\alpha_i - 1 + \frac{1}{p}}}{(p(\alpha_i - 1) + 1)^{\frac{1}{p}}} \|f_i\|_{L_q(g)}, \tag{6.36}$$

$x > a, i = 1, \ldots, m$.
So we got

$$\left|\left(I_{a+;g}^{\alpha_i} f_i\right)(x)\right| \leq \frac{(g(x) - g(a))^{\alpha_i - 1 + \frac{1}{p}}}{\Gamma(\alpha_i)(p(\alpha_i - 1) + 1)^{\frac{1}{p}}} \|f_i\|_{L_q(g)}, \tag{6.37}$$

$x > a, i = 1, \ldots, m$.
Hence

$$\prod_{i=1}^m \left|\left(I_{a+;g}^{\alpha_i} f_i\right)(x)\right|^p \leq \frac{(g(x) - g(a))^{p \sum\limits_{i=1}^m \alpha_i + m(1-p)}}{\prod\limits_{i=1}^m \left(\Gamma(\alpha_i)^p (p(\alpha_i - 1) + 1)\right)} \prod_{i=1}^m \|f_i\|_{L_q(g)}^p, \tag{6.38}$$

$x \in (a, b)$.
Consequently, we obtain

$$\int_a^b \left(\prod_{i=1}^m \left|\left(I_{a+;g}^{\alpha_i} f_i\right)(x)\right|^p\right) dg(x) \leq$$

$$\frac{\prod\limits_{i=1}^m \|f_i\|_{L_q(g)}^p \int_a^b (g(x) - g(a))^{p \sum\limits_{i=1}^m \alpha_i + m(1-p)} \, dg(x)}{\prod\limits_{i=1}^m \left(\Gamma(\alpha_i)^p (p(\alpha_i - 1) + 1)\right)}$$

$$= \prod_{i=1}^{m} \left[\frac{\|f_i\|_{L_q(g)}^{p}}{\left(\Gamma\left(\alpha_i\right)^p \left(p(\alpha_i - 1) + 1\right)\right)} \right] \frac{(g(b) - g(a))^{p \sum_{i=1}^{m} \alpha_i + m(1-p) + 1}}{\left(p \sum_{i=1}^{m} \alpha_i + m(1 - p) + 1 \right)}, \quad (6.39)$$

proving the claim. ∎

We also give

Theorem 6.7 *Let $p, q > 1$ such that $\frac{1}{p} + \frac{1}{q} = 1$; $\alpha_i > 0, i = 1, \ldots, m$; $r > 0$. Here $a, b \in \mathbb{R}$ and strictly increasing g with $I_{a+;g}^{\alpha_i}$ as in Definition 6.5, see (6.30). Let $f_i : (a, b) \to \mathbb{R}$, be Lebesgue measurable functions and $\|f_i\|_{L_q(g)}$ is finite, $i = 1, \ldots, m$. Then*

$$\left\| \prod_{i=1}^{m} \left(I_{a+;g}^{\alpha_i} f_i \right) \right\|_{L_r(g)} \leq$$

$$\frac{(g(b) - g(a))^{\sum_{i=1}^{m} \alpha_i - m + \frac{m}{p} + \frac{1}{r}}}{\left[\left(r \left(\sum_{i=1}^{m} \alpha_i - m + \frac{m}{p} \right) + 1 \right)^{\frac{1}{r}} \left(\prod_{i=1}^{m} \Gamma(\alpha_i) \left(p(\alpha_i - 1) + 1 \right)^{\frac{1}{p}} \right) \right]} \left(\prod_{i=1}^{m} \|f_i\|_{L_q(g)} \right).$$

$$(6.40)$$

Proof Using $r > 0$ and (6.37) we get

$$\left| \left(I_{a+;g}^{\alpha_i} f_i \right) (x) \right|^r \leq \frac{(g(x) - g(a))^{r\left(\alpha_i - 1 + \frac{1}{p}\right)}}{\Gamma\left(\alpha_i\right)^r \left(p(\alpha_i - 1) + 1 \right)^{\frac{r}{p}}} \|f_i\|_{L_q(g)}^r, \quad (6.41)$$

and

$$\prod_{i=1}^{m} \left| \left(I_{a+;g}^{\alpha_i} f_i \right) (x) \right|^r \leq \frac{(g(x) - g(a))^{r\left(\sum_{i=1}^{m} \alpha_i - m + \frac{m}{p}\right)}}{\left(\prod_{i=1}^{m} \Gamma\left(\alpha_i\right) \left(p(\alpha_i - 1) + 1 \right)^{\frac{1}{p}} \right)^r} \left(\prod_{i=1}^{m} \|f_i\|_{L_q(g)} \right)^r, \quad (6.42)$$

$x \in (a, b)$.

Consequently, it holds

$$\int_a^b \prod_{i=1}^{m} \left| \left(I_{a+;g}^{\alpha_i} f_i \right) (x) \right|^r dg(x) \leq \frac{\left(\int_a^b (g(x) - g(a))^{r\left(\sum_{i=1}^{m} \alpha_i - m + \frac{m}{p}\right)} dg(x) \right)}{\left(\prod_{i=1}^{m} \left(\Gamma\left(\alpha_i\right) \left(p(\alpha_i - 1) + 1 \right)^{\frac{1}{p}} \right) \right)^r}$$

$$\cdot \left(\prod_{i=1}^{m} \| f_i \|_{L_q(g)} \right)^r \qquad (6.43)$$

$$= \frac{(g(b) - g(a))^{r \left(\sum_{i=1}^{m} \alpha_i - m + \frac{m}{p} \right) + 1} \left(\prod_{i=1}^{m} \| f_i \|_{L_q(g)} \right)^r}{\left(r \left(\sum_{i=1}^{m} \alpha_i - m + \frac{m}{p} \right) + 1 \right) \left(\prod_{i=1}^{m} \left(\Gamma(\alpha_i) \left(p(\alpha_i - 1) + 1 \right)^{\frac{1}{p}} \right) \right)^r}. \qquad (6.44)$$

The claim is proved. ∎

We continue with

Theorem 6.8 *Let* $p, q > 1$ *such that* $\frac{1}{p} + \frac{1}{q} = 1$; $\alpha_i > 0, i = 1, \ldots, m$. *Here* $a, b \in \mathbb{R}$ *and strictly increasing* g *with* $I_{b-;g}^{\alpha_i}$ *as in Definition 6.5, see (6.31). Let* $f_i : (a, b) \to \mathbb{R}$, *be Lebesgue measurable functions and* $\| f_i \|_{L_q(g)}$ *is finite,* $i = 1, \ldots, m$. *Then*

$$\left\| \prod_{i=1}^{m} \left(I_{b-;g}^{\alpha_i} f_i \right) \right\|_{L_p(g)} \leq$$

$$\frac{(g(b) - g(a))^{\sum_{i=1}^{m} \alpha_i + m \left(\frac{1}{p} - 1 \right) + \frac{1}{p}}}{\left[\left(p \sum_{i=1}^{m} \alpha_i + m(1 - p) + 1 \right)^{\frac{1}{p}} \left(\prod_{i=1}^{m} \Gamma(\alpha_i) \left(p(\alpha_i - 1) + 1 \right)^{\frac{1}{p}} \right) \right]} \left(\prod_{i=1}^{m} \| f_i \|_{L_q(g)} \right).$$

$$\qquad (6.45)$$

Proof By (6.31) we have

$$\left(I_{b-;g}^{\alpha_i} f_i \right)(x) = \frac{1}{\Gamma(\alpha_i)} \int_x^b \frac{g'(t) f_i(t)}{(g(t) - g(x))^{1-\alpha_i}} dt, \qquad (6.46)$$

$x < b, i = 1, \ldots, m$.

We have that

$$\left| \left(I_{b-;g}^{\alpha_i} f_i \right)(x) \right| \leq \frac{1}{\Gamma(\alpha_i)} \int_x^b (g(t) - g(x))^{\alpha_i - 1} g'(t) | f_i(t) | dt$$

$$= \frac{1}{\Gamma(\alpha_i)} \int_x^b (g(t) - g(x))^{\alpha_i - 1} | f_i(t) | dg(t), \qquad (6.47)$$

$x < b, i = 1, \ldots, m$.

By Hölder's inequality we get

$$\left|\left(I_{b-;g}^{\alpha_i} f_i\right)(x)\right| \leq$$

$$\frac{1}{\Gamma(\alpha_i)} \left(\int_x^b (g(t) - g(x))^{p(\alpha_i-1)} \, dg(t)\right)^{\frac{1}{p}} \left(\int_x^b |f_i(t)|^q \, dg(t)\right)^{\frac{1}{q}}$$

$$\leq \frac{1}{\Gamma(\alpha_i)} \frac{(g(b) - g(x))^{\alpha_i-1+\frac{1}{p}}}{(p(\alpha_i-1)+1)^{\frac{1}{p}}} \left(\int_a^b |f_i(t)|^q \, dg(t)\right)^{\frac{1}{q}} \qquad (6.48)$$

$$= \frac{1}{\Gamma(\alpha_i)} \frac{(g(b) - g(x))^{\alpha_i-1+\frac{1}{p}}}{(p(\alpha_i-1)+1)^{\frac{1}{p}}} \|f_i\|_{L_q(g)}, \qquad (6.49)$$

$x < b, i = 1, \ldots, m$.
So we got

$$\left|\left(I_{b-;g}^{\alpha_i} f_i\right)(x)\right| \leq \frac{(g(b) - g(x))^{\alpha_i-1+\frac{1}{p}}}{\Gamma(\alpha_i)(p(\alpha_i-1)+1)^{\frac{1}{p}}} \|f_i\|_{L_q(g)}, \qquad (6.50)$$

$x < b, i = 1, \ldots, m$.
Hence

$$\prod_{i=1}^m \left|\left(I_{b-;g}^{\alpha_i} f_i\right)(x)\right|^p \leq \frac{(g(b) - g(x))^{p\sum\limits_{i=1}^m \alpha_i+m(1-p)}}{\prod\limits_{i=1}^m \left(\Gamma(\alpha_i)^p (p(\alpha_i-1)+1)\right)} \prod_{i=1}^m \|f_i\|_{L_q(g)}^p, \qquad (6.51)$$

$x \in (a, b)$.
Consequently, we obtain

$$\int_a^b \left(\prod_{i=1}^m \left|\left(I_{b-;g}^{\alpha_i} f_i\right)(x)\right|^p\right) dg(x) \leq$$

$$\frac{\prod\limits_{i=1}^m \|f_i\|_{L_q(g)}^p \left(\int_a^b (g(b) - g(x))^{p\sum\limits_{i=1}^m \alpha_i+m(1-p)} \, dg(x)\right)}{\prod\limits_{i=1}^m \left(\Gamma(\alpha_i)^p (p(\alpha_i-1)+1)\right)} =$$

$$\prod_{i=1}^{m} \left[\frac{\|f_i\|_{L_q(g)}^p}{(\Gamma(\alpha_i)^p (p(\alpha_i - 1) + 1))} \right] \frac{(g(b) - g(a))^{p \sum_{i=1}^{m} \alpha_i + m(1-p)+1}}{\left(p \sum_{i=1}^{m} \alpha_i + m(1 - p) + 1 \right)}, \qquad (6.52)$$

proving the claim. ∎

We also give

Theorem 6.9 *Let* $p, q > 1$ *such that* $\frac{1}{p} + \frac{1}{q} = 1$; $\alpha_i > 0, i = 1, \ldots, m$, $r > 0$. *Here* $a, b \in \mathbb{R}$ *and strictly increasing* g *with* $I_{b-;g}^{\alpha_i}$ *as in Definition 6.5, see (6.31). Let* $f_i : (a, b) \rightarrow \mathbb{R}$, *be Lebesgue measurable functions and* $\|f_i\|_{L_q(g)}$ *is finite,* $i = 1, \ldots, m$. *Then*

$$\left\| \prod_{i=1}^{m} \left(I_{b-;g}^{\alpha_i} f_i \right) \right\|_{L_r(g)} \leq$$

$$\frac{(g(b) - g(a))^{\sum_{i=1}^{m} \alpha_i - m + \frac{m}{p} + \frac{1}{r}}}{\left[\left(r \left(\sum_{i=1}^{m} \alpha_i - m + \frac{m}{p} \right) + 1 \right)^{\frac{1}{r}} \left(\prod_{i=1}^{m} \Gamma(\alpha_i) (p(\alpha_i - 1) + 1)^{\frac{1}{p}} \right) \right]} \left(\prod_{i=1}^{m} \|f_i\|_{L_q(g)} \right).$$

$$(6.53)$$

Proof Using $r > 0$ and (6.50) we get

$$\left| \left(I_{b-;g}^{\alpha_i} f_i \right)(x) \right|^r \leq \frac{(g(b) - g(x))^{r\left(\alpha_i - 1 + \frac{1}{p} \right)}}{\Gamma(\alpha_i)^r (p(\alpha_i - 1) + 1)^{\frac{r}{p}}} \|f_i\|_{L_q(g)}^r, \qquad (6.54)$$

and

$$\prod_{i=1}^{m} \left| \left(I_{b-;g}^{\alpha_i} f_i \right)(x) \right|^r \leq \frac{(g(b) - g(x))^{r\left(\sum_{i=1}^{m} \alpha_i - m + \frac{m}{p} \right)}}{\prod_{i=1}^{m} \left(\Gamma(\alpha_i) (p(\alpha_i - 1) + 1)^{\frac{1}{p}} \right)^r} \left(\prod_{i=1}^{m} \|f_i\|_{L_q(g)} \right)^r, \qquad (6.55)$$

$x \in (a, b)$.

Consequently, it holds

$$\int_a^b \prod_{i=1}^{m} \left| \left(I_{b-;g}^{\alpha_i} f_i \right)(x) \right|^r dg(x) \leq \frac{\left(\int_a^b (g(b) - g(x))^{r\left(\sum_{i=1}^{m} \alpha_i - m + \frac{m}{p} \right)} dg(x) \right)}{\left(\prod_{i=1}^{m} \left(\Gamma(\alpha_i) (p(\alpha_i - 1) + 1)^{\frac{1}{p}} \right) \right)^r}$$

$$\cdot \left(\prod_{i=1}^{m} \| f_i \|_{L_q(g)} \right)^r \tag{6.56}$$

$$= \frac{(g(b) - g(a))^{r \left(\sum_{i=1}^{m} \alpha_i - m + \frac{m}{p} \right) + 1} \left(\prod_{i=1}^{m} \| f_i \|_{L_q(g)} \right)^r}{\left(r \left(\sum_{i=1}^{m} \alpha_i - m + \frac{m}{p} \right) + 1 \right) \left(\prod_{i=1}^{m} \left(\Gamma(\alpha_i) \left(p(\alpha_i - 1) + 1 \right)^{\frac{1}{p}} \right) \right)^r}. \tag{6.57}$$

The claim is proved. ∎

We need

Definition 6.10 ([15]). Let $0 < a < b < \infty$, $\alpha > 0$. The left- and right-sided Hadamard fractional integrals of order α are given by

$$\left(J_{a+}^{\alpha} f \right)(x) = \frac{1}{\Gamma(\alpha)} \int_{a}^{x} \left(\ln \frac{x}{y} \right)^{\alpha - 1} \frac{f(y)}{y} dy, \quad x > a, \tag{6.58}$$

and

$$\left(J_{b-}^{\alpha} f \right)(x) = \frac{1}{\Gamma(\alpha)} \int_{x}^{b} \left(\ln \frac{y}{x} \right)^{\alpha - 1} \frac{f(y)}{y} dy, \quad x < b, \tag{6.59}$$

respectively.

Notice that the Hadamard fractional integrals of order α are special cases of left- and right-sided fractional integrals of a function f with respect to another function, here $g(x) = \ln x$ on $[a, b]$, $0 < a < b < \infty$.

Above f is a Lebesgue measurable function from (a, b) into $\mathbb{R}$, such that $\left(J_{a+}^{\alpha} (|f|) \right)(x)$ and/or $\left(J_{b-}^{\alpha} (|f|) \right)(x) \in \mathbb{R}$, $\forall x \in (a, b)$.

We present

Theorem 6.11 Let $p, q > 1$ such that $\frac{1}{p} + \frac{1}{q} = 1$; $\alpha_i > 0$, $i = 1, \ldots, m$. Here $0 < a < b < \infty$, and $J_{a+}^{\alpha_i}$ as in Definition 6.10, see (6.58). Let $f_i : (a, b) \to \mathbb{R}$, be Lebesgue measurable functions and $\| f_i \|_{L_q(ln)}$ is finite, $i = 1, \ldots, m$. Then

$$\left\| \prod_{i=1}^{m} \left(J_{a+}^{\alpha_i} f_i \right) \right\|_{L_p(ln)} \leq$$

$$\frac{\left(\ln(\frac{b}{a}) \right)^{\sum_{i=1}^{m} \alpha_i + m(\frac{1}{p} - 1) + \frac{1}{p}}}{\left(p \sum_{i=1}^{m} \alpha_i + m(1 - p) + 1 \right)^{\frac{1}{p}} \left(\prod_{i=1}^{m} \left(\Gamma(\alpha_i) \left(p(\alpha_i - 1) + 1 \right)^{\frac{1}{p}} \right) \right)} \left(\prod_{i=1}^{m} \| f_i \|_{L_q(ln)} \right). \tag{6.60}$$

Proof By Theorem 6.6, for $g(x) = \ln x$. ∎

We also have

Theorem 6.12 *Let* $p, q > 1$ *such that* $\frac{1}{p} + \frac{1}{q} = 1$; $\alpha_i > 0, i = 1, \ldots, m$; $r > 0$. *Here* $0 < a < b < \infty$, *and* $J_{a+}^{\alpha_i}$ *as in Definition 6.10, see (6.58). Let* $f_i : (a, b) \to \mathbb{R}$, *be Lebesgue measurable functions and* $\|f_i\|_{L_q(ln)}$ *is finite,* $i = 1, \ldots, m$. *Then*

$$\left\| \prod_{i=1}^{m} \left(J_{a+}^{\alpha_i} f_i \right) \right\|_{L_r(ln)} \leq$$

$$\frac{\left(\ln(\frac{b}{a}) \right)^{\sum_{i=1}^{m} \alpha_i - m + \frac{m}{p} + \frac{1}{r}}}{\left(r \left(\sum_{i=1}^{m} \alpha_i - m + \frac{m}{p} \right) + 1 \right)^{\frac{1}{r}} \left(\prod_{i=1}^{m} \left(\Gamma(\alpha_i) \left(p(\alpha_i - 1) + 1 \right)^{\frac{1}{p}} \right) \right)} \left(\prod_{i=1}^{m} \|f_i\|_{L_q(ln)} \right). \tag{6.61}$$

Proof By Theorem 6.7, for $g(x) = \ln x$. ∎

We continue with

Theorem 6.13 *Let* $p, q > 1$ *such that* $\frac{1}{p} + \frac{1}{q} = 1$; $\alpha_i > 0, i = 1, \ldots, m$. *Here* $0 < a < b < \infty$, *and* $J_{b-}^{\alpha_i}$ *as in Definition 6.10, see (6.59). Let* $f_i : (a, b) \to \mathbb{R}$, *be Lebesgue measurable functions and* $\|f_i\|_{L_q(ln)}$ *is finite,* $i = 1, \ldots, m$. *Then*

$$\left\| \prod_{i=1}^{m} \left(J_{b-}^{\alpha_i} f_i \right) \right\|_{L_p(ln)}$$

$$\leq \frac{\left(\ln(\frac{b}{a}) \right)^{\sum_{i=1}^{m} \alpha_i + m(\frac{1}{p} - 1) + \frac{1}{p}}}{\left(p \sum_{i=1}^{m} \alpha_i + m(1 - p) + 1 \right)^{\frac{1}{p}} \left(\prod_{i=1}^{m} \left(\Gamma(\alpha_i) \left(p(\alpha_i - 1) + 1 \right)^{\frac{1}{p}} \right) \right)} \left(\prod_{i=1}^{m} \|f_i\|_{L_q(ln)} \right). \tag{6.62}$$

Proof By Theorem 6.8, for $g(x) = \ln x$. ∎

We also have

Theorem 6.14 *Let* $p, q > 1$ *such that* $\frac{1}{p} + \frac{1}{q} = 1$; $\alpha_i > 0, i = 1, \ldots, m$; $r > 0$. *Here* $0 < a < b < \infty$, *and* $J_{b-}^{\alpha_i}$ *as in Definition 6.10, see (6.59). Let* $f_i : (a, b) \to \mathbb{R}$, *be Lebesgue measurable functions and* $\|f_i\|_{L_q(ln)}$ *is finite,* $i = 1, \ldots, m$. *Then*

$$\left\| \prod_{i=1}^{m} \left(J_{b-}^{\alpha_i} f_i \right) \right\|_{L_r(ln)} \leq$$

$$\frac{(\ln(\frac{b}{a}))^{\sum\limits_{i=1}^{m}\alpha_i-m+\frac{m}{p}+\frac{1}{r}}}{\left(r\left(\sum\limits_{i=1}^{m}\alpha_i-m+\frac{m}{p}\right)+1\right)^{\frac{1}{r}}\left(\prod\limits_{i=1}^{m}\left(\Gamma(\alpha_i)\,(p(\alpha_i-1)+1)^{\frac{1}{p}}\right)\right)}\left(\prod_{i=1}^{m}\|f_i\|_{L_q(\ln)}\right).$$

$$(6.63)$$

Proof By Theorem 6.9, for $g(x) = \ln x$. ∎

We need

Definition 6.15 ([17]). Let $(a, b), 0 \le a < b < \infty$; $\alpha, \sigma > 0$. We consider the left- and right-sided fractional integrals of order α as follows:

(1) for $\eta > -1$, we define

$$\left(I^{\alpha}_{a+;\sigma,\eta}f\right)(x) = \frac{\sigma x^{-\sigma(\alpha+\eta)}}{\Gamma(\alpha)}\int_a^x \frac{t^{\sigma\eta+\sigma-1}f(t)\,dt}{(x^\sigma-t^\sigma)^{1-\alpha}}, \qquad (6.64)$$

(2) for $\eta > 0$, we define

$$\left(I^{\alpha}_{b-;\sigma,\eta}f\right)(x) = \frac{\sigma x^{\sigma\eta}}{\Gamma(\alpha)}\int_x^b \frac{t^{\sigma(1-\eta-\alpha)-1}f(t)\,dt}{(t^\sigma-x^\sigma)^{1-\alpha}}. \qquad (6.65)$$

These are the Erdélyi-Kober type fractional integrals.

We present

Theorem 6.16 *Let $p, q > 1$ such that $\frac{1}{p} + \frac{1}{q} = 1$; $\alpha_i > 0, i = 1, \ldots, m$. Here $0 \le a < b < \infty, \sigma > 0, \eta > -1$, and $I^{\alpha_i}_{a+;\sigma,\eta}$ is as in Definition 6.15, see (6.64). Let $f_i : (a, b) \to \mathbb{R}$, be Lebesgue measurable functions and $\|x^{\sigma\eta}f_i(x)\|_{L_q(x^\sigma)}$ is finite, $i = 1, \ldots, m$. Then*

$$\left\|\prod_{i=1}^{m}\left(x^{\sigma(\alpha_i+\eta)}\left(I^{\alpha_i}_{a+;\sigma,\eta}f_i\right)(x)\right)\right\|_{L_p(x^\sigma)} \le \frac{(b^\sigma-a^\sigma)^{\sum\limits_{i=1}^{m}\alpha_i+m(\frac{1}{p}-1)+\frac{1}{p}}}{\left(p\sum\limits_{i=1}^{m}\alpha_i+m(1-p)+1\right)^{\frac{1}{p}}}$$

$$\cdot\frac{1}{\left(\prod\limits_{i=1}^{m}\left(\Gamma(\alpha_i)\,(p(\alpha_i-1)+1)^{\frac{1}{p}}\right)\right)}\left(\prod_{i=1}^{m}\|x^{\sigma\eta}f_i(x)\|_{L_q(x^\sigma)}\right). \qquad (6.66)$$

Proof By Definition 6.15, see (6.64), we have

$$\left(I^{\alpha_i}_{a+;\sigma,\eta}f_i\right)(x) = \frac{\sigma x^{-\sigma(\alpha_i+\eta)}}{\Gamma(\alpha_i)}\int_a^x \frac{t^{\sigma\eta+\sigma-1}f_i(t)\,dt}{(x^\sigma-t^\sigma)^{1-\alpha_i}}, \qquad (6.67)$$

$x > a$. We rewrite (6.67) as follows:

$$L_1(f_i)(x) := x^{\sigma(\alpha_i + \eta)} \left(I_{a+;\sigma,\eta}^{\alpha_i} f_i \right)(x)$$

$$= \frac{1}{\Gamma(\alpha_i)} \int_a^x (x^\sigma - t^\sigma)^{\alpha_i - 1} \left(t^{\sigma\eta} f_i(t) \right) dt^\sigma, \tag{6.68}$$

and by calling $F_{1i}(t) = t^{\sigma\eta} f_i(t)$, we have

$$L_1(f_i)(x) = \frac{1}{\Gamma(\alpha_i)} \int_a^x (x^\sigma - t^\sigma)^{\alpha_i - 1} F_{1i}(t) dt^\sigma, \tag{6.69}$$

$i = 1, \ldots, m$, $x > a$. Furthermore we notice that

$$|L_1(f_i)(x)| \leq \frac{1}{\Gamma(\alpha_i)} \int_a^x (x^\sigma - t^\sigma)^{\alpha_i - 1} |F_{1i}(t)| dt^\sigma, \tag{6.70}$$

$i = 1, \ldots, m$, $x > a$.

So that now we can act as in the proof of Theorem 6.6. ∎

We continue with

Theorem 6.17 Let $p, q > 1$ such that $\frac{1}{p} + \frac{1}{q} = 1$; $\alpha_i > 0$, $i = 1, \ldots, m$, $r > 0$. Here $0 \leq a < b < \infty$, $\sigma > 0$, $\eta > -1$, and $I_{a+;\sigma,\eta}^{\alpha_i}$ is as in Definition 6.15, see (6.64). Let $f_i : (a, b) \to \mathbb{R}$, be Lebesgue measurable functions and $\|x^{\sigma\eta} f_i(x)\|_{L_q(x^\sigma)}$ is finite, $i = 1, \ldots, m$. Then

$$\left\| \prod_{i=1}^m \left(x^{\sigma(\alpha_i + \eta)} \left(I_{a+;\sigma,\eta}^{\alpha_i} f_i \right)(x) \right) \right\|_{L_r(x^\sigma)} \leq \frac{(b^\sigma - a^\sigma)^{\sum\limits_{i=1}^m \alpha_i - m + \frac{m}{p} + \frac{1}{r}}}{\left(r \left(\sum\limits_{i=1}^m \alpha_i - m + \frac{m}{p} \right) + 1 \right)^{\frac{1}{r}}}$$

$$\cdot \frac{1}{\left(\prod\limits_{i=1}^m \left(\Gamma(\alpha_i) (p(\alpha_i - 1) + 1)^{\frac{1}{p}} \right) \right)} \left(\prod_{i=1}^m \left\| x^{\sigma\eta} f_i(x) \right\|_{L_q(x^\sigma)} \right). \tag{6.71}$$

Proof Based on the proof of Theorem 6.16, and similarly acting as in the proof of Theorem 6.7. ∎

We also have

Theorem 6.18 Let $p, q > 1$ such that $\frac{1}{p} + \frac{1}{q} = 1$; $\alpha_i > 0$, $i = 1, \ldots, m$. Here $0 \leq a < b < \infty$, $\sigma > 0$, $\eta > 0$, and $I_{b-;\sigma,\eta}^{\alpha_i}$ is as in Definition 6.15, see (6.65). Let $f_i : (a, b) \to \mathbb{R}$, be Lebesgue measurable functions and $\left\| x^{-\sigma(\eta + \alpha_i)} f_i(x) \right\|_{L_q(x^\sigma)}$ is finite, $i = 1, \ldots, m$. Then

$$\left\| \prod_{i=1}^{m} \left(x^{-\sigma \eta} \left(I_{b-;\sigma,\eta}^{\alpha_i} f_i \right)(x) \right) \right\|_{L_p(x^\sigma)} \leq \frac{(b^\sigma - a^\sigma)^{\sum\limits_{i=1}^{m} \alpha_i + m(\frac{1}{p}-1)+\frac{1}{p}}}{\left(p \sum\limits_{i=1}^{m} \alpha_i + m(1-p) + 1 \right)^{\frac{1}{p}}}$$

$$\cdot \frac{1}{\left(\prod\limits_{i=1}^{m} \left(\Gamma(\alpha_i) \left(p(\alpha_i - 1) + 1 \right)^{\frac{1}{p}} \right) \right)} \left(\prod_{i=1}^{m} \left\| x^{-\sigma(\eta+\alpha_i)} f_i(x) \right\|_{L_q(x^\sigma)} \right). \quad (6.72)$$

Proof By Definition 6.15, see (6.65) we have

$$\left(I_{b-;\sigma,\eta}^{\alpha_i} f_i \right)(x) = \frac{\sigma x^{\sigma \eta}}{\Gamma(\alpha_i)} \int_x^b \frac{t^{\sigma(1-\eta-\alpha_i)-1} f_i(t) \, dt}{(t^\sigma - x^\sigma)^{1-\alpha_i}}, \quad (6.73)$$

$x < b$. We rewrite (6.73) as follows:

$$L_2(f_i)(x) := x^{-\sigma \eta} \left(I_{b-;\sigma,\eta}^{\alpha_i} f_i \right)(x)$$

$$= \frac{1}{\Gamma(\alpha_i)} \int_x^b \left(t^\sigma - x^\sigma \right)^{\alpha_i - 1} \left(t^{-\sigma(\eta+\alpha_i)} f_i(t) \right) dt^\sigma, \quad (6.74)$$

and by calling $F_{2i}(t) = t^{-\sigma(\eta+\alpha_i)} f_i(t)$, we have

$$L_2(f_i)(x) = \frac{1}{\Gamma(\alpha_i)} \int_x^b \left(t^\sigma - x^\sigma \right)^{\alpha_i - 1} F_{2i}(t) dt^\sigma, \quad (6.75)$$

$i = 1, \ldots, m$, $x < b$. Furthermore we notice that

$$|L_2(f_i)(x)| \leq \frac{1}{\Gamma(\alpha_i)} \int_x^b \left(t^\sigma - x^\sigma \right)^{\alpha_i - 1} |F_{2i}(t)| \, dt^\sigma, \quad (6.76)$$

$i = 1, \ldots, m$, $x < b$.
So that now we can act as in the proof of Theorem 6.8. ∎

We continue with

Theorem 6.19 *Let $p, q > 1$ such that $\frac{1}{p} + \frac{1}{q} = 1$; $\alpha_i > 0$, $i = 1, \ldots, m$, $r > 0$. Here $0 \leq a < b < \infty$, $\sigma > 0$, $\eta > 0$, and $I_{b-;\sigma,\eta}^{\alpha_i}$ is as in Definition 6.15, see (6.65) Let $f_i : (a, b) \to \mathbb{R}$, be Lebesgue measurable functions and $\left\| x^{-\sigma(\eta+\alpha_i)} f_i(x) \right\|_{L_q(x^\sigma)}$ is finite, $i = 1, \ldots, m$. Then*

$$\left\|\prod_{i=1}^{m}\left(x^{-\sigma\eta}\left(I_{b-;\sigma,\eta}^{\alpha_i}f_i\right)(x)\right)\right\|_{L_r(x^\sigma)} \le \frac{(b^\sigma - a^\sigma)^{\sum\limits_{i=1}^{m}\alpha_i - m + \frac{m}{p} + \frac{1}{r}}}{\left(r\left(\sum\limits_{i=1}^{m}\alpha_i - m + \frac{m}{p}\right) + 1\right)^{\frac{1}{r}}}$$

$$\cdot \frac{1}{\left(\prod\limits_{i=1}^{m}\left(\Gamma(\alpha_i)\left(p(\alpha_i - 1) + 1\right)^{\frac{1}{p}}\right)\right)}\left(\prod_{i=1}^{m}\left\|x^{-\sigma(\eta+\alpha_i)}f_i(x)\right\|_{L_q(x^\sigma)}\right). \quad (6.77)$$

Proof Based on the proof of Theorem 6.18, and acting similarly as in the proof of Theorem 6.9. ∎

We make

Definition 6.20 Let $\prod\limits_{i=1}^{N}(a_i, b_i) \subset \mathbb{R}^N$, $N > 1$, $a_i < b_i$, $a_i, b_i \in \mathbb{R}$. Let $\alpha_i > 0$, $i = 1, \ldots, N$; $f \in L_1\left(\prod\limits_{i=1}^{N}(a_i, b_i)\right)$, and set $a = (a_1, \ldots, a_N)$, $b = (b_1, \ldots, b_N)$, $\alpha = (\alpha_1, \ldots, \alpha_N)$, $x = (x_1, \ldots, x_N)$, $t = (t_1, \ldots, t_N)$.

We define the left mixed Riemann-Liouville fractional multiple integral of order α (see also [17]):

$$\left(I_{a+}^{\alpha}f\right)(x) := \frac{1}{\prod\limits_{i=1}^{N}\Gamma(\alpha_i)}\int_{a_1}^{x_1}\cdots\int_{a_N}^{x_N}\prod_{i=1}^{N}(x_i - t_i)^{\alpha_i - 1}f(t_1, \ldots, t_N)\,dt_1\ldots dt_N,$$

$$(6.78)$$

with $x_i > a_i$, $i = 1, \ldots, N$.

We also define the right mixed Riemann-Liouville fractional multiple integral of order α (see also [14]):

$$\left(I_{b-}^{\alpha}f\right)(x) := \frac{1}{\prod\limits_{i=1}^{N}\Gamma(\alpha_i)}\int_{x_1}^{b_1}\cdots\int_{x_N}^{b_N}\prod_{i=1}^{N}(t_i - x_i)^{\alpha_i - 1}f(t_1, \ldots, t_N)\,dt_1\ldots dt_N,$$

$$(6.79)$$

with $x_i < b_i$, $i = 1, \ldots, N$.

Notice $I_{a+}^{\alpha}(|f|)$, $I_{b-}^{\alpha}(|f|)$ are finite if $f \in L_\infty\left(\prod\limits_{i=1}^{N}(a_i, b_i)\right)$.

We present

Theorem 6.21 *Let* $p, q > 1$ *such that* $\frac{1}{p} + \frac{1}{q} = 1$. *Here all as in Definition 6.20,* and *(6.78) for* I_{a+}^{α}. *Let* $f_j : \prod_{i=1}^{N} (a_i, b_i) \to \mathbb{R}$, $j = 1, \ldots, m$, *such that* $f_j \in L_q \left(\prod_{i=1}^{N} (a_i, b_i) \right)$. *Then it holds*

$$\left\| \prod_{j=1}^{m} I_{a+}^{\alpha} f_j \right\|_{p, \prod_{i=1}^{N} (a_i, b_i)} \leq$$

$$\prod_{i=1}^{N} \left(\frac{(b_i - a_i)^{\left(m\left((\alpha_i - 1) + \frac{1}{p} \right) + \frac{1}{p} \right)}}{(m (p(\alpha_i - 1) + 1) + 1)^{\frac{1}{p}} \left(\Gamma(\alpha_i) (p(\alpha_i - 1) + 1)^{\frac{1}{p}} \right)^{m}} \right) \qquad (6.80)$$

$$\cdot \left(\prod_{j=1}^{m} \| f_j \|_{q, \prod_{i=1}^{N} (a_i, b_i)} \right).$$

Proof By Definition 6.20, see (6.78), we have

$$\left(I_{a+}^{\alpha} f_j \right) (x) = \frac{1}{\prod_{i=1}^{N} \Gamma(\alpha_i)} \int_{a_1}^{x_1} \cdots \int_{a_N}^{x_N} \prod_{i=1}^{N} (x_i - t_i)^{\alpha_i - 1} f_j (t_1, \ldots, t_N) \, dt_1 \ldots dt_N,$$

$$(6.81)$$

furthermore it holds

$$\left| \left(I_{a+}^{\alpha} f_j \right) (x) \right| \leq \frac{1}{\prod_{i=1}^{N} \Gamma(\alpha_i)} \int_{a_1}^{x_1} \cdots \int_{a_N}^{x_N} \prod_{i=1}^{N} (x_i - t_i)^{\alpha_i - 1} \left| f_j (t_1, \ldots, t_N) \right| \, dt_1 \ldots dt_N,$$

$$(6.82)$$

$j = 1, \ldots, m$, $x \in \prod_{i=1}^{N} (a_i, b_i)$.

By Hölder's inequality we get

$$\left| \left(I_{a+}^{\alpha} f_j \right) (x) \right| \leq \frac{1}{\prod_{i=1}^{N} \Gamma(\alpha_i)} \left(\int_{a_1}^{x_1} \cdots \int_{a_N}^{x_N} \prod_{i=1}^{N} (x_i - t_i)^{p(\alpha_i - 1)} \, dt_1 \ldots dt_N \right)^{\frac{1}{p}}$$

$$\cdot \left(\int_{a_1}^{x_1} \cdots \int_{a_N}^{x_N} \left| f_j\left(t_1, \cdots, t_N\right) \right|^q dt_1 \ldots dt_N \right)^{\frac{1}{q}} \tag{6.83}$$

$$\le \frac{1}{\prod\limits_{i=1}^{N} \Gamma\left(\alpha_i\right)} \left(\prod_{i=1}^{N} \left(\int_{a_i}^{x_i} \left(x_i - t_i\right)^{p\left(\alpha_i - 1\right)} dt_i \right)^{\frac{1}{p}} \right) \left(\int_{\prod\limits_{i=1}^{N}\left(a_i, b_i\right)} \left| f_j\left(t\right) \right|^q dt \right)^{\frac{1}{q}}$$

$$\tag{6.84}$$

$$= \frac{1}{\prod\limits_{i=1}^{N} \Gamma\left(\alpha_i\right)} \left(\prod_{i=1}^{N} \left(\frac{\left(x_i - a_i\right)^{\left(\alpha_i - 1\right)+\frac{1}{p}}}{\left(p\left(\alpha_i - 1\right) + 1\right)^{\frac{1}{p}}} \right) \right) \left(\int_{\prod\limits_{i=1}^{N}\left(a_i, b_i\right)} \left| f_j\left(t\right) \right|^q dt \right)^{\frac{1}{q}}. \tag{6.85}$$

Hence

$$\prod_{j=1}^{m} \left| \left(I_{a+}^{\alpha} f_j \right)(x) \right|^p \le \frac{1}{\left(\prod\limits_{i=1}^{N} \Gamma\left(\alpha_i\right) \right)^{mp}} \left(\prod_{i=1}^{N} \frac{\left(x_i - a_i\right)^{\left(\alpha_i - 1\right)+\frac{1}{p}}}{\left(p\left(\alpha_i - 1\right) + 1\right)^{\frac{1}{p}}} \right)^{mp}$$

$$\cdot \prod_{j=1}^{m} \left(\int_{\prod\limits_{i=1}^{N}\left(a_i, b_i\right)} \left| f_j\left(t\right) \right|^q dt \right)^{\frac{p}{q}}, \tag{6.86}$$

for $x \in \prod\limits_{i=1}^{N} \left(a_i, b_i\right)$.

Consequently, we get

$$\int_{\prod\limits_{i=1}^{N}\left(a_i, b_i\right)} \prod_{j=1}^{m} \left| \left(I_{a+}^{\alpha} f_j \right)(x) \right|^p dx \le \frac{\left(\prod\limits_{j=1}^{m} \left(\int_{\prod\limits_{i=1}^{N}\left(a_i, b_i\right)} \left| f_j\left(t\right) \right|^q dt \right)^{\frac{p}{q}} \right)}{\left(\prod\limits_{i=1}^{N} \Gamma\left(\alpha_i\right) \right)^{mp} \left(\prod\limits_{i=1}^{N} \left(p\left(\alpha_i - 1\right) + 1\right)^m \right)}$$

$$\cdot \left(\int_{\prod\limits_{i=1}^{N}\left(a_i, b_i\right)} \prod_{i=1}^{N} \left(x_i - a_i\right)^{m\left(p\left(\alpha_i - 1\right)+1\right)} dx_1 \ldots dx_N \right) \tag{6.87}$$

$$= \prod_{i=1}^{N} \left(\frac{\left(b_i - a_i\right)^{m\left(p\left(\alpha_i - 1\right)+1\right)+1}}{\left(m\left(p\left(\alpha_i - 1\right) + 1\right) + 1\right) \left(\Gamma\left(\alpha_i\right)^p \left(p\left(\alpha_i - 1\right) + 1\right)\right)^m} \right)$$

$$\cdot \left(\prod_{j=1}^{m} \left(\int_{\prod_{i=1}^{N}(a_i,b_i)} |f_j(t)|^q \, dt \right)^{\frac{p}{q}} \right),$$
(6.88)

proving the claim. ∎

We have

Theorem 6.22 *Let $p, q > 1$ such that $\frac{1}{p} + \frac{1}{q} = 1$; $r > 0$. Here all as in Definition 6.20, and (6.78) for I_{a+}^{α}. Let $f_j : \prod_{i=1}^{N}(a_i,b_i) \to \mathbb{R}$, $j = 1, \ldots, m$, such that $f_j \in L_q\left(\prod_{i=1}^{N}(a_i,b_i)\right)$. Then*

$$\left\| \prod_{j=1}^{m} I_{a+}^{\alpha} f_j \right\|_{r, \prod_{i=1}^{N}(a_i,b_i)} \leq$$

$$\prod_{i=1}^{N} \left(\frac{(b_i - a_i)^{\left(m\left((\alpha_i - 1) + \frac{1}{p}\right) + \frac{1}{r}\right)}}{\left(mr\left((\alpha_i - 1) + \frac{1}{p}\right) + 1\right)^{\frac{1}{r}} \Gamma(\alpha_i)^m \left(p(\alpha_i - 1) + 1\right)^{\frac{m}{p}}} \right)$$
(6.89)

$$\cdot \left(\prod_{j=1}^{m} \|f_j\|_{q, \prod_{i=1}^{N}(a_i,b_i)} \right).$$

Proof We have

$$(I_{a+}^{\alpha} f_j)(x) = \frac{1}{\prod_{i=1}^{N} \Gamma(\alpha_i)} \int_{a_1}^{x_1} \cdots \int_{a_N}^{x_N} \prod_{i=1}^{N} (x_i - t_i)^{\alpha_i - 1} f_j(t_1, \ldots, t_N) \, dt_1 \ldots dt_N,$$
(6.90)

furthermore it holds

$$|(I_{a+}^{\alpha} f_j)(x)| \leq \frac{1}{\prod_{i=1}^{N} \Gamma(\alpha_i)} \int_{a_1}^{x_1} \cdots \int_{a_N}^{x_N} \prod_{i=1}^{N} (x_i - t_i)^{\alpha_i - 1} |f_j(t_1, \ldots, t_N)| \, dt_1 \ldots dt_N,$$
(6.91)

$$j = 1, \ldots, m, \ x \in \prod_{i=1}^{N}(a_i,b_i).$$

By using (6.85) of the proof of Theorem 6.21 and $r > 0$ we get

$$\prod_{j=1}^{m} \left| \left(I_{a+}^{\alpha} f_j \right)(x) \right|^{r} \leq \frac{1}{\left(\prod_{i=1}^{N} \Gamma(\alpha_i) \right)^{mr}} \left(\prod_{i=1}^{N} \left(\frac{(x_i - a_i)^{(\alpha_i - 1) + \frac{1}{p}}}{(p(\alpha_i - 1) + 1)^{\frac{1}{p}}} \right) \right)^{mr}$$

$$\cdot \prod_{j=1}^{m} \left(\int_{\prod_{i=1}^{N}(a_i,b_i)} |f_j(t)|^q \, dt \right)^{\frac{r}{q}}, \tag{6.92}$$

for $x \in \prod\limits_{i=1}^{N} (a_i, b_i)$.

Consequently, we get

$$\int_{\prod_{i=1}^{N}(a_i,b_i)} \prod_{j=1}^{m} \left| \left(I_{a+}^{\alpha} f_j \right)(x) \right|^{r} dx \leq$$

$$\frac{1}{\left(\prod_{i=1}^{N} \Gamma(\alpha_i) \right)^{mr}} \frac{\left(\prod_{j=1}^{m} \left(\int_{\prod_{i=1}^{N}(a_i,b_i)} |f_j(t)|^q \, dt \right)^{\frac{1}{q}} \right)^{r}}{\left(\prod_{i=1}^{N} (p(\alpha_i - 1) + 1)^{\frac{mr}{p}} \right)}$$

$$\cdot \left(\int_{\prod_{i=1}^{N}(a_i,b_i)} \prod_{i=1}^{N} (x_i - a_i)^{mr\left((\alpha_i - 1) + \frac{1}{p}\right)} dx \right) \tag{6.93}$$

$$= \prod_{i=1}^{N} \left(\frac{(b_i - a_i)^{mr\left((\alpha_i - 1) + \frac{1}{p}\right) + 1}}{\left(mr\left((\alpha_i - 1) + \frac{1}{p}\right) + 1 \right) \Gamma(\alpha_i)^{mr} (p(\alpha_i - 1) + 1)^{\frac{mr}{p}}} \right)$$

$$\cdot \left(\prod_{j=1}^{m} \|f_j\|_{q, \prod_{i=1}^{N}(a_i,b_i)} \right)^{r}, \tag{6.94}$$

proving the claim. ∎

We also give

Theorem 6.23 *Let $p, q > 1$ such that $\frac{1}{p} + \frac{1}{q} = 1$. Here all as in Definition 6.20, and (6.79) for I_{b-}^α. Let $f_j : \prod_{i=1}^{N} (a_i, b_i) \to \mathbb{R}$, $j = 1, \ldots, m$, such that $f_j \in L_q \left(\prod_{i=1}^{N} (a_i, b_i) \right)$. Then it holds*

$$\left\| \prod_{j=1}^{m} I_{b-}^\alpha f_j \right\|_{p, \prod_{i=1}^{N} (a_i, b_i)} \leq$$

$$\prod_{i=1}^{N} \left(\frac{(b_i - a_i)^{\left(m\left((\alpha_i - 1) + \frac{1}{p}\right) + \frac{1}{p}\right)}}{(m(p(\alpha_i - 1) + 1) + 1)^{\frac{1}{p}} \left(\Gamma(\alpha_i)(p(\alpha_i - 1) + 1)^{\frac{1}{p}} \right)^m} \right) \tag{6.95}$$

$$\cdot \left(\prod_{j=1}^{m} \| f_j \|_{q, \prod_{i=1}^{N} (a_i, b_i)} \right).$$

Proof By Definition 6.20, see (6.79), we have

$$\left(I_{b-}^\alpha f_j \right)(x) = \frac{1}{\prod_{i=1}^{N} \Gamma(\alpha_i)} \int_{x_1}^{b_1} \cdots \int_{x_N}^{b_N} \prod_{i=1}^{N} (t_i - x_i)^{\alpha_i - 1} f_j(t_1, \ldots, t_N) \, dt_1 \ldots dt_N,$$

$$\tag{6.96}$$

furthermore it holds

$$\left| \left(I_{b-}^\alpha f_j \right)(x) \right| \leq \frac{1}{\prod_{i=1}^{N} \Gamma(\alpha_i)} \int_{x_1}^{b_1} \cdots \int_{x_N}^{b_N} \prod_{i=1}^{N} (t_i - x_i)^{\alpha_i - 1} \left| f_j(t_1, \ldots, t_N) \right| dt_1 \ldots dt_N,$$

$$\tag{6.97}$$

$j = 1, \ldots, m$, $x \in \prod_{i=1}^{N} (a_i, b_i)$.

By Hölder's inequality we get

$$\left| \left(I_{b-}^\alpha f_j \right)(x) \right| \leq \frac{1}{\prod_{i=1}^{N} \Gamma(\alpha_i)} \left(\int_{x_1}^{b_1} \cdots \int_{x_N}^{b_N} \prod_{i=1}^{N} (t_i - x_i)^{p(\alpha_i - 1)} \, dt_1 \ldots dt_N \right)^{\frac{1}{p}}$$

$$\cdot \left(\int_{x_1}^{b_1} \cdots \int_{x_N}^{b_N} \left| f_j \left(t_1, \dots, t_N \right) \right|^q dt_1 \dots dt_N \right)^{\frac{1}{q}} \tag{6.98}$$

$$\leq \frac{1}{\prod\limits_{i=1}^{N} \Gamma\left(\alpha_i \right)} \left(\prod_{i=1}^{N} \left(\int_{x_i}^{b_i} \left(t_i - x_i \right)^{p(\alpha_i - 1)} dt_i \right)^{\frac{1}{p}} \right)$$

$$\cdot \left(\int_{\prod\limits_{i=1}^{N} (a_i, b_i)} \left| f_j \left(t \right) \right|^q dt \right)^{\frac{1}{q}} \tag{6.99}$$

$$= \frac{1}{\prod\limits_{i=1}^{N} \Gamma\left(\alpha_i \right)} \left(\prod_{i=1}^{N} \left(\frac{\left(b_i - x_i \right)^{(\alpha_i - 1) + \frac{1}{p}}}{\left(p\left(\alpha_i - 1 \right) + 1 \right)^{\frac{1}{p}}} \right) \right)$$

$$\cdot \left(\int_{\prod\limits_{i=1}^{N} (a_i, b_i)} \left| f_j \left(t \right) \right|^q dt \right)^{\frac{1}{q}}. \tag{6.100}$$

Hence

$$\prod_{j=1}^{m} \left| \left(I_{b-}^{\alpha} f_j \right) \left(x \right) \right|^p \leq \frac{1}{\left(\prod\limits_{i=1}^{N} \Gamma\left(\alpha_i \right) \right)^{mp}} \left(\prod_{i=1}^{N} \frac{\left(b_i - x_i \right)^{(\alpha_i - 1) + \frac{1}{p}}}{\left(p\left(\alpha_i - 1 \right) + 1 \right)^{\frac{1}{p}}} \right)^{mp}$$

$$\cdot \prod_{j=1}^{m} \left(\int_{\prod\limits_{i=1}^{N} (a_i, b_i)} \left| f_j \left(t \right) \right|^q dt \right)^{\frac{p}{q}}, \tag{6.101}$$

for $x \in \prod\limits_{i=1}^{N} \left(a_i, b_i \right)$.

Consequently, we get

$$\int_{\prod\limits_{i=1}^{N} (a_i, b_i)} \prod_{j=1}^{m} \left| \left(I_{b-}^{\alpha} f_j \right) \left(x \right) \right|^p dx \leq \frac{\left(\prod\limits_{j=1}^{m} \left(\int_{\prod\limits_{i=1}^{N} (a_i, b_i)} \left| f_j \left(t \right) \right|^q dt \right)^{\frac{p}{q}} \right)}{\left(\prod\limits_{i=1}^{N} \Gamma\left(\alpha_i \right) \right)^{mp} \left(\prod\limits_{i=1}^{N} \left(p\left(\alpha_i - 1 \right) + 1 \right)^m \right)}$$

$$\cdot \left(\int_{\prod\limits_{i=1}^{N}(a_i,b_i)} \prod_{i=1}^{N} (b_i - x_i)^{m(p(\alpha_i-1)+1)} \, dx_1 \ldots dx_N \right) \tag{6.102}$$

$$= \prod_{i=1}^{N} \left(\frac{(b_i - a_i)^{m(p(\alpha_i-1)+1)+1}}{(m(p(\alpha_i - 1) + 1) + 1)\left((\Gamma(\alpha_i))^p (p(\alpha_i - 1) + 1)\right)^m} \right)$$

$$\cdot \left(\prod_{j=1}^{m} \left(\int_{\prod\limits_{i=1}^{N}(a_i,b_i)} |f_j(t)|^q \, dt \right)^{\frac{p}{q}} \right), \tag{6.103}$$

proving the claim. ∎

We have

Theorem 6.24 *Let $p, q > 1$ such that $\frac{1}{p} + \frac{1}{q} = 1$; $r > 0$. Here all as in Definition 6.20, and (6.79) for I_{b-}^{α}. Let $f_j : \prod\limits_{i=1}^{N} (a_i, b_i) \to \mathbb{R}$, $j = 1, \ldots, m$, such that $f_j \in L_q\left(\prod\limits_{i=1}^{N} (a_i, b_i) \right)$. Then*

$$\left\| \prod_{j=1}^{m} I_{b-}^{\alpha} f_j \right\|_{r, \prod\limits_{i=1}^{N}(a_i,b_i)} \leq$$

$$\prod_{i=1}^{N} \left(\frac{(b_i - a_i)^{\left(m\left((\alpha_i-1)+\frac{1}{p}\right)+\frac{1}{r}\right)}}{\left(mr\left((\alpha_i - 1) + \frac{1}{p}\right) + 1\right)^{\frac{1}{r}} \Gamma(\alpha_i)^m (p(\alpha_i - 1) + 1)^{\frac{m}{p}}} \right)$$

$$\cdot \left(\prod_{j=1}^{m} \| f_j \|_{q, \prod\limits_{i=1}^{N}(a_i,b_i)} \right). \tag{6.104}$$

Proof We have

$$\left(I_{b-}^{\alpha} f_j \right)(x) = \frac{1}{\prod\limits_{i=1}^{N} \Gamma(\alpha_i)} \int_{x_1}^{b_1} \cdots \int_{x_N}^{b_N} \prod_{i=1}^{N} (t_i - x_i)^{\alpha_i - 1} f_j(t_1, \ldots, t_N) \, dt_1 \ldots dt_N,$$

$$\tag{6.105}$$

furthermore it holds

$$\left|\left(I_{b-}^{\alpha} f_j\right)(x)\right| \leq \frac{1}{\prod\limits_{i=1}^{N} \Gamma(\alpha_i)} \int_{x_1}^{b_1} \cdots \int_{x_N}^{b_N} \prod_{i=1}^{N} (t_i - x_i)^{\alpha_i - 1} \left|f_j(t_1, \ldots, t_N)\right| dt_1 \ldots dt_N,$$

(6.106)

$j = 1, \ldots, m, \ x \in \prod\limits_{i=1}^{N} (a_i, b_i).$

By using (6.100) of the proof of Theorem 6.23 and $r > 0$ we get

$$\prod_{j=1}^{m} \left|\left(I_{b-}^{\alpha} f_j\right)(x)\right|^r \leq \frac{1}{\left(\prod\limits_{i=1}^{N} \Gamma(\alpha_i)\right)^{mr}} \left(\prod_{i=1}^{N} \left(\frac{(b_i - x_i)^{(\alpha_i - 1) + \frac{1}{p}}}{(p(\alpha_i - 1) + 1)^{\frac{1}{p}}}\right)\right)^{mr}$$

$$\cdot \prod_{j=1}^{m} \left(\int_{\prod\limits_{i=1}^{N} (a_i, b_i)} |f_j(t)|^q \, dt\right)^{\frac{r}{q}},$$

(6.107)

for $x \in \prod\limits_{i=1}^{N} (a_i, b_i).$

Consequently, we get

$$\int_{\prod\limits_{i=1}^{N} (a_i, b_i)} \prod_{j=1}^{m} \left|\left(I_{b-}^{\alpha} f_j\right)(x)\right|^r dx \leq$$

$$\frac{1}{\left(\prod\limits_{i=1}^{N} \Gamma(\alpha_i)\right)^{mr}} \frac{\left(\prod\limits_{j=1}^{m} \left(\int_{\prod\limits_{i=1}^{N} (a_i, b_i)} |f_j(t)|^q \, dt\right)^{\frac{1}{q}}\right)^r}{\left(\prod\limits_{i=1}^{N} (p(\alpha_i - 1) + 1)^{\frac{mr}{p}}\right)}$$

$$\cdot \left(\int_{\prod\limits_{i=1}^{N} (a_i, b_i)} \prod_{i=1}^{N} (b_i - x_i)^{mr\left((\alpha_i - 1) + \frac{1}{p}\right)} dx\right)$$

(6.108)

$$= \prod_{i=1}^{N} \left(\frac{(b_i - a_i)^{mr\left((\alpha_i - 1) + \frac{1}{p}\right) + 1}}{\left(mr\left((\alpha_i - 1) + \frac{1}{p}\right) + 1\right) \Gamma(\alpha_i)^{mr} (p(\alpha_i - 1) + 1)^{\frac{mr}{p}}}\right)$$

$$\cdot \left(\prod_{j=1}^{m} \| f_j \|_{q, \prod_{i=1}^{N} (a_i, b_i)} \right)^r, \tag{6.109}$$

proving the claim. ∎

Definition 6.25 ([1], p. 448). The left generalized Riemann-Liouville fractional derivative of f of order $\beta > 0$ is given by

$$D_a^\beta f(x) = \frac{1}{\Gamma(n-\beta)} \left(\frac{d}{dx} \right)^n \int_a^x (x-y)^{n-\beta-1} f(y) \, dy, \tag{6.110}$$

where $n = [\beta] + 1$, $x \in [a, b]$.

For $a, b \in \mathbb{R}$, we say that $f \in L_1(a, b)$ has an L_∞ fractional derivative $D_a^\beta f$ ($\beta > 0$) in $[a, b]$, if and only if

(1) $D_a^{\beta-k} f \in C([a, b])$, $k = 2, \ldots, n = [\beta] + 1$,
(2) $D_a^{\beta-1} f \in AC([a, b])$
(3) $D_a^\beta f \in L_\infty(a, b)$.

Above we define $D_a^0 f := f$ and $D_a^{-\delta} f := I_{a+}^\delta f$, if $0 < \delta \le 1$.

From [1, p. 449] and [12] we mention and use

Lemma 6.26 *Let $\beta > \alpha \ge 0$ and let $f \in L_1(a, b)$ have an L_∞ fractional derivative $D_a^\beta f$ in $[a, b]$ and let $D_a^{\beta-k} f(a) = 0$, $k = 1, \ldots, [\beta] + 1$, then*

$$D_a^\alpha f(x) = \frac{1}{\Gamma(\beta-\alpha)} \int_a^x (x-y)^{\beta-\alpha-1} D_a^\beta f(y) \, dy, \tag{6.111}$$

for all $a \le x \le b$.

Here $D_a^\alpha f \in AC([a, b])$ for $\beta - \alpha \ge 1$, and $D_a^\alpha f \in C([a, b])$ for $\beta - \alpha \in (0, 1)$. Notice here that

$$D_a^\alpha f(x) = \left(I_{a+}^{\beta-\alpha} \left(D_a^\beta f \right) \right)(x), \quad a \le x \le b. \tag{6.112}$$

We present

Theorem 6.27 *Let $p, q > 1$ such that $\frac{1}{p} + \frac{1}{q} = 1$; $\beta_i > \alpha_i \ge 0$, $i = 1, \ldots, m$. Let $f_i \in L_1(a, b)$ have an L_∞ fractional derivative $D_a^{\beta_i} f_i$ in $[a, b]$ and let $D_a^{\beta_i - k_i} f_i(a) = 0$, $k_i = 1, \ldots, [\beta_i] + 1$. Then*

$$\left\| \prod_{i=1}^m \left(D_a^{\alpha_i} f_i \right) \right\|_p \le \frac{(b-a)^{\sum\limits_{i=1}^m (\beta_i - \alpha_i) + m\left(\frac{1}{p}-1\right) + \frac{1}{p}}}{\left(p \sum\limits_{i=1}^m (\beta_i - \alpha_i) + m(1-p) + 1 \right)^{\frac{1}{p}}}$$

$$\cdot \frac{1}{\left(\prod\limits_{i=1}^{m} \Gamma\left(\beta_i - \alpha_i\right) \left(p(\beta_i - \alpha_i - 1) + 1\right)^{\frac{1}{p}}\right)} \left(\prod\limits_{i=1}^{m} \left\| D_a^{\beta_i} f_i \right\|_q\right). \tag{6.113}$$

Proof Using Theorem 6.1, see (6.5), and Lemma 6.26, see (6.112). ∎

We also give

Theorem 6.28 *Let* $p, q > 1$ *such that* $\frac{1}{p} + \frac{1}{q} = 1$; $r > 0$, $\beta_i > \alpha_i \geq 0$, $i = 1, \ldots, m$. *Let* $f_i \in L_1(a, b)$ *have an* L_∞ *fractional derivative* $D_a^{\beta_i} f_i$ *in* $[a, b]$ *and let* $D_a^{\beta_i - k_i} f_i(a) = 0$, $k_i = 1, \ldots, [\beta_i] + 1$. *Then*

$$\left\| \prod_{i=1}^{m} \left(D_a^{\alpha_i} f_i\right) \right\|_r \leq \frac{(b-a)^{\sum\limits_{i=1}^{m}(\beta_i - \alpha_i) - m + \frac{m}{p} + \frac{1}{r}}}{\left(r\left(\sum\limits_{i=1}^{m}(\beta_i - \alpha_i) - m + \frac{m}{p}\right) + 1\right)^{\frac{1}{r}}}$$

$$\cdot \frac{1}{\left(\prod\limits_{i=1}^{m} \Gamma\left(\beta_i - \alpha_i\right) \left(p(\beta_i - \alpha_i - 1) + 1\right)^{\frac{1}{p}}\right)} \left(\prod\limits_{i=1}^{m} \left\| D_a^{\beta_i} f_i \right\|_q\right). \tag{6.114}$$

Proof Using Theorem 6.2, see (6.12), and Lemma 6.26, see (6.112). ∎

We need

Definition 6.29 ([9], p. 50, [1], p. 449). Let $\nu \geq 0$, $n := \lceil \nu \rceil$, $f \in AC^n([a, b])$. Then the left Caputo fractional derivative is given by

$$D_{*a}^{\nu} f(x) = \frac{1}{\Gamma(n-\nu)} \int_a^x (x-t)^{n-\nu-1} f^{(n)}(t)\, dt$$

$$= \left(I_{a+}^{n-\nu} f^{(n)}\right)(x), \tag{6.115}$$

and it exists almost everywhere for $x \in [a, b]$, in fact $D_{*a}^{\nu} f \in L_1(a, b)$, ([1], p. 394).

We have $D_{*a}^n f = f^{(n)}$, $n \in \mathbb{Z}_+$.

We also need

Theorem 6.30 ([4]). *Let* $\nu \geq \rho + 1$, $\rho > 0$, $\nu, \rho \notin \mathbb{N}$. *Call* $n := \lceil \nu \rceil$, $m^* := \lceil \rho \rceil$. *Assume* $f \in AC^n([a, b])$, *such that* $f^{(k)}(a) = 0$, $k = m^*, m^* + 1, \ldots, n - 1$, *and* $D_{*a}^{\nu} f \in L_\infty(a, b)$. *Then* $D_{*a}^{\rho} f \in AC([a, b])$ *(where* $D_{*a}^{\rho} f = \left(I_{a+}^{m^* - \rho} f^{(m^*)}\right)(x)$*), and*

$$D_{*a}^{\rho} f(x) = \frac{1}{\Gamma(\nu - \rho)} \int_a^x (x - t)^{\nu - \rho - 1} D_{*a}^{\nu} f(t) \, dt$$

$$= \left(I_{a+}^{\nu - \rho} \left(D_{*a}^{\nu} f \right) \right)(x), \tag{6.116}$$

$\forall \, x \in [a, b]$.

We present

Theorem 6.31 *Let* $p, q > 1$ *such that* $\frac{1}{p} + \frac{1}{q} = 1$; *and let* $\nu_i \geq \rho_i + 1$, $\rho_i > 0$, $\nu_i, \rho_i \notin \mathbb{N}$, $i = 1, \ldots, m$. *Call* $n_i := \lceil \nu_i \rceil$, $m_i^* := \lceil \rho_i \rceil$. *Suppose* $f_i \in AC^{n_i}([a, b])$, *such that* $f_i^{(k_i)}(a) = 0$, $k_i = m_i^*, m_i^* + 1, \ldots, n_i - 1$, *and* $D_{*a}^{\nu_i} f_i \in L_\infty(a, b)$. *Then*

$$\left\| \prod_{i=1}^m \left(D_{*a}^{\rho_i} f_i \right) \right\|_p \leq \frac{(b - a)^{\sum\limits_{i=1}^m (\nu_i - \rho_i) + m\left(\frac{1}{p} - 1\right) + \frac{1}{p}}}{\left(p \sum\limits_{i=1}^m (\nu_i - \rho_i) + m(1 - p) + 1 \right)^{\frac{1}{p}}}$$

$$\cdot \frac{1}{\left(\prod\limits_{i=1}^m \Gamma(\nu_i - \rho_i)(p(\nu_i - \rho_i - 1) + 1)^{\frac{1}{p}} \right)} \left(\prod_{i=1}^m \left\| D_{*a}^{\nu_i} f_i \right\|_q \right). \tag{6.117}$$

Proof Using Theorem 6.1, see (6.5), and Theorem 6.30, see (6.116). ∎

We also give

Theorem 6.32 *Let* $p, q > 1$ *such that* $\frac{1}{p} + \frac{1}{q} = 1$, $r > 0$; *and let* $\nu_i \geq \rho_i + 1$, $\rho_i > 0$, $\nu_i, \rho_i \notin \mathbb{N}$, $i = 1, \ldots, m$. *Call* $n_i := \lceil \nu_i \rceil$, $m_i^* := \lceil \rho_i \rceil$. *Suppose* $f_i \in AC^{n_i}([a, b])$, *such that* $f_i^{(k_i)}(a) = 0$, $k_i = m_i^*, m_i^* + 1, \ldots, n_i - 1$, *and* $D_{*a}^{\nu_i} f_i \in L_\infty(a, b)$. *Then*

$$\left\| \prod_{i=1}^m \left(D_{*a}^{\rho_i} f_i \right) \right\|_r \leq \frac{(b - a)^{\sum\limits_{i=1}^m (\nu_i - \rho_i) - m + \frac{m}{p} + \frac{1}{r}}}{\left(r \left(\sum\limits_{i=1}^m (\nu_i - \rho_i) - m + \frac{m}{p} \right) + 1 \right)^{\frac{1}{r}}}$$

$$\cdot \frac{1}{\left(\prod\limits_{i=1}^m \Gamma(\nu_i - \rho_i)(p(\nu_i - \rho_i - 1) + 1)^{\frac{1}{p}} \right)} \left(\prod_{i=1}^m \left\| D_{*a}^{\nu_i} f_i \right\|_q \right). \tag{6.118}$$

Proof Using Theorem 6.2, see (6.12), and Theorem 6.30, see (6.116). ∎

We need

Definition 6.33 ([2, 10, 11]). Let $\alpha \geq 0$, $n := \lceil \alpha \rceil$, $f \in AC^n([a, b])$. We define the right Caputo fractional derivative of order $\alpha \geq 0$, by

$$\overline{D}_{b-}^{\alpha} f(x) := (-1)^n I_{b-}^{n-\alpha} f^{(n)}(x), \tag{6.119}$$

we set $\overline{D}_-^0 f := f$, i.e.

$$\overline{D}_{b-}^{\alpha} f(x) = \frac{(-1)^n}{\Gamma(n-\alpha)} \int_x^b (J-x)^{n-\alpha-1} f^{(n)}(J) \, dJ. \tag{6.120}$$

Notice that $\overline{D}_{b-}^n f = (-1)^n f^{(n)}$, $n \in \mathbb{N}$.

We need

Theorem 6.34 ([4]). *Let* $f \in AC^n([a, b])$, $\alpha > 0$, $n \in \mathbb{N}$, $n := \lceil \alpha \rceil$, $\alpha \geq \rho + 1$, $\rho > 0$, $r = \lceil \rho \rceil$, $\alpha, \rho \notin \mathbb{N}$. *Assume* $f^{(k)}(b) = 0$, $k = r, r+1, \ldots, n-1$, *and* $\overline{D}_{b-}^{\alpha} f \in L_\infty([a, b])$. *Then*

$$\overline{D}_{b-}^{\rho} f(x) = \left(I_{b-}^{\alpha-\rho} \left(\overline{D}_{b-}^{\alpha} f \right) \right)(x) \in AC([a, b]), \tag{6.121}$$

that is

$$\overline{D}_{b-}^{\rho} f(x) = \frac{1}{\Gamma(\alpha-\rho)} \int_x^b (t-x)^{\alpha-\rho-1} \left(\overline{D}_{b-}^{\alpha} f \right)(t) \, dt, \tag{6.122}$$

$\forall \, x \in [a, b]$.

We present

Theorem 6.35 *Let* $p, q > 1$ *such that* $\frac{1}{p} + \frac{1}{q} = 1$; $\alpha_i \geq \rho_i + 1$, $\rho_i > 0$, $i = 1, \ldots, m$. *Suppose* $f_i \in AC^{n_i}([a, b])$, $n_i \in \mathbb{N}$, $n_i := \lceil \alpha_i \rceil$, $r_i = \lceil \rho_i \rceil$, $\alpha_i, \rho_i \notin \mathbb{N}$, *and* $f_i^{(k_i)}(b) = 0$, $k_i = r_i, r_i + 1, \ldots, n_i - 1$, *and* $\overline{D}_{b-}^{\alpha_i} f_i \in L_\infty([a, b])$, $i = 1, \ldots, m$. *Then*

$$\left\| \prod_{i=1}^m \left(\overline{D}_{b-}^{\rho_i} f_i \right) \right\|_p \leq \frac{(b-a)^{\sum_{i=1}^m (\alpha_i - \rho_i) + m\left(\frac{1}{p}-1\right) + \frac{1}{p}}}{\left(p \sum_{i=1}^m (\alpha_i - \rho_i) + m(1-p) + 1 \right)^{\frac{1}{p}}}$$

$$\cdot \frac{1}{\left(\prod_{i=1}^m \Gamma(\alpha_i - \rho_i) \left(p(\alpha_i - \rho_i - 1) + 1 \right)^{\frac{1}{p}} \right)} \left(\prod_{i=1}^m \left\| \overline{D}_{b-}^{\alpha_i} f_i \right\|_q \right). \tag{6.123}$$

Proof Using Theorem 6.3, see (6.17), and Theorem 6.34, see (6.121).   ■

We also give

Theorem 6.36 *Let $p, q > 1$ such that $\frac{1}{p} + \frac{1}{q} = 1, r > 0; \alpha_i \geq \rho_i + 1, \rho_i > 0, i = 1, \ldots, m$. Suppose $f_i \in AC^{n_i} ([a, b]), n_i \in \mathbb{N}, n_i := \lceil \alpha_i \rceil, r_i = \lceil \rho_i \rceil, \alpha_i, \rho_i \notin \mathbb{N}$, and $f_i^{(k_i)} (b) = 0, k_i = r_i, r_i + 1, \ldots, n_i - 1$, and $\overline{D}_{b-}^{\alpha_i} f_i \in L_\infty ([a, b]), i = 1, \ldots, m$. Then*

$$\left\| \prod_{i=1}^m \left(\overline{D}_{b-}^{\rho_i} f_i \right) \right\|_r \leq \frac{(b - a)^{\sum_{i=1}^m (\alpha_i - \rho_i) - m + \frac{m}{p} + \frac{1}{r}}}{\left(r \left(\sum_{i=1}^m (\alpha_i - \rho_i) - m + \frac{m}{p} \right) + 1 \right)^{\frac{1}{r}}}$$

$$\cdot \frac{1}{\left(\prod_{i=1}^m \Gamma (\alpha_i - \rho_i) (p(\alpha_i - \rho_i - 1) + 1)^{\frac{1}{p}} \right)} \left(\prod_{i=1}^m \left\| \overline{D}_{b-}^{\alpha_i} f_i \right\|_q \right). \qquad (6.124)$$

Proof Using Theorem 6.4, see (6.25), and Theorem 6.34, see (6.121).   ■

We need

Definition 6.37 Let $\nu > 0, n := [\nu], \alpha := \nu - n \ (0 \leq \alpha < 1)$. Let $a, b \in \mathbb{R}, a \leq x \leq b, f \in C ([a, b])$. We consider $C_a^\nu ([a, b]) := \{ f \in C^n ([a, b]) : I_{a+}^{1-\alpha} f^{(n)} \in C^1 ([a, b]) \}$. For $f \in C_a^\nu ([a, b])$, we define the left generalized ν-fractional derivative of f over $[a, b]$ as

$$\Delta_a^\nu f := \left(I_{a+}^{1-\alpha} f^{(n)} \right)', \qquad (6.125)$$

see [1], p. 24, and Canavati derivative in [8].

Notice here $\Delta_a^\nu f \in C ([a, b])$.
So that

$$\left(\Delta_a^\nu f \right) (x) = \frac{1}{\Gamma (1 - \alpha)} \frac{d}{dx} \int_a^x (x - t)^{-\alpha} f^{(n)} (t) \, dt, \qquad (6.126)$$

$\forall x \in [a, b]$.
Notice here that

$$\Delta_a^n f = f^{(n)}, \quad n \in \mathbb{Z}_+. \qquad (6.127)$$

We need

Theorem 6.38 ([4]). *Let $f \in C_a^\nu ([a, b]), n = [\nu]$, such that $f^{(i)} (a) = 0, i = r, r + 1, \ldots, n - 1$, where $r := [\rho]$, with $0 < \rho < \nu$. Then*

$$\left(\Delta_a^\rho f\right)(x) = \frac{1}{\Gamma(\nu - \rho)} \int_a^x (x - t)^{\nu - \rho - 1} \left(\Delta_a^\nu f\right)(t)\, dt, \tag{6.128}$$

i.e.

$$\left(\Delta_a^\rho f\right) = I_{a+}^{\nu - \rho} \left(\Delta_a^\nu f\right) \in C\left([a, b]\right). \tag{6.129}$$

Thus $f \in C_a^\rho\left([a, b]\right)$.

We present

Theorem 6.39 *Let* $p, q > 1$ *such that* $\frac{1}{p} + \frac{1}{q} = 1$; $\nu_i > \rho_i > 0$, $i = 1, \ldots, m$. *Let* $f_i \in C_a^{\nu_i}\left([a, b]\right)$, $n_i = [\nu_i]$, *such that* $f_i^{(k_i)}(a) = 0$, $k_i = r_i, r_i + 1, \ldots, n_i - 1$, *where* $r_i := [\rho_i]$, $i = 1, \ldots, m$. *Then*

$$\left\| \prod_{i=1}^m \left(\Delta_a^{\rho_i} f_i\right) \right\|_p \leq \frac{(b - a)^{\sum\limits_{i=1}^m (\nu_i - \rho_i) + m\left(\frac{1}{p} - 1\right) + \frac{1}{p}}}{\left(p \sum\limits_{i=1}^m (\nu_i - \rho_i) + m(1 - p) + 1\right)^{\frac{1}{p}}}$$

$$\cdot \frac{1}{\left(\prod\limits_{i=1}^m \Gamma(\nu_i - \rho_i)\left(p(\nu_i - \rho_i - 1) + 1\right)^{\frac{1}{p}}\right)} \left(\prod_{i=1}^m \left\| \Delta_a^{\nu_i} f_i \right\|_q\right). \tag{6.130}$$

Proof Using Theorem 6.1, see (6.5), and Theorem 6.38, see (6.129). ∎

We also give

Theorem 6.40 *Let* $p, q > 1$ *such that* $\frac{1}{p} + \frac{1}{q} = 1$, $r > 0$; $\nu_i > \rho_i > 0$, $i = 1, \ldots, m$. *Let* $f_i \in C_a^{\nu_i}\left([a, b]\right)$, $n_i = [\nu_i]$, *such that* $f_i^{(k_i)}(a) = 0$, $k_i = r_i, r_i + 1, \ldots, n_i - 1$, *where* $r_i := [\rho_i]$, $i = 1, \ldots, m$. *Then*

$$\left\| \prod_{i=1}^m \left(\Delta_a^{\rho_i} f_i\right) \right\|_r \leq \frac{(b - a)^{\sum\limits_{i=1}^m (\nu_i - \rho_i) - m + \frac{m}{p} + \frac{1}{r}}}{\left(r\left(\sum\limits_{i=1}^m (\nu_i - \rho_i) - m + \frac{m}{p}\right) + 1\right)^{\frac{1}{r}}}$$

$$\cdot \frac{1}{\left(\prod\limits_{i=1}^m \Gamma(\nu_i - \rho_i)\left(p(\nu_i - \rho_i - 1) + 1\right)^{\frac{1}{p}}\right)} \left(\prod_{i=1}^m \left\| \Delta_a^{\nu_i} f_i \right\|_q\right). \tag{6.131}$$

Proof Using Theorem 6.2, see (6.12), and Theorem 6.38, see (6.129). ∎

We need

Definition 6.41 ([2]). Let $\nu > 0$, $n := [\nu]$, $\alpha = \nu - n$, $0 < \alpha < 1$, $f \in C([a, b])$. Consider

$$C_{b-}^{\nu}([a, b]) := \{f \in C^n([a, b]) : I_{b-}^{1-\alpha} f^{(n)} \in C^1([a, b])\}. \tag{6.132}$$

Define the right generalized ν-fractional derivative of f over $[a, b]$, by

$$\Delta_{b-}^{\nu} f := (-1)^{n-1} \left(I_{b-}^{1-\alpha} f^{(n)}\right)'. \tag{6.133}$$

We set $\Delta_{b-}^0 f = f$. Notice that

$$\left(\Delta_{b-}^{\nu} f\right)(x) = \frac{(-1)^{n-1}}{\Gamma(1-\alpha)} \frac{d}{dx} \int_x^b (J - x)^{-\alpha} f^{(n)}(J) \, dJ, \tag{6.134}$$

and $\Delta_{b-}^{\nu} f \in C([a, b])$.

We also need

Theorem 6.42 ([4]). Let $f \in C_{b-}^{\nu}([a, b])$, $0 < \rho < \nu$. Assume $f^{(i)}(b) = 0$, $i = r, r+1, \ldots, n-1$, where $r := [\rho]$, $n := [\nu]$. Then

$$\Delta_{b-}^{\rho} f(x) = \frac{1}{\Gamma(\nu - \rho)} \int_x^b (J - x)^{\nu - \rho - 1} \left(\Delta_{b-}^{\nu} f\right)(J) \, dJ, \tag{6.135}$$

$\forall x \in [a, b]$, i.e.

$$\Delta_{b-}^{\rho} f = I_{b-}^{\nu - \rho} \left(\Delta_{b-}^{\nu} f\right) \in C([a, b]), \tag{6.136}$$

and $f \in C_{b-}^{\rho}([a, b])$.

We present

Theorem 6.43 Let $p, q > 1$ such that $\frac{1}{p} + \frac{1}{q} = 1$; $\nu_i > \rho_i > 0$, $i = 1, \ldots, m$. Let $f_i \in C_{b-}^{\nu_i}([a, b])$ such that $f_i^{(k_i)}(b) = 0$, $k_i = r_i, r_i + 1, \ldots, n_i - 1$, where $r_i := [\rho_i]$, $n_i := [\nu_i]$, $i = 1, \ldots, m$. Then

$$\left\| \prod_{i=1}^m \left(\Delta_{b-}^{\rho_i} f_i\right) \right\|_p \leq \frac{(b-a)^{\sum_{i=1}^m (\nu_i - \rho_i) + m\left(\frac{1}{p} - 1\right) + \frac{1}{p}}}{\left(p \sum_{i=1}^m (\nu_i - \rho_i) + m(1-p) + 1\right)^{\frac{1}{p}}}$$

$$\cdot \frac{1}{\left(\prod_{i=1}^m \Gamma(\nu_i - \rho_i)(p(\nu_i - \rho_i - 1) + 1)^{\frac{1}{p}}\right)} \left(\prod_{i=1}^m \left\| \Delta_{b-}^{\nu_i} f_i \right\|_q \right). \tag{6.137}$$

Proof Using Theorem 6.3, see (6.17), and Theorem 6.42, see (6.136). ∎

We also give

Theorem 6.44 *Let* $p, q > 1$ *such that* $\frac{1}{p} + \frac{1}{q} = 1$, $r > 0$; $\nu_i > \rho_i > 0$, $i =$
$1, \ldots, m$. *Let* $f_i \in C_{b-}^{\nu_i}([a, b])$ *such that* $f_i^{(k_i)}(b) = 0$, $k_i = r_i, r_i + 1, \ldots, n_i - 1$,
where $r_i := [\rho_i]$, $n_i := [\nu_i]$, $i = 1, \ldots, , m$. *Then*

$$\left\| \prod_{i=1}^{m} (\Delta_{b-}^{\rho_i} f_i) \right\|_r \leq \frac{(b-a)^{\sum_{i=1}^{m}(\nu_i - \rho_i) - m + \frac{m}{p} + \frac{1}{r}}}{\left(r \left(\sum_{i=1}^{m} (\nu_i - \rho_i) - m + \frac{m}{p} \right) + 1 \right)^{\frac{1}{r}}}$$

$$\cdot \frac{1}{\left(\prod_{i=1}^{m} \Gamma(\nu_i - \rho_i)(p(\nu_i - \rho_i - 1) + 1)^{\frac{1}{p}} \right)} \left(\prod_{i=1}^{m} \left\| \Delta_{b-}^{\nu_i} f_i \right\|_q \right). \qquad (6.138)$$

Proof Using Theorem 6.4, see (6.25), and Theorem 6.42, see (6.136). ∎

References

1. G.A. Anastassiou, *Fractional Differentiation Inequalities* (Research Monograph, Springer, New York, 2009)
2. G.A. Anastassiou, On right fractional calculus. Chaos, Solitons Fractals **42**, 365–376 (2009)
3. G.A. Anastassiou, Balanced fractional Opial inequalities. Chaos, Solitons Fractals **42**(3), 1523–1528 (2009)
4. G.A. Anastassiou, Fractional representation formulae and right fractional inequalities. Math. Comput. Model. **54**(11–12), 3098–3115 (2011)
5. G.A. Anastassiou, in *Univariate Hardy Type Fractional Inequalities*, ed. by G. Anastassiou, O. Duman. Proceedings of International Conference in Applied Mathematics and Approximation Theory 2012, Ankara, Turkey, 17–20 May 2012, Tobb Univrsity of Economics and Technology, Springer, New York, 2013
6. G.A. Anastassiou, Fractional integral inequalities involving convexity. Sarajevo J. Math. Spec. Issue Honoring 60th Birthday of M. Kulenovich **8**(21), 203-233 (2012)
7. G.A. Anastassiou, Basic Fractional Integral Inequalities. J. Appl. Funct. Anal. AMAT 2012 Proc. **8**(3–4), 267-300 (2013)
8. J.A. Canavati, The Riemann-Liouville Integral. Nieuw Archief Voor Wiskunde **5**(1), 53–75 (1987)
9. K. Diethelm, *The Analysis of Fractional Differential Equations*, Lecture Notes in Mathematics, vol. 2004, 1st edn. (Springer, New York, 2010)
10. A.M.A. El-Sayed, M. Gaber, On the finite Caputo and finite Riesz derivatives. Electron. J. Theor. Phys. **3**(12), 81–95 (2006)
11. R. Gorenflo, F. Mainardi, Essentials of Fractional Calculus, 2000, Maphysto Center, http://www.maphysto.dk/oldpages/events/LevyCAC2000/MainardiNotes/fm2k0a.ps
12. G.D. Handley, J.J. Koliha, J. Pečarić, Hilbert-Pachpatte type integral inequalities for fractional derivatives. Fract. Calc. Appl. Anal. **4**(1), 37–46 (2001)

13. H.G. Hardy, Notes on some points in the integral calculus. Messenger Math. **47**(10), 145–150 (1918)
14. S. Iqbal, K. Krulic, J. Pecaric, On an inequality of H.G. Hardy. J. Inequalities Appl. **264347**, p. 23 (2010)
15. A.A. Kilbas, H.M. Srivastava, J.J. Trujillo, *Theory and Applications of Fractional Differential Equations*, North-Holland Mathematics Studies, vol. 204 (Elsevier, New York, NY, USA, 2006)
16. T. Mamatov, S. Samko, Mixed fractional integration operators in mixed weighted Hölder spaces. Fract. Calc. Appl. Anal. **13**(3), 245–259 (2010)
17. S.G. Samko, A.A. Kilbas, O.I. Marichev, *Fractional Integral and Derivatives: Theory and Applications* (Gordon and Breach Science Publishers, Yverdon, 1993)

Chapter 7
Harmonic Multivariate Ostrowski and Grüss Inequalities Using Several Functions

Here we derive very general multivariate Ostrowski and Grüss type inequalities for several functions by involving harmonic polynomials. Estimates are with respect to all basic norms. We give applications. It follows [1].

7.1 Introduction

The problem of estimating the difference of a value of a function from its average is a paramount one. The answer to it are the Ostrowski type inequalities. Ostrowski type inequalities are very useful among others in Numerical Analysis for approximating integrals. The problem of estimating the difference between the average of a product of functions from the product of their averages is also a very important one. The answer to it are the Grüss type inequalities. Grüss type inequalities are very useful among others in Probability for estimating expected values, etc. There exists a vast literature on Ostrowski and Grüss type inequalities to all possible directions. Mathematical community is very much interested to these inequalities due to their applications. So here we derive very general Ostrowski and Grüss type inequalities for several multivariate functions, acting to all possible directions.

We are motivated by the following results.

Theorem 7.1 (1938, Ostrowski [8]) *Let* $f : [a, b] \to \mathbb{R}$ *be continuous on* $[a, b]$ *and differentiable on* (a, b) *whose derivative* $f' : (a, b) \to \mathbb{R}$ *is bounded on* (a, b), *i.e.,* $\|f'\|_\infty^{\sup} := \sup_{t \in (a,b)} |f'(t)| < +\infty$. *Then*

$$\left| \frac{1}{b-a} \int_a^b f(t)\, dt - f(x) \right| \le \left[\frac{1}{4} + \frac{\left(x - \frac{a+b}{2}\right)^2}{(b-a)^2} \right] \cdot (b-a) \left\| f' \right\|_\infty^{\sup}, \quad (7.1)$$

for any $x \in [a, b]$. *The constant* $\frac{1}{4}$ *is the best possible.*

© Springer International Publishing Switzerland 2016
G.A. Anastassiou, *Intelligent Comparisons: Analytic Inequalities*,
Studies in Computational Intelligence 609,
DOI 10.1007/978-3-319-21121-3_7

Theorem 7.2 (1935, Grüss [7]) *Let* f, g *be integrable functions from* $[a, b]$ *into* $\mathbb{R}$, *that satisfy the conditions*

$$m \leq f(x) \leq M, n \leq g(x) \leq N, \ x \in [a, b],$$

where $m, M, n, N \in \mathbb{R}$. *Then*

$$\left| \frac{1}{b-a} \int_a^b f(x) g(x) \, dx - \left(\frac{1}{b-a} \int_a^b f(x) \, dx \right) \left(\frac{1}{b-a} \int_a^b g(x) \, dx \right) \right| \quad (7.2)$$

$$\leq \frac{1}{4} (M - m)(N - n).$$

Theorem 7.3 (1998, Dragomir and Wang [5]) *Let* $f : [a, b] \to \mathbb{R}$ *is absolutely continuous function with* $f' \in L_p([a, b])$, $p > 1$, $\frac{1}{p} + \frac{1}{q} = 1$, $x \in [a, b]$. *Then*

$$\left| f(x) - \frac{1}{b-a} \int_a^b f(t) \, dt \right| \leq$$

$$\frac{1}{(q+1)^{\frac{1}{q}}} \left[\left(\frac{x-a}{b-a} \right)^{q+1} + \left(\frac{b-x}{b-a} \right)^{q+1} \right]^{\frac{1}{q}} (b-a)^{\frac{1}{q}} \left\| f' \right\|_p. \quad (7.3)$$

Theorem 7.4 (1882, Čebyšev [2]) *Let* $f, g : [a, b] \to \mathbb{R}$ *absolutely continuous functions with* $f', g' \in L_\infty([a, b])$. *Then*

$$\left| \frac{1}{b-a} \int_a^b f(x) g(x) \, dx - \left(\frac{1}{b-a} \int_a^b f(x) \, dx \right) \left(\frac{1}{b-a} \int_a^b g(x) \, dx \right) \right| \quad (7.4)$$

$$\leq \frac{1}{12} (b-a)^2 \left\| f' \right\|_\infty \left\| g' \right\|_\infty.$$

Above is also assumed that the involved integrals exist.

7.2 Background

Let $(P_n)_{n \in \mathbb{N}}$ be a harmonic sequence of polynomials, that is $P_n' = P_{n-1}$, $n \geq 1$, $P_0 = 1$. Furthermore, let $[a, b] \subset \mathbb{R}$, $a \neq b$, and $h : [a, b] \to \mathbb{R}$ be such that $h^{(n-1)}$ is absolutely continuous function for some fixed $n \geq 1$. We use the notation

$$L_n\left[h\left(x\right)\right] = \frac{1}{n}\left[h\left(x\right) + \sum_{k=1}^{n-1}\left(-1\right)^k P_k\left(x\right) h^{(k)}\left(x\right) + \right.$$

$$\left. \sum_{k=1}^{n-1} \frac{\left(-1\right)^k \left(n-k\right)}{b-a}\left[P_k\left(a\right) h^{(k-1)}\left(a\right) - P_k\left(b\right) h^{(k-1)}\left(b\right)\right]\right], \qquad (7.5)$$

$x \in [a, b]$, for convenience.

For $n = 1$ the above sums are defined to be zero, that is $L_1\left[h\left(x\right)\right] = h\left(x\right)$.

Dedic et al., see [3, 4], established the following identity,

$$L_n\left[h\left(x\right)\right] - \frac{1}{b-a}\int_a^b h\left(t\right) dt = \frac{\left(-1\right)^{n+1}}{n\left(b-a\right)}\int_a^b P_{n-1}\left(t\right) q\left(x,t\right) h^{(n)}\left(t\right) dt,$$
$$(7.6)$$

where

$$q\left(x, t\right) = \begin{cases} t - a, & \text{if } t \in [a, x], \\ t - b, & \text{if } t \in (x, b], \end{cases} \quad x \in [a, b]. \qquad (7.7)$$

For the harmonic sequence of polynomials $P_k\left(t\right) = \frac{\left(t-x\right)^k}{k!}$, $k \geq 0$, the identity (7.6) reduces to the Fink identity in [6], (see also [4], p. 177).

We rewrite (7.6) as follows:

$$h\left(x\right) + \sum_{k=1}^{n-1}\left(-1\right)^k P_k\left(x\right) h^{(k)}\left(x\right) + $$

$$\sum_{k=1}^{n-1} \frac{\left(-1\right)^k \left(n-k\right)}{b-a}\left[P_k\left(a\right) h^{(k-1)}\left(a\right) - P_k\left(b\right) h^{(k-1)}\left(b\right)\right] = $$

$$\frac{n}{b-a}\int_a^b h\left(t\right) dt + \frac{\left(-1\right)^{n+1}}{b-a}\int_a^b P_{n-1}\left(t\right) q\left(x,t\right) h^{(n)}\left(t\right) dt, \qquad (7.8)$$

$x \in [a, b]$.

That is the generalized Fink type representation formula.

$$h\left(x\right) = \sum_{k=1}^{n-1}\left(-1\right)^{k+1} P_k\left(x\right) h^{(k)}\left(x\right) + $$

$$\sum_{k=1}^{n-1} \frac{\left(-1\right)^k \left(n-k\right)}{b-a}\left[P_k\left(b\right) h^{(k-1)}\left(b\right) - P_k\left(a\right) h^{(k-1)}\left(a\right)\right] + $$

$$\frac{n}{b-a} \int_a^b h(t)\, dt + \frac{(-1)^{n+1}}{b-a} \int_a^b P_{n-1}(t)\, q(x,t)\, h^{(n)}(t)\, dt, \qquad (7.9)$$

$x \in [a, b]$, $n \geq 1$, when $n = 1$ the above sums are zero.

7.3 Main Results

Here $\prod_{i=1}^{m} [a_i, b_i] \subseteq \mathbb{R}^m$, $m, n \in \mathbb{N}$.

We make

General Assumption 7.5 Let $f : \prod_{i=1}^{m} [a_i, b_i] \to \mathbb{R}$. We assume

(1) for $j = 1, \ldots, m$ we have that $\frac{\partial^{n-1} f}{\partial x_j^{n-1}}(x_1, x_2, \ldots, x_{j-1}, s_j, x_{j+1}, \ldots, x_m)$ is absolutely continuous in s_j on $[a_j, b_j]$, for every $(x_1, x_2, \ldots, x_{j-1}, x_{j+1}, \ldots, x_m) \in \prod_{\substack{i=1 \\ i \neq j}}^{m} [a_i, b_i]$,

(2) for $j = 1, \ldots, m$ we have that $\frac{\partial^n f(s_1, \ldots, s_j, x_{j+1}, \ldots, x_m)}{\partial x_j^n}$ is continuous on $\prod_{i=1}^{j} [a_i, b_i]$, for every $(x_{j+1}, \ldots, x_m) \in \prod_{i=j+1}^{m} [a_i, b_i]$,

(3) for each $j = 1, \ldots, m$, and for every $l = 1, \ldots, n - 1$, we have that $\frac{\partial^l f}{\partial x_j^l}(s_1, s_2, \ldots, s_{j-1}, x_j, \ldots, x_m)$ is continuous on $\prod_{i=1}^{j-1} [a_i, b_i]$, for every $(x_j, \ldots, x_m) \in \prod_{i=j}^{m} [a_j, b_j]$.

(4) f is continuous on $\prod_{i=1}^{m} [a_i, b_i]$.

Brief Assumption 7.6 Let $f : \prod_{i=1}^{m} [a_i, b_i] \to \mathbb{R}$ with $\frac{\partial^l f}{\partial x_i^l}$ for $l = 0, 1, \ldots, n$; $i = 1, \ldots, m$, are continuous on $\prod_{i=1}^{m} [a_i, b_i]$.

Definition 7.7 We put

$$q(x_i, s_i) = \begin{cases} s_i - a_i, & \text{if } s_i \in [a_i, x_i], \\ s_i - b_i, & \text{if } s_i \in (x_i, b_i], \end{cases} \quad x_i \in [a_i, b_i], \tag{7.10}$$

$i = 1, \ldots, m.$

We present the following general representation result of Fink type.

Theorem 7.8 *Let f as in General Assumptions 7.5 or Brief Assumptions 7.6. Then*

$$f(x_1, \ldots, x_n) = \frac{n^m}{\displaystyle\prod_{i=1}^{m}(b_i - a_i)} \int_{\prod_{i=1}^{m}[a_i, b_i]} f(s_1, \ldots, s_m)\, ds_1 \ldots ds_m \tag{7.11}$$

$$+ \sum_{i=1}^{m} T_i(x_i, x_{i+1}, \ldots, x_m),$$

where

$$T_i(x_i, \ldots, x_m) := \frac{n^{i-1}}{\displaystyle\prod_{j=1}^{i-1}(b_j - a_j)} \cdot$$

$$\left[\sum_{k=1}^{n-1}(-1)^{k+1} P_k(x_i) \int_{a_1}^{b_1} \cdots \int_{a_{i-1}}^{b_{i-1}} \frac{\partial^k f(s_1, \ldots, s_{i-1}, x_i, \ldots, x_m)}{\partial x_i^k} ds_1 \ldots ds_{i-1} + \right.$$

$$\sum_{k=1}^{n-1} \frac{(-1)^k(n-k)}{b_i - a_i} \cdot$$

$$\left[P_k(b_i) \int_{a_1}^{b_1} \cdots \int_{a_{i-1}}^{b_{i-1}} \frac{\partial^{k-1} f(s_1, \ldots, s_{i-1}, b_i, x_{i+1}, \ldots, x_m)}{\partial x_i^{k-1}} ds_1 \ldots ds_{i-1} - \right.$$

$$P_k(a_i) \int_{a_1}^{b_1} \cdots \int_{a_{i-1}}^{b_{i-1}} \frac{\partial^{k-1} f(s_1, \ldots, s_{i-1}, a_i, x_{i+1}, \ldots, x_m)}{\partial x_i^{k-1}} ds_1 \ldots ds_{i-1} \left. \right] +$$

$$\frac{(-1)^{n+1}}{(b_i - a_i)} \int_{a_1}^{b_1} \cdots \int_{a_i}^{b_i} P_{n-1}(s_i)\, q(x_i, s_i) \frac{\partial^n f(s_1, \ldots, s_i, x_{i+1}, \ldots, x_m)}{\partial x_i^n} ds_1 \ldots ds_i \left. \right], \tag{7.12}$$

are continuous functions for all $i = 1, \ldots, m.$

Proof We apply (7.9) repeatedly.

We have

$$f(x_1, \ldots, x_m) = \sum_{k=1}^{n-1} (-1)^{k+1} P_k(x_1) \frac{\partial^k f(x_1, \ldots, x_m)}{\partial x_1^k} +$$

$$\sum_{k=1}^{n-1} \frac{(-1)^k (n-k)}{(b_1 - a_1)} \left[P_k(b_1) \frac{\partial^{k-1} f(b_1, x_2, \ldots, x_m)}{\partial x_1^{k-1}} - P_k(a_1) \frac{\partial^{k-1} f(a_1, x_2, \ldots, x_m)}{\partial x_1^{k-1}} \right]$$

$$+ \frac{n}{(b_1 - a_1)} \int_{a_1}^{b_1} f(s_1, x_2, \ldots, x_m) \, ds_1 +$$

$$\frac{(-1)^{n+1}}{(b_1 - a_1)} \int_{a_1}^{b_1} P_{n-1}(s_1) q(x_1, s_1) \frac{\partial^n f(s_1, x_2, \ldots, x_m)}{\partial x_1^n} ds_1, \qquad (7.13)$$

any $x_1 \in [a_1, b_1]$,

under the assumption that $\frac{\partial^{n-1} f(\cdot, x_2, \ldots, x_m)}{\partial x_1^{n-1}} \in AC([a_1, b_1])$.

Call

$$T_1(x_1, \ldots, x_m) := \sum_{k=1}^{n-1} (-1)^{k+1} P_k(x_1) \frac{\partial^k f(x_1, \ldots, x_m)}{\partial x_1^k} +$$

$$\sum_{k=1}^{n-1} \frac{(-1)^k (n-k)}{(b_1 - a_1)} \left[P_k(b_1) \frac{\partial^{k-1} f(b_1, x_2, \ldots, x_m)}{\partial x_1^{k-1}} - P_k(a_1) \frac{\partial^{k-1} f(a_1, x_2, \ldots, x_m)}{\partial x_1^{k-1}} \right]$$

$$+ \frac{(-1)^{n+1}}{(b_1 - a_1)} \int_{a_1}^{b_1} P_{n-1}(s_1) q(x_1, s_1) \frac{\partial^n f(s_1, x_2 \ldots, x_m)}{\partial x_1^n} ds. \qquad (7.14)$$

Hence it holds

$$f(x_1, \ldots, x_m) = \frac{n}{(b_1 - a_1)} \int_{a_1}^{b_1} f(s_1, x_2, \ldots, x_m) \, ds_1 + T_1(x_1, \ldots, x_m).$$
$$(7.15)$$

Next similarly we get

$$f(s_1, x_2, \ldots, x_m) = \sum_{k=1}^{n-1} (-1)^{k+1} P_k(x_2) \frac{\partial^k f(s_1, x_2, \ldots, x_m)}{\partial x_2^k} +$$

$$\sum_{k=1}^{n-1} \frac{(-1)^k (n-k)}{(b_2 - a_2)}.$$

$$\left[P_k(b_2) \frac{\partial^{k-1} f(s_1, b_2, x_3, \ldots, x_m)}{\partial x_2^{k-1}} - P_k(a_2) \frac{\partial^{k-1} f(s_1, a_2, x_3, \ldots, x_m)}{\partial x_2^{k-1}} \right] +$$

(7.16)

$$\frac{n}{(b_2 - a_2)} \int_{a_2}^{b_2} f(s_1, s_2, x_3, \ldots, x_m) \, ds_2 +$$

$$\frac{(-1)^{n+1}}{(b_2 - a_2)} \int_{a_2}^{b_2} P_{n-1}(s_2) q(x_2, s_2) \frac{\partial^n f(s_1, s_2, x_3, \ldots, x_m)}{\partial x_2^n} ds_2,$$

any $x_2 \in [a_2, b_2]$.

Hence it holds

$$f(x_1, \ldots, x_m) = \frac{n^2}{(b_1 - a_1)(b_2 - a_2)} \int_{a_1}^{b_1} \int_{a_2}^{b_2} f(s_1, s_2, x_3, \ldots, x_m) \, ds_1 ds_2 +$$

$$T_1(x_1, \ldots, x_m) + T_2(x_2, x_3, \ldots, x_m), \tag{7.17}$$

where

$$T_2(x_2, x_3, \ldots, x_m) := \frac{n}{(b_1 - a_1)} \left\{ \sum_{k=1}^{n-1} (-1)^{k+1} P_k(x_2) \int_{a_1}^{b_1} \frac{\partial^k f(s_1, x_2, \ldots, x_m)}{\partial x_2^k} ds_1 \right.$$

$$+ \sum_{k=1}^{n-1} \frac{(-1)^k (n-k)}{(b_2 - a_2)} \left[P_k(b_2) \int_{a_1}^{b_1} \frac{\partial^{k-1} f(s_1, b_2, x_3, \ldots, x_m)}{\partial x_2^{k-1}} ds_1 \right. \tag{7.18}$$

$$\left. - P_k(a_2) \int_{a_1}^{b_1} \frac{\partial^{k-1} f(s_1, a_2, x_3, \ldots, x_m)}{\partial x_2^{k-1}} ds_1 \right]$$

$$\left. + \frac{(-1)^{n+1}}{(b_2 - a_2)} \int_{a_1}^{b_1} \int_{a_2}^{b_2} P_{n-1}(s_2) q(x_2, s_2) \frac{\partial^n f(s_1, s_2, x_3, \ldots, x_m)}{\partial x_2^n} ds_1 ds_2 \right\}.$$

Next we see similarly that

$$f(s_1, s_2, x_3, \ldots, x_m) = \sum_{k=1}^{n-1} (-1)^{k+1} P_k(x_3) \frac{\partial^k f(s_1, s_2, x_3, \ldots, x_m)}{\partial x_3^k} +$$

$$\sum_{k=1}^{n-1} \frac{(-1)^k (n-k)}{(b_3 - a_3)} \left[P_k(b_3) \frac{\partial^{k-1} f(s_1, s_2, b_3, x_4, \ldots, x_m)}{\partial x_3^{k-1}} - \right.$$

$$P_k(a_3) \frac{\partial^{k-1} f(s_1, s_2, a_3, x_4, \ldots, x_m)}{\partial x_3^{k-1}}\Bigg] + \tag{7.19}$$

$$\frac{n}{(b_3 - a_3)} \int_{a_3}^{b_3} f(s_1, s_2, s_3, x_4, \ldots, x_m) \, ds_3 +$$

$$\frac{(-1)^{n+1}}{(b_3 - a_3)} \int_{a_3}^{b_3} P_{n-1}(s_3) q(x_3, s_3) \frac{\partial^n f(s_1, s_2, s_3, x_4, \ldots, x_m)}{\partial x_3^n} ds_3,$$

any $x_3 \in [a_3, b_3]$.
  So that we get

$$f(x_1, \ldots, x_m) = \frac{n^3}{\prod\limits_{j=1}^{3} (b_j - a_j)} \int_{a_1}^{b_1} \int_{a_2}^{b_2} \int_{a_3}^{b_3} f(s_1, s_2, s_3, x_4, \ldots, x_m) \, ds_1 ds_2 ds_3 +$$

$$T_1(x_1, \ldots, x_m) + T_2(x_2, \ldots, x_m) + T_3(x_3, \ldots, x_m), \tag{7.20}$$

where

$$T_3(x_3, \ldots, x_m) := \frac{n^2}{(b_1 - a_1)(b_2 - a_2)}.$$

$$\left[\sum_{k=1}^{n-1} (-1)^{k+1} P_k(x_3) \int_{a_1}^{b_1} \int_{a_2}^{b_2} \frac{\partial^k f(s_1, s_2, x_3, \ldots, x_m)}{\partial x_3^k} ds_1 ds_2 +\right.$$

$$\sum_{k=1}^{n-1} \frac{(-1)^k (n-k)}{(b_3 - a_3)} \left[P_k(b_3) \int_{a_1}^{b_1} \int_{a_2}^{b_2} \frac{\partial^{k-1} f(s_1, s_2, b_3, x_4, \ldots, x_m)}{\partial x_3^{k-1}} ds_1 ds_2 \right.$$

$$\left. -P_k(a_3) \int_{a_1}^{b_1} \int_{a_2}^{b_2} \frac{\partial^{k-1} f(s_1, s_2, a_3, x_4, \ldots, x_m)}{\partial x_3^{k-1}} ds_1 ds_2 \right] + \tag{7.21}$$

$$\left. \frac{(-1)^{n+1}}{(b_3 - a_3)} \int_{a_1}^{b_1} \int_{a_2}^{b_2} \int_{a_3}^{b_3} P_{n-1}(s_3) q(x_3, s_3) \frac{\partial^n f(s_1, s_2, s_3, x_4, \ldots, x_m)}{\partial x_3^n} ds_1 ds_2 ds_3 \right].$$

Furthermore we can write

$$f(s_1, s_2, s_3, x_4, \ldots, x_m) = \sum_{k=1}^{n-1} (-1)^{k+1} P_k(x_4) \frac{\partial^k f(s_1, s_2, s_3, x_4 \ldots, x_m)}{\partial x_4^k} +$$

$$\sum_{k=1}^{n-1} \frac{(-1)^k (n-k)}{(b_4 - a_4)} \left[P_k(b_4) \frac{\partial^{k-1} f(s_1, s_2, s_3, b_4, x_5, \ldots, x_m)}{\partial x_4^{k-1}} \right.$$

$$\left. - P_k(a_4) \frac{\partial^{k-1} f(s_1, s_2, s_3, a_4, x_5, \ldots, x_m)}{\partial x_4^{k-1}} \right] + \qquad (7.22)$$

$$\frac{n}{(b_4 - a_4)} \int_{a_4}^{b_4} f(s_1, s_2, s_3, s_4, x_5, \ldots, x_m) \, ds_4 +$$

$$\frac{(-1)^{n+1}}{(b_4 - a_4)} \int_{a_4}^{b_4} P_{n-1}(s_4) q(x_4, s_4) \frac{\partial^n f(s_1, s_2, s_3, s_4, x_5, \ldots, x_m)}{\partial x_4^n} ds_4,$$

any $x_4 \in [a_4, b_4]$.

Therefore it holds

$$f(x_1, \ldots, x_m) = \frac{n^4}{\prod\limits_{j=1}^{4}(b_j - a_j)} \cdot$$

$$\int_{a_1}^{b_1} \int_{a_2}^{b_2} \int_{a_3}^{b_3} \int_{a_4}^{b_4} f(s_1, s_2, s_3, s_4, x_5, \ldots, x_m) \, ds_1 ds_2 ds_3 ds_4 +$$

$$T_1(x_1, \ldots, x_m) + T_2(x_2, \ldots, x_m) + T_3(x_3, \ldots, x_m) + T_4(x_4, \ldots, x_m), \quad (7.23)$$

where

$$T_4(x_4, \ldots, x_m) := \frac{n^3}{\prod\limits_{j=1}^{3}(b_j - a_j)} \left[\sum_{k=1}^{n-1} (-1)^{k+1} P_k(x_4) \cdot \right.$$

$$\int_{a_1}^{b_1} \int_{a_2}^{b_2} \int_{a_3}^{b_3} \frac{\partial^k f(s_1, s_2, s_3, x_4, \ldots, x_m)}{\partial x_4^k} ds_1 ds_2 ds_3 +$$

$$\sum_{k=1}^{n-1} \frac{(-1)^k (n-k)}{(b_4 - a_4)} \left[P_k(b_4) \int_{a_1}^{b_1} \int_{a_2}^{b_2} \int_{a_3}^{b_3} \frac{\partial^{k-1} f(s_1, s_2, s_3, b_4, x_5, \ldots, x_m)}{\partial x_4^{k-1}} ds_1 ds_2 ds_3 \right.$$

$$-P_k(a_4) \int_{a_1}^{b_1} \int_{a_2}^{b_2} \int_{a_3}^{b_3} \frac{\partial^{k-1} f(s_1, s_2, s_3, a_4, x_5, \ldots, x_m)}{\partial x_4^{k-1}} ds_1 ds_2 ds_3 \Bigg] \qquad (7.24)$$

$$+ \frac{(-1)^{n+1}}{(b_4 - a_4)} \int_{a_1}^{b_1} \int_{a_2}^{b_2} \int_{a_3}^{b_3} \int_{a_4}^{b_4} P_{n-1}(s_4) q(x_4, s_4) \cdot$$

$$\frac{\partial^n f(s_1, s_2, s_3, s_4, x_5, \ldots, x_m)}{\partial x_4^n} ds_1 ds_2 ds_3 ds_4 \Bigg],$$

etc.

The theorem is proved. ∎

We make

Remark 7.9 Let f_λ, $\lambda = 1, \ldots, r \in \mathbb{N} - \{1\}$, as in Assumption 7.5 or Brief Assumption 7.6; $n_\lambda \in \mathbb{N}$ associated with f_λ. Here $x = (x_1, \ldots, x_m)$, $s = (s_1, \ldots, s_m) \in \prod_{i=1}^{m} [a_i, b_i]$. Then

$$f_\lambda(x) = \frac{n_\lambda^m}{\prod_{i=1}^{m}(b_i - a_i)} \int_{\prod_{i=1}^{m}[a_i, b_i]} f_\lambda(s)\, ds + \sum_{i=1}^{m} T_{i\lambda}(x_i, x_{i+1}, \ldots, x_m). \quad (7.25)$$

Here we have

$$T_{i\lambda}(x_i, \ldots, x_m) := \frac{n_\lambda^{i-1}}{\prod_{j=1}^{i-1}(b_j - a_j)} \left[\sum_{k=1}^{n_\lambda - 1} (-1)^{k+1} P_k(x_i) \cdot \right.$$

$$\int_{a_1}^{b_1} \ldots \int_{a_{i-1}}^{b_{i-1}} \frac{\partial^k f_\lambda(s_1, \ldots, s_{i-1}, x_i, \ldots, x_m)}{\partial x_i^k} ds_1 \ldots ds_{i-1} +$$

$$\sum_{k=1}^{n_\lambda - 1} \frac{(-1)^k (n_\lambda - k)}{b_i - a_i} \cdot$$

$$\left[P_k(b_i) \int_{a_1}^{b_1} \ldots \int_{a_{i-1}}^{b_{i-1}} \frac{\partial^{k-1} f_\lambda(s_1, \ldots, s_{i-1}, b_i, x_{i+1}, \ldots, x_m)}{\partial x_i^{k-1}} ds_1 \ldots ds_{i-1} \right.$$

$$\left. \left. -P_k(a_i) \int_{a_1}^{b_1} \ldots \int_{a_{i-1}}^{b_{i-1}} \frac{\partial^{k-1} f_\lambda(s_1, \ldots, s_{i-1}, a_i, x_{i+1}, \ldots, x_m)}{\partial x_i^{k-1}} ds_1 \ldots ds_{i-1} \right] + \right.$$

$$\qquad (7.26)$$

$$\frac{(-1)^{n_\lambda+1}}{(b_i - a_i)} \int_{a_1}^{b_1} \cdots \int_{a_i}^{b_i} P_{n_\lambda-1}(s_i) q(x_i, s_i) \frac{\partial^{n_\lambda} f_\lambda(s_1, \ldots, s_i, x_{i+1}, \ldots, x_m)}{\partial x_i^{n_\lambda}} ds_1 \ldots ds_i \Bigg],$$

are continuous functions, $i = 1, \ldots, m$; $\lambda = 1, \ldots, r$.

Hence it holds

$$\left(\prod_{\substack{\rho=1 \\ \rho \neq \lambda}}^{r} f_\rho(x) \right) f_\lambda(x) = \left(\frac{n_\lambda^m}{\prod_{i=1}^{m}(b_i - a_i)} \right) \left(\prod_{\substack{\rho=1 \\ \rho \neq \lambda}}^{r} f_\rho(x) \right) \left(\int_{\prod_{i=1}^{m}[a_i,b_i]} f_\lambda(s) \, ds \right) +$$

$$\left(\prod_{\substack{\rho=1 \\ \rho \neq \lambda}}^{r} f_\rho(x) \right) \left(\sum_{i=1}^{m} T_{i\lambda}(x_1, \ldots, x_m) \right).$$

(7.27)

Therefore we derive

$$\sum_{\lambda=1}^{r} \left(\left(\prod_{\substack{\rho=1 \\ \rho \neq \lambda}}^{r} f_\rho(x) \right) f_\lambda(x) \right) -$$

$$\frac{1}{\prod_{i=1}^{m}(b_i - a_i)} \left\{ \sum_{\lambda=1}^{r} \left(n_\lambda^m \left(\prod_{\substack{\rho=1 \\ \rho \neq \lambda}}^{r} f_\rho(x) \right) \left(\int_{\prod_{i=1}^{m}[a_i,b_i]} f_\lambda(s) \, ds \right) \right) \right\}$$

(7.28)

$$= \sum_{\lambda=1}^{r} \left(\left(\prod_{\substack{\rho=1 \\ \rho \neq \lambda}}^{r} f_\rho(x) \right) \left(\sum_{i=1}^{m} T_{i\lambda}(x_i, \ldots, x_m) \right) \right).$$

We notice that

$$T_{i\lambda}(x_i, \ldots, x_m) = A_{i\lambda}(x_i, \ldots, x_m) + B_{i\lambda}(x_i, \ldots, x_m),$$

(7.29)

$i = 1, \ldots, m$; where

$$A_{i\lambda}(x_i, \ldots, x_m) := \frac{n_\lambda^{i-1}}{\prod_{j=1}^{i-1}(b_j - a_j)} \left[\sum_{k=1}^{n_\lambda-1} (-1)^{k+1} P_k(x_i) \cdot \right.$$

(7.30)

$$\int_{a_1}^{b_1} \cdots \int_{a_{i-1}}^{b_{i-1}} \frac{\partial^k f_\lambda (s_1, \ldots, s_{i-1}, x_i, \ldots, x_m)}{\partial x_i^k} ds_1 \ldots ds_{i-1}$$

$$+ \sum_{k=1}^{n_\lambda - 1} \frac{(-1)^k (n_\lambda - k)}{b_i - a_i} \cdot$$

$$\left[P_k (b_i) \int_{a_1}^{b_1} \cdots \int_{a_{i-1}}^{b_{i-1}} \frac{\partial^{k-1} f_\lambda (s_1, \ldots, s_{i-1}, b_i, x_{i+1}, \ldots, x_m)}{\partial x_i^{k-1}} ds_1 \ldots ds_{i-1} - \right.$$

$$\left. P_k (a_i) \int_{a_1}^{b_1} \cdots \int_{a_{i-1}}^{b_{i-1}} \frac{\partial^{k-1} f_\lambda (s_1, \ldots, s_{i-1}, a_i, x_{i+1}, \ldots, x_m)}{\partial x_i^{k-1}} ds_1 \ldots ds_{i-1} \right] \right],$$

and

$$B_{i\lambda} (x_i, \ldots, x_m) := \frac{n_\lambda^{i-1} (-1)^{n_\lambda + 1}}{\prod\limits_{j=1}^{i} (b_j - a_j)} \cdot \tag{7.31}$$

$$\left[\int_{a_1}^{b_1} \cdots \int_{a_i}^{b_i} P_{n_\lambda - 1} (s_i) q (x_i, s_i) \frac{\partial^{n_\lambda} f_\lambda (s_1, \ldots, s_i, x_{i+1}, \ldots, x_m)}{\partial x_i^{n_\lambda}} ds_1 \ldots ds_i \right],$$

for all $i = 1, \ldots, m; \lambda = 1, \ldots, r$.

We call and have the identity

$$S (f_1, \ldots, f_r) (x) :=$$

$$\sum_{\lambda=1}^{r} \left\{ \left(\prod_{\substack{\rho=1 \\ \rho \neq \lambda}}^{r} f_\rho (x) \right) \left[f_\lambda (x) - \frac{n_\lambda^m}{\prod\limits_{i=1}^{m} (b_i - a_i)} \left(\int_{\prod\limits_{i=1}^{m} [a_i, b_i]} f_\lambda (s) \, ds \right) - \right. \right.$$

$$\left. \left. \sum_{i=1}^{m} A_{i\lambda} (x_i, \ldots, x_m) \right] \right\} = \sum_{\lambda=1}^{r} \left(\left(\prod_{\substack{\rho=1 \\ \rho \neq \lambda}}^{r} f_\rho (x) \right) \left(\sum_{i=1}^{m} B_{i\lambda} (x_i, \ldots, x_m) \right) \right),$$

$$\tag{7.32}$$

true for any fixed $x \in \prod\limits_{i=1}^{m} [a_i, b_i]$.

Then we have

$$|S(f_1, \ldots, f_r)(x)| \leq \sum_{\lambda=1}^{r} \left(\left(\prod_{\substack{\rho=1 \\ \rho \neq \lambda}}^{r} |f_\rho(x)| \right) \left(\sum_{i=1}^{m} |B_{i\lambda}(x_i, \ldots, x_m)| \right) \right). \quad (7.33)$$

We estimate the right hand side of (7.33).

We also make

Remark 7.10 We observe that

$$|B_{i\lambda}(x_i, \ldots, x_m)| \leq \frac{n_\lambda^{i-1}}{\prod_{j=1}^{i}(b_j - a_j)} \left[\int_{a_1}^{b_1} \ldots \int_{a_i}^{b_i} \left| P_{n_\lambda - 1}(s_i) \right| |q(x_i, s_i)| \cdot \right.$$

$$\left. \left| \frac{\partial^{n_\lambda} f_\lambda(s_1, \ldots, s_i, x_{i+1}, \ldots, x_m)}{\partial x_i^{n_\lambda}} \right| ds_1 \ldots ds_i \right] =: (\xi), \quad (7.34)$$

for all $i = 1, \ldots, m$; $\lambda = 1, \ldots, r$.
We know that

$$|q(x_i, s_i)| \leq \max(x_i - a_i, b_i - x_i) = \frac{(b_i - a_i) + |a_i + b_i - 2x_i|}{2}. \quad (7.35)$$

We have

$$(\xi) \overset{(7.35)}{\leq} \frac{n_\lambda^{i-1}}{\prod_{j=1}^{i}(b_j - a_j)} \left[\left\| P_{n_\lambda - 1} \right\|_{\infty, [a_i, b_i]} \left(\frac{(b_i - a_i) + |a_i + b_i - 2x_i|}{2} \right) \cdot \right.$$

$$\left. \left\| \frac{\partial^{n_\lambda} f_\lambda(\ldots, x_{i+1}, \ldots, x_m)}{\partial x_i^{n_\lambda}} \right\|_{L_1 \left(\prod_{j=1}^{i} [a_j, b_j] \right)} \right]. \quad (7.36)$$

Thus

$$|S(f_1,\ldots,f_r)(x)| \le \sum_{\lambda=1}^{r}\left(\left(\prod_{\substack{\rho=1\\\rho\neq\lambda}}^{r}|f_\rho(x)|\right)\left(\sum_{i=1}^{m}\frac{n_\lambda^{i-1}}{\prod_{j=1}^{i}(b_j-a_j)}\cdot\right.\right.$$

$$\left[\|P_{n_\lambda-1}\|_{\infty,[a_i,b_i]}\left(\frac{(b_i-a_i)+|a_i+b_i-2x_i|}{2}\right)\cdot\right.$$

$$\left.\left.\left.\left\|\frac{\partial^{n_\lambda}f_\lambda(\ldots,x_{i+1},\ldots,x_m)}{\partial x_i^{n_\lambda}}\right\|_{L_1\left(\prod_{j=1}^{i}[a_j,b_j]\right)}\right]\right)\right) =: \theta_1(x). \qquad (7.37)$$

Next let $p_{li} > 1 : \sum_{li=1}^{3}\frac{1}{p_{li}} = 1$. Then

$$|B_{i\lambda}(x_i,\ldots,x_m)| \overset{(7.34)}{\le} (\xi) \le \frac{n_\lambda^{i-1}}{\prod_{j=1}^{i}(b_j-a_j)}\cdot$$

$$\left[\left(\int_{a_1}^{b_1}\ldots\int_{a_i}^{b_i}|P_{n_\lambda-1}(s_i)|^{p_{1i}}\,ds_1\ldots ds_i\right)^{\frac{1}{p_{1i}}}\left(\int_{a_1}^{b_1}\ldots\int_{a_i}^{b_i}|q(x_i,s_i)|^{p_{2i}}\,ds_1\ldots ds_i\right)^{\frac{1}{p_{2i}}}\right.$$

$$\qquad (7.38)$$

$$\left.\cdot\left\|\frac{\partial^{n_\lambda}f_\lambda(\ldots,x_{i+1},\ldots,x_m)}{\partial x_i^{n_\lambda}}\right\|_{L_{p3i}\left(\prod_{j=1}^{i}[a_j,b_j]\right)}\right] =$$

$$\frac{n_\lambda^{i-1}}{\prod\limits_{j=1}^{i}(b_j - a_j)}\left[\left\|P_{n_\lambda-1}\right\|_{L_{p_{1i}}([a_i,b_i])}\left(\prod_{j=1}^{i-1}(b_j - a_j)\right)^{\frac{1}{p_{1i}}+\frac{1}{p_{2i}}}\cdot\right.$$

$$\left.\left(\frac{(b_i - x_i)^{p_{2i}+1} + (x_i - a_i)^{p_{2i}+1}}{p_{2i}+1}\right)^{\frac{1}{p_{2i}}}\left\|\frac{\partial^{n_\lambda}f_\lambda(...,x_{i+1},...,x_m)}{\partial x_i^{n_\lambda}}\right\|_{L_{p_{3i}}\left(\prod\limits_{j=1}^{i}[a_j,b_j]\right)}\right].$$

$$\tag{7.39}$$

Therefore we get

$$|S(f_1,...,f_r)(x)| \overset{(7.39)}{\leq} \sum_{\lambda=1}^{r}\left(\left(\prod_{\substack{\rho=1\\\rho\neq\lambda}}^{r}|f_\rho(x)|\right)\left(\sum_{i=1}^{m}\frac{n_\lambda^{i-1}}{\prod\limits_{j=1}^{i}(b_j - a_j)}\cdot\right.\right.$$

$$\left[\left\|P_{n_\lambda-1}\right\|_{L_{p_{1i}}([a_i,b_i])}\left(\prod_{j=1}^{i-1}(b_j - a_j)\right)^{\frac{1}{p_{1i}}+\frac{1}{p_{2i}}}\cdot\right.$$

$$\left(\frac{(b_i - x_i)^{p_{2i}+1} + (x_i - a_i)^{p_{2i}+1}}{p_{2i}+1}\right)^{\frac{1}{p_{2i}}}\cdot$$

$$\tag{7.40}$$

$$\left.\left.\left.\left\|\frac{\partial^{n_\lambda}f_\lambda(...,x_{i+1},...,x_m)}{\partial x_i^{n_\lambda}}\right\|_{L_{p_{3i}}\left(\prod\limits_{j=1}^{i}[a_j,b_j]\right)}\right]\right)\right) =: \theta_2(x).$$

We also have

$$
|B_{i\lambda}(x_i,\ldots,x_m)| \le (\xi) \le n_{\lambda}^{i-1} \left[\left\| P_{n_{\lambda}-1} \right\|_{\infty,[a_i,b_i]} \cdot \right.
$$

$$
\left\| \frac{\partial^{n_{\lambda}} f_{\lambda}(\ldots, x_{i+1},\ldots, x_m)}{\partial x_i^{n_{\lambda}}} \right\|_{\infty, \prod\limits_{j=1}^{i}[a_j,b_j]} \left(\frac{(b_i - x_i)^2 + (x_i - a_i)^2}{2(b_i - a_i)} \right) \right], \quad (7.41)
$$

$i = 1,\ldots, m;\ \lambda = 1,\ldots, r.$

Consequently we find

$$
|S(f_1,\ldots, f_r)(x)| \le \sum_{\lambda=1}^{r} \left(\left(\prod_{\substack{\rho=1 \\ \rho \ne \lambda}}^{r} |f_{\rho}(x)| \right) \left(\sum_{i=1}^{m} \left[n_{\lambda}^{i-1} \left(\frac{(b_i - x_i)^2 + (x_i - a_i)^2}{2(b_i - a_i)} \right) \right. \right. \right.
$$

$$
\left. \left. \left. \left\| P_{n_{\lambda}-1} \right\|_{\infty,[a_i,b_i]} \left\| \frac{\partial^{n_{\lambda}} f_{\lambda}(\ldots, x_{i+1},\ldots, x_m)}{\partial x_i^{n_{\lambda}}} \right\|_{\infty, \prod\limits_{j=1}^{i}[a_j,b_j]} \right] \right) \right) =: \theta_3(x). \quad (7.42)
$$

Finally we derive that

$$
|S(f_1,\ldots, f_r)(x)| \le \min\{\theta_1(x), \theta_2(x), \theta_3(x)\}. \quad (7.43)
$$

We have proved the following general multivariate Ostrowski type inequality for several functions.

Theorem 7.11 *Let f_{λ}, $\lambda = 1,\ldots, r \in \mathbb{N} - \{1\}$, as in Assumption 7.5 or Brief Assumption 7.6; $n_{\lambda} \in \mathbb{N}$ associated with f_{λ}, $x = (x_1,\ldots, x_m)$, $s = (s_1,\ldots, s_m) \in \prod\limits_{i=1}^{m} [a_i, b_i]$. Here $A_{i\lambda}(x_i,\ldots, x_m)$ as in (7.30), $i = 1,\ldots, m$. We put*

$$S(f_1, \ldots, f_r)(x) := \sum_{\lambda=1}^{r} \left\{ \left(\prod_{\substack{\rho=1 \\ \rho\neq\lambda}}^{r} f_\rho(x) \right) \left[f_\lambda(x) - \frac{n_\lambda^m}{\prod_{i=1}^{m}(b_i - a_i)} \cdot \right. \right.$$

$$\left. \left. \left(\int_{\prod_{i=1}^{m}[a_i,b_i]} f_\lambda(s)\, ds \right) - \sum_{i=1}^{m} A_{i\lambda}(x_i, \ldots, x_m) \right] \right\}. \qquad (7.44)$$

Here $\theta_1(x)$ *is as in (7.37). Let* $p_{li} > 1 : \sum_{li=1}^{3} \frac{1}{p_{li}} = 1, i = 1, \ldots, m,$ *and* $\theta_2(x)$ *as in (7.40). And* $\theta_3(x)$ *as in (7.42). Then*

$$|S(f_1, \ldots, f_r)(x)| \leq \min\{\theta_1(x), \theta_2(x), \theta_3(x)\}. \qquad (7.45)$$

We continue with

Remark 7.12 Additionally assume that $\frac{\partial^{n_\lambda} f_\lambda}{\partial x_i^{n_\lambda}}$ are continuous on $\prod_{j=1}^{m}[a_j, b_j]$ for all $i = 1, \ldots, m; \lambda = 1, \ldots, r.$

We define and observe

$$W := \int_{\prod_{j=1}^{m}[a_j,b_j]} S(f_1, \ldots, f_r)(x)\, dx = r \int_{\prod_{j=1}^{m}[a_j,b_j]} \left(\prod_{\substack{\rho=1 \\ \rho\neq\lambda}}^{r} f_\rho(x) \right) dx - \quad (7.46)$$

$$\frac{1}{\prod_{j=1}^{m}(b_j - a_j)} \sum_{\lambda=1}^{r} n_\lambda^m \left(\int_{\prod_{j=1}^{m}[a_j,b_j]} \left(\prod_{\substack{\rho=1 \\ \rho\neq\lambda}}^{r} f_\rho(x) \right) dx \right) \left(\int_{\prod_{i=1}^{m}[a_i,b_i]} f_\lambda(s)\, ds \right) -$$

$$\sum_{\lambda=1}^{r} \int_{\prod_{j=1}^{m}[a_j,b_j]} \left(\left(\prod_{\substack{\rho=1 \\ \rho\neq\lambda}}^{r} f_\rho(x) \right) \left(\sum_{i=1}^{m} A_{i\lambda}(x_i, \ldots, x_m) \right) \right) dx =$$

$$\sum_{\lambda=1}^{r}\left\{\int_{\prod\limits_{j=1}^{m}[a_j,b_j]}\left(\left(\prod_{\substack{\rho=1\\\rho\neq\lambda}}^{r}f_\rho(x)\right)\left(\sum_{i=1}^{m}B_{i\lambda}(x_i,\ldots,x_m)\right)\right)dx\right\}. \qquad (7.47)$$

Clearly here $B_{i\lambda}(x_i,\ldots,x_m)$ is a continuous function for all $i=1,\ldots,m;\;\lambda=1,\ldots,r$.

Hence

$$|W|\overset{(7.47)}{\le}\sum_{\lambda=1}^{r}\left\{\int_{\prod\limits_{j=1}^{m}[a_j,b_j]}\left\{\left(\prod_{\substack{\rho=1\\\rho\neq\lambda}}^{r}|f_\rho(x)|\right)\left(\sum_{i=1}^{m}|B_{i\lambda}(x_i,\ldots,x_m)|\right)\right\}dx\right\}$$
$$(7.48)$$

$$=:(\omega_1).$$

That is

$$|W|\le\sum_{\lambda=1}^{r}\left\{\left(\prod_{\substack{\rho=1\\\rho\neq\lambda}}^{r}\|f_\rho\|_{\infty,\prod\limits_{j=1}^{m}[a_j,b_j]}\right)\left(\sum_{i=1}^{m}\left(\int_{\prod\limits_{j=1}^{m}[a_j,b_j]}|B_{i\lambda}(x_i,\ldots,x_m)|\,dx\right)\right)\right\}.$$
$$(7.49)$$

From (7.41) we obtain

$$|B_{i\lambda}(x_i,\ldots,x_m)|\le$$

$$n_\lambda^{i-1}\|P_{n_\lambda-1}\|_{\infty,[a_i,b_i]}\left\|\frac{\partial^{n_\lambda}f_\lambda}{\partial x_i^{n_\lambda}}\right\|_{\infty,\prod\limits_{j=1}^{m}[a_j,b_j]}\left(\frac{(b_i-x_i)^2+(x_i-a_i)^2}{2(b_i-a_i)}\right), \qquad (7.50)$$

all $i=1,\ldots,m;\;\lambda=1,\ldots,r$.

Then

$$\int_{\prod\limits_{j=1}^{m}[a_j,b_j]}|B_{i\lambda}(x_i,\ldots,x_m)|\,dx\le$$

$$\frac{n_\lambda^{i-1}\|P_{n_\lambda-1}\|_{\infty,[a_i,b_i]}\left\|\frac{\partial^{n_\lambda}f_\lambda}{\partial x_i^{n_\lambda}}\right\|_{\infty,\prod\limits_{j=1}^{m}[a_j,b_j]}}{2(b_i-a_i)}.$$

$$\left(\prod_{\substack{j=1 \\ j \neq i}}^{m} (b_j - a_j) \right) \left(\int_{a_i}^{b_i} \left((b_i - x_i)^2 + (x_i - a_i)^2 \right) dx_i \right), \tag{7.51}$$

and consequently it holds,

$$\int_{\prod_{j=1}^{m} [a_j, b_j]} |B_{i\lambda}(x_i, \ldots, x_m)| \, dx \le$$

$$\frac{(b_i - a_i) \left(\prod_{j=1}^{m} (b_j - a_j) \right)}{3} n_\lambda^{i-1} \| P_{n_\lambda - 1} \|_{\infty, [a_i, b_i]} \left\| \frac{\partial^{n_\lambda} f_\lambda}{\partial x_i^{n_\lambda}} \right\|_{\infty, \prod_{j=1}^{m} [a_j, b_j]}, \tag{7.52}$$

for all $i = 1, \ldots, m$; $\lambda = 1, \ldots, r$.

Hence

$$\sum_{i=1}^{m} \left(\int_{\prod_{j=1}^{m} [a_j, b_j]} |B_{i\lambda}(x_i, \ldots, x_m)| \, dx \right) \le \left(\frac{\left(\prod_{j=1}^{m} (b_j - a_j) \right)}{3} \right) \cdot$$

$$\left(\sum_{i=1}^{m} \left[(b_i - a_i) \| P_{n_\lambda - 1} \|_{\infty, [a_i, b_i]} \left\| \frac{\partial^{n_\lambda} f_\lambda}{\partial x_i^{n_\lambda}} \right\|_{\infty, \prod_{j=1}^{m} [a_j, b_j]} n_\lambda^{i-1} \right] \right), \tag{7.53}$$

for $\lambda = 1, \ldots, r$.

Using (7.49) and (7.53) we obtain

$$|W| \le \left(\frac{\left(\prod_{j=1}^{m} (b_j - a_j) \right)}{3} \right) \left[\sum_{\lambda=1}^{r} \left\{ \left(\prod_{\substack{\rho=1 \\ \rho \neq \lambda}}^{r} \| f_\rho \|_{\infty, \prod_{j=1}^{m} [a_j, b_j]} \right) \cdot \right. \right.$$

$$\left(\sum_{i=1}^{m}\left[(b_i - a_i)\, n_\lambda^{i-1}\, \|P_{n_\lambda-1}\|_{\infty,[a_i,b_i]}\left\|\frac{\partial^{n_\lambda} f_\lambda}{\partial x_i^{n_\lambda}}\right\|_{\infty,\prod_{j=1}^{m}[a_j,b_j]}\right]\right)\right)\right] =: A_1.$$

(7.54)

Notice next that

$$(\omega_1) = \sum_{\lambda=1}^{r}\sum_{i=1}^{m}\left(\int_{\prod_{j=1}^{m}[a_j,b_j]}\left(\prod_{\substack{\rho=1\\\rho\neq\lambda}}^{r}|f_\rho(x)|\right)|B_{i\lambda}(x_i,\dots,x_m)|\,dx\right) =: (\omega_2).$$

(7.55)

Let $p, q > 1 : \frac{1}{p} + \frac{1}{q} = 1$. Then

$$(\omega_2) \le \sum_{\lambda=1}^{r}\sum_{i=1}^{m}\left\|\prod_{\substack{\rho=1\\\rho\neq\lambda}}^{r}f_\rho\right\|_{L_p\left(\prod_{j=1}^{m}[a_j,b_j]\right)}\|B_{i\lambda}\|_{L_q\left(\prod_{j=1}^{m}[a_j,b_j]\right)}\left(\prod_{j=1}^{i-1}(b_j - a_j)\right)^{\frac{1}{q}}.$$

(7.56)

We have also proved that

$$|W| \le \sum_{\lambda=1}^{r}\sum_{i=1}^{m}\left\|\prod_{\substack{\rho=1\\\rho\neq\lambda}}^{r}f_\rho\right\|_{L_p\left(\prod_{j=1}^{m}[a_j,b_j]\right)}\|B_{i\lambda}\|_{L_q\left(\prod_{j=i}^{m}[a_j,b_j]\right)}\left(\prod_{j=1}^{i-1}(b_j - a_j)\right)^{\frac{1}{q}}$$

(7.57)

$$=: A_2.$$

From (7.50) we get

$$|B_{i\lambda}(x_i,\dots,x_m)| \le n_\lambda^{i-1}\,\|P_{n_\lambda-1}\|_{\infty,[a_i,b_i]}\left\|\frac{\partial^{n_\lambda} f_\lambda}{\partial x_i^{n_\lambda}}\right\|_{\infty,\prod_{j=1}^{m}[a_j,b_j]}\left(\frac{b_i - a_i}{2}\right),$$

(7.58)

all $i = 1,\dots,m;\ \lambda = 1,\dots,r$.
  Thus it holds

$$\sum_{i=1}^{m}|B_{i\lambda}(x_i,\dots,x_m)| \le$$

$$\frac{1}{2}\left\{\sum_{i=1}^{m}\left[(b_i-a_i)\,n_\lambda^{i-1}\,\|P_{n_\lambda-1}\|_{\infty,[a_i,b_i]}\left\|\frac{\partial^{n_\lambda}f_\lambda}{\partial x_i^{n_\lambda}}\right\|_{\infty,\prod\limits_{j=1}^{m}[a_j,b_j]}\right]\right\},\qquad(7.59)$$

$\lambda=1,\dots,r.$

Using (7.48) and (7.59) we finally derive

$$|W|\le\frac{1}{2}\left\{\sum_{\lambda=1}^{r}\left\{\left\|\prod_{\substack{\rho=1\\\rho\ne\lambda}}^{r}f_\rho\right\|_{L_1\left(\prod\limits_{j=1}^{m}[a_j,b_j]\right)}\left[\sum_{i=1}^{m}\left[(b_i-a_i)\,n_\lambda^{i-1}\,\|P_{n_\lambda-1}\|_{\infty,[a_i,b_i]}\cdot\right.\right.\right.\right.$$

$$\left.\left.\left.\left.\left\|\frac{\partial^{n_\lambda}f_\lambda}{\partial x_i^{n_\lambda}}\right\|_{\infty,\prod\limits_{j=1}^{m}[a_j,b_j]}\right]\right]\right\}\right\}=:A_3.\qquad(7.60)$$

We have proved the following general multivariate Grüss inequality.

Theorem 7.13 *Let f_λ, $\lambda=1,\dots,r\in\mathbb{N}-\{1\}$, as in Assumptions 7.5 plus $\frac{\partial^{n_\lambda}f_\lambda}{\partial x_i^{n_\lambda}}$ are continuous on $\prod\limits_{j=1}^{m}[a_j,b_j]$ for all $i=1,\dots,m$; $\lambda=1,\dots,r$, or Brief Assumptions 7.6; $n_\lambda\in\mathbb{N}$ associated with f_λ. Here $A_{i\lambda}\,(x_i,\dots,x_m)$ as in (7.30), and $B_{i\lambda}\,(x_i,\dots,x_m)$ as in (7.31), $i=1,\dots,m$. We set*

$$W:=r\int_{\prod\limits_{j=1}^{m}[a_j,b_j]}\left(\prod_{\substack{\rho=1\\\rho\ne\lambda}}^{r}f_\rho\,(x)\right)dx-$$

$$\frac{1}{\prod\limits_{j=1}^{m}(b_j-a_j)}\sum_{\lambda=1}^{r}n_\lambda^{m}\left(\int_{\prod\limits_{j=1}^{m}[a_j,b_j]}\left(\prod_{\substack{\rho=1\\\rho\ne\lambda}}^{r}f_\rho\,(x)\right)dx\right)\left(\int_{\prod\limits_{i=1}^{m}[a_i,b_i]}f_\lambda\,(s)\,ds\right)-$$

$$\sum_{\lambda=1}^{r} \int_{\prod_{j=1}^{m}[a_j,b_j]} \left(\left(\prod_{\substack{\rho=1 \\ \rho \neq \lambda}}^{r} f_\rho(x) \right) \left(\sum_{i=1}^{m} A_{i\lambda}(x_i,\ldots,x_m) \right) \right) dx. \qquad (7.61)$$

Let A_1 as in (7.54); $p, q > 1 : \frac{1}{p} + \frac{1}{q} = 1$, A_2 as in (7.57), and A_3 as in (7.60).
   Then

$$|W| \leq \min\{A_1, A_2, A_3\}. \qquad (7.62)$$

7.4 Applications

We apply Theorems 7.8, 7.11 and 7.13 for the case of $n_1 = n_2 = \ldots = n_r = 1$.

   We simplify General Assumption 7.5 and Brief Assumption 7.6, respectively, as follows:

General Assumption 7.14 Let $f : \prod_{j=1}^{m}[a_j, b_j] \to \mathbb{R}$ satisfying:

(1) for $j = 1, \ldots, m$ we have that $f(x_1, x_2, \ldots, x_{j-1}, s_j, x_{j+1}, \ldots, x_m)$ is absolutely continuous in $s_j \in [a_j, b_j]$, for every $(x_1, x_2, \ldots, x_{j-1}, x_{j+1}, \ldots, x_m) \in \prod_{\substack{i=1 \\ i \neq j}}^{m}[a_i, b_i]$,

(2) for $j = 1, \ldots, m$ we have that $\frac{\partial f(s_1, \ldots, s_j, x_{j+1}, \ldots, x_m)}{\partial x_j}$ is continuous on $\prod_{i=1}^{j}[a_i, b_i]$, for every $(x_{j+1}, \ldots, x_m) \in \prod_{i=j+1}^{m}[a_i, b_i]$,

(3) f is continuous on $\prod_{i=1}^{m}[a_i, b_i]$.

Brief Assumption 7.15 Let $f : \prod_{j=1}^{m}[a_j, b_j] \to \mathbb{R}$ with $\frac{\partial^l f}{\partial x_i^l}$ for $l = 0, 1; j = 1, \ldots, m$, are continuous on $\prod_{j=1}^{m}[a_j, b_j]$.

   We give the following multivariate representation result which is an application of Theorem 7.8

Theorem 7.16 *Let f as in General Assumptions 7.14 or Brief Assumptions 7.15. Then*

$$f(x_1, \ldots, x_n) =$$

$$\frac{1}{\displaystyle\prod_{j=1}^{m}(b_j - a_j)} \int_{\prod_{j=1}^{m}[a_j,b_j]} f(s_1, \ldots, s_m)\, ds_1 \ldots ds_m + \sum_{i=1}^{m} T_i^*(x_i, \ldots, x_m), \quad (7.63)$$

where

$$T_i^*(x_i, \ldots, x_m) = \frac{1}{\displaystyle\prod_{j=1}^{i}(b_j - a_j)} \cdot$$

$$\int_{a_1}^{b_1} \ldots \int_{a_i}^{b_i} q(x_i, s_i)\, \frac{\partial f(s_1, \ldots, s_i, x_{i+1}, \ldots, x_m)}{\partial x_i}\, ds_1 \ldots ds_i, \quad (7.64)$$

are continuous functions for all $i = 1, \ldots, m$.

Next we make

Remark 7.17 Let f_λ as in Assumptions 7.14 or 7.15, $\lambda = 1, \ldots, r$. Then

$$f_\lambda(x) = \frac{1}{\displaystyle\prod_{j=1}^{m}(b_j - a_j)} \int_{\prod_{j=1}^{m}[a_j,b_j]} f_\lambda(s)\, ds + \sum_{i=1}^{m} T_{i\lambda}^*(x_i, \ldots, x_m). \quad (7.65)$$

Here the corresponding

$$A_{i\lambda}^*(x_i, \ldots, x_m) = 0,$$

and

$$B_{i\lambda}^*(x_i, \ldots, x_m) = \frac{1}{\displaystyle\prod_{j=1}^{i}(b_j - a_j)} \cdot$$

$$\int_{a_1}^{b_1} \ldots \int_{a_i}^{b_i} q(x_i, s_i)\, \frac{\partial f_\lambda(s_1, \ldots, s_i, x_{i+1}, \ldots, x_m)}{\partial x_i}\, ds_1 \ldots ds_i, \quad (7.66)$$

for all $i = 1, \ldots, m$; $\lambda = 1, \ldots, r$.

That is

$$T_{i\lambda}^* (x_i, \ldots, x_m) = B_{i\lambda}^* (x_i, \ldots, x_m), \tag{7.67}$$

$i = 1, \ldots, m; \lambda = 1, \ldots, r.$

We call and have the identity

$$S^* (f_1, \ldots, f_r) (x) :=$$

$$\sum_{\lambda=1}^{r} \left\{ \left(\prod_{\substack{\rho=1 \\ \rho \neq \lambda}}^{r} f_\rho (x) \right) \left[f_\lambda (x) - \frac{1}{\prod_{j=1}^{m} (b_j - a_j)} \left(\int_{\prod_{j=1}^{m} [a_j, b_j]} f_\lambda (s)\, ds \right) \right] \right\}$$

$$= \sum_{\lambda=1}^{r} \left(\left(\prod_{\substack{\rho=1 \\ \rho \neq \lambda}}^{r} f_\rho (x) \right) \left(\sum_{i=1}^{m} B_{i\lambda}^* (x_i, \ldots, x_m) \right) \right), \tag{7.68}$$

true for any fixed $x \in \prod_{j=1}^{m} [a_j, b_j]$.

Hence it holds

$$\left| S^* (f_1, \ldots, f_r) (x) \right| \leq \sum_{\lambda=1}^{r} \left(\left(\prod_{\substack{\rho=1 \\ \rho \neq \lambda}}^{r} \left| f_\rho (x) \right| \right) \left(\sum_{i=1}^{m} \left| B_{i\lambda}^* (x_i, \ldots, x_m) \right| \right) \right). \tag{7.69}$$

We obtain:

(1) From (7.37) we derive

$$\left| S^* (f_1, \ldots, f_r) (x) \right| \leq \sum_{\lambda=1}^{r} \left(\left(\prod_{\substack{\rho=1 \\ \rho \neq \lambda}}^{r} \left| f_\rho (x) \right| \right) \left(\sum_{i=1}^{m} \frac{1}{\prod_{j=1}^{i} (b_j - a_j)} \cdot \right. \right.$$

$$\left[\left(\frac{(b_i - a_i) + |a_i + b_i - 2x_i|}{2}\right)\left\|\frac{\partial f_\lambda\left(..., x_{i+1}, ..., x_m\right)}{\partial x_i}\right\|_{L_1\left(\prod_{j=1}^{i}[a_j,b_j]\right)}\right]\right)\right)$$

(7.70)

$$=: \theta_1^*(x).$$

(2) Let $p_{li} > 1 : \sum_{li=1}^{3}\frac{1}{p_{li}} = 1$. Then by (7.40) we derive

$$\left|S^*(f_1, ..., f_r)(x)\right| \le$$

$$\sum_{\lambda=1}^{r}\left(\left(\prod_{\substack{\rho=1\\\rho\neq\lambda}}^{r}|f_\rho(x)|\right)\left(\sum_{i=1}^{m}\frac{1}{\prod_{j=1}^{i}(b_j - a_j)}\left[(b_i - a_i)^{\frac{1}{p_{1i}}}\left(\prod_{j=1}^{i-1}(b_j - a_j)\right)^{\frac{1}{p_{1i}}+\frac{1}{p_{2i}}}\right.\right.\right.$$

$$\left(\frac{(b_i - x_i)^{p_{2i}+1} + (x_i - a_i)^{p_{2i}+1}}{p_{2i} + 1}\right)^{\frac{1}{p_{2i}}}.$$

$$\left.\left.\left.\left\|\frac{\partial f_\lambda\left(..., x_{i+1}, ..., x_m\right)}{\partial x_i}\right\|_{L_{p_{3i}}\left(\prod_{j=1}^{i}[a_j,b_j]\right)}\right]\right)\right) =: \theta_2^*(x).$$

(7.71)

(3) We get by (7.42) that

$$\left|S^*(f_1, ..., f_r)(x)\right| \le \sum_{\lambda=1}^{r}\left(\left(\prod_{\substack{\rho=1\\\rho\neq\lambda}}^{r}|f_\rho(x)|\right)\left(\sum_{i=1}^{m}\left(\frac{(b_i - x_i)^2 + (x_i - a_i)^2}{2(b_i - a_i)}\right).\right.\right.$$

$$\left.\left.\left\|\frac{\partial f_\lambda\left(..., x_{i+1}, ..., x_m\right)}{\partial x_i}\right\|_{\infty, \prod_{j=1}^{i}[a_j,b_j]}\right)\right] =: \theta_3^*(x).$$

(7.72)

So as an applications of Theorem 7.11 we give the following multivariate Ostrowski type inequality.

Theorem 7.18 *All as in Remark 7.17. Then*

$$\left| S^* \left(f_1, \ldots, f_r \right)(x) \right| \le \min \left\{ \theta_1^* \left(x \right), \theta_2^* \left(x \right), \theta_3^* \left(x \right) \right\}. \tag{7.73}$$

We also make

Remark 7.19 Here Assumption 7.14 hold and $\frac{\partial f_\lambda}{\partial x_i}$ are continuous on $\prod\limits_{j=1}^{m} [a_j, b_j]$

for all $i = 1, \ldots, m$; $\lambda = 1, \ldots, r$, or Assumption 7.15 are valid.

We set

$$W^* = \int_{\prod\limits_{j=1}^{m}[a_j,b_j]} S^* \left(f_1, \ldots, f_r \right)(x)\, dx = r \int_{\prod\limits_{j=1}^{m}[a_j,b_j]} \left(\prod_{\rho=1}^{r} f_\rho(x) \right) dx -$$

$$\tag{7.74}$$

$$\frac{1}{\prod\limits_{j=1}^{m}(b_j - a_j)} \sum_{\lambda=1}^{r} \left(\int_{\prod\limits_{j=1}^{m}[a_j,b_j]} \left(\prod_{\substack{\rho=1 \\ \rho\neq\lambda}}^{r} f_\rho(x) \right) dx \right) \left(\int_{\prod\limits_{i=1}^{m}[a_i,b_i]} f_\lambda(s)\, ds \right) =$$

$$\sum_{\lambda=1}^{r} \left\{ \int_{\prod\limits_{j=1}^{m}[a_j,b_j]} \left(\left(\prod_{\substack{\rho=1 \\ \rho\neq\lambda}}^{r} f_\rho(x) \right) \left(\sum_{i=1}^{m} B_{i\lambda}^* (x_i, \ldots, x_m) \right) \right) dx \right\}. \tag{7.75}$$

Here $B_{i\lambda}^*$ is a continuous for any $i = 1, \ldots, m$; $\lambda = 1, \ldots, r$.
Then

(1) following (7.49) we find

$$\left| W^* \right| \le \sum_{\lambda=1}^{r} \left\{ \left(\prod_{\substack{\rho=1 \\ \rho\neq\lambda}}^{r} \| f_\rho \|_{\infty, \prod\limits_{j=1}^{m}[a_j,b_j]} \right) \left(\sum_{i=1}^{m} \left(\int_{\prod\limits_{j=1}^{m}[a_j,b_j]} \left| B_{i\lambda}^* (x_i, \ldots, x_m) \right| dx \right) \right) \right\}. \tag{7.76}$$

We also get that

$$|W^*| \overset{(7.54)}{\leq} \left(\frac{\prod\limits_{j=1}^{m} (b_j - a_j)}{3} \right) \left[\sum_{\lambda=1}^{r} \left\{ \left(\prod_{\substack{\rho=1 \\ \rho \neq \lambda}}^{r} \|f_\rho\|_{\infty, \prod\limits_{j=1}^{m}[a_j,b_j]} \right) \right. \right.$$

$$\left. \left. \left(\sum_{i=1}^{m} (b_i - a_i) \left\| \frac{\partial f_\lambda}{\partial x_i} \right\|_{\infty, \prod\limits_{j=1}^{m}[a_j,b_j]} \right) \right\} \right] =: A_1^*. \tag{7.77}$$

(2) Let $p, q > 1 : \frac{1}{p} + \frac{1}{q} = 1$. Then

$$|W^*| \overset{(7.57)}{\leq} \sum_{\lambda=1}^{r} \sum_{i=1}^{m} \left\| \prod_{\substack{\rho=1 \\ \rho \neq \lambda}}^{r} f_\rho \right\|_{L_p \left(\prod\limits_{j=1}^{m}[a_j,b_j] \right)} \|B_{i\lambda}^*\|_{L_q \left(\prod\limits_{j=1}^{m}[a_j,b_j] \right)} \cdot \tag{7.78}$$

$$\left(\prod_{j=1}^{i-1} (b_j - a_j) \right)^{\frac{1}{q}} =: A_2^*.$$

(3) From (7.60) we obtain

$$|W^*| \leq \frac{1}{2} \left\{ \sum_{\lambda=1}^{r} \left\{ \left\| \prod_{\substack{\rho=1 \\ \rho \neq \lambda}}^{r} f_\rho \right\|_{L_1 \left(\prod\limits_{j=1}^{m}[a_j,b_j] \right)} \left[\sum_{i=1}^{m} \left[(b_i - a_i) \left\| \frac{\partial f_\lambda}{\partial x_i} \right\|_{\infty, \prod\limits_{j=1}^{m}[a_j,b_j]} \right] \right] \right\} \right\}$$

$$\tag{7.79}$$

$$=: A_3^*.$$

We have proved the following multivariate Grüss type inequality as an application of Theorem 7.13.

Theorem 7.20 *Here all as in Remark 7.19. We derive*

$$|W^*| \leq \min \left\{ A_1^*, A_2^*, A_3^* \right\}. \tag{7.80}$$

References

1. G.A. Anastassiou, Harmonic multivariate Ostrowski and Grüss type inequalities for several functions. Demonstr. Math. (Accepted 2014)
2. P.L. Čebyšev, Sur les expressions approximatives des intégrales définies par les aures prises entre les mêmes limites. Proc. Math. Soc. Charkov **2**, 93–98 (1882)
3. L.J. Dedic, M. Matic, J. Pečaric, On some generalizations of Ostrowski's inequality for Lipschitz functions and functions of bounded variation. Math. Inequal. Appl. **3**(1), 1–14 (2000)
4. L.J. Dedic, J. Pečaric, N. Ujevic, On generalizations of Ostrowski inequality and some related results. Czechoslovak Math. J. **53**(128), 173–189 (2003)
5. S.S. Dragomir, S. Wang, A new inequality of Ostrowski type in L_p norm. Indian J. Math. **40**, 299–304 (1998)
6. A.M. Fink, Bounds of the deviation of a function from its averages. Czechoslovak. Math. J. **42**(117), 289–310 (1992)
7. G. Grüss, Über das Maximum des absoluten Betrages von $\left[\left(\frac{1}{b-a} \right) \int_a^b f(x) g(x) dx - \left(\frac{1}{(b-a)^2} \int_a^b f(x) dx \int_a^b g(x) dx \right) \right]$. Math. Z. **39**, 215–226 (1935)
8. A. Ostrowski, Über die Absolutabweichung einer differentiebaren Funktion von ihrem Integralmittelwert. Comment. Math. Helv. **10**, 226–227 (1938)

Chapter 8
Fractional Ostrowski and Grüss Inequalities Using Several Functions

Using Caputo fractional left and right Taylor formulae we establish mixed fractional Ostrowski and Grüss type inequalities involving several functions. The estimates are with respect to all norms $\|\cdot\|_p$, $1 \le p \le \infty$. It follows [6].

8.1 Introduction

The following results motivate initially our chapter.

Theorem 8.1 (1938, Ostrowski [11]) *Let* $f : [a, b] \to \mathbb{R}$ *be continuous on* $[a, b]$ *and differentiable on* (a, b) *whose derivative* $f' : (a, b) \to \mathbb{R}$ *is bounded on* (a, b), *i.e.,* $\|f'\|_\infty^{\sup} := \sup\limits_{t \in (a,b)} |f'(t)| < +\infty$. *Then*

$$\left| \frac{1}{b-a} \int_a^b f(t)\, dt - f(x) \right| \le \left[\frac{1}{4} + \frac{\left(x - \frac{a+b}{2}\right)^2}{(b-a)^2} \right] (b-a) \|f'\|_\infty^{\sup}, \qquad (8.1)$$

for any $x \in [a, b]$. *The constant* $\frac{1}{4}$ *is the best possible.*

Ostrowski type inequalities have great applications to integration approximations in Numerical Analysis.

Theorem 8.2 (1882, Čebyšev [7]) *Let* $f, g : [a, b] \to \mathbb{R}$ *be absolutely continuous functions with* $f', g' \in L_\infty([a, b])$. *Then*

© Springer International Publishing Switzerland 2016
G.A. Anastassiou, *Intelligent Comparisons: Analytic Inequalities*,
Studies in Computational Intelligence 609,
DOI 10.1007/978-3-319-21121-3_8

$$\left| \frac{1}{b-a} \int_a^b f(x) g(x) \, dx - \left(\frac{1}{b-a} \int_a^b f(x) \, dx \right) \left(\frac{1}{b-a} \int_a^b g(x) \, dx \right) \right|$$

$$\le \frac{1}{12} (b-a)^2 \, \|f'\|_\infty \, \|g'\|_\infty . \tag{8.2}$$

The above integrals are assumed to exist.

Grüss type inequalities have applications to Probability.

We are also biggly inspired by the beautiful work of Pachpatte [12]. His article contains great ideas, but it is marred by many minor errors.

So here we present mixed fractional Ostrowski and Grüss type inequalities for several functions, acting to all possible directions. The estimates involve the left and right Caputo fractional derivatives. See also the monographs written by the author [1], Chaps. 24–26 and [3], Chaps. 2–6.

8.2 Background

Let $\nu \ge 0$; the operator I_{a+}^ν, defined for $f \in L_1[(a,b)]$ is given by

$$I_{a+}^\nu f(x) := \frac{1}{\Gamma(\nu)} \int_a^x (x-t)^{\nu-1} f(t) \, dt, \tag{8.3}$$

for $a \le x \le b$, is called the left Riemann-Liouville fractional integral operator of order ν. For $\nu = 0$, we set $I_{a+}^0 := I$, the identity operator, see [1], p. 392, also [8].

Let $\nu \ge 0$, $n := \lceil \nu \rceil$ ($\lceil \cdot \rceil$ ceiling of the number), $f \in AC^n([a,b])$ (it means $f^{(n-1)} \in AC([a,b])$, absolutely continuous functions).

Then the left Caputo fractional derivative is given by

$$D_{*a}^\nu f(x) := \frac{1}{\Gamma(n-\nu)} \int_a^x (x-t)^{n-\nu-1} f^{(n)}(t) \, dt = \left(I_{a+}^{n-\nu} f^{(n)} \right)(x), \tag{8.4}$$

and it exists almost everywhere for $x \in [a,b]$.

See Corollary 16.8, p. 394 of [1], and [8], pp. 49–50.

We need also the left Caputo fractional Taylor formula, see [1], p. 395 and [8], p. 54.

Theorem 8.3 *Let $\nu \ge 0$, $n := \lceil \nu \rceil$, $f \in AC^n([a,b])$. Then*

$$f(x) = \sum_{k=0}^{n-1} \frac{f^{(k)}(a)}{k!} (x-a)^k + I_{a+}^\nu D_{*a}^\nu f(x), \ \forall \, x \in [a,b]. \tag{8.5}$$

Here $I_{a+}^\nu D_{*a}^\nu f \in AC^n([a,b])$.

Let $f \in L_1([a, b])$, $\alpha > 0$. The right Riemann-Liouville fractional operator [2, 9, 10] of order α is denoted by

$$I_{b-}^{\alpha} f(x) := \frac{1}{\Gamma(\alpha)} \int_x^b (z - x)^{\alpha-1} f(z) \, dz, \quad \forall \quad x \in [a, b]. \tag{8.6}$$

We set $I_{b-}^0 := I$, the identity operator.

Let now $f \in AC^m([a, b])$, $m \in \mathbb{N}$, with $m := \lceil \alpha \rceil$.

We define the right Caputo fractional derivative of order $\alpha \geq 0$, by

$$D_{b-}^{\alpha} f(x) := (-1)^m I_{b-}^{m-\alpha} f^{(m)}(x), \tag{8.7}$$

we set $D_{b-}^0 f := f$, that is

$$D_{b-}^{\alpha} f(x) = \frac{(-1)^m}{\Gamma(m-\alpha)} \int_x^b (z - x)^{m-\alpha-1} f^{(m)}(z) \, dz. \tag{8.8}$$

We need the right Caputo fractional Taylor formula.

Theorem 8.4 ([2]) *Let $f \in AC^m([a, b])$, $x \in [a, b]$, $\alpha > 0$, $m := \lceil \alpha \rceil$. Then*

$$f(x) = \sum_{k=0}^{m-1} \frac{f^{(k)}(b)}{k!} (x - b)^k + \frac{1}{\Gamma(\alpha)} \int_x^z (z - x)^{\alpha-1} D_{b-}^{\alpha} f(z) \, dz. \tag{8.9}$$

We need

Proposition 8.5 ([4], p. 361) *Let $\alpha > 0$, $m = \lceil \alpha \rceil$, $f \in C^{m-1}([a, b])$, $f^{(m)} \in L_\infty([a, b])$; $x, x_0 \in [a, b] : x \geq x_0$. Then $D_{*x_0}^{\alpha} f(x)$ is continuous in x_0.*

Proposition 8.6 ([4], p. 361) *Let $\alpha > 0$, $m = \lceil \alpha \rceil$, $f \in C^{m-1}([a, b])$, $f^{(m)} \in L_\infty([a, b])$; $x, x_0 \in [a, b] : x \leq x_0$. Then $D_{x_0-}^{\alpha} f(x)$ is continuous in x_0.*

We also mention

Theorem 8.7 ([4], p. 362) *Let $f \in C^m([a, b])$, $m = \lceil \alpha \rceil$, $\alpha > 0$, $x, x_0 \in [a, b]$. Then $D_{*x_0}^{\alpha} f(x)$, $D_{x_0-}^{\alpha} f(x)$ are jointly continuous functions in (x, x_0) from $[a, b]^2$ into $\mathbb{R}$.*

Convention 8.8 *([4], p. 360) We suppose that*

$$D_{*x_0}^{\alpha} f(x) = 0, \text{ for } x < x_0, \tag{8.10}$$

and

$$D_{x_0-}^{\alpha} f(x) = 0, \text{ for } x > x_0, \tag{8.11}$$

for all $x, x_0 \in [a, b]$.

Finally we are motivated by the following mixed Caputo fractional Ostrowski type inequalities.

Theorem 8.9 ([3], p. 44) *Let* $[a, b] \subset \mathbb{R}$, $\alpha > 0$, $m = \lceil \alpha \rceil$, $f \in AC^m([a, b])$, *and* $\left\| D_{x_0-}^\alpha f \right\|_{\infty,[a,x_0]}$, $\left\| D_{*x_0}^\alpha f \right\|_{\infty,[x_0,b]} < \infty$, $x_0 \in [a, b]$. *Assume* $f^{(k)}(x_0) = 0$, $k = 1, \ldots, m - 1$. *Then*

$$\left| \frac{1}{b-a} \int_a^b f(x)\, dx - f(x_0) \right| \le \frac{1}{(b-a)\, \Gamma(\alpha+2)} \cdot$$

$$\left\{ \left\| D_{x_0-}^\alpha f \right\|_{\infty,[a,x_0]} (x_0 - a)^{\alpha+1} + \left\| D_{*x_0}^\alpha f \right\|_{\infty,[x_0,b]} (b - x_0)^{\alpha+1} \right\} \le \quad (8.12)$$

$$\frac{1}{\Gamma(\alpha+2)} \max \left\{ \left\| D_{x_0-}^\alpha f \right\|_{\infty,[a,x_0]}, \left\| D_{*x_0}^\alpha f \right\|_{\infty,[x_0,b]} \right\} (b-a)^\alpha. \quad (8.13)$$

Inequality (8.12) is sharp, infact it is attained.

Theorem 8.10 ([3], p. 45) *Le* $\alpha \ge 1$, $m = \lceil \alpha \rceil$, *and* $f \in AC^m([a, b])$. *Suppose that* $f^{(k)}(x_0) = 0$, $k = 1, \ldots, m - 1$, $x_0 \in [a, b]$ *and* $D_{x_0-}^\alpha f \in L_1([a, x_0])$, $D_{*x_0}^\alpha f \in L_1([x_0, b])$. *Then*

$$\left| \frac{1}{b-a} \int_a^b f(x)\, dx - f(x_0) \right| \le \frac{1}{(b-a)\, \Gamma(\alpha+1)} \cdot \quad (8.14)$$

$$\left\{ (x_0 - a)^\alpha \left\| D_{x_0-}^\alpha f \right\|_{L_1([a,x_0])} + (b - x_0)^\alpha \left\| D_{*x_0}^\alpha f \right\|_{L_1([x_0,b])} \right\} \le$$

$$\frac{1}{\Gamma(\alpha+1)} \max \left\{ \left\| D_{x_0-}^\alpha f \right\|_{L_1([a,x_0])}, \left\| D_{*x_0}^\alpha f \right\|_{L_1([x_0,b])} \right\} (b-a)^{\alpha-1}. \quad (8.15)$$

Theorem 8.11 ([3], p. 47) *Let* $p, q > 1 : \frac{1}{p} + \frac{1}{q} = 1$, $\alpha > \frac{1}{q}$, $m = \lceil \alpha \rceil$, $\alpha > 0$, *and* $f \in AC^m([a, b])$. *Suppose that* $f^{(k)}(x_0) = 0$, $k = 1, \ldots, m - 1$, $x_0 \in [a, b]$. *Assume* $D_{x_0-}^\alpha f \in L_q([a, x_0])$, *and* $D_{*x_0}^\alpha f \in L_q([x_0, b])$. *Then*

$$\left| \frac{1}{b-a} \int_a^b f(x)\, dx - f(x_0) \right| \le \frac{1}{(b-a)\, \Gamma(\alpha)\, (p(\alpha-1)+1)^{\frac{1}{p}} \left(\alpha + \frac{1}{p} \right)} \cdot$$

$$\quad (8.16)$$

$$\left\{ (x_0 - a)^{\alpha+\frac{1}{p}} \left\| D_{x_0-}^\alpha f \right\|_{L_q([a,x_0])} + (b - x_0)^{\alpha+\frac{1}{p}} \left\| D_{*x_0}^\alpha f \right\|_{L_q([x_0,b])} \right\} \le$$

$$\frac{1}{\Gamma(\alpha)\, (p(\alpha-1)+1)^{\frac{1}{p}} \left(\alpha + \frac{1}{p} \right)}.$$

$$\max \left\{ \left\| D_{x_0-}^{\alpha} f \right\|_{L_q([a,x_0])}, \left\| D_{*x_0}^{\alpha} f \right\|_{L_q([x_0,b])} \right\} (b-a)^{\alpha - \frac{1}{q}}. \tag{8.17}$$

In this chapter we generalize Theorems 8.9–8.11 for several functions. We also produce Caputo fractional Grüss type inequalities for several functions.

8.3 Main Results

We start with Caputo fractional mixed Ostrowski type inequalities involving several functions.

Theorem 8.12 *Let $x_0 \in [a, b] \subset \mathbb{R}$, $\alpha > 0$, $m = \lceil \alpha \rceil$, $f_i \in AC^m([a, b])$, $i = 1, \ldots, r \in \mathbb{N} - \{1\}$, with $f_i^{(k)}(x_0) = 0$, $k = 1, \ldots, m - 1$, $i = 1, \ldots, r$. Assume that $\left\| D_{x_0-}^{\alpha} f_i \right\|_{\infty,[a,x_0]}$, $\left\| D_{*x_0}^{\alpha} f_i \right\|_{\infty,[x_0,b]} < \infty$, $i = 1, \ldots, r$. Denote by*

$$\theta(f_1, \ldots, f_r)(x_0) := r \int_a^b \left(\prod_{k=1}^r f_k(x) \right) dx - \sum_{i=1}^r \left[f_i(x_0) \int_a^b \left(\prod_{\substack{j=1 \\ j \neq i}}^r f_j(x) \right) dx \right]. \tag{8.18}$$

Then

$$|\theta(f_1, \ldots, f_r)(x_0)| \leq \sum_{i=1}^r \left[\left[\left\| D_{x_0-}^{\alpha} f_i \right\|_{\infty,[a,x_0]} I_{a+}^{\alpha+1} \left(\prod_{\substack{j=1 \\ j \neq i}}^r |f_j(x_0)| \right) \right] \tag{8.19}$$

$$+ \left[\left\| D_{*x_0}^{\alpha} f_i \right\|_{\infty,[x_0,b]} I_{b-}^{\alpha+1} \left(\prod_{\substack{j=1 \\ j \neq i}}^r |f_j(x_0)| \right) \right] \right].$$

Inequality (8.19) is sharp, infact it is attained.

Theorem 8.13 *Let $\alpha \geq 1$, $m = \lceil \alpha \rceil$, and $f_i \in AC^m([a, b])$, $i = 1, \ldots, r \in \mathbb{N} - \{1\}$. Suppose that $f_i^{(k)}(x_0) = 0$, $k = 1, \ldots, m - 1$; $x_0 \in [a, b]$ and $D_{x_0-}^{\alpha} f_i \in L_1([a, x_0])$, $D_{*x_0}^{\alpha} f_i \in L_1([x_0, b])$, for all $i = 1, \ldots, r$. Then*

$$|\theta(f_1, \ldots, f_r)(x_0)| \leq \sum_{i=1}^r \left[\left[\left\| D_{x_0-}^{\alpha} f_i \right\|_{L_1([a,x_0])} I_{a+}^{\alpha} \left(\prod_{\substack{j=1 \\ j \neq i}}^r |f_j(x_0)| \right) \right] \tag{8.20}$$

$$+ \left[\left\| D^{\alpha}_{*x_0} f_i \right\|_{L_1([x_0,b])} I^{\alpha}_{b-} \left(\prod_{\substack{j=1 \\ j \neq i}}^{r} \left| f_j (x_0) \right| \right) \right] \right] .$$

Theorem 8.14 *Let* $p, q > 1 : \frac{1}{p} + \frac{1}{q} = 1$, $\alpha > \frac{1}{q}$, $m = \lceil \alpha \rceil$, $\alpha > 0$, *and* $f_i \in AC^m ([a,b])$, $i = 1, \ldots, r \in \mathbb{N} - \{1\}$. *Suppose that* $f_i^{(k)} (x_0) = 0$, $k = 1, \ldots, m - 1$, $x_0 \in [a,b]$; $i = 1, \ldots, r$. *Assume* $D^{\alpha}_{x_0-} f_i \in L_q ([a, x_0])$, *and* $D^{\alpha}_{*x_0} f_i \in L_q ([x_0, b])$, $i = 1, \ldots, r$. *Then*

$$\left| \theta (f_1, \ldots, f_r) (x_0) \right| \leq$$

$$\frac{\Gamma \left(\alpha + \frac{1}{p} \right)}{(p (\alpha - 1) + 1)^{\frac{1}{p}} \Gamma (\alpha)} \sum_{i=1}^{r} \left[\left[\left\| D^{\alpha}_{x_0-} f_i \right\|_{L_q([a, x_0])} I^{\alpha + \frac{1}{p}}_{a+} \left(\prod_{\substack{j=1 \\ j \neq i}}^{r} \left| f_j (x_0) \right| \right) \right] \right.$$

$$(8.21)$$

$$\left. + \left[\left\| D^{\alpha}_{*x_0} f_i \right\|_{L_q([x_0,b])} I^{\alpha + \frac{1}{p}}_{b-} \left(\prod_{\substack{j=1 \\ j \neq i}}^{r} \left| f_j (x_0) \right| \right) \right] \right] .$$

Proof of Theorems 8.12–8.14. Here $x_0 \in [a, b]$. Since $f_i^{(k)} (x_0) = 0$, $k = 1, \ldots, m - 1$; $i = 1, \ldots, r$, we have by Theorem 8.3 that

$$f_i (x) - f_i (x_0) = \frac{1}{\Gamma (\alpha)} \int_{x_0}^{x} (x - z)^{\alpha - 1} D^{\alpha}_{*x_0} f_i (z) \, dz, \quad \forall x \in [x_0, b], \quad (8.22)$$

and by Theorem 8.4 that

$$f_i (x) - f_i (x_0) = \frac{1}{\Gamma (\alpha)} \int_{x}^{x_0} (z - x)^{\alpha - 1} D^{\alpha}_{x_0-} f_i (z) \, dz, \quad \forall x \in [a, x_0]; \quad (8.23)$$

for all $i = 1, \ldots, r$.

Multiplying (8.22) and (8.23) by $\left(\prod_{\substack{j=1 \\ j \neq i}}^{r} f_j (x) \right)$ we get, respectively,

$$\prod_{k=1}^{r} f_k (x) - \left(\prod_{\substack{j=1 \\ j \neq i}}^{r} f_j (x) \right) f_i (x_0) =$$

$$\frac{\left(\prod_{\substack{j=1\\j\neq i}}^{r} f_j(x)\right)}{\Gamma(\alpha)} \int_{x_0}^{x} (x-z)^{\alpha-1} D_{*x_0}^{\alpha} f_i(z)\, dz, \quad \forall\, x \in [x_0, b], \tag{8.24}$$

and

$$\prod_{k=1}^{r} f_k(x) - \left(\prod_{\substack{j=1\\j\neq i}}^{r} f_j(x)\right) f_i(x_0) =$$

$$\frac{\left(\prod_{\substack{j=1\\j\neq i}}^{r} f_j(x)\right)}{\Gamma(\alpha)} \int_{x}^{x_0} (z-x)^{\alpha-1} D_{x_0-}^{\alpha} f_i(z)\, dz, \quad \forall\, x \in [a, x_0]; \tag{8.25}$$

for all $i = 1, \dots, r$.

Adding (8.24) and (8.25), separately, we obtain

$$r\left(\prod_{k=1}^{r} f_k(x)\right) - \sum_{i=1}^{r}\left[\left(\prod_{\substack{j=1\\j\neq i}}^{r} f_j(x)\right) f_i(x_0)\right] =$$

$$\frac{1}{\Gamma(\alpha)} \sum_{i=1}^{r}\left[\left(\prod_{\substack{j=1\\j\neq i}}^{r} f_j(x)\right)\int_{x_0}^{x} (x-z)^{\alpha-1} D_{*x_0}^{\alpha} f_i(z)\, dz\right], \quad \forall\, x \in [x_0, b],$$

$$\tag{8.26}$$

and

$$r\left(\prod_{k=1}^{r} f_k(x)\right) - \sum_{i=1}^{r}\left[\left(\prod_{\substack{j=1\\j\neq i}}^{r} f_j(x)\right) f_i(x_0)\right] =$$

$$\frac{1}{\Gamma(\alpha)} \sum_{i=1}^{r}\left[\left(\prod_{\substack{j=1\\j\neq i}}^{r} f_j(x)\right)\int_{x}^{x_0} (z-x)^{\alpha-1} D_{x_0-}^{\alpha} f_i(z)\, dz\right], \quad \forall\, x \in [a, x_0].$$

$$\tag{8.27}$$

Next we integrate (8.26) and (8.27) with respect to $x \in [a, b]$. We have

$$r\int_{x_0}^{b}\left(\prod_{k=1}^{r} f_k(x)\right) dx - \sum_{i=1}^{r}\left[f_i(x_0)\int_{x_0}^{b}\left(\prod_{\substack{j=1\\j\neq i}}^{r} f_j(x)\right) dx\right] =$$

$$\frac{1}{\Gamma(\alpha)} \sum_{i=1}^{r} \left[\int_{x_0}^{b} \left(\prod_{\substack{j=1 \\ j \neq i}}^{r} f_j(x) \right) \left(\int_{x_0}^{x} (x-z)^{\alpha-1} D_{*x_0}^{\alpha} f_i(z) \, dz \right) dx \right], \quad (8.28)$$

and

$$r \int_{a}^{x_0} \left(\prod_{k=1}^{r} f_k(x) \right) dx - \sum_{i=1}^{r} \left[f_i(x_0) \int_{a}^{x_0} \left(\prod_{\substack{j=1 \\ j \neq i}}^{r} f_j(x) \right) dx \right] =$$

$$\frac{1}{\Gamma(\alpha)} \sum_{i=1}^{r} \left[\int_{a}^{x_0} \left(\prod_{\substack{j=1 \\ j \neq i}}^{r} f_j(x) \right) \left(\int_{x}^{x_0} (z-x)^{\alpha-1} D_{x_0-}^{\alpha} f_i(z) \, dz \right) dx \right]. \quad (8.29)$$

Finally adding (8.28) and (8.29) we obtain the useful identity

$$\theta(f_1, \ldots, f_r)(x_0) :=$$

$$r \int_{a}^{b} \left(\prod_{k=1}^{r} f_k(x) \right) dx - \sum_{i=1}^{r} \left[f_i(x_0) \int_{a}^{b} \left(\prod_{\substack{j=1 \\ j \neq i}}^{r} f_j(x) \right) dx \right] =$$

$$\frac{1}{\Gamma(\alpha)} \sum_{i=1}^{r} \left[\left[\int_{a}^{x_0} \left(\prod_{\substack{j=1 \\ j \neq i}}^{r} f_j(x) \right) \left(\int_{x}^{x_0} (z-x)^{\alpha-1} D_{x_0-}^{\alpha} f_i(z) \, dz \right) dx \right] + \right.$$

$$\left. \left[\int_{x_0}^{b} \left(\prod_{\substack{j=1 \\ j \neq i}}^{r} f_j(x) \right) \left(\int_{x_0}^{x} (x-z)^{\alpha-1} D_{*x_0}^{\alpha} f_i(z) \, dz \right) dx \right] \right]. \quad (8.30)$$

Hence it holds

$$|\theta(f_1, \ldots, f_r)(x_0)| \leq$$

$$\frac{1}{\Gamma(\alpha)} \sum_{i=1}^{r} \left[\left[\int_{a}^{x_0} \left(\prod_{\substack{j=1 \\ j \neq i}}^{r} |f_j(x)| \right) \left(\int_{x}^{x_0} (z-x)^{\alpha-1} \left| D_{x_0-}^{\alpha} f_i(z) \right| dz \right) dx \right] + \right.$$

$$\left[\int_{x_0}^{b} \left(\prod_{\substack{j=1 \\ j \neq i}}^{r} |f_j(x)| \right) \left(\int_{x_0}^{x} (x-z)^{\alpha-1} |D_{*x_0}^{\alpha} f_i(z)| \, dz \right) dx \right] =: (*) . \quad (8.31)$$

We observe that

$$(*) \leq \frac{1}{\Gamma(\alpha+1)} \sum_{i=1}^{r} \left[\left[\|D_{x_0-}^{\alpha} f_i\|_{\infty,[a,x_0]} \int_{a}^{x_0} \left(\prod_{\substack{j=1 \\ j \neq i}}^{r} |f_j(x)| \right) (x_0-x)^{\alpha} \, dx \right] + \right.$$

$$\left. \left[\|D_{*x_0}^{\alpha} f_i\|_{\infty,[x_0,b]} \int_{x_0}^{b} \left(\prod_{\substack{j=1 \\ j \neq i}}^{r} |f_j(x)| \right) (x-x_0)^{\alpha} \, dx \right] \right] = \quad (8.32)$$

$$\sum_{i=1}^{r} \left[\left[\|D_{x_0-}^{\alpha} f_i\|_{\infty,[a,x_0]} I_{a+}^{\alpha+1} \left(\prod_{\substack{j=1 \\ j \neq i}}^{r} |f_j(x_0)| \right) \right] + \right.$$

$$\left. \left[\|D_{*x_0}^{\alpha} f_i\|_{\infty,[x_0,b]} I_{b-}^{\alpha+1} \left(\prod_{\substack{j=1 \\ j \neq i}}^{r} |f_j(x_0)| \right) \right] \right] . \quad (8.33)$$

Based on Proposition 15.114, p. 388, [1] we get that $I_{a+}^{\alpha+1} \left(\prod_{\substack{j=1 \\ j \neq i}}^{r} |f_j(x)| \right) \in$
$AC([a,b])$, so at x_0 is finite.

Also, based on [5] we get that $I_{b-}^{\alpha+1} \left(\prod_{\substack{j=1 \\ j \neq i}}^{r} |f_j(x)| \right) \in AC([a,b])$, so at x_0 is finite.

Next we prove that (8.19) is sharp, namely it is attained.
Set
$$\overline{f_i}(x) := \begin{cases} (x-x_0)^{\alpha}, & x \in [x_0,b], \\ (x_0-x)^{\alpha}, & x \in [a,x_0], \end{cases}$$

$\alpha > 0, a \leq x_0 \leq b, i = 1, \ldots, r.$

Observe that $\overline{f_i} \in AC^m([x_0,b])$, and in $AC^m([a,x_0])$. See that $\overline{f_{i-}}^{(k)}(x_0) = \overline{f_{i+}}^{(k)}(x_0) = 0, k = 0, 1, \ldots, m-1$. Hence there exists $\overline{f_i}^{(m-1)}$ at x_0, also $\overline{f_i}^{(m-1)} \in AC([a,b])$. That is $\overline{f_i} \in AC^m([a,b]), i = 1, \ldots, r.$

We find that

$$\left\| D_{x_0-}^{\alpha} \overline{f_i} \right\|_{\infty,[a,x_0]} = \Gamma(\alpha+1),$$

and

$$\left\| D_{*x_0}^{\alpha} \overline{f_i} \right\|_{\infty,[x_0,b]} = \Gamma(\alpha+1).$$

Consequently it holds

$$L.H.S.(8.19) = r \left[\frac{(b-x_0)^{\alpha r+1} + (x_0-a)^{\alpha r+1}}{\alpha r+1} \right] = R.H.S.(8.19),$$

proving optimality of (8.19).

Hence proving Theorem 8.12.

Next we notice, for $\alpha \geq 1$, that

$$(*) \leq \frac{1}{\Gamma(\alpha)} \sum_{i=1}^{r} \left[\left[\left\| D_{x_0-}^{\alpha} f_i \right\|_{L_1([a,x_0])} \int_a^{x_0} \left(\prod_{\substack{j=1 \\ j \neq i}}^{r} |f_j(x)| \right) (x_0-x)^{\alpha-1} dx \right] + \right.$$

$$\left[\left\| D_{*x_0}^{\alpha} f_i \right\|_{L_1([x_0,b])} \int_{x_0}^{b} \left(\prod_{\substack{j=1 \\ j \neq i}}^{r} |f_j(x)| \right) (x-x_0)^{\alpha-1} dx \right] \right] = \qquad (8.34)$$

$$\sum_{i=1}^{r} \left[\left[\left\| D_{x_0-}^{\alpha} f_i \right\|_{L_1([a,x_0])} I_{a+}^{\alpha} \left(\prod_{\substack{j=1 \\ j \neq i}}^{r} |f_j(x_0)| \right) \right] + \right.$$

$$\left[\left\| D_{*x_0}^{\alpha} f_i \right\|_{L_1([x_0,b])} I_{b-}^{\alpha} \left(\prod_{\substack{j=1 \\ j \neq i}}^{r} |f_j(x_0)| \right) \right] \right], \qquad (8.35)$$

proving Theorem 8.13.

Let now $p, q > 1 : \frac{1}{p} + \frac{1}{q} = 1$, with $\alpha > \frac{1}{q}$. Then

$$(*) \leq \frac{1}{\Gamma(\alpha)} \sum_{i=1}^{r} \left[\left[\int_a^{x_0} \left(\prod_{\substack{j=1 \\ j \neq i}}^{r} |f_j(x)| \right) \right. \right.$$

$$\left(\int_x^{x_0} (z-x)^{p(\alpha-1)}\,dz\right)^{\frac{1}{p}}\left(\int_x^{x_0}\left|D_{x_0-}^\alpha f_i(z)\right|^q dz\right)^{\frac{1}{q}} dx\right]+$$

$$\left[\int_{x_0}^b \left(\prod_{\substack{j=1\\j\neq i}}^r |f_j(x)|\right)\left(\int_{x_0}^x (x-z)^{p(\alpha-1)}\,dz\right)^{\frac{1}{p}}\left(\int_{x_0}^x\left|D_{*x_0}^\alpha f_i(z)\right|^q dz\right)^{\frac{1}{q}} dx\right]\right]\leq$$

$$(8.36)$$

$$\frac{1}{\Gamma(\alpha)}\sum_{i=1}^r\left[\left[\int_a^{x_0}\left(\prod_{\substack{j=1\\j\neq i}}^r |f_j(x)|\right)\frac{(x_0-x)^{\alpha-1+\frac{1}{p}}}{(p(\alpha-1)+1)^{\frac{1}{p}}}\left\|D_{x_0-}^\alpha f_i\right\|_{q,[a,x_0]} dx\right]+\right.$$

$$\left.\left[\int_{x_0}^b\left(\prod_{\substack{j=1\\j\neq i}}^r |f_j(x)|\right)\frac{(x-x_0)^{\alpha-1+\frac{1}{p}}}{(p(\alpha-1)+1)^{\frac{1}{p}}}\left\|D_{*x_0}^\alpha f_i\right\|_{q,[x_0,b]} dx\right]\right]= \quad (8.37)$$

$$\frac{1}{(p(\alpha-1)+1)^{\frac{1}{p}}\Gamma(\alpha)}\cdot$$

$$\sum_{i=1}^r\left[\left[\left\|D_{x_0-}^\alpha f_i\right\|_{q,[a,x_0]}\int_a^{x_0}(x_0-x)^{\alpha+\frac{1}{p}-1}\left(\prod_{\substack{j=1\\j\neq i}}^r |f_j(x)|\right)dx\right]+\right.$$

$$\left.\left[\left\|D_{*x_0}^\alpha f_i\right\|_{q,[x_0,b]}\int_{x_0}^b(x-x_0)^{\alpha+\frac{1}{p}-1}\left(\prod_{\substack{j=1\\j\neq i}}^r |f_j(x)|\right)dx\right]\right]= \quad (8.38)$$

$$\frac{\Gamma\left(\alpha+\frac{1}{p}\right)}{(p(\alpha-1)+1)^{\frac{1}{p}}\Gamma(\alpha)}\sum_{i=1}^r\left[\left[\left\|D_{x_0-}^\alpha f_i\right\|_{q,[a,x_0]}I_{a+}^{\alpha+\frac{1}{p}}\left(\prod_{\substack{j=1\\j\neq i}}^r |f_j(x_0)|\right)\right]+\right.$$

$$(8.39)$$

$$\left.\left[\left\|D_{*x_0}^\alpha f_i\right\|_{q,[x_0,b]}I_{b-}^{\alpha+\frac{1}{p}}\left(\prod_{\substack{j=1\\j\neq i}}^r |f_j(x_0)|\right)\right]\right],$$

that is proving Theorem 8.14. ∎

Next follow Caputo fractional Grüss type inequalities for several functions.

Theorem 8.15 *Let* $x_0 \in [a, b] \subset \mathbb{R}$, $0 < \alpha \leq 1$, $f_i \in AC([a, b])$, $i = 1, \ldots, r \in \mathbb{N} - \{1\}$. *Assume that* $\sup\limits_{x_0 \in [a,b]} \left\| D_{x_0-}^{\alpha} f_i \right\|_{\infty, [a,x_0]}$, $\sup\limits_{x_0 \in [a,b]} \left\| D_{*x_0}^{\alpha} f_i \right\|_{\infty, [x_0,b]} < \infty$, $i = 1, \ldots, r$. *Denote by*

$$\Delta(f_1, \ldots, f_r)(x_0) := r(b-a) \int_a^b \left(\prod_{k=1}^r f_k(x) \right) dx - \tag{8.40}$$

$$\sum_{i=1}^r \left[\left(\int_a^b f_i(x) \, dx \right) \left(\int_a^b \left(\prod_{\substack{j=1 \\ j \neq i}}^r f_j(x) \right) dx \right) \right].$$

Then

$$|\Delta(f_1, \ldots, f_r)| \leq (b-a) \cdot$$

$$\sum_{i=1}^r \left[\left[\sup_{x_0 \in [a,b]} \left\| D_{x_0-}^{\alpha} f_i \right\|_{\infty, [a,x_0]} \sup_{x_0 \in [a,b]} I_{a+}^{\alpha+1} \left(\prod_{\substack{j=1 \\ j \neq i}}^r |f_j(x_0)| \right) \right] + \right.$$

$$\left. \left[\sup_{x_0 \in [a,b]} \left\| D_{*x_0}^{\alpha} f_i \right\|_{\infty, [x_0,b]} \sup_{x_0 \in [a,b]} I_{b-}^{\alpha+1} \left(\prod_{\substack{j=1 \\ j \neq i}}^r |f_j(x_0)| \right) \right] \right]. \tag{8.41}$$

Theorem 8.16 *Let* $p, q > 1 : \frac{1}{p} + \frac{1}{q} = 1$, $\frac{1}{q} < \alpha \leq 1$, *and* $f_i \in AC([a, b])$, $i = 1, \ldots, r \in \mathbb{N} - \{1\}$, $x_0 \in [a, b]$. *Assume that* $\sup\limits_{x_0 \in [a,b]} \left\| D_{x_0-}^{\alpha} f_i \right\|_{L_q([a,x_0])}$, *and* $\sup\limits_{x_0 \in [a,b]} \left\| D_{*x_0}^{\alpha} f_i \right\|_{L_q([x_0,b])} < \infty$, $i = 1, \ldots, r$. *Then*

$$|\Delta(f_1, \ldots, f_r)| \leq \frac{(b-a) \Gamma\left(\alpha + \frac{1}{p}\right)}{(p(\alpha - 1) + 1)^{\frac{1}{p}} \Gamma(\alpha)} \cdot$$

$$\sum_{i=1}^r \left[\left[\sup_{x_0 \in [a,b]} \left\| D_{x_0-}^{\alpha} f_i \right\|_{L_q([a,x_0])} \sup_{x_0 \in [a,b]} I_{a+}^{\alpha + \frac{1}{p}} \left(\prod_{\substack{j=1 \\ j \neq i}}^r |f_j(x_0)| \right) \right] + \right.$$

$$\left[\sup_{x_0 \in [a,b]} \left\| D^\alpha_{*x_0} f_i \right\|_{L_q([x_0,b])} \sup_{x_0 \in [a,b]} I_{b-}^{\alpha+\frac{1}{p}} \left(\prod_{\substack{j=1 \\ j \neq i}}^r |f_j(x_0)| \right) \right] \right]. \tag{8.42}$$

Proof of Theorems 8.15, 8.16. Here $0 < \alpha \le 1$, i.e. $m = 1$, and $f_i \in AC([a,b])$, $i = 1, \ldots, r$. Now we are not tight up to any initial conditions, i.e. (8.22) and (8.23) are valid without initial conditions. Clearly here $\theta(f_1, \ldots, f_r)(x_0) \in AC([a,b])$. Integrating (8.30) over $[a, b]$ with respect to $x_0 \in [a, b]$ we derive

$$\Delta(f_1, \ldots f_r) := \int_a^b \theta(f_1, \ldots, f_r)(x_0)\, dx_0 = \tag{8.43}$$

$$r(b-a) \int_a^b \left(\prod_{k=1}^r f_k(x) \right) dx - \sum_{i=1}^r \left[\left(\int_a^b f_i(x)\, dx \right) \left(\int_a^b \left(\prod_{\substack{j=1 \\ j \neq i}}^r f_j(x) \right) dx \right) \right] =$$

$$\frac{1}{\Gamma(\alpha)} \int_a^b \left\{ \sum_{i=1}^r \left[\left[\int_a^{x_0} \left(\prod_{\substack{j=1 \\ j \neq i}}^r f_j(x) \right) \left(\int_x^{x_0} (z-x)^{\alpha-1} D^\alpha_{x_0-} f_i(z)\, dz \right) dx \right] + \right. \right.$$

$$\left. \left. \left[\int_{x_0}^b \left(\prod_{\substack{j=1 \\ j \neq i}}^r f_j(x) \right) \left(\int_{x_0}^x (x-z)^{\alpha-1} D^\alpha_{*x_0} f_i(z)\, dz \right) dx \right] \right] \right\} dx_0. \tag{8.44}$$

Hence it holds

$$|\Delta(f_1, \ldots, f_r)| \le$$

$$\frac{1}{\Gamma(\alpha)} \int_a^b \left| \sum_{i=1}^r \left[\left[\int_a^{x_0} \left(\prod_{\substack{j=1 \\ j \neq i}}^r f_j(x) \right) \left(\int_x^{x_0} (z-x)^{\alpha-1} D^\alpha_{x_0-} f_i(z)\, dz \right) dx \right] + \right. \right.$$

$$\left. \left. \left[\int_{x_0}^b \left(\prod_{\substack{j=1 \\ j \neq i}}^r f_j(x) \right) \left(\int_{x_0}^x (x-z)^{\alpha-1} D^\alpha_{*x_0} f_i(z)\, dz \right) dx \right] \right] \right| dx_0 =: (**). \tag{8.45}$$

Using (8.19) we have

$$(**) \le (b-a) \sum_{i=1}^{r} \left[\left[\sup_{x_0 \in [a,b]} \left\| D_{x_0-}^{\alpha} f_i \right\|_{\infty,[a,x_0]} \sup_{x_0 \in [a,b]} I_{a+}^{\alpha+1} \left(\prod_{\substack{j=1 \\ j \ne i}}^{r} |f_j(x_0)| \right) \right] \right.$$

$$\left. + \left[\sup_{x_0 \in [a,b]} \left\| D_{*x_0}^{\alpha} f_i \right\|_{\infty,[x_0,b]} \sup_{x_0 \in [a,b]} I_{b-}^{\alpha+1} \left(\prod_{\substack{j=1 \\ j \ne i}}^{r} |f_j(x_0)| \right) \right] \right], \tag{8.46}$$

proving Theorem 8.15.

When $p, q > 1 : \frac{1}{p} + \frac{1}{q} = 1$, with $\frac{1}{q} < \alpha \le 1$, by (8.21) we get

$$(**) \le \frac{(b-a) \, \Gamma \left(\alpha + \frac{1}{p} \right)}{(p(\alpha-1)+1)^{\frac{1}{p}} \, \Gamma(\alpha)} \cdot$$

$$\sum_{i=1}^{r} \left[\left[\sup_{x_0 \in [a,b]} \left\| D_{x_0-}^{\alpha} f_i \right\|_{q,[a,x_0]} \sup_{x_0 \in [a,b]} I_{a+}^{\alpha+\frac{1}{p}} \left(\prod_{\substack{j=1 \\ j \ne i}}^{r} |f_j(x_0)| \right) \right] + \right.$$

$$\left. \left[\sup_{x_0 \in [a,b]} \left\| D_{*x_0}^{\alpha} f_i \right\|_{q,[x_0,b]} \sup_{x_0 \in [a,b]} I_{b-}^{\alpha+\frac{1}{p}} \left(\prod_{\substack{j=1 \\ j \ne i}}^{r} |f_j(x_0)| \right) \right] \right], \tag{8.47}$$

proving Theorem 8.16.   ∎

References

1. G.A. Anastassiou, Fractional differentiation inequalities, *Research Monograph* (Springer, New York, 2009)
2. G.A. Anastassiou, On right fractional calculus. Chaos, Solitons Fractals **42**, 365–376 (2009)
3. G.A. Anastassiou, Advances on fractional inequalities, *Research Monograph* (Springer, New York, 2011)
4. G.A. Anastassiou, Intelligent mathematics: computational analysis, *Research Monograph* (Springer, Berlin, 2011)
5. G.A. Anastassiou, Fractional representation formulae and right fractional inequalities. Math. Comput. Model. **54**(11–12), 3098–3115 (2011)
6. G.A. Anastassiou, Fractional Ostrowski and Grüss type inequalities involving several functions. Panamerican Math. J. **24**(3), 1–14 (2014)

7. P.L. Čebyšev, Sur les expressions approximatives des intégrales définies par les aures prises entre les mêmes limites. Proc. Math. Soc. Charkov **2**, 93–98 (1882)
8. K. Diethelm, The Analysis of Fractional Differential Equations. Lecture Notes in Mathematics, vol. 2004, 1st edn. (Springer, New York, 2010)
9. A.M.A. El-Sayed, M. Gaber, On the finite Caputo and finite Riesz derivatives. Electron. J. Theor. Phys. **3**(12), 81–95 (2006)
10. R. Gorenflo, F. Mainardi, Essentials of fractional calculus. Maphysto Center (2000). http://www.maphysto.dk/oldpages/events/LevyCAC2000/MainardiNotes/fm2k0a.ps
11. A. Ostrowski, Über die Absolutabweichung einer differentiabaren Funcktion von ihrem Integralmittelwert. Comment. Math. Helv. **10**, 226–227 (1938)
12. D.B. Pachpatte, On right Caputo fractional Ostrowski Inequalities involving three functions. Int. J. Anal. **2013**, Article ID 127061, 5 p. doi:10.1155/2013/127061

Chapter 9
Further Interpretation of Some Fractional Ostrowski and Grüss Type Inequalities

We further interpret and simplify earlier produced fractional Ostrowski and Grüss type inequalities involving several functions. It follows [7].

9.1 Background

Let $\nu \geq 0$; the operator I_{a+}^{ν}, defined for $f \in L_1[(a,b)]$ is given by

$$I_{a+}^{\nu} f(x) := \frac{1}{\Gamma(\nu)} \int_a^x (x-t)^{\nu-1} f(t) \, dt, \tag{9.1}$$

for $a \leq x \leq b$, is called the left Riemann-Liouville fractional integral operator of order ν. For $\nu = 0$, we set $I_{a+}^0 := I$, the identity operator, see [1], p. 392, also [8].

Let $\nu \geq 0$, $n := \lceil \nu \rceil$ ($\lceil \cdot \rceil$ ceiling of the number), $f \in AC^n([a,b])$ (it means $f^{(n-1)} \in AC([a,b])$, absolutely continuous functions).

Then the left Caputo fractional derivative is given by

$$D_{*a}^{\nu} f(x) = \frac{1}{\Gamma(n-\nu)} \int_a^x (x-t)^{n-\nu-1} f^{(n)}(t) \, dt = \left(I_{a+}^{n-\nu} f^{(n)} \right)(x), \tag{9.2}$$

and it exists almost everywhere for $x \in [a,b]$.

Let $f \in L_1([a,b])$, $\alpha > 0$. The right Riemann-Liouville fractional operator [2, 9, 10] of order α is denoted by

$$I_{b-}^{\alpha} f(x) := \frac{1}{\Gamma(\alpha)} \int_x^b (z-x)^{\alpha-1} f(z) \, dz, \quad \forall x \in [a,b]. \tag{9.3}$$

We set $I_{b-}^0 := I$, the identity operator.

Let now $f \in AC^m([a,b])$, $m \in \mathbb{N}$, with $m := \lceil \alpha \rceil$.

© Springer International Publishing Switzerland 2016
G.A. Anastassiou, *Intelligent Comparisons: Analytic Inequalities*,
Studies in Computational Intelligence 609,
DOI 10.1007/978-3-319-21121-3_9

We define the right Caputo fractional derivative of order $\alpha \geq 0$, by

$$D^{\alpha}_{b-} f(x) := (-1)^m I^{m-\alpha}_{b-} f^{(m)}(x), \tag{9.4}$$

we set $D^0_{b-} f := f$, that is

$$D^{\alpha}_{b-} f(x) = \frac{(-1)^m}{\Gamma(m-\alpha)} \int_x^b (z-x)^{m-\alpha-1} f^{(m)}(z)\, dz. \tag{9.5}$$

We need

Proposition 9.1 ([4], p. 361) *Let $\alpha > 0$, $m = \lceil \alpha \rceil$, $f \in C^{m-1}([a, b])$, $f^{(m)} \in L_\infty([a, b])$; $x, x_0 \in [a, b] : x \geq x_0$. Then $D^{\alpha}_{*x_0} f(x)$ is continuous in x_0.*

Proposition 9.2 ([4], p. 361) *Let $\alpha > 0$, $m = \lceil \alpha \rceil$, $f \in C^{m-1}([a, b])$, $f^{(m)} \in L_\infty([a, b])$; $x, x_0 \in [a, b] : x \leq x_0$. Then $D^{\alpha}_{x_0-} f(x)$ is continuous in x_0.*

We also mention

Theorem 9.3 ([4], p. 362) *Let $f \in C^m([a, b])$, $m = \lceil \alpha \rceil$, $\alpha > 0$, $x, x_0 \in [a, b]$. Then $D^{\alpha}_{*x_0} f(x)$, $D^{\alpha}_{x_0-} f(x)$ are jointly continuous functions in (x, x_0) from $[a, b]^2$ into $\mathbb{R}$.*

Convention 9.4 ([4], p. 360) *We suppose that*

$$D^{\alpha}_{*x_0} f(x) = 0, \text{ for } x < x_0, \tag{9.6}$$

and

$$D^{\alpha}_{x_0-} f(x) = 0, \text{ for } x > x_0, \tag{9.7}$$

for all $x, x_0 \in [a, b]$.

9.2 Motivation

We mention some Caputo fractional mixed Ostrowski type inequalities involving several functions.

Theorem 9.5 ([6]) *Let $x_0 \in [a, b] \subset \mathbb{R}$, $\alpha > 0$, $m = \lceil \alpha \rceil$, $f_i \in AC^m([a, b])$, $i = 1, ..., r \in \mathbb{N} - \{1\}$, with $f_i^{(k)}(x_0) = 0$, $k = 1, ..., m-1$, $i = 1, ..., r$. Assume that $\left\| D^{\alpha}_{x_0-} f_i \right\|_{\infty,[a,x_0]}$, $\left\| D^{\alpha}_{*x_0} f_i \right\|_{\infty,[x_0,b]} < \infty$, $i = 1, ..., r$. Denote by*

$$\theta(f_1, ..., f_r)(x_0) := r \int_a^b \left(\prod_{k=1}^r f_k(x) \right) dx - \sum_{i=1}^r \left[f_i(x_0) \int_a^b \left(\prod_{\substack{j=1 \\ j \neq i}}^r f_j(x) \right) dx \right]. \tag{9.8}$$

Then

$$|\theta(f_1, ..., f_r)(x_0)| \leq \sum_{i=1}^{r} \left[\left[\|D_{x_0-}^{\alpha} f_i\|_{\infty,[a,x_0]} I_{a+}^{\alpha+1}\left(\prod_{\substack{j=1 \\ j \neq i}}^{r} |f_j(x_0)|\right)\right]\right. \tag{9.9}$$

$$\left. + \left[\|D_{*x_0}^{\alpha} f_i\|_{\infty,[x_0,b]} I_{b-}^{\alpha+1}\left(\prod_{\substack{j=1 \\ j \neq i}}^{r} |f_j(x_0)|\right)\right]\right].$$

Inequality (9.9) is sharp, infact it is attained.

Theorem 9.6 ([6]) *Let* $\alpha \geq 1$, $m = \lceil\alpha\rceil$, *and* $f_i \in AC^m([a,b])$, $i = 1, ..., r \in \mathbb{N} - \{1\}$. *Suppose that* $f_i^{(k)}(x_0) = 0$, $k = 1, ..., m-1$; $x_0 \in [a,b]$ *and* $D_{x_0-}^{\alpha} f_i \in L_1([a,x_0])$, $D_{*x_0}^{\alpha} f_i \in L_1([x_0,b])$, *for all* $i = 1, ..., r$. *Then*

$$|\theta(f_1, ..., f_r)(x_0)| \leq \sum_{i=1}^{r} \left[\left[\|D_{x_0-}^{\alpha} f_i\|_{L_1([a,x_0])} I_{a+}^{\alpha}\left(\prod_{\substack{j=1 \\ j \neq i}}^{r} |f_j(x_0)|\right)\right]\right. \tag{9.10}$$

$$\left. + \left[\|D_{*x_0}^{\alpha} f_i\|_{L_1([x_0,b])} I_{b-}^{\alpha}\left(\prod_{\substack{j=1 \\ j \neq i}}^{r} |f_j(x_0)|\right)\right]\right].$$

Theorem 9.7 ([6]) *Let* $p, q > 1 : \frac{1}{p} + \frac{1}{q} = 1$, $\alpha > \frac{1}{q}$, $m = \lceil\alpha\rceil$, $\alpha > 0$, *and* $f_i \in AC^m([a,b])$, $i = 1, ..., r \in \mathbb{N} - \{1\}$. *Suppose that* $f_i^{(k)}(x_0) = 0$, $k = 1, ..., m-1$, $x_0 \in [a,b]$; $i = 1, ..., r$. *Assume* $D_{x_0-}^{\alpha} f_i \in L_q([a,x_0])$, *and* $D_{*x_0}^{\alpha} f_i \in L_q([x_0,b])$, $i = 1, ..., r$. *Then*

$$|\theta(f_1, ..., f_r)(x_0)| \leq$$

$$\frac{\Gamma\left(\alpha + \frac{1}{p}\right)}{(p(\alpha-1)+1)^{\frac{1}{p}} \Gamma(\alpha)} \sum_{i=1}^{r} \left[\left[\|D_{x_0-}^{\alpha} f_i\|_{L_q([a,x_0])} I_{a+}^{\alpha+\frac{1}{p}}\left(\prod_{\substack{j=1 \\ j \neq i}}^{r} |f_j(x_0)|\right)\right]\right.$$

$$\tag{9.11}$$

$$\left. + \left[\|D_{*x_0}^{\alpha} f_i\|_{L_q([x_0,b])} I_{b-}^{\alpha+\frac{1}{p}}\left(\prod_{\substack{j=1 \\ j \neq i}}^{r} |f_j(x_0)|\right)\right]\right].$$

Next we mention some Caputo fractional Grüss type inequalities for several functions.

Theorem 9.8 ([6]) *Let* $x_0 \in [a, b] \subset \mathbb{R}$, $0 < \alpha \leq 1$, $f_i \in AC([a, b])$, $i = 1, ..., r \in \mathbb{N} - \{1\}$. *Assume that* $\sup\limits_{x_0 \in [a,b]} \left\| D^\alpha_{x_0-} f_i \right\|_{\infty,[a,x_0]}$, $\sup\limits_{x_0 \in [a,b]} \left\| D^\alpha_{*x_0} f_i \right\|_{\infty,[x_0,b]}$
$< \infty$, $i = 1, ..., r$. *Denote by*

$$\Delta(f_1, ..., f_r) := r(b - a) \int_a^b \left(\prod_{k=1}^r f_k(x) \right) dx - \tag{9.12}$$

$$\sum_{i=1}^r \left[\left(\int_a^b f_i(x)\, dx \right) \left(\int_a^b \left(\prod_{\substack{j=1 \\ j \neq i}}^r f_j(x) \right) dx \right) \right].$$

Then

$$|\Delta(f_1, ..., f_r)| \leq (b - a) \cdot$$

$$\sum_{i=1}^r \left[\left[\sup_{x_0 \in [a,b]} \left\| D^\alpha_{x_0-} f_i \right\|_{\infty,[a,x_0]} \sup_{x_0 \in [a,b]} I^{\alpha+1}_{a+} \left(\prod_{\substack{j=1 \\ j \neq i}}^r |f_j(x_0)| \right) \right] + \right.$$

$$\left. \left[\sup_{x_0 \in [a,b]} \left\| D^\alpha_{*x_0} f_i \right\|_{\infty,[x_0,b]} \sup_{x_0 \in [a,b]} I^{\alpha+1}_{b-} \left(\prod_{\substack{j=1 \\ j \neq i}}^r |f_j(x_0)| \right) \right] \right]. \tag{9.13}$$

Theorem 9.9 ([6]) *Let* $p, q > 1 : \frac{1}{p} + \frac{1}{q} = 1$, $\frac{1}{q} < \alpha \leq 1$, *and* $f_i \in AC([a, b])$, $i = 1, ..., r \in \mathbb{N} - \{1\}$, $x_0 \in [a, b]$. *Assume that* $\sup\limits_{x_0 \in [a,b]} \left\| D^\alpha_{x_0-} f_i \right\|_{L_q([a,x_0])}$, *and*
$\sup\limits_{x_0 \in [a,b]} \left\| D^\alpha_{*x_0} f_i \right\|_{L_q([x_0,b])} < \infty$, $i = 1, ..., r$. *Then*

$$|\Delta(f_1, ..., f_r)| \leq \frac{(b - a)\Gamma\left(\alpha + \frac{1}{p}\right)}{(p(\alpha - 1) + 1)^{\frac{1}{p}}\Gamma(\alpha)} \cdot$$

$$\sum_{i=1}^r \left[\left[\sup_{x_0 \in [a,b]} \left\| D^\alpha_{x_0-} f_i \right\|_{L_q([a,x_0])} \sup_{x_0 \in [a,b]} I^{\alpha+\frac{1}{p}}_{a+} \left(\prod_{\substack{j=1 \\ j \neq i}}^r |f_j(x_0)| \right) \right] + \right.$$

$$\left[\left[\sup_{x_0\in[a,b]}\left\|D_{*x_0}^\alpha f_i\right\|_{L_q([x_0,b])}\sup_{x_0\in[a,b]}I_{b-}^{\alpha+\frac{1}{p}}\left(\prod_{\substack{j=1\\j\neq i}}^r\left|f_j\left(x_0\right)\right|\right)\right]\right].$$ (9.14)

9.3 Main Results

We make

Remark 9.10 Let $g\in C\left([a,b]\right)$, $\alpha>0$, $x_0\in[a,b]\subset\mathbb{R}$. Notice that

$$I_{a+}^{\alpha+1}\left(g\right)\left(x_0\right)=\frac{1}{\Gamma\left(\alpha+1\right)}\int_a^{x_0}\left(x_0-z\right)^\alpha g\left(z\right)dz.$$ (9.15)

Hence

$$\left|I_{a+}^{\alpha+1}\left(g\right)\left(x_0\right)\right|\leq\frac{1}{\Gamma\left(\alpha+1\right)}\int_a^{x_0}\left(x_0-z\right)^\alpha\left|g\left(z\right)\right|dz\leq$$

$$\frac{\|g\|_{\infty,[a,x_0]}}{\Gamma\left(\alpha+1\right)}\int_a^{x_0}\left(x_0-z\right)^\alpha dz=\frac{\|g\|_{\infty,[a,x_0]}}{\Gamma\left(\alpha+1\right)}\frac{\left(x_0-a\right)^{\alpha+1}}{\left(\alpha+1\right)}$$

$$=\frac{\|g\|_{\infty,[a,x_0]}}{\Gamma\left(\alpha+2\right)}\left(x_0-a\right)^{\alpha+1}.$$ (9.16)

That is

$$\left|I_{a+}^{\alpha+1}\left(g\right)\left(x_0\right)\right|\leq\frac{\|g\|_{\infty,[a,x_0]}}{\Gamma\left(\alpha+2\right)}\left(x_0-a\right)^{\alpha+1}.$$ (9.17)

Similarly we have

$$I_{b-}^{\alpha+1}\left(g\right)\left(x_0\right)=\frac{1}{\Gamma\left(\alpha+1\right)}\int_{x_0}^b\left(z-x_0\right)^\alpha g\left(z\right)dz,$$ (9.18)

and

$$\left|I_{b-}^{\alpha+1}\left(g\right)\left(x_0\right)\right|\leq\frac{\|g\|_{\infty,[x_0,b]}}{\Gamma\left(\alpha+1\right)}\int_{x_0}^b\left(z-x_0\right)^\alpha dz$$

$$=\frac{\|g\|_{\infty,[x_0,b]}}{\Gamma\left(\alpha+1\right)}\frac{\left(b-x_0\right)^{\alpha+1}}{\left(\alpha+1\right)}=\frac{\|g\|_{\infty,[x_0,b]}}{\Gamma\left(\alpha+2\right)}\left(b-x_0\right)^{\alpha+1}.$$ (9.19)

That is

$$\left|I_{b-}^{\alpha+1}\left(g\right)\left(x_0\right)\right|\leq\frac{\|g\|_{\infty,[x_0,b]}}{\Gamma\left(\alpha+2\right)}\left(b-x_0\right)^{\alpha+1}.$$ (9.20)

Consequently we derive

$$I_{a+}^{\alpha+1}\left(\prod_{\substack{j=1\\j\neq i}}^{r}\left|f_j\left(x_0\right)\right|\right) \overset{(9.17)}{\leq} \frac{\left\|\prod_{\substack{j=1\\j\neq i}}^{r} f_j\right\|_{\infty,[a,x_0]}}{\Gamma\left(\alpha+2\right)} \left(x_0-a\right)^{\alpha+1}, \tag{9.21}$$

and

$$I_{b-}^{\alpha+1}\left(\prod_{\substack{j=1\\j\neq i}}^{r}\left|f_j\left(x_0\right)\right|\right) \overset{(9.20)}{\leq} \frac{\left\|\prod_{\substack{j=1\\j\neq i}}^{r} f_j\right\|_{\infty,[x_0,b]}}{\Gamma\left(\alpha+2\right)} \left(b-x_0\right)^{\alpha+1}. \tag{9.22}$$

Therefore it holds

$$\left|\theta\left(f_1,...,f_r\right)\left(x_0\right)\right| \overset{(9.9)}{\leq} \sum_{i=1}^{r}\left[\left[\left\|D_{x_0-}^{\alpha} f_i\right\|_{\infty,[a,x_0]} I_{a+}^{\alpha+1}\left(\prod_{\substack{j=1\\j\neq i}}^{r}\left|f_j\left(x_0\right)\right|\right)\right]+\right.$$

$$\left.\left[\left\|D_{*x_0}^{\alpha} f_i\right\|_{\infty,[x_0,b]} I_{b-}^{\alpha+1}\left(\prod_{\substack{j=1\\j\neq i}}^{r}\left|f_j\left(x_0\right)\right|\right)\right]\right] \overset{((9.21),(9.22))}{\leq}$$

$$\frac{1}{\Gamma\left(\alpha+2\right)}\sum_{i=1}^{r}\left[\left[\left\|D_{x_0-}^{\alpha} f_i\right\|_{\infty,[a,x_0]}\left\|\prod_{\substack{j=1\\j\neq i}}^{r} f_j\right\|_{\infty,[a,x_0]}\right]\left(x_0-a\right)^{\alpha+1}+\right.$$

$$\left.\left[\left\|D_{*x_0}^{\alpha} f_i\right\|_{\infty,[x_0,b]}\left\|\prod_{\substack{j=1\\j\neq i}}^{r} f_j\right\|_{\infty,[x_0,b]}\right]\left(b-x_0\right)^{\alpha+1}\right] =: \left(\xi_1\right). \tag{9.23}$$

Call

$$M_1\left(f_1,...,f_r\right)\left(x_0\right) := \max_{i=1,...,r}\left\{\left\|D_{x_0-}^{\alpha} f_i\right\|_{\infty,[a,x_0]},\left\|D_{*x_0}^{\alpha} f_i\right\|_{\infty,[x_0,b]}\right\}. \tag{9.24}$$

Then

$$
(\xi_1) \le \frac{M_1\,(f_1,\,...,\,f_r)\,(x_0)}{\Gamma\,(\alpha+2)} \sum_{i=1}^{r} \left[\left\| \prod_{\substack{j=1 \\ j \ne i}}^{r} f_j \right\|_{\infty,[a,x_0]} (x_0 - a)^{\alpha+1} + \right.
$$

$$
\left. \left\| \prod_{\substack{j=1 \\ j \ne i}}^{r} f_j \right\|_{\infty,[x_0,b]} (b - x_0)^{\alpha+1} \right] =: (\xi_2)\,. \tag{9.25}
$$

Call

$$
\psi_1\,(f_1,\,...,\,f_r)\,(x_0) := \max \left\{ \sum_{i=1}^{r} \left\| \prod_{\substack{j=1 \\ j \ne i}}^{r} f_j \right\|_{\infty,[a,x_0]}, \sum_{i=1}^{r} \left\| \prod_{\substack{j=1 \\ j \ne i}}^{r} f_j \right\|_{\infty,[x_0,b]} \right\}. \tag{9.26}
$$

So that

$$
(\xi_2) \le \frac{M_1\,(f_1,\,...,\,f_r)\,(x_0)\,\psi_1\,(f_1,\,...,\,f_r)\,(x_0)}{\Gamma\,(\alpha+2)} \left[(b - x_0)^{\alpha+1} + (x_0 - a)^{\alpha+1} \right] \le \tag{9.27}
$$

$$
\frac{M_1\,(f_1,\,...,\,f_r)\,(x_0)\,\psi_1\,(f_1,\,...,\,f_r)\,(x_0)}{\Gamma\,(\alpha+2)} (b - a)^{\alpha+1}\,.
$$

We have proved simpler interpretations of Caputo fractional mixed Ostrowski type inequalities involving several functions.

Theorem 9.11 *Here all as in Theorem 9.5, $M_1\,(f_1,\,...,\,f_r)\,(x_0)$ as in (9.24) and $\psi_1\,(f_1,\,...,\,f_r)\,(x_0)$ as in (9.26). Then*

$$
|\theta\,(f_1,\,...,\,f_r)\,(x_0)| \le
$$

$$
\frac{M_1\,(f_1,\,...,\,f_r)\,(x_0)\,\psi_1\,(f_1,\,...,\,f_r)\,(x_0)}{\Gamma\,(\alpha+2)} \left[(b - x_0)^{\alpha+1} + (x_0 - a)^{\alpha+1} \right] \le \tag{9.28}
$$

$$
\frac{M_1\,(f_1,\,...,\,f_r)\,(x_0)\,\psi_1\,(f_1,\,...,\,f_r)\,(x_0)}{\Gamma\,(\alpha+2)} (b - a)^{\alpha+1}\,. \tag{9.29}
$$

We make

Remark 9.12 Let $g \in C([a, b])$, $\alpha \geq 1$, $x_0 \in [a, b] \subset \mathbb{R}$. We have that

$$\left| I_{a+}^{\alpha}(g)(x_0) \right| \leq \frac{\|g\|_{\infty,[a,x_0]}}{\Gamma(\alpha+1)}(x_0 - a)^{\alpha}, \tag{9.30}$$

and

$$\left| I_{b-}^{\alpha}(g)(x_0) \right| \leq \frac{\|g\|_{\infty,[x_0,b]}}{\Gamma(\alpha+1)}(b - x_0)^{\alpha}. \tag{9.31}$$

Consequently we derive

$$I_{a+}^{\alpha}\left(\prod_{\substack{j=1 \\ j\neq i}}^{r} |f_j(x_0)| \right) \overset{(9.30)}{\leq} \frac{\left\| \prod_{\substack{j=1 \\ j\neq i}}^{r} f_j \right\|_{\infty,[a,x_0]}}{\Gamma(\alpha+1)}(x_0 - a)^{\alpha}, \tag{9.32}$$

$$I_{b-}^{\alpha}\left(\prod_{\substack{j=1 \\ j\neq i}}^{r} |f_j(x_0)| \right) \overset{(9.31)}{\leq} \frac{\left\| \prod_{\substack{j=1 \\ j\neq i}}^{r} f_j \right\|_{\infty,[x_0,b]}}{\Gamma(\alpha+1)}(b - x_0)^{\alpha}. \tag{9.33}$$

Therefore it holds

$$\left| \theta(f_1, \ldots, f_r)(x_0) \right| \overset{(9.10)}{\leq} \sum_{i=1}^{r} \left[\left[\left\| D_{x_0-}^{\alpha} f_i \right\|_{L_1([a,x_0])} I_{a+}^{\alpha}\left(\prod_{\substack{j=1 \\ j\neq i}}^{r} |f_j(x_0)| \right) \right] + \right.$$

$$\left. \left[\left\| D_{*x_0}^{\alpha} f_i \right\|_{L_1([x_0,b])} I_{b-}^{\alpha}\left(\prod_{\substack{j=1 \\ j\neq i}}^{r} |f_j(x_0)| \right) \right] \right] \overset{((9.32),(9.33))}{\leq}$$

$$\frac{1}{\Gamma(\alpha+1)} \sum_{i=1}^{r} \left[\left[\left\| D_{x_0-}^{\alpha} f_i \right\|_{L_1([a,x_0])} \left\| \prod_{\substack{j=1 \\ j\neq i}}^{r} f_j \right\|_{\infty,[a,x_0]} \right] (x_0 - a)^{\alpha} + \right.$$

$$\left[\left\| D^\alpha_{*x_0} f_i \right\|_{L_1([x_0,b])} \left\| \prod_{\substack{j=1 \\ j\neq i}}^{r} f_j \right\|_{\infty,[x_0,b]} (b-x_0)^\alpha \right] =: (\eta).$$ (9.34)

Call

$$M_2 (f_1, ..., f_r) (x_0) := \max_{i=1,...,r} \left\{ \left\| D^\alpha_{x_0-} f_i \right\|_{L_1([a,x_0])}, \left\| D^\alpha_{*x_0} f_i \right\|_{L_1([x_0,b])} \right\}.$$ (9.35)

Then

$$(\eta) \leq \frac{M_2 (f_1, ..., f_r) (x_0)}{\Gamma (\alpha + 1)}.$$

$$\sum_{i=1}^{r} \left[\left\| \prod_{\substack{j=1 \\ j\neq i}}^{r} f_j \right\|_{\infty,[a,x_0]} (x_0-a)^\alpha + \left\| \prod_{\substack{j=1 \\ j\neq i}}^{r} f_j \right\|_{\infty,[x_0,b]} (b-x_0)^\alpha \right]$$ (9.36)

$$\leq \frac{M_2 (f_1, ..., f_r) (x_0) \psi_1 (f_1, ..., f_r) (x_0)}{\Gamma (\alpha + 1)} \left[(b-x_0)^\alpha + (x_0-a)^\alpha \right]$$ (9.37)

$$\leq \frac{M_2 (f_1, ..., f_r) (x_0) \psi_1 (f_1, ..., f_r) (x_0)}{\Gamma (\alpha + 1)} (b-a)^\alpha.$$ (9.38)

We have proved

Theorem 9.13 *Let all as in Theorem 9.6, $M_2 (f_1, ..., f_r) (x_0)$ as in (9.35) and $\psi_1 (f_1, ..., f_r) (x_0)$ as in (9.26). Then*

$$|\theta (f_1, ..., f_r) (x_0)| \leq$$

$$\frac{M_2 (f_1, ..., f_r) (x_0) \psi_1 (f_1, ..., f_r) (x_0)}{\Gamma (\alpha + 1)} \left[(b-x_0)^\alpha + (x_0-a)^\alpha \right]$$ (9.39)

$$\leq \frac{M_2 (f_1, ..., f_r) (x_0) \psi_1 (f_1, ..., f_r) (x_0)}{\Gamma (\alpha + 1)} (b-a)^\alpha.$$ (9.40)

Similarly we obtain

Theorem 9.14 *Let all as in Theorem 9.7. Call*

$$M_3 (f_1, ..., f_r) (x_0) := \max_{i=1,...,r} \left\{ \left\| D^\alpha_{x_0-} f_i \right\|_{L_q([a,x_0])}, \left\| D^\alpha_{*x_0} f_i \right\|_{L_q([x_0,b])} \right\}.$$ (9.41)

Here $\psi_1 (f_1, ..., f_r) (x_0)$ as in (9.26). Then

$$|\theta (f_1, ..., f_r) (x_0)| \leq$$

$$\frac{M_3 (f_1, ..., f_r) (x_0) \psi_1 (f_1, ..., f_r) (x_0)}{\left(\alpha + \frac{1}{p}\right) (p (\alpha - 1) + 1)^{\frac{1}{p}} \Gamma (\alpha)} \left[(b - x_0)^{\alpha + \frac{1}{p}} + (x_0 - a)^{\alpha + \frac{1}{p}}\right] \leq \quad (9.42)$$

$$\frac{M_3 (f_1, ..., f_r) (x_0) \psi_1 (f_1, ..., f_r) (x_0)}{\left(\alpha + \frac{1}{p}\right) (p (\alpha - 1) + 1)^{\frac{1}{p}} \Gamma (\alpha)} (b - a)^{\alpha + \frac{1}{p}}. \quad (9.43)$$

Finally we give a simpler interpretation of Caputo fractional Grüss type inequalities (9.13), (9.14).

Theorem 9.15 *All as in Theorem 9.8. We define*

$$M_4 (f_1, ..., f_r) := \max_{i=1,...,r} \left\{ \sup_{x_0 \in [a,b]} \left\| D_{x_0-}^\alpha f_i \right\|_{\infty,[a,x_0]}, \sup_{x_0 \in [a,b]} \left\| D_{*x_0}^\alpha f_i \right\|_{\infty,[x_0,b]} \right\} \quad (9.44)$$

and

$$\psi_2 (f_1, ..., f_r) (x_0) :=$$

$$\max \left\{ \sum_{i=1}^r \sup_{x_0 \in [a,b]} \left\| \prod_{\substack{j=1 \\ j \neq i}}^r f_j \right\|_{\infty,[a,x_0]}, \sum_{i=1}^r \sup_{x_0 \in [a,b]} \left\| \prod_{\substack{j=1 \\ j \neq i}}^r f_j \right\|_{\infty,[x_0,b]} \right\}. \quad (9.45)$$

Then

$$|\Delta (f_1, ..., f_r)| \leq \frac{2M_4 (f_1, ..., f_r) \psi_2 (f_1, ..., f_r)}{\Gamma (\alpha + 2)} (b - a)^{\alpha + 2}. \quad (9.46)$$

Theorem 9.16 *All as in Theorem 9.9. We define*

$$M_5 (f_1, ..., f_r) := \max_{i=1,...,r} \left\{ \sup_{x_0 \in [a,b]} \left\| D_{x_0-}^\alpha f_i \right\|_{L_q([a,x_0])}, \sup_{x_0 \in [a,b]} \left\| D_{*x_0}^\alpha f_i \right\|_{L_q([x_0,b))} \right\} \quad (9.47)$$

Here ψ_2 is as in (9.45). Then

$$|\Delta (f_1, ..., f_r)| \leq \frac{2M_5 (f_1, ..., f_r) \psi_2 (f_1, ..., f_r)}{\left(\alpha + \frac{1}{p}\right) (p (\alpha - 1) + 1)^{\frac{1}{p}} \Gamma (\alpha)} (b - a)^{\alpha + \frac{1}{p} + 1}. \quad (9.48)$$

We finish with applications.

9.4 Applications

We apply above theory for $r = 2$. In that case

$$\theta\,(f_1,\,f_2)\,(x_0) = 2\int_a^b f_1\,(x)\,f_2\,(x)\,dx - f_1\,(x_0)\int_a^b f_2\,(x)\,dx - f_2\,(x_0)\int_a^b f_1\,(x)\,dx,$$
$$(9.49)$$

$x_0 \in [a,\,b]$,

$$M_1\,(f_1,\,f_2)\,(x_0) =$$

$$\max\left\{\left\|D_{x_0-}^\alpha f_1\right\|_{\infty,[a,x_0]},\left\|D_{x_0-}^\alpha f_2\right\|_{\infty,[a,x_0]},\left\|D_{*x_0}^\alpha f_1\right\|_{\infty,[x_0,b]},\left\|D_{*x_0}^\alpha f_2\right\|_{\infty,[x_0,b]}\right\},$$
$$(9.50)$$

$$\psi_1\,(f_1,\,f_2)\,(x_0) = \max\left\{\|f_1\|_{\infty,[a,x_0]} + \|f_2\|_{\infty,[a,x_0]},\|f_1\|_{\infty,[x_0,b]} + \|f_2\|_{\infty,[x_0,b]}\right\},$$
$$(9.51)$$

$$M_2\,(f_1,\,f_2)\,(x_0) = \max\left\{\left\|D_{x_0-}^\alpha f_1\right\|_{L_1([a,x_0])},\right.$$

$$\left.\left\|D_{x_0-}^\alpha f_2\right\|_{L_1([a,x_0])},\left\|D_{*x_0}^\alpha f_1\right\|_{L_1([x_0,b])},\left\|D_{*x_0}^\alpha f_2\right\|_{L_1([x_0,b])}\right\},$$
$$(9.52)$$

$$M_3\,(f_1,\,f_2)\,(x_0) := \max\left\{\left\|D_{x_0-}^\alpha f_1\right\|_{L_q([a,x_0])},\right.$$

$$\left.\left\|D_{x_0-}^\alpha f_2\right\|_{L_q([a,x_0])},\left\|D_{*x_0}^\alpha f_1\right\|_{L_q([x_0,b])},\left\|D_{*x_0}^\alpha f_2\right\|_{L_q([x_0,b])}\right\},$$
$$(9.53)$$

$$\Delta\,(f_1,\,f_2) = 2\left[(b-a)\int_a^b f_1\,(x)\,f_2\,(x)\,dx - \left(\int_a^b f_1\,(x)\,dx\right)\left(\int_a^b f_2\,(x)\,dx\right)\right],$$
$$(9.54)$$

$$M_4\,(f_1,\,f_2) = \max\left\{\sup_{x_0\in[a,b]}\left\|D_{x_0-}^\alpha f_1\right\|_{\infty,[a,x_0]},\sup_{x_0\in[a,b]}\left\|D_{x_0-}^\alpha f_2\right\|_{\infty,[a,x_0]},\right.$$

$$\left.\sup_{x_0\in[a,b]}\left\|D_{*x_0}^\alpha f_1\right\|_{\infty,[x_0,b]},\sup_{x_0\in[a,b]}\left\|D_{*x_0}^\alpha f_2\right\|_{\infty,[x_0,b]}\right\},$$
$$(9.55)$$

$$\psi_2\,(f_1,\,f_2) = \max\left\{\sup_{x_0\in[a,b]}\|f_1\|_{\infty,[a,x_0]} + \sup_{x_0\in[a,b]}\|f_2\|_{\infty,[a,x_0]},\right.$$

$$\left.\sup_{x_0\in[a,b]}\|f_1\|_{\infty,[x_0,b]} + \sup_{x_0\in[a,b]}\|f_2\|_{\infty,[x_0,b]}\right\},$$
$$(9.56)$$

and

$$M_5 (f_1, f_2) = \max \left\{ \sup_{x_0 \in [a,b]} \left\| D^{\alpha}_{x_0-} f_1 \right\|_{L_q([a,x_0])}, \sup_{x_0 \in [a,b]} \left\| D^{\alpha}_{x_0-} f_2 \right\|_{L_q([a,x_0])}, \right.$$

$$\left. \sup_{x_0 \in [a,b]} \left\| D^{\alpha}_{*x_0} f_1 \right\|_{L_q([x_0,b])}, \sup_{x_0 \in [a,b]} \left\| D^{\alpha}_{*x_0} f_2 \right\|_{L_q([x_0,b])} \right\}, \tag{9.57}$$

above $p, q > 1 : \frac{1}{p} + \frac{1}{q} = 1$.

Proposition 9.17 *Let* $x_0 \in [a, b] \subset \mathbb{R}$, $\alpha > 0$, $m = \lceil \alpha \rceil$, $f_1, f_2 \in AC^m ([a, b])$, *with* $f_1^{(k)} (x_0) = f_2^{(k)} (x_0) = 0$, $k = 1, ..., m - 1$. *Assume that* $\left\| D^{\alpha}_{x_0-} f_1 \right\|_{\infty, [a,x_0]}$, $\left\| D^{\alpha}_{x_0-} f_2 \right\|_{\infty, [a,x_0]}$, $\left\| D^{\alpha}_{*x_0} f_1 \right\|_{\infty, [x_0,b]}$, $\left\| D^{\alpha}_{*x_0} f_2 \right\|_{\infty, [x_0,b]} < \infty$. *Then*

$$|\theta (f_1, f_2) (x_0)| \leq \frac{M_1 (f_1, f_2) (x_0) \, \psi_1 (f_1, f_2) (x_0)}{\Gamma (\alpha + 2)} \left[(b - x_0)^{\alpha+1} + (x_0 - a)^{\alpha+1} \right] \tag{9.58}$$

$$\leq \frac{M_1 (f_1, f_2) (x_0) \, \psi_1 (f_1, f_2) (x_0)}{\Gamma (\alpha + 2)} (b - a)^{\alpha+1}. \tag{9.59}$$

Proof By Theorem 9.11. ∎

Proposition 9.18 *Let* $\alpha \geq 1$, $m = \lceil \alpha \rceil$, *and* $f_1, f_2 \in AC^m ([a, b])$. *Suppose that* $f_1^{(k)} (x_0) = f_2^{(k)} (x_0) = 0$, $k = 1, ..., m - 1$; $x_0 \in [a, b]$ *and* $D^{\alpha}_{x_0-} f_1, D^{\alpha}_{x_0-} f_2 \in L_1 ([a, x_0])$, $D^{\alpha}_{*x_0} f_1, D^{\alpha}_{*x_0} f_2 \in L_1 ([x_0, b])$. *Then*

$$|\theta (f_1, f_2) (x_0)| \leq \frac{M_2 (f_1, f_2) (x_0) \, \psi_1 (f_1, f_2) (x_0)}{\Gamma (\alpha + 1)} \left[(b - x_0)^{\alpha} + (x_0 - a)^{\alpha} \right] \tag{9.60}$$

$$\leq \frac{M_2 (f_1, f_2) (x_0) \, \psi_1 (f_1, f_2) (x_0)}{\Gamma (\alpha + 1)} (b - a)^{\alpha}. \tag{9.61}$$

Proof By Theorem 9.13. ∎

Proposition 9.19 *Let* $p, q > 1 : \frac{1}{p} + \frac{1}{q} = 1$, $\alpha > \frac{1}{q}$, $m = \lceil \alpha \rceil$, $\alpha > 0$, *and* $f_1, f_2 \in AC^m ([a, b])$. *Suppose that* $f_1^{(k)} (x_0) = f_2^{(k)} (x_0) = 0, k = 1, ..., m-1, x_0 \in [a, b]$. *Assume* $D^{\alpha}_{x_0-} f_1, D^{\alpha}_{x_0-} f_2 \in L_q ([a, x_0])$, *and* $D^{\alpha}_{*x_0} f_1, D^{\alpha}_{*x_0} f_2 \in L_q ([x_0, b])$. *Then*

$$|\theta (f_1, f_2) (x_0)| \leq \frac{M_3 (f_1, f_2) (x_0) \, \psi_1 (f_1, f_2) (x_0)}{\left(\alpha + \frac{1}{p} \right) (p (\alpha - 1) + 1)^{\frac{1}{p}} \Gamma (\alpha)} \left[(b - x_0)^{\alpha + \frac{1}{p}} + (x_0 - a)^{\alpha + \frac{1}{p}} \right] \tag{9.62}$$

$$\leq \frac{M_3 (f_1, f_2) (x_0) \, \psi_1 (f_1, f_2) (x_0)}{\left(\alpha + \frac{1}{p} \right) (p (\alpha - 1) + 1)^{\frac{1}{p}} \Gamma (\alpha)} (b - a)^{\alpha + \frac{1}{p}}. \tag{9.63}$$

Proof By Theorem 9.14. ∎

Proposition 9.20 *Let* $x_0 \in [a, b] \subset \mathbb{R}$, $0 < \alpha \leq 1$, $f_1, f_2 \in AC([a, b])$. *Assume that* $\sup_{x_0 \in [a,b]} \left\| D_{x_0-}^{\alpha} f_1 \right\|_{\infty, [a,x_0]}$, $\sup_{x_0 \in [a,b]} \left\| D_{x_0-}^{\alpha} f_2 \right\|_{\infty, [a,x_0]}$, $\sup_{x_0 \in [a,b]} \left\| D_{*x_0}^{\alpha} f_1 \right\|_{\infty, [x_0,b]}$, $\sup_{x_0 \in [a,b]} \left\| D_{*x_0}^{\alpha} f_2 \right\|_{\infty, [x_0,b]} < \infty$. *Then*

$$|\Delta(f_1, f_2)| \leq \frac{2M_4(f_1, f_2)\,\psi_2(f_1, f_2)}{\Gamma(\alpha + 2)}(b - a)^{\alpha + 2}. \tag{9.64}$$

Proof By Theorem 9.15. ∎

Proposition 9.21 *Let* $p, q > 1 : \frac{1}{p} + \frac{1}{q} = 1$, $\frac{1}{q} < \alpha \leq 1$, *and* $f_1, f_2 \in AC([a, b])$, $x_0 \in [a, b]$. *Assume that* $\sup_{x_0 \in [a,b]} \left\| D_{x_0-}^{\alpha} f_i \right\|_{L_q([a,x_0])}$, $\sup_{x_0 \in [a,b]} \left\| D_{*x_0}^{\alpha} f_i \right\|_{L_q([x_0,b])} < \infty$, $i = 1, 2$. *Then*

$$|\Delta(f_1, f_2)| \leq \frac{2M_5(f_1, f_2)\,\psi_2(f_1, f_2)}{\left(\alpha + \frac{1}{p}\right)(p(\alpha - 1) + 1)^{\frac{1}{p}}\,\Gamma(\alpha)}(b - a)^{\alpha + \frac{1}{p} + 1}. \tag{9.65}$$

Proof By Theorem 9.16. ∎

References

1. G.A. Anastassiou, *Fractional Differentiation Inequalities*, Research Monograph (Springer, New York, 2009)
2. G.A. Anastassiou, On right fractional calculus. Chaos Solitons Fractals **42**, 365–376 (2009)
3. G.A. Anastassiou, *Advances on Fractional Inequalities*, Research Monograph (Springer, New York, 2011)
4. G.A. Anastassiou, *Intelligent Mathematics: Computational Analysis*, Research Monograph (Springer, Berlin, 2011)
5. G.A. Anastassiou, Fractional representation formulae and right fractional inequalities. Math. Comput. Model. **54**(11–12), 3098–3115 (2011)
6. G.A. Anastassiou, Fractional Ostrowski and Grüss type inequalities involving several functions. Panam. Math. J. **24**(3), 1–14 (2014)
7. G.A. Anastassiou, Further interpretation of some fractional Ostrowski and Grüss type inequalities. J. Appl. Funct. Anal. **9**(3–4), 392–403 (2014)
8. K. Diethelm, *The Analysis of Fractional Differential Equations*, Lecture Notes in Mathematics, vol. 2004, 1st edn. (Springer, New York, 2010)
9. A.M.A. El-Sayed, M. Gaber, On the finite Caputo and finite Riesz derivatives. Electron. J. Theor. Phys. **3**(12), 81–95 (2006)
10. R. Gorenflo, F. Mainardi, Essentials of fractional calculus (Maphysto Center, 2000), http://www.maphysto.dk/oldpages/events/LevyCAC2000/MainardiNotes/fm2k0a.ps

Chapter 10
Multivariate Fractional Representation Formula and Ostrowski Inequality

Here we derive a multivariate fractional representation formula involving ordinary partial derivatives of first order. Then we prove a related multivariate fractional Ostrowski type inequality with respect to uniform norm. It follows [2].

10.1 Introduction

Let $f : [a, b] \to \mathbb{R}$ be differentiable on $[a, b]$, and $f' : [a, b] \to \mathbb{R}$ be integrable on $[a, b]$, then the following Montgomery identity holds [4]:

$$f(x) = \frac{1}{b-a} \int_a^b f(t) \, dt + \int_a^b P_1(x, t) f'(t) \, dt, \tag{10.1}$$

where $P_1(x, t)$ is the Peano kernel

$$P_1(x, t) = \begin{cases} \frac{t-a}{b-a}, & a \leq t \leq x, \\ \frac{t-b}{b-a}, & x < t \leq b, \end{cases} \tag{10.2}$$

The Riemann-Liouville integral operator of order $\alpha > 0$ with anchor point $a \in \mathbb{R}$ is defined by

$$J_a^\alpha f(x) := \frac{1}{\Gamma(\alpha)} \int_a^x (x - t)^{\alpha-1} f(t) \, dt, \tag{10.3}$$

$$J_a^0 f(x) := f(x), \quad x \in [a, b]. \tag{10.4}$$

Properties of the above operator can be found in [5].

When $\alpha = 1$, J_a^1 reduces to the classical integral.

In [1] we proved the following fractional representation formula of Montgomery identity type.

© Springer International Publishing Switzerland 2016
G.A. Anastassiou, *Intelligent Comparisons: Analytic Inequalities*,
Studies in Computational Intelligence 609,
DOI 10.1007/978-3-319-21121-3_10

Theorem 10.1 *Let $f : [a, b] \to \mathbb{R}$ be differentiable on $[a, b]$, and $f' : [a, b] \to \mathbb{R}$ be integrable on $[a, b]$, $\alpha \geq 1$, $x \in [a, b)$. Then*

$$f(x) =$$

$$(b - x)^{1-\alpha} \, \Gamma(\alpha) \left\{ \frac{J_a^\alpha f(b)}{b - a} - J_a^{\alpha-1}(P_1(x, b) f(b)) + J_a^\alpha (P_1(x, b) f'(b)) \right\}.$$

$$(10.5)$$

When $\alpha = 1$ the last (10.5) reduces to classic Montgomery identity (10.1).

We may rewrite (10.5) as follows

$$f(x) = (b - x)^{1-\alpha} \left[\frac{1}{b - a} \int_a^b (b - t)^{\alpha-1} f(t) \, dt - \right.$$

$$\left. (\alpha - 1) \int_a^b (b - t)^{\alpha-2} P_1(x, t) f(t) \, dt + \int_a^b (b - t)^{\alpha-1} P_1(x, t) f'(t) \, dt \right].$$

$$(10.6)$$

In this chapter based on (10.5), we establish a multivariate fractional representation formula for $f(x)$, $x \in \prod_{i=1}^m [a_i, b_i] \subset \mathbb{R}^m$, and from there we derive an interesting multivariate fractional Ostrowski type inequality.

10.2 Main Results

We make

Assumption 10.2 *Let $f \in C^1 \left(\prod_{i=1}^m [a_i, b_i] \right)$.*

Assumption 10.3 *Let $f : \prod_{i=1}^m [a_i, b_i] \to \mathbb{R}$ be measurable and bounded, such that there exist $\frac{\partial f}{\partial x_j} : \prod_{i=1}^m [a_i, b_i] \to \mathbb{R}$, and it is x_j-integrable for all $j = 1, \ldots, m$. Furthermore $\frac{\partial f}{\partial x_i}(t_1, \ldots, t_i, x_{i+1}, \ldots, x_m)$ it is integrable on $\prod_{j=1}^i [a_j, b_j]$, for all $i = 1, \ldots, m$, for any $(x_{i+1}, \ldots, x_m) \in \prod_{j=i+1}^m [a_j, b_j]$.*

Convention 10.4 *We set*

$$\prod_{j=1}^{0} \cdot = 1. \tag{10.7}$$

Notation 10.5 Here $x = \overrightarrow{x} = (x_1, \ldots, x_m) \in \mathbb{R}^m$, $m \in \mathbb{N} - \{1\}$. Likewise $t = \overrightarrow{t} = (t_1, \ldots, t_m)$, and $d\overrightarrow{t} = dt_1 dt_2 \ldots dt_m$. We denote the kernel

$$P_1(x_i, t_i) = \begin{cases} \frac{t_i - a_i}{b_i - a_i}, & a_i \leq t_i \leq x_i, \\ \frac{t_i - b_i}{b_i - a_i}, & x_i < t_i \leq b_i, \end{cases} \tag{10.8}$$

We need

Definition 10.6 (*see* [3]) Let $\prod_{i=1}^m [a_i, b_i] \subset \mathbb{R}^m$, $m \in \mathbb{N} - \{1\}$, $a_i < b_i$, $a_i, b_i \in \mathbb{R}$. Let $\alpha > 0$, $f \in L_1 \left(\prod_{i=1}^m [a_i, b_i] \right)$. We define the left mixed Riemann-Liouville fractional multiple integral of order α:

$$\left(I_{a+}^\alpha f \right)(x) := \frac{1}{(\Gamma(\alpha))^m} \int_{a_1}^{x_1} \cdots \int_{a_m}^{x_m} \left(\prod_{i=1}^m (x_i - t_i) \right)^{\alpha - 1} f(t_1, \ldots, t_m) \, dt_1 ... dt_m,$$

(10.9)

where $x_i \in [a_i, b_i]$, $i = 1, \ldots, m$, and $x = (x_1, \ldots, x_m)$, $a = (a_1, \ldots, a_m)$, $b = (b_1, \ldots, b_m)$.

We present the following multivariate fractional representation formula

Theorem 10.7 *Let f as in Assumptions 10.2 or 10.3, $\alpha \geq 1$, $x_i \in [a_i, b_i)$, $i = 1, \ldots, m$. Then*

$$f(x_1, \ldots, x_m) = \frac{\left(\prod_{i=1}^m (b_i - x_i) \right)^{1-\alpha} (\Gamma(\alpha))^m}{\prod_{i=1}^m (b_i - a_i)} \left(I_{a+}^\alpha f \right)(b) +$$

$$\sum_{i=1}^m A_i(x_1, \ldots, x_m) + \sum_{i=1}^m B_i(x_1, \ldots, x_m),$$

(10.10)

where for $i = 1, \ldots, m$:

$$A_i(x_1, \ldots, x_m) := \frac{-(\alpha - 1) \left(\prod_{j=1}^i (b_j - x_j) \right)^{1-\alpha}}{\prod_{j=1}^{i-1} (b_j - a_j)} \int_{\prod_{j=1}^i [a_j, b_j]} \left(\prod_{j=1}^{i-1} (b_j - t_j) \right)^{\alpha - 1} \cdot$$

(10.11)

$$(b_i - t_i)^{\alpha - 2} P_1(x_i, t_i) f(t_1, \ldots, t_i, x_{i+1}, \ldots, x_m) \, dt_1 ... dt_i,$$

and

$$B_i(x_1, \ldots, x_m) := \frac{\left(\prod_{j=1}^i (b_j - x_j) \right)^{1-\alpha}}{\prod_{j=1}^{i-1} (b_j - a_j)} \int_{\prod_{j=1}^i [a_j, b_j]} \left(\prod_{j=1}^i (b_j - t_j) \right)^{\alpha - 1} \cdot$$

(10.12)

$$P_1(x_i, t_i) \frac{\partial f}{\partial x_i} (t_1, \ldots, t_i, x_{i+1}, \ldots, x_m) \, dt_1 dt_2 ... dt_i.$$

Proof By (10.6) we have

$$f(x_1, \ldots, x_m) = (b_1 - x_1)^{1-\alpha} \left[\frac{1}{b_1 - a_1} \int_{a_1}^{b_1} (b_1 - t_1)^{\alpha - 1} f(t_1, x_2, \ldots, x_m) \, dt_1 \right.$$

$$- (\alpha - 1) \int_{a_1}^{b_1} (b_1 - t_1)^{\alpha - 2} P_1(x_1, t_1) f(t_1, x_2, \ldots, x_m) dt_1$$

$$+ \int_{a_1}^{b_1} (b_1 - t_1)^{\alpha - 1} P_1(x_1, t_1) \frac{\partial f}{\partial x_1} (t_1, x_2, \ldots, x_m) dt_1 \right], \tag{10.13}$$

and

$$f(t_1, x_2, \ldots, x_m) = (b_2 - x_2)^{1 - \alpha} \left[\frac{1}{b_2 - a_2} \int_{a_2}^{b_2} (b_2 - t_2)^{\alpha - 1} f(t_1, t_2, x_3, \ldots, x_m) dt_2 \right.$$

$$- (\alpha - 1) \int_{a_2}^{b_2} (b_2 - t_2)^{\alpha - 2} P_1(x_2, t_2) f(t_1, t_2, x_3, \ldots, x_m) dt_2$$

$$+ \int_{a_2}^{b_2} (b_2 - t_2)^{\alpha - 1} P_1(x_2, t_2) \frac{\partial f}{\partial x_2} (t_1, t_2, x_3, \ldots, x_m) dt_2 \right]. \tag{10.14}$$

We plug in (10.14) into (10.13). Hence

$$f(x_1, \ldots, x_m) = (b_1 - x_1)^{1 - \alpha} \left[\frac{1}{b_1 - a_1} \int_{a_1}^{b_1} (b_1 - t_1)^{\alpha - 1} (b_2 - x_2)^{1 - \alpha} \cdot \right.$$

$$\left[\frac{1}{b_2 - a_2} \int_{a_2}^{b_2} (b_2 - t_2)^{\alpha - 1} f(t_1, t_2, x_3, \ldots, x_m) dt_2 \right.$$

$$- (\alpha - 1) \int_{a_2}^{b_2} (b_2 - t_2)^{\alpha - 2} P_1(x_2, t_2) f(t_1, t_2, x_3, \ldots, x_m) dt_2 \tag{10.15}$$

$$+ \int_{a_2}^{b_2} (b_2 - t_2)^{\alpha - 1} P_1(x_2, t_2) \frac{\partial f}{\partial x_2} (t_1, t_2, x_3, \ldots, x_m) dt_2 \right] dt_1$$

$$- (\alpha - 1) \int_{a_1}^{b_1} (b_1 - t_1)^{\alpha - 2} P_1(x_1, t_1) f(t_1, x_2, \ldots, x_m) dt_1$$

$$+ \int_{a_1}^{b_1} (b_1 - t_1)^{\alpha - 1} P_1(x_1, t_1) \frac{\partial f}{\partial x_1} (t_1, x_2, \ldots, x_m) dt_1 \right].$$

That is we have so far

$$f(x_1, \ldots, x_m) = \frac{(b_1 - x_1)^{1 - \alpha} (b_2 - x_2)^{1 - \alpha}}{(b_1 - a_1)(b_2 - a_2)} \cdot$$

$$\int_{a_1}^{b_1} \int_{a_2}^{b_2} (b_1 - t_1)^{\alpha-1} (b_2 - t_2)^{\alpha-1} f(t_1, t_2, x_3, \ldots, x_m) \, dt_1 dt_2 - \quad (10.16)$$

$$\frac{(\alpha - 1)(b_1 - x_1)^{1-\alpha} (b_2 - x_2)^{1-\alpha}}{(b_1 - a_1)}.$$

$$\int_{a_1}^{b_1} \int_{a_2}^{b_2} (b_1 - t_1)^{\alpha-1} (b_2 - t_2)^{\alpha-1} P_1(x_2, t_2) f(t_1, t_2, x_3, \ldots, x_m) \, dt_1 dt_2$$

$$+ \frac{(b_1 - x_1)^{1-\alpha} (b_2 - x_2)^{1-\alpha}}{(b_1 - a_1)}.$$

$$\int_{a_1}^{b_1} \int_{a_2}^{b_2} (b_1 - t_1)^{\alpha-1} (b_2 - t_2)^{\alpha-1} P_1(x_2, t_2) \frac{\partial f}{\partial x_2}(t_1, t_2, x_3, \ldots, x_m) \, dt_1 dt_2$$

$$- (\alpha - 1)(b_1 - x_1)^{1-\alpha} \int_{a_1}^{b_1} (b_1 - t_1)^{\alpha-2} P_1(x_1, t_1) f(t_1, x_2, \ldots, x_m) \, dt_1$$

$$+ (b_1 - x_1)^{1-\alpha} \int_{a_1}^{b_1} (b_1 - t_1)^{\alpha-1} P_1(x_1, t_1) \frac{\partial f}{\partial x_1}(t_1, x_2, \ldots, x_m) \, dt_1.$$

Call

$$A_1(x_1, \ldots, x_m) := -(\alpha - 1)(b_1 - x_1)^{1-\alpha}.$$

$$\int_{a_1}^{b_1} (b_1 - t_1)^{\alpha-2} P_1(x_1, t_1) f(t_1, x_2, \ldots, x_m) \, dt_1, \quad (10.17)$$

$$B_1(x_1, \ldots, x_m) := (b_1 - x_1)^{1-\alpha}.$$

$$\int_{a_1}^{b_1} (b_1 - t_1)^{\alpha-1} P_1(x_1, t_1) \frac{\partial f}{\partial x_1}(t_1, x_2, \ldots, x_m) \, dt_1, \quad (10.18)$$

$$A_2(x_1, x_2, \ldots, x_m) := -\frac{(\alpha - 1)(b_1 - x_1)^{1-\alpha} (b_2 - x_2)^{1-\alpha}}{(b_1 - a_1)}. \quad (10.19)$$

$$\int_{a_1}^{b_1} \int_{a_2}^{b_2} (b_1 - t_1)^{\alpha-1} (b_2 - t_2)^{\alpha-1} P_1(x_2, t_2) f(t_1, t_2, x_3, \ldots, x_m) \, dt_1 dt_2,$$

and

$$B_2(x_1, x_2, \ldots, x_m) := \frac{(b_1 - x_1)^{1-\alpha} (b_2 - x_2)^{1-\alpha}}{(b_1 - a_1)}. \quad (10.20)$$

$$\int_{a_1}^{b_1} \int_{a_2}^{b_2} (b_1 - t_1)^{\alpha-1} (b_2 - t_2)^{\alpha-1} P_1(x_2, t_2) \frac{\partial f}{\partial x_2}(t_1, t_2, x_3, \dots, x_m) \, dt_1 dt_2.$$

We rewrite (10.16) as follows

$$f(x_1, \dots, x_m) = \frac{((b_1 - x_1)(b_2 - x_2))^{1-\alpha}}{(b_1 - a_1)(b_2 - a_2)} \cdot$$

$$\int_{a_1}^{b_1} \int_{a_2}^{b_2} ((b_1 - t_1)(b_2 - t_2))^{\alpha-1} f(t_1, t_2, x_3, \dots, x_m) \, dt_1 dt_2 + \qquad (10.21)$$

$$A_2(x_1, \dots, x_m) + B_2(x_1, \dots, x_m) + A_1(x_1, \dots, x_m) + B_1(x_1, \dots, x_m).$$

We continue with

$$f(t_1, t_2, x_3, \dots, x_m) \overset{(10.6)}{=} \frac{(b_3 - x_3)^{1-\alpha}}{b_3 - a_3} \int_{a_3}^{b_3} (b_3 - t_3)^{\alpha-1} f(t_1, t_2, t_3, x_4, \dots, x_m) \, dt_3$$

$$- (\alpha - 1)(b_3 - x_3)^{1-\alpha} \int_{a_3}^{b_3} (b_3 - t_3)^{\alpha-2} P_1(x_3, t_3) f(t_1, t_2, t_3, x_4, \dots, x_m) \, dt_3$$

$$(10.22)$$

$$+ (b_3 - x_3)^{1-\alpha} \int_{a_3}^{b_3} (b_3 - t_3)^{\alpha-1} P_1(x_3, t_3) \frac{\partial f}{\partial x_3}(t_1, t_2, t_3, x_4, \dots, x_m) \, dt_3.$$

Next plug (10.22) into (10.21). Hence it holds

$$f(x_1, \dots, x_m) = \frac{((b_1 - x_1)(b_2 - x_2)(b_3 - x_3))^{1-\alpha}}{(b_1 - a_1)(b_2 - a_2)(b_3 - a_3)} \cdot$$

$$\int_{a_1}^{b_1} \int_{a_2}^{b_2} \int_{a_3}^{b_3} ((b_1 - t_1)(b_2 - t_2)(b_3 - t_3))^{\alpha-1} f(t_1, t_2, t_3, x_4, \dots, x_m) \, dt_1 dt_2 dt_3$$

$$- \frac{(\alpha - 1)((b_1 - x_1)(b_2 - x_2)(b_3 - x_3))^{1-\alpha}}{(b_1 - a_1)(b_2 - a_2)} \cdot$$

$$\int_{a_1}^{b_1} \int_{a_2}^{b_2} \int_{a_3}^{b_3} ((b_1 - t_1)(b_2 - t_2))^{\alpha-1} (b_3 - t_3)^{\alpha-2} P_1(x_3, t_3) \cdot$$

$$f(t_1, t_2, t_3, x_4, \dots, x_m) \, dt_1 dt_2 dt_3$$

$$+ \frac{((b_1 - x_1)(b_2 - x_2)(b_3 - x_3))^{1-\alpha}}{(b_1 - a_1)(b_2 - a_2)} \cdot \qquad (10.23)$$

$$\int_{a_1}^{b_1} \int_{a_2}^{b_2} \int_{a_3}^{b_3} ((b_1 - t_1)(b_2 - t_2)(b_3 - t_3))^{\alpha-1} P_1(x_3, t_3) \cdot$$

$$\frac{\partial f}{\partial x_3}(t_1, t_2, t_3, x_4, \dots, x_m) \, dt_1 dt_2 dt_3$$

$$+A_1(x_1, \dots, x_m) + A_2(x_1, \dots, x_m) + B_1(x_1, \dots, x_m) + B_2(x_1, \dots, x_m).$$

Call

$$A_3(x_1, \dots, x_m) := -\frac{(\alpha - 1)((b_1 - x_1)(b_2 - x_2)(b_3 - x_3))^{1-\alpha}}{(b_1 - a_1)(b_2 - a_2)} . \qquad (10.24)$$

$$\int_{a_1}^{b_1} \int_{a_2}^{b_2} \int_{a_3}^{b_3} ((b_1 - t_1)(b_2 - t_2))^{\alpha-1}(b_3 - t_3)^{\alpha-2} P_1(x_3, t_3) \cdot$$

$$f(t_1, t_2, t_3, x_4, \dots, x_m) \, dt_1 dt_2 dt_3,$$

and

$$B_3(x_1, \dots, x_m) := \frac{((b_1 - x_1)(b_2 - x_2)(b_3 - x_3))^{1-\alpha}}{(b_1 - a_1)(b_2 - a_2)} . \qquad (10.25)$$

$$\int_{a_1}^{b_1} \int_{a_2}^{b_2} \int_{a_3}^{b_3} ((b_1 - t_1)(b_2 - t_2)(b_3 - t_3))^{\alpha-1} P_1(x_3, t_3) \cdot$$

$$\frac{\partial f}{\partial x_3}(t_1, t_2, t_3, x_4, \dots, x_m) \, dt_1 dt_2 dt_3.$$

Thus we have proved

$$f(x_1, \dots, x_m) = \frac{\left(\prod_{i=1}^{3}(b_i - x_i)\right)^{1-\alpha}}{\prod_{i=1}^{3}(b_i - a_i)} .$$

$$\int_{\prod_{i=1}^{3}[a_i, b_i]} \left(\prod_{i=1}^{3}(b_i - t_i)\right)^{\alpha-1} f(t_1, t_2, t_3, x_4, \dots, x_m) \, dt_1 dt_2 dt_3 +$$

$$\sum_{i=1}^{3} A_i(x_1, \dots, x_m) + \sum_{i=1}^{3} B_i(x_1, \dots, x_m) . \qquad (10.26)$$

Working similarly we finally obtain the fractional representation formula

$$f(x_1, \ldots, x_m) = \frac{\left(\prod_{i=1}^m (b_i - x_i)\right)^{1-\alpha}}{\prod_{i=1}^m (b_i - a_i)} \int_{\prod_{i=1}^m [a_i, b_i]} \left(\prod_{i=1}^m (b_i - t_i)\right)^{\alpha - 1} f\left(\overrightarrow{t}\right) d\overrightarrow{t}$$

$$+ \sum_{i=1}^m A_i(x_1, \ldots, x_m) + \sum_{i=1}^m B_i(x_1, \ldots, x_m). \qquad (10.27)$$

The proof of the theorem is now completed. ∎

We make

Remark 10.8 Let $f \in C^1\left(\prod_{i=1}^m [a_i, b_i]\right)$, $\alpha \geq 1$, $x_i \in [a_i, b_i)$, $i = 1, \ldots, m$. Denote by

$$\|f\|_\infty^{\sup} := \sup_{x \in \prod_{i=1}^m [a_i, b_i]} |f(x)|. \qquad (10.28)$$

We observe that

$$|B_i(x_1, \ldots, x_m)| \overset{(10.12)}{\leq} \frac{\left(\prod_{j=1}^i (b_j - x_j)\right)^{1-\alpha}}{\prod_{j=1}^{i-1} (b_j - a_j)} \left(\int_{\prod_{j=1}^i [a_j, b_j]} \left(\prod_{j=1}^i (b_j - t_j)\right)^{\alpha - 1} \cdot \right.$$
$$\qquad (10.29)$$

$$\left. |P_1(x_i, t_i)| \, dt_1 \ldots dt_i \right) \left\|\frac{\partial f}{\partial x_i}\right\|_\infty^{\sup} =$$

$$\frac{\left(\prod_{j=1}^i (b_j - x_j)\right)^{1-\alpha} \left\|\frac{\partial f}{\partial x_i}\right\|_\infty^{\sup}}{\prod_{j=1}^{i-1} (b_j - a_j)} \left(\prod_{j=1}^{i-1} \int_{a_j}^{b_j} (b_j - t_j)^{\alpha - 1} \, dt_j\right) \cdot$$

$$\left(\int_{a_i}^{b_i} (b_i - t_i)^{\alpha - 1} |P_1(x_i, t_i)| \, dt_i\right) = \qquad (10.30)$$

$$\frac{\left(\prod_{j=1}^i (b_j - x_j)\right)^{1-\alpha} \left\|\frac{\partial f}{\partial x_i}\right\|_\infty^{\sup}}{(b_i - a_i) \prod_{j=1}^{i-1} (b_j - a_j)} \frac{\left(\prod_{j=1}^{i-1} (b_j - a_j)\right)^\alpha}{\alpha^{i-1}} \cdot$$

$$\left[\int_{a_i}^{x_i} (b_i - t_i)^{\alpha - 1} (t_i - a_i) \, dt_i + \int_{x_i}^{b_i} (b_i - t_i)^{\alpha - 1} (b_i - t_i) \, dt_i\right] =$$

$$\frac{\left\|\frac{\partial f}{\partial x_i}\right\|_\infty^{\sup} \left(\prod_{j=1}^i (b_j - x_j)\right)^{1-\alpha} \left(\prod_{j=1}^{i-1} (b_j - a_j)\right)^{\alpha - 1}}{\alpha^{i-1} (b_i - a_i)}. \qquad (10.31)$$

$$\left[\int_{a_i}^{x_i} (b_i - t_i)^{\alpha-1} \left[(b_i - a_i) - (b_i - t_i) \right] dt_i + \frac{(b_i - x_i)^{\alpha+1}}{\alpha + 1} \right] =$$

$$\frac{\left\| \frac{\partial f}{\partial x_i} \right\|_\infty^{\sup} \left(\prod_{j=1}^{i} (b_j - x_j) \right)^{1-\alpha} \left(\prod_{j=1}^{i-1} (b_j - a_j) \right)^{\alpha-1}}{\alpha^{i-1} (b_i - a_i)}. \qquad (10.32)$$

$$\left[(b_i - a_i) \left[\frac{(b_i - a_i)^{\alpha}}{\alpha} - \frac{(b_i - x_i)^{\alpha}}{\alpha} \right] - \frac{(b_i - a_i)^{\alpha+1}}{\alpha + 1} + \frac{2 (b_i - x_i)^{\alpha+1}}{\alpha + 1} \right] =$$

$$\frac{\left\| \frac{\partial f}{\partial x_i} \right\|_\infty^{\sup} \left(\prod_{j=1}^{i} (b_j - x_j) \right)^{1-\alpha} \left(\prod_{j=1}^{i-1} (b_j - a_j) \right)^{\alpha-1}}{\alpha^{i-1} (b_i - a_i)}.$$

$$\left[\frac{(b_i - a_i)^{\alpha+1}}{\alpha (\alpha + 1)} + \frac{2 (b_i - x_i)^{\alpha+1}}{\alpha + 1} - (b_i - a_i) \frac{(b_i - x_i)^{\alpha}}{\alpha} \right]. \qquad (10.33)$$

We have proved for $i = 1, \ldots, m$, that

$$|B_i (x_1, \ldots, x_m)| \le \frac{\left\| \frac{\partial f}{\partial x_i} \right\|_\infty^{\sup} \left(\prod_{j=1}^{i} (b_j - x_j) \right)^{1-\alpha} \left(\prod_{j=1}^{i-1} (b_j - a_j) \right)^{\alpha-1}}{\alpha^{i-1}}.$$

$$\left[\frac{(b_i - a_i)^{\alpha}}{\alpha (\alpha + 1)} + \frac{2 (b_i - x_i)^{\alpha+1}}{(\alpha + 1) (b_i - a_i)} - \frac{(b_i - x_i)^{\alpha}}{\alpha} \right]. \qquad (10.34)$$

We have established the following multivariate fractional Ostrowski type inequality.

Theorem 10.9 Let $f \in C^1 \left(\prod_{i=1}^{m} [a_i, b_i] \right)$, $\alpha \ge 1$, $x_i \in [a_i, b_i)$, $i = 1, \ldots, m$. Then

$$\left| f (x_1, \ldots, x_m) - \frac{\left(\prod_{i=1}^{m} (b_i - x_i) \right)^{1-\alpha} (\Gamma (\alpha))^m (I_{a+}^{\alpha} f) (b)}{\prod_{i=1}^{m} (b_i - a_i)} - \sum_{i=1}^{m} A_i (x_1, \ldots, x_m) \right|$$

$$\le \sum_{i=1}^{m} \left\{ \frac{\left\| \frac{\partial f}{\partial x_i} \right\|_\infty^{\sup} \left(\prod_{j=1}^{i} (b_j - x_j) \right)^{1-\alpha} \left(\prod_{j=1}^{i-1} (b_j - a_j) \right)^{\alpha-1}}{\alpha^{i-1}} \right.$$

$$\left. \left[\frac{(b_i - a_i)^{\alpha}}{\alpha (\alpha + 1)} + \frac{2 (b_i - x_i)^{\alpha+1}}{(\alpha + 1) (b_i - a_i)} - \frac{(b_i - x_i)^{\alpha}}{\alpha} \right] \right\}. \qquad (10.35)$$

References

1. G. Anastassiou, M. Hooshmandasl, A. Ghasemi, F. Moftakharzadeh, Montgomery identities for fractional integrals and related fractional inequalities. J. Inequalities Pure Appl. Math. **10**(4), Article 97, 6 p. (2009)
2. G.A. Anastassiou, Multivariate fractional representation formula and Ostrowski type inequality. Sarajevo J. Math. **10**(22), 1–9 (2014)
3. T. Mamatov, S. Samko, Mixed fractional integration operators in mixed weighted Hölder spaces. Fract. Calc. Appl. Anal. **13**(3), 245–259 (2010)
4. D.S. Mitrinovic, J.E. Pecaric, A.M. Fink, *Inequalities for Functions and Their Integrals and Derivatives* (Kluwer Academic Publishers, Dordrecht, 1994)
5. S. Miller, B. Ross, *An Introduction to the Fractional Calculus and Fractional Differential Equations* (Wiley, New York, 1993)

Chapter 11
Multivariate Weighted Fractional Representation Formulae and Ostrowski Inequalities

Here we derive multivariate weighted fractional representation formulae involving ordinary partial derivatives of first order. Then we present related multivariate weighted fractional Ostrowski type inequalities with respect to uniform norm. It follows [2].

11.1 Introduction

Let $f : [a, b] \to \mathbb{R}$ be differentiable on $[a, b]$, and $f' : [a, b] \to \mathbb{R}$ be integrable on $[a, b]$. Suppose now that $w : [a, b] \to [0, \infty)$ is some probability density function, i.e. it is a nonnegative integrable function satisfying $\int_a^b w(t)\,dt = 1$, and $W(t) = \int_a^t w(x)\,dx$ for $t \in [a, b]$, $W(t) = 0$ for $t \leq a$ and $W(t) = 1$ for $t \geq b$. Then, the following identity (Pecarić [6]) is the weighted generalization of the Montgomery identity [5]

$$f(x) = \int_a^b w(t) f(t)\,dt + \int_a^b P_w(x, t) f'(t)\,dt, \qquad (11.1)$$

where the weighted Peano Kernel is

$$P_w(x, t) := \begin{cases} W(t), & a \leq t \leq x, \\ W(t) - 1, & x < t \leq b. \end{cases} \qquad (11.2)$$

In [1] we proved

Theorem 11.1 *Let $w : [a, b] \to [0, \infty)$ be a probability density function, i.e. $\int_a^b w(t)\,dt = 1$, and set $W(t) = \int_a^t w(x)\,dx$ for $a \leq t \leq b$, $W(t) = 0$ for $t \leq a$ and $W(t) = 1$ for $t \geq b$, $\alpha \geq 1$, and f is as in (11.1). Then the generalization of the weighted Montgomery identity for fractional integrals is the following*

© Springer International Publishing Switzerland 2016
G.A. Anastassiou, *Intelligent Comparisons: Analytic Inequalities*,
Studies in Computational Intelligence 609,
DOI 10.1007/978-3-319-21121-3_11

$$f(x) = (b - x)^{1-\alpha} \Gamma(\alpha) J_a^\alpha (w(b) f(b)) -$$

$$J_a^{\alpha-1} (Q_w(x, b) f(b)) + J_a^\alpha (Q_w(x, b) f'(b)), \tag{11.3}$$

where the weighted fractional Peano Kernel is

$$Q_w(x, t) := \begin{cases} (b - x)^{1-\alpha} \Gamma(\alpha) W(t), & a \le t \le x, \\ (b - x)^{1-\alpha} \Gamma(\alpha) (W(t) - 1), & x < t \le b, \end{cases} \tag{11.4}$$

i.e. $Q_w(x, t) = (b - x)^{1-\alpha} \Gamma(\alpha) P_w(x, t)$.

When $\alpha = 1$ then the weighted generalization of the Montgomery identity for fractional integrals in (11.3) reduces to the weighted generalization of the Montgomery identity for integrals in (11.1).

So for $\alpha \ge 1$ and $x \in [a, b)$ we can rewrite (11.3) as follows

$$f(x) = (b - x)^{1-\alpha} \int_a^b (b - t)^{\alpha-1} w(t) f(t) \, dt -$$

$$(b - x)^{1-\alpha} (\alpha - 1) \int_a^b (b - t)^{\alpha-2} P_w(x, t) f(t) \, dt + \tag{11.5}$$

$$(b - x)^{1-\alpha} \int_a^b (b - t)^{\alpha-1} P_w(x, t) f'(t) \, dt.$$

In this chapter based on (11.5), we establish a multivariate weighted general fractional representation formula for $f(x), x \in \prod_{i=1}^m [a_i, b_i] \subset \mathbb{R}^m$, and from there we derive an interesting multivariate weighted fractional Ostrowski type inequality. We finish with an application.

11.2 Main Results

We make

Assumption 11.2 Let $f \in C^1 \left(\prod_{i=1}^m [a_i, b_i] \right)$.

Assumption 11.3 Let $f : \prod_{i=1}^m [a_i, b_i] \to \mathbb{R}$ be measurable and bounded, such that there exist $\frac{\partial f}{\partial x_j} : \prod_{i=1}^m [a_i, b_i] \to \mathbb{R}$, and it is x_j-integrable for all $j = 1, \dots, m$. Furthermore $\frac{\partial f}{\partial x_i} (t_1, \dots, t_i, x_{i+1}, \dots, x_m)$ it is integrable on $\prod_{j=1}^i [a_j, b_j]$, for all $i = 1, \dots, m$, for any $(x_{i+1}, \dots, x_m) \in \prod_{j=i+1}^m [a_j, b_j]$.

Convention 11.4 *We set*

$$\prod_{j=1}^{0} \cdot = 1. \tag{11.6}$$

Notation 11.5 *Here* $x = \vec{x} = (x_1, \ldots, x_m) \in \mathbb{R}^m$, $m \in \mathbb{N} - \{1\}$. *Likewise* $t = \vec{t} = (t_1, \ldots, t_m)$, *and* $d\vec{t} = dt_1 dt_2 \ldots dt_m$. *Here* w_i, W_i *correspond to* $[a_i, b_i]$, $i = 1, \ldots, m$, *and are as* w, W *of Theorem 11.1.*

We need

Definition 11.6 (see [3] and [4]) *Let* $\prod_{i=1}^{m} [a_i, b_i] \subset \mathbb{R}^m$, $m \in \mathbb{N} - \{1\}$, $a_i < b_i$, $a_i, b_i \in \mathbb{R}$. *Let* $\alpha > 0$, $f \in L_1 \left(\prod_{i=1}^{m} [a_i, b_i] \right)$. *We define the left mixed Riemann-Liouville fractional multiple integral of order* α:

$$\left(I_{a+}^{\alpha} f \right)(x) := \frac{1}{(\Gamma(\alpha))^m} \int_{a_1}^{x_1} \cdots \int_{a_m}^{x_m} \left(\prod_{i=1}^{m} (x_i - t_i) \right)^{\alpha-1} f(t_1, \ldots, t_m) \, dt_1 \ldots dt_m, \tag{11.7}$$

where $x_i \in [a_i, b_i]$, $i = 1, \ldots, m$, *and* $x = (x_1, \ldots, x_m)$, $a = (a_1, \ldots, a_m)$, $b = (b_1, \ldots, b_m)$.

We present the following multivariate weighted fractional representation formula

Theorem 11.7 *Let* f *as in Assumptions 11.2 or 11.3,* $\alpha \geq 1$, $x_i \in [a_i, b_i)$, $i = 1, \ldots, m$. *Here* P_{w_i} *corresponds to* $[a_i, b_i]$, $i = 1, \ldots, m$, *and it is as in (11.2). The probability density function* w_j *is assumed to be bounded for all* $j = 1, \ldots, m$. *Then*

$$f(x_1, \ldots, x_m) = \left(\prod_{j=1}^{m} (b_j - x_j) \right)^{1-\alpha} (\Gamma(\alpha))^m \left(I_{a+}^{\alpha} \left(\prod_{j=1}^{m} w_j \right) f \right)(b) +$$

$$\sum_{i=1}^{m} A_i (x_1, \ldots, x_m) + \sum_{i=1}^{m} B_i (x_1, \ldots, x_m), \tag{11.8}$$

where for $i = 1, \ldots, m$:

$$A_i (x_1, \ldots, x_m) := -(\alpha - 1) \left(\prod_{j=1}^{i} (b_j - x_j) \right)^{1-\alpha} \int_{\prod_{j=1}^{i} [a_j, b_j]} \left(\prod_{j=1}^{i-1} (b_j - t_j) \right)^{\alpha-1} \cdot \tag{11.9}$$

$$(b_i - t_i)^{\alpha-2} \left(\prod_{j=1}^{i-1} w_j (t_j) \right) P_{w_i} (x_i, t_i) f(t_1, \ldots, t_i, x_{i+1}, \ldots, x_m) \, dt_1 \ldots dt_i,$$

and

$$B_i(x_1, \ldots, x_m) := \left(\prod_{j=1}^{i}(b_j - x_j)\right)^{1-\alpha} \int_{\prod_{j=1}^{i}[a_j, b_j]} \left(\prod_{j=1}^{i}(b_j - t_j)\right)^{\alpha-1} \cdot$$

$$\left(\prod_{j=1}^{i-1} w_j(t_j)\right) P_{w_i}(x_i, t_i) \frac{\partial f}{\partial x_i}(t_1, \ldots, t_i, x_{i+1}, \ldots, x_m) \, dt_1 \ldots dt_i.$$

(11.10)

Proof We have that

$$f(x_1, x_2, \ldots, x_m) \overset{(11.5)}{=} (b_1 - x_1)^{1-\alpha} \int_{a_1}^{b_1}(b_1 - t_1)^{\alpha-1} w_1(t_1) f(t_1, x_2, \ldots, x_m) \, dt_1 +$$

$$A_1(x_1, \ldots, x_m) + B_1(x_1, \ldots, x_m).$$

(11.11)

Similarly it holds

$$f(t_1, x_2, \ldots, x_m) \overset{(11.5)}{=} (b_2 - x_2)^{1-\alpha} \int_{a_2}^{b_2}(b_2 - t_2)^{\alpha-1} w_2(t_2) f(t_1, t_2, x_3, \ldots, x_m) \, dt_2$$

$$-(\alpha - 1)(b_2 - x_2)^{1-\alpha} \int_{a_2}^{b_2}(b_2 - t_2)^{\alpha-2} P_{w_2}(x_2, t_2) f(t_1, t_2, x_3, \ldots, x_m) \, dt_2 +$$

$$(b_2 - x_2)^{1-\alpha} \int_{a_2}^{b_2}(b_2 - t_2)^{\alpha-1} P_{w_2}(x_2, t_2) \frac{\partial f}{\partial x_2}(t_1, t_2, x_3, \ldots, x_m) \, dt_2. \quad (11.12)$$

Next we plug (11.12) into (11.11).

We get

$$f(x_1, \ldots, x_m) = ((b_1 - x_1)(b_2 - x_2))^{1-\alpha} \cdot$$

$$\int_{a_1}^{b_1} \int_{a_2}^{b_2} ((b_1 - t_1)(b_2 - t_2))^{\alpha-1} w_1(t_1) w_2(t_2) f(t_1, t_2, x_3, \ldots, x_m) \, dt_1 dt_2 +$$

(11.13)

$$A_2(x_1, \ldots, x_m) + B_2(x_1, \ldots, x_m) + A_1(x_1, \ldots, x_m) + B_1(x_1, \ldots, x_m).$$

We continue as above.

We also have

$$f(t_1, t_2, x_3, \ldots, x_m) \overset{(11.5)}{=} (b_3 - x_3)^{1-\alpha} \cdot$$

$$\int_{a_3}^{b_3}(b_3 - t_3)^{\alpha-1} w_3(t_3) f(t_1, t_2, t_3, x_4, \ldots, x_m) \, dt_3$$

$$- (\alpha - 1) (b_3 - x_3)^{1-\alpha} \int_{a_3}^{b_3} (b_3 - t_3)^{\alpha-2} P_{w_3}(x_3, t_3) f(t_1, t_2, t_3, x_4, \ldots, x_m) dt_3$$

$$\tag{11.14}$$

$$+ (b_3 - x_3)^{1-\alpha} \int_{a_3}^{b_3} (b_3 - t_3)^{\alpha-1} P_{w_3}(x_3, t_3) \frac{\partial f}{\partial x_3}(t_1, t_2, t_3, x_4, \ldots, x_m) dt_3.$$

We plug (11.14) into (11.13). Therefore it holds

$$f(x_1, \ldots, x_m) = \left(\prod_{j=1}^{3} (b_j - x_j)\right)^{1-\alpha} \int_{\prod_{j=1}^{3}[a_j, b_j]} \left(\prod_{j=1}^{3} (b_j - t_j)\right)^{\alpha-1} \cdot$$

$$\left(\prod_{j=1}^{3} w_j(t_j)\right) f(t_1, t_2, t_3, x_4, \ldots, x_m) dt_1 dt_2 dt_3 +$$

$$\sum_{j=1}^{3} A_j(x_1, \ldots, x_m) + \sum_{j=1}^{3} B_j(x_1, \ldots, x_m). \tag{11.15}$$

Continuing similarly we finally obtain

$$f(x_1, \ldots, x_m) = \left(\prod_{j=1}^{m} (b_j - x_j)\right)^{1-\alpha} \cdot$$

$$\int_{\prod_{j=1}^{m}[a_j, b_j]} \left(\prod_{j=1}^{m} (b_j - t_j)\right)^{\alpha-1} \left(\prod_{j=1}^{m} w_j(t_j)\right) f\left(\overrightarrow{t}\right) d\overrightarrow{t}$$

$$+ \sum_{i=1}^{m} A_i(x_1, \ldots, x_m) + \sum_{i=1}^{m} B_i(x_1, \ldots, x_m), \tag{11.16}$$

that is proving the claim. ∎

We make

Remark 11.8 Let $f \in C^1 \left(\prod_{i=1}^{m}[a_i, b_i]\right)$, $\alpha \geq 1$, $x_i \in [a_i, b_i]$, $i = 1, \ldots, m$. Denote by

$$\|f\|_{\infty}^{\sup} := \sup_{x \in \prod_{i=1}^{m}[a_i, b_i]} |f(x)|. \tag{11.17}$$

From (11.2) we get that

$$|P_w\left(x,t\right)| \le \begin{cases} W\left(x\right), & a \le t \le x, \\ 1 - W\left(x\right), & x < t \le b \end{cases}$$

$$\le \max\left\{W\left(x\right), 1 - W\left(x\right)\right\} = \frac{1 + |2W\left(x\right) - 1|}{2}. \tag{11.18}$$

That is

$$|P_w\left(x,t\right)| \le \frac{1 + |2W\left(x\right) - 1|}{2}, \tag{11.19}$$

for all $t \in [a, b]$, where $x \in [a, b]$ is fixed.

Consequently it holds

$$\left|P_{w_i}\left(x_i, t_i\right)\right| \le \frac{1 + |2W_i\left(x_i\right) - 1|}{2}, \quad i = 1, \ldots, m. \tag{11.20}$$

Assume here that

$$w_j\left(t_j\right) \le K_j, \tag{11.21}$$

for all $t_j \in \left[a_j, b_j\right]$, where $K_j > 0$, $j = 1, \ldots, m$.

Therefore we derive

$$|B_i\left(x_1, \ldots, x_m\right)| \le \left(\prod_{j=1}^{i}\left(b_j - x_j\right)\right)^{1-\alpha}\left(\prod_{j=1}^{i-1} K_j\right) \cdot$$

$$\left(\frac{1 + |2W_i\left(x_i\right) - 1|}{2}\right)\left\|\frac{\partial f}{\partial x_i}\right\|_{\infty}^{\sup} \prod_{j=1}^{i}\left(\int_{a_j}^{b_j}\left(b_j - t_j\right)^{\alpha-1} dt_j\right). \tag{11.22}$$

That is

$$|B_i\left(x_1, \ldots, x_m\right)| \le \left(\prod_{j=1}^{i}\left(b_j - x_j\right)\right)^{1-\alpha}\left(\frac{\prod_{j=1}^{i}\left(b_j - a_j\right)^{\alpha}}{\alpha^i}\right)\left(\prod_{j=1}^{i-1} K_j\right) \cdot$$

$$\left(\frac{1 + |2W_i\left(x_i\right) - 1|}{2}\right)\left\|\frac{\partial f}{\partial x_i}\right\|_{\infty}^{\sup}, \tag{11.23}$$

for all $i = 1, \ldots, m$.

Based on the above and Theorem 11.7 we have established the following multivariate weighted fractional Ostrowski type inequality.

Theorem 11.9 *Let* $f \in C^1 \left(\prod_{i=1}^m [a_i, b_i] \right)$, $\alpha \geq 1$, $x_i \in [a_i, b_i)$, $i = 1, \ldots, m$. *Here* P_{w_i} *corresponds to* $[a_i, b_i]$, $i = 1, \ldots, m$, *and it is as in (11.2). Assume that* $w_j \left(t_j \right) \leq K_j$, *for all* $t_j \in [a_j, b_j]$, *where* $K_j > 0$, $j = 1, \ldots, m$. *And* $A_i \left(x_1, \ldots, x_m \right)$ *is as in (11.9),* $i = 1, \ldots, m$. *Then*

$$
\left| f\left(x_1, \ldots, x_m\right) - \left(\prod_{j=1}^m \left(b_j - x_j \right) \right)^{1-\alpha} \left(\Gamma\left(\alpha\right) \right)^m \left(I_{a+}^{\alpha} \left(\prod_{j=1}^m w_j \right) f \right) (b) \right.
$$

$$
\left. - \sum_{i=1}^m A_i\left(x_1, \ldots, x_m\right) \right| \leq \sum_{i=1}^m \left\{ \left(\prod_{j=1}^i \left(b_j - x_j \right) \right)^{1-\alpha} \left(\frac{\prod_{j=1}^i \left(b_j - a_j \right)^{\alpha}}{\alpha^i} \right) \right.
$$

$$
\left. \left(\prod_{j=1}^{i-1} K_j \right) \left(\frac{1 + |2 W_i \left(x_i \right) - 1|}{2} \right) \left\| \frac{\partial f}{\partial x_i} \right\|_{\infty}^{\sup} \right\}. \tag{11.24}
$$

11.3 Application

Here we operate on $[0, 1]^m$, $m \in \mathbb{N} - \{1\}$. We notice that

$$
\int_0^1 \left(\frac{e^{-x}}{1 - e^{-1}} \right) dx = 1, \tag{11.25}
$$

and

$$
\frac{e^{-x}}{1 - e^{-1}} \leq \frac{1}{1 - e^{-1}}, \text{ for all } x \in [0, 1]. \tag{11.26}
$$

So here we choose as w_j the probability density function

$$
w_j^* \left(t_j \right) := \frac{e^{-t_j}}{1 - e^{-1}}, \tag{11.27}
$$

$j = 1, \ldots, m$, $t_j \in [0, 1]$.

So we have the corrsponding W_j as

$$
W_j^* \left(t_j \right) = \frac{1 - e^{-t_j}}{1 - e^{-1}}, \quad t_j \in [0, 1], \tag{11.28}
$$

and the corresponding P_{w_j} as

$$P_{w_j}^* \left(x_j, t_j\right) = \begin{cases} \frac{1-e^{-t_j}}{1-e^{-1}}, & 0 \le t_j \le x_j, \\ \frac{e^{-1}-e^{-t_j}}{1-e^{-1}}, & x_j < t_j \le 1, \end{cases} \tag{11.29}$$

$j = 1, \ldots, m$.

Set $\overrightarrow{0} = (0, \ldots, 0)$ and $\overrightarrow{1} = (1, \ldots, 1)$.

First we apply Theorem 11.7.

We have

Theorem 11.10 *Let* $f \in C^1 \left([0, 1]^m\right)$, $\alpha \ge 1$, $x_i \in [0, 1)$, $i = 1, \ldots, m$. *Then*

$$f(x_1, \ldots, x_m) = \left(\prod_{j=1}^m (1 - x_j) \right)^{1-\alpha} \left(\frac{\Gamma(\alpha)}{1 - e^{-1}} \right)^m \left(I_{\overrightarrow{0}+}^\alpha \left(e^{-\sum_{j=1}^m t_j} f(\cdot) \right) \right) \left(\overrightarrow{1} \right) \tag{11.30}$$

$$+ \sum_{i=1}^m A_i^* (x_1, \ldots, x_m) + \sum_{i=1}^m B_i^* (x_1, \ldots, x_m),$$

where for $i = 1, \ldots, m$:

$$A_i^* (x_1, \ldots, x_m) := \frac{-(\alpha - 1)}{\left(1 - e^{-1}\right)^{i-1}} \left(\prod_{j=1}^i (1 - x_j) \right)^{1-\alpha} \int_{[0,1]^i} \left(\prod_{j=1}^{i-1} (1 - t_j) \right)^{\alpha-1} \cdot \tag{11.31}$$

$$(1 - t_i)^{\alpha-2} e^{-\sum_{j=1}^{i-1} t_j} P_{w_i}^* (x_i, t_i) f (t_1, \ldots, t_i, x_{i+1}, \ldots, x_m) \, dt_1 \ldots dt_i,$$

and

$$B_i^* (x_1, \ldots, x_m) := \frac{\left(\prod_{j=1}^i (1 - x_j) \right)^{1-\alpha}}{\left(1 - e^{-1}\right)^{i-1}} \int_{[0,1]^i} \left(\prod_{j=1}^i (1 - t_j) \right)^{\alpha-1} \cdot \tag{11.32}$$

$$e^{-\sum_{j=1}^{i-1} t_j} P_{w_i}^* (x_i, t_i) \frac{\partial f}{\partial x_i} (t_1, \ldots, t_i, x_{i+1}, \ldots, x_m) \, dt_1 \ldots dt_i.$$

Above we set $\sum_{i=1}^0 \cdot = 0$.

Finally we apply Theorem 11.9.

Theorem 11.11 *Let* $f \in C^1 \left([0, 1]^m\right)$, $\alpha \ge 1$, $x_i \in [0, 1)$, $i = 1, \ldots, m$. *Here* $P_{w_i}^*$ *is as in (11.29) and* $A_i^* (x_1, \ldots, x_m)$ *as in (11.31),* $i = 1, \ldots, m$. *Then*

$$\left| f(x_1, \ldots, x_m) - \left(\prod_{j=1}^{m} (1 - x_j) \right)^{1-\alpha} \left(\frac{\Gamma(\alpha)}{1 - e^{-1}} \right)^m \left(I_{\overrightarrow{0}+}^\alpha \left(e^{-\Sigma_{j=1}^m t_j} f(\cdot) \right) \right) \left(\overrightarrow{1} \right) \right.$$

$$\tag{11.33}$$

$$\left. - \sum_{i=1}^{m} A_i^*(x_1, \ldots, x_m) \right| \leq \sum_{i=1}^{m} \left\{ \frac{\left(\prod_{j=1}^{i} (1 - x_j) \right)^{1-\alpha}}{\alpha^i \left(1 - e^{-1} \right)^{i-1}} \cdot \right.$$

$$\left. \left(\frac{1 + |2W_i^*(x_i) - 1|}{2} \right) \left\| \frac{\partial f}{\partial x_i} \right\|_\infty^{\text{sup}} \right\}.$$

References

1. G. Anastassiou, M. Hooshmandasl, A. Ghasemi, F. Moftakharzadeh, Montgomery identities for fractional integrals and related fractional inequalities. J. Inequalities Pure Appl. Math. **10**(4, 97), 6 (2009)
2. G.A. Anastassiou, Multivariate weighted fractional representation formulae and Ostrowski type inequalities. Studia Mathematica Babes Bolyai **59**(1), 3–10 (2014)
3. T. Mamatov, S. Samko, Mixed fractional integration operators in mixed weighted Hölder spaces. Fractional Calculus Appl. Anal. **13**(3), 245–259 (2010)
4. S. Miller, B. Ross, *An Introduction to the Fractional Calculus and Fractional Differential Equations* (Wiley, USA, 1993)
5. D.S. Mitrinovic, J.E. Pecaric, A.M. Fink, *Inequalities for Functions and their Integrals and Derivatives* (Kluwer Academic Publishers, Dordrecht, 1994)
6. J.E. Pečarić, *On the Čebyšev inequality*, Bul. Şti. Tehn. Inst. Politehn, "Traian Vuia" Timiş oara **25**(39, 1), 5–9 (1980)

Chapter 12
About Multivariate Lyapunov Inequalities

We transfer here basic univariate Lyapunov inequalities to the multivariate setting of a shell by using the polar method. It follows [1].

12.1 Background

Let A be a spherical shell $\subseteq \mathbb{R}^N$, $N > 1$, i.e. $A := B(0, R_2) - \overline{B(0, R_1)}$, $0 < R_1 < R_2$.

Here the ball

$$B(0, R) := \left\{ x \in \mathbb{R}^N : |x| < R \right\},$$

$R > 0$, where $|\cdot|$ is the Euclidean norm, also

$$S^{N-1} := \left\{ x \in \mathbb{R}^N : |x| = 1 \right\}$$

is the unit sphere in $\mathbb{R}^N$ with surface area

$$\omega_N := \frac{2\pi^{N/2}}{\Gamma(N/2)},$$

i.e.

$$\int_{S^{N-1}} d\omega = \frac{2\pi^{N/2}}{\Gamma(N/2)},$$

where Γ is the gamma function.

For $x \in \mathbb{R}^N - \{0\}$ one car write uniquely $x = r\omega$, where $r > 0$, $\omega \in S^{N-1}$, see [5], pp. 149–150. Here $r = |x| > 0$ and $\omega = \frac{x}{r} \in S^{N-1}$.

© Springer International Publishing Switzerland 2016
G.A. Anastassiou, *Intelligent Comparisons: Analytic Inequalities*,
Studies in Computational Intelligence 609,
DOI 10.1007/978-3-319-21121-3_12

For $F \in C\left(\overline{A}\right)$ we have

$$\int_A F(x)\, dx = \int_{S^{N-1}} \left(\int_{R_1}^{R_2} F(r\omega)\, r^{N-1} dr \right) d\omega. \qquad (12.1)$$

In particular $\overline{A} = [R_1, R_2] \times S^{N-1}$.

Here we deal with partial differential equations (PDE) involving radial derivatives of functions on $\overline{A}$, using the polar coordinates r, ω.

If $f \in C^n\left(\overline{A}\right)$, $n \in \mathbb{N}$, then for a fixed $\omega \in S^{N-1}$, $f(\cdot\omega) \in C^n([R_1, R_2])$.

12.2 Results

We mention the famous Lyapunov [3] inequality

Theorem 12.1 *If $x(t)$ is a nontrivial solution of*

$$x''(t) + q(t)\, x(t) = 0, \qquad (12.2)$$

where $q \in C([a, b])$, with $x(a) = x(b) = 0$, and $x(t) \neq 0$ for any $t \in (a, b)$, then

$$\int_a^b |q(s)|\, ds > \frac{4}{b-a}. \qquad (12.3)$$

We give our first multivariate Lyapunov inequality, the first in the literature

Theorem 12.2 *If $f \in C^2\left(\overline{A}\right)$ is a nontrivial solution of the PDE:*

$$\frac{\partial^2 f(x)}{\partial r^2} + q(x)\, f(x) = 0, \quad \forall x \in \overline{A}, \qquad (12.4)$$

where $q \in C\left(\overline{A}\right)$, with

$$f(\partial B(0, R_1)) = f(\partial B(0, R_2)) = 0,$$

and $f(x) \neq 0$ for any $x \in A$, then

$$\int_A |q(x)|\, dx > \frac{8\pi^{N/2} R_1^{N-1}}{\Gamma(N/2)(R_2 - R_1)}. \qquad (12.5)$$

Proof One can rewrite (12.4) as

$$\frac{\partial^2 f(r\omega)}{\partial r^2} + q(r\omega)\, f(r\omega) = 0, \quad \forall\, (r, \omega) \in [R_1, R_2] \times S^{N-1}, \qquad (12.6)$$

where $q(\cdot\omega) \in C([R_1, R_2])$, $\forall \omega \in S^{N-1}$, such that

$$f(R_1\omega) = f(R_2\omega) = 0, \forall \omega \in S^{N-1}. \tag{12.7}$$

Also $f(r\omega) \neq 0$, for any $r \in (R_1, R_2)$, $\forall \omega \in S^{N-1}$.

So for a fixed $\omega \in S^{N-1}$ by (12.3) we get

$$\frac{4}{R_2 - R_1} < \int_{R_1}^{R_2} |q(r\omega)| \, dr = \int_{R_1}^{R_2} r^{1-N} r^{N-1} |q(r\omega)| \, dr$$

$$\leq \left(\int_{R_1}^{R_2} r^{N-1} |q(r\omega)| \, dr \right) R_1^{1-N}. \tag{12.8}$$

That is

$$\int_{R_1}^{R_2} r^{N-1} |q(r\omega)| \, dr > \frac{4 R_1^{N-1}}{R_2 - R_1}, \tag{12.9}$$

$\forall \omega \in S^{N-1}$, and

$$\int_{S^{N-1}} \left(\int_{R_1}^{R_2} r^{N-1} |q(r\omega)| \, dr \right) d\omega > \left(\frac{4 R_1^{N-1}}{R_2 - R_1} \right) \left(\frac{2\pi^{N/2}}{\Gamma(N/2)} \right), \tag{12.10}$$

which by (12.1) proves (12.5). ∎

Remark 12.3 It was noticed by Wintner [6] that in (12.3) we can replace $|q(s)|$ by

$$q^+(t) = \max\{0, q(t)\} = \frac{q(t) + |q(t)|}{2},$$

the nonnegative part of $q(t)$.

The same can happen in (12.5), that is we have

$$\int_A q^+(x) \, dx > \frac{8\pi^{N/2} R_1^{N-1}}{\Gamma(N/2)(R_2 - R_1)}. \tag{12.11}$$

We need the 1999 Parhi and Panigrahi [4] result

Theorem 12.4 *Let $x(t)$ be a nontrivial solution of*

$$x''' + q(t) x = 0, \tag{12.12}$$

with $x(a) = x(b) = 0$, where $a, b \in \mathbb{R}$ with $a < b$ be consecutive zeros, and $x(t) \neq 0$ for $t \in (a, b)$. If there exists a $d \in (a, b)$ such that $x''(d) = 0$, then

$$\int_a^b |q(s)|\,ds > \frac{4}{(b-a)^2}. \tag{12.13}$$

We give

Theorem 12.5 *If $f \in C^3\left(\overline{A}\right)$ is a nontrivial solution of the PDE:*

$$\frac{\partial^3 f(x)}{\partial r^3} + q(x) f(x) = 0, \ \forall x \in \overline{A}, \tag{12.14}$$

where $q \in C\left(\overline{A}\right)$, with $f\left(\partial B\left(0, R_1\right)\right) = f\left(\partial B\left(0, R_2\right)\right) = 0$, and $f(x) \neq 0$ for any $x \in A$. If there exists a $d \in (R_1, R_2)$ such that $\frac{\partial^2 f(\partial B(0,d))}{\partial r^2} = 0$, then

$$\int_A |q(x)|\,dx > \frac{8\pi^{N/2} R_1^{N-1}}{\Gamma(N/2)(R_2 - R_1)^2}. \tag{12.15}$$

Proof One can rewrite (12.14) as

$$\frac{\partial^3 f(r\omega)}{\partial r^3} + q(r\omega) f(r\omega) = 0, \ \forall (r, \omega) \in [R_1, R_2] \times S^{N-1}, \tag{12.16}$$

where $q(\cdot\omega) \in C\left([R_1, R_2]\right),\ \forall \omega \in S^{N-1}$, such that

$$f(R_1\omega) = f(R_2\omega) = 0,\ \forall \omega \in S^{N-1}.$$

Also $f(r\omega) \neq 0$, for any $r \in (R_1, R_2),\ \forall \omega \in S^{N-1}$.

Furthermore there exists $d \in (R_1, R_2)$ such that $\frac{\partial^2 f(d\omega)}{\partial r^2} = 0,\ \forall \omega \in S^{N-1}$.

So for a fixed $\omega \in S^{N-1}$ by (12.13) we get

$$\frac{4}{(R_2 - R_1)^2} < \int_{R_1}^{R_2} |q(r\omega)|\,dr = \int_{R_1}^{R_2} r^{1-N} r^{N-1} |q(r\omega)|\,dr$$

$$\leq \left(\int_{R_1}^{R_2} r^{N-1} |q(r\omega)|\,dr\right) R_1^{1-N}. \tag{12.17}$$

That is

$$\int_{R_1}^{R_2} r^{N-1} |q(r\omega)|\,dr > \frac{4 R_1^{N-1}}{(R_2 - R_1)^2}, \tag{12.18}$$

$\forall \omega \in S^{N-1}$, and

$$\int_{S^{N-1}} \left(\int_{R_1}^{R_2} r^{N-1} |q(r\omega)|\,dr\right) d\omega > \left(\frac{4 R_1^{N-1}}{(R_2 - R_1)^2}\right)\left(\frac{2\pi^{N/2}}{\Gamma(N/2)}\right), \tag{12.19}$$

which by (12.1) proves (12.15). ■

We need the 2003 Yang's result [7],

Theorem 12.6 *Let $n \in \mathbb{N}$, $q(t) \in C([a,b])$. If there exists a $d \in (a,b)$ such that $x^{(2n)}(d) = 0$, where $x(t)$ is a solution of the following differential equation*

$$x^{(2n+1)} + q(t)x = 0, \tag{12.20}$$

satisfying $x^{(i)}(a) = x^{(i)}(b) = 0$, $i = 0, 1, \ldots, n-1$, $x(t) \neq 0$, $t \in (a,b)$, then

$$\int_a^b |q(s)| \, ds > \frac{n! 2^{n+1}}{(b-a)^{2n}}. \tag{12.21}$$

We give

Theorem 12.7 *Let $n \in \mathbb{N}$, $q \in C(\overline{A})$. If there exists a $d \in (R_1, R_2)$ such that $\frac{\partial^{2n} f}{\partial r^{2n}}(\partial B(0,d)) = 0$, where f is a solution of the PDE:*

$$\frac{\partial^{2n+1} f(x)}{\partial r^{2n+1}} + q(x)f(x) = 0, \; \forall x \in \overline{A}, \tag{12.22}$$

satifying $\frac{\partial^i f}{\partial r^i}(\partial B(0,R_1)) = \frac{\partial^i f}{\partial r^i}(\partial B(0,R_2)) = 0$, $i = 0, 1, \ldots, n-1$, and $f(x) \neq 0$ for any $x \in A$.
Then

$$\int_A |q(x)| \, dx > \frac{2^{n+2} \pi^{N/2} R_1^{N-1} n!}{\Gamma(N/2)(R_2 - R_1)^{2n}}. \tag{12.23}$$

Proof One can rewrite (12.22) as

$$\frac{\partial^{2n+1} f(r\omega)}{\partial r^{2n+1}} + q(r\omega)f(r\omega) = 0, \; \forall (r, \omega) \in [R_1, R_2] \times S^{N-1}, \tag{12.24}$$

where $q(\cdot\omega) \in C([R_1, R_2])$, $\forall \omega \in S^{N-1}$, such that for a $d \in (R_1, R_2)$ we have $\frac{\partial^{2n} f(d\omega)}{\partial r^{2n}} = 0$, $\forall \omega \in S^{N-1}$.

Furthermore it holds

$$\frac{\partial^i f}{\partial r^i}(R_1\omega) = \frac{\partial^i f}{\partial r^i}(R_2\omega) = 0, \forall \omega \in S^{N-1},$$

$i = 0, 1, \ldots, n-1$, with $f(r\omega) \neq 0$, for any $r \in (R_1, R_2)$, $\forall \omega \in S^{N-1}$.
So for a fixed $\omega \in S^{N-1}$ by (12.21) we get

$$\frac{n!2^{n+1}}{(R_2 - R_1)^{2n}} < \int_{R_1}^{R_2} |q\,(r\omega)|\,dr = \int_{R_1}^{R_2} r^{1-N} r^{N-1} |q\,(r\omega)|\,dr$$

$$\leq \left(\int_{R_1}^{R_2} r^{N-1} |q\,(r\omega)|\,dr \right) R_1^{1-N}. \tag{12.25}$$

That is

$$\int_{R_1}^{R_2} r^{N-1} |q\,(r\omega)|\,dr > \frac{n!2^{n+1} R_1^{N-1}}{(R_2 - R_1)^{2n}}, \tag{12.26}$$

$\forall \omega \in S^{N-1}$, and

$$\int_{S^{N-1}} \left(\int_{R_1}^{R_2} r^{N-1} |q\,(r\omega)|\,dr \right) d\omega > \left(\frac{n!2^{n+1} R_1^{N-1}}{(R_2 - R_1)^{2n}} \right) \left(\frac{2\pi^{N/2}}{\Gamma\,(N/2)} \right), \tag{12.27}$$

which by (12.1), proves (12.23). ■

We need the 2010 result of Cakmak [2], which follows

Theorem 12.8 *Let $n \in \mathbb{N}$, $n \geq 2$, $q\,(t) \in C\,([a, b])$. If the differential equation*

$$x^{(n)} + q\,(t)\,x = 0, \tag{12.28}$$

has a solution $x\,(t)$ satisfying the boundary value problem $x\,(a) = x\,(t_2) = \cdots = x\,(t_{n-1}) = x\,(b) = 0$, where $a = t_1 < t_2 < \cdots < t_{n-1} < t_n = b$ and $x\,(t) \neq 0$, for $t \in (t_k, t_{k+1})$, $k = 1, 2, \ldots, n - 1$, then

$$\int_a^b |q\,(s)|\,ds > \frac{(n - 2)!\,n^n}{(n - 1)^{n-1} (b - a)^{n-1}}. \tag{12.29}$$

We give

Theorem 12.9 *Let $n \in \mathbb{N}$, $n \geq 2$, $q \in C\,\left(\overline{A} \right)$. If the PDE*

$$\frac{\partial^n f\,(x)}{\partial r^n} + q\,(x)\,f\,(x) = 0, \ \forall x \in \overline{A}, \tag{12.30}$$

has a solution f satisfying the boundary value problem

$$f\,(\partial B\,(0, R_1)) = f\,(\partial B\,(0, t_2)) = \cdots = f\,(\partial B\,(0, t_{n-1})) = f\,(\partial B\,(0, R_2)) = 0, \tag{12.31}$$

where $R_1 = t_1 < t_2 < \cdots < t_{n-1} < t_n = R_2$, and $f\,(t\omega) \neq 0$, $\forall \omega \in S^{N-1}$, for any $t \in (t_k, t_{k+1})$, $k = 1, 2, \ldots, n - 1$.

Then

$$\int_A |q(x)| \, dx > \left(\frac{(n-2)! n^n R_1^{N-1}}{(n-1)^{n-1} (R_2 - R_1)^{n-1}} \right) \left(\frac{2\pi^{N/2}}{\Gamma(N/2)} \right). \tag{12.32}$$

Proof Similar to our earlier theorems in this chapter and based on Theorem 12.8. ■

We need the other 2010 result of Cakmak [2], which follows

Theorem 12.10 *Let us consider the following boundary value problem*

$$x^{(2n)} + q(t) x = 0, \tag{12.33}$$

$$x^{(2i)}(a) = x^{(2i)}(b) = 0, \ i = 0, 1, \ldots, n-1. \tag{12.34}$$

If $x(t)$ is a solution of (12.33) satisfying $x(t) \neq 0$ for any $t \in (a, b)$, then

$$\int_a^b |q(s)| \, ds > \frac{2^{2n}}{(b-a)^{2n-1}}. \tag{12.35}$$

We finish with

Theorem 12.11 *Let us consider on $\overline{A}$ the following boundary value problem*

$$\frac{\partial^{2n} f(x)}{\partial r^{2n}} + q(x) f(x) = 0, \ \forall x \in \overline{A}, \tag{12.36}$$

$$\frac{\partial^{2i} f}{\partial r^{2i}} (\partial (B(0, R_1))) = \frac{\partial^{2i} f}{\partial r^{2i}} (\partial (B(0, R_2))) = 0, \tag{12.37}$$

$i = 1, 2, \ldots, n-1.$
If $f \in C^{2n}(\overline{A})$ is a solution of (12.36) satisfying $f(x) \neq 0$ for any $x \in A$, then

$$\int_A |q(x)| \, dx > \left(\frac{2^{2n+1} R_1^{N-1}}{(R_2 - R_1)^{2n-1}} \right) \left(\frac{\pi^{N/2}}{\Gamma(N/2)} \right). \tag{12.38}$$

Proof Based on Theorem 12.10. ■

References

1. G.A. Anastassiou, Multivariate lyapunov inequalities. Appl. Math. Lett. **24**, 2167–2171 (2011)
2. D. Cakmak, Lyapunov-type integral inequalities for certain higher order differential equations. Appl. Math. Comput. **216**, 368–373 (2010)

3. A.M. Lyapunov, *Probleme General de la Stabilite du Mouvement*, (French Translation of a Russian paper dated 1893), Ann. Fac. Sci. Univ. Toulouse, 2 (1907) 27–247 (Reprinted as Ann. Math. Studies, No.17, Princeton, 1947)
4. N. Parhi, S. Panigrahi, On Liapunov-type inequality for third-order differential equations. J. Math. Anal. Appl. **233**(2), 445–460 (1999)
5. W. Rudin, *Real and Complex Analysis* (Mc. Graw Hill, New York, 1970)
6. A. Wintner, On the nonexistence of conjugate points. Amer. J. Math. **73**, 368–380 (1951)
7. X. Yang, On Lyapunov-type inequality for certain higher-order differential equations. Appl. Math. Comput. **134**(2–3), 307–317 (2003)

Chapter 13
Ostrowski Type Inequalities for Semigroups

Here we present Ostrowski type inequalities on Semigroups for various norms. We apply our results to the classical diffusion equation and its solution, the Gauss-Weierstrass singular integral. It follows [3].

13.1 Introduction

The main motivation here is the 1938, famous Ostrowski inequality, see [1, 9], which follows:

Theorem 13.1 *Let* $f : [a, b] \to \mathbb{R}$ *be continuous on* $[a, b]$ *and differentiable on* (a, b) *whose derivative* $f' : (a, b) \to \mathbb{R}$ *is bounded on* (a, b), *i.e.,* $\|f'\|_\infty :=$ $\sup_{t\epsilon(a,b)} |f'(t)| < \infty$. *Then*

$$\left| \frac{1}{b-a} \int_a^b f(t)\, dt - f(x) \right| \leq \left[\frac{1}{4} + \frac{\left(x - \frac{a+b}{2} \right)^2}{(b-a)^2} \right] (b-a) \|f'\|_\infty , \quad (13.1)$$

for any $x \in [a, b]$. *The constant* $\frac{1}{4}$ *is the best possible.*

We generalize here (13.1) to semigroups of linear operators and we expand to various directions. We give applications at the end.

13.2 Background

All this background comes from [4] (in general see also [6, 10]).

Let X a real or complex Banach space with elements $f, g, \ldots$ having norm $\|f\|, \|g\|, \ldots$ and let $\varepsilon(X)$ be the Banach algebra of endomorphisms of X.

© Springer International Publishing Switzerland 2016
G.A. Anastassiou, *Intelligent Comparisons: Analytic Inequalities*,
Studies in Computational Intelligence 609,
DOI 10.1007/978-3-319-21121-3_13

If $T \in \varepsilon(X)$, $\|T\|$ denotes the norm of T.

Definition 13.2 If $T(t)$ is an operator function on the non-negative real axis $0 \leq t < \infty$ to the Banach algebra $\varepsilon(X)$ satisfying the following conditions:

$$\begin{cases} (i)\ T(t_1 + t_2) = T(t_1)\,T(t_2)\,, \quad (t_1, t_2 \geq 0) \\ (ii)\ T(0) = I\ (I = identity\ operator)\,, \end{cases} \tag{13.2}$$

then $\{T(t)\,;\,0 \leq t < \infty\}$ is called a one-parameter semi-group of operators in $\varepsilon(X)$.

The semi-group $\{T(t)\,;\,0 \leq t < \infty\}$ is said to be of class C_0 if it satisfies the further property

$$(iii)\ s - \lim_{t \to 0+} T(t)\,f = f\ (f \in X) \tag{13.3}$$

referred to as the strong continuity of $T(t)$ at the origin.

In this chapter we shall assume that the family of bounded linear operators $\{T(t)\,;\,0 \leq t < \infty\}$ mapping X to itself is a semi-group of class C_0, thus all three conditions of the above definition are satisfied.

Proposition 13.3 (a) $\|T(t)\|$ is bounded on every finite subinterval of $[0, \infty)$.
(b) For each $f \in X$, the vector-valued function $T(t)\,f$ on $[0, \infty)$ is strongly continuous, thus vector-Riemann integrable on $[0, a]$, $a > 0$.

Definition 13.4 The infinitesimal generator A of the semi-group $\{T(t)\,;\,0 \leq t < \infty\}$ is defined by

$$Af = s - \lim_{\tau \to 0+} A_\tau f, \ A_\tau f = \frac{1}{\tau}[T(\tau) - I]\,f \tag{13.4}$$

whenever the limit exists; the domain of A, in symbols $D(A)$, is the set of elements f for which the limits exists.

Proposition 13.5 (a) $D(A)$ is a linear manifold in X and A is a linear operator.
(b) If $f \in D(A)$, then $T(t)\,f \in D(A)$ for each $t \geq 0$ and

$$\frac{d}{dt}T(t)\,f = AT(t)\,f = T(t)\,Af \quad (t \geq 0)\,; \tag{13.5}$$

furthermore,

$$T(t)\,f - f = \int_0^t T(u)\,Af\,du\ (t > 0)\,. \tag{13.6}$$

(c) $D(A)$ is dense in X, i.e., $\overline{D(A)} = X$, and A is a closed operator.

Definition 13.6 For $r = 0, 1, 2, \ldots$ the operator A^r is defined inductively by the relations $A^0 = I$, $A^1 = A$, and

$$D\left(A^r\right) = \left\{f; f \in D\left(A^{r-1}\right) \text{ and } A^{r-1}f \in D\left(A\right)\right\}$$

$$A^r f = A\left(A^{r-1}f\right) = s - \lim_{\tau \to 0+} A_\tau\left(A^{r-1}f\right)\left(f \in D\left(A^r\right)\right). \tag{13.7}$$

For the operator A^r and its domain $D\left(A^r\right)$ we have the following

Proposition 13.7 *(a)* $D\left(A^r\right)$ *is a linear subspace in* X *and* A^r *is a linear operator.* *(b) If* $f \in D\left(A^r\right)$, *so does* $T\left(t\right)f$ *for each* $t \geq 0$ *and*

$$\frac{d^r}{dt^r}T\left(t\right)f = A^r T\left(t\right)f = T\left(t\right)A^r f. \tag{13.8}$$

Furthermore

$$T\left(t\right)f - \sum_{k=0}^{r-1}\frac{t^k}{k!}A^k f = \frac{1}{(r-1)!}\int_0^t (t-u)^{r-1}T\left(u\right)A^r f\,du, \tag{13.9}$$

the Taylor's formula for semigroups.
Additionally it holds

$$|T\left(t\right) - I|^r = \int_0^t \int_0^t \cdots \int_0^t T\left(u_1 + u_2 + \ldots + u_r\right)A^r f\,du_1 du_2 \ldots du_r. \tag{13.10}$$

(c) $D\left(A^r\right)$ *is dense in* X *for* $r = 1, 2, \cdots$; *moreover,* $\bigcap_{r=1}^\infty D\left(A^r\right)$ *is dense in* X, A^r *is a closed operator.*

Integrals in (13.9) and (13.10) are vector valued Riemann integrals, see [4, 8].
Here we assume that $f \in D\left(A^r\right)$, $r \in \mathbb{N}$. Clearly here $\int_0^a T\left(t\right)f\,dt \in X$, where $a > 0$, see [10].

13.3 Ostrowski Type Inequalities for Semigroups

We present

Theorem 13.8 *Let* $f \in D\left(A\right)$, $a > 0$, *and denote*

$$\|\|T\left(\cdot\right)Af\|\|_{\infty,[0,a]} := \sup_{u \in [0,a]} \|T\left(u\right)Af\|. \tag{13.11}$$

Then

$$\left\|\frac{1}{a}\int_0^a T\left(t\right)f\,dt - T\left(t_0\right)f\right\| \leq \left(\frac{t_0^2 + (a - t_0)^2}{2a}\right)\|\|T\left(\cdot\right)Af\|\|_{\infty,[0,a]}, \tag{13.12}$$

for a fixed $t_0 \in [0, a]$.

Proof By (13.6) we get

$$T(t) f - f = \int_0^t T(u) A f du, \ t > 0, \tag{13.13}$$

and

$$T(t_0) f - f = \int_0^{t_0} T(u) A f du, \ t_0 > 0. \tag{13.14}$$

Therefore

$$E(t) f \ : \ = T(t) f - T(t_0) f$$
$$= \int_0^t T(u) A f du - \int_0^{t_0} T(u) A f du. \tag{13.15}$$

Hence in case of $t \geq t_0$ we obtain

$$E(t) f = \int_0^{t_0} T(u) A f du + \int_{t_0}^t T(u) A f du - \int_0^{t_0} T(u) A f du$$
$$= \int_{t_0}^t T(u) A f du. \tag{13.16}$$

In case of $t < t_0$ we derive

$$E(t) f = \int_0^t T(u) A f du - \int_0^t T(u) A f du - \int_t^{t_0} T(u) A f du$$
$$= -\int_t^{t_0} T(u) A f du = \int_{t_0}^t T(u) A f du. \tag{13.17}$$

Notice here $\| \|T(\cdot) A f\| \|_{\infty,[0,a]} < \infty$.

So if $t \geq t_0$ we get

$$\|E(t) f\| = \left\| \int_{t_0}^t T(u) A f du \right\|$$
$$\leq \int_{t_0}^t \|T(u) A f\| du \leq (t - t_0) \| \|T(\cdot) A f\| \|_{\infty,[0,a]}. \tag{13.18}$$

And in the case of $t < t_0$ we find that

$$\|E(t) f\| = \left\| \int_t^{t_0} T(u) A f du \right\|$$
$$\leq \int_t^{t_0} \|T(u) A f\| du \leq (t_0 - t) \| \|T(\cdot) A f\| \|_{\infty,[0,a]}. \tag{13.19}$$

That is

$$\| E(t) f \| \le |t - t_0| \, \| |T(\cdot) A f| \|_{\infty, [0,a]}, \tag{13.20}$$

$\forall t, t_0 \in [0, a]$.

It follows that

$$\left\| \frac{1}{a} \int_0^a T(t) f \, dt - T(t_0) f \right\| = \left\| \frac{1}{a} \int_0^a (T(t) f - T(t_0) f) \, dt \right\| \tag{13.21}$$

$$\le \frac{1}{a} \int_0^a \| (T(t) f - T(t_0) f) \| \, dt \le \frac{1}{a} \int_0^a \| E(t) f \| \, dt \tag{13.22}$$

$$\le \frac{\| |T(\cdot) A f| \|_{\infty, [0,a]}}{a} \int_0^a |t - t_0| \, dt$$

$$= \frac{\| |T(\cdot) A f| \|_{\infty, [0,a]}}{a} \left[\int_0^{t_0} (t_0 - t) \, dt + \int_{t_0}^a (t - t_0) \, dt \right]$$

$$= \frac{\| |T(\cdot) A f| \|_{\infty, [0,a]}}{a} \left[\frac{t_0^2}{2} + \frac{(a - t_0)^2}{2} \right] \tag{13.23}$$

$$= \| |T(\cdot) A f| \|_{\infty, [0,a]} \left(\frac{t_0^2 + (a - t_0)^2}{2a} \right),$$

proving (13.12). ∎

We continue with

Theorem 13.9 *Let* $f \in D(A^r)$, $r \in \mathbb{N}$, $a > 0$, *and denote*

$$\| |T(\cdot) A^r f| \|_{\infty, [0,a]} := \sup_{u \in [0,a]} \| T(u) A^r f \|, \tag{13.24}$$

$$\Delta(t) f := T(t) f - \sum_{k=1}^{r-1} \frac{t^k}{k!} A^k f, t > 0. \tag{13.25}$$

Then

$$\left\| \frac{1}{a} \int_0^a \Delta(t) f \, dt - \Delta(t_0) f \right\| \le \frac{1}{ar!} \left[\frac{\left(2r t_0^{r+1} + a^{r+1} \right)}{r+1} - t_0^r a \right]$$

$$\cdot \| |T(\cdot) A^r f| \|_{\infty, [0,a]}. \tag{13.26}$$

Proof By (13.9) we get

$$\Delta(t) f - f = \frac{1}{(r-1)!} \int_0^t (t - u)^{r-1} T(u) A^r f \, du, \tag{13.27}$$

and

$$\Delta\left(t_{0}\right)f-f=\frac{1}{(r-1)!}\int_{0}^{t_{0}}\left(t_{0}-u\right)^{r-1}T\left(u\right)A^{r}f du,$$

$\forall t, t_{0}\in[0,a]$.

Therefore we get

$$E\left(t\right)f:=\Delta\left(t\right)f-\Delta\left(t_{0}\right)f \tag{13.28}$$

$$=\frac{1}{(r-1)!}\left[\int_{0}^{t}\left(t-u\right)^{r-1}T\left(u\right)A^{r}f du-\int_{0}^{t_{0}}\left(t_{0}-u\right)^{r-1}T\left(u\right)A^{r}f du\right].$$

Case of $t\geq t_{0}$, we get

$$\left\|E\left(t\right)f\right\|=\frac{1}{(r-1)!}\left\|\int_{0}^{t_{0}}\left(t-u\right)^{r-1}T\left(u\right)A^{r}f du \right. \tag{13.29}$$

$$\left.+\int_{t_{0}}^{t}\left(t-u\right)^{r-1}T\left(u\right)A^{r}f du-\int_{0}^{t_{0}}\left(t_{0}-u\right)^{r-1}T\left(u\right)A^{r}f du\right\|$$

$$=\frac{1}{(r-1)!}\left\|\int_{0}^{t_{0}}\left[\left(t-u\right)^{r-1}-\left(t_{0}-u\right)^{r-1}\right]T\left(u\right)A^{r}f du\right.$$

$$\left.+\int_{t_{0}}^{t}\left(t-u\right)^{r-1}T\left(u\right)A^{r}f du\right\|$$

$$\leq\frac{\left\|\left\|T\left(\cdot\right)A^{r}f\right\|\right\|_{\infty,[0,a]}}{(r-1)!}\left[\int_{0}^{t_{0}}\left[\left(t-u\right)^{r-1}-\left(t_{0}-u\right)^{r-1}\right]du\right.$$

$$\left.+\int_{t_{0}}^{t}\left(t-u\right)^{r-1}du\right]$$

$$=\frac{\left\|\left\|T\left(\cdot\right)A^{r}f\right\|\right\|_{\infty,[0,a]}}{r!}\left(t^{r}-t_{0}^{r}\right). \tag{13.30}$$

So that

$$\left\|E\left(t\right)f\right\|\leq\frac{\left\|\left\|T\left(\cdot\right)A^{r}f\right\|\right\|_{\infty,[0,a]}}{r!}\left(t^{r}-t_{0}^{r}\right), \tag{13.31}$$

for $t\geq t_{0}$.

Case of $t<t_{0}$. We have

$$\left\|E\left(t\right)f\right\|=\frac{1}{(r-1)!}\left\|\int_{0}^{t}\left(t-u\right)^{r-1}T\left(u\right)A^{r}f du \right. \tag{13.32}$$

$$\left.-\int_{0}^{t}\left(t_{0}-u\right)^{r-1}T\left(u\right)A^{r}f du-\int_{t}^{t_{0}}\left(t_{0}-u\right)^{r-1}T\left(u\right)A^{r}f du\right\|$$

$$=\frac{1}{(r-1)!}\left\|\int_{0}^{t}\left[\left(t_{0}-u\right)^{r-1}-\left(t-u\right)^{r-1}\right]T\left(u\right)A^{r}f du\right.$$

$$+ \int_t^{t_0} (t_0 - u)^{r-1} T(u) A^r f du \Bigg\|$$

$$\leq \frac{\||T(\cdot) A^r f\||_{\infty,[0,a]}}{(r-1)!} \left[\int_0^t \left[(t_0 - u)^{r-1} - (t - u)^{r-1} \right] du \right.$$

$$+ \left. \int_t^{t_0} (t_0 - u)^{r-1} du \right] = \frac{\||T(\cdot) A^r f\||_{\infty,[0,a]}}{r!} (t_0^r - t^r).$$

So that

$$\|E(t) f\| \leq \frac{\||T(\cdot) A^r f\||_{\infty,[0,a]}}{r!} (t_0^r - t^r), \tag{13.33}$$

for $t < t_0$.

We have proved that

$$\|E(t) f\| \leq \frac{\||T(\cdot) A^r f\||_{\infty,[0,a]}}{r!} |t^r - t_0^r|, \tag{13.34}$$

$\forall t, t_0 \in [0, a]$.

Finally we see that

$$\left\| \frac{1}{a} \int_0^a \Delta(t) f dt - \Delta(t_0) f \right\| = \left\| \frac{1}{a} \int_0^a (\Delta(t) f - \Delta(t_0) f) dt \right\| \tag{13.35}$$

$$\leq \frac{1}{a} \int_0^a \|(\Delta(t) f - \Delta(t_0) f)\| dt = \frac{1}{a} \int_0^a \|E(t) f\| dt$$

$$\leq \frac{\||T(\cdot) A^r f\||_{\infty,[0,a]}}{ar!} \int_0^a |t^r - t_0^r| dt \tag{13.36}$$

$$= \frac{\||T(\cdot) A^r f\||_{\infty,[0,a]}}{ar!} \left[\int_0^{t_0} (t_0^r - t^r) dt + \int_{t_0}^a (t^r - t_0^r) dt \right]$$

$$= \frac{\||T(\cdot) A^r f\||_{\infty,[0,a]}}{ar!} \left[\frac{2rt_0^{r+1} + a^{r+1}}{r+1} - at_0^r \right],$$

proving (13.1). ∎

We give

Theorem 13.10 *Let* $f \in D(A)$, $a > 0$, $p, q > 1 : \frac{1}{p} + \frac{1}{q} = 1$. *Then*

$$\left\| \frac{1}{a} \int_0^a T(t) f dt - T(t_0) f \right\| \leq \left(\frac{t_0^2 + (a - t_0)^2}{2a} \right)^{\frac{1}{q}} \||T(\cdot) Af\||_{p,[0,a]}, \tag{13.37}$$

for a fixed $t_0 \in [0, a]$.

Proof We observe that

$$K := \frac{1}{a} \int_0^a T(t) f \, dt - T(t_0) f = \frac{1}{a} \int_0^a (T(t) f - T(t_0) f) \, dt$$

$$= \frac{1}{a} \left[\int_0^{t_0} (T(t) f - T(t_0) f) \, dt + \int_{t_0}^a (T(t) f - T(t_0) f) \, dt \right] \quad (13.38)$$

$$= \frac{1}{a} \left[\int_0^{t_0} \left(\int_{t_0}^t T(u) Af \, du \right) dt + \int_{t_0}^a \left(\int_{t_0}^t T(u) Af \, du \right) dt \right]$$

$$= \frac{1}{a} \int_0^a \left(\int_{t_0}^t T(u) Af \, du \right) dt. \quad (13.39)$$

That is we found

$$K = \frac{1}{a} \int_0^a \left(\int_{t_0}^t T(u) Af \, du \right) dt. \quad (13.40)$$

For $t \geq t_0$ we get

$$\left\| \int_{t_0}^t T(u) Af \, du \right\| \leq \int_{t_0}^t \| T(u) Af \| \, du$$

$$\leq (t - t_0)^{\frac{1}{q}} \left(\int_{t_0}^t \| T(u) Af \|^p \, du \right)^{\frac{1}{p}}. \quad (13.41)$$

For $t < t_0$ we obtain

$$\left\| \int_{t_0}^t T(u) Af \, du \right\| = \left\| \int_t^{t_0} T(u) Af \, du \right\|$$

$$\leq \int_t^{t_0} \| T(u) Af \| \, du \leq (t_0 - t)^{\frac{1}{q}} \left(\int_t^{t_0} \| T(u) Af \|^p \, du \right)^{\frac{1}{p}}. \quad (13.42)$$

Consequently we derive

$$\left\| \int_{t_0}^t T(u) Af \, du \right\| \leq |t - t_0|^{\frac{1}{q}} \left| \int_{t_0}^t \| T(u) Af \|^p \, du \right|^{\frac{1}{p}}, \quad (13.43)$$

$\forall t, t_0 \in [0, a]$.

Therefore

$$\|K\| \leq \frac{1}{a} \int_0^a \left\| \int_{t_0}^t T(u) \, Af \, du \right\| dt$$

$$\leq \frac{1}{a} \int_0^a |t - t_0|^{\frac{1}{q}} \left| \int_{t_0}^t \|T(u) \, Af\|^p \, du \right|^{\frac{1}{p}} dt$$

$$\leq \frac{1}{a} \left(\int_0^a |t - t_0| \, dt \right)^{\frac{1}{q}} \left(\int_0^a \left| \int_{t_0}^t \|T(u) \, Af\|^p \, du \right| dt \right)^{\frac{1}{p}}$$

$$=: (*) . \tag{13.44}$$

But it holds

$$\left| \int_{t_0}^t \|T(u) \, Af\|^p \, du \right| \leq \int_0^a \|T(u) \, Af\|^p \, du, \tag{13.45}$$

and

$$\left(\int_0^a \left| \int_{t_0}^t \|T(u) \, Af\|^p \, du \right| dt \right)^{\frac{1}{p}} \leq \left(\int_0^a \left(\int_0^a \|T(u) \, Af\|^p \, du \right) dt \right)^{\frac{1}{p}}$$

$$= a^{\frac{1}{p}} \|\|T(\cdot) \, Af\|\|_{p,[0,a]} . \tag{13.46}$$

Hence we have

$$(*) \leq \left(\frac{t_0^2 + (a - t_0)^2}{2} \right)^{\frac{1}{q}} a^{\frac{1}{p} - 1} \|\|T(\cdot) \, Af\|\|_{p,[0,a]}$$

$$= \left(\frac{t_0^2 + (a - t_0)^2}{2a} \right)^{\frac{1}{q}} \|\|T(\cdot) \, Af\|\|_{p,[0,a]} , \tag{13.47}$$

proving the claim. ∎

We also give

Theorem 13.11 *Let* $f \in D(A)$, $a > 0$. *Then*

$$\left\| \frac{1}{a} \int_0^a T(t) \, f \, dt - T(t_0) \, f \right\| \leq \|\|T(\cdot) \, Af\|\|_{1,[0,a]} . \tag{13.48}$$

Proof In (13.40) we found that

$$K \; := \; \frac{1}{a} \int_0^a T(t) f \, dt - T(t_0) f$$

$$= \frac{1}{a} \int_0^a \left(\int_{t_0}^t T(u) A f \, du \right) dt. \tag{13.49}$$

Hence

$$\|K\| \le \frac{1}{a} \int_0^a \left\| \int_{t_0}^t T(u) A f \, du \right\| dt \le \frac{1}{a} \int_0^a \left| \int_{t_0}^t \|T(u) A f\| \, du \right| dt$$

$$\le \frac{1}{a} \int_0^a \left(\int_0^a \|T(u) A f\| \, du \right) dt = \|\|T(\cdot) A f\|\|_{1,[0,a]}. \tag{13.50}$$

∎

We present

Theorem 13.12 Let $f \in D(A^r)$, $r \in \mathbb{N}$, $a > 0$, and denote

$$\Delta(t) f := T(t) f - \sum_{k=1}^{r-1} \frac{t^k}{k!} A^k f, t \ge 0.$$

Let $p, q > 1 : \frac{1}{p} + \frac{1}{q} = 1$. Then

$$\left\| \frac{1}{a} \int_0^a \Delta(t) f \, dt - \Delta(0) f \right\| = \left\| \frac{1}{a} \int_0^a \Delta(t) f \, dt - f \right\|$$

$$\le \frac{a^{r-\frac{1}{q}}}{(r-1)! \left(r + \frac{1}{p} \right) (p(r-1)+1)^{\frac{1}{p}}} \|\|T(\cdot) A^r f\|\|_{q,[0,a]}. \tag{13.51}$$

Proof By (13.9) we have $(t \ge 0)$

$$\|\Delta(t) f - \Delta(0) f\| = \|\Delta(t) f - f\|$$

$$= \left\| T(t) f - \sum_{k=0}^{r-1} \frac{t^k}{k!} A^k f \right\| = \frac{1}{(r-1)!} \left\| \int_0^t (t-u)^{r-1} T(u) A^r f \, du \right\|$$

$$\le \frac{1}{(r-1)!} \int_0^t (t-u)^{r-1} \|T(u) A^r f\| \, du \tag{13.52}$$

$$\le \frac{1}{(r-1)!} \left(\int_0^t (t-u)^{p(r-1)} \, du \right)^{\frac{1}{p}} \left(\int_0^t \|T(u) A^r f\|^q \, du \right)^{\frac{1}{q}}$$

$$\leq \frac{1}{(r-1)!} \frac{t^{(r-1)+\frac{1}{p}}}{(p(r-1)+1)^{\frac{1}{p}}} \left(\int_0^a \left\| T(u) A^r f \right\|^q du \right)^{\frac{1}{q}} \qquad (13.53)$$

$$= \frac{t^{r-\frac{1}{q}}}{(r-1)!(p(r-1)+1)^{\frac{1}{p}}} \left\| \| T(\cdot) A^r f \| \right\|_{q,[0,a]},$$

$0 \leq t \leq a$.

That is

$$\left\| \Delta(t) f - \Delta(0) f \right\| \leq \frac{t^{r-\frac{1}{q}}}{(r-1)!(p(r-1)+1)^{\frac{1}{p}}} \left\| \| T(\cdot) A^r f \| \right\|_{q,[0,a]}, \quad (13.54)$$

$0 \leq t \leq a$.

So we get

$$\left\| \frac{1}{a} \int_0^a \Delta(t) f dt - \Delta(0) f \right\| = \left\| \frac{1}{a} \int_0^a \Delta(t) f dt - f \right\|$$

$$= \frac{1}{a} \left\| \int_0^a (\Delta(t) f - f) dt \right\| \leq \frac{1}{a} \int_0^a \left\| \Delta(t) f - f \right\| dt \qquad (13.55)$$

$$\leq \frac{\left\| \| T(\cdot) A^r f \| \right\|_{q,[0,a]}}{a(r-1)!(p(r-1)+1)^{\frac{1}{p}}} \int_0^a \left(t^{r-\frac{1}{q}} \right) dt$$

$$= \frac{\left\| \| T(\cdot) A^r f \| \right\|_{q,[0,a]}}{(r-1)!(p(r-1)+1)^{\frac{1}{p}}} \frac{a^{r-\frac{1}{q}}}{\left(r-\frac{1}{q}+1 \right)} \qquad (13.56)$$

$$= \frac{\left\| \| T(\cdot) A^r f \| \right\|_{q,[0,a]}}{(r-1)!(p(r-1)+1)^{\frac{1}{p}}} \frac{a^{r-\frac{1}{q}}}{\left(r+\frac{1}{p} \right)},$$

proving the claim. ∎

We finally add

Theorem 13.13 *Let* $f \in D(A^r)$, $r \in \mathbb{N}$, $a > 0$. *Then*

$$\left\| \frac{1}{a} \int_0^a \Delta(t) f dt - f \right\| \leq \frac{a^{r-1}}{(r-1)!} \left\| \| T(\cdot) A^r f \| \right\|_{1,[0,a]}. \qquad (13.57)$$

Proof By (13.52) we get

$$\left\| \Delta(t) f - f \right\| \leq \frac{1}{(r-1)!} \int_0^t (t-u)^{r-1} \left\| T(u) A^r f \right\| du$$

$$\leq \frac{a^{r-1}}{(r-1)!} \int_0^a \left\| T(u) A^r f \right\| du = \frac{a^{r-1}}{(r-1)!} \left\| \| T(\cdot) A^r f \| \right\|_{1,[0,a]}. \qquad (13.58)$$

So that

$$\left\| \frac{1}{a} \int_0^a \Delta(t) f \, dt - f \right\| \leq \frac{1}{a} \int_0^a \| \Delta(t) f - f \| \, dt$$

$$\leq \frac{a^{r-1}}{(r-1)!} \left\| \| T(\cdot) A^r f \| \right\|_{1,[0,a]}, \qquad (13.59)$$

proving the claim. ■

Corollary 13.14 *Let* $f \in D(A^2)$, $a > 0$. *Then*

$$\left\| \frac{1}{a} \int_0^a (T(t) f - tAf) \, dt - f \right\| \leq \frac{2a^{3/2}}{5\sqrt{3}} \left\| \| T(\cdot) A^2 f \| \right\|_{2,[0,a]}. \qquad (13.60)$$

Proof Use (13.51) for $r = p = q = 2$. ■

13.4 Applications

Here see also [5]. It is known that the classical diffusion equation

$$\frac{\partial W}{\partial t} = \frac{\partial^2 W}{\partial x^2}, \quad -\infty < x < \infty, t > 0 \qquad (13.61)$$

with initial condition

$$\lim_{t \to 0+} W(x, t) = f(x), \qquad (13.62)$$

has under general conditions its solution given by

$$W(x, t, f) = [T(t) f](x) = \frac{1}{2\sqrt{\pi t}} \int_{-\infty}^{\infty} f(x + u) e^{-\frac{u^2}{4t}} \, du, \qquad (13.63)$$

the so called Gauss-Weierstrass singular integral.

The infinitesimal generator of the semigroup $\{T(t) ; 0 \leq t < \infty\}$ is $A = \partial^2/\partial x^2$ ([7], p. 578).

Here we suppose that $f, f^{(2k)}, k = 1, \ldots, r$, all belong to the Banach space $UCB(\mathbb{R})$, the space of bounded and uniformly continuous functions from $\mathbb{R}$ into itself, with norm

$$\| f \|_C := \sup_{x \in \mathbb{R}} | f(x) | . \qquad (13.64)$$

We define

$$\overline{\Delta}(t) f(x) := W(x, t, f) - \sum_{k=1}^{r-1} \frac{t^k}{k!} f^{(2k)}(x), \quad \text{for all } x \in \mathbb{R}. \tag{13.65}$$

In [2], p. 193 we found that

$$\left\| \left\| W\left(\cdot, t, f^{(2r)}\right) \right\|_C \right\|_\infty : = \sup_{t \in \mathbb{R}_+} \left\| W\left(\cdot, t, f^{(2r)}\right) \right\|_C \tag{13.66}$$

$$\leq \left\| f^{(2r)} \right\|_C < \infty.$$

First we apply Theorem 13.8 to obtain

Proposition 13.15 *Let $a > 0$. Then*

$$\left\| \frac{1}{a} \int_0^a W(\cdot, t, f) \, dt - W(\cdot, t_0, f) \right\|_C \leq \left(\frac{t_0^2 + (a - t_0)^2}{2a} \right) \left\| f^{(2)} \right\|_C, \tag{13.67}$$

for a fixed $t_0 \in [0, a]$.

Next we apply Theorem 13.9 to get

Proposition 13.16 *We have valid that*

$$\left\| \frac{1}{a} \int_0^a \overline{\Delta}(t) f \, dt - \overline{\Delta}(t_0) f \right\|_C \leq \frac{1}{ar!} \left[\left(\frac{2r t_0^{r+1} + a^{r+1}}{r+1} \right) - t_0^r a \right] \left\| f^{(2)} \right\|_C, \tag{13.68}$$

where $a > 0$.

An application of Theorem 13.10 follows.

Proposition 13.17 *Let $p, q > 1 : \frac{1}{p} + \frac{1}{q} = 1$, $a > 0$. Then*

$$\left\| \frac{1}{a} \int_0^a W(\cdot, t, f) \, dt - W(\cdot, t_0, f) \right\|_C \leq$$

$$\left(\frac{t_0^2 + (a - t_0)^2}{2a} \right)^{\frac{1}{q}} \left\| \left\| W\left(\cdot, \cdot, f^{(2)}\right) \right\| \right\|_{p, [0, a]}, \tag{13.69}$$

for a fixed $t_0 \in [0, a]$.

An application of Theorem 13.12 comes next

Proposition 13.18 *Let $p, q > 1 : \frac{1}{p} + \frac{1}{q} = 1$. Then*

$$\left\| \frac{1}{a} \int_0^a \overline{\Delta}(t)\, f\, dt - f \right\|_C \leq$$

$$\frac{a^{r-\frac{1}{q}}}{(r-1)!\left(r+\frac{1}{p}\right)(p(r-1)+1)^{\frac{1}{p}}} \left\| \left\| W\left(\cdot, \cdot, f^{(2r)}\right) \right\| \right\|_{q,[0,a]}, \qquad (13.70)$$

where $a > 0$.

We finish chapter with an application of Corollary 13.14.

Corollary 13.19 *We have valid that*

$$\left\| \frac{1}{a} \int_0^a \left(W(\cdot, t, f) - tf^{(2)} \right) dt - f \right\|_C \leq \frac{2a^{3/2}}{5\sqrt{3}} \left\| \left\| W\left(\cdot, \cdot, f^{(4)}\right) \right\| \right\|_{2,[0,a]}, \tag{13.71}$$

where $a > 0$.

References

1. G.A. Anastassiou, Ostrowski type inequalities. Proc. AMS **123**, 3775–3781 (1995)
2. G.A. Anastassiou, *Advanced Inequalities* (World Scientific, Singapore, 2011)
3. G.A. Anastassiou, Ostrowski inequalities for semigroups. Commun. Appl. Anal. **26**(4), 565–578 (2012)
4. P.L. Butzer, H. Berens, *Semi-Groups of Operators and Approximation* (Springer, New York, 1967)
5. P.L. Butzer, H.G. Tillmann, Approximation theorems for semi-groups of bounded linear transformations. Math. Ann. **140**, 256–262 (1960)
6. J.A. Goldstein, *Semigroups of Linear Operators and Applications* (Oxford University Press, Oxford, 1985)
7. E. Hille, R.S. Phillips, *Functional Analysis and Semigroups*, revised edn., pp. XII a. 808. American Mathematical Society Colloquium Publications, vol. 31. (American Mathematical Society Providence, RI, 1957)
8. Y. Katznelson, *An Introduction to Harmonic Analysis* (Dover, New York, 1976)
9. A. Ostrowski, Über die Absolutabweichung einer differentiebaren Funcktion von ihrem Integralmittelwert. Comments Math. Helv. **10**, 226–227 (1938)
10. G. Shilov, *Elementary Functional Analysis* (The MIT Press Cambridge, Massachusetts, 1974)

Chapter 14
About Ostrowski Inequalities for Cosine and Sine Operator Functions

Here we present Ostrowski type inequalities on Cosine and Sine Operator Functions for various norms. At the end we give applications. It follows [3].

14.1 Introduction

The main motivation here is the 1938, famous Ostrowski inequality, see [1, 2, 11], which follows:

Theorem 14.1 *Let* $f : [a, b] \to \mathbb{R}$ *be continuous on* $[a, b]$ *and differentiable on* (a, b) *whose derivative* $f' : (a, b) \to \mathbb{R}$ *is bounded on* (a, b), *i.e.* $\|f'\|_{\infty} :=$ $\sup\limits_{t \in (a,b)} |f'(t)| < \infty$. *Then*

$$\left| \frac{1}{b-a} \int_a^b f(t) \, dt - f(x) \right| \le \left[C + \frac{\left(x - \frac{a+b}{2}\right)^2}{(b-a)^2} \right] (b-a) \|f'\|_{\infty}, \quad (14.1)$$

for any $x \in [a, b]$, *where* C *is a positive constant. The constant* $C = \frac{1}{4}$ *is the best possible.*

We generalize here (14.1) to Cosine and Sine operator functions and we expand to various directions. At the end we give applications.

14.2 Background

(See [5, 7, 9, 10, 12]).

Let $(X, \|\cdot\|)$ be a real or complex Banach space. By definition, a cosine operator function is a family $\{C(t) ; t \in \mathbb{R}\}$ of bounded linear operators from X into itself, satisfying

© Springer International Publishing Switzerland 2016
G.A. Anastassiou, *Intelligent Comparisons: Analytic Inequalities*,
Studies in Computational Intelligence 609,
DOI 10.1007/978-3-319-21121-3_14

(i) $C(0) = I$, I the identity operator;

(ii) $C(t + s) + C(t - s) = 2C(t) C(s)$,   for all $t, s \in \mathbb{R}$;   (14.2)

     (the last product is composition)

(iii) $C(\cdot) f$ is continuous on $\mathbb{R}$,   for all $f \in X$.

Notice that $C(t) = C(-t)$, for all $t \in \mathbb{R}$.

The associated sine operator function $S(\cdot)$ is defined by

$$S(t) f := \int_0^t C(s) f ds, \quad \text{for all} \quad t \in \mathbb{R}, \quad \text{for all} \quad f \in X. \tag{14.3}$$

Notice $S(t) f \in X$ and is continuous in $t \in \mathbb{R}$.

The cosine operator function $C(\cdot)$ is such that $\|C(t)\| \le M e^{\omega t}$, for some $M \ge 1$, $\omega \ge 0$, for all $t \in \mathbb{R}_+$, here $\|\cdot\|$ is the norm of the operator.

The infinitesimal generator A of $C(\cdot)$ is the operator from X into itself defined as

$$Af := \lim_{t \to 0+} \frac{2}{t^2} (C(t) - I) f \tag{14.4}$$

with domain $D(A)$. The operator A is closed and $D(A)$ is dense in X, i.e. $\overline{D(A)} = X$, and satisfies

$$\int_0^t S(s) f ds \in D(A) \quad \text{and} \quad A \int_0^t S(s) f ds = C(t) f - f, \quad \text{for all} \quad f \in X. \tag{14.5}$$

Also it holds $A = C''(0)$, and $D(A)$ is the set of $f \in X : C(t) f$ is twice differentiable at $t = 0$; equivalently,

$$D(A) = \left\{ f \in X : C(\cdot) f \in C^2(\mathbb{R}, X) \right\}. \tag{14.6}$$

If $f \in D(A)$, then $C(t) f \in D(A)$, and $C''(t) f = C(t) Af = AC(t) f$, for all $t \in \mathbb{R}$; $C'(0) f = 0$, see [6, 13].

We define $A^0 = I$, $A^2 = A \circ A, \ldots, A^n = A \circ A^{n-1}$, $n \in \mathbb{N}$. Let $f \in D(A^n)$, then $C(t) f \in C^{2n}(\mathbb{R}, X)$, and $C^{(2n)}(t) f = C(t) A^n f = A^n C(t) f$, for all $t \in \mathbb{R}$, and $C^{(2k-1)}(0) f = 0$, $1 \le k \le n$, see [9].

For $f \in D(A^n)$, $t \in \mathbb{R}$, we have the cosine operator function's Taylor formula [9, 10] saying that

$$T_n(t) f := C(t) f - \sum_{k=0}^{n-1} \frac{t^{2k}}{(2k)!} A^k f = \int_0^t \frac{(t-s)^{2n-1}}{(2n-1)!} C(s) A^n f ds. \tag{14.7}$$

By integrating (14.7) we obtain the sine operator function's Taylor formula

$$M_n(t) f := S(t) f - ft - \frac{t^3}{3!} Af - \cdots - \frac{t^{2n-1}}{(2n-1)!} A^{n-1} f =$$

$$\int_0^t \frac{(t-s)^{2n}}{(2n)!} C(s) A^n f ds, \quad \text{for all } t \in \mathbb{R}, \tag{14.8}$$

all $f \in D(A^n)$.

Integrals in (14.7) and (14.8) are vector valued Riemann integrals, see [4, 8].

Here $f \in D(A^n)$, $n \in \mathbb{N}$.

Let $a > 0$ and $F \in C([0,a], X)$, then F is vector-Riemann integrable, see [12]. Clearly here $\int_0^a F(t) dt \in X$.

14.3 Ostrowski Type Inequalities

We first present results on the natural interval here $[0,a]$, $a > 0$.

Theorem 14.2 *Denote*

$$\|\|C(\cdot) A^n f\|\|_{\infty,[0,a]} := \sup_{t \in [0,a]} \|C(t) A^n f\|. \tag{14.9}$$

Here $t_0 \in [0,a]$.

Then

(i)

$$\left\| \frac{1}{a} \int_0^a T_n(t) f dt - T_n(t_0) f \right\| \leq \frac{\|\|C(\cdot) A^n f\|\|_{\infty,[0,a]}}{a(2n)!}.$$

$$\left[\left(\frac{4n t_0^{2n+1} + a^{2n+1}}{2n+1} \right) - a t_0^{2n} \right], \tag{14.10}$$

(ii)

$$\left\| \frac{1}{a} \int_0^a M_n(t) f dt - M_n(t_0) f \right\| \leq \frac{\|\|C(\cdot) A^n f\|\|_{\infty,[0,a]}}{a(2n+1)!}.$$

$$\left[\left(\frac{2(2n+1) t_0^{2(n+1)} + a^{2(n+1)}}{2(n+1)} \right) - a t_0^{2n+1} \right]. \tag{14.11}$$

Proof By (14.7) we have ($m := 2n - 1$)

$$T_n(t) f = \int_0^t \frac{(t-s)^m}{m!} C(s) A^n f ds,$$

and

$$T_n (t_0) f = \int_0^{t_0} \frac{(t_0 - s)^m}{m!} C (s) A^n f ds,$$

for any $t, t_0 \in [0, a]$.

We estimate

$$E_n (t) f := T_n (t) f - T_n (t_0) f.$$

Case of $t \geq t_0$: we have

$$\| E_n (t) f \| = \left\| \int_0^{t_0} \frac{(t - s)^m}{m!} C (s) A^n f ds + \right.$$

$$\left. \int_{t_0}^t \frac{(t - s)^m}{m!} C (s) A^n f ds - \int_0^{t_0} \frac{(t_0 - s)^m}{m!} C (s) A^n f ds \right\|$$

$$= \left\| \frac{1}{m!} \int_0^{t_0} \left((t - s)^m - (t_0 - s)^m \right) C (s) A^n f ds + \frac{1}{m!} \int_{t_0}^t (t - s)^m C (s) A^n f ds \right\|$$

$$\leq \frac{1}{m!} \left[\int_0^{t_0} \left((t - s)^m - (t_0 - s)^m \right) \| C (s) A^n f \| ds + \int_{t_0}^t (t - s)^m \| C (s) A^n f \| ds \right]$$

$$\leq \frac{\| \| C (\cdot) A^n f \| \|_{\infty,[0,a]}}{m!} \left[\int_0^{t_0} \left((t - s)^m - (t_0 - s)^m \right) ds + \int_{t_0}^t (t - s)^m ds \right]$$

$$= \frac{\| \| C (\cdot) A^n f \| \|_{\infty,[0,a]}}{(m + 1)!} \left[t^{m+1} - t_0^{m+1} \right].$$

That is

$$\| E_n (t) f \| \leq \frac{\| \| C (\cdot) A^n f \| \|_{\infty,[0,a]}}{(m + 1)!} \left[t^{m+1} - t_0^{m+1} \right],$$

for $t \geq t_0, t, t_0 \in [0, a]$.

Case of $t < t_0$: we similarly find that

$$\| E_n (t) f \| = \left\| \int_0^t \frac{(t - s)^m}{m!} C (s) A^n f ds - \int_0^{t_0} \frac{(t_0 - s)^m}{m!} C (s) A^n f ds \right\|$$

$$\leq \frac{\| \| C (\cdot) A^n f \| \|_{\infty,[0,a]}}{(m + 1)!} \left[t_0^{m+1} - t^{m+1} \right],$$

for $t < t_0, t, t_0 \in [0, a]$.

Therefore we get

$$\| E_n (t) f \| \leq \frac{\| \| C (\cdot) A^n f \| \|_{\infty,[0,a]}}{(m + 1)!} \left| t^{m+1} - t_0^{m+1} \right|,$$

for all $t, t_0 \in [0, a]$.

Next we observe

$$\left\| \frac{1}{a} \int_0^a T_n (t) f dt - T_n (t_0) f \right\| = \frac{1}{a} \left\| \int_0^a (T_n (t) f - T_n (t_0) f) dt \right\| \leq$$

$$\frac{1}{a} \int_0^a \| T_n (t) f - T_n (t_0) f \| dt = \frac{1}{a} \int_0^a \| E_n (t) f \| dt \leq$$

$$\frac{\| \| C (\cdot) A^n f \| \|_{\infty,[0,a]}}{(m + 1)! a} \int_0^a \left| t^{m+1} - t_0^{m+1} \right| dt =$$

$$\frac{\| \| C (\cdot) A^n f \| \|_{\infty,[0,a]}}{(m + 1)! a} \left[\left(\frac{2 (m + 1) t_0^{m+2} + a^{m+2}}{(m + 2)} \right) - a t_0^{m+1} \right].$$

So that

$$\left\| \frac{1}{a} \int_0^a T_n (t) f dt - T_n (t_0) f \right\| \leq$$

$$\frac{\| \| C (\cdot) A^n f \| \|_{\infty,[0,a]}}{(m + 1)! a} \left[\left(\frac{2 (m + 1) t_0^{m+2} + a^{m+2}}{(m + 2)} \right) - a t_0^{m+1} \right] =$$

$$\frac{\| \| C (\cdot) A^n f \| \|_{\infty,[0,a]}}{a (2n)!} \left[\left(\frac{4n t_0^{2n+1} + a^{2n+1}}{2n + 1} \right) - a t_0^{2n} \right],$$

proving (14.10).

Letting $m := 2n$, we similarly obtain

$$\left\| \frac{1}{a} \int_0^a M_n (t) f dt - M_n (t_0) f \right\| \leq$$

$$\frac{\| \| C (\cdot) A^n f \| \|_{\infty,[0,a]}}{(m + 1)! a} \left[\left(\frac{2 (m + 1) t_0^{m+2} + a^{m+2}}{(m + 2)} \right) - a t_0^{m+1} \right] =$$

$$\frac{\| \| C (\cdot) A^n f \| \|_{\infty,[0,a]}}{(2n + 1)! a} \left[\left(\frac{2 (2n + 1) t_0^{2(n+1)} + a^{2(n+1)}}{2 (n + 1)} \right) - a t_0^{2n+1} \right],$$

proving (14.11). ∎

When $n = 1$ we get

Proposition 14.3 *Let $f \in D(A)$, $t_0 \in [0, a]$, $a > 0$. We have valid*

(i)

$$\left\| \frac{1}{a} \int_0^a C(t) f \, dt - C(t_0) f \right\| \leq \frac{\| \| C(\cdot) Af \| \|_{\infty, [0,a]}}{2a} \left[\left(\frac{4t_0^3 + a^3}{3} \right) - a t_0^2 \right],$$

(14.12)

and

(ii)

$$\left\| \frac{1}{a} \int_0^a S(t) f \, dt - S(t_0) f + f \left(t_0 - \frac{a}{2} \right) \right\| \leq \frac{\| \| C(\cdot) Af \| \|_{\infty, [0,a]}}{6a}.$$

(14.13)

$$\left[\left(\frac{6t_0^4 + a^4}{4} \right) - a t_0^3 \right].$$

Next we call

$$T_n^*(t) f := C(t)(f) - \sum_{k=1}^{n-1} \frac{t^{2k}}{(2k)!} A^k f,$$

(14.14)

$t \in \mathbb{R}$.

Notice that

$$T_n^*(0) f = f,$$

(14.15)

furthermore we have

$$T_n^*(t) f - T_n^*(0) f = T_n^*(t) f - f = T_n(t) f,$$

(14.16)

for all $t \in \mathbb{R}$.

Next we give

Theorem 14.4 *Let $p, q > 1 : \frac{1}{p} + \frac{1}{q} = 1$. Then*

(i)

$$\left\| \frac{1}{a} \int_0^a T_n^*(t) f \, dt - f \right\| \leq$$

$$\frac{\| \| C(\cdot) A^n f \| \|_{q, [0,a]} \, a^{2n - \frac{1}{q}}}{(2n-1)! \, (p(2n-1) + 1)^{\frac{1}{p}} \left(2n + \frac{1}{p} \right)},$$

(14.17)

(ii)

$$\left\| \int_0^a M_n(t) \, f \, dt \right\| \le \frac{\| \|C(\cdot) A^n f\| \|_{q,[0,a]} \, a^{2n+\frac{1}{p}+1}}{(2n)! \, (2pn+1)^{\frac{1}{p}} \left(2n + \frac{1}{p} + 1\right)}. \tag{14.18}$$

Proof (i) Set $m = 2n - 1$, then

$$T_n(t) f = \int_0^t \frac{(t-s)^m}{m!} C(s) A^n f \, ds.$$

By Hölder's inequality we obtain

$$\|T_n(t) f\| = \left\| \int_0^t \frac{(t-s)^m}{m!} C(s) A^n f \, ds \right\| \le$$

$$\frac{1}{m!} \int_0^t (t-s)^m \, \|C(s) A^n f\| \, ds \le$$

$$\frac{1}{m!} \left(\int_0^t (t-s)^{pm} \, ds \right)^{\frac{1}{p}} \left(\int_0^t \|C(s) A^n f\|^q \, ds \right)^{\frac{1}{q}} \le$$

$$\frac{1}{m!} \left(\frac{t^{m+\frac{1}{p}}}{(pm+1)^{\frac{1}{p}}} \right) \left(\int_0^a \|C(s) A^n f\|^q \, ds \right)^{\frac{1}{q}} =$$

$$\frac{1}{m!} \left(\frac{t^{m+\frac{1}{p}}}{(pm+1)^{\frac{1}{p}}} \right) \| \|C(s) A^n f\| \|_{q,[0,a]},$$

for all $t \in [0, a]$.
That is we have

$$\|T_n(t) f\| \le \frac{1}{m!} \left(\frac{t^{m+\frac{1}{p}}}{(pm+1)^{\frac{1}{p}}} \right) \| \|C(s) A^n f\| \|_{q,[0,a]},$$

for all $t \in [0, a]$.
Therefore

$$\left\| \frac{1}{a} \int_0^a T_n^*(t) \, f \, dt - T_n^*(0) f \right\| = \left\| \frac{1}{a} \int_0^a \left(T_n^*(t) f - T_n^*(0) f \right) dt \right\| =$$

$$\left\| \frac{1}{a} \int_0^a T_n(t) \, f \, dt \right\| \le \frac{1}{a} \int_0^a \|T_n(t) f\| \, dt \le$$

$$\left(\frac{\||C\,(s)\,A^n f\||_{q,[0,a]}}{am!\,(pm+1)^{\frac{1}{p}}}\right)\int_0^a t^{m+\frac{1}{p}}\,dt = \frac{\||C\,(s)\,A^n f\||_{q,[0,a]}\,a^{m+\frac{1}{p}}}{m!\,(pm+1)^{\frac{1}{p}}\left(m+\frac{1}{p}+1\right)}.$$

We have proved that

$$\left\|\frac{1}{a}\int_0^a T_n^*\,(t)\,f\,dt - T_n^*\,(0)\,f\right\| \leq \frac{\||C\,(s)\,A^n f\||_{q,[0,a]}\,a^{m+\frac{1}{p}}}{m!\,(pm+1)^{\frac{1}{p}}\left(m+\frac{1}{p}+1\right)} =$$

$$\frac{\||C\,(s)\,A^n f\||_{q,[0,a]}\,a^{2n-\frac{1}{q}}}{(2n-1)!\,(p\,(2n-1)+1)^{\frac{1}{p}}\left(2n+\frac{1}{p}\right)},$$

establishing the claim.

(ii) Set $m = 2n$, then similarly we get

$$\left\|\frac{1}{a}\int_0^a M_n\,(t)\,f\,dt\right\| \leq \frac{\||C\,(s)\,A^n f\||_{q,[0,a]}\,a^{m+\frac{1}{p}}}{m!\,(pm+1)^{\frac{1}{p}}\left(m+\frac{1}{p}+1\right)} =$$

$$\frac{\||C\,(s)\,A^n f\||_{q,[0,a]}\,a^{2n+\frac{1}{p}}}{(2n)!\,(2pn+1)^{\frac{1}{p}}\left(2n+\frac{1}{p}+1\right)},$$

proving the claim. ■

Theorem 14.5 *We have valid that*

(i)

$$\left\|\frac{1}{a}\int_0^a T_n^*\,(t)\,f\,dt - f\right\| \leq \frac{a^{2n-1}}{(2n-1)!}\,\||C\,(\cdot)\,A^n f\||_{1,[0,a]}. \tag{14.19}$$

(ii)

$$\left\|\frac{1}{a}\int_0^a M_n\,(t)\,f\,dt\right\| \leq \frac{a^{2n+1}}{(2n)!}\,\||C\,(\cdot)\,A^n f\||_{1,[0,a]}. \tag{14.20}$$

Proof Set $m = 2n - 1$.

(i) We notice that

$$\|T_n\,(t)\,f\| = \left\|\int_0^t \frac{(t-s)^m}{m!}\,C\,(s)\,A^n f\,ds\right\| \leq$$

$$\frac{1}{m!}\int_0^t (t-s)^m\,\|C\,(s)\,A^n f\|\,ds \leq \frac{a^m}{m!}\int_0^t \|C\,(s)\,A^n f\|\,ds =$$

$$\frac{a^m}{m!} \int_0^a \|C(s) A^n f\| \, ds \le \frac{a^m}{m!} \|\|C(s) A^n f\|\|_{1,[0,a]},$$

for all $t \in [0, a]$,
i.e.

$$\|T_n(t) f\| \le \frac{a^m}{m!} \|\|C(s) A^n f\|\|_{1,[0,a]},$$

for all $t \in [0, a]$.
Therefore

$$\left\| \frac{1}{a} \int_0^a T_n(t) f \, dt \right\| \le \frac{1}{a} \int_0^a \|T_n(t) f\| \, dt \le \frac{a^m}{m!} \|\|C(s) A^n f\|\|_{1,[0,a]}$$

$$= \frac{a^{2n-1}}{(2n-1)!} \|\|C(s) A^n f\|\|_{1,[0,a]},$$

proving the claim.
(ii) Set $m = 2n$. Then similarly we get

$$\left\| \frac{1}{a} \int_0^a M_n(t) f \, dt \right\| \le \frac{a^{2n}}{(2n)!} \|\|C(s) A^n f\|\|_{1,[0,a]},$$

proving the claim. ∎

The case of $n = p = q = 2$ follows

Corollary 14.6 (to Theorem 14.4) *We have valid that*

(i)

$$\left\| \frac{1}{a} \int_0^a \left(C(t) f - \frac{t^2}{2} A f \right) dt - f \right\| \le \frac{\|\|C(\cdot) A^2 f\|\|_{2,[0,a]} a^{3.5}}{27\sqrt{7}}. \tag{14.21}$$

(ii)

$$\left\| \int_0^a \left(S(t) f - ft - \frac{t^3}{6} A f \right) dt \right\| \le \frac{\|\|C(\cdot) A^2 f\|\|_{2,[0,a]} a^{5.5}}{396}. \tag{14.22}$$

Next, we prove Ostrowski type inequality on the general interval $[a, b]$, $0 \le a < b$.

Theorem 14.7 *Let $t_0 \in [a, b]$. We have valid that*

(i)

$$\left\| \frac{1}{b-a} \int_a^b T_n(t) f \, dt - T_n(t_0) f \right\| \le \frac{\|\|C(\cdot) A^n f\|\|_{\infty,[0,b]}}{(2n)!(b-a)}.$$

$$\left[\frac{4nt_0^{2n+1}}{2n+1} - t_0^{2n}(a+b) + \left(\frac{a^{2n+1}+b^{2n+1}}{2n+1}\right)\right], \qquad (14.23)$$

(ii)

$$\left\|\frac{1}{b-a}\int_a^b M_n(t)\,fdt - M_n(t_0)\,f\right\| \leq \frac{\|\|C(\cdot)\,A^n f\|\|_{\infty,[0,b]}}{(2n+1)!\,(b-a)}\cdot$$

$$\left[\left(\frac{2(2n+1)\,t_0^{2(n+1)} + a^{2(n+1)} + b^{2(n+1)}}{2(n+1)}\right) - (a+b)\,t_0^{2n+1}\right]. \qquad (14.24)$$

Proof (i) Set $m := 2n - 1$, and

$$E_n(t)\,f := T_n(t)\,f - T_n(t_0)\,f,$$

where $t, t_0 \in [a, b]$, with $a \geq 0$.
As in the proof of Theorem 14.2 we obtain

$$\|E_n(t)\,f\| \leq \frac{\|\|C(\cdot)\,A^n f\|\|_{\infty,[0,b]}}{(m+1)!}\left|t^{m+1} - t_0^{m+1}\right|, \quad \text{for all} \quad t, t_0 \in [a, b].$$

Next we observe

$$\left\|\frac{1}{b-a}\int_a^b T_n(t)\,fdt - T_n(t_0)\,f\right\| = \frac{1}{b-a}\left\|\int_a^b (T_n(t)\,f - T_n(t_0)\,f)\,dt\right\| \leq$$

$$\frac{1}{b-a}\int_a^b \|T_n(t)\,f - T_n(t_0)\,f\|\,dt = \frac{1}{b-a}\int_a^b \|E_n(t)\,f\|\,dt \leq$$

$$\frac{\|\|C(\cdot)\,A^n f\|\|_{\infty,[0,b]}}{(m+1)!\,(b-a)}\int_a^b \left|t^{m+1} - t_0^{m+1}\right|\,dt =$$

$$\frac{\|\|C(\cdot)\,A^n f\|\|_{\infty,[0,b]}}{(m+1)!\,(b-a)}\left[\int_a^{t_0}\left(t_0^{m+1} - t^{m+1}\right)\,dt + \int_{t_0}^b\left(t^{m+1} - t_0^{m+1}\right)\,dt\right].$$

$$= \frac{\|\|C(\cdot)\,A^n f\|\|_{\infty,[0,b]}}{(m+1)!\,(b-a)}\left[t_0^{m+1}\,[2t_0 - a - b] + \left(\frac{a^{m+2} + b^{m+2} - 2t_0^{m+2}}{m+2}\right)\right] =$$

$$\frac{\|\|C(\cdot)\,A^n f\|\|_{\infty,[0,b]}}{(2n)!\,(b-a)}\left[t_0^{2n}\,[2t_0 - a - b] + \left(\frac{a^{2n+1} + b^{2n+1} - 2t_0^{2n+1}}{2n+1}\right)\right] =$$

$$\frac{\|\|C(\cdot)\,A^n f\|\|_{\infty,[0,b]}}{(2n)!\,(b-a)}\left[\frac{4nt_0^{2n+1}}{2n+1} - t_0^{2n}(a+b) + \left(\frac{a^{2n+1} + b^{2n+1}}{2n+1}\right)\right],$$

proving the claim.
(ii) Set $m = 2n$. Then similarly we find

$$\left\| \frac{1}{b-a} \int_a^b M_n(t) f \, dt - M_n(t_0) f \right\| \le$$

$$= \frac{\| \| C(\cdot) A^n f \| \|_{\infty,[0,b]}}{(m+1)!(b-a)} \left[t_0^{m+1} [2t_0 - a - b] + \left(\frac{a^{m+2} + b^{m+2} - 2t_0^{m+2}}{m+2} \right) \right] =$$

$$\frac{\| \| C(\cdot) A^n f \| \|_{\infty,[0,b]}}{(2n+1)!(b-a)} \left[\left(\frac{2(2n+1) t_0^{2(n+1)} + a^{2(n+1)} + b^{2(n+1)}}{2(n+1)} \right) - (a+b) t_0^{2n+1} \right],$$

proving the claim. ∎

When $n = 1$ we get

Proposition 14.8 *Let* $f \in D(A)$, $0 \le a < b$, $t_0 \in [0, a]$. *We have valid*

(i)

$$\left\| \frac{1}{b-a} \int_a^b C(t) f \, dt - C(t_0) f \right\| \le \frac{\| \| C(\cdot) A f \| \|_{\infty,[0,b]}}{2(b-a)}. \tag{14.25}$$

$$\left[\frac{4t_0^3}{3} - t_0^2 (a+b) + \left(\frac{a^3 + b^3}{3} \right) \right],$$

(ii)

$$\left\| \frac{1}{b-a} \int_a^b S(t) f \, dt - S(t_0) f + f \left(t_0 - \frac{a+b}{2} \right) \right\| \le \frac{\| \| C(\cdot) A f \| \|_{\infty,[0,b]}}{6(b-a)}. \tag{14.26}$$

$$\left[\left(\frac{6t_0^4 + a^4 + b^4}{4} \right) - (a+b) t_0^3 \right].$$

14.4 Applications

(See [7], p. 121).

Let X be the Banach space of odd, 2π-periodic real functions in the space of bounded uniformly continuous functions from $\mathbb{R}$ into itself: $BUC(\mathbb{R})$. Let $A := \frac{d^2}{dx^2}$ with $D(A^n) = \{f \in X : f^{(2k)} \in X, k = 1, \ldots, n\}$, $n \in \mathbb{N}$. A generates a Cosine function C^* given by

$$[C^*(t) f](x) = \frac{1}{2} [f(x+t) + f(x-t)], \quad \text{for all} \quad x, t \in \mathbb{R}. \tag{14.27}$$

The corresponding Sine function S^* is given by

$$[S^*(t) f](x) = \frac{1}{2}\left[\int_0^t f(x+s)\,ds + \int_0^t f(x-s)\,ds\right], \quad \text{for all} \quad x, t \in \mathbb{R}.$$
$$(14.28)$$

Here we consider $f \in D(A^n)$, $n \in \mathbb{N}$, as above. By (14.7) we obtain

$$\overline{T_n}(t) f := \frac{1}{2}[f(\cdot + t) + f(\cdot - t)] - \sum_{k=0}^{n-1} \frac{t^{2k}}{(2k)!} f^{(2k)}$$

$$= \int_0^t \frac{(t-s)^{2n-1}}{2(2n-1)!}\left[f^{(2n)}(\cdot + s) + f^{(2n)}(\cdot - s)\right] ds, \quad \text{for all} \quad t \in \mathbb{R}. \quad (14.29)$$

By (14.8) we get

$$\overline{M_n}(t) f := \frac{1}{2}\left[\int_0^t f(\cdot + s)\,ds + \int_0^t f(\cdot - s)\,ds\right] - \sum_{k=1}^{n} \frac{t^{2k-1}}{(2k-1)!} f^{(2(k-1))}$$

$$= \int_0^t \frac{(t-s)^{2n}}{2(2n)!}\left[f^{(2n)}(\cdot + s) + f^{(2n)}(\cdot - s)\right] ds, \quad \text{for all} \quad t \in \mathbb{R}. \quad (14.30)$$

Let $g \in BUC(\mathbb{R})$, we define $\|g\| = \|g\|_\infty := \sup_{x \in \mathbb{R}} |g(x)| < \infty$.

Notice also that

$$\left\|\left\|C^*(s) A^n f\right\|_\infty\right\|_\infty = \left\|\left\|C^*(s) f^{(2n)}\right\|_\infty\right\|_\infty$$

$$= \frac{1}{2}\left\|\left\|\left(f^{(2n)}(\cdot + s) + f^{(2n)}(\cdot - s)\right)\right\|_\infty\right\|_\infty$$

$$\leq \frac{1}{2}\left[\left\|\left\|f^{(2n)}(\cdot + s)\right\|_\infty\right\|_\infty + \left\|\left\|f^{(2n)}(\cdot - s)\right\|_\infty\right\|_\infty\right] \leq \left\|f^{(2n)}\right\|_\infty < \infty.$$
$$(14.31)$$

We have the following applications

Corollary 14.9 (to Theorem 14.7) *Let* $t_0 \in [a, b]$. *Then*

(i)

$$\left\|\frac{1}{b-a}\int_a^b \overline{T_n}(t) f\,dt - \overline{T_n}(t_0) f\right\|_\infty \leq \frac{\|f^{(2n)}\|_\infty}{(2n)!\,(b-a)}\cdot$$

$$\left[\frac{4nt_0^{2n+1}}{2n+1} - t_0^{2n}(a+b) + \left(\frac{a^{2n+1} + b^{2n+1}}{2n+1}\right)\right], \quad (14.32)$$

(ii)

$$\left\| \frac{1}{b-a} \int_a^b \overline{M_n}(t) f \, dt - \overline{M_n}(t_0) f \right\|_\infty \leq \frac{\|f^{(2n)}\|_\infty}{(2n+1)!(b-a)} \cdot$$

$$\left[\left(\frac{2(2n+1) t_0^{2(n+1)} + a^{2(n+1)} + b^{2(n+1)}}{2(n+1)} \right) - (a+b) t_0^{2n+1} \right]. \quad (14.33)$$

Corollary 14.10 *(to Proposition 14.8) Let $f \in X : f^{(2)} \in X, 0 \leq a < b, t_0 \in [a,b]$. Then*

(i)

$$\left\| \frac{1}{b-a} \int_a^b [f(\cdot + t) + f(\cdot - t)] \, dt - [f(\cdot + t_0) + f(\cdot - t_0)] \right\|_\infty \leq \quad (14.34)$$

$$\frac{\|f^{(2)}\|_\infty}{(b-a)} \left[\frac{4t_0^3}{3} - t_0^2(a+b) + \left(\frac{a^3 + b^3}{3} \right) \right],$$

(ii)

$$\left\| \frac{1}{b-a} \int_a^b S^*(t) f \, dt - S^*(t_0) f + f(\cdot) \cdot \left(t_0 - \frac{a+b}{2} \right) \right\|_\infty \leq \quad (14.35)$$

$$\frac{\|f^{(2)}\|_\infty}{6(b-a)} \left[\left(\frac{6t_0^4 + a^4 + b^4}{4} \right) - (a+b) t_0^3 \right].$$

We finish with

Corollary 14.11 (to Corollary 14.6) *We have valid that*

(i)

$$\left\| \frac{1}{2a} \int_0^a \left[f(\cdot + t) + f(\cdot - t) - t^2 f^{(2)}(\cdot) \right] dt - f(\cdot) \right\|_\infty \leq$$

$$\frac{\left\| \|f^{(4)}(\cdot + t) + f^{(4)}(\cdot - t)\|_\infty \right\|_{2,[0,a]} a^{3.5}}{54\sqrt{7}}, \quad (14.36)$$

(ii)

$$\left\| \int_0^a \left([S^*(t) f](x) - f(x) t - \frac{t^3}{6} f^{(2)}(x) \right) dt \right\|_{\infty,x} \leq$$

$$\frac{\left\| \|[C^*(t) f^{(4)}](x)\|_{\infty,x} \right\|_{2,[0,a],t} a^{5.5}}{396}. \quad (14.37)$$

References

1. G.A. Anastassiou, Ostrowski type inequalities. Proc. AMS **123**, 3775–3781 (1995)
2. G.A. Anastassiou, *Quantitative Approximations* (Chapman & Hall /CRC, Boca Raton, 2001)
3. G.A. Anastassiou, Ostrowski inequalities for cosine and sine operator functions. Mat. Vesnik **64**(4), 336–346 (2012)
4. P.L. Butzer, H. Berens, *Semi-Groups of Operators and Approximation* (Springer, New York, 1967)
5. D.-K. Chyan, S.-Y. Shaw, P. Piskarev, On maximal regularity and semivariation of cosine operator functions. J. Lond. Math. Soc. **2**(59), 1023–1032 (1999)
6. H.O. Fattorini, Ordinary differential equations in linear topological spaces. I. J. Diff. Equat. **5**, 72–105 (1968)
7. J.A. Goldstein, *Semigroups of Linear Operators and Applications* (Oxford University Press, Oxford, 1985)
8. Y. Katznelson, *An Introduction to Harmonic Analysis* (Dover, New York, 1976)
9. B. Nagy, On cosine operator functions in Banach spaces. Acta Scientarium Mathematicarum Szeged **36**, 281–289 (1974)
10. B. Nagy, Approximation theorems for Cosine operator functions. Acta Math. Acad. Scientarium Hung., Tomus **29**(1–2), 69–76 (1977)
11. A. Ostrowski, Über die Absolutabweichung einer differentiebaren Funktion von ihrem Integralmittelwert. Comment. Math. Helv. **10**, 226–227 (1938)
12. G. Shilov, *Elementary Functional Analysis* (The MIT Press Cambridge, Massachusetts, 1974)
13. M. Sova, *Cosine Operator Functions* (Rozprawy Matematyczne XLIX, Warszawa, 1966)

Chapter 15
About Hilbert-Pachpatte Inequalities for Semigroups, Cosine and Sine Operator Functions

Here we present Hilbert-Pachpatte type general L_p inequalities regarding Semigroups, Cosine and Sine Operator functions. We apply inequalities to specific cases of them. It follows [2].

15.1 Introduction

The results here are motivated by the original Hilbert double integral inequality.

Theorem 15.1 ([9, Theorem 316]) *If $p > 1$, $q = \frac{p}{p-1}$ and*

$$\int_0^\infty f^p(x)\,dx \le F, \int_0^\infty g^q(y)\,dy \le G,$$

then

$$\int_0^\infty \int_0^\infty \frac{f(x)g(y)}{x+y}dxdy < \frac{\pi}{\sin\frac{\pi}{p}}F^{\frac{1}{p}}G^{\frac{1}{q}}, \tag{15.1}$$

where f, g are nonnegative measurable functions, unless

$$f \equiv 0 \quad or \quad g \equiv 0.$$

The constant $\pi cosec \frac{\pi}{p}$ is the best possible in (15.1).

Also the results here are motivated by

Theorem 15.2 (Pachpatte [14, Theorem 1]) *Let $n \ge 1$ and $0 \le k \le n - 1$ be integers. Let $u \in C^n([0, x])$ and $v \in C^n([0, y])$, where $x > 0$, $y > 0$, and let $u^{(j)}(0) = v^{(j)}(0) = 0$ for $j \in \{0, \ldots, n-1\}$. Then*

© Springer International Publishing Switzerland 2016
G.A. Anastassiou, *Intelligent Comparisons: Analytic Inequalities*,
Studies in Computational Intelligence 609,
DOI 10.1007/978-3-319-21121-3_15

$$\int_0^x \int_0^y \frac{\left|u^{(k)}(s)\right|\left|v^{(k)}(t)\right|}{s^{2n-2k-1}+t^{2n-2k-1}}\,ds\,dt \leq M(n,k,x,y)\cdot \tag{15.2}$$

$$\left(\int_0^x (x-s)\left(u^{(n)}(s)\right)^2 ds\right)^{\frac{1}{2}}\left(\int_0^y (y-t)\left(v^{(n)}(t)\right)^2 dt\right)^{\frac{1}{2}},$$

where

$$M(n,k,x,y) := \frac{1}{2}\frac{\sqrt{xy}}{[(n-k-1)!]^2(2n-2k-1)}.$$

In this chapter we present Hilbert-Pachpatte inequalities for Semigroups, Cosine Operator functions and Sine Operator functions. At the end of each results section we give applications.

15.2 Semigroups Background

All this background comes from [3] (in general see also [8]).

Let X a real or complex Banach space with elements $f, g, \ldots$ having norm $\|f\|$, $\|g\|, \ldots$ and let $\varepsilon(x)$ be the Banach algebra of endomorphisms of X.

If $T \in \varepsilon(X)$, $\|T\|$ denotes the norm of T.

Definition 15.3 If $T(t)$ is an operator function on the non-negative real axis $0 \leq t < \infty$ to the Banach algebra $\varepsilon(X)$ satisfying the following conditions:

$$\begin{cases} (i)\, T(t_1+t_2) = T(t_1)\, T(t_2), & (t_1, t_2 \geq 0) \\ (ii)\, T(0) = I(I = \text{ identity operator}), \end{cases} \tag{15.3}$$

then $\{T(t)\,;\, 0 \leq t < \infty\}$ is called a one-parameter semi-group of operators in $\varepsilon(X)$.

The semi-group $\{T(t)\,;\, 0 \leq t < \infty\}$ is said to be of class C_0 if it satisfies the further property

$$(iii)\, s - \lim_{t\to 0+} T(t)\, f = f, \quad (f \in X) \tag{15.4}$$

referred to as the strong continuity of $T(t)$ at the origin.

In this chapter we shall assume that the family of bounded linear operators $\{T(t)\,;\, 0 \leq t < \infty\}$ mapping X to itself is a semi-group of class C_0, thus that all three conditions of the above definition are satisfied.

Proposition 15.4 *(a) $\|T(t)\|$ is bounded on every finite subinterval of $[0, \infty)$.*

(b) For each $f \in X$, the vector-valued function $T(t)\, f$ on $[0, \infty)$ is strongly continuous.

Definition 15.5 The infinitesimal generator A of the semi-group $\{T(t); 0 \le t < \infty\}$ is defined by

$$Af = s - \lim_{\tau \to 0+} A_\tau f, \quad A_\tau f = \frac{1}{\tau}[T(\tau) - I]f \tag{15.5}$$

whenever the limit exists; the domain of A, in symbols $D(A)$, is the set of elements f for which the limit exists.

Proposition 15.6 (a) $D(A)$ is a linear manifold in X and A is a linear operator.
(b) If $f \in D(A)$, then $T(t)f \in D(A)$ for each $t \ge 0$ and

$$\frac{d}{dt}T(t)f = AT(t)f = T(t)Af \quad (t \ge 0); \tag{15.6}$$

furthermore,

$$T(t)f - f = \int_0^t T(u)Af\,du \quad (t > 0). \tag{15.7}$$

(c) $D(A)$ is dense in X, i.e. $\overline{D(A)} = X$, and A is a closed operator.

Definition 15.7 For $\tau = 0, 1, 2, \ldots$ the operator A^τ is defined inductively by the relations $A^0 = I$, $A^1 = A$, and

$$D(A^r) = \left\{f; f \in D\left(A^{r-1}\right) \text{ and } A^{r-1}f \in D(A)\right\} \tag{15.8}$$

$$A^r f = A\left(A^{r-1}f\right) = s - \lim_{\tau \to 0+} A_\tau\left(A^{r-1}f\right) \quad (f \in D(A^r)).$$

For the operator A^r and its domain $D(A^r)$ we have the following

Proposition 15.8 (a) $D(A^r)$ is a linear subspace in X and A^r is a linear operator.
(b) If $f \in D(A^r)$, so does $T(t)f$ for each $t \ge 0$ and

$$\frac{d^r}{dt^r}T(t)f = A^r T(t)f = T(t)A^r f. \tag{15.9}$$

Moreover

$$\Delta_r(t)f := T(t)f - \sum_{k=0}^{r-1} \frac{t^k}{k!}A^k f = \frac{1}{(r-1)!}\int_0^t (t-u)^{r-1}T(u)A^r f\,du, \tag{15.10}$$

the Taylor's formula for semigroups.
(c) $D(A^r)$ is dense in X for $r = 1, 2, \ldots$; furthermore, $\cap_{r=1}^\infty D(A^r)$ is dense in X. A^r is a closed operator.

The integral in (15.10) is a vector valued Riemann integral, see [3, 11].

15.3 Hilbert-Pachpatte Inequalities for Semigroups

We give our first main result

Theorem 15.9 *Let $\varepsilon > 0$; $p, q > 1 : \frac{1}{p} + \frac{1}{q} = 1$. Let $\{T_i(t_i); 0 \le t_i < \infty\}$, $i = 1, 2$, mapping X into itself, semi-groups of class C_0. Here A_i, $i = 1, 2$, is the infinitesimal generator of T_i; we take $f_i \in D\left(A_i^{r_i}\right)$, where $r_1, r_2 \in \mathbb{N}$.*
    Denote

$$\Delta_{r_i}(t_i) f_i := T_i(t_i) f_i - \sum_{k_i=0}^{r_i-1} \frac{t^{k_i}}{k_i!} A_i^{k_i} f_i, \tag{15.11}$$

$i = 1, 2$; $t_i \in [0, a_i]$, $a_i > 0$.
    Then

$$\int_0^{a_1} \int_0^{a_2} \frac{\left\| \Delta_{r_1}(t_1) f_1 \right\| \left\| \Delta_{r_2}(t_2) f_2 \right\| dt_1 dt_2}{\left[\varepsilon + \frac{t_1^{p(r_1-1)+1}}{p(p(r_1-1)+1)} + \frac{t_2^{q(r_2-1)+1}}{q(q(r_2-1)+1)} \right]} \le \frac{a_1 a_2}{(r_1-1)!(r_2-1)!} \cdot \tag{15.12}$$

$$\left(\int_0^{a_1} \left\| T_1(u_1) A_1^{r_1} f_1 \right\|^q du_1 \right)^{\frac{1}{q}} \left(\int_0^{a_2} \left\| T_2(u_2) A_2^{r_2} f_2 \right\|^p du_2 \right)^{\frac{1}{p}}.$$

Proof We have

$$\Delta_{r_i}(t_i) f_i \; : \; = T_i(t_i) f_i - \sum_{k_i=0}^{r_i-1} \frac{t^{k_i}}{k_i!} A_i^{k_i} f_i \tag{15.13}$$

$$= \frac{1}{(r_i-1)!} \int_0^{t_i} (t_i - u_i)^{r_i-1} T_i(u_i) A_i^{r_i} f_i du_i,$$

$i = 1, 2$.
    That is

$$\left\| \Delta_{r_i}(t_i) f_i \right\| = \frac{1}{(r_i-1)!} \left\| \int_0^{t_i} (t_i - u_i)^{r_i-1} T_i(u_i) A_i^{r_i} f_i du_i \right\|$$

$$\le \frac{1}{(r_i-1)!} \int_0^{t_i} (t_i - u_i)^{r_i-1} \left\| T_i(u_i) A_i^{r_i} f_i \right\| du_i, \tag{15.14}$$

$i = 1, 2$.
    Consequently we obtain

$$\left\| \Delta_{r_1}(t_1) f_1 \right\| \le \frac{1}{(r_1-1)!} \cdot$$

$$\left(\int_0^{t_1} (t_1 - u_1)^{p(r_1-1)} \, du_1\right)^{\frac{1}{p}} \left(\int_0^{t_1} \left\|T_1(u_1) A_1^{r_1} f_1\right\|^q \, du_1\right)^{\frac{1}{q}}$$

$$= \frac{1}{(r_1 - 1)!} \frac{t_1^{r_1-\frac{1}{q}}}{(p(r_1-1)+1)^{\frac{1}{p}}} \left(\int_0^{t_1} \left\|T_1(u_1) A_1^{r_1} f_1\right\|^q \, du_1\right)^{\frac{1}{q}}, \qquad (15.15)$$

and

$$\left\|\Delta_{r_2}(t_2) f_2\right\| \le \frac{1}{(r_2 - 1)!} \cdot$$

$$\left(\int_0^{t_2} (t_2 - u_2)^{q(r_2-1)} \, du_2\right)^{\frac{1}{q}} \left(\int_0^{t_2} \left\|T_2(u_2) A_2^{r_2} f_2\right\|^p \, du_2\right)^{\frac{1}{p}}$$

$$= \frac{1}{(r_2 - 1)!} \frac{t_2^{r_2-\frac{1}{p}}}{(q(r_2-1)+1)^{\frac{1}{q}}} \left(\int_0^{t_2} \left\|T_2(u_2) A_2^{r_2} f_2\right\|^p \, du_2\right)^{\frac{1}{p}}. \qquad (15.16)$$

Therefore we get

$$\left\|\Delta_{r_1}(t_1) f_1\right\| \left\|\Delta_{r_2}(t_2) f_2\right\| \le \frac{1}{(r_1 - 1)!\,(r_2 - 1)!} \cdot$$

$$\frac{t_1^{\frac{p\left(r_1-\frac{1}{q}\right)}{p}} t_2^{\frac{q\left(r_2-\frac{1}{p}\right)}{q}}}{(p(r_1-1)+1)^{\frac{1}{p}} (q(r_2-1)+1)^{\frac{1}{q}}} \left(\int_0^{t_1} \left\|T_1(u_1) A_1^{r_1} f_1\right\|^q \, du_1\right)^{\frac{1}{q}} \cdot$$

$$\left(\int_0^{t_2} \left\|T_2(u_2) A_2^{r_2} f_2\right\|^p \, du_2\right)^{\frac{1}{p}} \qquad (15.17)$$

(using Young's inequality for $a, b \ge 0$, $a^{\frac{1}{p}} b^{\frac{1}{q}} \le \frac{a}{p} + \frac{b}{q}$)

$$\le \frac{1}{(r_1 - 1)!\,(r_2 - 1)!} \left[\frac{t_1^{p\left(r_1-\frac{1}{q}\right)}}{p(p(r_1-1)+1)} + \frac{t_2^{q\left(r_2-\frac{1}{p}\right)}}{q(q(r_2-1)+1)}\right] \cdot$$

$$\left(\int_0^{t_1} \left\|T_1(u_1) A_1^{r_1} f_1\right\|^q \, du_1\right)^{\frac{1}{q}} \left(\int_0^{t_2} \left\|T_2(u_2) A_2^{r_2} f_2\right\|^p \, du_2\right)^{\frac{1}{p}} \qquad (15.18)$$

$$\le \frac{1}{(r_1 - 1)!\,(r_2 - 1)!} \left[\varepsilon + \frac{t_1^{p(r_1-1)+1}}{p(p(r_1-1)+1)} + \frac{t_2^{q(r_2-1)+1}}{q(q(r_2-1)+1)}\right]$$

$$\left(\int_0^{t_1} \left\| T_1\left(u_1\right) A_1^{r_1} f_1 \right\|^q du_1\right)^{\frac{1}{q}} \left(\int_0^{t_2} \left\| T_2\left(u_2\right) A_2^{r_2} f_2 \right\|^p du_2\right)^{\frac{1}{p}}. \qquad (15.19)$$

So far we have proved

$$\frac{\left\| \Delta_{r_1}\left(t_1\right) f_1 \right\| \left\| \Delta_{r_2}\left(t_2\right) f_2 \right\|}{\left[\varepsilon + \frac{t_1^{p(r_1-1)+1}}{p(p(r_1-1)+1)} + \frac{t_2^{q(r_2-1)+1}}{q(q(r_2-1)+1)} \right]} \le \frac{1}{\left(r_1-1\right)!\left(r_2-1\right)!} \cdot$$

$$\left(\int_0^{t_1} \left\| T_1\left(u_1\right) A_1^{r_1} f_1 \right\|^q du_1\right)^{\frac{1}{q}} \left(\int_0^{t_2} \left\| T_2\left(u_2\right) A_2^{r_2} f_2 \right\|^p du_2\right)^{\frac{1}{p}}. \qquad (15.20)$$

Therefore we obtain

$$\int_0^{a_1} \int_0^{a_2} \frac{\left\| \Delta_{r_1}\left(t_1\right) f_1 \right\| \left\| \Delta_{r_2}\left(t_2\right) f_2 \right\| dt_1 dt_2}{\left[\varepsilon + \frac{t_1^{p(r_1-1)+1}}{p(p(r_1-1)+1)} + \frac{t_2^{q(r_2-1)+1}}{q(q(r_2-1)+1)} \right]} \le$$

$$\frac{1}{\left(r_1-1\right)!\left(r_2-1\right)!} \left(\int_0^{a_1} \left(\int_0^{t_1} \left\| T_1\left(u_1\right) A_1^{r_1} f_1 \right\|^q du_1\right)^{\frac{1}{q}} dt_1\right) \qquad (15.21)$$

$$\left(\int_0^{a_2} \left(\int_0^{t_2} \left\| T_2\left(u_2\right) A_2^{r_2} f_2 \right\|^p du_2\right)^{\frac{1}{p}} dt_2\right)$$

$$\le \frac{1}{\left(r_1-1\right)!\left(r_2-1\right)!} \left(\int_0^{a_1} \left(\int_0^{a_1} \left\| T_1\left(u_1\right) A_1^{r_1} f_1 \right\|^q du_1\right)^{\frac{1}{q}} dt_1\right)$$

$$\left(\int_0^{a_2} \left(\int_0^{a_2} \left\| T_2\left(u_2\right) A_2^{r_2} f_2 \right\|^p du_2\right)^{\frac{1}{p}} dt_2\right)$$

$$= \frac{a_1 a_2}{\left(r_1-1\right)!\left(r_2-1\right)!} \cdot$$

$$\left(\int_0^{a_1} \left\| T_1\left(u_1\right) A_1^{r_1} f_1 \right\|^q du_1\right)^{\frac{1}{q}} \left(\int_0^{a_2} \left\| T_2\left(u_2\right) A_2^{r_2} f_2 \right\|^p du_2\right)^{\frac{1}{p}}, \qquad (15.22)$$

proving the claim.   ■

The case of $p = q = r_1 = r_2 = 2$ follows

Corollary 15.10 (to Theorem 15.9) *Let* $\varepsilon > 0$; $\{T_i(t_i); 0 \le t_i < \infty\}$, $i = 1, 2$, *mapping* X *into itself, semi-groups of class* C_0. *Here* A_i, $i = 1, 2$, *is the infinitesimal generator of* T_i; *we take* $f_i \in D\left(A_i^2\right)$.
    Denote

$$\Delta_2^*(t_i) f_i := T_i(t_i) f_i - f_i - t_i A_i f_i, \tag{15.23}$$

$i = 1, 2$; $t_i \in [0, a_i]$, $a_i > 0$.
    Then

$$\int_0^{a_1} \int_0^{a_2} \frac{\left\| \Delta_2^*(t_1) f_1 \right\| \left\| \Delta_2^*(t_2) f_2 \right\| dt_1 dt_2}{\left[\varepsilon + \frac{t_1^3 + t_2^3}{6} \right]} \le \tag{15.24}$$

$$a_1 a_2 \left(\int_0^{a_1} \left\| T_1(u_1) A_1^2 f_1 \right\|^2 du_1 \right)^{\frac{1}{2}} \left(\int_0^{a_2} \left\| T_2(u_2) A_2^2 f_2 \right\|^2 du_2 \right)^{\frac{1}{2}}.$$

We finish this section with

Application 15.11 *(see also [4]) It is known the classical diffusion equation*

$$\frac{\partial W}{\partial t} = \frac{\partial^2 W}{\partial x^2}, \quad -\infty < x < \infty, \quad t > 0 \tag{15.25}$$

with initial condition

$$\lim_{t \to 0+} W(x, t) = f(x), \tag{15.26}$$

has under general conditions its solution given by

$$W(x, t, f) = [T(t) f](x) = \frac{1}{2\sqrt{\pi t}} \int_{-\infty}^{\infty} f(x + u) e^{-\frac{u^2}{4t}} du, \tag{15.27}$$

then so called Gauss-Weierstrass singular integral.

    The infinitesimal generator of the semigroup $\{T(t); 0 \le t < \infty\}$ *is* $A = \frac{\partial^2}{\partial x^2}$ *([10], p. 578).*

    Here we suppose that f, $f^{(2k)}$, $k = 1, \ldots, r$, *all belong to the Banach space* $UCB(\mathbb{R})$, *the space of bounded and uniformly continuous functions from* $\mathbb{R}$ *into itself, with norm*

$$\| f \|_C := \sup_{x \in \mathbb{R}} |f(x)|. \tag{15.28}$$

Here we define

$$\overline{\Delta_r}(t) f(x) := W(x, t, f) - \sum_{k=0}^{r-1} \frac{t^k}{k!} f^{(2k)}(x), \quad \text{for all } x \in \mathbb{R}. \tag{15.29}$$

From [1], pp. 247–248 we have

$$\left\| W\left(\cdot, t, f^{(2r)}\right) \right\|_C \le \left\| f^{(2r)} \right\|_C < \infty, \tag{15.30}$$

$\forall\, t \in \mathbb{R}.$

First we apply (15.12) of Theorem 15.9 to obtain

$$\int_0^{a_1} \int_0^{a_2} \frac{\left\| \overline{\Delta_r}\,(t_1)\, f_1 \right\|_C \left\| \overline{\Delta_r}\,(t_2)\, f_2 \right\|_C dt_1 dt_2}{\left[\varepsilon + \frac{t_1^{p(r-1)+1}}{p(p(r-1)+1)} + \frac{t_2^{q(r-1)+1}}{q(q(r-1)+1)} \right]} \le \frac{a_1 a_2}{((r-1)!)^2} \tag{15.31}$$

$$\left(\int_0^{a_1} \left\| W\left(\cdot, u_1, f_1^{(2r)}\right) \right\|_C^q du_1 \right)^{\frac{1}{q}} \left(\int_0^{a_2} \left\| W\left(\cdot, u_2, f_2^{(2r)}\right) \right\|_C^p du_2 \right)^{\frac{1}{p}},$$

where $f_1, f_2 \in UCB\,(\mathbb{R})$.

Consider the case of $p = q = r = 2$.
We apply (15.23) of Corollary 15.10.
Denote

$$\overline{\Delta_2^*}\,(t_i)\, f_i := W\,(\cdot, t_i, f_i) - f_i - t_i\, f_i^{(2)}, \tag{15.32}$$

$i = 1, 2;\ t_i \in [0, a_i],\ a_i > 0.$
Then

$$\int_0^{a_1} \int_0^{a_2} \frac{\left\| \overline{\Delta_2^*}\,(t_1)\, f_1 \right\|_C \left\| \overline{\Delta_2^*}\,(t_2)\, f_2 \right\|_C dt_1 dt_2}{\left[\varepsilon + \frac{t_1^3 + t_2^3}{6} \right]} \le \tag{15.33}$$

$$a_1 a_2 \left(\int_0^{a_1} \left\| W\left(\cdot, u_1, f_1^{(4)}\right) \right\|_C^2 du_1 \right)^{\frac{1}{2}} \left(\int_0^{a_2} \left\| W\left(\cdot, u_2, f_2^{(4)}\right) \right\|_C^2 du_2 \right)^{\frac{1}{2}},$$

where $f_1, f_2 \in UCB\,(\mathbb{R})$ for $r = 2$.

15.4 Cosine and Sine Operator Functions Background (see [5, 8, 12, 13])

Let $(X, \|\cdot\|)$ be a real or complex Banach space. By definition, a cosine operator function is a family $\{C\,(t)\,;\, t \in \mathbb{R}\}$ of bounded linear operators from X into itself, satisfying

$$(i)\, C\,(0) = I, I \text{ the identity operator;} \tag{15.34}$$

$$(ii)\, C\,(t+s) + C\,(t-s) = 2C\,(t)\,C\,(s), \quad \text{for all } t, s \in \mathbb{R};$$

(the last product is composition)

$$(iii) \, C \, (\cdot) \, f \text{ is continuous on } \mathbb{R}, \text{ for all } f \in X. \tag{15.35}$$

Notice that

$$C \, (t) = C \, (-t), \quad \text{for all } t \in \mathbb{R}.$$

The associated sine operator function $S \, (\cdot)$ is defined by

$$S \, (t) \, f := \int_0^t C \, (s) \, f ds, \text{ for all } t \in \mathbb{R}, \text{ for all } f \in X. \tag{15.36}$$

The cosine operator function $C \, (\cdot)$ is such that $\|C \, (t)\| \leq M e^{\omega |t|}$, for some $M \geq 1$, $\omega \geq 0$, for all $t \in \mathbb{R}$, here $\|\cdot\|$ is the norm of the operator.

The infinitesimal generator A of $C \, (\cdot)$ is the operator from X into itself defined as

$$Af := \lim_{t \to 0+} \frac{2}{t^2} \, (C \, (t) - I) \, f \tag{15.37}$$

with domain $D \, (A)$. The operator A is closed and $D \, (A)$ is dense in X, i.e. $\overline{D \, (A)} = X$, and one has

$$\int_0^t S \, (s) \, f ds \in D \, (A) \text{ and } A \int_0^t S \, (s) \, f ds = C \, (t) \, f - f, \text{ for all } f \in X. \tag{15.38}$$

Also one has $A = C'' \, (0)$, and $D \, (A)$ is the set of $f \in X$ such that $C \, (t) \, f$ is twice differentiable at $t = 0$; equivalently,

$$D \, (A) = \left\{ f \in X : C \, (\cdot) \, f \in C^2 \, (\mathbb{R}, X) \right\}. \tag{15.39}$$

If $f \in D \, (A)$, then $C \, (t) \, f \in D \, (A)$, and $C'' \, (t) \, f = C \, (t) \, Af = AC \, (t) \, f$, for all $t \in \mathbb{R}$; $C' \, (0) \, f = 0$, see [7, 15].

We define $A^0 = I, A^2 = A \circ A, \ldots, A^n = A \circ A^{n-1}, n \in \mathbb{N}$. Let $f \in D \, (A^n)$, then $C \, (t) \, f \in C^{2n} \, (\mathbb{R}, X)$, and $C^{(2n)} \, (t) \, f = C \, (t) \, A^n f = A^n C \, (t) \, f$, for all $t \in \mathbb{R}$, and $C^{(2k-1)} \, (0) \, f = 0, 1 \leq k \leq n$, see [12].

For $f \in D \, (A^n), t \in \mathbb{R}$, we have the cosine operator function's Taylor formula [12, 13] saying that

$$T_n \, (t) \, f := C \, (t) \, f - \sum_{k=0}^{n-1} \frac{t^{2k}}{(2k)!} A^k f = \int_0^t \frac{(t-s)^{2n-1}}{(2n-1)!} C \, (s) \, A^n f ds. \tag{15.40}$$

By integrating (15.40) we obtain the sine operator function's Taylor formula (see [1], p. 198)

$$M_n(t) f := S(t) f - ft - \frac{t^3}{3!} Af - \cdots - \frac{t^{2n-1}}{(2n-1)!} A^{n-1} f =$$

$$\int_0^t \frac{(t-s)^{2n}}{(2n)!} C(s) A^n f \, ds, \text{ for all } t \in \mathbb{R}, \tag{15.41}$$

all $f \in D(A^n)$.

The integrals in (15.40) and (15.41) are vector valued Riemann integrals, see [3, 11].

We give

Theorem 15.12 *Let $\varepsilon > 0$; $p, q > 1 : \frac{1}{p} + \frac{1}{q} = 1$. Let $\{C_i(t_i)\,; t_i \in \mathbb{R}\}$, $i = 1, 2$, mapping X into itself, cosine operator functions. Here A_i, $i = 1, 2$, is the infinitesimal generator of C_i; we take $f_i \in D\left(A_i^{n_i}\right)$, where $n_1, n_2 \in \mathbb{N}$.*
Denote

$$T_{n_i}(t_i) f_i := C_i(t_i) f_i - \sum_{k_i=0}^{n_i-1} \frac{t_i^{2k_i}}{(2k_i)!} A_i^{k_i} f_i, \tag{15.42}$$

$i = 1, 2$; $t_i \in [0, a_i]$, $a_i > 0$.
Then

$$\int_0^{a_1} \int_0^{a_2} \frac{\left\| T_{n_1}(t_1) f_1 \right\| \left\| T_{n_2}(t_2) f_2 \right\| dt_1 dt_2}{\left[\varepsilon + \frac{t_1^{p(2n_1-1)+1}}{p(p(2n_1-1)+1)} + \frac{t_2^{q(2n_2-1)+1}}{q(q(2n_2-1)+1)} \right]} \leq \frac{a_1 a_2}{(2n_1-1)!\,(2n_2-1)!} \cdot \tag{15.43}$$

$$\left(\int_0^{a_1} \left\| C_1(u_1) A_1^{n_1} f_1 \right\|^q du_1 \right)^{\frac{1}{q}} \left(\int_0^{a_2} \left\| C_2(u_2) A_2^{n_2} f_2 \right\|^p du_2 \right)^{\frac{1}{p}}.$$

Proof Similar to the proof of Theorem 15.9. ∎

The case of $p = q = n_1 = n_2 = 2$ follows

Corollary 15.13 (to Theorem 15.12) *Let $\varepsilon > 0$; $\{C_i(t_i)\,; t_i \in \mathbb{R}\}$, $i = 1, 2$, mapping X into itself, cosine operator functions. Here A_i, $i = 1, 2$, is the infinitesimal generator of C_i; we take $f_i \in D\left(A_i^2\right)$.*
Denote

$$T_2(t_i) f_i := C_i(t_i) f_i - f_i - \frac{t_i^2}{2} A_i f_i, \tag{15.44}$$

$i = 1, 2$; $t_i \in [0, a_i]$, $a_i > 0$.
Then

$$\int_0^{a_1} \int_0^{a_2} \frac{\left\| T_2(t_1) f_1 \right\| \left\| T_2(t_2) f_2 \right\| dt_1 dt_2}{\left[\varepsilon + \frac{t_1^7 + t_2^7}{14} \right]} \leq \tag{15.45}$$

$$\frac{a_1 a_2}{36} \left(\int_0^{a_1} \left\| C_1 \left(u_1 \right) A_1^2 f_1 \right\|^2 du_1 \right)^{\frac{1}{2}} \left(\int_0^{a_2} \left\| C_2 \left(u_2 \right) A_2^2 f_2 \right\|^2 du_2 \right)^{\frac{1}{2}}.$$

The corresponding result for sine operator functions follows

Theorem 15.14 *Let* $\varepsilon > 0$; $p, q > 1 : \frac{1}{p} + \frac{1}{q} = 1$. *Let* $\{C_i (t_i) ; t_i \in \mathbb{R}\}$, $i = 1, 2$, *mapping* X *into itself, cosine operator functions. Here* A_i, $i = 1, 2$, *is the infinitesimal generator of* C_i; *we take* $f_i \in D \left(A_i^{n_i} \right)$, *where* $n_1, n_2 \in \mathbb{N}$. *Let also the associated sine operator functions* $\{S_i (t_i) ; t_i \in \mathbb{R}\}$, $i = 1, 2$.*
    Denote

$$M_{n_i} (t_i) f_i := S_i (t_i) f_i - \sum_{k_i=1}^{n_i} \frac{t_i^{2k_i-1}}{(2k_i - 1)!} A_i^{k_i-1} f_i, \tag{15.46}$$

$i = 1, 2$; $t_i \in [0, a_i]$, $a_i > 0$.
    Then

$$\int_0^{a_1} \int_0^{a_2} \frac{\left\| M_{n_1} (t_1) f_1 \right\| \left\| M_{n_2} (t_2) f_2 \right\| dt_1 dt_2}{\left[\varepsilon + \frac{t_1^{(2pn_1+1)}}{p(2pn_1+1)} + \frac{t_2^{(2qn_2+1)}}{q(2qn_2+1)} \right]} \le \frac{a_1 a_2}{(2n_1)! \, (2n_2)!}. \tag{15.47}$$

$$\left(\int_0^{a_1} \left\| C_1 \left(u_1 \right) A_1^{n_1} f_1 \right\|^q du_1 \right)^{\frac{1}{q}} \left(\int_0^{a_2} \left\| C_2 \left(u_2 \right) A_2^{n_2} f_2 \right\|^p du_2 \right)^{\frac{1}{p}}.$$

Proof Similar to Theorem 15.9. ∎

The case of $p = q = n_1 = n_2 = 2$ follows

Corollary 15.15 (to Theorem 15.14) *Let* $\varepsilon > 0$ *and* $\{C_i (t_i) ; t_i \in \mathbb{R}\}$, $i = 1, 2$, *mapping* X *into itself, cosine operator functions. Here* A_i, $i = 1, 2$, *is the infinitesimal generator of* C_i; *we take* $f_i \in D \left(A_i^2 \right)$. *Let also the associated sine operator functions* $\{S_i (t_i) ; t_i \in \mathbb{R}\}$, $i = 1, 2$.*
    Denote

$$M_2 (t_i) f_i := S_i (t_i) f_i - f_i t_i - \frac{t_i^3}{6} A_i f_i, \tag{15.48}$$

$i = 1, 2$; $t_i \in [0, a_i]$, $a_i > 0$.
    Then

$$\int_0^{a_1} \int_0^{a_2} \frac{\left\| M_2 (t_1) f_1 \right\| \left\| M_2 (t_2) f_2 \right\| dt_1 dt_2}{\left[\varepsilon + \frac{t_1^9 + t_2^9}{18} \right]} \le \tag{15.49}$$

$$\frac{a_1 a_2}{576} \left(\int_0^{a_1} \left\| C_1 \left(u_1 \right) A_1^2 f_1 \right\|^2 du_1 \right)^{\frac{1}{2}} \left(\int_0^{a_2} \left\| C_2 \left(u_2 \right) A_2^2 f_2 \right\|^2 du_2 \right)^{\frac{1}{2}}.$$

Application 15.16 *(see [8], p. 121) Let X be the Banach space of odd, 2π -periodic real functions in the space of bounded uniformly continuous functions from $\mathbb{R}$ into itself: $BUC(\mathbb{R})$. Let $A := \frac{d^2}{dx^2}$ with $D(A^n) = \{f \in X : f^{(2k)} \in X, k = 1, \ldots, n\}$, $n \in \mathbb{N}$. A generates a cosine functions C^* given by*

$$C^*(t) f(x) = \frac{1}{2} [f(x+t) + f(x-t)], \ \forall x, t \in \mathbb{R}. \tag{15.50}$$

The corresponding sine function S^* is given by

$$S^*(t) f(x) = \frac{1}{2} \left[\int_0^t f(x+s) ds + \int_0^t f(x-s) ds \right], \ \forall x, t \in \mathbb{R}.$$

Here we consider $f \in D(A^n)$, $n \in \mathbb{N}$, as above. By (15.40) we get

$$T_n^*(t) f := \frac{1}{2} [f(\cdot + t) + f(\cdot - t)] - \sum_{k=0}^{n-1} \frac{t^{2k}}{(2k)!} f^{(2k)}$$

$$= \int_0^t \frac{(t-s)^{2n-1}}{2(2n-1)!} \left[f^{(2n)}(\cdot + s) + f^{(2n)}(\cdot - s) \right] ds, \ \forall t \in \mathbb{R}. \tag{15.51}$$

By (15.41) we get

$$M_n^*(t) f := \frac{1}{2} \left[\int_0^t f(\cdot + s) ds + \int_0^t f(\cdot - s) ds \right] - \sum_{k=1}^{n} \frac{t^{2k-1}}{(2k-1)!} f^{(2(k-1))}$$

$$= \int_0^t \frac{(t-s)^{2n}}{2(2n)!} \left[f^{(2n)}(\cdot + s) + f^{(2n)}(\cdot - s) \right] ds, \ \forall t \in \mathbb{R}. \tag{15.52}$$

Let $g \in BUC(\mathbb{R})$, we define $\|g\| = \|g\|_\infty := \sup_{x \in \mathbb{R}} |g(x)| < \infty$.

Notice also that

$$\left\| \left\| C^*(s) A^n f \right\|_\infty \right\|_\infty = \left\| \left\| C^*(s) f^{(2n)} \right\|_\infty \right\|_\infty = \tag{15.53}$$

$$\frac{1}{2} \left\| \left\| \left(f^{(2n)}(\cdot + s) + f^{(2n)}(\cdot - s) \right) \right\|_\infty \right\|_\infty \le$$

$$\frac{1}{2} \left[\left\| \left\| f^{(2n)}(\cdot + s) \right\|_\infty \right\|_\infty + \left\| \left\| f^{(2n)}(\cdot - s) \right\|_\infty \right\|_\infty \right] \le \left\| f^{(2n)} \right\|_\infty < \infty.$$

Let $a_1, a_2 > 0$.

First we apply (15.43) of Theorem 15.12 to obtain

$$\int_0^{a_1} \int_0^{a_2} \frac{\left\| T_n^*(t_1) f_1 \right\|_\infty \left\| T_n^*(t_2) f_2 \right\|_\infty dt_1 dt_2}{\left[\varepsilon + \frac{t_1^{p(2n-1)+1}}{p(p(2n-1)+1)} + \frac{t_2^{q(2n-1)+1}}{q(q(2n-1)+1)} \right]} \leq \frac{a_1 a_2}{((2n-1)!)^2} \cdot \tag{15.54}$$

$$\left(\int_0^{a_1} \left\| C^*(u_1) f_1^{(2n)} \right\|_\infty^q du_1 \right)^{\frac{1}{q}} \left(\int_0^{a_2} \left\| C^*(u_2) f_2^{(2n)} \right\|_\infty^p du_2 \right)^{\frac{1}{p}},$$

where $f_1, f_2 \in BUC(\mathbb{R})$.

Next we apply (15.45) of Corollary 15.13.
Here

$$T_2^*(t_i) f_i = C^*(t_i) f_i - f_i - \frac{t_i^2}{2} f_i^{(2)},$$

$i = 1, 2; t_i \in [0, a_i]$.
Then

$$\int_0^{a_1} \int_0^{a_2} \frac{\left\| T_2^*(t_1) f_1 \right\|_\infty \left\| T_2^*(t_2) f_2 \right\|_\infty dt_1 dt_2}{\left[\varepsilon + \frac{t_1^7 + t_2^7}{14} \right]} \leq \tag{15.55}$$

$$\frac{a_1 a_2}{36} \left(\int_0^{a_1} \left\| C^*(u_1) f_1^{(4)} \right\|_\infty^2 du_1 \right)^{\frac{1}{2}} \left(\int_0^{a_2} \left\| C^*(u_2) f_2^{(4)} \right\|_\infty^2 du_2 \right)^{\frac{1}{2}};$$

where $f_1, f_2 \in BUC(\mathbb{R})$ for $n = 2$.
Next we apply (15.47) of Theorem 15.14 to obtain

$$\int_0^{a_1} \int_0^{a_2} \frac{\left\| M_n^*(t_1) f_1 \right\|_\infty \left\| M_n^*(t_2) f_2 \right\|_\infty dt_1 dt_2}{\left[\varepsilon + \frac{t_1^{(2pn+1)}}{p(2pn+1)} + \frac{t_2^{(2qn+1)}}{q(2qn+1)} \right]} \leq \frac{a_1 a_2}{((2n)!)^2} \cdot \tag{15.56}$$

$$\left(\int_0^{a_1} \left\| C^*(u_1) f_1^{(2n)} \right\|_\infty^q du_1 \right)^{\frac{1}{q}} \left(\int_0^{a_2} \left\| C^*(u_2) f_2^{(2n)} \right\|_\infty^p du_2 \right)^{\frac{1}{p}},$$

where $f_1, f_2 \in BUC(\mathbb{R})$.
We finish with application of (15.49) of Corollary 15.15.
Notice here that

$$M_2^*(t_i) f_i = S(t_i) f_i - f_i t_i - \frac{t_i^3}{6} f_i^{(2)}, \tag{15.57}$$

$i = 1, 2; t_i \in [0, a_i]$.

We have

$$\int_0^{a_1} \int_0^{a_2} \frac{\left\| M_2^* (t_1) f_1 \right\|_\infty \left\| M_2^* (t_2) f_2 \right\|_\infty dt_1 dt_2}{\left[\varepsilon + \frac{t_1^9 + t_2^9}{18} \right]} \le \qquad (15.58)$$

$$\frac{a_1 a_2}{576} \left(\int_0^{a_1} \left\| C^* (u_1) f_1^{(4)} \right\|_\infty^2 du_1 \right)^{\frac{1}{2}} \left(\int_0^{a_2} \left\| C^* (u_2) f_2^{(4)} \right\|_\infty^2 du_2 \right)^{\frac{1}{2}},$$

where $f_1, f_2 \in BUC (\mathbb{R})$ for $n = 2$.

References

1. G.A. Anastassiou, *Advanced Inequalities* (World Scientific, Singapore, 2011)
2. G.A. Anastassiou, Hilbert-Pachpatte type inequalities for semigroups, cosine and sine operator functions. Appl. Math. Lett. **24**, 2172–2180 (2011)
3. P.L. Butzer, H. Berens, *Semi-Groups of Operators and Approximation* (Springer, New York, 1967)
4. P.L. Butzer, H.G. Tillman, Approximation theorerms for semi-groups of bounded linear transformations. Math. Ann. **140**, 256–262 (1960)
5. D.-K. Chyan, S.-Y. Shaw, P. Piskarev, On Maximal regularity and semivariation of Cosine operator functions. J. London Math. Soc. **59**(2), 1023–1032 (1999)
6. S.-K. Chua, R L. Wheeden, A note on sharp 1-dimensional Poincaré inequalities. Proc. AMS **134**(8), 2309–2316 (2006)
7. H.O. Fattorini, Ordinary differential equations in linear topological spaces. I. J. Diff. Equ. **5**, 72–105 (1968)
8. J.A. Goldstein, *Semigroups of Linear Operators and Applications* (Oxford University Press, Oxford, 1985)
9. G.H. Hardy, J.E. Littlewood, G. Polya, *Inequalities* (Cambridge University Press, Cambridge, 1934)
10. E. Hille, R.S. Phillips, *Functional Analysis and Semigroups*, revised edition, pp. XII a. 808, American Mathematical Society Colloquium Publications, vol. 31, American Mathematical Society, Providence (1957)
11. Y. Katznelson, *An introduction to Harmonic Analysis* (Dover, New York, 1976)
12. B. Nagy, On cosine operator functions in Banach spaces. Acta Scientarium Mathematicarum Szeged **36**, 281–289 (1974)
13. B. Nagy, Approximation theorems for Cosine operator functions, Acta Mathematica Academiae Scientarium Hungaricae. Tomus **29**(1–2), 69–76 (1977)
14. B.G. Pachpatte, Inequalities similar to the integral analogue of Hilbert's inequalities. Tamkang J. Math. **30**(1), 139–146 (1999)
15. M. Sova, *Cosine Operator Functions, Rozprawy Matematyczne*, vol. XLIX (Warszawa, New York, 1966)

Chapter 16
About Ostrowski and Landau Type Inequalities for Banach Space Valued Functions

Very general univariate Ostrowski type inequalities are presented regarding Banach space valued functions. They provide norm estimates for the deviation of the average of the function from its vector values, and they involve the nth order vector derivative, $n \geq 1$, with respect to $\|\cdot\|_p$, $1 \leq p \leq \infty$. On the way to prove our results we establish useful functional vector identities. Using our vector Ostrowski inequalities we derive the vector Landau inequalities. It follows [10].

16.1 Introduction

In 1938, Ostrowski [15] proved the following important inequality:

Theorem 16.1 *Let* $f : [a, b] \to \mathbb{R}$ *be continuous on* $[a, b]$ *and differentiable on* (a, b) *whose derivative* $f' : (a, b) \to \mathbb{R}$ *is bounded on* (a, b), *i.e.* $\|f'\|_\infty := \sup_{t \in (a,b)} |f'(t)| < \infty$. *Then*

$$\left| \frac{1}{b-a} \int_a^b f(t)\, dt - f(x) \right| \leq \left[\frac{1}{4} + \frac{\left(x - \frac{a+b}{2}\right)^2}{(b-a)^2} \right] (b-a) \|f'\|_\infty, \quad (16.1)$$

for any $x \in [a, b]$. *The constant* $\frac{1}{4}$ *is the best possible.*

Since then there has been a lot of activity around these inequalities with important applications to Numerical Analysis and Probability.

This chapter is also motivated by the following result.

Theorem 16.2 (see [4]) *Let* $f \in C^{n+1}([a, b])$, $n \in \mathbb{N}$ *and* $x \in [a, b]$ *be fixed, such that* $f^{(k)}(x) = 0$, $k = 1, \ldots, n$. *Then it holds*

© Springer International Publishing Switzerland 2016
G.A. Anastassiou, *Intelligent Comparisons: Analytic Inequalities*,
Studies in Computational Intelligence 609,
DOI 10.1007/978-3-319-21121-3_16

$$\left| \frac{1}{b-a} \int_a^b f(y) \, dy - f(x) \right| \leq \frac{\left\| f^{(n+1)} \right\|_\infty}{(n+2)!} \left(\frac{(x-a)^{n+2} + (b-x)^{n+2}}{b-a} \right).$$

$$(16.2)$$

*Inequality (16.2) is sharp. In particular, when n is odd is attained by $f^*(y) :=$
$(y-x)^{n+1} \cdot (b-a)$, while when n is even the optimal function is $\overline{f}(y) :=$
$|y-x|^{n+\alpha} \cdot (b-a)$, $\alpha > 1$.*

*Clearly inequality (16.2) generalizes inequality (16.1) for higher order derivatives
of f.*

This chapter is mostly motivated by [11].

Here we prove general Ostrowski type inequalities for Banach space valued functions. As applications we derive the vector Landau inequalities.

16.2 Background

(See [17], pp. 83–94).

Let $f(t)$ be a function defined on $[a, b] \subseteq \mathbb{R}$ taking values in a real or complex normed linear space $(X, \|\cdot\|)$. Then $f(t)$ is said to be differentiable at a point $t_0 \in [a, b]$ if the limit

$$f'(t_0) := \lim_{h \to 0} \frac{f(t_0 + h) - f(t_0)}{h} \tag{16.3}$$

exists in X, the convergence is in $\|\cdot\|$. This is called the derivative of $f(t)$ at $t = t_0$.

We call $f(t)$ differentiable on $[a, b]$, iff there exists $f'(t) \in X$ for all $t \in [a, b]$.

Similarly and inductively are defined higher order derivatives of f, denoted f'', $f^{(3)}, \ldots, f^{(k)}$, $k \in \mathbb{N}$, just as for numerical functions.

For all the properties of derivatives see [17], pp. 83–86.

Let now $(X, \|\cdot\|)$ be a Banach space, and $f : [a, b] \to X$.

We define the vector valued Riemann integral $\int_a^b f(t) \, dt \in X$ as the limit of the vector valued Riemann sums in X, convergence is in $\|\cdot\|$. The definition is as for the numerical valued functions.

If $\int_a^b f(t) \, dt \in X$ we call f integrable on $[a, b]$. If $f \in C([a, b], X)$, then f is integrable, [17], p. 87.

For all the properties of vector valued Riemann integrals see [17], pp. 86–91.

We define the space $C^n([a, b], X)$, $n \in \mathbb{N}$, of n-times continuously differentiable functions from $[a, b]$ into X; here continuity is with respect to $\|\cdot\|$ and defined in the usual way as for numerical functions.

Let $(X, \|\cdot\|)$ be a Banach space and $f \in C^n([a, b], X)$, then we have the vector valued Taylor's formula, see [17], pp. 93–94, and also [16], (IV, 9; 47).

It holds

$$f(y) - f(x) - f'(x)(y-x) - \frac{1}{2} f''(x)(y-x)^2 - \ldots - \frac{1}{(n-1)!} f^{(n-1)}(x)(y-x)^{n-1}$$

$$= \frac{1}{(n-1)!} \int_x^y (y-t)^{n-1} f^{(n)}(t)\,dt, \quad \forall\, x, y \in [a, b]. \tag{16.4}$$

In particular (16.4) is true when $X = \mathbb{R}^m$, $\mathbb{C}^m$, $m \in \mathbb{N}$, etc.

In case of some $x_0 \in [a, b]$ such that $f^{(k)}(x_0) = 0$, $k = 0, 1, \ldots, n-1$, then

$$f(y) = \frac{1}{(n-1)!} \int_{x_0}^y (y-t)^{n-1} f^{(n)}(t)\,dt, \quad \forall\, y \in [a, b], \tag{16.5}$$

see also [7].

A function $f(t)$ with values in a normed linear space X is said to be piecewise continuous (see [17], p. 85) on the interval $a \le t \le b$ if there exists a partition $a = t_0 < t_1 < t_2 < \cdots < t_n = b$ such that $f(t)$ is continuous on every open interval $t_k < t < t_{k+1}$ and has finite limits $f(t_0+0)$, $f(t_1-0)$, $f(t_1+0)$, $f(t_2-0)$, $f(t_2+0), \ldots, f(t_n-0)$.

Here $f(t_k-0) = \lim_{t \uparrow t_k} f(t)$, $f(t_k+0) = \lim_{t \downarrow t_k} f(t)$.

The values of $f(t)$ at the points t_k can be arbitrary on even undefined.

A function $f(t)$ with values in normed linear space X is said to be piecewise smooth on $[a, b]$, if it is continuous on $[a, b]$ and has a derivative $f'(t)$ at all but a finite number of points of $[a, b]$, and if $f'(t)$ is piecewise continuous on $[a, b]$ (see [17], p. 85).

Let $u(t)$ and $v(t)$ be two piecewise smooth functions on $[a, b]$, one a numerical function and the other a vector function with values in Banach space X. Then we have the following integration by parts formula

$$\int_a^b u(t)\,dv(t) = u(t)\,v(t)\,|_a^b - \int_a^b v(t)\,du(t), \tag{16.6}$$

see [17], p. 93.

We need also the mean value theorem for Banach space valued functions.

Theorem 16.3 (see [13], p. 3) *Let $f \in C([a, b], X)$, where X is a Banach space. Assume f' exists on $[a, b]$ and $\|f'(t)\| \le K$, $a < t < b$, then*

$$\|f(b) - f(a)\| \le K(b-a). \tag{16.7}$$

Notice also that

$$\left(\frac{1}{4} + \frac{\left(x - \frac{a+b}{2}\right)^2}{(b-a)^2} \right)(b-a) = \frac{(x-a)^2 + (b-x)^2}{2(b-a)}, \quad \forall\, x \in [a, b]. \tag{16.8}$$

16.3 Main Results

First we present the vector space analog of Theorem 22.1, p. 498 of [5], see also [4, 11].

Theorem 16.4 *Let* $(X, \|\cdot\|)$ *be a Banach space and* $f \in C^1([a, b], X)$, *with* $x \in [a, b]$ *be fixed. Here* i_0 *is a fixed unit vector in* X. *Denote* $\||f'|\|_\infty :=$ $\sup_{x \in [a,b]} \|f'(x)\|$. *Then*

$$\left\| \frac{1}{b-a} \int_a^b f(t)\, dt - f(x) \right\| \leq \left(\frac{(x-a)^2 + (b-x)^2}{2(b-a)} \right) \||f'|\|_\infty \qquad (16.9)$$

$$\leq \frac{(b-a)}{2} \||f'|\|_\infty . \qquad (16.10)$$

Inequality (16.9) is sharp.
In particular the optimal function is

$$f^*(y) := |y - x|^\alpha \cdot (b - a) \cdot i_0, \quad \alpha > 1. \qquad (16.11)$$

Proof Here $\int_a^b f(t)\, dt \in X$. We observe that

$$\left\| \frac{1}{b-a} \int_a^b f(t)\, dt - f(x) \right\| = \left\| \frac{1}{b-a} \int_a^b (f(t) - f(x))\, dt \right\| \leq \qquad (16.12)$$

$$\frac{1}{b-a} \int_a^b \|f(t) - f(x)\|\, dt \leq$$

(by Theorem 16.3)

$$\frac{1}{b-a} \left(\int_a^b |t - x|\, dt \right) \||f'|\|_\infty = \frac{\||f'|\|_\infty}{b-a} \left[\int_a^x (x - t)\, dt + \int_x^b (t - x)\, dt \right]$$

$$= \frac{\||f'|\|_\infty}{b-a} \left(\frac{(x-a)^2 + (b-x)^2}{2} \right) \qquad (16.13)$$

$$\leq \frac{\||f'|\|_\infty}{b-a} \left(\frac{(b-a)^2}{2} \right) = \frac{(b-a)\||f'|\|_\infty}{2},$$

proving (16.9) and (16.10).

Notice that $f^*(x) = 0$. Then the left hand side of (16.9) (L.H.S. (16.9)),

$$L.H.S.\,(16.9) = \left\| \frac{1}{b-a} \int_a^b |y-x|^\alpha \cdot (b-a) \cdot i_0 dy \right\|$$

$$= \int_a^b |y-x|^\alpha \, dy = \frac{(x-a)^{\alpha+1} + (b-x)^{\alpha+1}}{\alpha+1}, \qquad (16.14)$$

and

$$\lim_{\alpha \to 1} L.H.S.\,(16.9) = \frac{(x-a)^2 + (b-x)^2}{2}. \qquad (16.15)$$

Notice that

$$f^{*\prime}(y) = \alpha |y-x|^{\alpha-1} \cdot sign\,(y-x) \cdot (b-a) \cdot i_0, \qquad (16.16)$$

hence

$$\left\| f^{*\prime}(y) \right\| = \alpha \cdot |y-x|^{\alpha-1} \cdot (b-a),$$

and

$$\left\| \left\| f^{*\prime} \right\| \right\|_\infty = \alpha\,(b-a)\,(\max\,(b-x,\,x-a))^{\alpha-1}. \qquad (16.17)$$

So that the right hand side of (16.9) (R.H.S. (16.9)),

$$R.H.S.\,(16.9) = \left(\frac{(x-a)^2 + (b-x)^2}{2} \right) \alpha\,(\max\,(b-x,\,x-a))^{\alpha-1}, \qquad (16.18)$$

and

$$\lim_{\alpha \to 1} R.H.S.\,(16.9) = \frac{(x-a)^2 + (b-x)^2}{2}. \qquad (16.19)$$

That is

$$\lim_{\alpha \to 1} L.H.S.\,(16.9) = \lim_{\alpha \to 1} R.H.S.\,(16.9), \qquad (16.20)$$

establishing sharpness of (16.9).

When $x = a$ or $x = b$, then (16.9) is attained by $f_a(y) := (y-a)(b-a) i_0$, or $f_b(y) := (y-b)(b-a) i_0$, respectively, then both sides of (16.9) equal $\frac{(b-a)^2}{2}$. ∎

Next we present a vector Montgomery identity.

Theorem 16.5 *Let* $(X, \|\cdot\|)$ *be a Banach space and* $f \in C^1([a, b], X)$. *Let* $x \in [a, b]$ *be fixed and define*

$$P(x, t) := \begin{cases} \frac{t-a}{b-a}, & a \leq t \leq x, \\ \frac{t-b}{b-a}, & x < t \leq b. \end{cases} \tag{16.21}$$

Then

$$f(x) = \frac{1}{b-a} \int_a^b f(t)\,dt + \int_a^b P(x, t)\,f'(t)\,dt. \tag{16.22}$$

Proof We use integration by part, see (16.6). We have

$$\int_a^x (t - a)\,f'(t)\,dt = \int_a^x (t - a)\,df(t) =$$

$$(t - a)\,f(t)\,|_a^x - \int_a^x f(t)\,d(t - a) = (x - a)\,f(x) - \int_a^x f(t)\,dt, \tag{16.23}$$

and

$$\int_x^b (t - b)\,f'(t)\,dt = \int_x^b (t - b)\,df(t) =$$

$$(t - b)\,f(t)\,|_x^b - \int_x^b f(t)\,d(t - b) = (b - x)\,f(x) - \int_x^b f(t)\,dt. \tag{16.24}$$

Therefore we get

$$\int_a^x (t - a)\,f'(t)\,dt + \int_x^b (t - b)\,f'(t)\,dt =$$

$$(x - a)\,f(x) + (b - x)\,f(x) - \int_a^x f(t)\,dt - \int_x^b f(t)\,dt$$

$$= (b - a)\,f(x) - \int_a^b f(t)\,dt, \tag{16.25}$$

and

$$f(x) - \frac{1}{b-a} \int_a^b f(t)\,dt = \int_a^x \frac{(t - a)}{(b - a)}\,f'(t)\,dt + \int_x^b \frac{(t - b)}{(b - a)}\,f'(t)\,dt, \tag{16.26}$$

proving the claim. ∎

Next we give and L_p, $p > 1$, Ostrowski basic inequality, see also [11].

Theorem 16.6 *Let $f \in C^1([a, b], X)$, where $(X, \|\cdot\|)$ is a Banach space. Let $p, q > 1 : \frac{1}{p} + \frac{1}{q} = 1$ Denote*

$$\|f\|_p := \left(\int_a^b \|f(t)\|^p \, dt \right)^{\frac{1}{p}}.$$

Here $x \in [a, b]$ is fixed.
Then

$$\left\| \frac{1}{b-a} \int_a^b f(t) \, dt - f(x) \right\| \leq$$

$$\frac{1}{(b-a)(q+1)^{\frac{1}{q}}} \left((x-a)^{q+1} + (b-x)^{q+1} \right)^{\frac{1}{q}} \|f'\|_p \leq \tag{16.27}$$

$$\frac{1}{(q+1)^{\frac{1}{q}}} (b-a)^{\frac{1}{q}} \|f'\|_p. \tag{16.28}$$

Proof By (16.22) we get

$$\left\| f(x) - \frac{1}{b-a} \int_a^b f(t) \, dt \right\| = \left\| \int_a^b P(x, t) f'(t) \, dt \right\| \leq$$

$$\int_a^b |P(x, t)| \, \|f'(t)\| \, dt \leq \left(\int_a^b |P(x, t)|^q \, dt \right)^{\frac{1}{q}} \left(\int_a^b \|f'(t)\|^p \, dt \right)^{\frac{1}{p}} = \tag{16.29}$$

$$\frac{\left(\int_a^x (t-a)^q \, dt + \int_x^b (b-t)^q \, dt \right)^{\frac{1}{q}}}{b-a} \|f'\|_p =$$

$$\frac{1}{(b-a)(q+1)^{\frac{1}{q}}} \left((x-a)^{q+1} + (b-x)^{q+1} \right)^{\frac{1}{q}} \|f'\|_p$$

(notice here $(x-a)^{q+1} + (b-x)^{q+1} \leq (b-a)^{q+1}$)

$$\leq \frac{(b-a)^{\frac{1}{q}}}{(q+1)^{\frac{1}{q}}} \|f'\|_p, \tag{16.30}$$

proving the claim. ∎

It follows the L_1 corresponding basic Ostrowski inequality, see also [11].

Theorem 16.7 *Let $f \in C^1([a, b], X)$, where $(X, \|\cdot\|)$ is a Banach space. Denote*

$$\|\|f\|\|_1 := \int_a^b \|f(t)\| \, dt. \tag{16.31}$$

Here $x \in [a, b]$ is fixed. Then

$$\left\| \frac{1}{b-a} \int_a^b f(t) \, dt - f(x) \right\| \leq \left[\frac{1}{2} + \left| \frac{x - \left(\frac{a+b}{2} \right)}{b-a} \right| \right] \|\|f'\|\|_1 \tag{16.32}$$

$$\leq \|\|f'\|\|_1. \tag{16.33}$$

Proof We have

$$\left\| f(x) - \frac{1}{b-a} \int_a^b f(t) \, dt \right\| \leq \int_a^b |P(x, t)| \, \|f'(t)\| \, dt = \tag{16.34}$$

$$\frac{1}{b-a} \left[\int_a^x (t - a) \|f'(t)\| \, dt + \int_x^b (b - t) \|f'(t)\| \, dt \right]$$

$$\leq \frac{1}{b-a} \left[(x - a) \int_a^x \|f'(t)\| \, dt + (b - x) \int_x^b \|f'(t)\| \, dt \right]$$

$$\leq \frac{1}{b-a} \max(x - a, b - x) \left(\int_a^b \|f'(t)\| \, dt \right) \tag{16.35}$$

$$= \frac{1}{b-a} \left[\frac{1}{2}(b - a) + \left| x - \left(\frac{a+b}{2} \right) \right| \right] \|\|f'\|\|_1 \tag{16.36}$$

$$= \left[\frac{1}{2} + \frac{|x - \left(\frac{a+b}{2} \right)|}{b-a} \right] \|\|f'\|\|_1$$

$$\leq \left[\frac{1}{2} + \frac{(b - a)}{2(b - a)} \right] \|\|f'\|\|_1 = \|\|f'\|\|_1, \tag{16.37}$$

proving the claim. ■

We make the following

Remark 16.8 Again here $(X, \|\cdot\|)$ is a Banach space. Let $f \in C^{n+1}([a, b], X)$, $n \in \mathbb{N}$, $x \in [a, b]$ be fixed.

Then by vector Taylor's theorem (16.4) we obtain

$$f(y) - f(x) = \sum_{k=1}^{n} \frac{f^{(k)}(x)}{k!}(y-x)^k + R_n(x,y),$$ (16.38)

where

$$R_n(x,y) := \int_x^y \left(f^{(n)}(t) - f^{(n)}(x) \right) \frac{(y-t)^{n-1}}{(n-1)!} dt,$$ (16.39)

here y can be $\geq x$ or $\leq x$.

Let $y \geq x$, then it holds

$$\|R_n(x,y)\| \leq \int_x^y \left\| f^{(n)}(t) - f^{(n)}(x) \right\| \frac{(y-t)^{n-1}}{(n-1)!} dt$$

$$\leq \left\| \left| f^{(n+1)} \right| \right\|_\infty \int_x^y (t-x) \frac{(y-t)^{n-1}}{(n-1)!} dt = \left\| \left| f^{(n+1)} \right| \right\|_\infty \frac{(y-x)^{n+1}}{(n+1)!}.$$

That is,

$$\|R_n(x,y)\| \leq \frac{\left\| \left| f^{(n+1)} \right| \right\|_\infty}{(n+1)!}(y-x)^{n+1}, \quad y \geq x.$$ (16.40)

Now let $x \geq y$, then it holds

$$\|R_n(x,y)\| = \left\| \int_y^x \left(f^{(n)}(t) - f^{(n)}(x) \right) \frac{(y-t)^{n-1}}{(n-1)!} dt \right\|$$

$$\leq \int_y^x \left\| f^{(n)}(t) - f^{(n)}(x) \right\| \frac{|y-t|^{n-1}}{(n-1)!} dt$$

$$\leq \frac{\left\| \left| f^{(n+1)} \right| \right\|_\infty}{(n-1)!} \int_y^x (x-t)(t-y)^{n-1} dt = \frac{\left\| \left| f^{(n+1)} \right| \right\|_\infty}{(n+1)!}(x-y)^{n+1}.$$

That is,

$$\|R_n(x,y)\| \leq \frac{\left\| \left| f^{(n+1)} \right| \right\|_\infty}{(n+1)!}(x-y)^{n+1}, \quad x \geq y.$$ (16.41)

By (16.40) and (16.41) we have that

$$\|R_n(x,y)\| \leq \frac{\left\| \left| f^{(n+1)} \right| \right\|_\infty}{(n+1)!}|y-x|^{n+1}, \quad \text{for all } x,y \in [a,b].$$ (16.42)

In the following we treat

$$\left\| \frac{1}{b-a} \int_a^b f(y)\,dy - f(x) \right\| = \frac{1}{b-a} \left\| \int_a^b (f(y) - f(x))\,dy \right\|$$

$$= \frac{1}{b-a} \left\| \int_a^b \left[\sum_{k=1}^n \frac{f^{(k)}(x)}{k!} (y-x)^k + R_n(x,y) \right] dy \right\|$$

$$= \frac{1}{b-a} \left\| \sum_{k=1}^n \frac{f^{(k)}(x)}{k!} \int_a^b (y-x)^k \, dy + \int_a^b R_n(x,y)\,dy \right\|$$

$$= \frac{1}{b-a} \left\| \sum_{k=1}^n \frac{f^{(k+1)}(x)}{(k+1)!} \left[(b-x)^{k+1} - (a-x)^{k+1} \right] + \int_a^b R_n(x,y)\,dy \right\| \le$$

from (16.42)

$$\frac{1}{b-a} \left[\sum_{k=1}^n \frac{\left\| f^{(k)}(x) \right\|}{(k+1)!} \left| (b-x)^{k+1} - (a-x)^{k+1} \right| + \frac{\left\| f^{(n+1)} \right\|_\infty}{(n+1)!} \int_a^b |y-x|^{n+1}\,dy \right].$$

That is, we have established

$$\left\| \frac{1}{b-a} \int_a^b f(y)\,dy - f(x) \right\| \le$$

$$\frac{1}{b-a} \left[\sum_{k=1}^n \frac{\left\| f^{(k)}(x) \right\|}{(k+1)!} \left| (b-x)^{k+1} - (a-x)^{k+1} \right| + \right.$$

$$\left. \frac{\left\| f^{(n+1)} \right\|_\infty}{(n+2)!} \left((x-a)^{n+2} + (b-x)^{n+2} \right) \right], \tag{16.43}$$

where $f \in C^{n+1}([a,b], X)$, $n \in \mathbb{N}$, $x \in [a,b]$ is fixed.
If we pick $x = \frac{a+b}{2}$, then

$$b - x = x - a = \frac{b-a}{2}.$$

Hence

$$\left\| \frac{1}{b-a} \int_a^b f(y)\,dy - f\left(\frac{a+b}{2} \right) \right\| \le$$

$$\frac{1}{b-a}\left[\sum_{1\leq k \text{ even} \leq n}\frac{\left\|f^{(k)}\left(\frac{a+b}{2}\right)\right\|(b-a)^{k+1}}{(k+1)!}\cdot\frac{1}{2^k}+\right.$$

$$\left.\frac{\left\|f^{(n+1)}\right\|_\infty(b-a)^{n+2}}{(n+2)!}\frac{1}{2^{n+1}}\right],\tag{16.44}$$

where $f \in C^{n+1}([a,b],X)$, $n \in \mathbb{N}$.

The above considerations and the proved inequalities (16.43) and (16.44), lead to the next results.

(See also [5], p. 502 and [4]).

Theorem 16.9 *Let* $f \in C^{n+1}([a,b],X)$, *where* $(X, \|\cdot\|)$ *is a Banach space,* $n \in \mathbb{N}$, $x \in [a,b]$ *be fixed, such that* $f^{(k)}(x) = 0$, $k = 1, \ldots, n$. *Also* i_0 *is a unit vector in* X. *Then*

$$\left\|\frac{1}{b-a}\int_a^b f(y)\,dy - f(x)\right\| \leq$$

$$\frac{\left\|f^{(n+1)}\right\|_\infty}{(n+2)!}\left(\frac{(x-a)^{n+2}+(b-x)^{n+2}}{b-a}\right).\tag{16.45}$$

Inequality (16.45) is sharp. In particular, when n *is odd it is attained by* $f^*(y) := (y-x)^{n+1}(b-a)i_0$, *while when* n *is even the optimal function is* $\widetilde{f}(y) := |y-x|^{n+\alpha}(b-a)i_0$, $\alpha > 1$.

Proof Inequality (16.45) comes directly from (16.43). In the following we establish the sharpness of (16.45).

When n is odd: see that $f^{*(k)}(x) = 0$, $k = 0, 1, \ldots, n$, and $f^{*(n+1)}(y) = (n+1)!(b-a)i_0$. Thus $\left\|f^{*(n+1)}\right\|_\infty = (n+1)!(b-a)$.

Plugging f^* into (16.45) we obtain that

$$L.H.S.(16.45) = \frac{(b-x)^{n+2}+(x-a)^{n+2}}{n+2}.\tag{16.46}$$

Also, it holds

$$R.H.S.(16.45) = \frac{(x-a)^{n+2}+(b-x)^{n+2}}{n+2}.\tag{16.47}$$

From (16.46) and (16.47), when n is odd, inequality (16.45) is established sharp, in particular, it is attained by f^*.

When n is even: Observe that $\overline{f}^{(k)}(x) = 0$, $k = 0, 1, \ldots, n$, and

$$\overline{f}^{(n+1)}(y) = (n + \alpha)(n + \alpha - 1) \ldots (\alpha + 1) \cdot \alpha \cdot |y - x|^{\alpha - 1} \cdot sign\,(y - x) \cdot (b - a) \cdot i_0.$$

Thus

$$\left\| \overline{f}^{(n+1)}(y) \right\| = \left(\prod_{j=0}^{n} (n + \alpha - j) \right) \cdot |y - x|^{\alpha - 1} \cdot (b - a)$$

and

$$\left\| \left\| \overline{f}^{(n+1)} \right\| \right\|_\infty = \left(\prod_{j=0}^{n} (n + \alpha - j) \right) \cdot (\max\,(b - x, x - a))^{\alpha - 1} \cdot (b - a).$$

Therefore we have

$$R.H.S.\,(16.45) = \frac{\left(\displaystyle\prod_{j=0}^{n} (n + \alpha - j) \right) \cdot (\max\,(b - x, x - a))^{\alpha - 1}}{(n + 2)!}.$$

$$\left((x - a)^{n+2} + (b - x)^{n+2} \right), \quad \alpha > 1.$$

Hence

$$\lim_{\alpha \to 1} R.H.S.\,(16.45) = \frac{(x - a)^{n+2} + (b - x)^{n+2}}{n + 2} \tag{16.48}$$

and

$$L.H.S.\,(16.45) = \frac{(x - a)^{n+\alpha+1} + (b - x)^{n+\alpha+1}}{n + \alpha + 1}.$$

Consequently

$$\lim_{\alpha \to 1} L.H.S.\,(16.45) = \frac{(x - a)^{n+2} + (b - x)^{n+2}}{n + 2}. \tag{16.49}$$

From (16.48) and (16.49) we get the sharpness of (16.45) when n is even. ■

Notice that when $x = a$, or $x = b$ and n is even, inequality (16.45) can be attained by $\widetilde{f}_a(y) := (y - a)^{n+1} \cdot (b - a) \cdot i_0$, or $\widetilde{f}_b(y) := (y - b)^{n+1} \cdot (b - a) \cdot i_0$, respectively (then both sides of (16.45) equal $\frac{(b-a)^{n+2}}{(n+2)}$).

When $x = \frac{a+b}{2}$, we have a case of special interest next.

Theorem 16.10 *Let $f \in C^{n+1}\left([a, b], X\right)$, where $(X, \|\cdot\|)$ is a Banach space, $n \in \mathbb{N}$, such that $f^{(k)}\left(\frac{a+b}{2}\right) = 0$, all k even $\in \{1, \ldots, n\}$, i_0 is a unit vector in X. Then*

$$\left\| \frac{1}{b-a} \int_a^b f\left(y\right) dy - f\left(\frac{a+b}{2}\right) \right\| \leq \frac{\left\| \left| f^{(n+1)} \right| \right\|_\infty}{(n+2)!} \frac{(b-a)^{n+1}}{2^{n+1}}. \quad (16.50)$$

Inequality (16.50) is sharp. More precisely, when n is odd it is attained by $f^*\left(y\right) := \left(y - \frac{a+b}{2}\right)^{n+1} (b-a) i_0$, while when n is even the optimal function is $\widetilde{f}\left(y\right) := \left| y - \frac{a+b}{2} \right|^{n+\alpha} (b-a) i_0$, $\alpha > 1$.

Corollary 16.11 (to Theorem 16.10, case of $n = 1$). *Let $f \in C^2\left([a, b], X\right)$, such that $f''\left(\frac{a+b}{2}\right) = 0$. Then*

$$\left\| \frac{1}{b-a} \int_a^b f\left(y\right) dy - f\left(\frac{a+b}{2}\right) \right\| \leq \left\| f'' \right\|_\infty \frac{(b-a)^2}{24}, \quad (16.51)$$

which is sharp, it is attained by $f^*\left(y\right) := \left(y - \frac{a+b}{2}\right)^2 (b-a) i_0$.

We need

Theorem 16.12 ([17], p. 92) *Let $G\left(t\right)$ be a function with values in a Banach space X, and suppose $G\left(t\right)$ is differentiable on $[a, b]$, with a piecewise continuous derivative. Then*

$$G\left(t\right) = G\left(a\right) + \int_a^t G'\left(t\right) dt, \quad (16.52)$$

for every $t \in [a, b]$.

We also need (see also [8], p. 6 and [6]).

Theorem 16.13 *Let $f \in C^n\left([a, b], X\right)$, $n \in \mathbb{N}$, $(X, \|\cdot\|)$ is a Banach space and $x \in [a, b]$. Define the kernel*

$$P\left(r, s\right) := \begin{cases} \frac{s-a}{b-a}, & a \leq s \leq r, \\ \frac{s-b}{b-a}, & r < s \leq b. \end{cases} \quad (16.53)$$

where $r, s \in [a, b]$. Then

$$\theta_{1,n} := f\left(x\right) - \frac{1}{b-a} \int_a^b f\left(s_1\right) ds_1 - \quad (16.54)$$

$$\sum_{k=0}^{n-2} \left(\frac{f^{(k)}\left(b\right) - f^{(k)}\left(a\right)}{b-a} \right) \cdot \underbrace{\int_a^b \cdots \int_a^b}_{(k+1)th\text{-}integral} P\left(x, s_1\right) \prod_{i=1}^{k} P\left(s_i, s_{i+1}\right) ds_1 ds_2 \ldots ds_{k+1}$$

$$= \int_a^b \cdots \int_a^b P(x, s_1) \prod_{i=1}^{n-1} P(s_i, s_{i+1}) f^{(n)}(s_n) \, ds_1 ds_2 \ldots ds_n =: \theta_{2,n}.$$

We make the conventions that $\prod_{k=0}^{-1} \cdot = 0$, $\prod_{i=1}^{0} \cdot = 1$.

Proof Here we use repeatedly Theorems 16.5 and 16.12.

We have by (16.22) that

$$f(x) = \frac{1}{b-a} \int_a^b f(s_1) \, ds_1 + \int_a^b P(x, s_1) f'(s_1) \, ds_1.$$

Doing the same for the derivative of f we get

$$f'(s_1) = \frac{1}{b-a} \int_a^b f'(s_2) \, ds_2 + \int_a^b P(s_1, s_2) f''(s_2) \, ds_2.$$

That is

$$f'(s_1) = \frac{f(b) - f(a)}{b-a} + \int_a^b P(s_1, s_2) f''(s_2) \, ds_2.$$

Similarly for f'' we obtain

$$f''(s_2) = \frac{f'(b) - f'(a)}{b-a} + \int_a^b P(s_2, s_3) f'''(s_3) \, ds_3.$$

And in general we have

$$f^{(n-1)}(s_{n-1}) = \frac{f^{(n-2)}(b) - f^{(n-2)}(a)}{b-a} + \int_a^b P(s_{n-1}, s_n) f^{(n)}(s_n) \, ds_n.$$

We see that

$$\int_a^b P(x, s_1) f'(s_1) \, ds_1 = \left(\frac{f(b) - f(a)}{b-a} \right) \int_a^b P(x, s_1) \, ds_1 +$$

$$\int_a^b \int_a^b P(x, s_1) P(s_1, s_2) f''(s_2) \, ds_2 ds_1.$$

That is

$$f(x) = \frac{1}{b-a} \int_a^b f(s_1)\, ds_1 + \left(\frac{f(b) - f(a)}{b-a} \right) \int_a^b P(x, s_1)\, ds_1 +$$

$$\int_a^b \int_a^b P(x, s_1)\, P(s_1, s_2)\, f''(s_2)\, ds_2 ds_1.$$

Now we observe that

$$\int_a^b \int_a^b P(x, s_1)\, P(s_1, s_2)\, f''(s_2)\, ds_2 ds_1 =$$

$$\left(\frac{f'(b) - f'(a)}{b-a} \right) \int_a^b \int_a^b P(x, s_1)\, P(s_1, s_2)\, ds_2 ds_1 +$$

$$\int_a^b \int_a^b \int_a^b P(x, s_1)\, P(s_1, s_2)\, P(s_2, s_3)\, f'''(s_3)\, ds_3 ds_2 ds_1.$$

Hence we get

$$f(x) = \frac{1}{b-a} \int_a^b f(s_1)\, ds_1 + \left(\frac{f(b) - f(a)}{b-a} \right) \int_a^b P(x, s_1)\, ds_1 +$$

$$\left(\frac{f'(b) - f'(a)}{b-a} \right) \int_a^b \int_a^b P(x, s_1)\, P(s_1, s_2)\, ds_2 ds_1 +$$

$$\int_a^b \int_a^b \int_a^b P(x, s_1)\, P(s_1, s_2)\, P(s_2, s_3)\, f'''(s_3)\, ds_3 ds_2 ds_1.$$

Therefore in general we derive

$$f(x) = \frac{1}{b-a} \int_a^b f(s_1)\, ds_1 + \left(\frac{f(b) - f(a)}{b-a} \right) \int_a^b P(x, s_1)\, ds_1 +$$

$$\left(\frac{f'(b) - f'(a)}{b-a} \right) \int_a^b \int_a^b P(x, s_1)\, P(s_1, s_2)\, ds_2 ds_1 +$$

$$\left(\frac{f''(b) - f''(a)}{b-a} \right) \int_a^b \int_a^b \int_a^b P(x, s_1)\, P(s_1, s_2)\, P(s_2, s_3)\, ds_3 ds_2 ds_1 +$$

$$\left(\frac{f'''(b) - f'''(a)}{b-a} \right).$$

$$\int_a^b \int_a^b \int_a^b \int_a^b P(x, s_1) P(s_1, s_2) P(s_2, s_3) P(s_3, s_4) ds_4 ds_3 ds_2 ds_1 + \ldots +$$

$$\left(\frac{f^{(n-2)}(b) - f^{(n-2)}(a)}{b-a} \right) \underbrace{\int_a^b \cdots \int_a^b}_{(n-1)\text{th-integral}} P(x, s_1) \prod_{i=1}^{n-2} P(s_i, s_{i+1}) ds_{n-1} \ldots ds_1 +$$

$$\underbrace{\int_a^b \cdots \int_a^b}_{n\text{th-integral}} P(x, s_1) \prod_{i=1}^{n-1} P(s_i, s_{i+1}) f^{(n)}(s_n) ds_n \ldots ds_1.$$

The last equality is written in short as follows

$$f(x) = \frac{1}{b-a} \int_a^b f(s_1) ds_1 + \sum_{k=0}^{n-2} \left(\frac{f^{(k)}(b) - f^{(k)}(a)}{b-a} \right) \cdot$$

$$\left(\underbrace{\int_a^b \cdots \int_a^b}_{(k+1)\text{th-integral}} P(x, s_1) \prod_{i=1}^{k} P(s_i, s_{i+1}) ds_1 ds_2 \ldots ds_{k+1} \right)$$

$$+ \int_a^b \cdots \int_a^b P(x, s_1) \prod_{i=1}^{n-1} P(s_i, s_{i+1}) f^{(n)}(s_n) ds_1 ds_2 \ldots ds_n. \qquad (16.55)$$

So we have proved (16.54). ■

A special very common case follows.

Corollary 16.14 *Under the assumptions and notations of Theorem 16.13 additionally suppose that*

$$f^{(k)}(a) = f^{(k)}(b), \quad k = 0, 1, \ldots, n-2; \ \text{when } n \geq 2.$$

Then

$$\theta_{1,n} = f(x) - \frac{1}{b-a} \int_a^b f(s_1) ds_1 = \qquad (16.56)$$

$$= \int_a^b \cdots \int_a^b P(x, s_1) \prod_{i=1}^{n-1} P(s_i, s_{i+1}) f^{(n)}(s_n) ds_1 \ldots ds_n = \theta_{2,n},$$

for $n \in \mathbb{N}$, $x \in [a, b]$.

Proof Directly from (16.54). ∎

We present the following Ostrowski type inequalities using Theorem 16.13.

Theorem 16.15 *All as in Theorem 16.13. Then*

$$\|\theta_{1,n}\| \leq \left\|\left|f^{(n)}\right|\right\|_{\infty} \cdot \int_a^b \cdots \int_a^b |P(x, s_1)| \cdot \prod_{i=1}^{n-1} |P(s_i, s_{i+1})| \, ds_1 \ldots ds_n =: A_1.$$

(16.57)

Theorem 16.16 *All as in Theorem 16.13. Let* $p, q > 1 : \frac{1}{p} + \frac{1}{q} = 1$. *Then*

$$\|\theta_{1,n}\| \leq \left\|\left|f^{(n)}\right|\right\|_p \cdot \int_a^b \cdots \int_a^b |P(x, s_1)| \cdot \left(\prod_{i=1}^{n-2} |P(s_i, s_{i+1})|\right) \cdot$$

(16.58)

$$\|P(s_{n-1}, \cdot)\|_q \, ds_1 ds_2 \ldots ds_{n-1} =: A_2.$$

Theorem 16.17 *All as in Theorem 16.13. Then*

$$\|\theta_{1,n}\| \leq \left\|\left|f^{(n)}\right|\right\|_1 \cdot \left(\int_a^b \cdots \int_a^b |P(x, s_1)| \cdot \left(\prod_{i=1}^{n-2} |P(s_i, s_{i+1})|\right) \cdot\right.$$

(16.59)

$$\left. \|P(s_{n-1}, \cdot)\|_{\infty} \, ds_1 ds_2 \ldots ds_{n-1}\right) =: A_3.$$

Corollary 16.18 *All as in Theorem 16.13, and* $f^{(k)}(a) = f^{(k)}(b), k = 0, 1, \ldots, n-2$; *when* $\mathbb{N} \ni n \geq 2$, $x \in [a, b]$. *Then*

$$\left\| f(x) - \frac{1}{b-a} \int_a^b f(s_1) \, ds_1 \right\| \leq \min\{A_1, A_2, A_3\}.$$

(16.60)

Proof Use of Corollary 16.14. ∎

We make

Remark 16.19 Let $f \in C^n([a, b], X)$, $n \in \mathbb{N}$, $(X, \|\cdot\|)$ a Banach space. By vector Taylor formula (16.4) we get

$$f(x) = f(y) + \sum_{k=1}^{n-1} \frac{f^{(k)}(y)(x-y)^k}{k!} +$$

$$\frac{1}{(n-1)!} \int_y^x (x-t)^{n-1} f^{(n)}(t) \, dt,$$

(16.61)

$\forall \, x, y \in [a, b]$.

We integrate (16.61) with respect to y to get (see also [12])

$$Ef(x) := f(x)(b-a) - \int_a^b f(y)\,dy - \sum_{k=1}^{n-1} \frac{1}{k!} \int_a^b f^{(k)}(y)(x-y)^k\,dy$$

$$= \frac{1}{(n-1)!} \int_a^b \left(\int_y^x (x-t)^{n-1} f^{(n)}(t)\,dt \right) dy. \tag{16.62}$$

We observe the following

$$\left\| \int_a^b \left(\int_y^x (x-t)^{n-1} f^{(n)}(t)\,dt \right) dy \right\| =$$

$$\left\| \int_a^x \left(\int_y^x (x-t)^{n-1} f^{(n)}(t)\,dt \right) dy + \int_x^b \left(\int_y^x (x-t)^{n-1} f^{(n)}(t)\,dt \right) dy \right\| \le$$
$$\tag{16.63}$$

$$\left\| \int_a^x \left(\int_y^x (x-t)^{n-1} f^{(n)}(t)\,dt \right) dy \right\| + \left\| \int_x^b \left(\int_x^y (x-t)^{n-1} f^{(n)}(t)\,dt \right) dy \right\| \le$$

$$\int_a^x \left(\int_y^x (x-t)^{n-1} \left\| f^{(n)}(t) \right\| dt \right) dy + \int_x^b \left(\int_x^y |x-t|^{n-1} \left\| f^{(n)}(t) \right\| dt \right) dy =$$
$$\tag{16.64}$$

$$\int_a^x \left(\int_a^t (x-t)^{n-1} \left\| f^{(n)}(t) \right\| dy \right) dt + \int_x^b \left(\int_t^b |x-t|^{n-1} \left\| f^{(n)}(t) \right\| dy \right) dt =$$

$$\int_a^x (x-t)^{n-1} \left\| f^{(n)}(t) \right\| (t-a)\,dt + \int_x^b |x-t|^{n-1} \left\| f^{(n)}(t) \right\| (b-t)\,dt =$$
$$\tag{16.65}$$

(calling

$$K(t,x) := \begin{cases} t-a, & a \le t \le x, \\ t-b, & x < t \le b, \end{cases}) \tag{16.66}$$

$$\int_a^b |x-t|^{n-1} \left\| f^{(n)}(t) \right\| |K(t,x)|\,dt.$$

So that we have proved

$$\frac{1}{(n-1)!} \left\| \int_a^b \left(\int_y^x (x-t)^{n-1} f^{(n)}(t)\,dt \right) dy \right\| \le$$

$$\frac{1}{(n-1)!} \int_a^b |x-t|^{n-1} |K(t,x)| \left\| f^{(n)}(t) \right\| dt. \tag{16.67}$$

We call (as in [12])

$$F_k(x) := \left(\frac{n-k}{k!}\right)\left(\frac{f^{(k-1)}(a)(x-a)^k - f^{(k-1)}(b)(x-b)^k}{b-a}\right), \qquad (16.68)$$

$k = 1, \ldots, n-1$.

We also call (as in [12])

$$I_k(x) := \int_a^b \frac{f^{(k)}(y)(x-y)^k}{k!} dy, \qquad (16.69)$$

$k = 1, \ldots, n-1$.

Using the integration by parts formula (16.6) we obtain (see also [12])

$$(n-k)(I_k(x) - I_{k-1}(x)) = -(b-a)F_k(x), \qquad (16.70)$$

$1 \le k \le n-1$.

Hence it holds

$$\sum_{k=1}^{n-1}(n-k)(I_k(x) - I_{k-1}(x)) = -(b-a)\sum_{k=1}^{n-1}F_k(x). \qquad (16.71)$$

Simplification of the last formula results to

$$\sum_{k=1}^{n-1}I_k(x) = (n-1)I_0 - (b-a)\sum_{k=1}^{n-1}F_k(x), \qquad (16.72)$$

where

$$I_0 := \int_a^b f(y)\, dy. \qquad (16.73)$$

Consequently we find

$$Ef(x) = f(x)(b-a) - I_0 - (n-1)I_0 + (b-a)\sum_{k=1}^{n-1}F_k(x) = \qquad (16.74)$$

$$f(x)(b-a) - nI_0 + (b-a)\sum_{k=1}^{n-1}F_k(x),$$

and

$$\frac{Ef(x)}{n(b-a)} = \left(\frac{f(x) + \sum_{k=1}^{n-1} F_k(x)}{n}\right) - \frac{I_0}{b-a}. \tag{16.75}$$

We established that (by (16.67) and (16.75))

$$\left\|\frac{1}{n}\left(f(x) + \sum_{k=1}^{n-1} F_k(x)\right) - \frac{1}{b-a}\int_a^b f(y)\,dy\right\| \leq \tag{16.76}$$

$$\frac{1}{n!(b-a)}\int_a^b |x-t|^{n-1}|K(t,x)|\left\|f^{(n)}(t)\right\|dt.$$

The last inequality is the vector analog of equality (16.10) in [12].

We have proved

Theorem 16.20 *Let* $f \in C^n([a,b], X)$, $n \in \mathbb{N}$, $(X, \|\cdot\|)$ *a Banach space. Set*

$$K(t,x) := \begin{cases} t-a, & a \leq t \leq x, \\ t-b, & x < t \leq b \end{cases} \tag{16.77}$$

and

$$F_k(x) := \left(\frac{n-k}{k!}\right)\left(\frac{f^{(k-1)}(a)(x-a)^k - f^{(k-1)}(b)(x-b)^k}{b-a}\right), \tag{16.78}$$

$k = 1, \ldots, n-1$.
Then
(1)

$$\left\|\frac{1}{n}\left(f(x) + \sum_{k=1}^{n-1} F_k(x)\right) - \frac{1}{b-a}\int_a^b f(y)\,dy\right\| \leq \tag{16.79}$$

$$\frac{1}{n!(b-a)}\int_a^b |x-t|^{n-1}|K(t,x)|\left\|f^{(n)}(t)\right\|dt =: J_n(f).$$

When $n = 1$ *we take* $\sum_{k=1}^{0} \cdot = 0$.

(2) *In case of $f^{(i)}(a) = f^{(i)}(b) = 0$, for $i = 0, 1, \ldots, n - 2$, we have*

$$\left\| \frac{f(x)}{n} - \frac{1}{b-a} \int_a^b f(y) \, dy \right\| \le J_n(f). \tag{16.80}$$

We make

Remark 16.21 (continuation of Remark 16.19)

(1) Let $p, q > 1 : \frac{1}{p} + \frac{1}{q} = 1$. Then

$$J := \int_a^b |x - t|^{n-1} |K(t, x)| \left\| f^{(n)}(t) \right\| dt \le$$

$$\left\| f^{(n)} \right\|_p \left(\int_a^b \left(|x - t|^{n-1} |K(t, x)| \right)^q dt \right)^{\frac{1}{q}}, \tag{16.81}$$

with equality when

$$\left\| f^{(n)}(t) \right\| = \lambda \left(|x - t|^{n-1} |K(t, x)| \right)^{q-1}. \tag{16.82}$$

As in [12] we have

$$\left(\int_a^b \left(|x - t|^{n-1} |K(t, x)| \right)^q dt \right)^{\frac{1}{q}} = (B(q + 1, (n - 1)q + 1))^{\frac{1}{q}} \cdot \tag{16.83}$$

$$\left[(x - a)^{nq+1} + (b - x)^{nq+1} \right]^{\frac{1}{q}},$$

where B is the beta function

$$B(x, y) = \int_0^1 t^{x-1} (1 - t)^{y-1} \, dt. \tag{16.84}$$

Consequently we have

$$J \le \left\| f^{(n)} \right\|_p (B(q + 1, (n - 1)q + 1))^{\frac{1}{q}} \left[(x - a)^{nq+1} + (b - x)^{nq+1} \right]^{\frac{1}{q}}. \tag{16.85}$$

(2) As in [12] we find easily

$$J \le \left\| f^{(n)} \right\|_\infty \left(\frac{(x - a)^{n+1} + (b - x)^{n+1}}{n(n + 1)} \right). \tag{16.86}$$

(3) We also have

$$J \leq \left\| \left| f^{(n)} \right| \right\|_1 \cdot \sup_{a \leq x \leq b} \left(|x - t|^{n-1} \, |K(t, x)| \right) = \tag{16.87}$$

$$\left\| \left| f^{(n)} \right| \right\|_1 \cdot \sup_{a \leq x \leq b} \left(\left| (x - t)^{n-1} \, K(t, x) \right| \right).$$

As in [12], one finds easily that

$$\sup_{a \leq x \leq b} \left(\left| (x - t)^{n-1} \, K(t, x) \right| \right) = \frac{1}{n} \left(\frac{n-1}{n} \right)^{n-1} \max\{ (x - a)^n, (b - x)^n \}. \tag{16.88}$$

Therefore we get

$$J \leq \left\| \left| f^{(n)} \right| \right\|_1 \cdot \frac{1}{n} \left(\frac{n-1}{n} \right)^{n-1} \max\{ (x - a)^n, (b - x)^n \}. \tag{16.89}$$

We have established (see also (16.79))

Theorem 16.22 Let $f \in C^n([a, b], X)$, $n \in \mathbb{N}$, $(X, \|\cdot\|)$ a Banach space, $p, q > 1 : \frac{1}{p} + \frac{1}{q} = 1$ Define

$$F_k(x) := \left(\frac{n-k}{k!} \right) \left(\frac{f^{(k-1)}(a)(x-a)^k - f^{(k-1)}(b)(x-b)^k}{b-a} \right), \tag{16.90}$$

$k = 1, \ldots, n - 1$. Then

(1)

$$\left\| \frac{1}{n} \left(f(x) + \sum_{k=1}^{n-1} F_k(x) \right) - \frac{1}{b-a} \int_a^b f(y) \, dy \right\| \leq$$

$$\frac{1}{n!(b-a)} \min \left\{ \left\| \left| f^{(n)} \right| \right\|_p (B(q+1, (n-1)q+1))^{\frac{1}{q}} \left[(x-a)^{nq+1} + (b-x)^{nq+1} \right]^{\frac{1}{q}}, \right.$$

$$\left\| \left| f^{(n)} \right| \right\|_\infty \left(\frac{(x-a)^{n+1} + (b-x)^{n+1}}{n(n+1)} \right),$$

$$\left. \left\| \left| f^{(n)} \right| \right\|_1 \cdot \frac{(n-1)^{n-1}}{n^n} \max\{ (x-a)^n, (b-x)^n \} \right\} =: \theta, \tag{16.91}$$

(2) In case of $f^{(i)}(a) = f^{(i)}(b) = 0$, for $i = 0, 1, \ldots, n-2$, we have

$$\left\| \frac{f(x)}{n} - \frac{1}{b-a} \int_a^b f(y)\, dy \right\| \leq \theta. \tag{16.92}$$

We need

Remark 16.23 Let $f \in C^n([a, b], X)$, $n \in \mathbb{N}$, $(X, \|\cdot\|)$ a Banach space. Then by (16.62) and (16.74) we get

$$f(x) = \frac{n \int_a^b f(y)\, dy}{b-a} - \sum_{k=1}^{n-1} F_k(x) +$$

$$\frac{1}{(n-1)!\,(b-a)} \int_a^b \left(\int_y^x (x-t)^{n-1} f^{(n)}(t)\, dt \right) dy. \tag{16.93}$$

Let $w : [a, b] \to [0, \infty)$ such that w is a continuous weight function with $\int_a^b w(t)\, dt = 1$. Then

$$\int_a^b f(x)\, w(x)\, dx = \frac{n \int_a^b f(y)\, dy}{b-a} - \sum_{k=1}^{n-1} \int_a^b F_k(x)\, w(x)\, dx + \tag{16.94}$$

$$\frac{1}{(n-1)!\,(b-a)} \int_a^b w(x) \left(\int_a^b \left(\int_y^x (x-t)^{n-1} f^{(n)}(t)\, dt \right) dy \right) dx.$$

Therefore by substracting (16.94) from (16.93) we obtain

$$f(x) - \int_a^b f(x)\, w(x)\, dx = -\sum_{k=1}^{n-1} F_k(x) + \sum_{k=1}^{n-1} \int_a^b F_k(x)\, w(x)\, dx +$$

$$\frac{1}{(n-1)!\,(b-a)} \int_a^b \left(\int_y^x (x-t)^{n-1} f^{(n)}(t)\, dt \right) dy -$$

$$\frac{1}{(n-1)!\,(b-a)} \int_a^b w(x) \left(\int_a^b \left(\int_y^x (x-t)^{n-1} f^{(n)}(t)\, dt \right) dy \right) dx.$$

So we have derived the representation formula

$$f(x) = \int_a^b f(x)\, w(x)\, dx - \sum_{k=1}^{n-1} F_k(x) + \sum_{k=1}^{n-1} \int_a^b F_k(x)\, w(x)\, dx +$$

$$\frac{1}{(n-1)!\,(b-a)} \int_a^b \left(\int_y^x (x-t)^{n-1}\, f^{(n)}\,(t)\,dt \right) dy -$$ (16.95)

$$\frac{1}{(n-1)!\,(b-a)} \int_a^b w\,(x) \left(\int_a^b \left(\int_y^x (x-t)^{n-1}\, f^{(n)}\,(t)\,dt \right) dy \right) dx.$$

Using (16.95) one can prove various Ostrowski type inequalities, integral means inequalities, as well as to have applications in Numerical Functional Analysis: trapezoid and midpoint rules, but to keep chapter short we omit it, see also [3].

16.4 Applications to Landau Inequalities

We establish here vector Landau inequalities.

Theorem 16.24 *Let* $f \in C^2\,(I, X)$, $(X, \|\cdot\|)$ *is a Banach space and* $I = \mathbb{R}_+, \mathbb{R}$. *We assume here*

$$\||f\||_{\infty,I}, \, \||f''\||_{\infty,I} < \infty.$$ (16.96)

Then

$$\||f'\||_{\infty,I} \le 2\sqrt{\||f\||_{\infty,I}\, \||f''\||_{\infty,I}},$$ (16.97)

where

$$\||f\||_{\infty,I} := \sup_{t \in I} \| f\,(t) \|.$$

Proof Here $f \in C^2\,(I, X)$, $(X, \|\cdot\|)$ is a Banach space, while $I = \mathbb{R}_+, \mathbb{R}$, and $[a, b] \subset I$, $x \in [a, b]$.
Then we apply (16.10) for f' to get

$$\left\| \frac{1}{b-a} \int_a^b f'\,(t)\,dt - f'\,(x) \right\| \le \frac{(b-a)}{2} \||f''\||_{\infty,[a,b]}$$ (16.98)

$$\le \frac{(b-a)}{2} \||f''\||_{\infty,I}.$$

That is

$$\left\| \frac{f\,(b) - f\,(a)}{b-a} - f'\,(x) \right\| \le \frac{(b-a)}{2} \||f''\||_{\infty,I}$$ (16.99)

and

$$\left\| f'(x) \right\| - \frac{\| f(b) - f(a) \|}{b - a} \le \frac{(b - a)}{2} \left\| \left\| f'' \right\| \right\|_{\infty, I} . \tag{16.100}$$

Therefore

$$\left\| f'(x) \right\| \le \frac{1}{b - a} \left\| f(b) - f(a) \right\| + \frac{(b - a)}{2} \left\| \left\| f'' \right\| \right\|_{\infty, I} \le$$

$$\frac{2}{(b - a)} \left\| \left\| f \right\| \right\|_{\infty, I} + \frac{(b - a)}{2} \left\| \left\| f'' \right\| \right\|_{\infty, I} \tag{16.101}$$

(under the assumptions $\left\| \left\| f \right\| \right\|_{\infty, I}$, $\left\| \left\| f'' \right\| \right\|_{\infty, I}$ both are finite).

Consequently it holds

$$\left\| \left\| f' \right\| \right\|_{\infty, I} \le \frac{2 \left\| \left\| f \right\| \right\|_{\infty, I}}{t} + \frac{t}{2} \left\| \left\| f'' \right\| \right\|_{\infty, I}, \ \forall \, t > 0. \tag{16.102}$$

The function

$$y(t) = \frac{2}{t} \left\| \left\| f \right\| \right\|_{\infty, I} + \frac{t}{2} \left\| \left\| f'' \right\| \right\|_{\infty, I} \tag{16.103}$$

attains minimal value only for

$$t_{\min} = 2 \sqrt{\frac{\left\| \left\| f \right\| \right\|_{\infty, I}}{\left\| \left\| f'' \right\| \right\|_{\infty, I}}} \tag{16.104}$$

which

$$y_{\min} = 2 \sqrt{\left\| \left\| f \right\| \right\|_{\infty, I} \left\| \left\| f'' \right\| \right\|_{\infty, I}}. \tag{16.105}$$

Consequently

$$\left\| \left\| f' \right\| \right\|_{\infty, I} \le 2 \sqrt{\left\| \left\| f \right\| \right\|_{\infty, I} \left\| \left\| f'' \right\| \right\|_{\infty, I}}, \tag{16.106}$$

proving the claim. See also [1, 2]. ■

Corollary 16.25 *Let* $f \in C^2 (\mathbb{R}_+, X)$, $(X, \|\cdot\|)$ *is a Banach space. We assume here*

$$\left\| \left\| f \right\| \right\|_{\infty, \mathbb{R}_+}, \ \left\| \left\| f'' \right\| \right\|_{\infty, \mathbb{R}_+} < \infty.$$

Then

$$\left\| \left\| f' \right\| \right\|_{\infty, \mathbb{R}_+} \le 2 \sqrt{\left\| \left\| f \right\| \right\|_{\infty, \mathbb{R}_+} \left\| \left\| f'' \right\| \right\|_{\infty, \mathbb{R}_+}}, \tag{16.107}$$

When $X = \mathbb{R}$, Landau proved in 1913, see [14], that 2 is the best possible constant in (16.107). He proved also the optimal inequality [14]

$$\||f'\||_{\infty,\mathbb{R}} \leq \sqrt{2 \||f\||_{\infty,\mathbb{R}} \||f''\||_{\infty,\mathbb{R}}} \tag{16.108}$$

where $f \in C^2 (\mathbb{R}, \mathbb{R})$ and $\||f\||_{\infty,\mathbb{R}}$, $\||f''\||_{\infty,\mathbb{R}} < \infty$. The constant $\sqrt{2}$ is the best possible in (16.108).

We need

Lemma 16.26 (see [9]) *Let* $y(t) := \frac{\mu}{t} + \theta t^{\nu}$, $0 < \nu \leq 1$, $t > 0$, $\mu, \theta > 0$. *Then* y *has only one critical number on* $(0, \infty)$, $t_0 = \left(\frac{\mu}{\nu\theta}\right)^{\frac{1}{(\nu+1)}}$ *so that* $y'(t_0) = 0$, *also* $y''(t_0) > 0$, *and the global minimum of* y *is*

$$y(t_0) = \left(\theta\mu^{\nu}\right)^{\frac{1}{(\nu+1)}} (\nu + 1) \nu^{-\left(\frac{\nu}{\nu+1}\right)}. \tag{16.109}$$

The function y *has no other extrema.*

We give the following L_p vector Landau inequality. Here the vector improper integrals are defined as in [17], pp. 90–91, similar to numerical ones.

Theorem 16.27 *Let* $f \in C^2 (I, X)$, $(X, \|\cdot\|)$ *is a Banach space and* $I = \mathbb{R}_+, \mathbb{R}$. *Let also* $p, q > 1 : \frac{1}{p} + \frac{1}{q} = 1$. *Assume* $\||f\||_{\infty,I}$, $\||f''\||_{p,I} < \infty$, *where*

$$\||f''\||_{p,I} := \left(\int_I \|f''(t)\|^p \, dt\right)^{\frac{1}{p}}.$$

Then

$$\||f'\||_{\infty,I} \leq \left(2 \||f\||_{\infty,I}\right)^{\frac{1}{(q+1)}} \left(\left(\frac{q+1}{q}\right) \||f''\||_{p,I}\right)^{\left(\frac{q}{q+1}\right)}. \tag{16.110}$$

Proof Here $f \in C^2 (I, X)$, $(X, \|\cdot\|)$ is a Banach space, while $I = \mathbb{R}_+, \mathbb{R}$, and $[a, b] \subset I$, $x \in [a, b]$.

Then we apply (16.28) for f' to get

$$\left\| \frac{1}{b-a} \int_a^b f'(t) \, dt - f'(x) \right\| \leq (b-a)^{\frac{1}{q}} \frac{\||f''\||_{p,[a,b]}}{(q+1)^{\frac{1}{q}}}$$

$$\leq (b-a)^{\frac{1}{q}} \frac{\||f''\||_{p,I}}{(q+1)^{\frac{1}{q}}}. \tag{16.111}$$

That is

$$\left\| \frac{f(b) - f(a)}{b - a} - f'(x) \right\| \leq (b - a)^{\frac{1}{q}} \frac{\left\| \left\| f'' \right\| \right\|_{p,I}}{(q+1)^{\frac{1}{q}}}, \tag{16.112}$$

and

$$\left\| f'(x) \right\| - \left\| \frac{f(b) - f(a)}{b - a} \right\| \leq (b - a)^{\frac{1}{q}} \frac{\left\| \left\| f'' \right\| \right\|_{p,I}}{(q+1)^{\frac{1}{q}}}. \tag{16.113}$$

Therefore

$$\left\| f'(x) \right\| \leq \frac{2}{(b - a)} \left\| \left\| f \right\| \right\|_{\infty,I} + (b - a)^{\frac{1}{q}} \frac{\left\| \left\| f'' \right\| \right\|_{p,I}}{(q+1)^{\frac{1}{q}}} \tag{16.114}$$

(under the assumptions $\left\| \left\| f \right\| \right\|_{\infty,I}$, $\left\| \left\| f'' \right\| \right\|_{p,I}$ both are finite).

Consequently it holds

$$\left\| \left\| f' \right\| \right\|_{\infty,I} \leq \frac{2}{t} \left\| \left\| f \right\| \right\|_{\infty,I} + \frac{t^{\frac{1}{q}} \left\| \left\| f'' \right\| \right\|_{p,I}}{(q+1)^{\frac{1}{q}}}, \; \forall \, t > 0. \tag{16.115}$$

The function

$$y(t) = \frac{2}{t} \left\| \left\| f \right\| \right\|_{\infty,I} + \frac{t^{\frac{1}{q}} \left\| \left\| f'' \right\| \right\|_{p,I}}{(q+1)^{\frac{1}{q}}} \tag{16.116}$$

can be written as $y(t) = \frac{\mu}{t} + \theta t^{\nu}$, where

$$\mu := 2 \left\| \left\| f \right\| \right\|_{\infty,I}, \; \theta := \frac{\left\| \left\| f'' \right\| \right\|_{p,I}}{(q+1)^{\frac{1}{q}}}, \; \nu = \frac{1}{q} \in (0, 1). \tag{16.117}$$

By Lemma 16.26, y has global minimum only as an extremum, which is

$$y(t_0) = \left(\frac{\left\| \left\| f'' \right\| \right\|_{p,I}}{(q+1)^{\frac{1}{q}}} \left(2 \left\| \left\| f \right\| \right\|_{\infty,I} \right)^{\frac{1}{q}} \right)^{\frac{1}{\left(\frac{1}{q} + 1 \right)}} \left(\frac{1}{q} + 1 \right) \left(\frac{1}{q} \right)^{-\left(\frac{\frac{1}{q}}{\frac{1}{q} + 1} \right)}$$

$$= \left(\frac{q+1}{q} \right)^{\left(\frac{q}{q+1} \right)} \left(2 \left\| \left\| f \right\| \right\|_{\infty,I} \right)^{\frac{1}{q+1}} \left\| \left\| f'' \right\| \right\|_{p,I}^{\left(\frac{q}{q+1} \right)}, \tag{16.118}$$

where here the only critical number is

$$t_0 = \left(\frac{2q \, (q+1)^{\frac{1}{q}} \, \| \| f \| \|_{\infty, I}}{\| \| f'' \| \|_{p, I}} \right)^{\left(\frac{q}{q+1} \right)}. \tag{16.119}$$

We have proved that

$$\| \| f' \| \|_{\infty, I} \leq \left(\frac{q+1}{q} \right)^{\left(\frac{q}{q+1} \right)} \left(2 \, \| \| f \| \|_{\infty, I} \right)^{\frac{1}{(q+1)}} \| \| f'' \| \|_{p, I}^{\left(\frac{q}{q+1} \right)}, \tag{16.120}$$

establishing the claim. ∎

We continue with a new type of vector Landau inequality

Theorem 16.28 *Let* $f \in C^{n+1}([a, b], X), n \in \mathbb{N}, (X, \| \cdot \|)$ *a Banach space. Assume the boundary conditions* $f^{(i)}(a) = f^{(i)}(b) = 0, for \, i = 1, 2, \ldots, n-1$ *when* $n \geq 2$. *Let* $p, q > 1 : \frac{1}{p} + \frac{1}{q} = 1$. *Then*

$$\| \| f' \| \|_{\infty, [a, b]} \leq \frac{2n \, \| \| f \| \|_{\infty, [a, b]}}{b - a} +$$

$$\frac{1}{(n-1)!} \min \left\{ \left\| \left\| f^{(n+1)} \right\| \right\|_p (B \, (q+1, (n-1) \, q + 1))^{\frac{1}{q}} (b-a)^{n - \frac{1}{p}}, \right.$$

$$\left\| \left\| f^{(n+1)} \right\| \right\|_\infty \frac{(b-a)^n}{n \, (n+1)}, \left\| \left\| f^{(n+1)} \right\| \right\|_1 \cdot \frac{(n-1)^{n-1}}{n^n} (b-a)^{n-1} \right\}. \tag{16.121}$$

Proof Here $f \in C^{n+1}([a, b], X), n \in \mathbb{N}$, with $(X, \| \cdot \|)$ a Banach space. Then $f' \in C^n([a, b], X)$. Additionally assume that $f^{(i)}(a) = f^{(i)}(b) = 0$, for $i = 1, 2, \ldots, n - 1$. That is $(f')^{(i)}(a) = (f')^{(i)}(b) = 0$ for $i = 0, 1, \ldots, n - 2$. Let also $p, q > 1 : \frac{1}{p} + \frac{1}{q} = 1, x \in [a, b]$.

Then by Theorem 16.22, (16.92) we get

$$\left\| \frac{f'(x)}{n} - \frac{1}{b-a} \int_a^b f'(y) \, dy \right\| \leq \frac{1}{n! \, (b-a)} \cdot$$

$$\min \left\{ \left\| \left\| f^{(n+1)} \right\| \right\|_p (B \, (q+1, (n-1) \, q + 1))^{\frac{1}{q}} \left[(x-a)^{nq+1} + (b-x)^{nq+1} \right]^{\frac{1}{q}}, \right.$$

$$\left\| \left\| f^{(n+1)} \right\| \right\|_\infty \left(\frac{(x-a)^{n+1} + (b-x)^{n+1}}{n \, (n+1)} \right), \tag{16.122}$$

$$\left. \left\|f^{(n+1)}\right\|_1 \cdot \frac{(n-1)^{n-1}}{n^n} \max\{(x-a)^n, (b-x)^n\} \right\} =: T.$$

Therefore

$$\left\|\frac{f'(x)}{n} - \frac{f(b)-f(a)}{b-a}\right\| \leq T,$$

and

$$\left\|\frac{f'(x)}{n}\right\| - \left\|\frac{f(b)-f(a)}{b-a}\right\| \leq T, \tag{16.123}$$

leading to

$$\left\|\frac{f'(x)}{n}\right\| \leq \left\|\frac{f(b)-f(a)}{b-a}\right\| + T \tag{16.124}$$

$$\leq \frac{2\,\|f\|_{\infty,[a,b]}}{b-a} + T,$$

and

$$\left\|f'(x)\right\| \leq \frac{2n\,\|f\|_{\infty,[a,b]}}{b-a} + nT. \tag{16.125}$$

We notice that

$$T \leq \frac{1}{n!\,(b-a)} \min\left\{ \left\|f^{(n+1)}\right\|_p (B(q+1,(n-1)q+1))^{\frac{1}{q}} (b-a)^{n+\frac{1}{q}}, \right.$$

$$\left. \left\|f^{(n+1)}\right\|_\infty \frac{(b-a)^{n+1}}{n(n+1)}, \left\|f^{(n+1)}\right\|_1 \cdot \frac{(n-1)^{n-1}}{n^n} (b-a)^n \right\} =: M, \tag{16.126}$$

(by $(x-a)^\alpha + (b-x)^\alpha \leq (b-a)^\alpha$, $\alpha \geq 1$).

Consequently we derive

$$\left\|f'(x)\right\| \leq \frac{2n\,\|f\|_{\infty,[a,b]}}{b-a} + nM, \tag{16.127}$$

leading to

$$\left\|f'\right\|_{\infty,[a,b]} \leq \frac{2n\,\|f\|_{\infty,[a,b]}}{b-a} + nM, \tag{16.128}$$

and proving the claim. ■

Here also see the related results of [1, 2].
We finish with

Corollary 16.29 (to Theorem 16.27) *Let* $f \in C^2 (I, X)$, $(X, \|\cdot\|)$ *a Banach space and* $I = \mathbb{R}_+, \mathbb{R}$. *Assume* $\||f|\|_{\infty, I}$, $\||f''|\|_{2, I} < \infty$. *Then*

$$\||f'|\|_{\infty, I} \leq \left(2 \||f|\|_{\infty, I}\right)^{\frac{1}{3}} \left(\frac{3}{2} \||f''|\|_{2, I}\right)^{\frac{2}{3}}. \tag{16.129}$$

Corollary 16.30 (to Theorem 16.28) *Let* $f \in C^4 ([a, b], X)$, $(X, \|\cdot\|)$ *a Banach space. Assume the boundary conditions* $f'(a) = f'(b) = f''(a) = f''(b) = 0$. *Then*

$$\||f'|\|_{\infty, [a,b]} \leq \frac{6 \||f|\|_{\infty, [a,b]}}{b - a} +$$

$$\frac{1}{2} \min \left\{ \||f^{(4)}|\|_2 \frac{(b - a)^{\frac{5}{2}}}{\sqrt{105}}, \ \||f^{(4)}|\|_\infty \frac{(b - a)^3}{12}, \ \||f^{(4)}|\|_1 \cdot \frac{4 (b - a)^2}{27} \right\}. \tag{16.130}$$

References

1. A. Aglic Aljinovic, L.J. Marangunic, J. Pecaric, On Landau type inequalities via extension of Montgomery identity, Euler and Fink identities. Nonlinear Funct. Anal. & Appl. **10**(2), 273-283 (2005)
2. A. Aglic Aljinovic, L.J. Marangunic, J. Pecaric, On Landau type inequalities via Ostrowski inequalities. Nonlinear Funct. Anal. & Appl. **10**(4), 565-579 (2005)
3. A. Aglic Aljinovic, J. Pecaric, A. Vukelic, The extension of montgomery identity via Fink identity with applications. J. Inequalities Appl. **1**, 67-80 (2005)
4. G.A. Anastassiou, Ostrowski type inequalities. Proc. AMS **123**, 3775–3781 (1995)
5. G.A. Anastassiou, *Quantitative Approximations* (Chapman and Hall/CRC, Boca Raton, 2001)
6. G.A. Anastassiou, Univariate Ostrowski inequalities, revisited. Manatshefte Math. **135**, 175–189 (2002)
7. G.A. Anastassiou, Opial type inequalities for vector valued functions. Bull. Hellenic Math. Soc. **55**, 1–8 (2008)
8. G.A. Anastassiou, *Advanced Inequalities* (World Scientific, Singapore, 2011)
9. G.A. Anastassiou, Left Caputo fractional $\|\cdot\|_\infty$-Landau inequalities. Appl. Math. Lett. **24**(7), 1149–1154 (2011)
10. G.A. Anastassiou, Ostrowski and Landau inequalities for Banach space valued functions. Math. Comput. Modell. **55**, 312–329 (2012)
11. N.S. Barnett, C. Buse, P. Cerone, S. Dragomir, Ostrowski's Inequality for vector-valued functions and applications. Comput. Math. Appl. **44**, 559–572 (2002)
12. A.M. Fink, Bounds on the deviation of a function from its averages. Czech. Math. J. **42**(2), 289–310 (1992)
13. G. Ladas, V. Laksmikantham, *Differential Equations in Abstract Spaces* (Academic Press, New York, 1972)

14. E. Landau, Einige Ungleichungen für zweimal differenzierbaren Fuktionen. Proc. Lond. Math. Soc. Ser. **2**(13), 43-49 (1913)
15. A. Ostrowski, Über die Absolutabweichung einer differentiebaren Funktion von ihrem Integralmittelwert. Comment. Math. Helv. **10**, 226–227 (1938)
16. L. Schwartz, *Analyse Mathematique* (Hermann, Paris, 1967)
17. G. Shilov, *Elementary Functional Analysis* (The MIT Press Cambridge, Massachusetts, 1974)

Chapter 17
Multidimensional Ostrowski Type Inequalities for Banach Space Valued Functions

Here we are dealing with smooth functions from a real box to a Banach space. For these we establish vector multivariate sharp Ostrowski type inequalities to all possible directions. In establishing them we prove interesting multivariate vector identities using integration by parts and other basic analytical methods. It follows [9].

17.1 Introduction

In 1938, Ostrowski [12] proved the following inequality:

Theorem 17.1 *Let* $f : [a, b] \to \mathbb{R}$ *be continuous on* $[a, b]$ *and differentiable on* (a, b) *whose derivative* $f' : (a, b) \to \mathbb{R}$ *is bounded on* (a, b)*, i.e.,* $\|f'\|_\infty :=$ $\sup_{t \in (a,b)} |f'(t)| < \infty$*. Then*

$$\left| \frac{1}{b-a} \int_a^b f(t)\, dt - f(x) \right| \leq \left[\frac{1}{4} + \frac{\left(x - \frac{a+b}{2}\right)^2}{(b-a)^2} \right] (b-a)\, \|f'\|_\infty, \quad (17.1)$$

for any $x \in [a, b]$*. The constant* $\frac{1}{4}$ *is the best possible.*

Since then there has been a lot of activity around these inequalities with important applications to Numerical Analysis and Probability.

This chapter is also greatly motivated by the following result:

Theorem 17.2 (see [1]) *Let* $f \in C^1 \left(\prod_{i=1}^k [a_i, b_i] \right)$*, where* $a_i < b_i$*;* $a_i, b_i \in \mathbb{R}$*,* $i = 1, \ldots, k$*, and let* $\vec{x_0} := (x_{01}, \ldots, x_{0k}) \in \prod_{i=1}^k [a_i, b_i]$ *be fixed. Then*

© Springer International Publishing Switzerland 2016
G.A. Anastassiou, *Intelligent Comparisons: Analytic Inequalities*,
Studies in Computational Intelligence 609,
DOI 10.1007/978-3-319-21121-3_17

$$\left| \frac{1}{\prod_{i=1}^{k} (b_i - a_i)} \int_{a_1}^{b_1} \cdots \int_{a_i}^{b_i} \cdots \int_{a_k}^{b_k} f(z_1, \ldots, z_k) \, dz_1 \ldots dz_k - f(\vec{x_0}) \right| \leq$$

$$(17.2)$$

$$\sum_{i=1}^{k} \left(\frac{(x_{0i} - a_i)^2 + (b_i - x_{0i})^2}{2(b_i - a_i)} \right) \left\| \frac{\partial f}{\partial z_i} \right\|_{\infty}.$$

Inequality (17.2) is sharp, here the optimal function is

$$f^*(z_1, \ldots, z_k) := \sum_{i=1}^{k} |z_i - x_{0i}|^{\alpha_i}, \quad \alpha_i > 1.$$

Clearly inequality (17.2) generalizes inequality (17.1) to multidimension.

In this chapter we establish multivariate Ostrowski inequalities for smooth functions from a real box to a Banach space. These involve the norms $\|\cdot\|_p$, $1 \leq p \leq \infty$. Some of these inequalities are sharp.

17.2 Background

(see [14], pp. 83–94)

Let $f(t)$ be a function defined on $[a, b] \subseteq \mathbb{R}$ taking values in a real or complex normed linear space $(X, \|\cdot\|)$. Then $f(t)$ is said to be differentiable at a point $t_0 \in [a, b]$ if the limit

$$f'(t_0) := \lim_{h \to 0} \frac{f(t_0 + h) - f(t_0)}{h} \qquad (17.3)$$

exists in X, the convergence is in $\|\cdot\|$. This is called the derivative of $f(t)$ at $t = t_0$.

We call $f(t)$ differentiable on $[a, b]$, if there exists $f'(t) \in X$ for all $t \in [a, b]$.

Similarly and inductively are defined higher order derivatives of f, denoted f'', $f^{(3)}, \ldots, f^{(k)}$, $k \in \mathbb{N}$, just as for numerical functions.

For all the properties of derivatives see [14], pp. 83–86.

Let now $(X, \|\cdot\|)$ be a Banach space, and $f : [a, b] \to X$.

We define the vector valued Riemann integral $\int_a^b f(t) \, dt \in X$ as the limit of the vector valued Riemann sums in X, convergence is in $\|\cdot\|$. The definition is as for the numerical valued functions.

If $\int_a^b f(t) \, dt \in X$ we call f integrable on $[a, b]$. If $f \in C([a, b], X)$, then f is integrable, [14], p. 87.

For all the properties of vector valued Riemann integrals see [14], pp. 86–91.

We define the space $C^n([a, b], X)$, $n \in \mathbb{N}$, of n-times continuously differentiable functions from $[a, b]$ into X; here continuity is with respect to $\|\cdot\|$ and defined in the usual way as for numerical functions.

Let $(X, \|\cdot\|)$ be a Banach space and $f \in C^n([a, b], X)$, then we have the vector valued Taylor's formula, see [14], pp. 93–94, and also [13], (IV, 9; 47).

It holds

$$f(y) - f(x) - f'(x)(y - x) - \frac{1}{2}f''(x)(y - x)^2 - \cdots - \frac{1}{(n-1)!}f^{(n-1)}(x)(y - x)^{n-1}$$

$$= \frac{1}{(n-1)!} \int_x^y (y - t)^{n-1} f^{(n)}(t) \, dt, \ \forall \, x, y \in [a, b]. \tag{17.4}$$

In particular (17.4) is true when $X = \mathbb{R}^m, \mathbb{C}^m, m \in \mathbb{N}$, etc.

A function $f(t)$ with values in a normed linear space X is said to be piecewise continuous (see [14], p. 85) on the interval $a \leq t \leq b$ if there exists a partition $a = t_0 < t_1 < t_2 < \cdots < t_n = b$ such that $f(t)$ is continuous on every open interval $t_k < t < t_{k+1}$ and has finite limits $f(t_0 + 0), f(t_1 - 0), f(t_1 + 0), f(t_2 - 0),$ $f(t_2 + 0), \ldots, f(t_n - 0)$.

Here $f(t_k - 0) = \lim_{t \uparrow t_k} f(t), f(t_k + 0) = \lim_{t \downarrow t_k} f(t)$.

The values of $f(t)$ at the points t_k can be arbitrary or even undefined.

A function $f(t)$ with values in normed linear space X is said to be piecewise smooth on $[a, b]$, if it is continuous on $[a, b]$ and has a derivative $f'(t)$ at all but a finite number of points of $[a, b]$, and if $f'(t)$ is piecewise continuous on $[a, b]$ (see [14], p. 85).

Let $u(t)$ and $v(t)$ be two piecewise smooth functions on $[a, b]$, one a numerical function and the other a vector function with values in Banach space X. Then we have the following integration by parts formula

$$\int_a^b u(t) \, dv(t) = u(t) \, v(t) \, |_a^b - \int_a^b v(t) \, du(t), \tag{17.5}$$

see [14], p. 93.

We mention also the mean value theorem for Banach space valued functions.

Theorem 17.3 (see [11], p. 3) *Let $f \in C([a, b], X)$, where X is a Banach space. Assume f' exists on $[a, b]$ and $\|f'(z)\| \leq K$, $a < t < b$, then*

$$\|f(b) - f(a)\| \leq K(b - a). \tag{17.6}$$

Here the multiple Riemann integral of a function from a real box to a Banach space is defined similarly to numerical one however convergence is with respect to $\|\cdot\|$. Similarly are defined the vector valued partial derivatives as in the numerical case.

We mention the equality of vector valued mixed partial derivatives.

Proposition 17.4 (see Proposition 4.11 of [10], p. 90) *Let* $Q = (a, b) \times (c, d) \subseteq \mathbb{R}^2$ *and* $f \in C(Q, X)$, *where* $(X, \|\cdot\|)$ *is a Banach space. Assume that* $\frac{\partial}{\partial t} f(s, t)$, $\frac{\partial}{\partial s} f(s, t)$ *and* $\frac{\partial^2}{\partial t \partial s} f(s, t)$ *eixst and are continuous for* $(s, t) \in Q$, *then* $\frac{\partial^2}{\partial s \partial t} f(s, t)$ *exists for* $(s, t) \in Q$ *and*

$$\frac{\partial^2}{\partial s \partial t} f(s, t) = \frac{\partial^2}{\partial t \partial s} f(s, t), \text{ for } (s, t) \in Q.$$

Notice also that

$$\left(\frac{1}{4} + \frac{\left(x - \frac{a+b}{2}\right)^2}{(b-a)^2} \right)(b-a) = \frac{(x-a)^2 + (b-x)^2}{2(b-a)}, \ \forall \, x \in [a, b]. \tag{17.7}$$

17.3 Main Results

Here we present the first vector multivariate Ostrowski type inequality, see also the real analog p. 507, Theorem 23.1 of [1, 2].

Theorem 17.5 *Let* $(X, \|\cdot\|)$ *be a Banach space and* $f \in C^1 \left(\prod_{i=1}^{k} [a_i, b_i], X \right)$, $a_i < b_i$; $a_i, b_i \in \mathbb{R}$, $i = 1, \ldots, k$, *and let* $\vec{x_0} := (x_{01}, \ldots, x_{0k}) \in \prod_{i=1}^{k} [a_i, b_i]$ *be fixed. Here* i_0 *is a fixed unit vector in* X. *Denote*

$$\left\| \frac{\partial f}{\partial z_i} \right\|_\infty := \sup_{x \in \prod_{i=1}^{k} [a_i, b_i]} \left\| \frac{\partial f(x)}{\partial z_i} \right\|.$$

Then

$$\left\| \frac{1}{\prod_{i=1}^{k} (b_i - a_i)} \int_{a_1}^{b_1} \cdots \int_{a_i}^{b_i} \cdots \int_{a_k}^{b_k} f(z_1, \ldots, z_k) \, dz_1 \ldots dz_k - f(\vec{x_0}) \right\| \leq$$

$$\tag{17.8}$$

$$\sum_{i=1}^{k} \left(\frac{(x_{0i} - a_i)^2 + (b_i - x_{0i})^2}{2(b_i - a_i)} \right) \left\| \frac{\partial f}{\partial z_i} \right\|_\infty.$$

Inequality (17.8) is sharp, the optimal function is

$$f^*(z_1, \ldots, z_k) := \left(\sum_{i=1}^{k} |z_i - x_{0i}|^{\alpha_i} \right) \cdot i_0, \ \alpha_i > 1.$$

Proof Set $\vec{z} := (z_1, \ldots, z_k)$. Consider $g_{\vec{z}}(t) := f(\vec{x_0} + t(\vec{z} - \vec{x_0}))$, $t \geq 0$. Note that $g_{\vec{z}}(0) = f(\vec{x_0})$, $g_{\vec{z}}(1) = f(\vec{z})$. Hence

$$\left\| f(\vec{z}) - f(\vec{x_0}) \right\| = \left\| g_{\vec{z}}(1) - g_{\vec{z}}(0) \right\| \leq$$

$$\left\| g'_{\vec{z}}(\xi) \right\|_{\infty,(0,1)} (1 - 0) = \left\| g'_{\vec{z}}(\xi) \right\|_{\infty,(0,1)}.$$

Since

$$g'_{\vec{z}}(\xi) = (z_1 - x_{01}) \frac{\partial f}{\partial z_1} (x_0 + \xi(z - x_0)) + \cdots + (z_k - x_{0k}) \frac{\partial f}{\partial z_k} (x_0 + \xi(z - x_0))$$

we get

$$\left\| f(\vec{z}) - f(\vec{x_0}) \right\| \leq \sum_{i=1}^{k} |z_i - x_{0i}| \left\| \frac{\partial f}{\partial z_i} (\vec{x_0} + \xi(\vec{z} - \vec{x_0})) \right\|_{\infty,(0,1)}$$

$$\leq \sum_{i=1}^{k} |z_i - x_{0i}| \left\| \frac{\partial f}{\partial z_i} \right\|_{\infty}. \tag{17.9}$$

Next we see that

$$\left\| \frac{1}{\prod_{i=1}^{k} (b_i - a_i)} \int_{a_1}^{b_1} \cdots \int_{a_i}^{b_i} \cdots \int_{a_k}^{b_k} f(z_1, \ldots, z_k) \, dz_1 \ldots dz_k - f(\vec{x_0}) \right\|$$

$$= \frac{1}{\prod_{i=1}^{k} (b_i - a_i)} \left\| \int_{a_1}^{b_1} \cdots \int_{a_i}^{b_i} \cdots \int_{a_k}^{b_k} (f(\vec{z}) - f(\vec{x_0})) \, d\vec{z} \right\|$$

$$\leq \frac{1}{\prod_{i=1}^{k} (b_i - a_i)} \int_{a_1}^{b_1} \cdots \int_{a_i}^{b_i} \cdots \int_{a_k}^{b_k} \left\| f(\vec{z}) - f(\vec{x_0}) \right\| d\vec{z}$$

$$\overset{(17.9)}{\leq} \frac{1}{\prod_{i=1}^{k} (b_i - a_i)} \int_{a_1}^{b_1} \cdots \int_{a_i}^{b_i} \cdots \int_{a_k}^{b_k} \left(\sum_{i=1}^{k} |z_i - x_{0i}| \left\| \frac{\partial f}{\partial z_i} \right\|_{\infty} \right) dz_1 \ldots dz_k$$

$$= \frac{1}{\prod_{i=1}^{k} (b_i - a_i)} \left[\sum_{i=1}^{k} \left(\int_{a_1}^{b_1} \cdots \int_{a_i}^{b_i} \cdots \int_{a_k}^{b_k} |z_i - x_{0i}| \, dz_1 \ldots dz_k \right) \left\| \frac{\partial f}{\partial z_i} \right\|_{\infty} \right] =: (*).$$

Here notice that

$$\int_{a_i}^{b_i} |z_i - x_{0i}| \, dz_i = \frac{(x_{0i} - a_i)^2 + (b_i - x_{0i})^2}{2}, \quad i = 1, \ldots, k. \tag{17.10}$$

Therefore, by (17.10) we obtain

$$(*) = \frac{1}{\prod_{j=1}^{k} (b_j - a_j)} \cdot$$

$$\left[\sum_{i=1}^{k} \left\| \frac{\partial f}{\partial z_i} \right\|_{\infty} \left(\frac{(x_{0i} - a_i)^2 + (b_i - x_{0i})^2}{2} \right) \left(\prod_{\substack{j=1 \\ j \neq i}}^{k} (b_j - a_j) \right) \right]$$

$$= \sum_{i=1}^{k} \left(\frac{(x_{0i} - a_i)^2 + (b_i - x_{0i})^2}{2 (b_i - a_i)} \right) \left\| \frac{\partial f}{\partial z_i} \right\|_{\infty},$$

so that we establish inequality (17.8).

In the following we prove the sharpness of (17.8): Notice that $f^* \left(\overrightarrow{x_0} \right) = 0$ and

$$\frac{\partial f^*}{\partial z_i} \left(\overrightarrow{z} \right) = \alpha_i |z_i - x_0|^{\alpha_i - 1} \cdot sgn \, (z_i - x_{0i}) \cdot i_0, \ \alpha_i > 1.$$

In particular we find

$$\left\| \frac{\partial f^* \left(\overrightarrow{x} \right)}{\partial z_i} \right\| = \alpha_i |z_i - x_{0i}|^{\alpha_i - 1},$$

and $(a_i \leq z_i \leq b_i)$

$$\left\| \frac{\partial f^*}{\partial z_i} \right\|_{\infty} = \alpha_i \, (\max \, (b_i - x_{0i}, x_{0i} - a_i))^{\alpha_i - 1}.$$

Consequently, we observe

$$R.H.S. \, (17.8) = \sum_{i=1}^{k} \left(\frac{(x_{0i} - a_i)^2 + (b_i - x_{0i})^2}{2 (b_i - a_i)} \right) \left\| \frac{\partial f^*}{\partial z_i} \right\|_{\infty} =$$

$$\sum_{i=1}^{k} \left(\frac{(x_{0i} - a_i)^2 + (b_i - x_{0i})^2}{2 (b_i - a_i)} \right) \alpha_i \, (\max \, (b_i - x_{0i}, x_{0i} - a_i))^{\alpha_i - 1},$$

and

$$\lim_{\substack{\alpha_i \to 1 \\ i=1,\dots,k}} R.H.S. \, (17.8) = \sum_{i=1}^{k} \left(\frac{(x_{0i} - a_i)^2 + (b_i - x_{0i})^2}{2 (b_i - a_i)} \right). \qquad (17.11)$$

Moreover, we get that

$$L.H.S.\,(17.8) = \frac{1}{\prod_{i=1}^{k}(b_i - a_i)} \cdot$$

$$\left(\int_{a_1}^{b_1} \cdots \int_{a_i}^{b_i} \cdots \int_{a_k}^{b_k} \left(\sum_{i=1}^{k} |z_i - x_{0i}|^{\alpha_i} \right) dz_1 \ldots dz_k \right) \|i_0\|$$

$$= \frac{1}{\prod_{i=1}^{k}(b_i - a_i)} \left[\sum_{i=1}^{k} \left(\int_{a_1}^{b_1} \cdots \int_{a_i}^{b_i} \cdots \int_{a_k}^{b_k} |z_i - x_{0i}|^{\alpha_i} \, dz_1 \ldots dz_k \right) \right]$$

$$= \frac{1}{\prod_{i=1}^{k}(b_i - a_i)} \left[\sum_{i=1}^{k} \left(\frac{(x_{0i} - a_i)^{\alpha_i+1} + (b_i - x_{0i})^{\alpha_i+1}}{\alpha_i + 1} \right) \left(\prod_{\substack{j=1 \\ j \neq i}}^{k} (b_j - a_j) \right) \right]$$

$$= \sum_{i=1}^{k} \left(\frac{(x_{0i} - a_i)^{\alpha_i+1} + (b_i - x_{0i})^{\alpha_i+1}}{(\alpha_i + 1)(b_i - a_i)} \right),$$

and

$$\lim_{\substack{\alpha_i \to 1 \\ i=1,\ldots,k}} L.H.S.\,(17.8) = \sum_{i=1}^{k} \left(\frac{(x_{0i} - a_i)^2 + (b_i - x_{0i})^2}{2(b_i - a_i)} \right). \tag{17.12}$$

At the end from (17.11) and (17.12) we obtain that

$$\lim_{\substack{\alpha_i \to 1 \\ i=1,\ldots,k}} L.H.S.\,(17.8) = \lim_{\substack{\alpha_i \to 1 \\ i=1,\ldots,k}} R.H.S.\,(17.8),$$

proving the inequality (17.8) is sharp. ∎

Regarding vector higher order derivatives we give the following results:

Theorem 17.6 *Let* $(X, \|\cdot\|)$ *be a Banach space and* $f \in C^{n+1}\left(\prod_{i=1}^{k} [a_i, b_i], X \right)$, $n \in \mathbb{N}$ *and fixed* $\overrightarrow{x_0} \in \prod_{i=1}^{k} [a_i, b_i]$, $k \geq 1$, *such that all vector partial derivatives* $f_\alpha := \frac{\partial^\alpha f}{\partial z^\alpha}$, *where* $\alpha = (\alpha_1, \ldots, \alpha_k)$, $\alpha_i \in \mathbb{Z}^+$, $i = 1, \ldots, k$, $|\alpha| = \sum_{i=1}^{k} \alpha_i = j$, $j = 1, \ldots, n$ *fulfill* $f_\alpha\left(\overrightarrow{x_0} \right) = 0$. *Then*

$$\left\| \frac{1}{\prod_{i=1}^{k} (b_i - a_i)} \int_{\prod_{i=1}^{k} [a_i, b_i]} f(\overrightarrow{z}) \, d\overrightarrow{z} - f(\overrightarrow{x_0}) \right\| \le$$

$$\left(\frac{D_{n+1}(f)}{(n+1)! \prod_{i=1}^{k} (b_i - a_i)} \right) \int_{\prod_{i=1}^{k} [a_i, b_i]} \left(\left\| \overrightarrow{z} - \overrightarrow{x_0} \right\|_{l_1} \right)^{n+1} d\overrightarrow{z}, \qquad (17.13)$$

where

$$D_{n+1}(f) := \max_{\alpha : |\alpha| = n+1} \|\|f\|\|_\infty, \qquad (17.14)$$

and

$$\left\| \overrightarrow{z} - \overrightarrow{x_0} \right\|_{l_1} := \sum_{i=1}^{k} |z_i - x_{0i}|. \qquad (17.15)$$

Proof Take $g_{\overrightarrow{z}}(t) := f(\overrightarrow{x_0} + t(\overrightarrow{z} - \overrightarrow{x_0}))$, $0 \le t \le 1$. Notice that $g_{\overrightarrow{z}}(0) = f(\overrightarrow{x_0})$ and $g_{\overrightarrow{z}}(1) = f(\overrightarrow{z})$. The jth derivative of $g_{\overrightarrow{z}}(t)$, based on Proposition 17.4, is given by

$$g_{\overrightarrow{z}}^{(j)}(t) = \left[\left(\sum_{i=1}^{k} (z_i - x_{0i}) \frac{\partial}{\partial z_i} \right)^j f \right] (x_{01} + t(z_1 - x_{01}), \ldots, x_{0k} + t(z_k - x_{0k}))$$

$$(17.16)$$

and

$$g_{\overrightarrow{z}}^{(j)}(0) = \left[\left(\sum_{i=1}^{k} (z_i - x_{0i}) \frac{\partial}{\partial z_i} \right)^j f \right] (\overrightarrow{x_0}), \qquad (17.17)$$

for $j = 1, \ldots, n+1$.

Let f_α be a partial derivative of $f \in C^{n+1} \left(\prod_{i=1}^{k} [a_i, b_i] \right)$. Because by assumption of the theorem we have $f_\alpha(\overrightarrow{x_0}) = 0$ for all $\alpha : |\alpha| = j$, $j = 1, \ldots, n$, we find that

$$g_{\overrightarrow{z}}^{(j)}(0) = 0, \quad j = 1, \ldots, n.$$

Hence by vector Taylor's theorem (17.4) we see that

$$f(\overrightarrow{z}) - f(\overrightarrow{x_0}) = \sum_{j=1}^{n} \frac{g_{\overrightarrow{z}}^{(j)}(0)}{j!} + R_n(\overrightarrow{z}, 0) = R_n(\overrightarrow{z}, 0), \qquad (17.18)$$

where

$$R_n\left(\overrightarrow{z},0\right) := \int_0^1 \left(\int_0^{t_1} \cdots \left(\int_0^{t_{n-1}} \left(g_{\overrightarrow{z}}^{(n)}\left(t_n\right) - g_{\overrightarrow{z}}^{(n)}\left(0\right)\right) dt_n\right) \cdots\right) dt_1. \quad (17.19)$$

Therefore,

$$\left\|R_n\left(\overrightarrow{z},0\right)\right\| \le \int_0^1 \left(\int_0^{t_1} \cdots \left(\int_0^{t_{n-1}} \left\|g_{\overrightarrow{z}}^{(n+1)}\left(\xi\left(t_n\right)\right)\right\|_{\infty} t_n dt_n\right) \cdots\right) dt_1, \quad (17.20)$$

by the vector mean value Theorem 17.3 applied on $g_{\overrightarrow{z}}^{(n)}$ over $(0, t_n)$. Moreover, we get

$$\left\|R_n\left(\overrightarrow{z},0\right)\right\| \le \left\|g_{\overrightarrow{z}}^{(n+1)}\right\|_{\infty,[0,1]} \int_0^1 \int_0^{t_1} \cdots \left(\int_0^{t_{n-1}} t_n dt_n\right) \ldots dt_1$$

$$= \frac{\left\|g_{\overrightarrow{z}}^{(n+1)}\right\|_{\infty,[0,1]}}{(n+1)!}. \quad (17.21)$$

However, there exists a $t_0 \in [0, 1]$ such that $\left\|g_{\overrightarrow{z}}^{(n+1)}\right\|_{\infty,[0,1]} = \left\|g_{\overrightarrow{z}}^{(n+1)}\left(t_0\right)\right\|$. That is,

$$\left\|g_{\overrightarrow{z}}^{(n+1)}\right\|_{\infty,[0,1]} = \left\|\left[\left(\sum_{i=1}^{k}\left(z_i - x_{0i}\right)\frac{\partial}{\partial z_i}\right)^{n+1} f\right]\left(\overrightarrow{x_0} + t_0\left(\overrightarrow{z} - \overrightarrow{x_0}\right)\right)\right\|$$

$$\le \left[\left(\sum_{i=1}^{k}\left|z_i - x_{0i}\right|\left\|\frac{\partial}{\partial z_i}\right\|\right)^{n+1} f\right]\left(\overrightarrow{x_0} + t_0\left(\overrightarrow{z} - \overrightarrow{x_0}\right)\right).$$

I.e.,

$$\left\|g_{\overrightarrow{z}}^{(n+1)}\right\|_{\infty,[0,1]} \le \left[\left(\sum_{i=1}^{k}\left|z_i - x_{0i}\right|\left\|\frac{\partial}{\partial z_i}\right\|_{\infty}\right)^{n+1} f\right]. \quad (17.22)$$

Hence by (17.22) we get

$$\left\|f\left(\overrightarrow{z}\right) - f\left(\overrightarrow{x_0}\right)\right\| = \left\|R_n\left(\overrightarrow{z},0\right)\right\|$$

$$\le \frac{\left[\left(\sum_{i=1}^{k}\left|z_i - x_{0i}\right|\left\|\frac{\partial}{\partial z_i}\right\|_{\infty}\right)^{n+1} f\right]}{(n+1)!}. \quad (17.23)$$

In the following we observe

$$\left\| \frac{1}{\prod_{i=1}^{k} (b_i - a_i)} \int_{\prod_{i=1}^{k} [a_i, b_i]} f\left(\vec{z}\right) d\vec{z} - f\left(\vec{x_0}\right) \right\| =$$

$$\frac{1}{\prod_{i=1}^{k} (b_i - a_i)} \left\| \int_{\prod_{i=1}^{k} [a_i, b_i]} \left(f\left(\vec{z}\right) - f\left(\vec{x_0}\right)\right) d\vec{z} \right\| \leq$$

$$\frac{1}{\prod_{i=1}^{k} (b_i - a_i)} \int_{\prod_{i=1}^{k} [a_i, b_i]} \left\| f\left(\vec{z}\right) - f\left(\vec{x_0}\right) \right\| d\vec{z} \leq$$

(by (17.23))

$$\frac{1}{(n+1)! \prod_{i=1}^{k} (b_i - a_i)} \int_{\prod_{i=1}^{k} [a_i, b_i]} \left(\left(\sum_{i=1}^{k} |z_i - x_{0i}| \left\| \frac{\partial}{\partial z_i} \right\|_{\infty} \right)^{n+1} f \right) d\vec{z} \leq$$

$$\left(\frac{D_{n+1}(f)}{(n+1)! \prod_{i=1}^{k} (b_i - a_i)} \right) \int_{\prod_{i=1}^{k} [a_i, b_i]} \left(\left\| \vec{z} - \vec{x_0} \right\|_{l_1} \right)^{n+1} d\vec{z}. \qquad (17.24)$$

This establishes inequality (17.13). ∎

Corollary 17.7 (to Theorem 17.6) *Under the assumptions of Theorem 17.6 we find that*

$$\left\| \frac{1}{\prod_{i=1}^{k} (b_i - a_i)} \int_{\prod_{i=1}^{k} [a_i, b_i]} f\left(\vec{z}\right) d\vec{z} - f\left(\vec{x_0}\right) \right\| \leq$$

$$\frac{1}{(n+1)! \prod_{i=1}^{k} (b_i - a_i)} \int_{\prod_{i=1}^{k} [a_i, b_i]} \left(\left(\sum_{i=1}^{k} |z_i - x_{0i}| \left\| \frac{\partial}{\partial z_i} \right\|_{\infty} \right)^{n+1} f \right) d\vec{z}. \qquad (17.25)$$

Furthermore, (17.25) is sharp: when n is odd it is attained by

$$f^*(z_1, \ldots, z_k) := \left(\sum_{i=1}^{k} (z_i - x_{0i})^{n+1} \right) \cdot i_0 \qquad (17.26)$$

while when n is even the optimal function is

$$\overline{f}(z_1, \ldots, z_k) := \left(\sum_{i=1}^{k} |z_i - x_{0i}|^{n+\alpha_i} \right) \cdot i_0, \ \alpha_i > 1, \qquad (17.27)$$

where $i_0 \in X : \|i_0\| = 1$.

Proof Inequality (17.25) comes directly from (17.24).

Next we prove the sharpness of (17.25).

(i) When n is odd: Notice that $f^* (\overrightarrow{x_0}) = 0$ and

$$\left\| \frac{\partial^{n+1} f^*}{\partial z_i^{n+1}} \right\|_{\infty} = (n+1)!,$$

furthermore any mixed partial of f^* equals zero. Thus by plugging f^* into (17.25) we observe that

$$R.H.S. (17.25) =$$

$$\frac{1}{(n+1)! \prod_{i=1}^{k} (b_i - a_i)} \int_{\prod_{i=1}^{k} [a_i, b_i]} \left[\sum_{i=1}^{k} |z_i - x_{0i}|^{n+1} (n+1)! \right] d\overrightarrow{z}$$

$$= \frac{1}{\prod_{i=1}^{k} (b_i - a_i)} \int_{\prod_{i=1}^{k} [a_i, b_i]} \left(\sum_{i=1}^{k} (z_i - x_{0i})^{n+1} \right) d\overrightarrow{z} \qquad (17.28)$$

$$= \frac{1}{\prod_{i=1}^{k} (b_i - a_i)} \int_{\prod_{i=1}^{k} [a_i, b_i]} f^* (\overrightarrow{z}) d\overrightarrow{z} = L.H.S. (17.25),$$

proving the sharpness of (17.25) when n is odd.

(ii) When n is even: Notice that $\overline{f} (\overrightarrow{x_0}) = 0$ and any mixed partial of $\overline{f}$ equals zero. Especially we observe that

$$\left\| \frac{\partial^{n+1} \overline{f} (\overrightarrow{z})}{\partial z_i^{n+1}} \right\| = \left(\prod_{j=0}^{n} (n + \alpha_i - j) \right) |z_i - x_{0i}|^{\alpha_i - 1}, \quad \alpha_i > 1, \qquad (17.29)$$

and

$$\left\| \frac{\partial^{n+1} \overline{f}}{\partial z_i^{n+1}} \right\|_{\infty} = \left(\prod_{j=0}^{n} (n + \alpha_i - j) \right) \|z_i - x_{0i}\|_{\infty}^{\alpha_i - 1} \qquad (17.30)$$

(here $\|z_i - x_{0i}\|_{\infty} < +\infty$), all $i = 1, \ldots, k$. Hence by plugging $\overline{f}$ into (17.25) we obtain

$$\lim_{\text{all } \alpha_i \to 1} R.H.S. (17.25) = \frac{1}{(n+1)! \prod_{i=1}^{k} (b_i - a_i)} \cdot$$

$$\lim_{\text{all } \alpha_i \to 1} \int_{\prod_{i=1}^{k} [a_i, b_i]} \left[\sum_{i=1}^{k} |z_i - x_{0i}|^{n+1} \left(\prod_{j=0}^{n} (n + \alpha_i - j) \right) \|z_i - x_{0i}\|_{\infty}^{\alpha_i - 1} \right] d\overrightarrow{z}$$

$$= \frac{1}{\prod_{i=1}^{k} (b_i - a_i)} \int_{\prod_{i=1}^{k} [a_i, b_i]} \left(\sum_{i=1}^{k} |z_i - x_{0i}|^{n+1} \right) d\vec{z}. \tag{17.31}$$

Furthermore,

$$\lim_{\text{all } \alpha_i \to 1} L.H.S. (17.25) =$$

$$\frac{1}{\prod_{i=1}^{k} (b_i - a_i)} \lim_{\text{all } \alpha_i \to 1} \int_{\prod_{i=1}^{k} [a_i, b_i]} \left(\sum_{i=1}^{k} |z_i - x_{0i}|^{n+\alpha_i} \right) d\vec{z} = \tag{17.32}$$

$$\frac{1}{\prod_{i=1}^{k} (b_i - a_i)} \int_{\prod_{i=1}^{k} [a_i, b_i]} \left(\sum_{i=1}^{k} |z_i - x_{0i}|^{n+1} \right) d\vec{z}.$$

So we found that

$$\lim_{\text{all } \alpha_i \to 1} R.H.S. (17.25) = \lim_{\text{all } \alpha_i \to 1} L.H.S. (17.25), \tag{17.33}$$

proving the sharpness of (17.25) when n is even. ∎

Corollary 17.8 (to Corollary 17.7) *Let* $f \in C^{n+1} ([a_1, b_1] \times [a_2, b_2], X)$, $n \in \mathbb{N}$ *where* $a_1 < b_1$, $a_2 < b_2$; $a_1, a_2, b_1, b_2 \in \mathbb{R}$ *and let* $\vec{x_0} = (x_{01}, x_{02}) \in [a_1, b_1] \times [a_2, b_2]$ *be fixed. We suppose here that all vector partial derivatives* $f_\alpha := \frac{\partial^\alpha f}{\partial z^\alpha}$, *where* $\alpha = (\alpha_1, \alpha_2)$, $\alpha_1, \alpha_2 \in \mathbb{Z}^+$, $|\alpha| = \alpha_1 + \alpha_2 = j$, $j = 1, \ldots, n$ *fulfill* $f_\alpha (\vec{x_0}) = 0$. *Then*

$$\left\| \frac{1}{(b_1 - a_1)(b_2 - a_2)} \int_{a_1}^{b_1} \int_{a_2}^{b_2} f(z_1, z_2) \, dz_1 dz_2 - f(\vec{x_0}) \right\| \leq$$

$$\sum_{l=0}^{n+1} \left\{ \frac{\left[(x_{01} - a_1)^{n+2-l} + (b_1 - x_{01})^{n+2-l} \right] \left[(x_{02} - a_2)^{l+1} + (b_2 - x_{02})^{l+1} \right]}{(n + 2 - l)! \, (l + 1)! \, (b_1 - a_1)(b_2 - a_2)} \right\}$$

$$\left\| \left\| \frac{\partial^{n+1} f}{\partial z_1^{n+1-l} \partial z_2^l} \right\| \right\|_\infty. \tag{17.34}$$

Inequality (17.34) is sharp, exactly the same manner as inequality (17.25).

Corollary 17.9 *Let* $f \in C^2 ([a_1, b_1] \times [a_2, b_2], X)$, *where* $a_1 < b_1$, $a_2 < b_2$; $a_1, a_2, b_1, b_2 \in \mathbb{R}$ *and let* $\vec{x_0} = (x_{01}, x_{02}) \in [a_1, b_1] \times [a_2, b_2]$ *be fixed. We suppose that* $\frac{\partial f}{\partial z_1} (\vec{x_0}) = \frac{\partial f}{\partial z_2} (\vec{x_0}) = 0$. *Then*

$$\left\| \frac{1}{(b_1 - a_1)(b_2 - a_2)} \int_{a_1}^{b_1} \int_{a_2}^{b_2} f(z_1, z_2) \, dz_1 dz_2 - f(\vec{x_0}) \right\| \leq$$

$$\left(\frac{(x_{01} - a_1)^3 + (b_1 - x_{01})^3}{6\,(b_1 - a_1)}\right)\left\|\frac{\partial^2 f}{\partial z_1^2}\right\|_\infty +$$

$$\left[\frac{\left((x_{01} - a_1)^2 + (b_1 - x_{01})^2\right)\left((x_{02} - a_2)^2 + (b_2 - x_{02})^2\right)}{4\,(b_1 - a_1)\,(b_2 - a_2)}\right]\left\|\frac{\partial^2 f}{\partial z_1 \partial z_2}\right\|_\infty +$$

$$\left(\frac{(x_{02} - a_2)^3 + (b_2 - x_{02})^3}{6\,(b_2 - a_2)}\right)\left\|\frac{\partial^2 f}{\partial z_2^2}\right\|_\infty. \tag{17.35}$$

Inequality (17.35) is sharp; in fact it is attained by

$$f^*\,(z_1, z_2) = \left((z_1 - x_{01})^2 + (z_2 - x_{02})^2\right) \cdot i_0, \tag{17.36}$$

where $i_0 \in X : \|i_0\| = 1$.

Proof Application of Corollary 17.8 when $n = 1$. ∎

Corollary 17.10 *Let $f \in C^2\left(\prod\limits_{i=1}^{3} [a_i, b_i], X\right)$, where $a_i < b_i$, $i = 1, 2, 3$; $a_i, b_i \in$*
$\mathbb{R}$ and let $\overrightarrow{x_0} = (x_{01}, x_{02}, x_{03}) \in \prod\limits_{i=1}^{3} [a_i, b_i]$ is fixed. We suppose here that $\frac{\partial f}{\partial z_i}\left(\overrightarrow{x_0}\right) =$
0; $i = 1, 2, 3$. Then

$$\left\|\frac{1}{\prod_{i=1}^{3}(b_i - a_i)}\int_{a_1}^{b_1}\int_{a_2}^{b_2}\int_{a_3}^{b_3} f\,(z_1, z_2, z_3)\,dz_1 dz_2 dz_3 - f\left(\overrightarrow{x_0}\right)\right\| \leq$$

$$\left\{\sum_{i=1}^{3}\left(\frac{(x_{0i} - a_i)^3 + (b_i - x_{0i})^3}{6\,(b_i - a_i)}\right)\left\|\frac{\partial^2 f}{\partial z_i^2}\right\|_\infty +\right.$$

$$\sum_{i=1}^{2}\frac{\left((x_{0i} - a_i)^2 + (b_i - x_{0i})^2\right)\left((x_{0,i+1} - a_{i+1})^2 + (b_{i+1} - x_{0,i+1})^2\right)}{4\,(b_i - a_i)\,(b_{i+1} - a_{i+1})}$$

$$\cdot\left\|\frac{\partial^2 f}{\partial z_i \partial z_{i+1}}\right\|_\infty +$$

$$\left.\frac{\left((x_{03} - a_3)^2 + (b_3 - x_{03})^2\right)\left((x_{0,1} - a_1)^2 + (b_1 - x_{01})^2\right)}{4\,(b_3 - a_3)\,(b_1 - a_1)}\left\|\frac{\partial^2 f}{\partial z_3 \partial z_1}\right\|_\infty\right\}. \tag{17.37}$$

Inequality (17.37) is sharp and it is attained by

$$f^* (z_1, z_2, z_3) = \left(\sum_{i=1}^{3} (z_i - x_{0i})^2 \right) \cdot i_0, \qquad (17.38)$$

where $i_0 \in X : \|i_0\| = 1$.

Proof Use of Corollary 17.7 for $k = 3$ and $n = 1$. ∎

We further need

Theorem 17.11 *Here $(X, \|\cdot\|)$ is a Banach space. Let $f : [a, A] \times [b, B] \times [c, C] \to X$ be a mapping three times continuously differentiable. Let also $(x, y, z) \in [a, A] \times [b, B] \times [c, C]$ be fixed. We define the kernels $p : [a, A]^2 \to \mathbb{R}$, $q : [b, B]^2 \to \mathbb{R}$, and $\theta : [c, C]^2 \to \mathbb{R}$:*

$$p (x, s) := \begin{cases} s - a, & s \in [a, x], \\ s - A, & s \in (x, A], \end{cases}$$

$$q (y, t) := \begin{cases} t - b, & t \in [b, y], \\ t - B, & t \in (y, B], \end{cases}$$

and

$$\theta (z, r) := \begin{cases} r - c, & r \in [c, z], \\ r - C, & r \in (z, C]. \end{cases}$$

Then

$$\theta_{1,3} := \int_a^A \int_b^B \int_c^C p (x, s) q (y, t) \theta (z, t) f_{s,t,r}''' (s, t, r) \, ds \, dt \, dr =$$

$$\{(A - a)(B - b)(C - c) f (x, y, z)\} - \Bigg[(B - b)(C - c) \int_a^A f (s, y, z) \, ds$$

$$+ (A - a)(C - c) \int_b^B f (x, t, z) \, dt + (A - a)(B - b) \int_c^C f (x, y, r) \, dr \Bigg]$$

$$+ \Bigg[(C - c) \int_a^A \int_b^B f (s, t, z) \, ds \, dt + (B - b) \int_a^A \int_c^C f (s, y, r) \, ds \, dr \quad (17.39)$$

$$+ (A - a) \int_b^B \int_c^C f (x, t, r) \, dt \, dr \Bigg] - \int_a^A \int_b^B \int_c^C f (s, t, r) \, ds \, dt \, dr =: \theta_{2,3}.$$

Proof Similar to [4], see also [7], p. 82, using integration by parts several times. ∎

In general we state

Theorem 17.12 *Here* $(X, \|\cdot\|)$ *is a Banach space. Let* $f : \times_{i=1}^{n} [a_i, b_i] \to X$ *be a mapping* n *times continuously differentiable,* $n > 1$. *Let also* $(x_1, \ldots, x_n)$ $\in \times_{i=1}^{n} [a_i, b_i]$ *be fixed. We define the kernels* $p_i : [a_i, b_i]^2 \to \mathbb{R}$:

$$p_i(x_i, s_i) := \begin{cases} s_i - a_i, & s_i \in [a_i, x_i], \\ s_i - b_i, & s_i \in (x_i, b_i], \end{cases} \quad \text{for all } i = 1, \ldots, n.$$

Then

$$\theta_{1,n} := \int_{\times_{i=1}^{n}[a_i,b_i]} \prod_{i=1}^{n} p_i(x_i, s_i) \frac{\partial^n f(s_1, \ldots, s_n)}{\partial s_1 \ldots \partial s_n} ds_1 \ldots ds_n =$$

$$\left\{ \left(\prod_{i=1}^{n} (b_i - a_i) \right) f(x_1, \ldots, x_n) \right\} -$$

$$\left[\sum_{i=1}^{\binom{n}{1}} \left(\prod_{\substack{j=1 \\ j \neq i}}^{n} (b_j - a_j) \int_{a_i}^{b_i} f(x_1, \ldots, s_i, \ldots, x_n) \, ds_i \right) \right] +$$

$$\left[\sum_{i=1}^{\binom{n}{2}} \left(\prod_{\substack{k=1 \\ k \neq i, j}}^{n} (b_k - a_k) \left(\int_{a_i}^{b_i} \int_{a_j}^{b_j} f(x_1, \ldots, s_i, \ldots, s_j, \ldots, x_n) \, ds_i ds_j \right) \right) \right]_{(l)}$$

$$- + \cdots - + \cdots + \tag{17.40}$$

$$(-1)^{n-1} \left[\sum_{i=1}^{\binom{n}{n-1}} (b_j - a_j) \int_{\substack{\times_{i=1}^{n}[a_i,b_i] \\ i \neq j}} f(s_1, \ldots, x_j, \ldots, s_n) \, ds_1 \ldots \widehat{ds_j} \ldots ds_n \right]$$

$$+ (-1)^n \int_{\times_{i=1}^{n}[a_i,b_i]} f(s_1, \ldots, s_n) \, ds_1 \ldots ds_n =: \theta_{2,n}.$$

The above l *counts all the* $(i, j)'s$ $i < j$ *and* $i, j = 1, \ldots, n$. *Also* $\widehat{ds_j}$ *means* ds_j *is missing.*

Proof Similar to Theorem 17.11, see also [4]. ∎

We present the following Ostrowski type inequalities

Theorem 17.13 *Under the notations and assumptions of Theorem 17.11, we obtain*

$$\|\theta_{2,3}\| \le \frac{\||f_{s,t,r}'''\|\|_\infty}{8} \cdot \left\{\left((x-a)^2 + (A-x)^2\right) \cdot \left((y-b)^2 + (B-y)^2\right)\right.$$

$$\left. \cdot \left((z-c)^2 + (C-z)^2\right)\right\}, \text{ for all } (x, y, z) \in [a, A] \times [b, B] \times [c, C]. \quad (17.41)$$

Proof Notice that

$$\|\theta_{2,3}\| = \|\theta_{1,3}\| \le$$

$$\||f_{s,t,r}'''\|\|_\infty \left(\int_a^A |p(x, s)| \, ds\right) \left(\int_b^B |q(y, t)| \, dt\right) \left(\int_c^C |\theta(z, r)| \, dr\right).$$

Also see that

$$\int_a^A |p(x, s)| \, ds = \frac{1}{2} \left\{(x-a)^2 + (A-x)^2\right\},$$

etc. ∎

The counterpart of the last theorem is

Theorem 17.14 *Under the notations and assumptions of Theorem 17.12, we derive*

$$\|\theta_{2,n}\| \le \frac{\left\|\left\|\frac{\partial^n f(s_1,\dots,s_n)}{\partial s_1 \dots \partial s_n}\right\|\right\|_\infty}{2^n} \left\{\prod_{i=1}^{n}\left[(x_i - a_i)^2 + (b_i - x_i)^2\right]\right\}, \quad (17.42)$$

for all $(x_1, \dots, x_n) \in \times_{i=1}^{n} [a_i, b_i]$.

Proof As in Theorem 17.13. ∎

It follows

Theorem 17.15 *Let* $p, q > 1 : \frac{1}{p} + \frac{1}{q} = 1$. *Under the notations and assumptions of Theorem 17.11 we get*

$$\|\theta_{2,3}\| \le \frac{\||f_{s,t,r}'''\|\|_p}{(q+1)^{\frac{3}{q}}} \cdot \left\{\left[\left((x-a)^{q+1} + (A-x)^{q+1}\right) \cdot \left((y-b)^{q+1} + (B-y)^{q+1}\right)\right.\right.$$

$$\cdot \left((z - c)^{q+1} + (C - z)^{q+1} \right) \right]^{\frac{1}{q}} \right\}, \text{ for all } (x, y, z) \in [a, A] \times [b, B] \times [c, C]. \quad (17.43)$$

Here

$$\|f_{s,t,r}'''\|_p := \left(\int_a^A \int_b^B \int_c^C \|f_{s,t,r}''' (s, t, r)\|^p \, ds \, dt \, dr \right)^{\frac{1}{p}}.$$

Proof Notice that

$$\|\theta_{2,3}\| = \|\theta_{1,3}\| = \left\| \int_a^A \int_b^B \int_c^C p(x, s) \, q(y, t) \, \theta(z, r) \, f_{s,t,r}''' (s, t, r) \, ds \, dt \, dr \right\|$$

$$\leq \int_a^A \int_b^B \int_c^C |p(x, s)| \, |q(y, t)| \, |\theta(z, r)| \, \|f_{s,t,r}''' (s, t, r)\| \, ds \, dt \, dr$$

(by Hölder's inequality)

$$\leq \left(\int_a^A \int_b^B \int_c^C \|f_{s,t,r}''' (s, t, r)\|^p \, ds \, dt \, dr \right)^{\frac{1}{p}} \cdot$$

$$\left(\int_a^A \int_b^B \int_c^C (|p(x, s)| \, |q(y, t)| \, |\theta(z, r)|)^q \, ds \, dt \, dr \right)^{\frac{1}{q}}$$

$$= \|f_{s,t,r}'''\|_p \left\{ \left(\int_a^A |p(x, s)|^q \, ds \right) \left(\int_b^B |q(y, t)|^q \, dt \right) \left(\int_c^C |\theta(z, r)|^q \, dr \right) \right\}^{\frac{1}{q}}$$

$$= \|f_{s,t,r}'''\|_p \left\{ \left(\frac{(x - a)^{q+1} + (A - x)^{q+1}}{q + 1} \right) \cdot \right.$$

$$\left(\frac{(y - b)^{q+1} + (B - y)^{q+1}}{q + 1} \right) \left(\frac{(z - c)^{q+1} + (C - z)^{q+1}}{q + 1} \right) \right\}^{\frac{1}{q}}.$$

∎

The corresponding general L_p-case follows.

Theorem 17.16 *Let* $p, q > 1 : \frac{1}{p} + \frac{1}{q} = 1$. *Under the notations and assumptions of Theorem 17.12 we find*

$$\left\| \theta_{2,n} \right\| \le \frac{\left\| \left| \frac{\partial^n f(x_1,\ldots,x_n)}{\partial x_1 \ldots \partial x_n} \right| \right\|_p}{(q+1)^{\frac{n}{q}}} \left\{ \prod_{i=1}^{n} \left[(x_i - a_i)^{q+1} + (b_i - x_i)^{q+1} \right] \right\}^{\frac{1}{q}}, \quad (17.44)$$

for any $(x_1, \ldots, x_n) \in \times_{i=1}^{n} [a_i, b_i]$.

Proof Similar to Theorem 17.15. ∎

Remark 17.17 Equalities (17.39) and (17.40) can simplify alot, if for instance we assume in Theorem 17.11 that there exists an $(x_0, y_0, z_0) \in [a, A] \times [b, B] \times [c, C]$ such that

$$f(x_0, \cdot, \cdot) = f(\cdot, y_0, \cdot) = f(\cdot, \cdot, z_0) = 0. \quad (17.45)$$

Also in Theorem 17.12 we may assume that there exists an $\left(x_1^0, x_2^0, \ldots, x_n^0 \right) \in \times_{i=1}^{n} [a_i, b_i]$ such that

$$f\left(x_1^0, x_2, \ldots, x_n \right) = f\left(x_1, x_2^0, x_3, \ldots, x_n \right) = \cdots = f\left(x_1, \ldots, x_{n-1}, x_n^0 \right) = 0, \quad (17.46)$$

for any $(x_1, \ldots, x_n) \in \times_{i=1}^{n} [a_i, b_i]$. So in these particular cases we obtain that

$$\theta_{2,3}(x_0, y_0, z_0) = -\int_a^A \int_b^B \int_c^C f(s, t, r)\, ds dt dr, \quad (17.47)$$

and

$$\theta_{2,n}\left(x_1^0, x_2^0, \ldots, x_n^0 \right) = (-1)^n \int_{\times_{i=1}^{n}[a_i,b_i]} f(s_1, \ldots, s_n)\, ds_1 \ldots ds_n. \quad (17.48)$$

Hence in these cases we obtain

$$\left\| \theta_{2,3}(x_0, y_0, z_0) \right\| = \left\| \int_a^A \int_b^B \int_c^C f(s, t, r)\, ds dt dr \right\|, \quad (17.49)$$

and

$$\left\| \theta_{2,n}\left(x_1^0, x_2^0, \ldots, x_n^0 \right) \right\| = \left\| \int_{\times_{i=1}^{n}[a_i,b_i]} f(s_1, \ldots, s_n)\, ds_1 \ldots ds_n \right\|. \quad (17.50)$$

So (17.41)–(17.44) simplify alot and they become very interesting inequalities.

Theorem 17.18 *Inequalities (17.41) and (17.42) are sharp.*

Proof It is enough to prove that (17.42) is sharp. Let $i_0 \in X : \|i_0\| = 1$. Here the optimal function will be

$$f^* (s_1, \ldots, s_n) := \prod_{i=1}^{n} \left| s_i - x_i^0 \right|^{\alpha} (b_i - a_i) i_0, \ \alpha > 1, \tag{17.51}$$

where $(x_1^0, x_2^0, \ldots, x_n^0)$ is fixed in $\times_{i=1}^{n} [a_i, b_i]$, $i_0 \in X$, $\|i_0\| = 1$. Notice here that

$$f^* \left(s_1, \ldots, x_j^0, \ldots, s_n \right) = 0, \text{ for all } j = 1, \ldots, n, \text{ and any } (s_1, \ldots, s_n) \in \times_{i=1}^{n} [a_i, b_i].$$

Therefore by Remark 17.17 we have (17.50). We observe that

$$\frac{\partial^n f^* (s_1, \ldots, s_n)}{\partial s_1 \ldots \partial s_n} = \alpha^n \left(\prod_{i=1}^{n} (b_i - a_i) \right) \left(\prod_{i=1}^{n} \left| s_i - x_i^0 \right|^{\alpha-1} sign \left(s_i - x_i^0 \right) \right) i_0,$$

$$\tag{17.52}$$

and

$$\left\| \frac{\partial^n f^* (s_1, \ldots, s_n)}{\partial s_1 \ldots \partial s_n} \right\| = \alpha^n \left(\prod_{i=1}^{n} (b_i - a_i) \right) \left(\prod_{i=1}^{n} \left| s_i - x_i^0 \right|^{\alpha-1} \right). \tag{17.53}$$

Consequently we find

$$\left\| \frac{\partial^n f^* (s_1, \ldots, s_n)}{\partial s_1 \ldots \partial s_n} \right\|_{\infty} = \alpha^n \left(\prod_{i=1}^{n} (b_i - a_i) \right) \left(\prod_{i=1}^{n} \left(\max \left(b_i - x_i^0, x_i^0 - a_i \right) \right)^{\alpha-1} \right). \tag{17.54}$$

First we calculate the left-hand side of corresponding inequality (17.42). We have

$$\left\| \int_{\times_{i=1}^{n} [a_i, b_i]} f^* (s_1, \ldots, s_n) \, ds_1 \ldots ds_n \right\| =$$

$$\int_{\times_{i=1}^{n} [a_i, b_i]} \left(\prod_{i=1}^{n} (b_i - a_i) \right) \left(\prod_{i=1}^{n} \left| s_i - x_i^0 \right|^{\alpha} \right) ds_1 \ldots ds_n =$$

$$\left(\prod_{i=1}^{n} (b_i - a_i) \right) \int_{\times_{i=1}^{n} [a_i, b_i]} \left(\prod_{i=1}^{n} \left| s_i - x_i^0 \right|^{\alpha} \right) ds_1 \ldots ds_n =$$

$$\left(\prod_{i=1}^{n} (b_i - a_i) \right) \left(\prod_{i=1}^{n} \left(\int_{a_i}^{b_i} \left| s_i - x_i^0 \right|^{\alpha} ds_i \right) \right) =$$

$$\left(\prod_{i=1}^{n}(b_i - a_i)\right)\left(\prod_{i=1}^{n}\left(\frac{(x_i^0 - a_i)^{\alpha+1} + (b_i - x_i^0)^{\alpha+1}}{\alpha + 1}\right)\right). \qquad (17.55)$$

That is,

$$L.H.S.\,(17.42) = \frac{\left(\prod_{i=1}^{n}(b_i - a_i)\right)}{(\alpha + 1)^n}\left(\prod_{i=1}^{n}\left(\left(x_i^0 - a_i\right)^{\alpha+1} + \left(b_i - x_i^0\right)^{\alpha+1}\right)\right). \qquad (17.56)$$

And next we see that

$$R.H.S.\,(17.42) = \frac{\alpha^n\left(\prod_{i=1}^{n}(b_i - a_i)\right)\left(\prod_{i=1}^{n}\left(\max\left(b_i - x_i^0, x_i^0 - a_i\right)\right)^{\alpha-1}\right)}{2^n} \qquad (17.57)$$

$$\cdot\left\{\prod_{i=1}^{n}\left(\left(x_i^0 - a_i\right)^2 + \left(b_i - x_i^0\right)^2\right)\right\}.$$

Now let $\alpha \to 1$. We find

$$\lim_{\alpha \to 1} L.H.S.\,(17.42) = \frac{\left(\prod_{i=1}^{n}(b_i - a_i)\right)}{2^n}\left(\prod_{i=1}^{n}\left(\left(x_i^0 - a_i\right)^2 + \left(b_i - x_i^0\right)^2\right)\right), \qquad (17.58)$$

and

$$\lim_{\alpha \to 1} R.H.S.\,(17.42) = \frac{\left(\prod_{i=1}^{n}(b_i - a_i)\right)}{2^n}\left(\prod_{i=1}^{n}\left(\left(x_i^0 - a_i\right)^2 + \left(b_i - x_i^0\right)^2\right)\right). \qquad (17.59)$$

That is,

$$\lim_{\alpha \to 1} L.H.S.\,(17.42) = \lim_{\alpha \to 1} R.H.S.\,(17.42), \qquad (17.60)$$

hence proving the sharpness of (17.42). ∎

Remark 17.19 Anorther interesting case for (17.39) and (17.40) is to assume that for specific (x, y, z) $((x_1, \ldots, x_n)$, respectively) all the marginal integrals of f are equal to zero. Then we get

$$\theta_{2,3} = (A - a)(B - b)(C - c) f(x, y, z) - \int_a^A \int_b^B \int_c^C f(s, t, r) \, ds \, dt \, dr,$$
(17.61)

and

$$\theta_{2,n} = \left(\prod_{i=1}^n (b_i - a_i) \right) f(x_1, \ldots, x_n) + (-1)^n \int_{\times_{i=1}^n [a_i, b_i]} f(s_1, \ldots, s_n) \, ds_1 \ldots ds_n.$$
(17.62)

Hence inequalities (17.41)–(17.44) become again alot simpler.

Next we mention a vector Montgomery identity derived by applying twice integration by parts.

Theorem 17.20 ([8]) *Let* $(X, \|\cdot\|)$ *be a Banach space and* $f \in C^1([a, b], X)$. *Let* $x \in [a, b]$ *be fixed and define*

$$P(x, t) := \begin{cases} \frac{t-a}{b-a}, & a \le t \le x, \\ \frac{t-b}{b-a}, & x < t \le b. \end{cases}$$
(17.63)

Then

$$f(x) = \frac{1}{b - a} \int_a^b f(t) \, dt + \int_a^b P(x, t) f'(t) \, dt.$$
(17.64)

We present a vector multivariate Montgomery identity, see also the real analog in [3]. We have the representation

Theorem 17.21 *Let* $(X, \|\cdot\|)$ *is a Banach space and* $f \in C^3 \left(\times_{i=1}^3 [a_i, b_i], X \right)$. *Let* $(x_1, x_2, x_3) \in \times_{i=1}^3 [a_i, b_i]$. *Define the kernels* $p_i : [a_i, b_i]^2 \to \mathbb{R}$:

$$p_i(x_i, s_i) := \begin{cases} s_i - a_i, & s_i \in [a_i, x_i], \\ s_i - b_i, & s_i \in (x_i, b_i], \end{cases}$$
(17.65)

for $i = 1, 2, 3$.
    Then

$$f(x_1, x_2, x_3) = \frac{1}{\prod_{i=1}^3 (b_i - a_i)} \left\{ \int_{a_1}^{b_1} \int_{a_2}^{b_2} \int_{a_3}^{b_3} f(s_1, s_2, s_3) \, ds_3 \, ds_2 \, ds_1 \right.$$

$$+ \sum_{j=1}^3 \left(\int_{a_1}^{b_1} \int_{a_2}^{b_2} \int_{a_3}^{b_3} p_j(x_j, s_j) \frac{\partial f(s_1, s_2, s_3)}{\partial s_j} \, ds_3 \, ds_2 \, ds_1 \right)$$
(17.66)

$$+ \sum_{\substack{l=1 \\ j<k}}^{3} \left(\int_{a_1}^{b_1} \int_{a_2}^{b_2} \int_{a_3}^{b_3} p_j\left(x_j, s_j\right) p_k\left(x_k, s_k\right) \frac{\partial^2 f\left(s_1, s_2, s_3\right)}{\partial s_k \partial s_j} ds_3 ds_2 ds_1 \right)_{(l)}$$

$$+ \int_{a_1}^{b_1} \int_{a_2}^{b_2} \int_{a_3}^{b_3} \left(\prod_{i=1}^{3} p_i\left(x_i, s_i\right) \right) \frac{\partial^3 f\left(s_1, s_2, s_3\right)}{\partial s_3 \partial s_2 \partial s_1} ds_3 ds_2 ds_1 \Biggr\} .$$

Above l counts (j, k) : $j < k$; $j, k \in \{1, 2, 3\}$.

Proof Multiple use of (17.64), see also [3] and [6], p. 16. ∎

A generalization of Theorem 17.21 to any $n \in \mathbb{N}$ follows.

Theorem 17.22 *Let $(X, \|\cdot\|)$ is a Banach space and $f \in C^n\left(\times_{i=1}^{n} [a_i, b_i], X\right)$. Let $(x_1, \ldots, x_n) \in \times_{i=1}^{n} [a_i, b_i]$. Define the kernels $p_i : [a_i, b_i]^2 \to \mathbb{R}$:*

$$p_i\left(x_i, s_i\right) := \begin{cases} s_i - a_i, & s_i \in [a_i, x_i], \\ s_i - b_i, & s_i \in (x_i, b_i], \end{cases} \tag{17.67}$$

for $i = 1, 2, \ldots, n$.
 Then

$$f\left(x_1, x_2, \ldots, x_n\right) = \frac{1}{\prod\limits_{i=1}^{n} (b_i - a_i)} \Biggl\{ \int_{\times_{i=1}^{n}[a_i, b_i]} f\left(s_1, s_2, \ldots, s_n\right) ds_n ds_{n-1} \ldots ds_1$$

$$+ \sum_{j=1}^{n} \left(\int_{\times_{i=1}^{n}[a_i, b_i]} p_j\left(x_j, s_j\right) \frac{\partial f\left(s_1, s_2, \ldots, s_n\right)}{\partial s_j} ds_n \ldots ds_1 \right) \tag{17.68}$$

$$+ \sum_{\substack{l_1=1 \\ j<k}}^{\binom{n}{2}} \left(\int_{\times_{i=1}^{n}[a_i, b_i]} p_j\left(x_j, s_j\right) p_k\left(x_k, s_k\right) \frac{\partial^2 f\left(s_1, s_2, \ldots, s_n\right)}{\partial s_k \partial s_j} ds_n \ldots ds_1 \right)_{(l_1)}$$

$$\sum_{\substack{l_2=1 \\ j<k<r}}^{\binom{n}{3}} \left(\int_{\times_{i=1}^{n}[a_i, b_i]} p_j\left(x_j, s_j\right) p_k\left(x_k, s_k\right) p_r\left(x_r, s_r\right) \frac{\partial^3 f\left(s_1, \ldots, s_n\right)}{\partial s_r \partial s_k \partial s_j} ds_n \ldots ds_1 \right)_{(l_2)}$$

$$+ \cdots + \sum_{l=1}^{\binom{n}{n-1}} \left(\int_{\times_{i=1}^{n}[a_i, b_i]} p_1\left(x_1, s_1\right) \ldots \widehat{p_l\left(x_l, s_l\right)} \ldots p_n\left(x_n, s_n\right) \right.$$

$$\frac{\partial^{n-1} f(s_1, \ldots, s_n)}{\partial s_n \ldots \widehat{\partial s_l} \ldots \partial s_1} ds_n \ldots ds_1 \bigg)$$

$$+ \int_{\times_{i=1}^n [a_i, b_i]} \left(\prod_{i=1}^n p_i(x_i, s_i) \right) \frac{\partial^n f(s_1, \ldots, s_n)}{\partial s_n \ldots \partial s_1} ds_n \ldots ds_1 \bigg\}.$$

Above l_1 counts $(j, k) : j < k$; $j, k \in \{1, 2, \ldots, n\}$, also l_2 counts $(j, k, r) : j < k < r$; $j, k, r \in \{1, 2, \ldots, n\}$, etc. Also $p_l(x_l, s_l)$, $\widehat{\partial s_l}$ mean that $p_l(x_l, s_l)$, ∂s_l are missing, respectively.

Proof Similar to Theorem 17.21, see also [3]. ∎

Next we obtain the following Ostrowski type inequalities.

Theorem 17.23 *Let $f : \times_{i=1}^3 [a_i, b_i] \to X$ as in Theorem 17.21. Then*

$$\left\| f(x_1, x_2, x_3) - \frac{1}{\prod_{i=1}^3 (b_i - a_i)} \int_{a_1}^{b_1} \int_{a_2}^{b_2} \int_{a_3}^{b_3} f(s_1, s_2, s_3) \, ds_3 ds_2 ds_1 \right\|$$

$$\leq \frac{(M_{1,3} + M_{2,3} + M_{3,3})}{\prod_{i=1}^3 (b_i - a_i)}. \tag{17.69}$$

Here we have

$$M_{1,3} := \min \begin{cases} \sum_{j=1}^3 \left(\left\| \left\| \frac{\partial f}{\partial s_j} \right\| \right\|_\infty \left(\prod_{\substack{i=1 \\ i \neq j}}^3 (b_i - a_i) \right) \left(\frac{(x_j - a_j)^2 + (b_j - x_j)^2}{2} \right) \right), \\[3ex] \sum_{j=1}^3 \left(\left\| \left\| \frac{\partial f}{\partial s_j} \right\| \right\|_{p_j} \left(\prod_{\substack{i=1 \\ i \neq j}}^3 (b_i - a_i)^{\frac{1}{q_j}} \right) \right. \\[3ex] \left. \left[\frac{(x_j - a_j)^{q_j+1} + (b_j - x_j)^{q_j+1}}{q_j + 1} \right]^{\frac{1}{q_j}} \right), \\[2ex] \text{when } p_j, q_j > 1 : \frac{1}{p_j} + \frac{1}{q_j} = 1, \text{ for } j = 1, 2, 3; \\[2ex] \sum_{j=1}^3 \left(\left\| \left\| \frac{\partial f}{\partial s_j} \right\| \right\|_1 \left(\frac{b_j - a_j}{2} + \left| x_j - \frac{a_j + b_j}{2} \right| \right) \right). \end{cases}$$

$$\tag{17.70}$$

Also we have

$$M_{2,3} := \min \left\{ \begin{array}{l} \sum_{\substack{l=1 \\ j<k}}^{3} \left(\left\| \left\| \frac{\partial^2 f}{\partial s_k \partial s_j} \right\| \right\|_\infty \left(\prod_{\substack{i=1 \\ i \neq j,k}}^{3} (b_i - a_i) \right) \right. \\ \left. \frac{\left((x_j-a_j)^2+(b_j-x_j)^2\right)\left((x_k-a_k)^2+(b_k-x_k)^2\right)}{4} \right)_{(l)}, \\[2em] \sum_{\substack{l=1 \\ j<k}}^{3} \left(\left\| \left\| \frac{\partial^2 f}{\partial s_k \partial s_j} \right\| \right\|_{p_{kj}} \left(\prod_{\substack{i=1 \\ i \neq j,k}}^{3} (b_i - a_i)^{\frac{1}{q_{kj}}} \right) \cdot \right. \\ \left. \left[\frac{(x_j-a_j)^{q_{kj}+1}+(b_j-x_j)^{q_{kj}+1}}{q_{kj}+1} \right]^{\frac{1}{q_{kj}}} \left[\frac{(x_k-a_k)^{q_{kj}+1}+(b_k-x_k)^{q_{kj}+1}}{q_{kj}+1} \right]^{\frac{1}{q_{kj}}} \right)_{(l)}, \\[1em] \text{when } p_{kj}, q_{kj} > 1 : \frac{1}{p_{kj}} + \frac{1}{q_{kj}} = 1, \text{ for } j,k \in \{1,2,3\}; \\[1em] \sum_{\substack{l=1 \\ j<k}}^{3} \left(\left\| \left\| \frac{\partial^2 f}{\partial s_k \partial s_j} \right\| \right\|_1 \left(\frac{b_j-a_j}{2} + \left| x_j - \frac{a_j+b_j}{2} \right| \right) \right. \\ \left. \left(\frac{b_k-a_k}{2} + \left| x_k - \frac{a_k+b_k}{2} \right| \right) \right)_{(l)} . \end{array} \right. \tag{17.71}$$

And finally

$$M_{3,3} := \min \left\{ \begin{array}{l} \left\| \left\| \frac{\partial^3 f(s_1,s_2,s_3)}{\partial s_3 \partial s_2 \partial s_1} \right\| \right\|_\infty \frac{\prod_{j=1}^{3}\left((x_j-a_j)^2+(b_j-x_j)^2\right)}{8}, \\[1.5em] \left\| \left\| \frac{\partial^3 f(s_1,s_2,s_3)}{\partial s_3 \partial s_2 \partial s_1} \right\| \right\|_{p_{123}} \prod_{i=1}^{3} \left(\frac{(x_j-a_j)^{q_{123}+1}+(b_j-x_j)^{q_{123}+1}}{q_{123}+1} \right)^{\frac{1}{q_{123}}}, \\[1em] \text{when } p_{123}, q_{123} > 1 : \frac{1}{p_{123}} + \frac{1}{q_{123}} = 1; \\[1em] \left\| \left\| \frac{\partial^3 f(s_1,s_2,s_3)}{\partial s_3 \partial s_2 \partial s_1} \right\| \right\|_1 \prod_{i=1}^{3} \left(\frac{b_j-a_j}{2} + \left| x_j - \frac{a_j+b_j}{2} \right| \right) . \end{array} \right. \tag{17.72}$$

Inequality (17.69) is true for any $(x_1, x_2, x_3) \in \times_{i=1}^{3} [a_i, b_i]$, *where* $\|\cdot\|_p$ $(1 \leq p \leq \infty)$ *are the usual* L_p-*norms on* $\times_{i=1}^{3} [a_i, b_i]$.

Proof We have by (17.66) that

$$\left\| f\left(x_{1}, x_{2}, x_{3}\right)-\frac{1}{\prod_{i=1}^{3}\left(b_{i}-a_{i}\right)} \int_{a_{1}}^{b_{1}} \int_{a_{2}}^{b_{2}} \int_{a_{3}}^{b_{3}} f\left(s_{1}, s_{2}, s_{3}\right) d s_{3} d s_{2} d s_{1} \right\|$$

$$\leq \frac{1}{\prod_{i=1}^{3}\left(b_{i}-a_{i}\right)}\left(\sum_{j=1}^{3}\left(\int_{a_{1}}^{b_{1}} \int_{a_{2}}^{b_{2}} \int_{a_{3}}^{b_{3}}\left|p_{j}\left(x_{j}, s_{j}\right)\right|\left\|\frac{\partial f\left(s_{1}, s_{2}, s_{3}\right)}{\partial s_{j}}\right\| d s_{3} d s_{2} d s_{1}\right)\right.$$

$$+\sum_{\substack{l=1 \\ j<k}}^{3}\left(\int_{a_{1}}^{b_{1}} \int_{a_{2}}^{b_{2}} \int_{a_{3}}^{b_{3}}\left|p_{j}\left(x_{j}, s_{j}\right)\right|\left|p_{k}\left(x_{k}, s_{k}\right)\right|\left\|\frac{\partial^{2} f\left(s_{1}, s_{2}, s_{3}\right)}{\partial s_{k} \partial s_{j}}\right\| d s_{3} d s_{2} d s_{1}\right)_{(l)}$$

$$+\int_{a_{1}}^{b_{1}} \int_{a_{2}}^{b_{2}} \int_{a_{3}}^{b_{3}}\left(\prod_{i=1}^{3}\left|p_{i}\left(x_{i}, s_{i}\right)\right|\right)\left\|\frac{\partial^{3} f\left(s_{1}, s_{2}, s_{3}\right)}{\partial s_{3} \partial s_{2} \partial s_{1}}\right\| d s_{3} d s_{2} d s_{1}\right).$$

We notice the following ($j = 1, 2, 3$)

$$\int_{a_{1}}^{b_{1}} \int_{a_{2}}^{b_{2}} \int_{a_{3}}^{b_{3}}\left|p_{j}\left(x_{j}, s_{j}\right)\right|\left\|\frac{\partial f\left(s_{1}, s_{2}, s_{3}\right)}{\partial s_{j}}\right\| d s_{3} d s_{2} d s_{1} \leq$$

$$\min \left\{\begin{array}{l} \left\|\left\|\frac{\partial f}{\partial s_{j}}\right\|\right\|_{\infty} \int_{a_{1}}^{b_{1}} \int_{a_{2}}^{b_{2}} \int_{a_{3}}^{b_{3}}\left|p_{j}\left(x_{j}, s_{j}\right)\right| d s_{3} d s_{2} d s_{1}, \\ \left\|\left\|\frac{\partial f}{\partial s_{j}}\right\|\right\|_{p_{j}}\left(\int_{a_{1}}^{b_{1}} \int_{a_{2}}^{b_{2}} \int_{a_{3}}^{b_{3}}\left|p_{j}\left(x_{j}, s_{j}\right)\right|^{q_{j}} d s_{3} d s_{2} d s_{1}\right)^{\frac{1}{q_{j}}}, \\ p_{j}, q_{j}>1: \frac{1}{p_{j}}+\frac{1}{q_{j}}=1 ; \\ \left\|\left\|\frac{\partial f}{\partial s_{j}}\right\|\right\|_{1} \sup_{s_{j} \in\left[a_{j}, b_{j}\right]}\left|p_{j}\left(x_{j}, s_{j}\right)\right| . \end{array}\right.$$

Furthermore we see that

$$\int_{a_{1}}^{b_{1}} \int_{a_{2}}^{b_{2}} \int_{a_{3}}^{b_{3}}\left|p_{j}\left(x_{j}, s_{j}\right)\right| d s_{3} d s_{2} d s_{1}=$$

$$\left(b_{1}-a_{1}\right)\left(\widehat{b_{j}-a_{j}}\right)\left(b_{3}-a_{3}\right)\left(\frac{\left(x_{j}-a_{2}\right)^{2}+\left(b_{j}-x_{j}\right)^{2}}{2}\right) ;$$

$\widehat{b_j - a_j}$ means $(b_j - a_j)$ is missing, for $j = 1, 2, 3$. Also we find

$$\left(\int_{a_1}^{b_1} \int_{a_2}^{b_2} \int_{a_3}^{b_3} |p_j (x_j, s_j)|^{q_j} ds_3 ds_2 ds_1 \right)^{\frac{1}{q_j}} =$$

$$(b_1 - a_1)^{\frac{1}{q_j}} \left(\widehat{b_j - a_j} \right)^{\frac{1}{q_j}} (b_3 - a_3)^{\frac{1}{q_j}} \left(\frac{(x_j - a_2)^{q_j + 1} + (b_j - x_j)^{q_j + 1}}{q_j + 1} \right)^{\frac{1}{q_j}};$$

$\left(\widehat{b_j - a_j} \right)^{\frac{1}{q_j}}$ means $(b_j - a_j)^{\frac{1}{q_j}}$ is missing, for $j = 1, 2, 3$. Also

$$\sup_{s_j \in [a_j, b_j]} |p_j (x_j, s_j)| = \max\{x_j - a_j, b_j - x_j\} = \frac{b_j - a_j}{2} + \left| x_j - \frac{a_j + b_j}{2} \right|,$$

for $j = 1, 2, 3$.

Putting things together we get

$$\sum_{j=1}^{3} \left(\int_{a_1}^{b_1} \int_{a_2}^{b_2} \int_{a_3}^{b_3} |p_j (x_j, s_j)| \left\| \frac{\partial f (s_1, s_2, s_3)}{\partial s_j} \right\| ds_3 ds_2 ds_1 \right) \leq$$

$$\min \left\{ \begin{array}{l} \sum_{j=1}^{3} \left(\left\| \left\| \frac{\partial f}{\partial s_j} \right\| \right\|_{\infty} \left(\prod_{\substack{i=1 \\ i \neq j}}^{3} (b_i - a_i) \right) \left(\frac{(x_j - a_j)^2 + (b_j - x_j)^2}{2} \right) \right), \\[2em] \sum_{j=1}^{3} \left(\left\| \left\| \frac{\partial f}{\partial s_j} \right\| \right\|_{p_j} \left(\prod_{\substack{i=1 \\ i \neq j}}^{3} (b_i - a_i)^{\frac{1}{q_j}} \right) \cdot \right. \\[1em] \left. \left[\frac{(x_j - a_j)^{q_j + 1} + (b_j - x_j)^{q_j + 1}}{q_j + 1} \right]^{\frac{1}{q_j}} \right), \\[1.5em] \text{when } p_j, q_j > 1 : \frac{1}{p_j} + \frac{1}{q_j} = 1, \text{ for } j = 1, 2, 3; \\[1em] \sum_{j=1}^{3} \left(\left\| \left\| \frac{\partial f}{\partial s_j} \right\| \right\|_{1} \left(\frac{b_j - a_j}{2} + \left| x_j - \frac{a_j + b_j}{2} \right| \right) \right). \end{array} \right. \tag{17.73}$$

Similarly work, next we find that

$$\sum_{\substack{l=1 \\ j<k}}^{3} \left(\int_{a_1}^{b_1} \int_{a_2}^{b_2} \int_{a_3}^{b_3} |p_j (x_j, s_j)| |p_k (x_k, s_k)| \left\| \frac{\partial^2 f (s_1, s_2, s_3)}{\partial s_k \partial s_j} \right\| ds_3 ds_2 ds_1 \right)_{(l)} \leq$$

$$
\min \left\{
\begin{array}{l}
\sum_{\substack{l=1 \\ j<k}}^{3} \left(\left\| \left\| \frac{\partial^2 f}{\partial s_k \partial s_j} \right\| \right\|_\infty \left(\prod_{\substack{i=1 \\ i \neq j,k}}^{3} (b_i - a_i) \right) \right. \\
\left. \frac{\left((x_j - a_j)^2 + (b_j - x_j)^2 \right)\left((x_k - a_k)^2 + (b_k - x_k)^2 \right)}{4} \right)_{(l)}, \\[3ex]
\sum_{\substack{l=1 \\ j<k}}^{3} \left(\left\| \left\| \frac{\partial^2 f}{\partial s_k \partial s_j} \right\| \right\|_{p_{kj}} \left(\prod_{\substack{i=1 \\ i \neq j,k}}^{3} (b_i - a_i)^{\frac{1}{q_{kj}}} \right) \cdot \right. \\
\left. \left[\frac{(x_j - a_j)^{q_{kj}+1} + (b_j - x_j)^{q_{kj}+1}}{q_{kj}+1} \right]^{\frac{1}{q_{kj}}} \left[\frac{(x_k - a_k)^{q_{kj}+1} + (b_k - x_k)^{q_{kj}+1}}{q_{kj}+1} \right]^{\frac{1}{q_{kj}}} \right)_{(l)}, \\[3ex]
\text{when } p_{kj}, q_{kj} > 1 : \frac{1}{p_{kj}} + \frac{1}{q_{kj}} = 1, \text{ for } j,k \in \{1,2,3\}; \\[2ex]
\sum_{\substack{l=1 \\ j<k}}^{3} \left(\left\| \left\| \frac{\partial^2 f}{\partial s_k \partial s_j} \right\| \right\|_1 \left(\frac{b_j - a_j}{2} + \left| x_j - \frac{a_j + b_j}{2} \right| \right) \right. \\
\left. \left(\frac{b_k - a_k}{2} + \left| x_k - \frac{a_k + b_k}{2} \right| \right) \right)_{(l)}.
\end{array}
\right.
$$

$$(17.74)$$

Finally we get that

$$
\int_{a_1}^{b_1} \int_{a_2}^{b_2} \int_{a_3}^{b_3} \left(\prod_{i=1}^{3} |p_i(x_i, s_i)| \right) \left\| \frac{\partial^3 f(s_1, s_2, s_3)}{\partial s_3 \partial s_2 \partial s_1} \right\| ds_3 ds_2 ds_1 \leq
$$

$$
\min \left\{
\begin{array}{l}
\left\| \left\| \frac{\partial^3 f(s_1,s_2,s_3)}{\partial s_3 \partial s_2 \partial s_1} \right\| \right\|_\infty \dfrac{\prod_{j=1}^{3} \left((x_j - a_j)^2 + (b_j - x_j)^2 \right)}{8}, \\[3ex]
\left\| \left\| \frac{\partial^3 f(s_1,s_2,s_3)}{\partial s_3 \partial s_2 \partial s_1} \right\| \right\|_{p_{123}} \prod_{i=1}^{3} \left(\dfrac{(x_j-a_j)^{q_{123}+1} + (b_j - x_j)^{q_{123}+1}}{q_{123}+1} \right)^{\frac{1}{q_{123}}}, \\[3ex]
\text{when } p_{123}, q_{123} > 1 : \frac{1}{p_{123}} + \frac{1}{q_{123}} = 1; \\[2ex]
\left\| \left\| \frac{\partial^3 f(s_1,s_2,s_3)}{\partial s_3 \partial s_2 \partial s_1} \right\| \right\|_1 \prod_{i=1}^{3} \left(\frac{b_j - a_j}{2} + \left| x_j - \frac{a_j + b_j}{2} \right| \right).
\end{array}
\right.
$$

$$(17.75)$$

Taking into account (17.73)–(17.75) we have completed the proof of (17.69). ■

A generalization of Theorem 17.23 follows.

Theorem 17.24 *Let* $f : \times_{i=1}^{n} [a_i, b_i] \to X$ *as in Theorem 17.22,* $n \in \mathbb{N}$, *and* $(x_1, \ldots, x_n) \in \times_{i=1}^{n} [a_i, b_i]$. *Here* $\|\cdot\|_p$ $(1 \leq p \leq \infty)$ *is the usual* L_p-*norm on* $\times_{i=1}^{n} [a_i, b_i]$. *Then*

$$\left\| f(x_1, x_2, \ldots, x_n) - \frac{1}{\prod\limits_{i=1}^{n} (b_i - a_i)} \int_{\times_{i=1}^{n}[a_i,b_i]} f(s_1, s_2, \ldots, s_n) \, ds_n \ldots ds_1 \right\|$$

$$\leq \frac{\left(\sum\limits_{i=1}^{n} M_{1,n} \right)}{\prod\limits_{i=1}^{n} (b_i - a_i)}. \tag{17.76}$$

Here we have

$$M_{1,n} := \min \begin{cases} \sum\limits_{j=1}^{n} \left(\left\| \frac{\partial f}{\partial s_j} \right\|_{\infty} \left(\prod\limits_{\substack{i=1 \\ i \neq j}}^{n} (b_i - a_i) \right) \left(\frac{(x_j - a_j)^2 + (b_j - x_j)^2}{2} \right) \right), \\[2em] \sum\limits_{j=1}^{n} \left(\left\| \frac{\partial f}{\partial s_j} \right\|_p \left(\prod\limits_{\substack{i=1 \\ i \neq j}}^{n} (b_i - a_i)^{\frac{1}{q}} \right) \cdot \left[\frac{(x_j - a_j)^{q+1} + (b_j - x_j)^{q+1}}{q+1} \right]^{\frac{1}{q}} \right), \\[2em] \quad when \ p, q > 1 : \frac{1}{p} + \frac{1}{q} = 1, for \ j = 1, 2, \ldots, n; \\[1em] \sum\limits_{j=1}^{n} \left(\left\| \frac{\partial f}{\partial s_j} \right\|_1 \left(\frac{b_j - a_j}{2} + \left| x_j - \frac{a_j + b_j}{2} \right| \right) \right). \end{cases}$$

$$\tag{17.77}$$

Also we have for $j, k \in \{1, 2, \ldots, n\}$ *that*

$$M_{2,n} := \min \begin{cases} \sum_{\substack{l_1=1 \\ j<k}}^{\binom{n}{2}} \left(\left\| \left| \frac{\partial^2 f}{\partial s_k \partial s_j} \right| \right\|_\infty \left(\prod_{\substack{i=1 \\ i\neq j,k}}^{n} (b_i - a_i) \right) \right. \\ \qquad\qquad \left. \frac{\left((x_j-a_j)^2 + (b_j-x_j)^2 \right)\left((x_k-a_k)^2 + (b_k-x_k)^2 \right)}{4} \right)_{(l_1)}, \\[2mm] \sum_{\substack{l_1=1 \\ j<k}}^{\binom{n}{2}} \left(\left\| \left| \frac{\partial^2 f}{\partial s_k \partial s_j} \right| \right\|_p \left(\prod_{\substack{i=1 \\ i\neq j,k}}^{n} (b_i - a_i)^{\frac{1}{q}} \right) \cdot \right. \\ \qquad \left. \left[\frac{(x_j-a_j)^{q+1} + (b_j-x_j)^{q+1}}{q+1} \right]^{\frac{1}{q}} \left[\frac{(x_k-a_k)^{q+1} + (b_k-x_k)^{q+1}}{q+1} \right]^{\frac{1}{q}} \right)_{(l_1)}, \\ \text{when } p,q > 1 : \frac{1}{p} + \frac{1}{q} = 1; \\[2mm] \sum_{\substack{l_1=1 \\ j<k}}^{\binom{n}{2}} \left(\left\| \left| \frac{\partial^2 f}{\partial s_k \partial s_j} \right| \right\|_1 \left(\frac{b_j-a_j}{2} + \left| x_j - \frac{a_j+b_j}{2} \right| \right) \right. \\ \qquad \left. \left(\frac{b_k-a_k}{2} + \left| x_k - \frac{a_k+b_k}{2} \right| \right) \right)_{(l_1)}. \end{cases} \tag{17.78}$$

And for $j, k, r \in \{1, \dots, n\}$ that

$$M_{3,n} := \min \begin{cases} \sum_{\substack{l_2=1 \\ j<k}}^{\binom{n}{3}} \left(\left\| \left| \frac{\partial^3 f}{\partial s_r \partial s_k \partial s_j} \right| \right\|_\infty \left(\prod_{\substack{i=1 \\ i\neq j,k,r}}^{n} (b_i - a_i) \right) \right. \\ \qquad\qquad \left. \prod_{m\in\{j,k,r\}} \cdot \frac{\left((x_m-a_m)^2 + (b_m-x_m)^2 \right)}{2} \right)_{(l_2)}, \\[2mm] \sum_{\substack{l_2=1 \\ j<k}}^{\binom{n}{3}} \left(\left\| \left| \frac{\partial^3 f}{\partial s_r \partial s_k \partial s_j} \right| \right\|_p \left(\prod_{\substack{i=1 \\ i\neq j,k,r}}^{n} (b_i - a_i)^{\frac{1}{q}} \right) \cdot \right. \\ \qquad \left. \prod_{m\in\{j,k,r\}} \left[\frac{(x_m-a_m)^{q+1} + (b_m-x_m)^{q+1}}{q+1} \right]^{\frac{1}{q}} \right)_{(l_2)}, \\ \text{when } p,q > 1 : \frac{1}{p} + \frac{1}{q} = 1; \\[2mm] \sum_{\substack{l_2=1 \\ j<k}}^{\binom{n}{3}} \left(\left\| \left| \frac{\partial^3 f}{\partial s_r \partial s_k \partial s_j} \right| \right\|_1 \prod_{m\in\{j,k,r\}} \left(\frac{b_m-a_m}{2} + \left| x_m - \frac{a_m+b_m}{2} \right| \right) \right)_{(l_2)}. \end{cases}$$

$$\tag{17.79}$$

And for $l = 1, \ldots, n$ that

$$
M_{n-1,n} := \min
\begin{cases}
\binom{n}{n-1} \\
\displaystyle\sum_{l=1} \left(\left\| \frac{\partial^{n-1} f}{\partial s_n \ldots \widehat{\partial s_l} \ldots \partial s_1} \right\|_{\infty} (b_l - a_l) \frac{\displaystyle\prod_{\substack{m=1 \\ m \neq l}}^{n} \left((x_m - a_m)^2 + (b_m - x_m)^2 \right)}{2^{n-1}} \right), \\[2em]
\binom{n}{n-1} \\
\displaystyle\sum_{l=1} \left(\left\| \frac{\partial^{n-1} f}{\partial s_n \ldots \widehat{\partial s_l} \ldots \partial s_1} \right\|_{p} (b_l - a_l)^{\frac{1}{q}} \cdot \right. \\
\left. \prod_{\substack{m=1 \\ m \neq l}}^{n} \left[\frac{(x_m - a_m)^{q+1} + (b_m - x_m)^{q+1}}{q+1} \right]^{\frac{1}{q}} \right), \\
\text{when } p, q > 1 : \frac{1}{p} + \frac{1}{q} = 1; \\[1em]
\binom{n}{n-1} \\
\displaystyle\sum_{l=1} \left(\left\| \frac{\partial^{n-1} f}{\partial s_n \ldots \widehat{\partial s_l} \ldots \partial s_1} \right\|_{1} \prod_{\substack{m=1 \\ m \neq l}}^{n} \left(\frac{b_m - a_m}{2} + \left| x_m - \frac{a_m + b_m}{2} \right| \right) \right).
\end{cases}
\tag{17.80}
$$

Finally we have

$$
M_{n,n} := \min
\begin{cases}
\left\| \frac{\partial^n f}{\partial s_n \ldots \partial s_1} \right\|_{\infty} \frac{\displaystyle\prod_{j=1}^{n} \left((x_j - a_j)^2 + (b_j - x_j)^2 \right)}{2^n}, \\[1.5em]
\left\| \frac{\partial^n f}{\partial s_n \ldots \partial s_1} \right\|_{p} \displaystyle\sum_{j=1}^{n} \left(\frac{(x_j - a_j)^{q+1} + (b_j - x_j)^{q+1}}{q+1} \right)^{\frac{1}{q}}, \\
\text{when } p, q > 1 : \frac{1}{p} + \frac{1}{q} = 1; \\[1em]
\left\| \frac{\partial^n f}{\partial s_n \ldots \partial s_1} \right\|_{1} \displaystyle\prod_{i=1}^{n} \left(\frac{b_j - a_j}{2} + \left| x_j - \frac{a_j + b_j}{2} \right| \right).
\end{cases}
\tag{17.81}
$$

Proof Similar to Theorem 17.23. ■

We need from [8]

Lemma 17.25 *Let $f \in C^n([a, b], X)$, $n \in \mathbb{N}$, $(X, \|\cdot\|)$ a Banach space, $x \in [a, b]$. Then*

$$
f(x) = \frac{n \int_a^b f(y) \, dy}{b - a} + \sum_{k=1}^{n-1} \left(\frac{n-k}{k!} \right) \left(\frac{f^{(k-1)}(b)(x-b)^k - f^{(k-1)}(a)(x-a)^k}{b - a} \right)
$$

$$
+ \frac{1}{(n-1)!(b-a)} \int_a^b \left(\int_y^x (x-t)^{n-1} f^{(n)}(t) \, dt \right) dy.
\tag{17.82}
$$

We present the representation result

Theorem 17.26 *Let* $(X, \|\cdot\|)$ *be a Banach space and* $f \in C^n\left(\prod_{j=1}^{m}[a_j, b_j], X\right)$,

$m, n \in \mathbb{N}$. *Let also* $(x_1, \ldots, x_m) \in \prod_{j=1}^{m}[a_j, b_j]$ *be fixed. Then*

$$f(x_1, \ldots, x_n) = \frac{n^m}{\prod\limits_{j=1}^{m}(b_j - a_j)} \int_{\prod\limits_{j=1}^{m}[a_j, b_j]} f(s_1, \ldots, s_m)\, ds_1 \ldots ds_m$$

$$+ \sum_{i=1}^{m} T_i, \tag{17.83}$$

where for $i = 1, \ldots, m$ *we put*

$$T_i := T_i(x_i, \ldots, x_m) := \left(\frac{n^{i-1}}{\prod\limits_{j=1}^{i}(b_j - a_j)}\right) \left\{\sum_{k=1}^{n-1}\left(\frac{n-k}{k!}\right)\right.$$

$$\left\{\int_{\prod\limits_{j=1}^{i-1}[a_j, b_j]} \left(\frac{\partial^{k-1} f}{\partial x_i^{k-1}}(s_1, s_2, \ldots, s_{i-1}, b_i, x_{i+1}, \ldots, x_m)(x_i - b_i)^k -\right.\right.$$

$$\tag{17.84}$$

$$\left.\left.\frac{\partial^{k-1} f}{\partial x_i^{k-1}}(s_1, s_2, \ldots, s_{i-1}, a_i, x_{i+1}, \ldots, x_m)(x_i - a_i)^k\right) ds_1 ds_2 \ldots ds_{i-1}\right\}\right\}$$

$$+ \left(\frac{n^{i-1}}{(n-1)! \prod\limits_{j=1}^{i}(b_j - a_j)}\right) \left(\int_{\prod\limits_{j=1}^{i}[a_j, b_j]} \left(\int_{s_i}^{x_i}(x_i - t_i)^{n-1} \cdot\right.\right.$$

$$\frac{\partial^n f}{\partial x_i^n}\left(s_1, s_2, s_3, \ldots, s_{i-1}, t_i, x_{i+1}, \ldots, x_m\right) dt_i\Bigg) ds_i ds_{i-1} ds_{i-2} \cdots ds_2 ds_1 \Bigg).$$

When $n = 1$ the $\sum\limits_{k=1}^{n-1}$ in (17.84) is zero. (For the real analog see [5] and [7], p. 367.)

Proof Here $(X, \|\cdot\|)$ is a Banach space and $f \in C^n\left(\prod\limits_{j=1}^m [a_j, b_j], X\right)$, $m, n \in \mathbb{N}$. Hence by Lemma 17.25 we have

$$f(x_1, \ldots, x_m) = \frac{n}{b_1 - a_1} \int_{a_1}^{b_1} f(s_1, x_2, \ldots, x_m)\, ds_1 + T_1(x_1, \ldots, x_m), \quad (17.85)$$

where

$$T_1(x_1, \ldots, x_m) := \frac{1}{(b_1 - a_1)} \sum_{k=1}^{n-1} \left(\frac{n-k}{k!}\right). \qquad (17.86)$$

$$\left(\frac{\partial^{k-1} f(b_1, x_2, \ldots, x_m)}{\partial x_1^{k-1}}(x_1 - b_1)^k - \frac{\partial^{k-1} f(a_1, x_2, \ldots, x_m)}{\partial x_1^{k-1}}(x_1 - a_1)^k\right)$$

$$+ \frac{1}{(n-1)!\,(b_1 - a_1)} \int_{a_1}^{b_1} \left(\int_{s_1}^{x_1} (x_1 - t_1)^{n-1} \frac{\partial^n f}{\partial x_1^n}(t_1, x_2, \ldots, x_m)\, dt_1\right) ds_1.$$

But it holds

$$f(s_1, x_2, \ldots, x_m) = \frac{n}{b_2 - a_2} \int_{a_2}^{b_2} f(s_1, s_2, x_3 \ldots, x_m)\, ds_2 + \qquad (17.87)$$

$$\frac{1}{(b_2 - a_2)} \sum_{k=1}^{n-1} \left(\frac{n-k}{k!}\right).$$

$$\left(\frac{\partial^{k-1} f(s_1, b_2, x_3 \ldots, x_m)}{\partial x_2^{k-1}}(x_2 - b_2)^k - \frac{\partial^{k-1} f(s_1, a_2, x_3, \ldots, x_m)}{\partial x_2^{k-1}}(x_2 - a_2)^k\right)$$

$$+ \frac{1}{(n-1)!\,(b_2 - a_2)} \int_{a_2}^{b_2} \left(\int_{s_2}^{x_2} (x_2 - t_2)^{n-1} \frac{\partial^n f}{\partial x_2^n}(s_1, t_2, x_3, \ldots, x_m)\, dt_2\right) ds_2.$$

Combining (17.85) and (17.87) we obtain

$$f(x_1, \ldots, x_m) = \frac{n^2}{(b_1 - a_1)(b_2 - a_2)} \int_{a_1}^{b_1} \int_{a_2}^{b_2} f(s_1, s_2, x_3, \ldots, x_m)\, ds_2 ds_1$$

$$+ T_2(x_2, x_3, \ldots, x_m) + T_1(x_1, \ldots, x_m), \qquad (17.88)$$

where

$$T_2(x_2, x_3, \ldots, x_m) := \frac{n}{(b_1 - a_1)(b_2 - a_2)} \sum_{k=1}^{n-1} \left(\frac{n-k}{k!}\right) \cdot$$

$$\int_{a_1}^{b_1} \left(\frac{\partial^{k-1} f(s_1, b_2, x_3 \ldots, x_m)}{\partial x_2^{k-1}} (x_2 - b_2)^k - \frac{\partial^{k-1} f(s_1, a_2, x_3, \ldots, x_m)}{\partial x_2^{k-1}} (x_2 - a_2)^k \right)$$

$$\cdot ds_1 + \frac{n}{(n-1)!\,(b_1 - a_1)(b_2 - a_2)} \cdot$$

$$\int_{a_1}^{b_1} \left(\int_{a_2}^{b_2} \left(\int_{s_2}^{x_2} (x_2 - t_2)^{n-1} \frac{\partial^n f}{\partial x_2^n}(s_1, t_2, x_3, \ldots, x_m)\, dt_2 \right) ds_2 \right) ds_1.$$

Next we see that

$$f(s_1, s_2, x_3, \ldots, x_m) = \frac{n}{b_3 - a_3} \int_{a_3}^{b_3} f(s_1, s_2, s_3, x_4 \ldots, x_m)\, ds_3 +$$

$$\frac{1}{(b_3 - a_3)} \sum_{k=1}^{n-1} \left(\frac{n-k}{k!}\right) \cdot \qquad (17.89)$$

$$\left(\frac{\partial^{k-1} f(s_1, s_2, b_3, x_4, \ldots, x_m)}{\partial x_3^{k-1}} (x_3 - b_3)^k - \frac{\partial^{k-1} f(s_1, s_2, a_3, x_4, \ldots, x_m)}{\partial x_3^{k-1}} (x_3 - a_3)^k \right)$$

$$+ \frac{1}{(n-1)!\,(b_3 - a_3)} \int_{a_3}^{b_3} \left(\int_{s_3}^{x_3} (x_3 - t_3)^{n-1} \frac{\partial^n f}{\partial x_3^n}(s_1, s_2, t_3, x_4, \ldots, x_m)\, dt_3 \right) ds_3.$$

Combining (17.88) and (17.89) we get

$$f(x_1, \ldots, x_m) = \frac{n^3}{(b_1 - a_1)(b_2 - a_2)(b_3 - a_3)} \cdot$$

$$\int_{a_1}^{b_1} \int_{a_2}^{b_2} \int_{a_3}^{b_3} f(s_1, s_2, s_3, x_4 \ldots, x_m)\, ds_3 ds_2 ds_1 +$$

$$+ T_3 (x_3, x_4, \ldots, x_m) + T_1 + T_2, \qquad (17.90)$$

where

$$T_3 (x_3, x_4, \ldots, x_m) := \frac{n^2}{3 \atop \prod_{i=1} (b_i - a_i)} \sum_{k=1}^{n-1} \left(\frac{n-k}{k!} \right)$$

$$\left(\int_{a_1}^{b_1} \left(\int_{a_2}^{b_2} \left(\frac{\partial^{k-1} f (s_1, s_2, b_3, x_4, \ldots, x_m)}{\partial x_3^{k-1}} (x_3 - b_3)^k - \right. \right. \right.$$

$$\left. \left. \left. - \frac{\partial^{k-1} f (s_1, s_2, a_3, x_4, \ldots, x_m)}{\partial x_3^{k-1}} (x_3 - a_3)^k \right) ds_2 \right) ds_1 \right)$$

$$+ \frac{n^2}{(n-1)! \prod_{i=1}^{3} (b_i - a_i)} \int_{a_1}^{b_1} \left(\int_{a_2}^{b_2} \left(\int_{a_3}^{b_3} \right. \right.$$

$$\left. \left. \left(\int_{s_3}^{x_3} (x_3 - t_3)^{n-1} \frac{\partial^n f}{\partial x_3^n} (s_1, s_2, t_3, x_4, \ldots, x_m) \, dt_3 \right) ds_3 \right) ds_2 \right) ds_1. \qquad (17.91)$$

We also observe that

$$f (s_1, s_2, s_3, x_4, \ldots, x_m) = \frac{n}{b_4 - a_4} \int_{a_4}^{b_4} f (s_1, s_2, s_3, s_4, x_5, \ldots, x_m) \, ds_4 + \qquad (17.92)$$

$$\frac{1}{(b_4 - a_4)} \sum_{k=1}^{n-1} \left(\frac{n-k}{k!} \right) \cdot \left(\frac{\partial^{k-1} f (s_1, s_2, s_3, b_4, x_5, \ldots, x_m)}{\partial x_4^{k-1}} (x_4 - b_4)^k \right.$$

$$\left. - \frac{\partial^{k-1} f (s_1, s_2, s_3, a_4, x_5, \ldots, x_m)}{\partial x_4^{k-1}} (x_4 - a_4)^k \right) +$$

$$\frac{1}{(n-1)! (b_4 - a_4)} \int_{a_4}^{b_4} \left(\int_{s_4}^{x_4} (x_4 - t_4)^{n-1} \frac{\partial^n f}{\partial x_4^n} (s_1, s_2, s_3, t_4, x_5, \ldots, x_m) \, dt_4 \right) ds_4.$$

Combining (17.90) and (17.92) we get

$$f (x_1, x_2, \ldots, x_m) = \frac{n^4}{\prod_{j=1}^{4} (b_j - a_j)} \cdot$$

$$\int_{\prod\limits_{j=1}^{4}[a_j,b_j]} f\left(s_1,s_2,s_3,s_4,x_5,\ldots,x_m\right) ds_1 ds_2 ds_3 ds_4 + \sum_{j=1}^{4} T_j, \qquad (17.93)$$

where

$$T_4\left(x_4,x_5,\ldots,x_m\right) := \frac{n^3}{\prod\limits_{j=1}^{4}\left(b_j-a_j\right)}\left\{\sum_{k=1}^{n-1}\left(\frac{n-k}{k!}\right)\right. \qquad (17.94)$$

$$\left(\int_{\prod\limits_{j=1}^{3}[a_j,b_j]}\left(\frac{\partial^{k-1}f\left(s_1,s_2,s_3,b_4,x_5,\ldots,x_m\right)}{\partial x_4^{k-1}}\left(x_4-b_4\right)^k-\right.\right.$$

$$\left.\left.-\frac{\partial^{k-1}f\left(s_1,s_2,s_3,a_4,x_5,\ldots,x_m\right)}{\partial x_4^{k-1}}\left(x_4-a_4\right)^k ds_1 ds_2 ds_3\right)\right\}$$

$$+\frac{n^3}{(n-1)!\prod\limits_{i=1}^{4}\left(b_i-a_i\right)}\left(\int_{\prod\limits_{j=1}^{4}[a_j,b_j]}\right.$$

$$\left.\left(\int_{s_4}^{x_4}\left(x_4-t_4\right)^{n-1}\frac{\partial^n f}{\partial x_4^n}\left(s_1,s_2,s_3,t_4,x_5,\ldots,x_m\right) dt_4\right) ds_4 ds_3 ds_2 ds_1\right).$$

Etc. doing the same procedure m times, so proving the claim. ∎

Remark 17.27 We call for $i=1,\ldots,m$,

$$A_i := A_i\left(x_i,\ldots,x_m\right) :=$$

$$\left(\frac{n^{i-1}}{\prod\limits_{j=1}^{i}\left(b_j-a_j\right)}\right)\left\{\sum_{k=1}^{n-1}\left(\frac{n-k}{k!}\right)\right.$$

$$\left\{ \int_{\prod_{j=1}^{i-1}[a_j,b_j]} \left(\frac{\partial^{k-1} f}{\partial x_i^{k-1}}(s_1, s_2, \ldots, s_{i-1}, b_i, x_{i+1}, \ldots, x_m)(x_i - b_i)^k - \right. \right.$$

$$\left. \left. \frac{\partial^{k-1} f}{\partial x_i^{k-1}}(s_1, s_2, \ldots, s_{i-1}, a_i, x_{i+1}, \ldots, x_m)(x_i - a_i)^k \right) ds_1 ds_2 \cdots ds_{i-1} \right\} \right\}, \tag{17.95}$$

when $n = 1$, then $A_i = 0$ all $i = 1, \ldots, m$.

And for $i = 1, \ldots, m$, we call

$$B_i := B_i(x_i, \ldots, x_m) :=$$

$$\left(\frac{n^{i-1}}{(n-1)! \prod_{j=1}^{i}(b_j - a_j)} \right) \left(\int_{\prod_{j=1}^{i}[a_j, b_j]} \left(\int_{s_i}^{t_i} (x_i - t_i)^{n-1} \cdot \right. \right. \tag{17.96}$$

$$\left. \left. \frac{\partial^n f}{\partial x_i^n}(s_1, s_2, s_3, \ldots, s_{i-1}, t_i, x_{i+1}, \ldots, x_m) dt_i \right) ds_i ds_{i-1} ds_{i-2} \cdots ds_2 ds_1 \right).$$

So that we have

$$T_i = A_i + B_i, \tag{17.97}$$

$i = 1, \ldots, m$.

We set

$$K(t_i, x_i) := \begin{cases} t_i - a_i, & a_i \le t_i \le x_i, \\ t_i - b_i, & x_i < t_i \le b_i, \end{cases} \tag{17.98}$$

for $i = 1, \ldots, m$.

Call

$$g^{(n)}(t_i) := \frac{\partial^n f}{\partial x_i^n}(s_1, s_2, s_3, \ldots, s_{i-1}, t_i, x_{i+1}, \ldots, x_m), \tag{17.99}$$

$i = 1, \ldots, m$.

Then as in [8] one can prove that

$$\left\| \int_{a_i}^{b_i} \left(\int_{s_i}^{t_i} (x_i - t_i)^{n-1} g^{(n)} (t_i) \, dt_i \right) ds_i \right\| \le$$

$$\int_{a_i}^{b_i} |x_i - t_i|^{n-1} |K (t_i, x_i)| \left\| g^{(n)} (t_i) \right\| dt_i, \qquad (17.100)$$

for $i = 1, \ldots, m$.

The last gives

$$\|B_i\| \le \left(\frac{n^{i-1}}{(n-1)! \prod_{j=1}^{i} (b_j - a_j)} \right) \left(\int_{\prod_{j=1}^{i} [a_j, b_j]} \left(|x_i - s_i|^{n-1} |K (s_i, x_i)| \cdot \right. \right.$$

$$\qquad (17.101)$$

$$\left. \left. \left\| \frac{\partial^n f}{\partial x_i^n} (s_1, s_2, s_3, \ldots, s_{i-1}, s_i, x_{i+1}, \ldots, x_m) \right\| ds_i ds_{i-1} \cdots ds_1 \right) \right),$$

for $i = 1, \ldots, m$.

Thus by (17.83) we get

$$E_f (x_1, \ldots, x_m) := f (x_1, \ldots, x_m) -$$

$$\frac{n^m}{\prod_{j=1}^{m} (b_j - a_j)} \int_{\prod_{j=1}^{m} [a_j, b_j]} f (s_1, \ldots, s_m) \, ds_1 \cdots ds_m - \sum_{i=1}^{m} A_i = \sum_{i=1}^{m} B_i.$$

$$\qquad (17.102)$$

Hence

$$\left\| E_f (x_1, \ldots, x_n) \right\| \le \sum_{i=1}^{m} \|B_i\| . \qquad (17.103)$$

Next we estimate E_f via some Ostrowski type inequalities

Theorem 17.28 *Assume all as in Theorem 17.26. Then*

$$\left\| E_f\left(x_1, \ldots, x_m\right) \right\| \leq \frac{\displaystyle\sum_{i=1}^{m}}{(n-1)!} \left\{ \left\| \left\| \frac{\partial^n f}{\partial x_i^n} \left(\overbrace{\cdots}^{-i-}, x_{i+1}, \ldots, x_m \right) \right\| \right\|_\infty \right.$$

$$\frac{n^{i-1}}{(b_i - a_i)} \left[\left(\frac{b_i\,(b_i - x_i)^n - a_i\,(x_i - a_i)^n}{n} \right) + \right.$$

$$\sum_{\lambda=0}^{n-1} \binom{n-1}{\lambda} (-1)^{n-\lambda-1} \left\{ \frac{(n+3)\,x_i^{n+1}}{(\lambda+2)\,(n-\lambda+1)} - \right.$$

$$\left. \left. \left. \left(\frac{x_i^\lambda a_i^{n-\lambda+1}}{(n-\lambda+1)} + \frac{x_i^{n-\lambda-1} b_i^{\lambda+2}}{(\lambda+2)} \right) \right\} \right] \right\}, \qquad (17.104)$$

$$\forall\, (x_1, \ldots, x_m) \in \prod_{j=1}^{m} [a_j, b_j].$$

Proof We observe by (17.101) that

$$\| B_i \| \leq \left(\frac{n^{i-1}}{(n-1)! \displaystyle\prod_{j=1}^{i} (b_j - a_j)} \right) \left(\int_{\prod_{j=1}^{i} [a_j, b_j]} \right.$$

$$\left. \left(|x_i - s_i|^{n-1} |K(s_i, x_i)| \left\| \frac{\partial^n f}{\partial x_i^n} (s_1, s_2, s_3, \ldots, s_i, x_{i+1}, \ldots, x_m) \right\| ds_1 \cdots ds_i \right) \right)$$

$$\leq \frac{n^{i-1}}{(n-1)!\,(b_i - a_i)} \left\| \left\| \frac{\partial^n f}{\partial x_i^n} \left(\overbrace{\cdots}^{-i-}, x_{i+1}, \ldots, x_m \right) \right\| \right\|_\infty$$

$$\times \int_{a_i}^{b_i} |x_i - s_i|^{n-1} |K(s_i, x_i)|\, ds_i =: (*). \qquad (17.105)$$

We find that

$$\int_{a_i}^{b_i} |x_i - s_i|^{n-1} |K(s_i, x_i)| \, ds_i =$$

$$\int_{a_i}^{x_i} |x_i - s_i|^{n-1} |K(s_i, x_i)| \, ds_i + \int_{x_i}^{b_i} |x_i - s_i|^{n-1} |K(s_i, x_i)| \, ds_i =$$

$$\int_{a_1}^{x_1} (x_i - s_i)^{n-1} (s_i - a_i) \, ds_i + \int_{x_1}^{b_1} (s_i - x_i)^{n-1} (b_i - s_i) \, ds_i =$$

$$\left[\left(\frac{b_i (b_i - x_i)^n - a_i (x_i - a_i)^n}{n} \right) + \sum_{\lambda=0}^{n-1} \binom{n-1}{\lambda} (-1)^{n-\lambda-1} \cdot \right.$$

$$\left. \left\{ \frac{(n+3) x_i^{n+1}}{(\lambda+2)(n-\lambda+1)} - \left(\frac{x_i^\lambda a_i^{n-\lambda+1}}{(n-\lambda+1)} + \frac{x_i^{n-\lambda-1} b_i^{\lambda+2}}{(\lambda+2)} \right) \right\} \right] =: \gamma_i. \quad (17.106)$$

Therefore we obtain

$$(*) \le \frac{n^{i-1}}{(n-1)! (b_i - a_i)} \left\| \frac{\partial^n f}{\partial x_i^n} \left(\overset{-i-}{\overset{\frown}{\cdots}}, x_{i+1}, \dots, x_m \right) \right\|_\infty \gamma_i. \quad (17.107)$$

Finally we get

$$\| E_f (x_1, \dots, x_m) \| \le \sum_{i=1}^{m} \frac{n^{i-1}}{(n-1)! (b_i - a_i)} \left\| \frac{\partial^n f}{\partial x_i^n} \left(\overset{-i-}{\overset{\frown}{\cdots}}, x_{i+1}, \dots, x_m \right) \right\|_\infty \gamma_i, \quad (17.108)$$

proving the claim. ∎

We continue with

Theorem 17.29 *Assume all as in Theorem 17.26. Then*

$$\| E_f (x_1, x_2 \dots, x_m) \| \le \frac{\sum_{i=1}^{m}}{(n-1)!} \left\| \frac{\partial^n f}{\partial x_i^n} \right\|_\infty \times$$

$$\left\{ \frac{n^{i-1}}{(b_i - a_i)} \left[\left(\frac{b_i (b_i - x_i)^n - a_i (x_i - a_i)^n}{n} \right) + \right. \right.$$

$$\left. \sum_{\lambda=0}^{n-1} \binom{n-1}{\lambda} (-1)^{n-\lambda-1} \left\{ \frac{(n+3) x_i^{n+1}}{(\lambda+2)(n-\lambda+1)} - \right. \right.$$

$$\left(\frac{x_i^{\lambda} a_i^{n-\lambda+1}}{(n-\lambda+1)} + \frac{x_i^{n-\lambda-1} b_i^{\lambda+2}}{(\lambda+2)} \right) \Bigg] \Bigg] \Bigg\}, \tag{17.109}$$

$$\forall \, (x_1, \ldots, x_m) \in \prod_{j=1}^{m} [a_j, b_j].$$

Proof From Theorem 17.28. ∎

Next we give

Theorem 17.30 *Assume all as in Theorem 17.26, and let* $p, q > 1$ *such that* $\frac{1}{p} + \frac{1}{q} = 1$. *Then*

$$\|E_f(x_1, \ldots, x_m)\| \le \sum_{i=1}^{m} \frac{1}{(n-1)!} \left\{ \left(\frac{n^{i-1}}{(b_i - a_i) \left(\prod_{j=1}^{i-1} (b_j - a_j) \right)^{\frac{1}{p}}} \right) \times \right.$$

$$\left\{ \left((x_i - a_i)^{nq+1} + (b_i - x_i)^{nq+1} \right) B\left((n-1)q + 1, q + 1 \right) \right\}^{\frac{1}{q}} \times$$

$$\left. \left\| \frac{\partial^n f}{\partial x_i^n} \left(\overset{-i-}{\overset{\frown}{\cdots}}, x_{i+1}, \ldots, x_m \right) \right\|_p \right\}, \tag{17.110}$$

$$\forall \, (x_1, \ldots, x_m) \in \prod_{j=1}^{m} [a_j, b_j].$$

Proof We notice that

$$\|B_i\| \le \left(\frac{n^{i-1}}{(n-1)! \prod_{j=1}^{i} (b_j - a_j)} \right) \left\| \frac{\partial^n f}{\partial x_i^n} \left(\overset{-i-}{\overset{\frown}{\cdots}}, x_{i+1}, \ldots, x_m \right) \right\|_p$$

$$\times \left(\int_{a_i}^{b_i} |x_i - s_i|^{q(n-1)} |K(s_i, x_i)|^q \, ds_i \right)^{\frac{1}{q}} \left(\prod_{j=1}^{i-1} (b_j - a_j) \right)^{\frac{1}{q}}$$

(by [15], p. 256)

$$= \left(\frac{n^{i-1}}{(n-1)! \, (b_i - a_i) \left(\prod_{j=1}^{i-1} (b_j - a_j) \right)^{\frac{1}{p}}} \right) \times$$

$$\left\{ \left((x_i - a_i)^{nq+1} + (b_i - x_i)^{nq+1} \right) B \left((n-1) q + 1, q + 1 \right) \right\}^{\frac{1}{q}}$$

$$\times \left\| \frac{\partial^n f}{\partial x_i^n} \left(\overbrace{\ldots}^{-i-}, x_{i+1}, \ldots, x_m \right) \right\|_p, \tag{17.111}$$

proving the claim.   ∎

We also present

Theorem 17.31 *Assume all as in Theorem 17.26. Then*

$$\| E_f (x_1, \ldots, x_m) \| \leq \frac{\sum_{i=1}^{m}}{(n-1)!} \left\{ \frac{n^{i-1}}{\prod_{j=1}^{i} (b_j - a_j)} \times \right.$$

$$(\max (x_i - a_i, b_i - x_i))^n \left\| \frac{\partial^n f}{\partial x_i^n} \left(\overbrace{\ldots}^{-i-}, x_{i+1}, \ldots, x_m \right) \right\|_1, \tag{17.112}$$

$$\forall \, (x_1, \ldots, x_m) \in \prod_{j=1}^{m} [a_j, b_j].$$

Proof We see that

$$
\|B_i\| \leq \left(\frac{n^{i-1}}{(n-1)! \displaystyle\prod_{j=1}^{i} (b_j - a_j)} \right) \left(\sup_{s_i \in [a_i, b_i]} |x_i - s_i|^{n-1} |K(s_i, x_i)| \right)
$$

$$
\times \left\| \left\| \frac{\partial^n f}{\partial x_i^n} \left(\overset{-i-}{\underset{\cdots}{\frown}}, x_{i+1}, \ldots, x_m \right) \right\| \right\|_1
$$

$$
\leq \left(\frac{n^{i-1}}{(n-1)! \displaystyle\prod_{j=1}^{i} (b_j - a_j)} \right) \left(\max (x_i - a_i, b_i - x_i) \right)^n
$$

$$
\times \left\| \left\| \frac{\partial^n f}{\partial x_i^n} \left(\overset{-i-}{\underset{\cdots}{\frown}}, x_{i+1}, \ldots, x_m \right) \right\| \right\|_1, \tag{17.113}
$$

proving the claim. ∎

We give

Corollary 17.32 *Assume all as in Theorem 17.26. Then*

$$
\| E_f (x_1, \ldots, x_m) \| \leq \min \{ R.H.S. \, (17.104), R.H.S. \, (17.110), R.H.S. \, (17.112) \}, \tag{17.114}
$$

where $p, q > 1 : \frac{1}{p} + \frac{1}{q} = 1, \forall (x_1, \ldots, x_m) \in \displaystyle\prod_{j=1}^{m} [a_j, b_j]$.

Proof By (17.104), (17.110) and (17.112). ∎

We further give

Corollary 17.33 *Let $(X, \|\cdot\|)$ be a Banach space and $f \in C^1 ([a_1, b_1] \times [a_2, b_2], X)$. Let $(x_1, x_2) \in [a_1, b_1] \times [a_2, b_2]$. Then*

$$
f(x_1, x_2) = \frac{1}{(b_1 - a_1)(b_2 - a_2)} \int_{[a_1, b_1] \times [a_2, b_2]} f(s_1, s_2) \, ds_1 ds_2 + T_1 + T_2, \tag{17.115}
$$

where

$$
T_1 = T_1 (x_1, x_2) = \frac{1}{(b_1 - a_1)} \int_{a_1}^{b_1} K(s_1, x_1) \frac{\partial f}{\partial x_1} (s_1, s_2) \, ds_1
$$

$$= B_1 (x_1, x_2), \tag{17.116}$$

$$T_2 = T_2 (x_2) = \frac{1}{(b_1 - a_1) (b_2 - a_2)} \int_{a_1}^{b_1} \int_{a_2}^{b_2} K (s_2, x_2) \frac{\partial f}{\partial x_2} (s_1, s_2) \, ds_1 ds_2$$

$$= B_2 (x_2), \tag{17.117}$$

Proof By Theorem 17.26. ∎

We need

Remark 17.34 Denote here

$$E_f (x_1, x_2) = f (x_1, x_2) - \frac{1}{(b_1 - a_1) (b_2 - a_2)} \int_{a_1}^{b_1} \int_{a_2}^{b_2} f (s_1, s_2) \, ds_1 ds_2.$$
$$\tag{17.118}$$

Hence

$$\left\| E_f (x_1, x_2) \right\| \leq \| B_1 \| + \| B_2 \|. \tag{17.119}$$

We give the following special Ostrowski type inequalities.

Corollary 17.35 *Assume all as in Corollary 17.33. Then*

$$\left\| E_f (x_1, x_2) \right\| = \left\| f (x_1, x_2) - \frac{1}{(b_1 - a_1) (b_2 - a_2)} \int_{a_1}^{b_1} \int_{a_2}^{b_2} f (s_1, s_2) \, ds_1 ds_2 \right\|$$

$$\leq \left\| \frac{\partial f}{\partial x_1} (\cdot, x_2) \right\|_{\infty, [a_1, b_1]} \frac{[a_1^2 + b_1^2 - x_1 (a_1 + b_1)]}{(b_1 - a_1)} +$$

$$\left\| \frac{\partial f}{\partial x_2} (\cdot, \cdot) \right\|_{\infty, [a_1, b_1] \times [a_2, b_2]} \frac{[a_2^2 + b_2^2 - x_2 (a_2 + b_2)]}{(b_2 - a_2)}, \tag{17.120}$$

$\forall (x_1, x_2) \in [a_1, b_1] \times [a_2, b_2]$.

Proof By Theorem 17.28. ∎

We continue with

Corollary 17.36 *Assume all as in Corollary 17.33.*

$$\left| E_f (x_1, x_2) \right| \leq \max \left\{ \left\| \frac{\partial f}{\partial x_1} \right\|_\infty, \left\| \frac{\partial f}{\partial x_2} \right\|_\infty \right\} \times$$

$$\left\{\left(\frac{a_1^2 + b_1^2 - x_1\,(a_1 + b_1)}{b_1 - a_1}\right) + \left(\frac{a_2^2 + b_2^2 - x_2\,(a_2 + b_2)}{b_2 - a_2}\right)\right\},\qquad (17.121)$$

$\forall\,(x_1, x_2) \in [a_1, b_1] \times [a_2, b_2]$.

Proof By (17.120). ∎

Next we have

Corollary 17.37 *Assume all as in Corollary 17.33. Then*

$$\left\|f\,(x_1, x_2) - \frac{1}{(b_1 - a_1)\,(b_2 - a_2)} \int_{a_1}^{b_1} \int_{a_2}^{b_2} f\,(s_1, s_2)\,ds_1 ds_2\right\| \le$$

$$\frac{1}{(b_1 - a_1)} \sqrt{\frac{(x_1 - a_1)^3 + (b_1 - x_1)^3}{3}} \left\|\frac{\partial f}{\partial x_1}\,(\cdot, x_2)\right\|_{2,[a_1,b_1]} +$$

$$\frac{1}{(b_2 - a_2)\,\sqrt{b_1 - a_1}} \sqrt{\frac{(x_2 - a_2)^3 + (b_2 - x_2)^3}{3}} \left\|\frac{\partial f}{\partial x_2}\,(\cdot, \cdot)\right\|_{2,[a_1,b_1]\times[a_2,b_2]},$$
$$(17.122)$$

$\forall\,(x_1, x_2) \in [a_1, b_1] \times [a_2, b_2]$.

Proof By (17.110). ∎

We present

Corollary 17.38 *Assume all as in Corollary 17.33. Then*

$$\left\|f\,(x_1, x_2) - \frac{1}{(b_1 - a_1)\,(b_2 - a_2)} \int_{a_1}^{b_1} \int_{a_2}^{b_2} f\,(s_1, s_2)\,ds_1 ds_2\right\| \le$$

$$\frac{1}{(b_1 - a_1)} \max\,(x_1 - a_1, b_1 - x_1) \left\|\frac{\partial f}{\partial x_1}\,(\cdot, x_2)\right\|_{1,[a_1,b_1]} +$$

$$\frac{1}{(b_1 - a_1)\,(b_2 - a_2)} \max\,(x_2 - a_2, b_2 - x_2) \left\|\frac{\partial f}{\partial x_2}\,(\cdot, \cdot)\right\|_{1,[a_1,b_1]\times[a_2,b_2]},$$
$$(17.123)$$

$\forall\,(x_1, x_2) \in [a_1, b_1] \times [a_2, b_2]$.

Proof By (17.112). ∎

We end Ostrowski type inequality applications for $n = 1, m = 2$ with

Corollary 17.39 *Assume all as in Corollary 17.33. Then*

$$\left\| f(x_1, x_2) - \frac{1}{(b_1 - a_1)(b_2 - a_2)} \int_{a_1}^{b_1} \int_{a_2}^{b_2} f(s_1, s_2) \, ds_1 ds_2 \right\| \le$$

$$\min \{ R.H.S. (17.120), R.H.S. (17.122), R.H.S. (17.123) \}, \tag{17.124}$$

$\forall (x_1, x_2) \in [a_1, b_1] \times [a_2, b_2]$.

We continue with

Corollary 17.40 *Let $(X, \|\cdot\|)$ be a Banach space and $f \in C^2 \left(\prod_{j=1}^{3} [a_j, b_j], X \right)$.*

Let also $(x_1, x_2, x_3) \in \prod_{j=1}^{3} [a_j, b_j]$. Then

$$f(x_1, x_2, x_3) = \frac{8}{\prod_{j=1}^{3} (b_j - a_j)} \int_{\prod_{j=1}^{3} [a_j, b_j]} f(s_1, s_2, s_3) \, ds_1 ds_2 ds_3 + \sum_{i=1}^{3} T_i.$$

$$\tag{17.125}$$

Here for $i = 1, 2, 3$ we have

$$T_1 = T_1(x_1, x_2, x_3) = \frac{1}{(b_1 - a_1)} \times$$

$$\left\{ (f(b_1, x_2, x_3)(x_1 - b_1) - f(a_1, x_2, x_3)(x_1 - a_1)) + \right.$$

$$\left. \int_{a_1}^{b_1} (x_1 - s_1) K(s_1, x_1) \frac{\partial^2 f}{\partial x_1^2}(s_1, x_2, x_3) \, ds_1 \right\}, \tag{17.126}$$

$$T_2 = T_2(x_2, x_3) = \frac{2}{(b_1 - a_1)(b_2 - a_2)} \times$$

$$\left\{ \int_{a_1}^{b_1} (f(s_1, b_2, x_3)(x_2 - b_2) - f(s_1, a_2, x_3)(x_2 - a_2)) \, ds_1 + \right.$$

$$\int_{\prod\limits_{j=1}^{2}[a_j,b_j]} (x_2 - s_2)\, K\,(s_2, x_2)\, \frac{\partial^2 f}{\partial x_2^2}\,(s_1, s_2, x_3)\, ds_1 ds_2 \Bigg\} , \qquad (17.127)$$

and

$$T_3 = T_3\,(x_3) = \frac{4}{\prod\limits_{j=1}^{3} (b_j - a_j)} \times$$

$$\Bigg\{ \int_{\prod\limits_{j=1}^{2}[a_j,b_j]} (f\,(s_1, s_2, b_3)\,(x_3 - b_3) - f\,(s_1, s_2, a_3)\,(x_3 - a_3))\, ds_1 ds_2 +$$

$$\int_{\prod\limits_{j=1}^{3}[a_j,b_j]} (x_3 - s_3)\, K\,(s_3, x_3)\, \frac{\partial^2 f}{\partial x_3^2}\,(s_1, s_2, s_3)\, ds_1 ds_2 ds_3 \Bigg\} . \qquad (17.128)$$

We need

Remark 17.41   Here we have

$$A_1 = A_1\,(x_1, x_2, x_3) = \frac{f\,(b_1, x_2, x_3)\,(x_1 - b_1) - f\,(a_1, x_2, x_3)\,(x_1 - a_1)}{b_1 - a_1},$$
$$(17.129)$$

$$\|B_1\| = \|B_1\,(x_1, x_2, x_3)\| \le$$

$$\frac{1}{(b_1 - a_1)} \int_{a_1}^{b_1} |x_1 - s_1|\, |K\,(s_1, x_1)| \left\| \frac{\partial^2 f}{\partial x_1^2}\,(s_1, x_2, x_3) \right\| ds_1, \qquad (17.130)$$

also

$$A_2 = A_2\,(x_2, x_3) = \frac{2}{(b_1 - a_1)\,(b_2 - a_2)} \times$$

$$\int_{a_1}^{b_1} (f\,(s_1, b_2, x_3)\,(x_2 - b_2) - f\,(s_1, a_2, x_3)\,(x_2 - a_2))\, ds_1, \qquad (17.131)$$

$$\|B_2\| = \|B_2(x_2, x_3)\| \leq \frac{2}{(b_1 - a_1)(b_2 - a_2)} \times$$

$$\int_{\prod_{j=1}^{2} [a_j, b_j]} |x_2 - s_2| |K(s_2, x_2)| \left\| \frac{\partial^2 f}{\partial x_2^2}(s_1, s_2, x_3) \right\| ds_1 ds_2, \qquad (17.132)$$

and

$$A_3 = A_3(x_3) = \frac{4}{\prod_{j=1}^{3}(b_j - a_j)} \times$$

$$\int_{\prod_{j=1}^{2} [a_j, b_j]} (f(s_1, s_2, b_3)(x_3 - b_3) - f(s_1, s_2, a_3)(x_3 - a_3)) \, ds_1 ds_2,$$

$$(17.133)$$

$$\|B_3\| = \|B_3(x_3)\| = \frac{4}{\prod_{j=1}^{3}(b_j - a_j)} \times$$

$$\int_{\prod_{j=1}^{3} [a_j, b_j]} |x_3 - s_3| |K(s_3, x_3)| \left\| \frac{\partial^2 f}{\partial x_3^2}(s_1, s_2, s_3) \right\| ds_1 ds_2 ds_3. \qquad (17.134)$$

Notice here

$$T_i = A_i + B_i, \quad i = 1, 2, 3. \qquad (17.135)$$

We also denote

$$E_f(x_1, x_2, x_3) := f(x_1, x_2, x_3) - \frac{8}{\prod_{j=1}^{3}(b_j - a_j)} \times$$

$$\int_{\prod_{j=1}^{3} [a_j, b_j]} f(s_1, s_2, s_3) \, ds_1 ds_2 ds_3 - \sum_{i=1}^{3} A_i. \qquad (17.136)$$

Hence

$$\left\| E_f\left(x_1, x_2, x_3\right)\right\| \le \sum_{i=1}^{3} \left\| B_i\right\|. \tag{17.137}$$

We give

Corollary 17.42 *Assume all as in Corollary 17.40. Then*

$$\left\| E_f\left(x_1, x_2, x_3\right)\right\| \le \sum_{i=1}^{3} \left\{ \left\| \frac{\partial^2 f}{\partial x_i^2}\left(\overbrace{\underset{\cdots}{-i-}}, x_3\right)\right\|_{\infty} \frac{2^{i-1}}{\left(b_i - a_i\right)} \right.$$

$$\left[\left(\frac{b_i\left(b_i - x_i\right)^2 - a_i\left(x_i - a_i\right)^2}{2}\right) + \sum_{\lambda=0}^{1}(-1)^{1-\lambda}\right.$$

$$\left.\left.\left\{\frac{5x_i^3}{(\lambda+2)(3-\lambda)} - \left(\frac{x_i^{\lambda} a_i^{3-\lambda}}{3-\lambda} + \frac{x_i^{1-\lambda} b_i^{\lambda+2}}{\lambda+2}\right)\right\}\right]\right\}, \tag{17.138}$$

$$\forall\,(x_1, x_2, x_3) \in \prod_{j=1}^{3} [a_j, b_j].$$

Proof By (17.104). ∎

We continue with

Corollary 17.43 *Assume all as in Corollary 17.40. Then*

$$\left\| E_f\left(x_1, x_2, x_3\right)\right\| \le \sum_{i=1}^{3} \left\| \frac{\partial^2 f}{\partial x_i^2}\right\|_{\infty} \left\{ \frac{2^{i-1}}{\left(b_i - a_i\right)} \times \right.$$

$$\left[\left(\frac{b_i\left(b_i - x_i\right)^2 - a_i\left(x_i - a_i\right)^2}{2}\right) + \sum_{\lambda=0}^{1}(-1)^{1-\lambda} \times\right.$$

$$\left.\left.\left\{\frac{5x_i^3}{(\lambda+2)(3-\lambda)} - \left(\frac{x_i^{\lambda} a_i^{3-\lambda}}{3-\lambda} + \frac{x_i^{1-\lambda} b_i^{\lambda+2}}{\lambda+2}\right)\right\}\right]\right\}, \tag{17.139}$$

$$\forall\,(x_1, x_2, x_3) \in \prod_{j=1}^{3} [a_j, b_j].$$

Proof By (17.109). ∎

Next we give

Corollary 17.44 *Assume all as in Corollary 17.40. Then*

$$
\left\| E_f\left(x_1, x_2, x_3\right) \right\| \leq \sum_{i=1}^{3} \left\{ \frac{2^{i-1}}{(b_i - a_i)\left(\prod_{j=1}^{i-1} (b_j - a_j)\right)^{\frac{1}{2}}} \times \right.
$$

$$
\left. \left(\sqrt{\frac{(x_i - a_i)^5 + (b_i - x_i)^5}{30}} \right) \left\| \frac{\partial^2 f}{\partial x_i^2}\left(\overset{-i-}{\cdots}, x_3\right) \right\|_2 \right\}, \qquad (17.140)
$$

$$
\forall \, (x_1, x_2, x_3) \in \prod_{j=1}^{3} [a_j, b_j].
$$

Proof By (17.110). ∎

We also present

Corollary 17.45 *Assume all as in Corollary 17.40. Then*

$$
\left\| E_f\left(x_1, x_2, x_3\right) \right\| \leq \sum_{i=1}^{3} \left\{ \frac{2^{i-1}}{\prod_{j=1}^{i} (b_j - a_j)} \times \right.
$$

$$
\left. \left(\max\left(x_i - a_i, b_i - x_i\right)\right)^2 \left\| \frac{\partial^2 f}{\partial x_i^2}\left(\overset{-i-}{\cdots}, x_3\right) \right\|_1 \right\}, \qquad (17.141)
$$

$$
\forall \, (x_1, x_2, x_3) \in \prod_{j=1}^{3} [a_j, b_j].
$$

Proof By (17.112). ∎

We finish with

Corollary 17.46 *Assume all as in Corollary 17.40. Then*

$$\left\| E_f \left(x_1, x_2, x_3 \right) \right\| \leq$$

$$\min \left\{ R.H.S. \, (17.138), \, R.H.S. \, (17.140), \, R.H.S. \, (17.141) \right\}, \qquad (17.142)$$

$$\forall \, (x_1, x_2, x_3) \in \prod_{j=1}^{3} \left[a_j, b_j \right].$$

Proof By Corollary 17.32. ∎

References

1. G.A. Anastassiou, Multivariate Ostrowski type inequalities. Acta Math. Hung. **76**(4), 267–278 (1997)
2. G.A. Anastassiou, *Quantitative Approximations* (Chapman & Hall/CRC, Boca Raton, 2001)
3. G.A. Anastassiou, Multivariate montgomery identities and Ostrowski inequalities. Numer. Funct. Anal. Opt. **23**(3–4), 247–263 (2002)
4. G.A. Anastassiou, Multidimensional Ostrowski inequalities, revisited. Acta Math. Hung. **97**(4), 339–353 (2002)
5. G.A. Anastassiou, Multivariate fink type identity and multivariate Ostrowski, comparison of means and Grüss type inequalities. Math. Comput. Modell. **46**, 351–374 (2007)
6. G.A. Anastassiou, *Probabilistic Inequalities* (World Scientific, Singapore, 2010)
7. G.A. Anastassiou, *Advanced Inequalities* (World Scientific, Singapore, 2011)
8. G.A. Anastassiou, Ostrowski and Landau inequalities for Banach space valued functions. Math. Comput. Modell. **55**, 312–329 (2012)
9. G.A. Anastassiou, Multidimensional Ostrowski inequalities for Banach space valued functions. J. Nonlinear Evol. Eqn. Appl. **2**, 23-56 (2011)
10. B. Driver, *Analysis Tools with Applications* (Springer, New York, 2003)
11. G. Ladas, V. Laksmikantham, *Differential Equations in Abstract Spaces* (Academic Press, New York, 1972)
12. A. Ostrowski, Über die Absolutabweichung einer differentiebaren Funktion von ihrem integralmittelwert. Comment. Math. Helv. **10**, 226–227 (1938)
13. L. Schwartz, *Analyse Mathematique* (Hermann, Paris, 1967)
14. G. Shilov, *Elementary Functional Analysis* (The MIT Press Cambridge, Massachusetts, 1974)
15. E.T. Whittaker, G.N. Watson, *A Course in Modern Analysis* (University Press, Cambridge, 1927)

Chapter 18
About Fractional Representation Formulae and Right Fractional Inequalities

Here we prove fractional representation formulae involving generalized fractional derivatives, Caputo fractional derivatives and Riemann-Liouville fractional derivatives. Then we establish Poincaré, Sobolev, Hilbert-Pachpatte and Opial type fractional inequalities, involving the right versions of the above mentioned fractional derivatives. It follows [7].

18.1 Introduction

This chapter also continues author's monograph [2], and articles [3–6].

Among others we establish fractional representation formulae for left and right generalized fractional derivatives, left and right Caputo fractional derivatives, and right Riemann-Liouville fractional derivatives.

Then, based on these formulae, we prove new fractional differentiation inequalities of Poincaré type, Sobolev type, Hilbert-Pachpatte type, and Opial type.

These inequalities involve all right fractional derivatives mentioned above.

To give a flavor of our work we mention:

let $f \in AC^m([a, b])$ (means $f^{(m-1)}$ is absolutely continuous), $m \in \mathbb{N}$, $m := \lceil \alpha \rceil$ (ceiling of the number), $\alpha \notin \mathbb{N}$, $\alpha \geq \gamma + 1$, $\gamma \geq 0$, $n := \lceil \gamma \rceil$. Assume $f^{(k)}(b) = 0$, $k = n, n + 1, \ldots, m - 1$, $\overline{D}_{b_-}^{\alpha} f \in L_\infty(a, b)$ (where $\overline{D}_{b_-}^{\alpha} f$ is the right Caputo fractional derivative), see Theorem 18.15.

Then we have the representation formula

$$\overline{D}_{b_-}^{\gamma} f(x) = \frac{1}{\Gamma(\alpha - \gamma)} \int_x^b (t - x)^{\alpha - \gamma - 1} \left(\overline{D}_{b_-}^{\alpha} f\right)(t)\, dt, \qquad (18.1)$$

$\forall x \in [a, b]$, Γ stands for gamma function.

© Springer International Publishing Switzerland 2016
G.A. Anastassiou, *Intelligent Comparisons: Analytic Inequalities*,
Studies in Computational Intelligence 609,
DOI 10.1007/978-3-319-21121-3_18

Using (18.1) and $p, q > 1 : \frac{1}{p} + \frac{1}{q} = 1$, we prove the following Poincaré type right Caputo fractional inequality, see Theorem 18.21,

$$\int_a^b \left| \overline{D_{b-}^{\gamma}} f(x) \right|^q dx \leq \frac{(b-a)^{q(\alpha-\gamma)}}{\left\{ (\Gamma(\alpha-\gamma))^q (p(\alpha-\gamma-1)+1)^{(q/p)} q(\alpha-\gamma) \right\}}$$

$$\cdot \left(\int_a^b \left| \overline{D_{b-}^{\alpha}} f(\varsigma) \right|^q d\varsigma \right). \tag{18.2}$$

18.2   Representation Results

We start with

Background 18.1 *Let $\nu > 0$, $n := [\nu]$ (integral part of ν), and $\alpha := \nu - n$ ($0 < \alpha < 1$). The gamma function Γ is given by $\Gamma(\nu) = \int_0^\infty e^{-t} t^{\nu-1} dt$. Here $[a, b] \subseteq \mathbb{R}$, $x, x_0 \in [a, b]$ such that $x \geq x_0$, where x_0 is fixed. Let $f \in C([a, b])$ and define the Riemann-Liouville integral*

$$\left(J_\nu^{x_0} f \right)(x) := \frac{1}{\Gamma(\nu)} \int_{x_0}^x (x-t)^{\nu-1} f(t) dt, \tag{18.3}$$

$x_0 \leq x \leq b$. *We define the subspace $C_{x_0}^\nu([a, b])$ of $C^n([a, b])$:*

$$C_{x_0}^\nu([a, b]) := \left\{ f \in C^n([a, b]) : J_{1-\alpha}^{x_0} f^{(n)} \in C^1([x_0, b]) \right\}. \tag{18.4}$$

For $f \in C_{x_0}^\nu([a, b])$, we define the *(left) generalized ν-fractional derivative of* f *over* $[x_0, b]$ as

$$D_{x_0}^\nu f := \left(J_{1-\alpha}^{x_0} f^{(n)} \right)', \tag{18.5}$$

see [2], p. 24, and Canavati derivative in [8].
    We need the (left) fractional Taylor theorem see [2], p. 25 which follows

Theorem 18.2 *Let $f \in C_{x_0}^\nu([a, b])$, $x_0 \in [a, b]$ fixed.*

(i) If $\nu \geq 1$ then

$$f(x) = f(x_0) + f'(x_0)(x-x_0) + f''(x_0) \frac{(x-x_0)^2}{2}$$

$$+ \cdots + f^{(n-1)}(x_0) \frac{(x-x_0)^{n-1}}{(n-1)!} + \left(J_\nu^{x_0} D_{x_0}^\nu f \right)(x), \tag{18.6}$$

for all $x \in [a, b] : x \geq x_0$.

(ii) If $0 < \nu < 1$ then

$$f(x) = \left(J_\nu^{x_0} D_{x_0}^\nu f\right)(x),\tag{18.7}$$

for all $x \in [a, b] : x \geq x_0$.

For convenience we call

$$\left(T_{n-1, x_0} f\right)(x) := f(x_0) + f'(x_0)(x - x_0) + f''(x_0)\frac{(x - x_0)^2}{2}$$

$$+ \cdots + f^{(n-1)}(x_0)\frac{(x - x_0)^{n-1}}{(n-1)!},\tag{18.8}$$

and we have

$$\left(J_\nu^{x_0} D_{x_0}^\nu f\right)(x) = \frac{1}{\Gamma(\nu)} \int_{x_0}^x (x - t)^{\nu-1} \left(D_{x_0}^\nu f\right)(t)\, dt.\tag{18.9}$$

We need from [1], p. 543.

Lemma 18.3 *Let $f \in C([a, b])$, $\mu, \nu > 0$. Then*

$$J_\mu^{x_0}\left(J_\nu^{x_0} f\right) = J_{\mu+\nu}^{x_0}(f).\tag{18.10}$$

We present our first main result, a representation theorem for (left) generalized ν−fractional derivatives.

Theorem 18.4 *Let $f \in C_{x_0}^\nu([a, b])$ such that $f^{(i)}(x_0) = 0, i = m, m+1, \ldots, n-1$, where $m := [\gamma]$, with $0 < \gamma < \nu$. Then*

$$\left(D_{x_0}^\gamma f\right)(x) = \frac{1}{\Gamma(\nu - \gamma)} \int_{x_0}^x (x - t)^{\nu-\gamma-1} \left(D_{x_0}^\nu f\right)(t)\, dt,$$

i.e.

$$\left(D_{x_0}^\gamma f\right)(x) = J_{\nu-\gamma}^{x_0}\left(D_{x_0}^\nu f\right) \in C([x_0, b]).\tag{18.11}$$

Consequently $f \in C_{x_0}^\gamma([a, b])$.

This theorem was first proved in [7], however the proof is very complicated and not the most natural one. Our proof is elementary, basic and constructive. Theorem 18.4 improves many results from [2], starting from Remark 3.4, p. 26 there.

Proof Call $\rho = \gamma - m$, notice $m \le n$.

Case $\nu \ge 1$.

If $m = n$, then $\left(T_{n-1,x_0} f\right)^{(m)} = 0$.

If $m \le n - 1$, then

$$\left(T_{n-1,x_0} f\right)^{(m)} (x) = f^{(m)} (x_0) + f^{(m+1)} (x_0) (x - x_0) + f^{(m+2)} (x_0) \frac{(x - x_0)^2}{2}$$

$$+ \cdots + f^{(n-1)} (x_0) \frac{(x - x_0)^{n-m-1}}{(n - m - 1)!} \tag{18.12}$$

$$= 0.$$

Hence $\left(T_{n-1,x_0} f\right)^{(m)} = 0$, and

$$f^{(m)} (x) = \left(\left(J_\nu^{x_0} D_{x_0}^\nu f\right) (x)\right)^{(m)}. \tag{18.13}$$

If $0 < \nu < 1$, then $m = 0$, and

$$f (x) = \left(J_\nu^{x_0} D_{x_0}^\nu f\right) (x), \ x \ge x_0. \tag{18.14}$$

So in the case of $\nu \ge 1$ we have

$$f^{(m)} (x) = \frac{1}{\Gamma (\nu)} \left(\int_{x_0}^x (x - t)^{\nu-1} \left(D_{x_0}^\nu f\right) (t) \, dt\right)^{(m)}. \tag{18.15}$$

Here we have $D_{x_0}^\nu f \in C ([x_0, b])$, also $m < \nu$, that is $\nu - m > 0$.

We apply Theorem 7.7, p. 117 of [2], which says: Let $0 \le s \le x$ and $f \in L_\infty ([0, x])$, $r > 0$, define $F (s) := \int_0^s (s - t)^r f (t) \, dt$, then $F' (s) = r \int_0^s (s - t)^{r-1} f (t) \, dt$, $\forall s \in [0, x]$.

If $\nu = 1$ then $m = 0$, therefore of interest is $\nu > 1$.

Also we use $\Gamma (p + 1) = p \Gamma (p)$, $p > 0$.

So we have

$$f^{(m)} (x) = \frac{1}{\Gamma (\nu - m)} \int_{x_0}^x (x - t)^{\nu-m-1} \left(D_{x_0}^\nu f\right) (t) \, dt, \tag{18.16}$$

that is

$$f^{(m)} (x) = \left(J_{\nu-m}^{x_0} \left(D_{x_0}^\nu f\right)\right) (x), \ \forall x \in [x_0, b]. \tag{18.17}$$

Proposition 15.144, p. 388, [2], says that:

Let $r > 0$, $F \in L_\infty (a, b)$, and $G (s) := \int_0^s (s - t)^{r-1} F (t) \, dt$, all $s \in [a, b]$. Then $G \in AC ([a, b])$ (absolutely continuous functions) for $r \ge 1$ and $G \in C ([a, b])$, when $r \in (0, 1)$.

Hence $f^{(m)} \in C([x_0, b])$.

By Lemma 18.3 we observe

$$J_{1-\rho}^{x_0}\left(f^{(m)}\right) = J_{1-\rho}^{x_0}\left(J_{\nu-m}^{x_0}\left(D_{x_0}^{\nu} f\right)\right) = \left(J_{1-\rho}^{x_0} \circ J_{\nu-m}^{x_0}\right)\left(D_{x_0}^{\nu} f\right)$$

$$= J_{1-\rho+\nu-m}^{x_0}\left(D_{x_0}^{\nu} f\right) = J_{(\nu-\gamma)+1}^{x_0}\left(D_{x_0}^{\nu} f\right). \tag{18.18}$$

That is

$$J_{1-\rho}^{x_0}\left(f^{(m)}\right) = J_{(\nu-\gamma)+1}^{x_0}\left(D_{x_0}^{\nu} f\right), \tag{18.19}$$

i.e.

$$J_{1-\rho}^{x_0}\left(f^{(m)}\right)(x) = \frac{1}{\Gamma(\nu-\gamma+1)} \int_{x_0}^{x} (x-t)^{\nu-\gamma}\left(D_{x_0}^{\nu} f\right)(t)\, dt. \tag{18.20}$$

Consequently we get

$$D_{x_0}^{\gamma} f(x) = \left(J_{1-\rho}^{x_0}\left(f^{(m)}\right)(x)\right)' \tag{18.21}$$

$$= \frac{1}{\Gamma(\nu-\gamma)} \int_{x_0}^{x} (x-t)^{\nu-\gamma-1}\left(D_{x_0}^{\nu} f\right)(t)\, dt \in C([x_0, b]).$$

The case $0 < \nu < 1$ is treated totally the same way with the same result. ■

We need and prove the following result

Theorem 18.5 *Let $r > 0$, $F \in L_\infty(a, b)$, and*

$$G(s) := \int_{s}^{b} (t-s)^{r-1} F(t)\, dt, \tag{18.22}$$

all $s \in [a, b]$. Then $G \in AC([a, b])$ (absolutely continuous functions) for $r \geq 1$ and $G \in C([a, b])$, when $r \in (0, 1)$.

Proof (1) Case $r \geq 1$. We use the definition of absolute continuity. So for every $\epsilon > 0$ we need $\delta > 0$: whenever (a_i, b_i), $i = 1, \ldots, n$, are disjoint subintervals of $[a, b]$, then

$$\sum_{i=1}^{n} (b_i - a_i) < \delta \Rightarrow \sum_{i=1}^{n} |G(b_i) - G(a_i)| < \epsilon. \tag{18.23}$$

If $\|F\|_\infty = 0$, then $G(s) = 0$, for all $s \in [a, b]$, the trivial case and all fulfilled.

So we assume $\|F\|_\infty \neq 0$. Hence we have

$$
G(b_i) - G(a_i) = \int_{b_i}^{b} (t - b_i)^{r-1} F(t)\, dt - \int_{a_i}^{b} (t - a_i)^{r-1} F(t)\, dt
$$

$$
= \int_{b_i}^{b} (t - b_i)^{r-1} F(t)\, dt - \int_{a_i}^{b_i} (t - a_i)^{r-1} F(t)\, dt - \int_{b_i}^{b} (t - a_i)^{r-1} F(t)\, dt
$$

$$
= \int_{b_i}^{b} \left((t - b_i)^{r-1} - (t - a_i)^{r-1} \right) F(t)\, dt - \int_{a_i}^{b_i} (t - a_i)^{r-1} F(t)\, dt.
$$

$$(18.24)$$

Call

$$
I_i := \int_{b_i}^{b} \left| (t - b_i)^{r-1} - (t - a_i)^{r-1} \right| dt. \tag{18.25}
$$

Thus

$$
|G(b_i) - G(a_i)| \leq \left[I_i + \frac{(b_i - a_i)^r}{r} \right] \|F\|_\infty =: T_i. \tag{18.26}
$$

If $r = 1$, then $I_i = 0$, and

$$
|G(b_i) - G(a_i)| \leq \|F\|_\infty (b_i - a_i), \tag{18.27}
$$

for all $i = 1, \ldots, n$.

If $r > 1$, then because $\left[(t - a_i)^{r-1} - (t - b_i)^{r-1} \right] \geq 0$, for all $t \in [b_i, b]$, we find

$$
I_i = \int_{b_i}^{b} \left((t - a_i)^{r-1} - (t - b_i)^{r-1} \right) dt
$$

$$
= \frac{(b - a_i)^r - (b_i - a_i)^r - (b - b_i)^r}{r}
$$

$$
= \frac{r(b - \xi)^{r-1}(b_i - a_i) - (b_i - a_i)^r}{r}, \tag{18.28}
$$

for some $\xi \in (a_i, b_i)$. Therefore it holds

$$
I_i \leq \frac{r(b - a)^{r-1}(b_i - a_i) - (b_i - a_i)^r}{r}, \tag{18.29}
$$

and

$$
\left(I_i + \frac{(b_i - a_i)^r}{r} \right) \leq (b - a)^{r-1}(b_i - a_i). \tag{18.30}
$$

That is

$$T_i \leq \|F\|_\infty (b-a)^{r-1} (b_i - a_i), \tag{18.31}$$

so that

$$|G(b_i) - G(a_i)| \leq \|F\|_\infty (b-a)^{r-1} (b_i - a_i),$$

for all $i = 1, \ldots, n$.

So in the case of $r = 1$, and by choosing $\delta := \epsilon / \|F\|_\infty$, we get

$$\sum_{i=1}^n |G(b_i) - G(a_i)| \leq \|F\|_\infty \left(\sum_{i=1}^n (b_i - a_i) \right)$$
$$\leq \|F\|_\infty \, \delta = \epsilon, \tag{18.32}$$

proving for $r = 1$ that G is absolutely continuous. In the case of $r > 1$, and by choosing $\delta := \epsilon / \left(\|F\|_\infty (b-a)^{r-1} \right)$, we get

$$\sum_{i=1}^n |G(b_i) - G(a_i)| \leq \|F\|_\infty (b-a)^{r-1} \left(\sum_{i=1}^n (b_i - a_i) \right)$$
$$\leq \|F\|_\infty (b-a)^{r-1} \delta = \epsilon, \tag{18.33}$$

proving for $r > 1$ that G is absolutely continuous again.

(2) Case of $0 < r < 1$u. Let $a_{i_*}, b_{i_*} \in [a,b] : a_{i_*} \leq b_{i_*}$. Then $\left(t - a_{i_*} \right)^{r-1} \leq \left(t - b_{i_*} \right)^{r-1}$, for all $t \in [b_i, b]$. Then

$$I_{i_*} = \int_{b_{i_*}}^b \left(\left(t - b_{i_*} \right)^{r-1} - \left(t - a_{i_*} \right)^{r-1} \right) dt$$
$$= \frac{\left(b - b_{i_*} \right)^r}{r} - \left(\frac{\left(b - a_{i_*} \right)^r - \left(b_{i_*} - a_{i_*} \right)^r}{r} \right)$$
$$\leq \frac{\left(b_{i_*} - a_{i_*} \right)^r}{r}, \tag{18.34}$$

by $\left(b - b_{i_*} \right)^r - \left(b - a_{i_*} \right)^r < 0$. Therefore

$$I_{i_*} \leq \frac{\left(b_{i_*} - a_{i_*} \right)^r}{r}, \tag{18.35}$$

and

$$T_{i_*} \leq \frac{2 \left(b_{i_*} - a_{i_*}\right)^r}{r} \, \|F\|_\infty,$$
(18.36)

proving that

$$\left|G\left(b_{i_*}\right) - G\left(a_{i_*}\right)\right| \leq \left(\frac{2\,\|F\|_\infty}{r}\right) \left(b_{i_*} - a_{i_*}\right)^r,$$
(18.37)

which proves that G is continuous.

Taking the special case of $b = 0$, $F(t) = 1$, for all $t \in [a, 0]$ we get that

$$G(s) = \frac{(-s)^r}{r}, \text{ all } s \in [a, 0],$$
(18.38)

for $0 < r < 1$.

The last is a Lipschitz function of order $r \in (0, 1)$, which is not absolutely continuous. Consequently G for $r \in (0, 1)$ in general, cannot be absolutely continuous. That completes the proof. ∎

Background 18.6 (*see also* [3]).

Let $\nu > 0$, $n := [\nu]$, $\alpha = \nu - n$, $0 < \alpha < 1$, $f \in C\left([a, b]\right)$, *call the right Riemann-Liouville fractional integral operator by*

$$\left(J_{b-}^\nu f\right)(x) := \frac{1}{\Gamma(\nu)} \int_x^b (\zeta - x)^{\nu-1} f(\zeta)\,d\zeta,$$
(18.39)

$x \in [a, b]$, *see also [10], [11]. Define the subspace of functions*

$$C_{b-}^\nu \left([a, b]\right) := \left\{ f \in C^n \left([a, b]\right) : J_{b-}^{1-\alpha} f^{(n)} \in C^1 \left([a, b]\right) \right\}.$$

Define the *(right) generalized ν–fractional derivative of f over $[a, b]$* as

$$D_{b-}^\nu f := (-1)^{n-1} \left(J_{b-}^{1-\alpha} f^{(n)}\right)'.$$
(18.40)

We set $D_{b-}^0 f = f$.

We need the (right) fractional Taylor Theorem [3].

Theorem 18.7 *Let $f \in C_{b-}^\nu \left([a, b]\right)$, $\nu > 0$, $n := [\nu]$. Then*

(i) *If $\nu \geq 1$, we get*

$$f(x) = \sum_{k=0}^{n-1} \frac{f^{(k)}(b)}{k!} (x - b)^k + \left(J_{b-}^\nu D_{b-}^\nu f\right)(x),$$
(18.41)

$\forall x \in [a, b]$.

(ii) If $0 < \nu < 1$, *we get*

$$f(x) = J^{\nu}_{b-} D^{\nu}_{b-} f(x), \tag{18.42}$$

$\forall x \in [a, b]$. *That is here*

$$\left(J^{\nu}_{b-} D^{\nu}_{b-} f\right)(x) = \frac{1}{\Gamma(\nu)} \int_x^b (\zeta - x)^{\nu-1} \left(D^{\nu}_{b-} f\right)(\zeta) \, d\zeta. \tag{18.43}$$

We present the following representation theorem for right generalized $\nu-$ fractional derivatives.

Theorem 18.8 *Let* $f \in C^{\nu}_{b-}([a, b])$, $0 < \gamma < \nu$. *Assume* $f^{(i)}(b) = 0$, $i = m, m+1, \ldots, n-1$, *where* $m := [\gamma]$. *Then*

$$D^{\gamma}_{b-} f(x) = \frac{1}{\Gamma(\nu - \gamma)} \int_x^b (\zeta - x)^{\nu-\gamma-1} \left(D^{\nu}_{b-} f\right)(\zeta) \, d\zeta, \tag{18.44}$$

$\forall x \in [a, b]$, *i.e.,*

$$D^{\gamma}_{b-} f = J^{\nu-\gamma}_{b-} \left(D^{\nu}_{b-} f\right) \in C([a, b]), \tag{18.45}$$

and $f \in C^{\gamma}_{b-}([a, b])$.

Proof We notice that $\nu > m$ and $\nu - m > 0$. Of course $m \leq n$.
Set $\rho := \gamma - m$.
Case $\nu \geq 1$.
Call

$$\left(T_{n-1,b} f\right)(x) = \sum_{k=0}^{n-1} \frac{f^{(k)}(b)}{k!} (x - b)^k. \tag{18.46}$$

If $m = n$, then $\left(T_{n-1,b} f\right)^{(m)} = 0$. If $m \leq n - 1$, then

$$\left(T_{n-1,b} f\right)^{(m)}(x) = f^{(m)}(b) + f^{(m+1)}(b)(x - b) + f^{(m+2)}(b) \frac{(x - b)^2}{2}$$

$$+ \cdots + f^{(n-1)}(b) \frac{(x - b)^{n-m-1}}{(n - m - 1)!} = 0, \tag{18.47}$$

by the assumption.
Hence always it holds

$$\left(T_{n-1,b} f\right)^{(m)} = 0. \tag{18.48}$$

So that we have

$$f^{(m)}(x) = \left(\left(J_{b-}^{\nu} D_{b-}^{\nu} f\right)(x)\right)^{(m)}. \tag{18.49}$$

If $0 < \nu < 1$, then $m = 0$, and

$$f(x) = \left(\left(J_{b-}^{\nu} D_{b-}^{\nu} f\right)(x)\right), \; \forall x \in [a, b]. \tag{18.50}$$

Thus in the case of $\nu \geq 1$ we have

$$f^{(m)}(x) = \frac{1}{\Gamma(\nu)} \left(\int_x^b (\zeta - x)^{\nu-1} \left(D_{b-}^{\nu} f\right)(\zeta) \, d\zeta\right)^{(m)}, \tag{18.51}$$

where $D_{b-}^{\nu} f \in C([a, b])$.

Case of $\nu = 1$, then $m = 0$ and

$$f(x) = \left(J_{b-}^{\nu} D_{b-}^{\nu} f\right)(x). \tag{18.52}$$

Of interest is the case of $\nu > 1$ and $\nu - 1 > 0$.

So by Theorem 7, [6], we get

$$f^{(m)}(x) = \frac{(-1)^m}{\Gamma(\nu - m)} \int_x^b (\zeta - x)^{\nu-m-1} \left(D_{b-}^{\nu} f\right)(\zeta) \, d\zeta. \tag{18.53}$$

That is

$$f^{(m)}(x) = (-1)^m \left(J_{b-}^{\nu-m} D_{b-}^{\nu} f\right)(x). \tag{18.54}$$

Here $f^{(m)} \in C([a, b])$, by Theorem 18.5.

We observe that

$$J_{b-}^{1-\rho} f^{(m)} = J_{b-}^{1-\rho} (-1)^m J_{b-}^{\nu-m} \left(D_{b-}^{\nu} f\right)$$

$$= (-1)^m J_{b-}^{1-\rho+\nu-m} \left(D_{b-}^{\nu} f\right) = (-1)^m J_{b-}^{(\nu-\gamma)+1} \left(D_{b-}^{\nu} f\right),$$

by Theorem 4 of [3].

That is

$$J_{b-}^{1-\rho} f^{(m)} = (-1)^m J_{b-}^{(\nu-\gamma)+1} \left(D_{b-}^{\nu} f\right), \tag{18.55}$$

which means

$$\left(J_{b-}^{1-\rho} f^{(m)}\right)(x) = \frac{(-1)^m}{\Gamma(\nu - \gamma + 1)} \int_x^b (\zeta - x)^{\nu-\gamma} \left(D_{b-}^{\nu} f\right)(\zeta) \, d\zeta. \tag{18.56}$$

Therefore

$$D_{b-}^{\gamma} f(x) (-1)^{m+1} = \frac{D_{b-}^{\gamma} f(x)}{(-1)^{m-1}} = \left(\left(J_{b-}^{1-\rho} f^{(m)} \right)(x) \right)' \tag{18.57}$$

$$= \frac{(-1)^{m+1}}{\Gamma(\nu - \gamma)} \int_x^b (\zeta - x)^{\nu - \gamma - 1} \left(D_{b-}^{\nu} f \right)(\zeta) \, d\zeta.$$

I.e.

$$D_{b-}^{\gamma} f(x) = \frac{1}{\Gamma(\nu - \gamma)} \int_x^b (\zeta - x)^{\nu - \gamma - 1} \left(D_{b-}^{\nu} f \right)(\zeta) \, d\zeta$$

$$\in C([a, b]). \tag{18.58}$$

The case $0 < \nu < 1$ is treated totally the same way. ∎

We need

Background 18.9 Let $\nu \geq 0$; the operator J_a^{ν}, defined on $L_1[a, b]$ is given by

$$J_a^{\nu} f(x) := \frac{1}{\Gamma(\nu)} \int_a^x (x - t)^{\nu - 1} f(t) \, dt, \tag{18.59}$$

for $a \leq x \leq b$, is called the left Riemann-Liouville fractional integral operator of order ν. For $\nu = 0$, we set $J_a^0 := I$, the identity operator, see [2], p. 392, also [9].

Let $\nu \geq 0$, $n := \lceil \nu \rceil$ ($\lceil \cdot \rceil$ ceiling of the number), $f \in AC^n([a, b])$ (it means $f^{(n-1)} \in AC([a, b])$). Then the *left Caputo fractional derivative* is given by

$$D_{*a}^{\nu} f(x) = \frac{1}{\Gamma(n - \nu)} \int_a^x (x - t)^{n - \nu - 1} f^{(n)}(t) \, dt \tag{18.60}$$

$$= \left(J_a^{n - \nu} f^{(n)} \right)(x)$$

and it exists almost everywhere for x in $[a, b]$.

See Corollary 16.8, p. 394, of [2], and [9], pp. 49–50.

We need also the left Caputo fractional Taylor formula, see [2], p. 395 and [9], p. 54.

Theorem 18.10 Let $\nu \geq 0$, $n := \lceil \nu \rceil$, $f \in AC^n([a, b])$. Then

$$f(x) = \sum_{k=0}^{n-1} \frac{f^{(k)}(a)}{k!} (x - a)^k + J_a^{\nu} D_{*a}^{\nu} f(x), \tag{18.61}$$

$\forall x \in [a, b]$. Here $J_a^{\nu} D_{*a}^{\nu} f \in AC^n([a, b])$.

We give the following left Caputo fractional derivatives representation theorem.

Theorem 18.11 *Let* $\nu \geq \gamma + 1$, $\gamma > 0$, $\nu, \gamma \notin \mathbb{N}$. *Call* $n := \lceil \nu \rceil$, $m := \lceil \gamma \rceil$. *Assume* $f \in AC^n ([a, b])$, *such that* $f^{(k)} (a) = 0$, $k = m, m + 1, ..., n - 1$, *and* $D_{*a}^{\nu} f \in L_{\infty} (a, b)$. *Then* $D_{*a}^{\gamma} f \in AC ([a, b])$ *(where* $D_{*a}^{\gamma} f := J_a^{m-\gamma} f^{(m)} (x)$*)*,

$$D_{*a}^{\gamma} f (x) = \frac{1}{\Gamma (\nu - \gamma)} \int_a^x (x - t)^{\nu - \gamma - 1} D_{*a}^{\nu} f (t) \, dt, \qquad (18.62)$$

$\forall x \in [a, b]$.

The last theorem improves Theorem 16.16, pp. 395–396 of [2] and consequently the related results from [2].

Proof Notice that

$$\left(\sum_{k=0}^{n-1} \frac{f^{(k)} (a)}{k!} (x - a)^k \right)^{(m)} = 0, \qquad (18.63)$$

by $f^{(k)} (a) = 0, k = m, m + 1, ..., n - 1$.

Here by Lemma 16.15, p. 395 of [2], $m \leq n - 1$.

Since $f \in AC^n ([a, b])$, then $f \in C^{n-1} ([a, b])$ and $f \in C^m ([a, b])$.

Furthermore $f \in AC^n ([a, b])$ implies $f \in AC^m ([a, b])$, so that $D_{*a}^{\gamma} f$ makes sense and is continuous on $[a, b]$, see Proposition 15.114, p. 388 of [2].

Therefore

$$f^{(m)}(x) = \left(J_a^{\nu} D_{*a}^{\nu} f (x) \right)^{(m)}$$
$$\text{(by Theorem 7.7, p. 117 of [2])}$$
$$= \left(J_a^{\nu - m} D_{*a}^{\nu} f \right) (x), \, \forall x \in [a, b], \qquad (18.64)$$

notice here $\nu \geq \gamma + 1 > \lceil \gamma \rceil =: m$, hence $\nu - m > 0$.

It follows

$$D_{*a}^{\gamma} f = J_a^{m-\gamma} f^{(m)}$$
$$= J_a^{m-\gamma} J_a^{\nu - m} \left(D_{*a}^{\nu} f \right)$$
$$\text{(by [10])}$$
$$= J_a^{\nu - \gamma} \left(D_{*a}^{\nu} f \right), \, \text{on } [a, b]. \qquad (18.65)$$

That is proving (18.62), and Proposition 15.114, p. 388 of [2], since $\nu - \gamma \geq 1$, implies $D_{*a}^{\gamma} f \in AC ([a, b])$. ∎

We need

Background 18.12 *(see also [3, 11, 12]) Let* $f \in L_1([a, b])$, $\alpha > 0$. *The right Riemann-Liouville fractional operator of order* α *is also denoted by*

$$I_{b-}^{\alpha} f(x) := \frac{1}{\Gamma(\alpha)} \int_x^b (\zeta - x)^{\alpha-1} f(\zeta) \, d\zeta, \tag{18.66}$$

$\forall x \in [a, b]$. *We set* $I_{b-}^0 := I$, *the identity operator.*

Let now $f \in AC^m([a, b])$, $m \in \mathbb{N}$, with $m := \lceil \alpha \rceil$.
We define the *right Caputo fractional derivative of order* $\alpha \geq 0$, by

$$\overline{D}_{b-}^{\alpha} f(x) := (-1)^m I_{b-}^{m-\alpha} f^{(m)}(x), \tag{18.67}$$

we set $\overline{D}_{b-}^0 f := f$,
i.e.

$$\overline{D}_{b-}^{\alpha} f(x) = \frac{(-1)^m}{\Gamma(m-\alpha)} \int_x^b (\zeta - x)^{m-\alpha-1} f^{(m)}(\zeta) \, d\zeta. \tag{18.68}$$

We use the right Caputo fractional Taylor formula.

Theorem 18.13 ([3]). *Let* $f \in AC^m([a, b])$, $x \in [a, b]$, $\alpha > 0$, $m := \lceil \alpha \rceil$. *Then*

$$f(x) = \sum_{k=0}^{m-1} \frac{f^{(k)}(b)}{k!} (x - b)^k + \frac{1}{\Gamma(\alpha)} \int_x^b (\zeta - x)^{\alpha-1} \overline{D}_{b-}^{\alpha} f(\zeta) \, d\zeta. \tag{18.69}$$

We need also

Theorem 18.14 ([3]). *Let* $\alpha, \beta \geq 0$, $f \in L_1([a, b])$. *Then*

$$I_{b-}^{\alpha} I_{b-}^{\beta} f = I_{b-}^{\alpha+\beta} f = I_{b-}^{\beta} I_{b-}^{\alpha} f, \tag{18.70}$$

valid almost everywhere on $[a, b]$. *If additionally* $f \in C([a, b])$ *or* $\alpha + \beta \geq 1$, *then (18.70) is true on all of* $[a, b]$.

We present the following representation theorem regarding right Caputo fractional derivatives.

Theorem 18.15 *Let* $f \in AC^m([a, b])$, $m \in \mathbb{N}$, $m := \lceil \alpha \rceil$, $\alpha \notin \mathbb{N}$, $\alpha \geq \gamma + 1$, $\gamma > 0$, $n := \lceil \gamma \rceil$. *Assume* $f^{(k)}(b) = 0$, $k = n, n+1, ..., m+1$ *and* $\overline{D}_{b-}^{\alpha} f \in L_{\infty}([a, b])$. *Then*

$$\overline{D}_{b-}^{\gamma} f(x) = \left(I_{b-}^{\alpha-\gamma} \left(\overline{D}_{b-}^{\alpha} f \right) \right)(x) \tag{18.71}$$

$$\in AC([a, b]),$$

i.e.

$$\overline{D}_{b-}^{\gamma} f(x) = \frac{1}{\Gamma(\alpha - \gamma)} \int_{x}^{b} (t - x)^{\alpha - \gamma - 1} \left(\overline{D}_{b-}^{\alpha} f\right)(t) \, dt, \tag{18.72}$$

$\forall x \in [a, b]$.

Proof Notice that $n = \lceil \gamma \rceil < \gamma + 1$, $\alpha > n$ and $m > n$. As in earlier proofs, here using (18.69) and Theorem 7 of [6], we derive

$$f^{(n)}(x) = \frac{(-1)^{n}}{\Gamma(\alpha - n)} \int_{x}^{b} (\xi - x)^{\alpha - n - 1} \overline{D}_{b-}^{\alpha} f(\xi) \, d\xi, \tag{18.73}$$

i.e.

$$f^{(n)}(x) = (-1)^{n} I_{b-}^{\alpha - n} \left(\overline{D}_{b-}^{\alpha} f\right)(x). \tag{18.74}$$

Also since $f \in AC^{m}([a, b])$, we get $f \in AC^{n}([a, b])$. By definition

$$
\begin{aligned}
\overline{D}_{b-}^{\gamma} f(x) &= (-1)^{n} I_{b-}^{n - \gamma} f^{(n)}(x) \\
&\overset{(18.74)}{=} (-1)^{n} I_{b-}^{n - \gamma} (-1)^{n} I_{b-}^{\alpha - n} \left(\overline{D}_{b-}^{\alpha} f\right)(x) \\
&= I_{b-}^{n - \gamma + \alpha - n} \left(\overline{D}_{b-}^{\alpha} f\right)(x) \\
&= I_{b-}^{\alpha - \gamma} \left(\overline{D}_{b-}^{\alpha} f\right)(x), \tag{18.75}
\end{aligned}
$$

(notice $\alpha - \gamma \geq 1$ and by Theorem 18.14) proving (18.71).
    We observe that $\overline{D}_{b-}^{\gamma} f \in AC([a, b])$, by application of Theorem 18.5. ∎

    We need (see [3, 11, 12])

Definition 18.16 Let $\alpha > 0$, $m := \lceil \alpha \rceil$, $f \in AC^{m}([a, b])$. We define the right Riemann-Liouville fractional derivative by

$$\mathcal{D}_{b-}^{\alpha} f(x) := \frac{(-1)^{m}}{\Gamma(m - \alpha)} \left(\frac{d}{dx}\right)^{m} \int_{x}^{b} (t - x)^{m - \alpha - 1} f(t) \, dt, \tag{18.76}$$

we set $\mathcal{D}_{b-}^{0} f := f$.

Remark 18.17 Let now $f \in C^{m}([a, b])$, $m := \lceil \alpha \rceil$, $\alpha > 0$. In [5] we proved that $\overline{D}_{b-}^{\alpha} f \in C([a, b])$. Since $C^{m}([a, b]) \subset AC^{m}([a, b])$, we have $f \in AC^{m}([a, b])$. Thus by Theorem 9 of [3], we obtain that also $\mathcal{D}_{b-}^{\alpha} f$ exists on $[a, b]$. Furthermore if $f^{(k)}(b) = 0$, $k = 0, 1, ..., m - 1$, we get

$$\overline{D}_{b-}^{\alpha} f(x) = \mathcal{D}_{b-}^{\alpha} f(x), \quad \forall x \in [a, b], \tag{18.77}$$

hence

$$\mathcal{D}_{b-}^{\alpha} f \in C\left([a, b]\right).$$

Let $\gamma > 0 : \alpha \geq \gamma + 1, n := \lceil \gamma \rceil$. Then by Theorem 18.15 we have

$$\overline{D}_{b-}^{\gamma} f(x) = \frac{1}{\Gamma(\alpha - \gamma)} \int_{x}^{b} (t - x)^{\alpha - \gamma - 1} \left(\mathcal{D}_{b-}^{\alpha} f\right)(t) \, dt. \tag{18.78}$$

Since $n \leq m - 1$ (by Lemma 16.15, p. 395 of [2]); if $f \in C^{m}\left([a, b]\right)$, then $f \in C^{n}\left([a, b]\right)$.

Also we have $f^{(k)}(b) = 0, k = 0, 1, \ldots, n - 1$, and of course $f \in AC^{n}\left([a, b]\right)$. Here again we have $\overline{D}_{b-}^{\gamma} f \in AC\left([a, b]\right)$. Hence similarly it holds $\mathcal{D}_{b-}^{\gamma} f(x) = \overline{D}_{b-}^{\gamma} f(x), \forall x \in [a, b]$. Then we obtain

$$\mathcal{D}_{b-}^{\gamma} f(x) = \frac{1}{\Gamma(\alpha - \gamma)} \int_{x}^{b} (t - x)^{\alpha - \gamma - 1} \left(\mathcal{D}_{b-}^{\alpha} f\right)(t) \, dt. \tag{18.79}$$

We have established the following representation theorem regarding right Riemann-Liouville fractional derivatives.

Theorem 18.18 Let $f \in C^{m}\left([a, b]\right), m \in \mathbb{N}, m := \lceil \alpha \rceil, \alpha \notin \mathbb{N}, \alpha \geq \gamma + 1, \gamma > 0,$ $f^{(k)}(b) = 0, k = 0, 1, \ldots, m - 1$. Then

$$\mathcal{D}_{b-}^{\gamma} f(x) = \frac{1}{\Gamma(\alpha - \gamma)} \int_{x}^{b} (t - x)^{\alpha - \gamma - 1} \left(\mathcal{D}_{b-}^{\alpha} f\right)(t) \, dt, \tag{18.80}$$

$\forall x \in [a, b]$.

Also $\mathcal{D}_{b-}^{\gamma} f \in AC\left([a, b]\right)$.

18.3 Right Fractional Inequalities

We present the following right fractional Poincaré type inequality involving right generalized fractional derivatives.

Theorem 18.19 Let $f \in C_{b-}^{\nu}\left([a, b]\right), 0 \leq \gamma < \nu$. Assume $f^{(i)}(b) = 0, i = m,$ $m + 1, \ldots, n - 1$, where $m := \lceil \gamma \rceil$. Let $p, q > 1 : \frac{1}{p} + \frac{1}{q} = 1$ so that $\nu - \gamma > \frac{1}{q}$. Then

$$\int_a^b \left| D_{b-}^\gamma f(x) \right|^q dx \le \frac{(b-a)^{q(\nu-\gamma)}}{\left[(\Gamma(\nu-\gamma))^q (p(\nu-\gamma-1)+1)^{(q/p)} q(\nu-\gamma) \right]}$$
$$\cdot \left(\int_a^b \left| D_{b-}^\nu f(\zeta) \right|^q d\zeta \right). \tag{18.81}$$

Proof When $0 < \gamma < \nu$ by (18.44) we have

$$\left| D_{b-}^\gamma f(x) \right| \le \frac{1}{\Gamma(\nu-\gamma)} \int_x^b (\zeta-x)^{\nu-\gamma-1} \left| D_{b-}^\nu f(\zeta) \right| d\zeta$$
$$\text{(by the Holder's inequality)}$$

$$\le \frac{1}{\Gamma(\nu-\gamma)} \left(\int_x^b (\zeta-x)^{p(\nu-\gamma-1)} d\zeta \right)^{1/p} \left(\int_x^b \left| D_{b-}^\nu f(\zeta) \right|^q d\zeta \right)^{1/q} \tag{18.82}$$

$$= \frac{1}{\Gamma(\nu-\gamma)} \frac{(b-x)^{(\nu-\gamma-1)+\frac{1}{p}}}{(p(\nu-\gamma-1)+1)^{\frac{1}{p}}} \left(\int_x^b \left| D_{b-}^\nu f(\zeta) \right|^q d\zeta \right)^{1/q}$$
$$\le \frac{1}{\Gamma(\nu-\gamma)} \frac{(b-x)^{(\nu-\gamma-1)+\frac{1}{p}}}{(p(\nu-\gamma-1)+1)^{\frac{1}{p}}} \left(\int_a^b \left| D_{b-}^\nu f(\zeta) \right|^q d\zeta \right)^{1/q}. \tag{18.83}$$

That is, we have

$$\left| D_{b-}^\gamma f(x) \right| \le \frac{1}{\Gamma(\nu-\gamma)} \frac{(b-x)^{(\nu-\gamma-1)+\frac{1}{p}}}{(p(\nu-\gamma-1)+1)^{\frac{1}{p}}} \left(\int_a^b \left| D_{b-}^\nu f(\zeta) \right|^q d\zeta \right)^{1/q},$$
$$\tag{18.84}$$

$\forall x \in [a, b]$.

Consequently we find

$$\left| D_{b-}^\gamma f(x) \right|^q \le \left(\frac{(b-x)^{q(\nu-\gamma-1)+\frac{q}{p}}}{(\Gamma(\nu-\gamma))^q (p(\nu-\gamma-1)+1)^{\frac{q}{p}}} \right)$$
$$\cdot \left(\int_a^b \left| D_{b-}^\nu f(\zeta) \right|^q d\zeta \right), \tag{18.85}$$

$\forall x \in [a, b]$.

Hence we obtain

$$\int_a^b \left| D_{b-}^\gamma f(x) \right|^q dx \leq \frac{(b-a)^{q(\nu-\gamma-1)+\frac{q}{p}+1}}{(\Gamma(\nu-\gamma))^q (p(\nu-\gamma-1)+1)^{\frac{q}{p}}}$$

$$\cdot \frac{1}{\left(q(\nu-\gamma-1)+\frac{q}{p}+1\right)} \left(\int_a^b \left| D_{b-}^\nu f(\zeta) \right|^q d\zeta \right) \qquad (18.86)$$

$$= \frac{(b-a)^{q(\nu-\gamma)}}{(\Gamma(\nu-\gamma))^q (p(\nu-\gamma-1)+1)^{\frac{q}{p}}} \frac{1}{q(\nu-\gamma)} \left(\int_a^b \left| D_{b-}^\nu f(\zeta) \right|^q d\zeta \right), \qquad (18.87)$$

proving the claim. ∎

When $\gamma = 0$ we act similarly.

Case $\gamma = 0$ follows.

Corollary 18.20 (to Theorem 18.19). Let $f \in C_{b-}^\nu ([a, b])$. Assume $f^{(i)}(b) = 0$, $i = 0, \ldots, n-1$. Let $p, q > 1 : \frac{1}{p} + \frac{1}{q} = 1$ so that $\nu > \frac{1}{q}$. Then

$$\int_a^b |f(x)|^q dx \leq \frac{(b-a)^{q\nu}}{(\Gamma(\nu))^q (p(\nu-1)+1)^{\frac{q}{p}} q\nu}$$

$$\cdot \left(\int_a^b \left| D_{b-}^\nu f(\zeta) \right|^q d\zeta \right). \qquad (18.88)$$

Similarly we present a Poincaré type inequality regarding right Caputo fractional derivatives

Theorem 18.21 Let $f \in AC^m ([a, b]), m \in \mathbb{N}, m := \lceil \alpha \rceil, \alpha \notin \mathbb{N}, \alpha \geq \gamma+1, \gamma \geq 0$, $n := \lceil \gamma \rceil$. Assume $f^{(k)}(b) = 0, k = n, n+1, \ldots, m-1$, and $\overline{D}_{b-}^\alpha f \in L_\infty ([a, b])$. Let $p, q > 1 : \frac{1}{p} + \frac{1}{q} = 1$. Then

$$\int_a^b \left| \overline{D}_{b-}^\gamma f(x) \right|^q dx \leq \frac{(b-a)^{q(\alpha-\gamma)}}{(\Gamma(\alpha-\gamma))^q (p(\alpha-\gamma-1)+1)^{\frac{q}{p}} q(\alpha-\gamma)}$$

$$\cdot \left(\int_a^b \left| \overline{D}_{b-}^\alpha f(\zeta) \right|^q d\zeta \right). \qquad (18.89)$$

We also give a Poincaré type inequality for right Riemann-Liouville fractional derivatives

Theorem 18.22 *Let* $f \in C^m$ $([a, b])$, $m \in \mathbb{N}$, $m := \lceil \alpha \rceil$, $\alpha \notin \mathbb{N}$, $\alpha \geq \gamma + 1$, $\gamma \geq 0$, $f^{(k)}$ $(b) = 0$, $k = 0, 1, ..., m - 1$. *Let* $p, q > 1 : \frac{1}{p} + \frac{1}{q} = 1$. *Then*

$$\int_a^b |\mathcal{D}_{b-}^\gamma f(x)|^q \, dx \leq \frac{(b-a)^{q(\alpha-\gamma)}}{(\Gamma(\alpha-\gamma))^q (p(\alpha-\gamma-1)+1)^{\frac{q}{p}} q(\alpha-\gamma)}$$

$$\cdot \left(\int_a^b |\mathcal{D}_{b-}^\alpha f(\zeta)|^q \, d\zeta \right). \tag{18.90}$$

Next we present the following right fractional Sobolev type inequality involving right generalized fractional derivatives.

Theorem 18.23 *Let* $f \in C_{b-}^\nu$ $([a, b])$, $0 \leq \gamma < \nu$, $r \geq 1$. *Assume* $f^{(i)}$ $(b) = 0$, $i = m, m+1, ..., n-1$, *where* $m := [\gamma]$. *Let* $p, q > 1 : \frac{1}{p} + \frac{1}{q} = 1$ *so that* $\nu - \gamma > \frac{1}{q}$. *Then*

(i)

$$\left\| D_{b-}^\gamma f \right\|_{L_r(a,b)} \leq \frac{(b-a)^{\nu-\gamma-\frac{1}{q}+\frac{1}{r}}}{\Gamma(\nu-\gamma)(p(\nu-\gamma-1)+1)^{\frac{1}{p}}} \tag{18.91}$$

$$\cdot \frac{1}{\left(r \left(\nu - \gamma - \frac{1}{q} \right) + 1 \right)^{\frac{1}{r}}} \left\| D_{b-}^\nu f \right\|_{L_q(a,b)}.$$

(ii) When $\gamma = 0$ *and* $\nu > \frac{1}{q}$, *we get*

$$\| f \|_{L_r(a,b)} \leq \frac{(b-a)^{\nu-\frac{1}{q}+\frac{1}{r}}}{\Gamma(\nu)(p(\nu-1)+1)^{\frac{1}{p}} \left(r \left(\nu - \frac{1}{q} \right) + 1 \right)^{\frac{1}{r}}} \left\| D_{b-}^\nu f \right\|_{L_q(a,b)}. \tag{18.92}$$

Proof As in the proof of Theorem 18.19, (18.84), we get that

$$|D_{b-}^\gamma f(x)| \leq \frac{1}{\Gamma(\nu-\gamma)} \frac{(b-x)^{(\nu-\gamma-1)+\frac{1}{p}}}{(p(\nu-\gamma-1)+1)^{\frac{1}{p}}} \left(\int_a^b |D_{b-}^\nu f(\zeta)|^q \, d\zeta \right)^{\frac{1}{q}}, \tag{18.93}$$

$\forall x \in [a, b]$.

That is

$$|D_{b-}^\gamma f(x)| \leq \frac{(b-x)^{\left(\nu-\gamma-\frac{1}{q} \right)}}{\Gamma(\nu-\gamma)(p(\nu-\gamma-1)+1)^{\frac{1}{p}}} \left\| D_{b-}^\nu f \right\|_{L_q(a,b)}. \tag{18.94}$$

Hence, by $r \geq 1$ we obtain

$$\left| D_{b-}^{\gamma} f(x) \right|^r \leq \frac{(b-x)^{r\left(\nu-\gamma-\frac{1}{q}\right)}}{(\Gamma(\nu-\gamma))^r (p(\nu-\gamma-1)+1)^{\frac{r}{p}}} \left\| D_{b-}^{\nu} f \right\|_{L_q(a,b)}^r, \quad (18.95)$$

$\forall x \in [a, b]$.

Consequently it holds

$$\int_a^b \left| D_{b-}^{\gamma} f(x) \right|^r dx \leq \frac{(b-a)^{r\left(\nu-\gamma-\frac{1}{q}\right)+1}}{(\Gamma(\nu-\gamma))^r (p(\nu-\gamma-1)+1)^{\frac{r}{p}}} \quad (18.96)$$
$$\cdot \frac{1}{\left(r\left(\nu-\gamma-\frac{1}{q}\right)+1\right)} \left\| D_{b-}^{\nu} f \right\|_{L_q(a,b)}^r.$$

That is

$$\left\| D_{b-}^{\gamma} f \right\|_{L_r(a,b)} \leq \frac{(b-a)^{\nu-\gamma-\frac{1}{q}+\frac{1}{r}}}{(\Gamma(\nu-\gamma))(p(\nu-\gamma-1)+1)^{\frac{1}{p}}} \quad (18.97)$$
$$\cdot \frac{1}{\left(r\left(\nu-\gamma-\frac{1}{q}\right)+1\right)^{\frac{1}{r}}} \left\| D_{b-}^{\nu} f \right\|_{L_q(a,b)},$$

proving the claim. ∎

We continue with a Sobolev type inequality regarding right Caputo fractional derivatives.

Theorem 18.24 *Let* $f \in AC^m([a,b])$, $m \in \mathbb{N}$, $m := \lceil \alpha \rceil$, $\alpha \notin \mathbb{N}$, $\alpha \geq \gamma + 1$, $\gamma \geq 0$, $n := \lceil \gamma \rceil$, $r \geq 1$. *Assume* $f^{(k)}(b) = 0$, $k = n, n+1, \ldots, m-1$, *and* $\overline{D}_{b-}^{\alpha} f \in L_\infty([a,b])$. *Let* $p, q > 1 : \frac{1}{p} + \frac{1}{q} = 1$. *Then*

(i)

$$\left\| \overline{D}_{b-}^{\gamma} f \right\|_{L_r(a,b)} \leq \frac{(b-a)^{\alpha-\gamma+\frac{1}{r}-\frac{1}{q}} \left\| \overline{D}_{b-}^{\alpha} f \right\|_{L_q(a,b)}}{\Gamma(\alpha-\gamma)(p(\alpha-\gamma-1)+1)^{\frac{1}{p}} \left(r\left(\alpha-\gamma-\frac{1}{q}\right)+1\right)^{\frac{1}{r}}}. \quad (18.98)$$

(ii) When $\gamma = 0$ *we get*

$$\|f\|_{L_r(a,b)} \leq \frac{(b-a)^{\alpha-\frac{1}{q}+\frac{1}{r}} \left\| \overline{D}_{b-}^{\alpha} f \right\|_{L_q(a,b)}}{\Gamma(\alpha)(p(\alpha-1)+1)^{\frac{1}{p}} \left(r\left(\alpha-\frac{1}{q}\right)+1\right)^{\frac{1}{r}}}. \quad (18.99)$$

We also give a Sobolev type inequality for right Riemann-Liouville fractional derivatives

Theorem 18.25 *Let* $f \in C^m([a, b])$, $m \in \mathbb{N}$, $m := \lceil \alpha \rceil$, $\alpha \notin \mathbb{N}$, $\alpha \geq \gamma + 1$, $\gamma \geq 0$, $r \geq 1$, $f^{(k)}(b) = 0$, $k = 0, 1,..., m - 1$. *Let* $p, q > 1 : \frac{1}{p} + \frac{1}{q} = 1$. *Then*

(i)

$$\left\| \mathcal{D}_{b-}^{\gamma} f \right\|_{L_r(a,b)} \leq \frac{(b-a)^{\alpha - \gamma + \frac{1}{r} - \frac{1}{q}} \left\| \mathcal{D}_{b-}^{\alpha} f \right\|_{L_q(a,b)}}{\Gamma(\alpha - \gamma)(p(\alpha - \gamma - 1) + 1)^{\frac{1}{p}} \left(r\left(\alpha - \gamma - \frac{1}{q}\right) + 1 \right)^{\frac{1}{r}}}.$$

(18.100)

(ii) *When* $\gamma = 0$ *we get*

$$\left\| f \right\|_{L_r(a,b)} \leq \frac{(b-a)^{\alpha + \frac{1}{r} - \frac{1}{q}} \left\| \mathcal{D}_{b-}^{\alpha} f \right\|_{L_q(a,b)}}{\Gamma(\alpha)(p(\alpha - 1) + 1)^{\frac{1}{p}} \left(r\left(\alpha - \frac{1}{q}\right) + 1 \right)^{\frac{1}{r}}}.$$

(18.101)

Next we prove a Hilbert-Pachpatte type fractional inequality regarding right generalized fractional derivatives.

Theorem 18.26 *Let* $i = 1, 2$; $f_i \in C_{b_i-}^{\nu_i}([a_i, b_i])$, $0 \leq \gamma_i < \nu_i$. *Let* $m_i = \lceil \gamma_i \rceil$, $n_i := \lceil \nu_i \rceil$, *and assume* $f_i^{(j_i)}(b_i) = 0$, $j_i = m_i, m_i + 1,..., n_i - 1$, $i = 1, 2$. *Let* $p, q > 1 : \frac{1}{p} + \frac{1}{q} = 1$ *and assume* $\nu_1 - \gamma_1 > \frac{1}{q}$, $\nu_2 - \gamma_2 > \frac{1}{p}$. *Then*

$$\int_{a_1}^{b_1} \int_{a_2}^{b_2} \frac{\left| D_{b_1-}^{\gamma_1} f_1(x_1) \right| \left| D_{b_2-}^{\gamma_2} f_2(x_2) \right| dx_1 dx_2}{\left[\left(\frac{(b_1 - x_1)^{p(\nu_1 - \gamma_1 - 1) + 1}}{p(p(\nu_1 - \gamma_1 - 1) + 1)} \right) + \left(\frac{(b_2 - x_2)^{q(\nu_2 - \gamma_2 - 1) + 1}}{q(q(\nu_2 - \gamma_2 - 1) + 1)} \right) \right]}$$

$$\leq \frac{(b_1 - a_1)(b_2 - a_2)}{\Gamma(\nu_1 - \gamma_1) \Gamma(\nu_2 - \gamma_2)} \left(\int_{a_1}^{b_1} \left| D_{b_1-}^{\nu_1} f_1(\zeta_1) \right|^q d\zeta_1 \right)^{\frac{1}{q}}$$

$$\cdot \left(\int_{a_2}^{b_2} \left| D_{b_2-}^{\nu_2} f_2(\zeta_2) \right|^p d\zeta_2 \right)^{\frac{1}{p}}.$$

(18.102)

Proof By Theorem 18.8 we get

$$D_{b_i-}^{\gamma_i} f_i(x_i) = \frac{1}{\Gamma(\nu_i - \gamma_i)} \int_{x_i}^{b_i} (\zeta_i - x_i)^{\nu_i - \gamma_i - 1} \left(D_{b_i-}^{\nu_i} f_i \right)(\zeta_i) d\zeta_i,$$

(18.103)

$\forall x_i \in [a_i, b_i]$; $i = 1, 2$.

Hence by Hölder's inequality we obtain

$$
\left| D_{b_1-}^{\gamma_1} f_1(x_1) \right| \le \frac{1}{\Gamma(\nu_1 - \gamma_1)} \left(\int_{x_1}^{b_1} (\zeta_1 - x_1)^{p(\nu_1-\gamma_1-1)} d\zeta_1 \right)^{\frac{1}{p}}
$$
$$
\cdot \left(\int_{x_1}^{b_1} \left| D_{b_1-}^{\nu_1} f_1(\zeta_1) \right|^q d\zeta_1 \right)^{\frac{1}{q}}, \tag{18.104}
$$

and

$$
\left| D_{b_2-}^{\gamma_2} f_2(x_2) \right| \le \frac{1}{\Gamma(\nu_2 - \gamma_2)} \left(\int_{x_2}^{b_2} (\zeta_2 - x_2)^{q(\nu_2-\gamma_2-1)} d\zeta_2 \right)^{\frac{1}{q}}
$$
$$
\cdot \left(\int_{x_2}^{b_2} \left| D_{b_2-}^{\nu_2} f_2(\zeta_2) \right|^p d\zeta_2 \right)^{\frac{1}{p}}. \tag{18.105}
$$

So we have

$$
\left| D_{b_1-}^{\gamma_1} f_1(x_1) \right| \le \frac{1}{\Gamma(\nu_1 - \gamma_1)} \left(\frac{(b_1 - x_1)^{p(\nu_1-\gamma_1-1)+1}}{(p(\nu_1 - \gamma_1 - 1) + 1)} \right)^{\frac{1}{p}}
$$
$$
\left(\int_{x_1}^{b_1} \left| D_{b_1-}^{\nu_1} f_1(\zeta_1) \right|^q d\zeta_1 \right)^{\frac{1}{q}}, \tag{18.106}
$$

and

$$
\left| D_{b_2-}^{\gamma_2} f_2(x_2) \right| \le \frac{1}{\Gamma(\nu_2 - \gamma_2)} \left(\frac{(b_2 - x_2)^{q(\nu_2-\gamma_2-1)+1}}{(q(\nu_2 - \gamma_2 - 1) + 1)} \right)^{\frac{1}{q}}
$$
$$
\left(\int_{x_2}^{b_2} \left| D_{b_2-}^{\nu_2} f_2(\zeta_2) \right|^p d\zeta_2 \right)^{\frac{1}{p}}. \tag{18.107}
$$

Hence

$$
\left| D_{b_1-}^{\gamma_1} f_1(x_1) \right| \left| D_{b_2-}^{\gamma_2} f_2(x_2) \right| \le \frac{1}{\Gamma(\nu_1 - \gamma_1) \Gamma(\nu_2 - \gamma_2)} \tag{18.108}
$$
$$
\cdot \left(\frac{(b_1 - x_1)^{p(\nu_1-\gamma_1-1)+1}}{(p(\nu_1 - \gamma_1 - 1) + 1)} \right)^{\frac{1}{p}} \left(\frac{(b_2 - x_2)^{q(\nu_2-\gamma_2-1)+1}}{(q(\nu_2 - \gamma_2 - 1) + 1)} \right)^{\frac{1}{q}}
$$
$$
\cdot \left(\int_{x_1}^{b_1} \left| D_{b_1-}^{\nu_1} f_1(\zeta_1) \right|^q d\zeta_1 \right)^{\frac{1}{q}} \left(\int_{x_2}^{b_2} \left| D_{b_2-}^{\nu_2} f_2(\zeta_2) \right|^p d\zeta_2 \right)^{\frac{1}{p}}
$$

(using Young's inequality for $a, b \ge 0$, $a^{\frac{1}{p}} b^{\frac{1}{q}} \le \frac{a}{p} + \frac{b}{q}$)

$$\leq \frac{1}{\Gamma\left(\nu_1 - \gamma_1\right)\Gamma\left(\nu_2 - \gamma_2\right)} \left(\frac{(b_1 - x_1)^{p(\nu_1 - \gamma_1 - 1) + 1}}{p\left(p\left(\nu_1 - \gamma_1 - 1\right) + 1\right)} \right. \tag{18.109}$$

$$\left. + \left(\frac{(b_2 - x_2)^{q(\nu_2 - \gamma_2 - 1) + 1}}{q\left(q\left(\nu_2 - \gamma_2 - 1\right) + 1\right)} \right) \right)$$

$$\cdot \left(\int_{x_1}^{b_1} \left| D_{b_1-}^{\nu_1} f_1\left(\zeta_1\right) \right|^q d\zeta_1 \right)^{\frac{1}{q}} \left(\int_{x_2}^{b_2} \left| D_{b_2-}^{\nu_2} f_2\left(\zeta_2\right) \right|^p d\zeta_2 \right)^{\frac{1}{p}}.$$

So far we have proved that

$$\frac{\left| D_{b_1-}^{\gamma_1} f_1\left(x_1\right) \right| \left| D_{b_2-}^{\gamma_2} f_2\left(x_2\right) \right|}{\left(\frac{(b_1 - x_1)^{p(\nu_1 - \gamma_1 - 1) + 1}}{p(p(\nu_1 - \gamma_1 - 1) + 1)} + \left(\frac{(b_2 - x_2)^{q(\nu_2 - \gamma_2 - 1) + 1}}{q(q(\nu_2 - \gamma_2 - 1) + 1)} \right) \right)}$$

$$\leq \frac{1}{\Gamma\left(\nu_1 - \gamma_1\right)\Gamma\left(\nu_2 - \gamma_2\right)} \tag{18.110}$$

$$\cdot \left(\int_{x_1}^{b_1} \left| D_{b_1-}^{\nu_1} f_1\left(\zeta_1\right) \right|^q d\zeta_1 \right)^{\frac{1}{q}} \left(\int_{x_2}^{b_2} \left| D_{b_2-}^{\nu_2} f_2\left(\zeta_2\right) \right|^p d\zeta_2 \right)^{\frac{1}{p}}.$$

The denominator in (18.110) can be zero only when $x_1 = b_1$ and $x_2 = b_2$.
   Therefore we obtain

$$\int_{a_1}^{b_1} \int_{a_2}^{b_2} \frac{\left| D_{b_1-}^{\gamma_1} f_1\left(x_1\right) \right| \left| D_{b_2-}^{\gamma_2} f_2\left(x_2\right) \right| dx_1 dx_2}{\left[\left(\frac{(b_1 - x_1)^{p(\nu_1 - \gamma_1 - 1) + 1}}{p(p(\nu_1 - \gamma_1 - 1) + 1)} \right) + \left(\frac{(b_2 - x_2)^{q(\nu_2 - \gamma_2 - 1) + 1}}{q(q(\nu_2 - \gamma_2 - 1) + 1)} \right) \right]}$$

$$\leq \frac{1}{\Gamma\left(\nu_1 - \gamma_1\right)\Gamma\left(\nu_2 - \gamma_2\right)}$$

$$\cdot \left(\int_{a_1}^{b_1} \left(\int_{x_1}^{b_1} \left| D_{b_1-}^{\nu_1} f_1\left(\zeta_1\right) \right|^q d\zeta_1 \right)^{\frac{1}{q}} dx_1 \right)$$

$$\cdot \left(\int_{a_2}^{b_2} \left(\int_{x_2}^{b_2} \left| D_{b_2-}^{\nu_2} f_2\left(\zeta_2\right) \right|^p d\zeta_2 \right)^{\frac{1}{p}} dx_2 \right) \tag{18.111}$$

$$\leq \frac{1}{\Gamma\left(\nu_1 - \gamma_1\right)\Gamma\left(\nu_2 - \gamma_2\right)}$$

$$\cdot \left(\int_{a_1}^{b_1} \left(\int_{a_1}^{b_1} \left| D_{b_1-}^{\nu_1} f_1\left(\zeta_1\right) \right|^q d\zeta_1 \right)^{\frac{1}{q}} dx_1 \right)$$

$$\cdot \left(\int_{a_2}^{b_2} \left(\int_{a_2}^{b_2} \left| D_{b_2-}^{\nu_2} f_2 (\zeta_2) \right|^p d\zeta_2 \right)^{\frac{1}{p}} dx_2 \right)$$

$$= \frac{(b_1 - a_1)(b_2 - a_2)}{\Gamma(\nu_1 - \gamma_1) \Gamma(\nu_2 - \gamma_2)} \left(\int_{a_1}^{b_1} \left| D_{b_1-}^{\nu_1} f_1 (\zeta_1) \right|^q d\zeta_1 \right)^{\frac{1}{q}}$$

$$\cdot \left(\int_{a_2}^{b_2} \left| D_{b_2-}^{\nu_2} f_2 (\zeta_2) \right|^p d\zeta_2 \right)^{\frac{1}{p}}.$$
(18.112)

proving the claim. ∎

We continue with a Hilbert-Pachpatte type fractional inequality for right Caputo fractional derivatives.

Theorem 18.27 *Let* $i = 1, 2$; $f_i \in AC^{m_i}([a_i, b_i])$, $\mathbb{N} \ni m_i := \lceil \alpha_i \rceil$, $\alpha_i \notin \mathbb{N}$, $\alpha_i \geq \gamma_i + 1$, $\gamma_i \geq 0$, $n_i = \lceil \gamma_i \rceil$. *Assume* $f_i^{(k_i)}(b_i) = 0$, $k_i = n_i, n_i + 1, ..., m_i - 1$, *and* $\overline{D}_{b_i-}^{\alpha_i} f_i \in L_\infty (a_i, b_i)$. *Let* $p, q > 1 : \frac{1}{p} + \frac{1}{q} = 1$. *Then*

$$\int_{a_1}^{b_1} \int_{a_2}^{b_2} \frac{\left| \overline{D}_{b_1-}^{\gamma_1} f_1 (x_1) \right| \left| \overline{D}_{b_2-}^{\gamma_2} f_2 (x_2) \right| dx_1 dx_2}{\left[\left(\frac{(b_1-x_1)^{p(\alpha_1-\gamma_1-1)+1}}{p(p(\alpha_1-\gamma_1-1)+1)} \right) + \left(\frac{(b_2-x_2)^{q(\alpha_2-\gamma_2-1)+1}}{q(q(\alpha_2-\gamma_2-1)+1)} \right) \right]} \leq$$

$$\frac{(b_1 - a_1)(b_2 - a_2)}{\Gamma(\alpha_1 - \gamma_1) \Gamma(\alpha_2 - \gamma_2)} \left(\int_{a_1}^{b_1} \left| D_{b_1-}^{\alpha_1} f_1 (\zeta_1) \right|^q d\zeta_1 \right)^{\frac{1}{q}} \left(\int_{a_2}^{b_2} \left| D_{b_2-}^{\alpha_2} f_2 (\zeta_2) \right|^p d\zeta_2 \right)^{\frac{1}{p}}$$
(18.113)

We continue with a Hilbert-Pachpatte type fractional inequality for right Riemann-Liouville fractional derivatives

Theorem 18.28 *Let* $i = 1, 2$; $f_i \in C^{m_i}([a_i, b_i])$, $m_i \in \mathbb{N}$, $m_i := \lceil \alpha_i \rceil$, $\alpha_i \notin \mathbb{N}$, $\alpha_i \geq \gamma_i + 1$, $\gamma_i \geq 0$, $f_i^{(k_i)}(b_i) = 0$, $k_i = 0, 1, ..., m_i - 1$. *Let* $p, q > 1 : \frac{1}{p} + \frac{1}{q} = 1$. *Then*

$$\int_{a_1}^{b_1} \int_{a_2}^{b_2} \frac{\left| \mathcal{D}_{b_1-}^{\gamma_1} f_1 (x_1) \right| \left| \mathcal{D}_{b_2-}^{\gamma_2} f_2 (x_2) \right| dx_1 dx_2}{\left[\left(\frac{(b_1-x_1)^{p(\alpha_1-\gamma_1-1)+1}}{p(p(\alpha_1-\gamma_1-1)+1)} \right) + \left(\frac{(b_2-x_2)^{q(\alpha_2-\gamma_2-1)+1}}{q(q(\alpha_2-\gamma_2-1)+1)} \right) \right]}$$

$$\leq \frac{(b_1 - a_1)(b_2 - a_2)}{\Gamma(\alpha_1 - \gamma_1) \Gamma(\alpha_2 - \gamma_2)} \left(\int_{a_1}^{b_1} \left| \mathcal{D}_{b_1-}^{\alpha_1} f_1 (\zeta_1) \right|^q d\zeta_1 \right)^{\frac{1}{q}}$$

$$\cdot \left(\int_{a_2}^{b_2} \left| \mathcal{D}_{b_2-}^{\alpha_2} f_2 (\zeta_2) \right|^p d\zeta_2 \right)^{\frac{1}{p}}.$$
(18.114)

Next we present a basic fractional Opial type inequality for right Caputo fractional derivatives.

Theorem 18.29 *Let* $f \in AC^m ([a, b]), m \in \mathbb{N}, m := \lceil \alpha \rceil, \alpha \notin \mathbb{N}, \alpha \geq \gamma + 1, \gamma \geq 0,$ $n := \lceil \gamma \rceil$ *Assume* $f^{(k)} (b) = 0, k = n, n + 1, ..., m - 1,$ *and* $\overline{D}_{b-}^{\alpha} f \in L_{\infty} (a, b);$ *Let* $p, q > 1 : \frac{1}{p} + \frac{1}{q} = 1.$ *Then*

$$
\int_x^b \left| \overline{D}_{b-}^{\gamma} f (w) \right| \left| \overline{D}_{b-}^{\alpha} f (w) \right| dw
$$

$$
\leq \frac{(b - x)^{\left((\alpha - \gamma - 1) + \frac{2}{p} \right)}}{2^{\frac{1}{q}} \Gamma (\alpha - \gamma) [(p (\alpha - \gamma - 1) + 1) (p (\alpha - \gamma - 1) + 2)]^{\frac{1}{p}}}
$$

$$
\cdot \left(\int_x^b \left| \overline{D}_{b-}^{\alpha} f (w) \right|^q dw \right)^{\frac{2}{q}} , \tag{18.115}
$$

$\forall x \in [a, b].$

Proof We apply Theorem 18.15 and Hölder's inequality to obtain

$$
\left| \overline{D}_{b-}^{\gamma} f (x) \right| \leq \frac{1}{\Gamma (\alpha - \gamma)} \int_x^b (t - x)^{\alpha - \gamma - 1} \left| \left(\overline{D}_{b-}^{\alpha} f \right) (t) \right| dt
$$

$$
\leq \frac{1}{\Gamma (\alpha - \gamma)} \left(\int_x^b \left((t - x)^{\alpha - \gamma - 1} \right)^p dt \right)^{\frac{1}{p}} \left(\int_x^b \left| \overline{D}_{b-}^{\alpha} f (t) \right|^q dt \right)^{\frac{1}{q}}
$$

$$
= \frac{1}{\Gamma (\alpha - \gamma)} \frac{(b - x)^{(p(\alpha - \gamma - 1) + 1)/p}}{(p (\alpha - \gamma - 1) + 1)^{\frac{1}{p}}} \left(\int_x^b \left| \overline{D}_{b-}^{\alpha} f (t) \right|^q dt \right)^{1/q} . \tag{18.116}
$$

Set

$$
\lambda (x) := \int_x^b \left| \overline{D}_{b-}^{\alpha} f (t) \right|^q dt = - \int_b^x \left| \overline{D}_{b-}^{\alpha} f (t) \right|^q dt, \tag{18.117}
$$

$\lambda (b) = 0.$
   Then

$$
\lambda' (x) = - \left| \overline{D}_{b-}^{\alpha} f (x) \right|^q , \text{ a.e. in } (a, b), \tag{18.118}
$$

and

$$
\left| \overline{D}_{b-}^{\alpha} f (x) \right| = (-\lambda' (x))^{\frac{1}{q}} , \text{ a.e. in } (a, b). \tag{18.119}
$$

Therefore we get

$$\left|\overline{D}_{b-}^{\gamma} f(w)\right| \left|\overline{D}_{b-}^{\alpha} f(w)\right| \le \frac{1}{\Gamma(\alpha-\gamma)} \frac{(b-w)^{(p(\alpha-\gamma-1)+1)/p}}{(p(\alpha-\gamma-1)+1)^{\frac{1}{p}}} \left(-\lambda(w)\,\lambda'(w)\right)^{\frac{1}{q}},$$
(18.120)

a.e. on (x,b).

Consequently we find

$$\int_{x}^{b} \left|\overline{D}_{b-}^{\gamma} f(w)\right| \left|\overline{D}_{b-}^{\alpha} f(w)\right| dw \le \frac{1}{\Gamma(\alpha-\gamma)(p(\alpha-\gamma-1)+1)^{\frac{1}{p}}}$$
(18.121)

$$\cdot \int_{x}^{b} (b-w)^{(p(\alpha-\gamma-1)+1)/p} \left(-\lambda(w)\,\lambda'(w)\right)^{\frac{1}{q}} dw$$

(by Holder's inequality)

$$\le \frac{1}{\Gamma(\alpha-\gamma)(p(\alpha-\gamma-1)+1)^{\frac{1}{p}}}$$

$$\cdot \left(\int_{x}^{b} (b-w)^{p(\alpha-\gamma-1)+1} dw\right)^{\frac{1}{p}} \left(\int_{x}^{b} \left(-\lambda(w)\,\lambda'(w)\right) dw\right)^{\frac{1}{q}}$$

$$= \frac{(b-x)^{\left((\alpha-\gamma-1)+\frac{2}{p}\right)}}{\Gamma(\alpha-\gamma)[(p(\alpha-\gamma-1)+1)(p(\alpha-\gamma-1)+2)]^{\frac{1}{p}}} \frac{(\lambda(x))^{\frac{2}{q}}}{2^{\frac{1}{q}}},$$
(18.122)

proving the claim. ∎

Corollary 18.30 (*to Theorem* 18.29) *Case of* $p = q = 2$. *Then*

$$\int_{x}^{b} \left|\overline{D}_{b-}^{\gamma} f(w)\right| \left|\overline{D}_{b-}^{\alpha} f(w)\right| dw \le \frac{(b-x)^{(\alpha-\gamma)}}{2\Gamma(\alpha-\gamma)[(2(\alpha-\gamma)-1)(\alpha-\gamma)]^{\frac{1}{2}}}$$

$$\cdot \left(\int_{x}^{b} \left|\overline{D}_{b-}^{\alpha} f(w)\right|^{2} dw\right), \forall x \in [a,b].$$
(18.123)

It follows a basic fractional Opial type inequality for right Riemann-Liouville fractional derivatives

Theorem 18.31 *Let* $f \in C^{m}([a,b])$, $m \in \mathbb{N}$, $m := \lceil \alpha \rceil$, $\alpha \notin \mathbb{N}$, $\alpha \ge \gamma + 1$, $\gamma \ge 0$, $f^{(k)}(b) = 0$, $k = 0, 1, \dots, m-1$. *Let* $p, q > 1 : \frac{1}{p} + \frac{1}{q} = 1$. *Then*

$$\int_x^b \left|\mathcal{D}_{b-}^\gamma f\left(w\right)\right| \left|\mathcal{D}_{b-}^\alpha f\left(w\right)\right| dw \le \frac{(b-x)^{\left((\alpha-\gamma-1)+\frac{2}{p}\right)}}{\left[\left(p\left(\alpha-\gamma-1\right)+1\right)\left(p\left(\alpha-\gamma-1\right)+2\right)\right]^{\frac{1}{p}}}$$

$$\cdot \frac{1}{2^{\frac{1}{q}}\Gamma\left(\alpha-\gamma\right)} \left(\int_x^b \left|\mathcal{D}_{b-}^\alpha f\left(w\right)\right|^q dw\right)^{\frac{2}{q}}, \ \forall x \in [a,b]. \tag{18.124}$$

Next we present a basic fractional Opial type inequality for right generalized fractional derivatives.

Theorem 18.32 *Let* $f \in C_{b-}^\nu\left([a,b]\right)$, $0 \le \gamma < \nu$. *Assume* $f^{(i)}\left(b\right) = 0$, $i = m$, $m+1,\ldots, n-1$, *where* $m := [\gamma]$. *Let* $p,q > 1 : \frac{1}{p}+\frac{1}{q} = 1$ *so that* $\nu-\gamma > \frac{1}{q}$. *Then*

$$\int_x^b \left|D_{b-}^\gamma f\left(w\right)\right| \left|D_{b-}^\nu f\left(w\right)\right| dw \le \frac{(b-x)^{\left((\nu-\gamma-1)+\frac{2}{p}\right)}}{\left[\left(p\left(\nu-\gamma-1\right)+1\right)\left(p\left(\nu-\gamma-1\right)+2\right)\right]^{\frac{1}{p}}} \tag{18.125}$$

$$\cdot \frac{1}{2^{\frac{1}{q}}\Gamma\left(\nu-\gamma\right)} \left(\int_x^b \left|D_{b-}^\nu f\left(w\right)\right|^q dw\right)^{\frac{2}{q}}, \ \forall x \in [a,b].$$

We continue with a more general Opial type inequality regarding right Caputo fractional derivatives.

Theorem 18.33 *Let* $f \in AC^m\left([a,b]\right)$, $m \in \mathbb{N}, m := \lceil\alpha\rceil$, $\alpha \notin \mathbb{N}$, $\alpha \ge \gamma_j + 1$, $\gamma_j \ge 0$, $j = 1,2,\ldots, r \in \mathbb{N}$, $n_j := \lceil\gamma_j\rceil$. *Call* $n_0 := \min\{n_1, n_2, \ldots, n_r\}$. *Assume* $f^{(k)}\left(b\right) = 0$, $k = n_0, n_0+1,\ldots, m-1$, *and* $\overline{D}_{b-}^\alpha f \in L_\infty\left([a,b]\right)$. *Let* $p,q > 1 : \frac{1}{p}+\frac{1}{q} = 1$. *Then*

$$\int_x^b \prod_{j=1}^r \left|\overline{D}_{b-}^{\gamma_j} f\left(w\right)\right| dw \le \left\{\prod_{j=1}^r \frac{1}{\Gamma\left(\alpha-\gamma_j\right)\left(p\left(\alpha-\gamma_j-1\right)+1\right)^{\frac{1}{p}}}\right\}$$

$$\cdot \frac{(b-x)^{\left(r\alpha-\left(\sum_{j=1}^r \gamma_j\right)-r+\frac{r}{p}+1\right)}}{\left(r\alpha-\left(\sum_{j=1}^r \gamma_j\right)-r+\frac{r}{p}+1\right)} \left(\int_x^b \left|\overline{D}_{b-}^\alpha f\left(w\right)\right|^q dw\right)^{\frac{r}{q}}, \tag{18.126}$$

$\forall x \in [a,b]$.

Proof We observe that

$$\int_x^b \prod_{j=1}^r \left|\overline{D}_{b-}^{\gamma_j} f(w)\right| dw$$

$$\overset{(18.72)}{\leq} \int_x^b \prod_{j=1}^r \frac{1}{\Gamma(\alpha - \gamma_j)} \left(\int_w^b (t-w)^{\alpha - \gamma_j - 1} \left|\left(\overline{D}_{b-}^\alpha f\right)(t)\right| dt\right) dw$$

(by Holder's inequality)

$$\leq \left(\prod_{j=1}^r \frac{1}{\Gamma(\alpha - \gamma_j)}\right) \int_x^b \prod_{j=1}^r \left(\int_w^b (t-w)^{p(\alpha - \gamma_j - 1)} dt\right)^{\frac{1}{p}}$$

$$\cdot \prod_{j=1}^r \left(\int_w^b \left|\overline{D}_{b-}^\alpha f(t)\right|^q dt\right)^{\frac{1}{q}} dw \qquad (18.127)$$

$$\leq \left(\int_x^b \left|\overline{D}_{b-}^\alpha f(t)\right|^q dt\right)^{\frac{r}{q}} \left(\prod_{j=1}^r \frac{1}{\Gamma(\alpha - \gamma_j)}\right)$$

$$\cdot \int_x^b \prod_{j=1}^r \left(\frac{(b-w)^{\left((\alpha - \gamma_j - 1) + \frac{1}{p}\right)}}{(p(\alpha - \gamma_j - 1) + 1)^{\frac{1}{p}}}\right) dw \qquad (18.128)$$

$$= \left(\int_x^b \left|\overline{D}_{b-}^\alpha f(t)\right|^q dt\right)^{\frac{r}{q}} \left(\prod_{j=1}^r \frac{1}{\Gamma(\alpha - \gamma_j)\left(p(\alpha - \gamma_j - 1) + 1\right)^{\frac{1}{p}}}\right)$$

$$\cdot \int_x^b (b-w)^{r\alpha - \left(\sum_{j=1}^r \gamma_j\right) - r + \frac{r}{p}} dw$$

$$= \left\{\prod_{j=1}^r \frac{1}{\Gamma(\alpha - \gamma_j)\left(p(\alpha - \gamma_j - 1) + 1\right)^{\frac{1}{p}}}\right\}$$

$$\cdot \frac{(b-x)^{\left(r\alpha - \left(\sum_{j=1}^r \gamma_j\right) - r + \frac{r}{p} + 1\right)}}{\left(r\alpha - \left(\sum_{j=1}^r \gamma_j\right) - r + \frac{r}{p} + 1\right)} \left(\int_x^b \left|\overline{D}_{b-}^\alpha f(t)\right|^q dt\right)^{\frac{r}{q}}, \qquad (18.129)$$

proving the claim. ∎

We give a more general Opial type inequality for right Riemann-Liouville fractional derivatives

Theorem 18.34 *Let* $f \in C^m([a,b])$, $m \in \mathbb{N}$, $m := \lceil \alpha \rceil$, $\alpha \notin \mathbb{N}$, $\alpha \geq \gamma_j + 1$, $\gamma_j \geq 0$, $j = 1, \ldots r \in \mathbb{N}$, $f^{(k)}(b) = 0$, $k = 0, 1, \ldots, m - 1$. *Let* $p, q > 1$: $\frac{1}{p} + \frac{1}{q} = 1$. *Then*

$$\int_x^b \prod_{j=1}^r \left| \mathcal{D}_{b-}^{\gamma_j} f(w) \right| dw \leq \left\{ \prod_{j=1}^r \frac{1}{\Gamma(\alpha - \gamma_j)(p(\alpha - \gamma_j - 1) + 1)^{\frac{1}{p}}} \right\}$$

$$\cdot \frac{(b-x)^{\left(r\alpha - \left(\sum_{j=1}^r \gamma_j \right) - r + \frac{r}{p} + 1 \right)}}{\left(r\alpha - \left(\sum_{j=1}^r \gamma_j \right) - r + \frac{r}{p} + 1 \right)} \left(\int_x^b \left| \mathcal{D}_{b-}^{\alpha} f(w) \right|^q dw \right)^{\frac{r}{q}}, \quad (18.130)$$

$\forall x \in [a, b]$.

We finish with a more general Opial type inequality involving the right generalized fractional derivatives.

Theorem 18.35 *Let* $f \in C_{b-}^{\nu}([a,b])$, $0 \leq \gamma_i < \nu$, $j = 1, \ldots, r \in \mathbb{N}$. *Denote* $m_j := \lceil \gamma_j \rceil$, *and* $m_0 := \min\{m_1, \ldots, m_r\}$. *Assume* $f^{(i)}(b) = 0$, $i = m_0$, $m_0 + 1, \ldots, n - 1$, *where* $n := \lceil \nu \rceil$. *Let* $p, q > 1$: $\frac{1}{p} + \frac{1}{q} = 1$ *so that* $\nu - \gamma_j > \frac{1}{q}$, $j = 1, \ldots, r$. *Then*

$$\int_x^b \prod_{j=1}^r \left| D_{b-}^{\gamma_j} f(w) \right| dw \leq \left\{ \prod_{j=1}^r \frac{1}{\Gamma(\nu - \gamma_j)(p(\nu - \gamma_j - 1) + 1)^{\frac{1}{p}}} \right\}$$

$$\cdot \frac{(b-x)^{\left(r\nu - \left(\sum_{j=1}^r \gamma_j \right) - r + \frac{r}{p} + 1 \right)}}{\left(r\nu - \left(\sum_{j=1}^r \gamma_j \right) - r + \frac{r}{p} + 1 \right)} \left(\int_x^b \left| D_{b-}^{\nu} f(w) \right|^q dw \right)^{\frac{r}{q}}, \quad (18.131)$$

$\forall x \in [a, b]$.

References

1. G.A. Anastassiou, *Quantitative Approximations* (Chapman & Hall/CRC, New York, 2001)
2. G.A. Anastassiou, *Fractional Differentiation Inequalities* (Springer, New York, 2009)
3. G.A. Anastassiou, On right fractional calculus. Chaos, Solitons Fractals **42**, 365–376 (2009)
4. G.A. Anastassiou, Balanced fractional opial inequalities. Chaos, Solitons Fractals **42**(3), 1523–1528 (2009)
5. G.A. Anastassiou, Fractional Korovkin theory. Chaos, Solitons Fractals **42**, 2080–2094 (2009)
6. G.A. Anastassiou, Opial-type inequalities for functions and their ordinary and balanced fractional derivatives. J. Comput. Anal. Appl. **14**(5), 862–879 (2012)

7. G.A. Anastassiou, Fractional representation formulae and right fractional inequalities. Math. Comput. Modell. **54**(11–12), 3098–3115 (2011)
8. M. Andric, J.E. Pecaric, I. Peric, *Improvements of composition rule for the Canavati Fractional derivatives and applications to Opial-type inequalities, Dynamical Systems and Applications* (2011) (Accepted)
9. J.A. Canavati, The Riemann-Liouville integral. Nieuw Archief Voor Wiskunde **5**(1), 53–75 (1987)
10. K. Diethelm, *The Analysis of Fractional Differential Equations*, 1st edn. Lecture Notes in Mathematics (Springer, New York, 2010)
11. A.M.A. El-Sayed, M. Gaber, On the finite Caputo and finite Riesz derivatives. Electron. J. Theoret. Phys. **3**(12), 81–95 (2006)
12. R. Gorenflo, F. Mainardi, *Essentials of Fractional Calculus*. Maphysto Center. http://www. maphysto.dk/oldpages/events/LevyCAC2000/MainardiNotes/fm2k0a.ps (2000)

Chapter 19
About Canavati Fractional Ostrowski Inequalities

Here we present Ostrowski type inequalities involving left and right Canavati type generalised fractional derivatives. Combining these we obtain fractional Ostrowski type inequalities of mixed form. Then we establish Ostrowski type inequalities for ordinary and fractional derivatives involving complex valued functions defined on the unit circle. It follows [6].

19.1 Introduction

In 1938, Ostrowski [13] proved the following important inequality:

Theorem 19.1 *Let* $f : [a, b] \to \mathbb{R}$ *be continuous on* $[a, b]$ *and differentiable on* (a, b) *whose derivative* $f' : (a, b) \to \mathbb{R}$ *is bounded on* (a, b), *i.e.,* $\|f'\|_\infty :=$ $\sup_{t \in (a,b)} |f'(t)| < +\infty$. *Then*

$$\left| \frac{1}{b-a} \int_a^b f(t)\, dt - f(x) \right| \le \left[\frac{1}{4} + \frac{\left(x - \frac{a+b}{2}\right)^2}{(b-a)^2} \right] \cdot (b-a) \|f'\|_\infty, \quad (19.1)$$

for any $x \in [a, b]$. *The constant* $\frac{1}{4}$ *is the best possible.*

Since then there has been a lot of activity around these inequalities with important applications to numerical analysis and probability.

In this chapter we present various general Ostrowski type inequalities involving fractional derivatives of Canavati type.

At the end we give applications to complex valued functions defined on the unit circle.

© Springer International Publishing Switzerland 2016
G.A. Anastassiou, *Intelligent Comparisons: Analytic Inequalities*,
Studies in Computational Intelligence 609,
DOI 10.1007/978-3-319-21121-3_19

19.2 Background

Let $\nu > 0$, $n := [\nu]$ (integral part of ν), and $\alpha := \nu - n$ $(0 < \alpha < 1)$. The gamma function Γ is given by $\Gamma(\nu) = \int_0^\infty e^{-t} t^{\nu-1} dt$. Here $[a, b] \subseteq \mathbb{R}$, $x, x_0 \in [a, b]$ such that $x \geq x_0$, where x_0 is fixed. Let $f \in C([a, b])$ and define the left Riemann-Liouville integral

$$\left(J_\nu^{x_0} f\right)(x) := \frac{1}{\Gamma(\nu)} \int_{x_0}^x (x - t)^{\nu-1} f(t) dt, \tag{19.2}$$

$x_0 \leq x \leq b$. We define the subspace $C_{x_0}^\nu([a, b])$ of $C^n([a, b])$:

$$C_{x_0}^\nu([a, b]) := \left\{ f \in C^n([a, b]) : J_{1-\alpha}^{x_0} f^{(n)} \in C^1([x_0, b]) \right\}. \tag{19.3}$$

For $f \in C_{x_0}^\nu([a, b])$, we define the left generalized ν-fractional derivative of f over $[x_0, b]$ as

$$D_{x_0}^\nu f := \left(J_{1-\alpha}^{x_0} f^{(n)} \right)', \tag{19.4}$$

see [4], p. 24 and Canavati derivative in [7].

Notice that $D_{x_0}^\nu f \in C([x_0, b])$.

We need the following generalization of Taylor's formula at the fractional level, see [4], pp. 8–10, and [7].

Theorem 19.2 Let $f \in C_{x_0}^\nu([a, b])$, $x_0 \in [a, b]$ fixed.

(i) *If $\nu \geq 1$ then*

$$f(x) = f(x_0) + f'(x_0)(x - x_0) + f''(x_0) \frac{(x - x_0)^2}{2} + \cdots + \tag{19.5}$$

$$f^{(n-1)}(x_0) \frac{(x - x_0)^{n-1}}{(n-1)!} + \left(J_\nu^{x_0} D_{x_0}^\nu f \right)(x), \quad all \ x \in [a, b] : x \geq x_0.$$

(ii) *If $0 < \nu < 1$ we get*

$$f(x) = \left(J_\nu^{x_0} D_{x_0}^\nu f \right)(x), \quad all \ x \in [a, b] : x \geq x_0 \tag{19.6}$$

We will use (19.5).

Furthermore we need:

Let $\alpha > 0$, $m = [\alpha]$, $\beta = \alpha - m$, $0 < \beta < 1$, $f \in C([a, b])$, call the right Riemann-Liouville fractional integral operator by

$$\left(J_{b-}^\alpha f\right)(x) := \frac{1}{\Gamma(\alpha)} \int_x^b (J - x)^{\alpha-1} f(J) dJ, \tag{19.7}$$

$x \in [a, b]$, see also [5, 10–12, 14]. Define the subspace of functions

$$C_{b-}^{\alpha} ([a, b]) := \left\{ f \in C^m ([a, b]) : J_{b-}^{1-\beta} f^{(m)} \in C^1 ([a, b]) \right\}. \quad (19.8)$$

Define the right generalized α-fractional derivative of f over $[a, b]$ as

$$D_{b-}^{\alpha} f := (-1)^{m-1} \left(J_{b-}^{1-\beta} f^{(m)} \right)', \quad (19.9)$$

see [5]. We set $D_{b-}^0 f = f$. Notice that $D_{b-}^{\alpha} f \in C ([a, b])$.

From [5], we need the following Taylor fractional formula.

Theorem 19.3 *Let* $f \in C_{b-}^{\alpha} ([a, b])$, $\alpha > 0$, $m := [\alpha]$. *Then*

(1) If $\alpha \geq 1$, *we get*

$$f (x) = \sum_{k=0}^{m-1} \frac{f^{(k)} (b_-)}{k!} (x - b)^k + \left(J_{b-}^{\alpha} D_{b-}^{\alpha} f \right) (x), \quad \forall x \in [a, b]. \quad (19.10)$$

(2) If $0 < \alpha < 1$, *we get*

$$f (x) = J_{b-}^{\alpha} D_{b-}^{\alpha} f (x), \quad \forall x \in [a, b]. \quad (19.11)$$

We will use (19.10).

In [4], pp. 589–594, and [3], we proved the first fractional Ostrowski inequality.

Theorem 19.4 *Let* $a \leq x_0 < b$, x_0 *is fixed. Let* $f \in C_{x_0}^{\nu} ([a, b])$, $\nu \geq 1$, $n := [\nu]$. *Assume* $f^{(i)} (x_0) = 0$, $i = 1, \ldots, n - 1$. *Then*

$$\left| \frac{1}{b - x_0} \int_{x_0}^b f (y) \, dy - f (x_0) \right| \leq \frac{\left\| D_{x_0}^{\nu} f \right\|_{\infty, [x_0, b]}}{\Gamma (\nu + 2)} \cdot (b - x_0)^{\nu}. \quad (19.12)$$

Inequality (19.12) is sharp, namely it is attained by

$$f (x) := (x - x_0)^{\nu}, \quad \nu \geq 1, \quad x \in [a, b]. \quad (19.13)$$

When $1 \leq \nu < 2$ *the assumption* $f^{(i)} (x_0) = 0$, $i = 1, \ldots, n - 1$ *is void.*

19.3 Main Results

We give

Theorem 19.5 *Same assumptions as in Theorem 19.4. Then*

$$\left| \frac{1}{b - x_0} \int_{x_0}^b f(y)\, dy - f(x_0) \right| \leq \frac{\left\| D_{x_0}^\nu f \right\|_{L_1([x_0,b])} (b - x_0)^{\nu-1}}{\Gamma(\nu + 1)}. \qquad (19.14)$$

Proof By (19.5) we get

$$f(y) - f(x_0) = \frac{1}{\Gamma(\nu)} \int_{x_0}^y (y - w)^{\nu-1} \left(D_{x_0}^\nu f \right)(w)\, dw, \quad \forall\, y \geq x_0. \qquad (19.15)$$

Hence

$$|f(y) - f(x_0)| \leq \frac{1}{\Gamma(\nu)} \int_{x_0}^y (y - w)^{\nu-1} \left| \left(D_{x_0}^\nu f \right)(w) \right| dw$$

$$\leq \frac{(y - x_0)^{\nu-1}}{\Gamma(\nu)} \int_{x_0}^y \left| \left(D_{x_0}^\nu f \right)(w) \right| dw \leq \frac{(y - x_0)^{\nu-1}}{\Gamma(\nu)} \int_{x_0}^b \left| \left(D_{x_0}^\nu f \right)(w) \right| dw$$

$$= \frac{(y - x_0)^{\nu-1}}{\Gamma(\nu)} \left\| D_{x_0}^\nu f \right\|_{L_1([x_0,b])}.$$

i.e.

$$|f(y) - f(x_0)| \leq \frac{(y - x_0)^{\nu-1}}{\Gamma(\nu)} \left\| D_{x_0}^\nu f \right\|_{L_1([x_0,b])}, \quad \forall\, y \in [x_0, b]. \qquad (19.16)$$

Therefore we get

$$\left| \frac{1}{b - x_0} \int_{x_0}^b f(y)\, dy - f(x_0) \right| = \frac{1}{b - x_0} \left| \int_{x_0}^b (f(y) - f(x_0))\, dy \right| \leq$$

$$\frac{1}{b - x_0} \int_{x_0}^b |f(y) - f(x_0)|\, dy \overset{(19.16)}{\leq}$$

$$\frac{1}{b - x_0} \left(\int_{x_0}^b (y - x_0)^{\nu-1}\, dy \right) \frac{\left\| D_{x_0}^\nu f \right\|_{L_1([x_0,b])}}{\Gamma(\nu)} = \frac{(b - x_0)^{\nu-1}}{\nu} \frac{\left\| D_{x_0}^\nu f \right\|_{L_1([x_0,b])}}{\Gamma(\nu)}$$

$$= \frac{(b - x_0)^{\nu-1}}{\Gamma(\nu + 1)} \left\| D_{x_0}^\nu f \right\|_{L_1([x_0,b])},$$

proving the claim. ∎

We continue with

Theorem 19.6 *Same assumptions as in Theorem 19.4. Let $p, q > 1 : \frac{1}{p} + \frac{1}{q} = 1$.*
Then

$$\left| \frac{1}{b - x_0} \int_{x_0}^{b} f(y) \, dy - f(x_0) \right| \leq \frac{\left\| D_{x_0}^{\nu} f \right\|_{L_q([x_0,b])}}{\Gamma(\nu) \, (p(\nu - 1) + 1)^{\frac{1}{p}} \left(\nu + \frac{1}{p}\right)} (b - x_0)^{\nu - 1 + \frac{1}{p}}.$$

(19.17)

Proof We notice that

$$|f(y) - f(x_0)| \leq \frac{1}{\Gamma(\nu)} \int_{x_0}^{y} (y - w)^{\nu - 1} \left|(D_{x_0}^{\nu} f)(w)\right| dw$$

$$\leq \frac{1}{\Gamma(\nu)} \left(\int_{x_0}^{y} (y - w)^{p(\nu - 1)} \, dw \right)^{\frac{1}{p}} \left(\int_{x_0}^{y} \left|(D_{x_0}^{\nu} f)(w)\right|^q dw \right)^{\frac{1}{q}} \leq$$

$$\frac{1}{\Gamma(\nu)} \left(\frac{(y - x_0)^{p(\nu - 1) + 1}}{p(\nu - 1) + 1} \right)^{\frac{1}{p}} \left(\int_{x_0}^{b} \left|(D_{x_0}^{\nu} f)(w)\right|^q dw \right)^{\frac{1}{q}} =$$

$$\frac{1}{\Gamma(\nu)} \frac{(y - x_0)^{(\nu - 1) + \frac{1}{p}}}{(p(\nu - 1) + 1)^{\frac{1}{p}}} \left\| (D_{x_0}^{\nu} f) \right\|_{L_q([x_0,b])}.$$

That is

$$|f(y) - f(x_0)| \leq \frac{\left\| D_{x_0}^{\nu} f \right\|_{L_q([x_0,b])}}{\Gamma(\nu)} \frac{(y - x_0)^{(\nu - 1) + \frac{1}{p}}}{(p(\nu - 1) + 1)^{\frac{1}{p}}}, \quad \forall \, y \in [x_0, b]. \quad (19.18)$$

Consequently we obtain

$$\left| \frac{1}{b - x_0} \int_{x_0}^{b} f(y) \, dy - f(x_0) \right| \leq \frac{1}{b - x_0} \int_{x_0}^{b} |f(y) - f(x_0)| \, dy \overset{(19.18)}{\leq}$$

$$\left(\frac{1}{b - x_0} \right) \frac{\left\| D_{x_0}^{\nu} f \right\|_{L_q([x_0,b])}}{\Gamma(\nu) \, (p(\nu - 1) + 1)^{\frac{1}{p}}} \int_{x_0}^{b} (y - x_0)^{(\nu - 1) + \frac{1}{p}} \, dy =$$

$$\frac{\left\| D_{x_0}^{\nu} f \right\|_{L_q([x_0,b])}}{\Gamma(\nu) \, (p(\nu - 1) + 1)^{\frac{1}{p}}} \frac{(b - x_0)^{\nu + \frac{1}{p} - 1}}{\left(\nu + \frac{1}{p} \right)},$$

(19.19)

proving the claim. ∎

Combining Theorems 19.4–19.6 we derive

Proposition 19.7 *Let all as in Theorem 19.6. Then*

$$\left| \frac{1}{b - x_0} \int_{x_0}^{b} f(y)\,dy - f(x_0) \right| \leq \min \left\{ \frac{\left\| D_{x_0}^{\nu} f \right\|_{\infty,[x_0,b]}}{\Gamma(\nu + 2)} (b - x_0)^{\nu}, \right. \qquad (19.20)$$

$$\left. \frac{\left\| D_{x_0}^{\nu} f \right\|_{L_1([x_0,b])} (b - x_0)^{\nu-1}}{\Gamma(\nu + 1)}, \frac{\left\| D_{x_0}^{\nu} f \right\|_{L_q([x_0,b])} (b - x_0)^{\nu-1+\frac{1}{p}}}{\Gamma(\nu)(p(\nu - 1) + 1)^{\frac{1}{p}} \left(\nu + \frac{1}{p}\right)} \right\}.$$

We continue with right Canavati fractional Ostrowski inequalities.

Theorem 19.8 *Let* $\alpha \geq 1$, $m = \lceil \alpha \rceil$, $f \in C_{b-}^{\alpha}([a, b])$. *Assume* $f^{(k)}(b_-) = 0$, $k = 1, \ldots, m - 1$; *which is void when* $1 \leq \alpha < 2$. *Then*

$$\left| \frac{1}{b - a} \int_{a}^{b} f(x)\,dx - f(b) \right| \leq \frac{\left\| D_{b-}^{\alpha} f \right\|_{\infty,[a,b]}}{\Gamma(\alpha + 2)} (b - a)^{\alpha}. \qquad (19.21)$$

Proof Let $x \in [a, b]$. By (19.10) we get

$$f(x) - f(b) = \frac{1}{\Gamma(\alpha)} \int_{x}^{b} (J - x)^{\alpha-1} D_{b-}^{\alpha} f(J)\,dJ. \qquad (19.22)$$

Then, as before, we get

$$|f(x) - f(b)| \leq \frac{(b - x)^{\alpha}}{\Gamma(\alpha + 1)} \left\| D_{b-}^{\alpha} f \right\|_{\infty,[a,b]}, \quad \forall x \in [a, b]. \qquad (19.23)$$

Hence it holds

$$\left| \frac{1}{b - a} \int_{a}^{b} f(x)\,dx - f(b) \right| \leq \frac{1}{b - a} \int_{a}^{b} |f(x) - f(b)|\,dx \overset{(19.23)}{\leq}$$

$$\frac{\left\| D_{b-}^{\alpha} f \right\|_{\infty,[a,b]}}{\Gamma(\alpha + 1)(b - a)} \int_{a}^{b} (b - x)^{\alpha}\,dx = \frac{\left\| D_{b-}^{\alpha} f \right\|_{\infty,[a,b]}}{(\Gamma(\alpha + 1))(\alpha + 1)} (b - a)^{\alpha}$$

$$= \frac{\left\| D_{b-}^{\alpha} f \right\|_{\infty,[a,b]}}{\Gamma(\alpha + 2)} (b - a)^{\alpha}, \qquad (19.24)$$

proving the claim. ∎

We continue with

Theorem 19.9 *Same assumptions as in Theorem 19.8. Then*

$$\left| \frac{1}{b - a} \int_{a}^{b} f(x)\,dx - f(b) \right| \leq \frac{\left\| D_{b-}^{\alpha} f \right\|_{L_1([a,b])} (b - a)^{\alpha-1}}{\Gamma(\alpha + 1)}. \qquad (19.25)$$

Proof We have again

$$|f(x) - f(b)| \le \frac{1}{\Gamma(\alpha)} \int_x^b (J - x)^{\alpha-1} \left| D_{b-}^\alpha f(J) \right| dJ \qquad (19.26)$$

$$\le \frac{(b - x)^{\alpha-1}}{\Gamma(\alpha)} \left\| D_{b-}^\alpha f \right\|_{L_1([a,b])}, \quad \forall x \in [a, b].$$

Hence

$$\left| \frac{1}{b-a} \int_a^b f(x) \, dx - f(b) \right| \le \frac{1}{b-a} \int_a^b |f(x) - f(b)| \, dx \overset{(19.26)}{\le}$$

$$\frac{\left\| D_{b-}^\alpha f \right\|_{L_1([a,b])}}{(\Gamma(\alpha))(b-a)} \int_a^b (b-x)^{\alpha-1} \, dx = \frac{\left\| D_{b-}^\alpha f \right\|_{L_1([a,b])}}{\Gamma(\alpha+1)} (b-a)^{\alpha-1}, \quad (19.27)$$

proving the claim. ∎

We also have

Theorem 19.10 *Same assumptions as in Theorem 19.8. Let* $p, q > 1 : \frac{1}{p} + \frac{1}{q} = 1$ *Then*

$$\left| \frac{1}{b-a} \int_a^b f(x) \, dx - f(b) \right| \le \frac{\left\| D_{b-}^\alpha f \right\|_{L_q([a,b])}}{\Gamma(\alpha)(p(\alpha-1)+1)^{\frac{1}{p}} \left(\alpha + \frac{1}{p}\right)} (b-a)^{\alpha-1+\frac{1}{p}}.$$

$$(19.28)$$

Proof As before we obtain

$$|f(x) - f(b)| \le \frac{\left\| D_{b-}^\alpha f \right\|_{L_q([a,b])}}{\Gamma(\alpha)(p(\alpha-1)+1)^{\frac{1}{p}}} (b-x)^{\alpha-1+\frac{1}{p}}, \quad \forall x \in [a, b].$$

$$(19.29)$$

Hence

$$\left| \frac{1}{b-a} \int_a^b f(x) \, dx - f(b) \right| \le \frac{1}{b-a} \int_a^b |f(x) - f(b)| \, dx \overset{(19.29)}{\le}$$

$$\frac{\left\| D_{b-}^\alpha f \right\|_{L_q([a,b])}}{\Gamma(\alpha)(p(\alpha-1)+1)^{\frac{1}{p}}(b-a)} \left(\int_a^b (b-x)^{\alpha-1+\frac{1}{p}} \, dx \right) =$$

$$\frac{\left\| D_{b-}^\alpha f \right\|_{L_q([a,b])}}{(p(\alpha-1)+1)^{\frac{1}{p}} \Gamma(\alpha)} \frac{(b-a)^{\alpha-1+\frac{1}{p}}}{\left(\alpha + \frac{1}{p}\right)}, \qquad (19.30)$$

proving the claim. ∎

Combining Theorems 19.8–19.10 we derive

Proposition 19.11 *Here all as in Theorem 19.10. Then*

$$\left| \frac{1}{b-a} \int_a^b f(x)\,dx - f(b) \right| \leq \min \left\{ \frac{\left\| D_{b-}^\alpha f \right\|_{\infty,[a,b]}}{\Gamma(\alpha+2)} (b-a)^\alpha, \right. \tag{19.31}$$

$$\left. \frac{\left\| D_{b-}^\alpha f \right\|_{L_1([a,b])}}{\Gamma(\alpha+1)} (b-a)^{\alpha-1}, \frac{\left\| D_{b-}^\alpha f \right\|_{L_q([a,b])}}{\Gamma(\alpha)\,(p(\alpha-1)+1)^{\frac{1}{p}} \left(\alpha+\frac{1}{p}\right)} (b-a)^{\alpha-1+\frac{1}{p}} \right\}.$$

We also give optimality of (19.21).

Proposition 19.12 *Inequality (19.21) is sharp, namely it is attained by*

$$f_*(J) = (b-J)^\alpha, \quad \alpha \geq 1, \quad J \in [a,b], \tag{19.32}$$

$m := [\alpha]$.

Proof We have that

$$f_*^{(m)}(J) = (-1)^m \alpha(\alpha-1)(\alpha-2)\ldots(\alpha-m+2)(\alpha-m+1)(b-J)^{\alpha-m}. \tag{19.33}$$

We also notice

$$\left(J_{b-}^{1-(\alpha-m)} f_*^{(m)} \right)(x) \overset{(19.7)}{=} \frac{1}{\Gamma(1-\alpha+m)} \int_x^b (J-x)^{-\alpha+m} f_*^{(m)}(J)\,dJ$$

$$= \frac{1}{\Gamma(1-\alpha+m)} \int_x^b (J-x)^{m-\alpha}(-1)^m \alpha(\alpha-1)(\alpha-2)\ldots(\alpha-m+2)$$

$$\cdot (\alpha-m+1)(b-J)^{\alpha-m}\,dJ$$

$$= \frac{(-1)^m \alpha(\alpha-1)(\alpha-2)\ldots(\alpha-m+2)(\alpha-m+1)}{\Gamma(1-\alpha+m)} \cdot$$

$$\int_x^b (b-J)^{(\alpha-m+1)-1}(J-x)^{(1+m-\alpha)-1}\,dJ$$

$$= \frac{(-1)^m \alpha(\alpha-1)(\alpha-2)\ldots(\alpha-m+2)(\alpha-m+1)}{\Gamma(1-\alpha+m)} \cdot$$

$$\frac{\Gamma(\alpha - m + 1)\Gamma(1 + m - \alpha)}{\Gamma(2)}(b - x)$$

$$= (-1)^m \alpha (\alpha - 1)(\alpha - 2)\ldots(\alpha - m + 2)(\alpha - m + 1)\Gamma(\alpha - m + 1)(b - x)$$

$$= (-1)^m \alpha (\alpha - 1)(\alpha - 2)\ldots(\alpha - m + 2)\Gamma(\alpha - m + 2)(b - x)$$

$$= \cdots = (-1)^m \Gamma(\alpha + 1)(b - x). \tag{19.34}$$

That is

$$\left(J_{b-}^{1-\alpha+m} f_*^{(m)}\right)(x) = (-1)^m \Gamma(\alpha + 1)(b - x). \tag{19.35}$$

Therefore it holds

$$\left(D_{b-}^{\alpha} f_*\right)(x) \overset{(19.9)}{=} (-1)^{m-1}(-1)^m \Gamma(\alpha + 1)(-1) = \Gamma(\alpha + 1). \tag{19.36}$$

So that

$$\left\| D_{b-}^{\alpha} f_* \right\|_{\infty,[a,b]} = \Gamma(\alpha + 1). \tag{19.37}$$

We also notice that $f_*^{(k)}(b_-) = 0$, $k = 1, \ldots, m - 1$, and $f_*(b) = 0$.
    We observe further that

$$L.H.S.\,(19.21) = \frac{(b - a)^{\alpha}}{\alpha + 1}, \tag{19.38}$$

and

$$R.H.S.\,(19.21) = \frac{\Gamma(\alpha + 1)}{\Gamma(\alpha + 2)}(b - a)^{\alpha} = \frac{(b - a)^{\alpha}}{\alpha + 1}, \tag{19.39}$$

proving the claim. ∎

Next we present mixed Canavati fractional Ostrowski type inequalities.

Theorem 19.13 *Let $\alpha \geq 1$, $m = \lceil \alpha \rceil$, $x \in [a, b]$ fixed, $f \in C([a, b])$ with $f \in C_{x-}^{\alpha}([a, x])$ and $f \in C_x^{\alpha}([x, b])$. Assume that $f^{(k)}(x) = 0$, $k = 1, \ldots, m - 1$, which is void when $1 \leq \alpha < 2$. Then*

$$\left| \frac{1}{b - a} \int_a^b f(y)\,dy - f(x) \right| \leq \tag{19.40}$$

$$\frac{1}{(b - a)\Gamma(\alpha + 2)}\left\{ \left\| D_{x-}^{\alpha} f \right\|_{\infty,[a,x]}(x - a)^{\alpha+1} + \left\| D_x^{\alpha} f \right\|_{\infty,[x,b]}(b - x)^{\alpha+1} \right\} \leq$$

$$\left(\frac{(b - x)^{\alpha+1} + (x - a)^{\alpha+1}}{(b - a)\Gamma(\alpha + 2)} \right)\max\left\{ \left\| D_{x-}^{\alpha} f \right\|_{\infty,[a,x]}, \left\| D_x^{\alpha} f \right\|_{\infty,[x,b]} \right\}. \tag{19.41}$$

Proof Let $x \in [a, b]$. By (19.10) we get

$$f(y) - f(x) = \frac{1}{\Gamma(\alpha)} \int_y^x (J - y)^{\alpha - 1} D_{x-}^{\alpha} f(J) \, dJ, \quad \forall \, y \in [a, x]. \quad (19.42)$$

Hence

$$|f(y) - f(x)| \leq \frac{1}{\Gamma(\alpha)} \int_y^x (J - y)^{\alpha - 1} \left| D_{x-}^{\alpha} f(J) \right| dJ \leq \frac{(x - y)^{\alpha}}{\Gamma(\alpha + 1)} \left\| D_{x-}^{\alpha} f \right\|_{\infty, [a, x]},$$
$$(19.43)$$

$\forall \, y \in [a, x]$.

Similarly, by (19.5) we get

$$f(y) - f(x) = \frac{1}{\Gamma(\alpha)} \int_x^y (y - w)^{\alpha - 1} \left(D_x^{\alpha} f \right)(w) \, dw, \quad \forall \, y \in [x, b]. \quad (19.44)$$

Hence

$$|f(y) - f(x)| \leq \frac{1}{\Gamma(\alpha)} \int_x^y (y - w)^{\alpha - 1} \left| D_x^{\alpha} f(w) \right| dw \leq \left\| D_x^{\alpha} f \right\|_{\infty, [x, b]} \frac{(y - x)^{\alpha}}{\Gamma(\alpha + 1)},$$
$$(19.45)$$

$\forall \, y \in [x, b]$.

We observe that

$$\left| \frac{1}{b - a} \int_a^b f(y) \, dy - f(x) \right| \leq \frac{1}{b - a} \int_a^b |f(y) - f(x)| \, dy = \quad (19.46)$$

$$\frac{1}{b - a} \left\{ \int_a^x |f(y) - f(x)| \, dy + \int_x^b |f(y) - f(x)| \, dy \right\} \overset{\text{by ((19.43), (19.45))}}{\leq}$$

$$\frac{1}{b - a} \left\{ \frac{\left\| D_{x-}^{\alpha} f \right\|_{\infty, [a, x]}}{\Gamma(\alpha + 1)} \int_a^x (x - y)^{\alpha} \, dy + \frac{\left\| D_x^{\alpha} f \right\|_{\infty, [x, b]}}{\Gamma(\alpha + 1)} \int_x^b (y - x)^{\alpha} \, dy \right\} =$$
$$(19.47)$$

$$\frac{1}{(b - a) \, \Gamma(\alpha + 2)} \left\{ \left\| D_{x-}^{\alpha} f \right\|_{\infty, [a, x]} (x - a)^{\alpha + 1} + \left\| D_x^{\alpha} f \right\|_{\infty, [x, b]} (b - x)^{\alpha + 1} \right\},$$
$$(19.48)$$

proving the claim.   ■

We continue with the optimality of Theorem 19.13.

Proposition 19.14 *Inequalities (19.40), (19.41) are sharp, namely are attained by*

$$\overline{f}(J) = \begin{cases} (x - J)^{\alpha}, & J \in [a, x], \\ (J - x)^{\alpha}, & J \in [x, b], \end{cases} \tag{19.49}$$

where $\alpha \geq 1$, $x \in [a, b]$ is fixed.
See that $\overline{f}^{(k)}(x_-) = \overline{f}^{(k)}(x_+) = 0$, $k = 0, 1, \ldots, m - 1$.
We have that

$$\left\| D_{x-}^{\alpha} \overline{f} \right\|_{\infty, [a,x]} = \left\| D_x^{\alpha} \overline{f} \right\|_{\infty, [x,b]} = \Gamma(\alpha + 1). \tag{19.50}$$

Furthermore we notice

$$L.H.S. (19.40) = \frac{1}{(\alpha + 1)(b - a)} \left\{ (b - x)^{\alpha+1} + (x - a)^{\alpha+1} \right\}, \tag{19.51}$$

and

$$R.H.S. (19.41) = \frac{\left((b - x)^{\alpha+1} + (x - a)^{\alpha+1}\right)}{(b - a)\,\Gamma(\alpha + 2)} \Gamma(\alpha + 1)$$

$$= \frac{\left((b - x)^{\alpha+1} + (x - a)^{\alpha+1}\right)}{(\alpha + 1)(b - a)}, \tag{19.52}$$

proving the claim.

We continue with

Theorem 19.15 *All as in Theorem 19.13. Then ($x \in [a, b]$)*

$$\left| \frac{1}{b - a} \int_a^b f(y)\, dy - f(x) \right| \leq \tag{19.53}$$

$$\frac{1}{(b - a)\,\Gamma(\alpha + 1)} \left\{ \left\| D_{x-}^{\alpha} f \right\|_{L_1([a,x])} (x - a)^{\alpha} + \left\| D_x^{\alpha} f \right\|_{L_1([x,b])} (b - x)^{\alpha} \right\} \leq$$

$$\left(\frac{(b - x)^{\alpha} + (x - a)^{\alpha}}{(b - a)\,\Gamma(\alpha + 1)} \right) \max \left\{ \left\| D_{x-}^{\alpha} f \right\|_{L_1([a,x])}, \left\| D_x^{\alpha} f \right\|_{L_1([x,b])} \right\}. \tag{19.54}$$

Proof Let $x \in [a, b]$. From (19.42) we get ($y \in [a, x]$)

$$|f(y) - f(x)| \leq \frac{1}{\Gamma(\alpha)} \int_y^x (J - y)^{\alpha-1} \left| D_{x-}^{\alpha} f(J) \right| dJ \leq \tag{19.55}$$

$$\frac{1}{\Gamma(\alpha)} (x - y)^{\alpha-1} \int_y^x \left| D_{x-}^{\alpha} f(J) \right| dJ \leq \frac{(x - y)^{\alpha-1}}{\Gamma(\alpha)} \int_a^x \left| D_{x-}^{\alpha} f(J) \right| dJ. \tag{19.56}$$

That is

$$|f(y) - f(x)| \leq (x-y)^{\alpha-1} \frac{\left\| D_{x-}^{\alpha} f \right\|_{L_1([a,x])}}{\Gamma(\alpha)}, \quad \forall\, y \in [a, x]. \tag{19.57}$$

Similarly from (19.44) we get

$$|f(y) - f(x)| \leq \frac{1}{\Gamma(\alpha)} \int_x^y (y-w)^{\alpha-1} \left| (D_x^{\alpha} f)(w) \right| dw \tag{19.58}$$

$$\leq \frac{(y-x)^{\alpha-1}}{\Gamma(\alpha)} \left\| D_x^{\alpha} f \right\|_{L_1([x,b])}, \quad \forall\, y \in [x, b]. \tag{19.59}$$

From (19.46) we obtain

$$\left| \frac{1}{b-a} \int_a^b f(y)\, dy - f(x) \right| \leq$$

$$\frac{1}{b-a} \left\{ \frac{\left\| D_{x-}^{\alpha} f \right\|_{L_1([a,x])}}{\Gamma(\alpha)} \int_a^x (x-y)^{\alpha-1}\, dy + \frac{\left\| D_x^{\alpha} f \right\|_{L_1([x,b])}}{\Gamma(\alpha)} \int_x^b (y-x)^{\alpha-1}\, dy \right\} \tag{19.60}$$

$$= \frac{1}{(b-a)\,\Gamma(\alpha+1)} \left\{ \left\| D_{x-}^{\alpha} f \right\|_{L_1([a,x])} (x-a)^{\alpha} + \left\| D_x^{\alpha} f \right\|_{L_1([x,b])} (b-x)^{\alpha} \right\} \leq$$

$$\left(\frac{(b-x)^{\alpha} + (x-a)^{\alpha}}{(b-a)\,\Gamma(\alpha+1)} \right) \max \left\{ \left\| D_{x-}^{\alpha} f \right\|_{L_1([a,x])}, \left\| D_x^{\alpha} f \right\|_{L_1([x,b])} \right\}, \tag{19.61}$$

proving the claim. ∎

We also give

Theorem 19.16 *All as in Theorem 19.13. Let* $p, q > 1 : \frac{1}{p} + \frac{1}{q} = 1$. *Then* $(x \in [a, b])$

$$\left| \frac{1}{b-a} \int_a^b f(y)\, dy - f(x) \right| \leq \frac{1}{(b-a)\,\Gamma(\alpha)\,(p(\alpha-1)+1)^{\frac{1}{p}} \left(\alpha + \frac{1}{p} \right)} \tag{19.62}$$

$$\left\{ (x-a)^{\alpha+\frac{1}{p}} \left\| D_{x-}^{\alpha} f \right\|_{L_q([a,x])} + (b-x)^{\alpha+\frac{1}{p}} \left\| D_x^{\alpha} f \right\|_{L_q([x,b])} \right\} \leq$$

$$\left(\frac{(b-x)^{\alpha+\frac{1}{p}} + (x-a)^{\alpha+\frac{1}{p}}}{(b-a)\,\Gamma(\alpha)\,(p(\alpha-1)+1)^{\frac{1}{p}}\left(\alpha+\frac{1}{p}\right)} \right) \cdot$$

$$\max\left\{ \left\| D_{x-}^{\alpha} f \right\|_{L_q([a,x])}, \left\| D_{x*}^{\alpha} f \right\|_{L_q([x,b])} \right\}. \tag{19.63}$$

Proof By (19.55) we get

$$|f(y) - f(x)| \leq \frac{1}{\Gamma(\alpha)} \left(\int_y^x (J-y)^{p(\alpha-1)} \, dJ \right)^{\frac{1}{p}} \left\| D_{x-}^{\alpha} f \right\|_{L_q([a,x])}$$

$$= \frac{1}{\Gamma(\alpha)} \frac{(x-y)^{\alpha-1+\frac{1}{p}}}{(p(\alpha-1)+1)^{\frac{1}{p}}} \left\| D_{x-}^{\alpha} f \right\|_{L_q([a,x])}, \quad \forall y \in [a, x]. \tag{19.64}$$

Similarly from (19.58) we derive

$$|f(y) - f(x)| \leq \frac{1}{\Gamma(\alpha)} \left(\int_x^y (y-w)^{p(\alpha-1)} \, dw \right)^{\frac{1}{p}} \left\| D_{x*}^{\alpha} f \right\|_{L_q([x,b])}$$

$$= \frac{1}{\Gamma(\alpha)} \frac{(y-x)^{\alpha-1+\frac{1}{p}}}{(p(\alpha-1)+1)^{\frac{1}{p}}} \left\| D_{x*}^{\alpha} f \right\|_{L_q([x,b])}, \quad \forall y \in [x, b]. \tag{19.65}$$

By (19.46) we derive

$$\left| \frac{1}{b-a} \int_a^b f(y) \, dy - f(x) \right| \leq \frac{1}{\Gamma(\alpha)\,(b-a)\,(p(\alpha-1)+1)^{\frac{1}{p}}}$$

$$\left\{ \left(\int_a^x (x-y)^{\alpha-1+\frac{1}{p}} \, dy \right) \left\| D_{x-}^{\alpha} f \right\|_{L_q([a,x])} + \right.$$

$$\left. \left(\int_x^b (y-x)^{\alpha-1+\frac{1}{p}} \, dy \right) \left\| D_{x*}^{\alpha} f \right\|_{L_q([x,b])} \right\} \tag{19.66}$$

$$= \frac{1}{(b-a)\,\Gamma(\alpha)\,(p(\alpha-1)+1)^{\frac{1}{p}}\left(\alpha+\frac{1}{p}\right)} \cdot$$

$$\left\{ (x-a)^{\alpha+\frac{1}{p}} \left\| D_{x-}^{\alpha} f \right\|_{L_q([a,x])} + (b-x)^{\alpha+\frac{1}{p}} \left\| D_{x*}^{\alpha} f \right\|_{L_q([x,b])} \right\} \leq \tag{19.67}$$

$$\left(\frac{(b-x)^{\alpha+\frac{1}{p}} + (x-a)^{\alpha+\frac{1}{p}}}{(b-a)\,\Gamma\,(\alpha)\,(p\,(\alpha-1)+1)^{\frac{1}{p}}\left(\alpha+\frac{1}{p}\right)} \right) \max \left\{ \left\| D^{\alpha}_{x-} f \right\|_{L_q([a,x])}, \left\| D^{\alpha}_{x} f \right\|_{L_q([x,b])} \right\},$$

$$(19.68)$$

proving the claim. ■

Corollary 19.17 *All as in Theorem 19.13. Then*

$$\left| \frac{1}{b-a} \int_a^b f(y)\,dy - f(x) \right| \le \tag{19.69}$$

$$\left(\frac{(b-x)^{\alpha+\frac{1}{2}} + (x-a)^{\alpha+\frac{1}{2}}}{(b-a)\,\Gamma\,(\alpha)\,\sqrt{2\alpha-1}\,(\alpha+\frac{1}{2})} \right) \max \left\{ \left\| D^{\alpha}_{x-} f \right\|_{L_2([a,x])}, \left\| D^{\alpha}_{x} f \right\|_{L_2([x,b])} \right\}.$$

Proof By Theorem 19.16. ■

Combining Theorems 19.13, 19.15 and 19.16 we derive

Theorem 19.18 *Here all as in Theorem 19.13. Let any $p, q > 1 : \frac{1}{p} + \frac{1}{q} = 1$. Then*

$$\left| \frac{1}{b-a} \int_a^b f(y)\,dy - f(x) \right| \le$$

$$\min \left\{ \frac{1}{(b-a)\,\Gamma\,(\alpha+2)} \left\{ \left\| D^{\alpha}_{x-} f \right\|_{\infty,[a,x]} (x-a)^{\alpha+1} + \left\| D^{\alpha}_{x} f \right\|_{\infty,[x,b]} (b-x)^{\alpha+1} \right\}, \right.$$

$$\frac{1}{(b-a)\,\Gamma\,(\alpha+1)} \left\{ \left\| D^{\alpha}_{x-} f \right\|_{L_1([a,x])} (x-a)^{\alpha} + \left\| D^{\alpha}_{x} f \right\|_{L_1([x,b])} (b-x)^{\alpha} \right\},$$

$$(19.70)$$

$$\frac{1}{(b-a)\,\Gamma\,(\alpha)\,(p\,(\alpha-1)+1)^{\frac{1}{p}}\left(\alpha+\frac{1}{p}\right)}\cdot$$

$$\left\{ (x-a)^{\alpha+\frac{1}{p}} \left\| D^{\alpha}_{x-} f \right\|_{L_q([a,x])} + (b-x)^{\alpha+\frac{1}{p}} \left\| D^{\alpha}_{x} f \right\|_{L_q([x,b])} \right\} \right\} \le$$

$$\min \left\{ \left(\frac{(b-x)^{\alpha+1} + (x-a)^{\alpha+1}}{(b-a)\,\Gamma\,(\alpha+2)} \right) \max \left\{ \left\| D^{\alpha}_{x-} f \right\|_{\infty,[a,x]}, \left\| D^{\alpha}_{x} f \right\|_{\infty,[x,b]} \right\}, \right.$$

$$\left(\frac{(b-x)^{\alpha} + (x-a)^{\alpha}}{(b-a)\,\Gamma\,(\alpha+1)} \right) \max \left\{ \left\| D^{\alpha}_{x-} f \right\|_{L_1([a,x])}, \left\| D^{\alpha}_{x} f \right\|_{L_1([x,b])} \right\}, \tag{19.71}$$

$$\left(\frac{(b-x)^{\alpha+\frac{1}{p}} + (x-a)^{\alpha+\frac{1}{p}}}{(b-a)\,\Gamma(\alpha)\,(p\,(\alpha-1)+1)^{\frac{1}{p}}\left(\alpha+\frac{1}{p}\right)} \right) \cdot$$

$$\max\left\{ \left\| D_{x-}^{\alpha} f \right\|_{L_q([a,x])},\ \left\| D_x^{\alpha} f \right\|_{L_q([x,b])} \right\} .$$

19.4 Applications

Inequalities for complex valued functions defined on the unit circle were studied extensively by Dragomir, see [8, 9].

We give here our version for these functions involved in Ostrowski type inequalities, by applying results of this chapter.

Let $t \in [a, b] \subseteq [0, 2\pi)$, the unit circle arc $A = \left\{ z \in \mathbb{C} : z = e^{it},\ t \in [a, b] \right\}$, and $f : A \to \mathbb{C}$ be a continuous function. Clearly here there exist functions $u, v : A \to \mathbb{R}$ continuous, the real and the complex part of f, respectively, such that

$$f\left(e^{it}\right) = u\left(e^{it}\right) + iv\left(e^{it}\right). \tag{19.72}$$

So that f is continuous, iff u, v are continuous.

Call $g(t) = f\left(e^{it}\right)$, $l_1(t) = u\left(e^{it}\right)$, $l_2(t) = v\left(e^{it}\right)$, $t \in [a, b]$; so that $g : [a, b] \to \mathbb{C}$ and $l_1, l_2 : [a, b] \to \mathbb{R}$ are continuous functions in t.

If g has a derivative with respect to t, then l_1, l_2 have also derivatives with respect to t. In that case

$$f_t\left(e^{it}\right) = u_t\left(e^{it}\right) + iv_t\left(e^{it}\right), \tag{19.73}$$

(i.e. $g'(t) = l_1'(t) + il_2'(t)$), which means

$$f_t(\cos t + i\sin t) = u_t(\cos t + i\sin t) + iv_t(\cos t + i\sin t). \tag{19.74}$$

Let us call $x = \cos t$, $y = \sin t$. Then

$$u_t\left(e^{it}\right) = u_t(\cos t + i\sin t) = u_t(x + iy) = u_t(x, y) =$$

$$\frac{\partial u}{\partial x}\frac{\partial x}{\partial t} + \frac{\partial u}{\partial y}\frac{\partial y}{\partial t} = \frac{\partial u\left(e^{it}\right)}{\partial x}(-\sin t) + \frac{\partial u\left(e^{it}\right)}{\partial y}\cos t. \tag{19.75}$$

Similarly we find that

$$v_t\left(e^{it}\right) = \frac{\partial v\left(e^{it}\right)}{\partial x}(-\sin t) + \frac{\partial v\left(e^{it}\right)}{\partial y}\cos t. \tag{19.76}$$

So that

$$\left\| u_t\left(e^{it}\right) \right\|_{\infty,[a,b]} \le \left\| \frac{\partial u\left(e^{it}\right)}{\partial x} \right\|_{\infty,[a,b]} + \left\| \frac{\partial u\left(e^{it}\right)}{\partial y} \right\|_{\infty,[a,b]}, \qquad (19.77)$$

and

$$\left\| v_t\left(e^{it}\right) \right\|_{\infty,[a,b]} \le \left\| \frac{\partial v\left(e^{it}\right)}{\partial x} \right\|_{\infty,[a,b]} + \left\| \frac{\partial v\left(e^{it}\right)}{\partial y} \right\|_{\infty,[a,b]}, \qquad (19.78)$$

Consequently it holds

$$\left\| f_t\left(e^{it}\right) \right\|_{\infty,[a,b]} \le$$

$$\left\| \frac{\partial u\left(e^{it}\right)}{\partial x} \right\|_{\infty,[a,b]} + \left\| \frac{\partial v\left(e^{it}\right)}{\partial x} \right\|_{\infty,[a,b]} + \left\| \frac{\partial u\left(e^{it}\right)}{\partial y} \right\|_{\infty,[a,b]} + \left\| \frac{\partial v\left(e^{it}\right)}{\partial y} \right\|_{\infty,[a,b]}.$$
$$(19.79)$$

Since g is continuous on $[a, b]$, then $\int_a^b f\left(e^{it}\right) dt$ exists. Furthermore it holds

$$\int_a^b f\left(e^{it}\right) dt = \int_a^b u\left(e^{it}\right) dt + i \int_a^b v\left(e^{it}\right) dt. \qquad (19.80)$$

Let now $t_0 \in [a, b]$. We observe that

$$\left| \frac{1}{b-a} \int_a^b f\left(e^{it}\right) dt - f\left(e^{it_0}\right) \right| =$$

$$\left| \frac{1}{b-a} \int_a^b u\left(e^{it}\right) dt + i\frac{1}{b-a} \int_a^b v\left(e^{it}\right) dt - u\left(e^{it_0}\right) - iv\left(e^{it_0}\right) \right| \le \quad (19.81)$$

$$\left| \frac{1}{b-a} \int_a^b u\left(e^{it}\right) dt - u\left(e^{it_0}\right) \right| + \left| \frac{1}{b-a} \int_a^b v\left(e^{it}\right) dt - v\left(e^{it_0}\right) \right| \overset{\text{(by (19.1))}}{\le}$$

$$\left[\frac{1}{4} + \frac{\left(t_0 - \frac{a+b}{2}\right)^2}{(b-a)^2} \right] (b-a) \left[\left\| u_t\left(e^{it}\right) \right\|_{\infty,[a,b]} + \left\| v_t\left(e^{it}\right) \right\|_{\infty,[a,b]} \right]. \qquad (19.82)$$

We have proved the following version of Ostrowski inequality for complex functions.

Theorem 19.19 *Let $f \in C\left(A, \mathbb{C}\right)$ with its real and complex part $u\left(e^{it}\right)$, $v\left(e^{it}\right) \in C^1\left([a, b]\right)$ as functions of t, where $t_0 \in [a, b] \subseteq [0, 2\pi)$. Then*

$$\left| \frac{1}{b-a} \int_a^b f\left(e^{it}\right) dt - f\left(e^{it_0}\right) \right| \le$$

$$\left[\frac{1}{4} + \frac{\left(t_0 - \frac{a+b}{2}\right)^2}{(b-a)^2}\right](b-a)\left[\left\|u_t\left(e^{it}\right)\right\|_{\infty,[a,b]} + \left\|v_t\left(e^{it}\right)\right\|_{\infty,[a,b]}\right] \leq \quad (19.83)$$

$$\left[\frac{1}{4} + \frac{\left(t_0 - \frac{a+b}{2}\right)^2}{(b-a)^2}\right](b-a) \cdot \quad (19.84)$$

$$\left[\left[\left\|\frac{\partial u\left(e^{it}\right)}{\partial x}\right\|_{\infty,[a,b]} + \left\|\frac{\partial v\left(e^{it}\right)}{\partial x}\right\|_{\infty,[a,b]} + \left\|\frac{\partial u\left(e^{it}\right)}{\partial y}\right\|_{\infty,[a,b]} + \left\|\frac{\partial v\left(e^{it}\right)}{\partial y}\right\|_{\infty,[a,b]}\right]\right].$$

Inequality (19.83) is sharp.

An explanation follows next.

For $z \in \mathbb{C} - \{0\}$ we call principal value of $\log(z)$ the complex valued function

$$Log(z) := \ln|z| + i\,Arg(z), \quad (19.85)$$

where $0 \leq Arg(z) < 2\pi$.

For $t \in [0, 2\pi)$ we have that

$$Log\left(e^{it}\right) = it. \quad (19.86)$$

Let here $a = 0 < b < 2\pi$, and $t_0 = 0$. Here $l_1(t) = 0$ and $l_2(t) = t$, with $l_2'(t) = 1$. Notice in general that

$$\left(\frac{1}{4} + \frac{\left(t_0 - \frac{a+b}{2}\right)^2}{(b-a)^2}\right)(b-a) = \frac{(b-t_0)^2 + (t_0-a)^2}{2(b-a)}, \quad (19.87)$$

for any $t_0 \in [a, b]$, see [10], p. 498 and [1].

Hence we have

$$L.H.S. (19.84) = \left|\frac{1}{b}\int_0^b Log\left(e^{it}\right)dt\right| = \left|\frac{1}{b}\int_0^b itdt\right| = \frac{1}{b}\int_0^b tdt = \frac{b}{2}. \quad (19.88)$$

Furthermore it holds

$$R.H.S. (19.84) = \frac{b}{2} \cdot 1 = \frac{b}{2}. \quad (19.89)$$

By (19.88) and (19.89) we conclude that inequality (19.83) is attained by *Log* at $t_0 = 0$ on $[0, b]$, that is a sharp inequality.

We now move at the fractional level.

Let $t_0 \in [a, b]$, we rewrite (19.81) as follows

$$\left| \frac{1}{b-a} \int_a^b f\left(e^{it}\right) dt - f\left(e^{it_0}\right) \right| \leq$$

$$\left| \frac{1}{b-a} \int_a^b l_1(t) dt - l_1(t_0) \right| + \left| \frac{1}{b-a} \int_a^b l_2(t) dt - l_2(t_0) \right|. \tag{19.90}$$

By applying Theorem 19.18 to each of the last two summands we derive the following complex fractional Ostrowski inequality.

Theorem 19.20 *Let $f \in C(A, \mathbb{C})$, $t, t_0 \in [a, b] \subseteq [0, 2\pi)$; any $p, q > 1 : \frac{1}{p} + \frac{1}{q} = 1$. Let $\alpha \geq 1$, $m = [\alpha]$, with $l_1, l_2 \in C_{t_0-}^\alpha([a, t_0])$ and $l_1, l_2 \in C_{t_0}^\alpha([t_0, b])$. Assume that $l_1^{(k)}(t_0) = l_2^{(k)}(t_0) = 0$, $k = 1, \ldots, m-1$, which is void when $1 \leq \alpha < 2$. Then*

$$\left| \frac{1}{b-a} \int_a^b f\left(e^{it}\right) dt - f\left(e^{it_0}\right) \right| \leq$$

$$\min\left\{ \frac{1}{(b-a)\,\Gamma(\alpha+2)} \left\{ \left\| D_{t_0-}^\alpha l_1 \right\|_{\infty,[a,t_0]} (t_0-a)^{\alpha+1} + \left\| D_{t_0}^\alpha l_1 \right\|_{\infty,[t_0,b]} (b-t_0)^{\alpha+1} \right\}, \right.$$

$$\frac{1}{(b-a)\,\Gamma(\alpha+1)} \left\{ \left\| D_{t_0-}^\alpha l_1 \right\|_{L_1([a,t_0])} (t_0-a)^\alpha + \left\| D_{t_0}^\alpha l_1 \right\|_{L_1([t_0,b])} (b-t_0)^\alpha \right\},$$

$$\frac{1}{(b-a)\,\Gamma(\alpha)\,(p(\alpha-1)+1)^{\frac{1}{p}}\left(\alpha+\frac{1}{p}\right)} .$$

$$\left\{ (t_0-a)^{\alpha+\frac{1}{p}} \left\| D_{t_0-}^\alpha l_1 \right\|_{L_q([a,t_0])} + (b-t_0)^{\alpha+\frac{1}{p}} \left\| D_{t_0}^\alpha l_1 \right\|_{L_q([t_0,b])} \right\} \right\} +$$

$$\min\left\{ \frac{1}{(b-a)\,\Gamma(\alpha+2)} \left\{ \left\| D_{t_0-}^\alpha l_2 \right\|_{\infty,[a,t_0]} (t_0-a)^{\alpha+1} + \left\| D_{t_0}^\alpha l_2 \right\|_{\infty,[t_0,b]} (b-t_0)^{\alpha+1} \right\}, \right.$$

$$\frac{1}{(b-a)\,\Gamma(\alpha+1)} \left\{ \left\| D_{t_0-}^\alpha l_2 \right\|_{L_1([a,t_0])} (t_0-a)^\alpha + \left\| D_{t_0}^\alpha l_2 \right\|_{L_1([t_0,b])} (b-t_0)^\alpha \right\},$$

$$\frac{1}{(b-a)\,\Gamma(\alpha)\,(p(\alpha-1)+1)^{\frac{1}{p}}\left(\alpha+\frac{1}{p}\right)} .$$

$$\left\{ (t_0-a)^{\alpha+\frac{1}{p}} \left\| D_{t_0-}^\alpha l_2 \right\|_{L_q([a,t_0])} + (b-t_0)^{\alpha+\frac{1}{p}} \left\| D_{t_0}^\alpha l_2 \right\|_{L_q([t_0,b])} \right\} \right\} \leq \tag{19.91}$$

$$\min\left\{\left(\frac{(b-t_0)^{\alpha+1}+(t_0-a)^{\alpha+1}}{(b-a)\,\Gamma\,(\alpha+2)}\right)\max\left\{\left\|D^{\alpha}_{t_0-}l_1\right\|_{\infty,[a,t_0]},\,\left\|D^{\alpha}_{t_0}l_1\right\|_{\infty,[t_0,b]}\right\},\right.$$

$$\left(\frac{(b-t_0)^{\alpha}+(t_0-a)^{\alpha}}{(b-a)\,\Gamma\,(\alpha+1)}\right)\max\left\{\left\|D^{\alpha}_{t_0-}l_1\right\|_{L_1([a,t_0])},\,\left\|D^{\alpha}_{t_0}l_1\right\|_{L_1([t_0,b])}\right\},$$

$$\left(\frac{(b-t_0)^{\alpha+\frac{1}{p}}+(t_0-a)^{\alpha+\frac{1}{p}}}{(b-a)\,\Gamma\,(\alpha)\,(p\,(\alpha-1)+1)^{\frac{1}{p}}\left(\alpha+\frac{1}{p}\right)}\right)\cdot$$

$$\left.\max\left\{\left\|D^{\alpha}_{t_0-}l_1\right\|_{L_q([a,t_0])},\,\left\|D^{\alpha}_{t_0}l_1\right\|_{L_q([t_0,b])}\right\}\right\}+$$

$$\min\left\{\left(\frac{(b-t_0)^{\alpha+1}+(t_0-a)^{\alpha+1}}{(b-a)\,\Gamma\,(\alpha+2)}\right)\max\left\{\left\|D^{\alpha}_{t_0-}l_2\right\|_{\infty,[a,t_0]},\,\left\|D^{\alpha}_{t_0}l_2\right\|_{\infty,[t_0,b]}\right\},\right.$$

$$\left(\frac{(b-t_0)^{\alpha}+(t_0-a)^{\alpha}}{(b-a)\,\Gamma\,(\alpha+1)}\right)\max\left\{\left\|D^{\alpha}_{t_0-}l_2\right\|_{L_1([a,t_0])},\,\left\|D^{\alpha}_{t_0}l_2\right\|_{L_1([t_0,b])}\right\},$$

$$\left(\frac{(b-t_0)^{\alpha+\frac{1}{p}}+(t_0-a)^{\alpha+\frac{1}{p}}}{(b-a)\,\Gamma\,(\alpha)\,(p\,(\alpha-1)+1)^{\frac{1}{p}}\left(\alpha+\frac{1}{p}\right)}\right)\cdot$$

$$\left.\max\left\{\left\|D^{\alpha}_{t_0-}l_2\right\|_{L_q([a,t_0])},\,\left\|D^{\alpha}_{t_0}l_2\right\|_{L_q([t_0,b])}\right\}\right\}. \tag{19.92}$$

References

1. G.A. Anastassiou, Ostrowski type inequalities. Proc. AMS **123**, 3775–3781 (1995)
2. G. Anastassiou, *Quantitative Approximation* (Chapman & Hall/CRC, Boca Raton, 2001)
3. G. Anastassiou, Fractional Ostrowski type inequalities. Commun. Appl. Anal. **7**(2), 203–208 (2003)
4. G.A. Anastassiou, *Fractional Differentiation Inequalities Research Monograph* (Springer, New York, 2009)
5. G.A. Anastassiou, On right fractional calculus. Chaos, Solitons and Fractals **42**, 365–376 (2009)
6. G.A. Anastassiou, Canavati fractional Ostrowski type inequalities. Commun. Appl. Anal. **17**(2), 157–172 (2013)
7. J.A. Canavati, The Riemann-Liouville integral. Nieuw Archief Voor Wiskunde **5**(1), 53–75 (1987)
8. S.S. Dragomir, Ostrowski's type inequalities for complex functions defined on unit circle with applications for unitary operators in hilbert spaces. 1RGMIA. Res. Rep. Coll. **6**(6) (2013). http://rgmia.org/v16.php
9. S.S. Dragomir, Generalized trapezoidal type inequalities for complex functions defined on unit circle with applications for unitary operators in hilbert spaces. RGMIA. Res. Rep. Coll. **16**(9) (2013). http://rgmia.org/v16.php

10. A.M.A. El-Sayed, M. Gaber, On the finite Caputo and finite Riesz derivatives. Electron. J. Theor. Phys. **3**(12), 81–95 (2006)
11. G.S. Frederico, D.F.M. Torres, Fractional optimal control in the sense of caputo and the fractional Noether's theorem. Int. Math. forum **3**(10), 479–493 (2008)
12. R. Gorenflo, F. Mainardi, *Essentials of Fractional Calculus*, Maphysto Center (2000). http://www.maphysto.dk/oldpages/events/LevyCAC2000/MainardiNotes/fm2k0a.ps
13. A. Ostrowski, Über die Absolutabweichung einer differentiabaren Funcktion von ihrem Integralmittelwert. Comment. Math. Helv. **10**, 226–227 (1938)
14. S.G. Samko, A.A. Kilbas, O.I. Marichev, *Fractional Integrals and Derivatives, Theory and Applications*, (Gordon and Breach, Amsterdam, 1993) [English translation from the Russian, Integrals and Derivatives of Fractional Order and Some of Their Applications (Nauka i Tekhnika, Minsk, 1987)]

Chapter 20
The Most General Fractional Representation Formula for Functions and Consequences

Here we present the most general fractional representation formulae for a function in terms of the most general fractional integral operators due to Kalla [4–6]. The last include most of the well-known fractional integrals such as of Riemann-Liouville, Erdé lyi-Kober and Saigo, etc. Based on these we derive very general fractional Ostrowski type inequalities. It follows [2].

20.1 Introduction

Let $f : [a, b] \rightarrow \mathbb{R}$ be differentiable on $[a, b]$, and $f' : [a, b] \rightarrow \mathbb{R}$ be integrable on $[a, b]$, then the following Montgomery identity holds [11]:

$$f(x) = \frac{1}{b-a} \int_a^b f(t)\, dt + \int_a^b P_1(x, t)\, f'(t)\, dt, \qquad (20.1)$$

where $P_1(x, t)$ is the Peano kernel

$$P_1(x, t) = \begin{cases} \frac{t-a}{b-a}, & a \leq t \leq x, \\ \frac{t-b}{b-a}, & x < t \leq b, \end{cases} \qquad (20.2)$$

The Riemann-Liouville integral operator of order $\alpha > 0$ with anchor point $a \in \mathbb{R}$ is defined by

$$J_a^\alpha f(x) := \frac{1}{\Gamma(\alpha)} \int_a^x (x - t)^{\alpha-1} f(t)\, dt, \qquad (20.3)$$

$$J_a^0 f(x) := f(x), x \in [a, b]. \qquad (20.4)$$

Properties of the above operator can be found in [10].

When $\alpha = 1$, J_a^1 reduces to the classical integral.

© Springer International Publishing Switzerland 2016
G.A. Anastassiou, *Intelligent Comparisons: Analytic Inequalities*,
Studies in Computational Intelligence 609,
DOI 10.1007/978-3-319-21121-3_20

In [1] we proved the following fractional representation formula of Montgomery identity type.

Theorem 20.1 *Let $f : [a, b] \to \mathbb{R}$ be differentiable on $[a, b]$, and $f' : [a, b] \to \mathbb{R}$ be integrable on $[a, b]$, $\alpha \geq 1$, $x \in [a, b)$. Then*

$$f(x) =$$

$$(b - x)^{1-\alpha} \, \Gamma(\alpha) \left\{ \frac{J_a^\alpha f(b)}{b - a} - J_a^{\alpha - 1} \left(P_1(x, b) f(b) \right) + J_a^\alpha \left(P_1(x, b) f'(b) \right) \right\}. \tag{20.5}$$

When $\alpha = 1$ the last (20.5) reduces to classic Montgomery identity (20.1).

Motivated by (20.5), here we establish a very general fractional representation formula based on the most general fractional integral due to Kalla [4–6]. The last integral includes almost all other fractional integrals as special cases. We then establish a very general fractional Ostrowski type inequality.

We finish with applications.

20.2 Main Results

Here let $f : \mathbb{R}_+ \to \mathbb{R}$ differentiable with $f' : \mathbb{R}_+ \to \mathbb{R}$ be integrable. Let also $\Phi : [0, 1] \to \mathbb{R}_+$ a general kernel function, which is differentiable with $\Phi' : [0, 1] \to \mathbb{R}_+$ being integrable too. For z *in* $(0, 1)$ we assume $\Phi(z) > 0$.

Let here the parameters γ, δ be such that $\gamma > -1$ and $\delta \in \mathbb{R}$. Set $\varepsilon := \delta - \gamma - 1$, that is $\delta = \varepsilon + \gamma + 1$.

The most general fractional integral operator was defined by Kalla [4–6], see also [8], as follows:

$$I_\Phi^{\gamma, \delta} f(x) := x^\delta \int_0^1 \Phi(\sigma) \, \sigma^\gamma f(x\sigma) \, d\sigma, \tag{20.6}$$

for any $x > 0$, with $I_\Phi^{\gamma, \delta} f(0) := 0$.

Here we consider $b > 0$ fixed, and $0 < x < b$. We operate on $[0, b]$.

By convenient change of variable we can rewrite $I_\Phi^{\gamma, \delta} f(x)$ as follows:

$$I_\Phi^{\gamma, \varepsilon} f(x) := x^\varepsilon \int_0^x \Phi\left(\frac{w}{x}\right) w^\gamma f(w) \, dw. \tag{20.7}$$

That is

$$I_\Phi^{\gamma, \varepsilon} f(x) = I_\Phi^{\gamma, \delta} f(x), \text{ for any } x > 0. \tag{20.8}$$

We take $\gamma > 0$ from now on.

We present the following most general fractional representation formula.

Theorem 20.2 *All as above described. Then*

$$f(x) = b^{\gamma+1-\delta} x^{-\gamma} \left(\Phi\left(\frac{x}{b}\right) \right)^{-1} \left[\frac{1}{b} I_\Phi^{\gamma,\delta} f(b) + \gamma I_\Phi^{\gamma-1,\delta} \left(P_1(x,b) f(b) \right) \right.$$

$$\left. + \frac{1}{b} I_{\Phi'}^{\gamma,\delta} \left(P_1(x,b) f(b) \right) + I_\Phi^{\gamma,\delta} \left(P_1(x,b) f'(b) \right) \right]. \qquad (20.9)$$

Proof We observe that

$$I_\Phi^{\gamma,\varepsilon} \left(P_1(x,b) f'(b) \right) = b^\varepsilon \int_0^b \Phi\left(\frac{w}{b}\right) w^\gamma P_1(x,w) f'(w)\, dw = \qquad (20.10)$$

$$b^\varepsilon \left[\int_0^x \Phi\left(\frac{w}{b}\right) w^\gamma \frac{w}{b} f'(w)\, dw + \int_x^b \Phi\left(\frac{w}{b}\right) w^\gamma \left(\frac{w-b}{b}\right) f'(w)\, dw \right] =$$
$$(20.11)$$

$$b^{\varepsilon-1} \left[\int_0^x \Phi\left(\frac{w}{b}\right) w^{\gamma+1} f'(w)\, dw + \int_x^b \Phi\left(\frac{w}{b}\right) \left(w^{\gamma+1} - bw^\gamma \right) f'(w)\, dw \right] =$$

$$b^{\varepsilon-1} \left[\Phi\left(\frac{x}{b}\right) x^{\gamma+1} f(x) - \int_0^x f(w)\, d\left(\Phi\left(\frac{w}{b}\right) w^{\gamma+1} \right) - \right.$$

$$\left. \Phi\left(\frac{x}{b}\right) \left(x^{\gamma+1} - bx^\gamma \right) f(x) - \int_x^b f(w)\, d\left(\Phi\left(\frac{w}{b}\right) \left(w^{\gamma+1} - bw^\gamma \right) \right) \right] =$$

$$b^{\varepsilon-1} \left[bx^\gamma \Phi\left(\frac{x}{b}\right) f(x) - \int_0^x f(w) \left[\frac{1}{b} \Phi'\left(\frac{w}{b}\right) w^{\gamma+1} + (\gamma+1) \Phi\left(\frac{w}{b}\right) w^\gamma \right] dw - \right.$$
$$(20.12)$$

$$\left. \int_x^b f(w) \left[\frac{1}{b} \Phi'\left(\frac{w}{b}\right) \left(w^{\gamma+1} - bw^\gamma \right) + \Phi\left(\frac{w}{b}\right) \left((\gamma+1) w^\gamma - b\gamma w^{\gamma-1} \right) \right] dw \right] =$$

$$b^{\varepsilon-1} \left[bx^\gamma \Phi\left(\frac{x}{b}\right) f(x) - \frac{1}{b} \int_0^x f(w) \Phi'\left(\frac{w}{b}\right) w^{\gamma+1} dw - \right.$$

$$(\gamma+1) \int_0^x f(w) \Phi\left(\frac{w}{b}\right) w^\gamma dw - \int_0^b f(w) \left[\frac{1}{b} \Phi'\left(\frac{w}{b}\right) \left(w^{\gamma+1} - bw^\gamma \right) + \right.$$

$$\left. \Phi\left(\frac{w}{b}\right) \left((\gamma+1) w^\gamma - b\gamma w^{\gamma-1} \right) \right] dw + \int_0^x f(w) \left[\frac{1}{b} \Phi'\left(\frac{w}{b}\right) \left(w^{\gamma+1} - bw^\gamma \right) + \right.$$
$$(20.13)$$

$$\left. \Phi\left(\frac{w}{b}\right) \left((\gamma+1) w^\gamma - b\gamma w^{\gamma-1} \right) \right] dw \right] =$$

$$b^{\varepsilon-1}\left[bx^\gamma\Phi\left(\frac{x}{b}\right)f(x)-\frac{1}{b}\int_0^b f(w)\,\Phi'\left(\frac{w}{b}\right)w^{\gamma+1}dw+\int_0^b f(w)\,\Phi'\left(\frac{w}{b}\right)w^\gamma dw-\right.$$

$$(\gamma+1)\int_0^b f(w)\,\Phi\left(\frac{w}{b}\right)w^\gamma dw+b\gamma\int_0^b f(w)\,\Phi\left(\frac{w}{b}\right)w^{\gamma-1}dw-$$

$$\left.\int_0^x f(w)\,\Phi'\left(\frac{w}{b}\right)w^\gamma dw-b\gamma\int_0^x f(w)\,\Phi\left(\frac{w}{b}\right)w^{\gamma-1}dw\right]=:(\eta).\quad(20.14)$$

We notice that

$$-\frac{1}{b}\int_0^b f(w)\,\Phi'\left(\frac{w}{b}\right)w^{\gamma+1}dw=-\left[\int_0^x f(w)\,\Phi'\left(\frac{w}{b}\right)\frac{w}{b}w^\gamma dw+\right.$$

$$\left.\int_x^b f(w)\,\Phi'\left(\frac{w}{b}\right)\frac{(w-b)}{b}w^\gamma dw+\int_x^b f(w)\,\Phi'\left(\frac{w}{b}\right)w^\gamma dw\right]=\quad(20.15)$$

$$-\int_0^b f(w)\,\Phi'\left(\frac{w}{b}\right)P_1(x,w)\,w^\gamma dw-\int_0^b f(w)\,\Phi'\left(\frac{w}{b}\right)w^\gamma dw$$

$$+\int_0^x f(w)\,\Phi'\left(\frac{w}{b}\right)w^\gamma dw.$$

Furthermore we have

$$-\gamma\int_0^b f(w)\,\Phi\left(\frac{w}{b}\right)w^\gamma dw=-\gamma\left[b\int_0^x f(w)\,\Phi\left(\frac{w}{b}\right)\frac{w}{b}w^{\gamma-1}dw+\right.$$

$$\left.b\int_x^b f(w)\,\Phi\left(\frac{w}{b}\right)\frac{(w-b)}{b}w^{\gamma-1}dw+b\int_x^b f(w)\,\Phi\left(\frac{w}{b}\right)w^{\gamma-1}dw\right]=$$

$$\text{(20.16)}$$

$$-b\gamma\int_0^b f(w)\,\Phi\left(\frac{w}{b}\right)P_1(x,w)\,w^{\gamma-1}dw-b\gamma\int_0^b f(w)\,\Phi\left(\frac{w}{b}\right)w^{\gamma-1}dw$$

$$+b\gamma\int_0^x f(w)\,\Phi\left(\frac{w}{b}\right)w^{\gamma-1}dw.$$

Putting together (20.10), (20.14)–(20.16) we obtain

$$I_\Phi^{\gamma,\varepsilon}\left(P_1(x,b)\,f'(b)\right)=(\eta)=$$

$$b^{\varepsilon-1}\left[bx^\gamma\Phi\left(\frac{x}{b}\right)f(x)-\int_0^b f(w)\,\Phi'\left(\frac{w}{b}\right)P_1(x,w)\,w^\gamma dw-\right.\qquad(20.17)$$

$$\int_0^b f(w) \, \Phi\left(\frac{w}{b}\right) w^\gamma dw - b\gamma \int_0^b f(w) \, \Phi\left(\frac{w}{b}\right) P_1(x, w) \, w^{\gamma-1} dw \Bigg] =$$

$$b^{\varepsilon-1}\left[bx^\gamma \Phi\left(\frac{x}{b}\right) f(x) - \frac{1}{b^\varepsilon} I_{\Phi'}^{\gamma,\varepsilon}(P_1(x, b) \, f(b))\right.$$

$$\left. - \frac{1}{b^\varepsilon} I_\Phi^{\gamma,\varepsilon} f(b) - \gamma b^{1-\varepsilon} I_\Phi^{\gamma-1,\varepsilon}(P_1(x, b) \, f(b))\right] = \qquad (20.18)$$

$$b^\varepsilon x^\gamma \Phi\left(\frac{x}{b}\right) f(x) - \frac{1}{b} I_{\Phi'}^{\gamma,\varepsilon}(P_1(x, b) \, f(b)) - \frac{1}{b} I_\Phi^{\gamma,\varepsilon} f(b) - \gamma I_\Phi^{\gamma-1,\varepsilon}(P_1(x, b) \, f(b)).$$

That is

$$I_\Phi^{\gamma,\varepsilon}(P_1(x, b) \, f'(b)) = b^\varepsilon x^\gamma \Phi\left(\frac{x}{b}\right) f(x) -$$

$$\frac{1}{b} I_{\Phi'}^{\gamma,\varepsilon}(P_1(x, b) \, f(b)) - \frac{1}{b} I_\Phi^{\gamma,\varepsilon} f(b) - \gamma I_\Phi^{\gamma-1,\varepsilon}(P_1(x, b) \, f(b)). \qquad (20.19)$$

Solving the last (20.19) for $f(x)$ we get

$$f(x) = b^{-\varepsilon} x^{-\gamma} \left(\Phi\left(\frac{x}{b}\right)\right)^{-1}\left[\frac{1}{b} I_\Phi^{\gamma,\varepsilon} f(b) + \gamma I_\Phi^{\gamma-1,\varepsilon}(P_1(x, b) \, f(b)) +\right.$$

$$\left.\frac{1}{b} I_{\Phi'}^{\gamma,\varepsilon}(P_1(x, b) \, f(b)) + I_\Phi^{\gamma,\varepsilon}(P_1(x, b) \, f'(b))\right], \qquad (20.20)$$

proving the claim.   ∎

Next we establish a very general fractional Ostrowski type inequality.

Theorem 20.3 *Here all as in Theorem 20.2. Then*

$$\left|f(x) - b^{\gamma+1-\delta} x^{-\gamma} \left(\Phi\left(\frac{x}{b}\right)\right)^{-1}\left[\frac{1}{b} I_\Phi^{\gamma,\delta} f(b) +\right.\right.$$

$$\left.\left.\gamma I_\Phi^{\gamma-1,\delta}(P_1(x, b) \, f(b)) + \frac{1}{b} I_{\Phi'}^{\gamma,\delta}(P_1(x, b) \, f(b))\right]\right| \le$$

$$b^{-1} x^{-\gamma} \left(\Phi\left(\frac{x}{b}\right)\right)^{-1} \|\Phi\|_{\infty,[0,1]} \|f'\|_{\infty,[0,b]} \left[\frac{\left(2x^{\gamma+2} - b^{\gamma+2}\right)}{\gamma+2} + \frac{b\left(b^{\gamma+1} - x^{\gamma+1}\right)}{\gamma+1}\right].$$
$$(20.21)$$

Proof We observe that

$$\left|I_\Phi^{\gamma,\delta}(P_1(x, b) \, f'(b))\right| = \left|I_\Phi^{\gamma,\varepsilon}(P_1(x, b) \, f'(b))\right| = \qquad (20.22)$$

$$b^\varepsilon \left| \int_0^b \Phi\left(\frac{w}{b}\right) w^\gamma P_1(x, w) f'(w) dw \right| \le b^\varepsilon \int_0^b \Phi\left(\frac{w}{b}\right) w^\gamma |P_1(x, w)| |f'(w)| dw \le$$

$$b^\varepsilon \|\Phi\|_{\infty,[0,1]} \|f'\|_{\infty,[0,b]} \int_0^b w^\gamma |P_1(x, w)| dw = \qquad (20.23)$$

$$b^\varepsilon \|\Phi\|_{\infty,[0,1]} \|f'\|_{\infty,[0,b]} \left[\frac{1}{b} \int_0^x w^{\gamma+1} dw + \frac{1}{b} \int_x^b w^\gamma (b - w) dw \right] =$$

$$b^{\varepsilon-1} \|\Phi\|_{\infty,[0,1]} \|f'\|_{\infty,[0,b]} \left[\frac{2x^{\gamma+2}}{\gamma+2} + \frac{b}{\gamma+1}\left(b^{\gamma+1} - x^{\gamma+1}\right) - \frac{b^{\gamma+2}}{\gamma+2}\right].$$
$$\qquad (20.24)$$

That is we derived

$$\left| I_\Phi^{\gamma,\delta}\left(P_1(x, b) f'(b)\right) \right| \le$$

$$b^{\delta-\gamma-2} \|\Phi\|_{\infty,[0,1]} \|f'\|_{\infty,[0,b]} \left[\frac{\left(2x^{\gamma+2} - b^{\gamma+2}\right)}{\gamma+2} + \frac{b\left(b^{\gamma+1} - x^{\gamma+1}\right)}{\gamma+1}\right]. \quad (20.25)$$

The claim is proved.   ∎

20.3 Applications

We mention

Definition 20.4 Let $\alpha > 0$, $\beta, \eta \in \mathbb{R}$, then the Saigo fractional integral $I_{0,t}^{\alpha,\beta,\eta}$ of order α for $f \in C(\mathbb{R}_+)$ is defined by ([13], see also [7, p.19], [12]):

$$I_{0,t}^{\alpha,\beta,\eta}\{f(t)\} = \frac{t^{-\alpha-\beta}}{\Gamma(\alpha)} \int_0^t (t - \tau)^{\alpha-1} {}_2F_1\left(\alpha + \beta, -\eta; \alpha; 1 - \frac{\tau}{t}\right) f(\tau) d\tau,$$
$$\qquad (20.26)$$

where the function ${}_2F_1$ in (20.26) is the Gaussian hypergeometric function defined by

$${}_2F_1(a, b; c; t) = \sum_{n=0}^\infty \frac{(a)_n (b)_n}{(c)_n} \frac{t^n}{n!}, \qquad (20.27)$$

and $(a)_n$ is the Pochhammer symbol $(a)_n = a(a+1)\ldots(a+n-1)$, $(a)_0 = 1$; where $c \ne 0, -1, -2, \ldots$.

Note 20.5 *Given that $a + b < c$, ${}_2F_1$ converges on $[-1, 1]$, see [3].*

Furthermore we have

$$\frac{d\,_2F_1\,(a,b;\,c;\,t)}{dt} = \left(\frac{ab}{c}\right)\,_2F_1\,(a+1,b+1;\,c+1;\,t)\,, \qquad (20.28)$$

which converges on $[-1, 1]$ *when* $1 + a + b < c$*. So when* $1 + a + b < c$*, then both (20.27) and (20.28) converge on* $[-1, 1]$*. Therefore when* $\eta > 1 + \beta$ *we get that both* $_2F_1\left(\alpha + \beta, -\eta; \alpha; 1 - \frac{\tau}{t}\right)$ *and its derivative with respect to* τ : $\left(\frac{(\alpha+\beta)\eta}{t\alpha}\right)$ $_2F_1\left(\alpha + \beta + 1, -\eta + 1; \alpha + 1; 1 - \frac{\tau}{t}\right)$*, converge on* $[0, 1]$*; notice here* $0 \le 1 - \frac{\tau}{t} \le 1$*,* $t > 0$*.*

Remark 20.6 The integral operator $I_{0,t}^{\alpha,\beta,\eta}$ includes both the Riemann-Liouville and the Erdélyi-Kober fractional integral operators given by

$$J_0^\alpha\{f(x)\} = I_{0,t}^{\alpha,-\alpha,\eta}\{f(t)\} = \frac{1}{\Gamma(\alpha)} \int_0^t (t-\tau)^{\alpha-1} f(\tau)\, d\tau \, (\alpha > 0), \qquad (20.29)$$

and

$$I^{\alpha,\eta}\{f(t)\} = I_{0,t}^{\alpha,0,\eta}\{f(t)\} = \frac{t^{-\alpha-\eta}}{\Gamma(\alpha)} \int_0^t (t-\tau)^{\alpha-1} \tau^\eta f(\tau)\, d\tau (\alpha > 0, \eta \in \mathbb{R}).$$
$$\qquad (20.30)$$

Remark 20.7 By a simple change of variable $(w = \frac{\tau}{t})$ we get

$$I_{0,t}^{\alpha,\beta,\eta}\{f(t)\} = \frac{t^{-\beta}}{\Gamma(\alpha)} \int_0^1 (1-w)^{\alpha-1}\,_2F_1\,(\alpha + \beta, -\eta; \alpha; 1 - w)\, f(tw)\, dw. \qquad (20.31)$$

Similarly we find

$$J_0^\alpha\{f(t)\} = \frac{t^\alpha}{\Gamma(\alpha)} \int_0^1 (1-w)^{\alpha-1} f(tw)\, dw, \qquad (20.32)$$

and

$$I^{\alpha,\eta}\{f(t)\} = \frac{1}{\Gamma(\alpha)} \int_0^1 (1-w)^{\alpha-1} w^\eta f(tw)\, dw. \qquad (20.33)$$

Remark 20.8 [9] The above Saigo fractional integral (20.26) and its special cases of Riemann-Liouville and Erdélyi-Kober fractional integrals (20.29), (20.30), are all examples of the Kalla [6] generalized fractional integral in the reduced form

$$K_\Phi^\gamma f(x) = x^{-\gamma-1} \int_0^x \Phi\left(\frac{w}{x}\right) w^\gamma f(w)\, dw = \int_0^1 \Phi(\sigma) \sigma^\gamma f(x\sigma)\, d\sigma, \qquad (20.34)$$

where $x > 0$, $\gamma > -1$ and Φ continuous arbitrary Kernel function.

Notice that (by (20.6) and (20.34))

$$I_\Phi^{\gamma,\delta} f(x) = x^\delta K_\Phi^\gamma f(x),$$ (20.35)

for any $x > 0$, where $\gamma > -1$ and $\delta \in \mathbb{R}$.

So for $b > 0$ we get

$$I_\Phi^{\gamma,\delta} f(b) = b^\delta K_\Phi^\gamma f(b).$$ (20.36)

Next we restrict ourselves to $\gamma > 0$. By Theorem 20.2 and (20.36) we obtain the following general fractional representation formula

Theorem 20.9 *It holds*

$$f(x) = b^{\gamma+1-\delta} x^{-\gamma} \left(\Phi\left(\frac{x}{b}\right) \right)^{-1} \left[b^{\delta-1} K_\Phi^\gamma f(b) + \gamma b^\delta K_\Phi^{\gamma-1} (P_1(x,b) f(b)) + \right.$$

$$\left. b^{\delta-1} K_{\Phi'}^\gamma (P_1(x,b) f(b)) + b^\delta K_\Phi^\gamma \left(P_1(x,b) f'(b) \right) \right].$$ (20.37)

We finish the following very general fractional Ostrowski type inequality, a direct application of (20.21) and (20.36).

Theorem 20.10 *All as in Theorem 20.3. Then*

$$\left| f(x) - b^{\gamma+1-\delta} x^{-\gamma} \left(\Phi\left(\frac{x}{b}\right) \right)^{-1} \left[b^{\delta-1} K_\Phi^\gamma f(b) + \right. \right.$$ (20.38)

$$\left. \left. \gamma b^\delta K_\Phi^{\gamma-1} (P_1(x,b) f(b)) + b^{\delta-1} K_{\Phi'}^\gamma (P_1(x,b) f(b)) \right] \right| \le$$

$$b^{-1} x^{-\gamma} \left(\Phi\left(\frac{x}{b}\right) \right)^{-1} \|\Phi\|_{\infty,[0,1]} \|f'\|_{\infty,[0,b]} \left[\frac{\left(2x^{\gamma+2} - b^{\gamma+2} \right)}{\gamma+2} + \frac{b\left(b^{\gamma+1} - x^{\gamma+1} \right)}{\gamma+1} \right].$$

Comment 20.11 *One can apply (20.37) and (20.38) for the Riemann-Liouville and Erdélyi-Kober fractional integrals, as well as many other fractional integrals. To keep chapter short we omit this task.*

References

1. G. Anastassiou, M. Hooshmandasl, A. Ghasemi, F. Moftakharzadeh, Montgomery identities for fractional integrals and related fractional inequalities. J. Inequal. Pure Appl. Math. **10**(4), article no 97, 6 (2009)
2. G.A. Anastassiou, Most general fractional representation formula for functions and implications. Serelica Math. **40**, 89–98 (2014)

3. E.A. Chistova (originator), Hypergeometric function, Encyclopedia of Mathematics, http://www.encyclopediaofmath.org/index.php?title=Hypergeometric_function&oldid=12873, ISBN 1402006098
4. S.L. Kalla, On operators of fractional integration. I. Mat. Notae **22**, 89–93 (1970)
5. S.L. Kalla, On operators of fractional integration. II Mat. Notae **25**, 29–35 (1976)
6. S.L. Kalla, Operators of fractional integration. In: *Proceedings of Conference on Analytic Functions, Kozubnik 1979*, Lecture Notes of Mathematic, vol. 798, pp. 258–280 (Springer, Berlin, 1980)
7. V.S. Kiryakova, *Generalized Fractional Calculus and Applications*, Pitman Research Notes in Mathematics Series, vol. 301 (Longman Scientific & Technical, Harlow, 1994)
8. V.S. Kiryakova, All the special functions are fractional differintegrals of elementary functions. J. Phys. A: Math. Gen. **30**, 5085–5103 (1997)
9. V.S. Kiryakova, On two Saigo's fractional integral operators in the class of univalent functions. Fractional Calculus & Applied Analysis **9**(2), 159–176 (2006)
10. S. Miller, B. Ross, *An Introduction to the Fractional Calculus and Fractional Differential Equations* (Wiley, USA, 1993)
11. D.S. Mitrinovic, J.E. Pecaric, A.M. Fink, *Inequalities for functions and their integrals and derivatives* (Kluwer Academic Publishers, Dordrecht, 1994)
12. R.K. Raina, Solution of Abel-type integral equation involving the Appell hypergeometric function. Integr. Transform. Spec. Funct. **21**(7), 515–522 (2010)
13. M. Saigo, A remark on integral operators involving the Gauss hypergeometric functions. Math. Rep. Kyushu Univ. **11**, 135–143 (1978)

Chapter 21
Rational Inequalities for Integral Operators Using Convexity

Here we present integral inequalities for convex and increasing functions applied to products of ratios of functions and other important mixtures. As applications we derive a wide range of fractional inequalities of Hardy type. They involve the left and right Riemann-Liouville fractional integrals and their generalizations, in particular the Hadamard fractional integrals. Also inequalities for Riemann-Liouville, Caputo, Canavati and their generalizations fractional derivatives. These application inequalities are of L_p type, $p \geq 1$, exponential type and of other general forms. It follows [5].

21.1 Introduction

Let $(\Omega_1, \Sigma_1, \mu_1)$ and $(\Omega_2, \Sigma_2, \mu_2)$ be measure spaces with positive σ-finite measures, and let $k_i : \Omega_1 \times \Omega_2 \to \mathbb{R}$ be nonnegative measurable functions, $k_i(x, \cdot)$ measurable on Ω_2 and

$$K_i(x) = \int_{\Omega_2} k_i(x, y) \, d\mu_2(y), \quad \text{for any } x \in \Omega_1, \tag{21.1}$$

$i = 1, \ldots, m$. We assume that $K_i(x) > 0$ a.e. on Ω_1, and the weight functions are nonnegative measurable functions on the related set.

We consider measurable functions $g_i : \Omega_1 \to \mathbb{R}$ with the representation

$$g_i(x) = \int_{\Omega_2} k_i(x, y) f_i(y) \, d\mu_2(y), \tag{21.2}$$

where $f_i : \Omega_2 \to \mathbb{R}$ are measurable functions, $i = 1, \ldots, m$.

Here u stands for a weight function on Ω_1.

In [1] we proved the following general result.

© Springer International Publishing Switzerland 2016
G.A. Anastassiou, *Intelligent Comparisons: Analytic Inequalities*,
Studies in Computational Intelligence 609,
DOI 10.1007/978-3-319-21121-3_21

Theorem 21.1 *Let $j \in \{1, \ldots, m\}$ be fixed. Assume that the function*

$$x \mapsto \left(\frac{u(x) \prod\limits_{i=1}^{m} k_i(x, y)}{\prod\limits_{i=1}^{m} K_i(x)} \right)$$

is integrable on Ω_1 for each fixed $y \in \Omega_2$. Define λ_m on Ω_2 by

$$\lambda_m(y) := \int_{\Omega_1} \left(\frac{u(x) \prod\limits_{i=1}^{m} k_i(x, y)}{\prod\limits_{i=1}^{m} K_i(x)} \right) d\mu_1(x) < \infty. \tag{21.3}$$

If $\Phi_i : \mathbb{R}_+ \to \mathbb{R}_+$, $i = 1, \ldots, m$, are convex and increasing functions.
Then

$$\int_{\Omega_1} u(x) \prod_{i=1}^{m} \Phi_i \left(\left| \frac{g_i(x)}{K_i(x)} \right| \right) d\mu_1(x) \leq \tag{21.4}$$

$$\left(\prod_{\substack{i=1 \\ (i \neq j)}}^{m} \int_{\Omega_2} \Phi_i(|f_i(y)|) d\mu_2(y) \right) \left(\int_{\Omega_2} \Phi_j(|f_j(y)|) \lambda_m(y) d\mu_2(y) \right),$$

true for all measurable functions, $i = 1, \ldots, m$, $f_i : \Omega_2 \to \mathbb{R}$ such that:

(i) f_i, $\Phi_i(|f_i|)$ are both $k_i(x, y) d\mu_2(y)$ -integrable, μ_1 -a.e. in $x \in \Omega_1$,

(ii) $\lambda_m \Phi_j(|f_j|)$; $\Phi_1(|f_1|)$, $\Phi_2(|f_2|)$, $\Phi_3(|f_3|)$, $\ldots$, $\widetilde{\Phi_j(|f_j|)}$, $\ldots$, $\Phi_m(|f_m|)$ are all μ_2 -integrable,

and for all corresponding functions g_i given by (21.2). Above $\widetilde{\Phi_j(|f_j|)}$ means missing item.

Here $\mathbb{R}^* := \mathbb{R} \cup \{\pm\infty\}$. Let $\varphi : \mathbb{R}^{*2} \to \mathbb{R}^*$ be a Borel measurable function. Let $f_{1i}, f_{2i} : \Omega_2 \to \mathbb{R}$ be measurable functions, $i = 1, \ldots, m$.

The function $\varphi(f_{1i}(y), f_{2i}(y))$, $y \in \Omega_2$, $i = 1, \ldots, m$, is Σ_2-measurable. In this chapter we assume that $0 < \varphi(f_{1i}(y), f_{2i}(y)) < \infty$, a.e., $i = 1, \ldots, m$.

We consider

$$f_{3i}(y) := \frac{f_{1i}(y)}{\varphi(f_{1i}(y), f_{2i}(y))}, \tag{21.5}$$

$i = 1, \ldots, m$, $y \in \Omega_2$, which is a measurable function.

We also consider here

$$k_i^* (x, y) := k_i (x, y) \varphi (f_{1i} (y), f_{2i} (y)), \qquad (21.6)$$

$y \in \Omega_2, i = 1, \ldots, m$, which is a nonnegative a.e. measurable function on $\Omega_1 \times \Omega_2$. We have that $k_i^* (x, \cdot)$ is measurable on $\Omega_2, i = 1, \ldots, m$.
Denote by

$$K_i^* (x) := \int_{\Omega_2} k_i^* (x, y) \, d\mu_2 (y) \qquad (21.7)$$

$$= \int_{\Omega_2} k_i (x, y) \varphi (f_{1i} (y), f_{2i} (y)) \, d\mu_2 (y), \quad i = 1, \ldots, m.$$

We assume that $K_i^* (x) > 0$, a.e. on Ω_1.
So here the function

$$g_{1i} (x) = \int_{\Omega_2} k_i (x, y) f_{1i} (y) \, d\mu_2 (y)$$

$$= \int_{\Omega_2} k_i (x, y) \varphi (f_{1i} (y), f_{2i} (y)) \left(\frac{f_{1i} (y)}{\varphi (f_{1i} (y), f_{2i} (y))} \right) d\mu_2 (y)$$

$$= \int_{\Omega_2} k_i^* (x, y) f_{3i} (y) \, d\mu_2 (y), \quad i = 1, \ldots, m. \qquad (21.8)$$

A typical example is when

$$\varphi (f_{1i} (y), f_{2i} (y)) = f_{2i} (y), \quad i = 1, \ldots, m, y \in \Omega_2. \qquad (21.9)$$

In that case we have that

$$f_{3i} (y) = \frac{f_{1i} (y)}{f_{2i} (y)}, \quad i = 1, \ldots, m, \ y \in \Omega_2. \qquad (21.10)$$

Tha latter case was studied in [14], for $i = 1$, which is an article with interesting ideas however containing several mistakes.
In the special case (21.10) we get that

$$K_i^* (x) = g_{2i} (x) := \int_{\Omega_2} k_i (x, y) f_{2i} (y) \, d\mu_2 (y), \quad i = 1, \ldots, m. \qquad (21.11)$$

In this chapter we get first general results by applying Theorem 21.1 for (f_{3i}, g_{1i}), $i = 1, \ldots, m$, and on other various important settings, then we give wide applications to Fractional Calculus.

21.2 Main Results

We present

Theorem 21.2 *Let $j \in \{1, \ldots, m\}$ be fixed. Assume that the function*

$$x \mapsto \left(\frac{u(x) \prod\limits_{i=1}^{m} k_i(x, y)\, \varphi(f_{1i}(y), f_{2i}(y))}{\prod\limits_{i=1}^{m} K_i^*(x)} \right)$$

is integrable on Ω_1, for each $y \in \Omega_2$. Define λ_m^* on Ω_2 by

$$\lambda_m^*(y) := \left(\prod_{i=1}^{m} \varphi(f_{1i}(y), f_{2i}(y)) \right) \int_{\Omega_1} \left(\frac{u(x) \prod\limits_{i=1}^{m} k_i(x, y)}{\prod\limits_{i=1}^{m} K_i^*(x)} \right) d\mu_1(x) < \infty.$$

$$(21.12)$$

Here $\Phi_i : \mathbb{R}_+ \to \mathbb{R}_+$, $i = 1, \ldots, m$, are convex and increasing functions.
   Then

$$\int_{\Omega_1} u(x) \prod_{i=1}^{m} \Phi_i \left(\left| \frac{g_{1i}(x)}{K_i^*(x)} \right| \right) d\mu_1(x) \leq \tag{21.13}$$

$$\left(\prod_{\substack{i=1 \\ i \neq j}}^{m} \int_{\Omega_2} \Phi_i \left(\left| \frac{f_{1i}(y)}{\varphi(f_{1i}(y), f_{2i}(y))} \right| \right) d\mu_2(y) \right) \cdot$$

$$\left(\int_{\Omega_2} \Phi_j \left(\left| \frac{f_{1j}(y)}{\varphi(f_{1j}(y), f_{2j}(y))} \right| \right) \lambda_m^*(y)\, d\mu_2(y) \right),$$

true for all measurable functions, $i = 1, \ldots, m$, $f_{1i}, f_{2i} : \Omega_2 \to \mathbb{R}$ such that

(i) $\left(\frac{f_{1i}(y)}{\varphi(f_{1i}(y), f_{2i}(y))} \right)$, $\Phi_i \left(\left| \frac{f_{1i}(y)}{\varphi(f_{1i}(y), f_{2i}(y))} \right| \right)$ *are both $k_i(x, y)\, \varphi(f_{1i}(y), f_{2i}(y))$*
   $d\mu_2(y)$ *-integrable, μ_1 -a.e. in $x \in \Omega_1$,*

(ii) $\lambda_m^* \Phi_j \left(\left| \frac{f_{1j}(y)}{\varphi(f_{1j}(y), f_{2j}(y))} \right| \right)$ *and $\Phi_i \left(\left| \frac{f_{1i}(y)}{\varphi(f_{1i}(y), f_{2i}(y))} \right| \right)$, for $i \in \{1, \ldots, m\} - \{j\}$*
   are all μ_2 -integrable

and for all corresponding functions g_{1i} given by (21.8).

Proof Direct application of Theorem 21.1 on the setting described at introduction. ∎

In the special case of (21.9)–(21.11) we derive

Theorem 21.3 *Here* $0 < f_{2i}(y) < \infty$, *a.e.*, $i = 1, \ldots, m$. *Let* $j \in \{1, \ldots, m\}$ *be fixed. Assume that the function*

$$x \mapsto \left(\frac{u(x) \prod\limits_{i=1}^{m} k_i(x, y) f_{2i}(y)}{\prod\limits_{i=1}^{m} g_{2i}(x)} \right)$$

is integrable on Ω_1, *for each* $y \in \Omega_2$. *Define* λ_m^{**} *on* Ω_2 *by*

$$\lambda_m^{**}(y) := \left(\prod_{i=1}^{m} f_{2i}(x) \right) \int_{\Omega_1} \left(\frac{u(x) \prod\limits_{i=1}^{m} k_i(x, y)}{\prod\limits_{i=1}^{m} g_{2i}(x)} \right) d\mu_1(x) < \infty.$$

Here $\Phi_i : \mathbb{R}_+ \to \mathbb{R}_+$, $i = 1, \ldots, m$, *are convex and increasing functions. Then*

$$\int_{\Omega_1} u(x) \prod_{i=1}^{m} \Phi_i \left(\left| \frac{g_{1i}(x)}{g_{2i}(x)} \right| \right) d\mu_1(x) \leq \qquad (21.14)$$

$$\left(\prod_{\substack{i=1 \\ i \neq j}}^{m} \int_{\Omega_2} \Phi_i \left(\left| \frac{f_{1i}(y)}{f_{2i}(y)} \right| \right) d\mu_2(y) \right) \left(\int_{\Omega_2} \Phi_j \left(\left| \frac{f_{1j}(y)}{f_{2j}(y)} \right| \right) \lambda_m^{**}(y) d\mu_2(y) \right),$$

true for all measurable functions, $i = 1, \ldots, m$, $f_{1i}, f_{2i} : \Omega_2 \to \mathbb{R}$ *such that*

(i) $\frac{f_{1i}(y)}{f_{2i}(y)}$, $\Phi_i \left(\left| \frac{f_{1i}(y)}{f_{2i}(y)} \right| \right)$ *are both* $k_i(x, y) f_{2i}(y) d\mu_2(y)$ *-integrable,* μ_1 *-a.e. in* $x \in \Omega_1$,

(ii) $\lambda_m^{**} \Phi_j \left(\left| \frac{f_{1j}(y)}{f_{2j}(y)} \right| \right)$, *and* $\Phi_i \left(\left| \frac{f_{1i}(y)}{f_{2i}(y)} \right| \right)$, *for* $i \in \{1, \ldots, m\} - \{j\}$, *are all* μ_2 *-integrable*

and for all corresponding functions g_{1i} *given by (21.8), and* g_{2i} *given by (21.11).*

Proof By Theorem 21.2. ∎

Theorem 21.3 generalizes and fixes Theorem 1.2 of [14], which inspired the current chapter.

Next we consider the case of $\varphi(s_i, t_i) = |a_{1i} s_i + a_{2i} t_i|^r$, where $r \in \mathbb{R}$; $s_i, t_i \in \mathbb{R}^*$; $a_{1i}, a_{2i} \in \mathbb{R}$, $i = 1, \ldots, m$. We assume here that

$$0 < |a_{1i} f_{1i}(y) + a_{2i} f_{2i}(y)|^r < \infty, \tag{21.15}$$

a.e., $i = 1, \ldots, m$.

We further assume that

$$\overline{K_i^*}(x) := \int_{\Omega_2} k_i(x, y) |a_{1i} f_{1i}(y) + a_{2i} f_{2i}(y)|^r \, d\mu_2(y) > 0, \tag{21.16}$$

a.e. on Ω_1, $i = 1, \ldots, m$.

Here we have

$$f_{3i}(y) = \frac{f_{1i}(y)}{|a_{1i} f_{1i}(y) + a_{2i} f_{2i}(y)|^r}, \tag{21.17}$$

$i = 1, \ldots, m$, $y \in \Omega_2$.

Denote by

$$\overline{k_i^*}(x, y) := k_i(x, y) |a_{1i} f_{1i}(y) + a_{2i} f_{2i}(y)|^r, \tag{21.18}$$

$i = 1, \ldots, m$.

By Theorem 21.2 we obtain

Theorem 21.4 *Let $j \in \{1, \ldots, m\}$ be fixed. Assume that the function*

$$x \mapsto \left(\frac{u(x) \prod_{i=1}^{m} k_i(x, y) |a_{1i} f_{1i}(y) + a_{2i} f_{2i}(y)|^r}{\prod_{i=1}^{m} \overline{K_i^*}(x)} \right)$$

is integrable on Ω_1, for each $y \in \Omega_2$. Define $\overline{\lambda_m^*}$ on Ω_2 by

$$\overline{\lambda_m^*}(y) := \left(\prod_{i=1}^{m} |a_{1i} f_{1i}(y) + a_{2i} f_{2i}(y)|^r \right) \int_{\Omega_1} \left(\frac{u(x) \prod_{i=1}^{m} k_i(x, y)}{\prod_{i=1}^{m} \overline{K_i^*}(x)} \right) d\mu_1(x) < \infty. \tag{21.19}$$

Here $\Phi_i : \mathbb{R}_+ \to \mathbb{R}_+$, $i = 1, \ldots, m$, are convex and increasing functions.
Then

$$\int_{\Omega_1} u(x) \prod_{i=1}^{m} \Phi_i \left(\left| \frac{g_{1i}(x)}{\overline{K_i^*}(x)} \right| \right) d\mu_1(x) \leq \tag{21.20}$$

$$\left(\prod_{\substack{i=1 \\ i \neq j}}^{m} \int_{\Omega_2} \Phi_i \left(\left| \frac{f_{1i}(y)}{(a_{1i} f_{1i}(y) + a_{2i} f_{2i}(y))^r} \right| \right) d\mu_2(y) \right).$$

$$\left(\int_{\Omega_2} \Phi_j\left(\left|\frac{f_{1j}(y)}{\left(a_{1j}f_{1j}(y)+a_{2j}f_{2j}(y)\right)^r}\right|\right)\overline{\lambda_m^*}(y)\,d\mu_2(y)\right),$$

true for all measurable functions, $i = 1, \ldots, m$, f_{1i}, $f_{2i} : \Omega_2 \to \mathbb{R}$ such that

(i) $\left(\frac{f_{1i}(y)}{|a_{1i}f_{1i}(y)+a_{2i}f_{2i}(y)|^r}\right)$, $\Phi_i\left(\left|\frac{f_{1i}(y)}{\left(a_{1i}f_{1i}(y)+a_{2i}f_{2i}(y)\right)^r}\right|\right)$ *are both*
$k_i(x, y)\,|a_{1i}f_{1i}(y)+a_{2i}f_{2i}(y)|^r\,d\mu_2(y)$ *-integrable, μ_1 -a.e. in $x \in \Omega_1$,*

(ii) $\overline{\lambda_m^*}\,\Phi_j\left(\left|\frac{f_{1j}(y)}{\left(a_{1j}f_{1j}(y)+a_{2j}f_{2j}(y)\right)^r}\right|\right)$ *and* $\Phi_i\left(\left|\frac{f_{1i}(y)}{\left(a_{1i}f_{1i}(y)+a_{2i}f_{2i}(y)\right)^r}\right|\right)$, *for $i \in \{1,$*
$\ldots, m\} - \{j\}$ are all μ_2 -integrable

and for all corresponding functions g_{1i} given by (21.8).

In Theorem 21.4 of great interest is the case of $r \in \mathbb{Z} - \{0\}$ and $a_{1i} = a_{2i} = 1$, all $i = 1, \ldots, m$; or $a_{1i} = 1$, $a_{2i} = -1$, all $i = 1, \ldots, m$.

Another interesting case arises when

$$\varphi\left(f_{1i}(y), f_{2i}(y)\right) := |f_{1i}(y)|^{r_1}\,|f_{2i}(y)|^{r_2}, \tag{21.21}$$

$i = 1, \ldots, m$, where $r_1, r_2 \in \mathbb{R}$. We assume that

$$0 < |f_{1i}(y)|^{r_1}\,|f_{2i}(y)|^{r_2} < \infty, a.e., i = 1, \ldots, m. \tag{21.22}$$

In this case

$$f_{3i}(y) = \frac{f_{1i}(y)}{|f_{1i}(y)|^{r_1}\,|f_{2i}(y)|^{r_2}}, \tag{21.23}$$

$i = 1, \ldots, m$, $y \in \Omega_2$, also

$$k_i^*(x, y) = k_{pi}^*(x, y) := k_i(x, y)\,|f_{1i}(y)|^{r_1}\,|f_{2i}(y)|^{r_2}, \tag{21.24}$$

$y \in \Omega_2, i = 1, \ldots, m$.

We have

$$K_i^*(x) = K_{pi}^*(x) := \int_{\Omega_2} k_i(x, y)\,|f_{1i}(y)|^{r_1}\,|f_{2i}(y)|^{r_2}\,d\mu_2(y), \tag{21.25}$$

$i = 1, \ldots, m$.

We assume that $K_{pi}^* > 0$, a.e. on Ω_1.

By Theorem 21.2 we derive

Theorem 21.5 *Let $j \in \{1, \ldots, m\}$ be fixed. Assume that the function*

$$x \mapsto \left(\frac{u(x)\prod_{i=1}^{m} k_i(x, y)\,|f_{1i}(y)|^{r_1}\,|f_{2i}(y)|^{r_2}}{\prod_{i=1}^{m} K_{pi}^*(x)}\right)$$

is integrable on Ω_1, for each $y \in \Omega_2$. Define λ^*_{pm} on Ω_2 by

$$\lambda^*_{pm}(y) := \left(\prod_{i=1}^{m} |f_{1i}(y)|^{r_1} |f_{2i}(y)|^{r_2}\right) \int_{\Omega_1} \left(\frac{u(x) \prod_{i=1}^{m} k_i(x, y)}{\prod_{i=1}^{m} K^*_{pi}(x)}\right) d\mu_1(x) < \infty.$$

$$(21.26)$$

Here $\Phi_i : \mathbb{R}_+ \to \mathbb{R}_+$, $i = 1, \ldots, m$, are convex and increasing functions. Then

$$\int_{\Omega_1} u(x) \prod_{i=1}^{m} \Phi_i \left(\left|\frac{g_{1i}(x)}{K^*_{pi}(x)}\right|\right) d\mu_1(x) \le \qquad (21.27)$$

$$\left(\prod_{\substack{i=1 \\ i \neq j}}^{m} \int_{\Omega_2} \Phi_i \left(|f_{1i}(y)|^{1-r_1} |f_{2i}(y)|^{-r_2}\right) d\mu_2(y)\right) \cdot$$

$$\left(\int_{\Omega_2} \Phi_j \left(|f_{1j}(y)|^{1-r_1} |f_{2j}(y)|^{-r_2}\right) \lambda^*_{pm}(y) \, d\mu_2(y)\right),$$

true for all measurable functions, $i = 1, \ldots, m$, $f_{1i}, f_{2i} : \Omega_2 \to \mathbb{R}$ such that

(i) $\left(\frac{f_{1i}(y)}{|f_{1i}(y)|^{r_1} |f_{2i}(y)|^{r_2}}\right)$, $\Phi_i \left(|f_{1i}(y)|^{1-r_1} |f_{2i}(y)|^{-r_2}\right)$ *are both*
$k_i(x, y) |f_{1i}(y)|^{r_1} |f_{2i}(y)|^{r_2} d\mu_2(y)$ *-integrable, μ_1 -a.e. in $x \in \Omega_1$,*

(ii) $\lambda^*_{pm} \Phi_j \left(|f_{1j}(y)|^{1-r_1} |f_{2j}(y)|^{-r_2}\right)$ *and $\Phi_i \left(|f_{1i}(y)|^{1-r_1} |f_{2i}(y)|^{-r_2}\right)$, for $i \in$*
$\{1, \ldots, m\} - \{j\}$ *are all μ_2 -integrable*

and for all corresponding functions g_{1i} given by (21.8).

In Theorem 21.5 of interest will be the case of $r_1 = 1 - n$, $r_2 = -n$, $n \in \mathbb{N}$. In that case $|f_{1i}(y)|^{1-r_1} |f_{2i}(y)|^{-r_2} = |f_{1i}(y) f_{2i}(y)|^n$, etc.

Next we apply Theorem 21.2 for specific convex functions.

Theorem 21.6 *Let $j \in \{1, \ldots, m\}$ be fixed. Assume that the function*

$$x \mapsto \left(\frac{u(x) \prod_{i=1}^{m} k_i(x, y) \, \varphi(f_{1i}(y), f_{2i}(y))}{\prod_{i=1}^{m} K^*_i(x)}\right)$$

is integrable on Ω_1, for each $y \in \Omega_2$. Define λ_m^* on Ω_2 by

$$\lambda_m^* (y) := \left(\prod_{i=1}^{m} \varphi \left(f_{1i} (y), f_{2i} (y) \right) \right) \int_{\Omega_1} \left(\frac{u (x) \prod_{i=1}^{m} k_i (x, y)}{\prod_{i=1}^{m} K_i^* (x)} \right) d\mu_1 (x) < \infty.$$

(21.28)

Then

$$\int_{\Omega_1} u (x) e^{\sum_{i=1}^{m} \left| \frac{g_{1i} (x)}{K_i^* (x)} \right|} d\mu_1 (x) \leq$$

(21.29)

$$\left(\prod_{\substack{i=1 \\ i \neq j}}^{m} \int_{\Omega_2} e^{\left| \frac{f_{1i} (y)}{\varphi (f_{1i} (y), f_{2i} (y))} \right|} d\mu_2 (y) \right) \left(\int_{\Omega_2} e^{\left| \frac{f_{1j} (y)}{\varphi (f_{1j} (y), f_{2j} (y))} \right|} \lambda_m^* (y) d\mu_2 (y) \right),$$

true for all measurable functions, $i = 1, \ldots, m$, $f_{1i}, f_{2i} : \Omega_2 \to \mathbb{R}$ such that

(i) $\frac{f_{1i} (y)}{\varphi (f_{1i} (y), f_{2i} (y))}$, $e^{\left| \frac{f_{1i} (y)}{\varphi (f_{1i} (y), f_{2i} (y))} \right|}$ *are both $k_i (x, y) \varphi \left(f_{1i} (y), f_{2i} (y) \right) d\mu_2 (y)$*
-integrable, μ_1 -a.e. in $x \in \Omega_1$,

(ii) $\lambda_m^* e^{\left| \frac{f_{1j} (y)}{\varphi (f_{1j} (y), f_{2j} (y))} \right|}$ *and* $e^{\left| \frac{f_{1i} (y)}{\varphi (f_{1i} (y), f_{2i} (y))} \right|}$, *for $i \in \{1, \ldots, m\} - \{j\}$ are all μ_2*
-integrable

and for all corresponding functions g_{1i} given by (21.8).

We continue with

Theorem 21.7 *Let $j \in \{1, \ldots, m\}$ be fixed. Assume that the function*

$$x \mapsto \left(\frac{u (x) \prod_{i=1}^{m} k_i (x, y) \varphi \left(f_{1i} (y), f_{2i} (y) \right)}{\prod_{i=1}^{m} K_i^* (x)} \right)$$

is integrable on Ω_1, for each $y \in \Omega_2$. Define λ_m^* on Ω_2 by

$$\lambda_m^* (y) := \left(\prod_{i=1}^{m} \varphi \left(f_{1i} (y), f_{2i} (y) \right) \right) \int_{\Omega_1} \left(\frac{u (x) \prod_{i=1}^{m} k_i (x, y)}{\prod_{i=1}^{m} K_i^* (x)} \right) d\mu_1 (x) < \infty.$$

(21.30)

Let $p_i \geq 1, i = 1, \ldots, m$.
  Then

$$\int_{\Omega_1} u(x) \prod_{i=1}^m \left| \frac{g_{1i}(x)}{K_i^*(x)} \right|^{p_i} d\mu_1(x) \leq \qquad (21.31)$$

$$\left(\prod_{\substack{i=1 \\ i \neq j}}^m \int_{\Omega_2} \left| \frac{f_{1i}(y)}{\varphi(f_{1i}(y), f_{2i}(y))} \right|^{p_i} d\mu_2(y) \right) \cdot$$

$$\left(\int_{\Omega_2} \left| \frac{f_{1j}(y)}{\varphi(f_{1j}(y), f_{2j}(y))} \right|^{p_j} \lambda_m^*(y) d\mu_2(y) \right),$$

true for all measurable functions, $i = 1, \ldots, m$, $f_{1i}, f_{2i} : \Omega_2 \to \mathbb{R}$ such that

(i) $\left| \frac{f_{1i}(y)}{\varphi(f_{1i}(y), f_{2i}(y))} \right|^{p_i}$ *is $k_i(x, y) \varphi(f_{1i}(y), f_{2i}(y)) d\mu_2(y)$ -integrable, μ_1 -a.e. in $x \in \Omega_1$,*

(ii) $\lambda_m^* \left| \frac{f_{1j}(y)}{\varphi(f_{1j}(y), f_{2j}(y))} \right|^{p_j}$ *and* $\left| \frac{f_{1i}(y)}{\varphi(f_{1i}(y), f_{2i}(y))} \right|^{p_i}$, *for $i \in \{1, \ldots, m\} - \{j\}$ are all μ_2 -integrable*

and for all corresponding functions g_{1i} given by (21.8).

We continue as follows:
Choosing $r_1 = 0, r_2 = -1, i = 1, \ldots, m$, on (21.21) we have that

$$\varphi(f_{1i}(y), f_{2i}(y)) = |f_{2i}(y)|^{-1}. \qquad (21.32)$$

We assume that
$$0 < |f_{2i}(y)|^{-1} < \infty, a.e., i = 1, \ldots, m, \qquad (21.33)$$

which is the same as

$$0 < |f_{2i}(y)| < \infty, \ a.e., i = 1, \ldots, m. \qquad (21.34)$$

In this case
$$f_{3i}(y) = f_{1i}(y) |f_{2i}(y)|, \qquad (21.35)$$

$i = 1, \ldots, m, y \in \Omega_2$, also it is

$$k_{pi}^*(x, y) = \overline{k_{pi}^*}(x, y) := \frac{k_i(x, y)}{|f_{2i}(y)|}, \qquad (21.36)$$

$y \in \Omega_2, i = 1, \ldots, m.$

We have that

$$K_{pi}^*(x) = \overline{K_{pi}^*}(x) := \int_{\Omega_2} \frac{k_i(x, y)}{|f_{2i}(y)|} d\mu_2(y),$$ (21.37)

$i = 1, \ldots, m$.

We assume that $\overline{K_{pi}^*}(x) > 0$, a.e. on Ω_1.

By Theorem 21.5 we obtain

Corollary 21.8 *Let* $j \in \{1, \ldots, m\}$ *be fixed. Assume that the function*

$$x \mapsto \left(\frac{u(x) \prod\limits_{i=1}^m k_i(x, y)}{\prod\limits_{i=1}^m \overline{K_{pi}^*}(x) |f_{2i}(y)|} \right)$$

is integrable on Ω_1, *for each* $y \in \Omega_2$. *Define* $\overline{\lambda_{pm}^*}$ *on* Ω_2 *by*

$$\overline{\lambda_{pm}^*}(y) := \left(\frac{1}{\prod\limits_{i=1}^m |f_{2i}(y)|} \right) \int_{\Omega_1} \left(\frac{u(x) \prod\limits_{i=1}^m k_i(x, y)}{\prod\limits_{i=1}^m \overline{K_{pi}^*}(x)} \right) d\mu_1(x) < \infty. \quad (21.38)$$

Here $\Phi_i : \mathbb{R}_+ \to \mathbb{R}_+$, $i = 1, \ldots, m$, *are convex and increasing functions.*
Then

$$\int_{\Omega_1} u(x) \prod_{i=1}^m \Phi_i \left(\left| \frac{g_{1i}(x)}{K_{pi}^*(x)} \right| \right) d\mu_1(x) \leq \quad (21.39)$$

$$\left(\prod_{\substack{i=1 \\ (i \neq j)}}^m \int_{\Omega_2} \Phi_i(|f_{1i}(y) f_{2i}(y)|) d\mu_2(y) \right) \left(\int_{\Omega_2} \Phi_j(|f_{1j}(y) f_{2j}(y)|) \overline{\lambda_{pm}^*}(y) d\mu_2(y) \right)$$

true for all measurable functions, $i = 1, \ldots, m$, $f_{1i}, f_{2i} : \Omega_2 \to \mathbb{R}$ *such that*

(i) $f_{1i}(y) f_{2i}(y)$, $\Phi_i(|f_{1i}(y) f_{2i}(y)|)$ *are both* $\frac{k_i(x,y)}{|f_{2i}(y)|} d\mu_2(y)$ *-integrable,* μ_1 *-a.e. in* $x \in \Omega_1$,

(ii) $\overline{\lambda_{pm}^*} \Phi_j(|f_{1j}(y) f_{2j}(y)|)$ *and* $\Phi_i(|f_{1i}(y) f_{2i}(y)|)$, *for* $i \in \{1, \ldots, m\} - \{j\}$ *are all* μ_2 *-integrable*

and for all corresponding functions g_{1i} *given by (21.8).*

To keep exposition short, in the rest of this chapter we give only applications of Theorem 21.3 to Fractional Calculus. We need the following:

Let $a < b$, $a, b \in \mathbb{R}$. By $C^N([a, b])$, we denote the space of all functions on $[a, b]$ which have continuous derivatives up to order N, and $AC([a, b])$ is the space of all absolutely continuous functions on $[a, b]$. By $AC^N([a, b])$, we denote the space of all functions g with $g^{(N-1)} \in AC([a, b])$. For any $\alpha \in \mathbb{R}$, we denote by $[\alpha]$ the integral part of α (the integer k satisfying $k \leq \alpha < k + 1$), and $\lceil \alpha \rceil$ is the ceiling of α ($\min\{n \in \mathbb{N}, n \geq \alpha\}$). By $L_1(a, b)$, we denote the space of all functions integrable on the interval (a, b), and by $L_\infty(a, b)$ the set of all functions measurable and essentially bounded on (a, b). Clearly, $L_\infty(a, b) \subset L_1(a, b)$.

We give the definition of the Riemann-Liouville fractional integrals, see [15]. Let $[a, b]$, $(-\infty < a < b < \infty)$ be a finite interval on the real axis $\mathbb{R}$.

The Riemann-Liouville fractional integrals $I_{a+}^\alpha f$ and $I_{b-}^\alpha f$ of order $\alpha > 0$ are defined by

$$\left(I_{a+}^\alpha f\right)(x) = \frac{1}{\Gamma(\alpha)} \int_a^x f(t)(x - t)^{\alpha-1} \, dt, \quad (x > a), \tag{21.40}$$

$$\left(I_{b-}^\alpha f\right)(x) = \frac{1}{\Gamma(\alpha)} \int_x^b f(t)(t - x)^{\alpha-1} \, dt, \quad (x < b), \tag{21.41}$$

respectively. Here $\Gamma(\alpha)$ is the Gamma function. These integrals are called the left-sided and the right-sided fractional integrals.

Let f_{1i}, f_{2i} be Lebesgue measurable functions from (a, b) into $\mathbb{R}$, such that $\left(I_{a+}^{\alpha_i}(|f_{1i}|)\right)(x)$, $\left(I_{a+}^{\alpha_i}(|f_{2i}|)\right)(x) \in \mathbb{R}$, $\forall\, x \in (a, b)$, $\alpha_i > 0$, $i = 1, \ldots, m$, e.g. when f_{1i}, $f_{2i} \in L_\infty(a, b)$.

Assume $0 < f_{2i}(y) < \infty$, a.e., $i = 1, \ldots, m$.

Consider

$$g_{1i}(x) = \left(I_{a+}^{\alpha_i} f_{1i}\right)(x), \tag{21.42}$$

$$g_{2i}(x) = \left(I_{a+}^{\alpha_i} f_{2i}\right)(x), \tag{21.43}$$

$x \in (a, b)$, $i = 1, \ldots, m$.

Notice that $g_{1i}(x)$, $g_{2i}(x) \in \mathbb{R}$ and they are Lebesgue measurable. We pick $\Omega_1 = \Omega_2 = (a, b)$, $d\mu_1(x) = dx$, $d\mu_2(y) = dy$, the Lebesgue measure.

We see that

$$\left(I_{a+}^{\alpha_i} f_{1i}\right)(x) = \int_a^b \frac{\chi_{(a,x]}(t)(x - t)^{\alpha_i-1}}{\Gamma(\alpha_i)} f_{1i}(t) \, dt, \tag{21.44}$$

$$\left(I_{a+}^{\alpha_i} f_{2i}\right)(x) = \int_a^b \frac{\chi_{(a,x]}(t)(x - t)^{\alpha_i-1}}{\Gamma(\alpha_i)} f_{2i}(t) \, dt, \tag{21.45}$$

where χ stands for the characteristic function, $x > a$.

So here it is

$$k_i(x, t) := \frac{\chi_{(a,x]}(t)(x-t)^{\alpha_i-1}}{\Gamma(\alpha_i)}, \quad i = 1, \ldots, m. \quad (21.46)$$

In fact

$$k_i(x, y) = \begin{cases} \frac{(x-y)^{\alpha_i-1}}{\Gamma(\alpha_i)}, & a < y \le x, \\ 0, & x < y < b. \end{cases} \quad (21.47)$$

Let $j \in \{1, \ldots, m\}$ be fixed.

Assume that the function

$$x \to \frac{\left(\prod_{i=1}^{m} f_{2i}(y)\right) u(x) \chi_{(a,x]}(y)(x-y)^{\left(\sum_{i=1}^{m} \alpha_i\right)-m}}{\left(\prod_{i=1}^{m}(I_{a+}^{\alpha_i} f_{2i})(x)\right)\left(\prod_{i=1}^{m}\Gamma(\alpha_i)\right)} \quad (21.48)$$

is integrable on (a, b), for each $y \in (a, b)$.

Here we have

$$\lambda_m^{**}(y) = \theta_m(y) := \left(\prod_{i=1}^{m}\left(\frac{f_{2i}(y)}{\Gamma(\alpha_i)}\right)\right)\int_y^b u(x)\left(\frac{(x-y)^{\left(\sum_{i=1}^{m}\alpha_i\right)-m}}{\prod_{i=1}^{m}(I_{a+}^{\alpha_i} f_{2i})(x)}\right) dx < \infty, \quad (21.49)$$

for any $y \in (a, b)$.

Here $\Phi_i : \mathbb{R}_+ \to \mathbb{R}_+$, $i = 1, \ldots, m$, are convex and increasing functions.

We get

Proposition 21.9 *Here all as above. It holds*

$$\int_a^b u(x) \prod_{i=1}^{m} \Phi_i\left(\left|\frac{I_{a+}^{\alpha_i} f_{1i}(x)}{I_{a+}^{\alpha_i} f_{2i}(x)}\right|\right) dx \le \quad (21.50)$$

$$\left(\prod_{\substack{i=1 \\ i\ne j}}^{m}\int_a^b \Phi_i\left(\left|\frac{f_{1i}(y)}{f_{2i}(y)}\right|\right) dy\right)\left(\int_a^b \Phi_j\left(\left|\frac{f_{1j}(y)}{f_{2j}(y)}\right|\right)\theta_m(y)\, dy\right),$$

true for all measurable functions, $i = 1, \ldots, m$, $f_{1i}, f_{2i} : (a, b) \to \mathbb{R}$ such that

(i) $\frac{f_{1i}(y)}{f_{2i}(y)}$, $\Phi_i\left(\left|\frac{f_{1i}(y)}{f_{2i}(y)}\right|\right)$ are both $\frac{\chi_{(a,x]}(y)(x-y)^{\alpha_i-1}}{\Gamma(\alpha_i)} f_{2i}(y)\,dy$ -integrable, a.e. in $x \in (a,b)$,

(ii) $\theta_m(y)\,\Phi_j\left(\left|\frac{f_{1j}(y)}{f_{2j}(y)}\right|\right)$; and $\Phi_i\left(\left|\frac{f_{1i}(y)}{f_{2i}(y)}\right|\right)$, for $i \in \{1,\dots,m\} - \{j\}$ are all Lebesgue integrable.

Corollary 21.10 *It holds*

$$\int_a^b u(x)\, e^{\left(\sum_{i=1}^m \left|\frac{I_{a+}^{\alpha_i} f_{1i}(x)}{I_{a+}^{\alpha_i} f_{2i}(x)}\right|\right)}\,dx \le \tag{21.51}$$

$$\left(\prod_{\substack{i=1\\i\ne j}}^m \int_a^b e^{\left|\frac{f_{1i}(y)}{f_{2i}(y)}\right|}\,dy\right)\left(\int_a^b e^{\left|\frac{f_{1j}(y)}{f_{2j}(y)}\right|}\theta_m(y)\,dy\right),$$

true for all measurable functions, $i = 1,\dots,m$, $f_{1i}, f_{2i} : (a,b) \to \mathbb{R}$ such that

(i) $\frac{f_{1i}(y)}{f_{2i}(y)}$, $e^{\left|\frac{f_{1i}(y)}{f_{2i}(y)}\right|}$ are both $\frac{\chi_{(a,x]}(y)(x-y)^{\alpha_i-1}}{\Gamma(\alpha_i)} f_{2i}(y)\,dy$ -integrable, a.e. in $x \in (a,b)$,

(ii) $\theta_m(y)\,e^{\left|\frac{f_{1j}(y)}{f_{2j}(y)}\right|}$; and $e^{\left|\frac{f_{1i}(y)}{f_{2i}(y)}\right|}$, for $i \in \{1,\dots,m\}-\{j\}$ are all Lebesgue integrable.

Corollary 21.11 *Let $p_i \ge 1$, $i = 1,\dots,m$. It holds*

$$\int_a^b u(x)\left(\prod_{i=1}^m \left|\frac{I_{a+}^{\alpha_i} f_{1i}(x)}{I_{a+}^{\alpha_i} f_{2i}(x)}\right|^{p_i}\right)dx \le \tag{21.52}$$

$$\left(\prod_{\substack{i=1\\i\ne j}}^m \int_a^b \left|\frac{f_{1i}(y)}{f_{2i}(y)}\right|^{p_i}\,dy\right)\left(\int_a^b \left|\frac{f_{1j}(y)}{f_{2j}(y)}\right|^{p_j}\theta_m(y)\,dy\right),$$

true for all measurable functions, $i = 1,\dots,m$, $f_{1i}, f_{2i} : (a,b) \to \mathbb{R}$ such that

(i) $\left|\frac{f_{1i}(y)}{f_{2i}(y)}\right|^{p_i}$ is $\frac{\chi_{(a,x]}(y)(x-y)^{\alpha_i-1}}{\Gamma(\alpha_i)} f_{2i}(y)\,dy$ -integrable, a.e. in $x \in (a,b)$,

(ii) $\theta_m(y)\left|\frac{f_{1j}(y)}{f_{2j}(y)}\right|^{p_j}$; and $\left|\frac{f_{1i}(y)}{f_{2i}(y)}\right|^{p_i}$, for $i \in \{1,\dots,m\} - \{j\}$ are all Lebesgue integrable.

Let us assume that $0 < f_{2i}(x) < \infty$, a.e. in $x \in (a,b)$, and we choose $u(x) = \prod_{i=1}^m \left(I_{a+}^{\alpha_i} f_{2i}\right)(x)$. Then

$$\theta_m(y) = \left(\prod_{i=1}^{m}\left(\frac{f_{2i}(y)}{\Gamma(\alpha_i)}\right)\right)\frac{(b-y)^{\left(\sum\limits_{i=1}^{m}\alpha_i-m+1\right)}}{\left(\sum\limits_{i=1}^{m}\alpha_i-m+1\right)}, \tag{21.53}$$

given that $\sum\limits_{i=1}^{m}\alpha_i > m-1$.

Corollary 21.12 Let $p_i \geq 1$, $i = 1,\ldots,m$, and $\sum\limits_{i=1}^{m}\alpha_i > m - 1$. Assume $0 < f_{2i}(x) < \infty$, a.e., $i = 1,\ldots,m$. It holds

$$\int_a^b \prod_{i=1}^{m}\left(I_{a+}^{\alpha_i}f_{1i}(x)\right)^{p_i}\left(I_{a+}^{\alpha_i}f_{2i}(x)\right)^{1-p_i}dx \leq \tag{21.54}$$

$$\frac{1}{\left(\prod\limits_{i=1}^{m}\Gamma(\alpha_i)\right)\left(\sum\limits_{i=1}^{m}\alpha_i-m+1\right)}\left(\prod_{\substack{i=1\\i\neq j}}^{m}\int_a^b\left(\frac{f_{1i}(y)}{f_{2i}(y)}\right)^{p_i}dy\right)\cdot$$

$$\left(\int_a^b\left(f_{1j}(y)\right)^{p_j}\left(f_{2j}(y)\right)^{1-p_j}\left(\prod_{\substack{i=1\\i\neq j}}^{m}f_{2i}(y)\right)(b-y)^{\left(\sum\limits_{i=1}^{m}\alpha_i-m+1\right)}dy\right) \leq$$

$$\frac{(b-a)^{\left(\sum\limits_{i=1}^{m}\alpha_i-m+1\right)}}{\left(\sum\limits_{i=1}^{m}\alpha_i-m+1\right)\left(\prod\limits_{i=1}^{m}\Gamma(\alpha_i)\right)}\cdot$$

$$\left(\prod_{\substack{i=1\\i\neq j}}^{m}\int_a^b\left(\frac{f_{1i}(y)}{f_{2i}(y)}\right)^{p_i}dy\right)\left(\int_a^b\left(f_{1j}(y)\right)^{p_j}\left(f_{2j}(y)\right)^{1-p_j}\left(\prod_{\substack{i=1\\i\neq j}}^{m}f_{2i}(y)\right)dy\right),$$

true for all measurable functions, $i = 1,\ldots,m$, $f_{1i}, f_{2i} : (a,b) \to \mathbb{R}$ such that

(i) $\left(\frac{f_{1i}(y)}{f_{2i}(y)}\right)^{p_i}$ *is* $\frac{\chi_{(a,x]}(y)(x-y)^{\alpha_i-1}}{\Gamma(\alpha_i)}f_{2i}(y)\,dy$ *-integrable, a.e. in* $x \in (a,b)$, $i = 1,\ldots,m$,

(ii) $\left(f_{1j}(y)\right)^{p_j}\left(f_{2j}(y)\right)^{1-p_j}\left(\prod_{\substack{i=1\\i\neq j}}^{m}f_{2i}(y)\right)$; and $\left(\frac{f_{1i}(y)}{f_{2i}(y)}\right)^{p_i}$, for $i\in\{1,\dots,$

$m\}-\{j\}$ are all Lebesgue integrable.

Let f_{1i},f_{2i} be Lebesgue measurable functions from (a,b) into $\mathbb{R}$, such that $\left(I_{b-}^{\alpha_i}(|f_{1i}|)\right)(x),\left(I_{b-}^{\alpha_i}(|f_{2i}|)\right)(x)\in\mathbb{R}$, $\forall\,x\in(a,b),\alpha_i>0,i=1,\dots,m$, e.g. when $f_{1i},f_{2i}\in L_\infty(a,b)$.

Assume $0<f_{2i}(y)<\infty$, a.e., $i=1,\dots,m$.

Consider

$$g_{1i}(x)=\left(I_{b-}^{\alpha_i}f_{1i}\right)(x),\tag{21.55}$$

$$g_{2i}(x)=\left(I_{b-}^{\alpha_i}f_{2i}\right)(x),\tag{21.56}$$

$x\in(a,b),i=1,\dots,m.$

Notice that $g_{1i}(x),g_{2i}(x)\in\mathbb{R}$ and they are Lebesgue measurable. We pick $\Omega_1=\Omega_2=(a,b),d\mu_1(x)=dx,d\mu_2(y)=dy$, the Lebesgue measure.

We see that

$$\left(I_{b-}^{\alpha_i}f_{1i}\right)(x)=\int_a^b\frac{\chi_{[x,b)}(t)(t-x)^{\alpha_i-1}}{\Gamma(\alpha_i)}f_{1i}(t)\,dt,\tag{21.57}$$

$$\left(I_{b-}^{\alpha_i}f_{2i}\right)(x)=\int_a^b\frac{\chi_{[x,b)}(t)(t-x)^{\alpha_i-1}}{\Gamma(\alpha_i)}f_{2i}(t)\,dt,\tag{21.58}$$

$x<b.$

So here it is

$$k_i(x,t):=\frac{\chi_{[x,b)}(t)(t-x)^{\alpha_i-1}}{\Gamma(\alpha_i)},\quad i=1,\dots,m.\tag{21.59}$$

In fact here

$$k_i(x,y)=\begin{cases}\frac{(y-x)^{\alpha_i-1}}{\Gamma(\alpha_i)},&x\leq y<b,\\0,&a<y<x.\end{cases}\tag{21.60}$$

Let $j\in\{1,\dots,m\}$ be fixed.

Assume that the function

$$x\to\frac{\left(\prod_{i=1}^{m}f_{2i}(y)\right)u(x)\chi_{[x,b)}(y)(y-x)^{\left(\sum_{i=1}^{m}\alpha_i\right)-m}}{\left(\prod_{i=1}^{m}\left(I_{b-}^{\alpha_i}f_{2i}\right)(x)\right)\left(\prod_{i=1}^{m}\Gamma(\alpha_i)\right)}\tag{21.61}$$

is integrable on (a, b), for each $y \in (a, b)$.

Here we have

$$\lambda_m^{**}(y) = \psi_m(y) := \left(\prod_{i=1}^m \left(\frac{f_{2i}(y)}{\Gamma(\alpha_i)}\right)\right) \int_a^y u(x) \left(\frac{(y-x)^{\left(\sum_{i=1}^m \alpha_i\right) - m}}{\prod_{i=1}^m \left(I_{b-}^{\alpha_i} f_{2i}\right)(x)}\right) dx < \infty,$$

$$(21.62)$$

for any $y \in (a, b)$.

Here $\Phi_i : \mathbb{R}_+ \to \mathbb{R}_+, i = 1, \ldots, m$, are convex and increasing functions.
We get

Proposition 21.13 *Here all as above. It holds*

$$\int_a^b u(x) \prod_{i=1}^m \Phi_i \left(\left|\frac{I_{b-}^{\alpha_i} f_{1i}(x)}{I_{b-}^{\alpha_i} f_{2i}(x)}\right|\right) dx \leq \qquad (21.63)$$

$$\left(\prod_{\substack{i=1 \\ i \neq j}}^m \int_a^b \Phi_i \left(\left|\frac{f_{1i}(y)}{f_{2i}(y)}\right|\right) dy\right) \left(\int_a^b \Phi_j \left(\left|\frac{f_{1j}(y)}{f_{2j}(y)}\right|\right) \psi_m(y) \, dy\right),$$

true for all measurable functions, $i = 1, \ldots, m$, $f_{1i}, f_{2i} : (a, b) \to \mathbb{R}$ such that

(i) $\frac{f_{1i}(y)}{f_{2i}(y)}$, $\Phi_i \left(\left|\frac{f_{1i}(y)}{f_{2i}(y)}\right|\right)$ *are both* $\frac{\chi_{[x,b)}(y)(y-x)^{\alpha_i - 1}}{\Gamma(\alpha_i)} f_{2i}(y) \, dy$ *-integrable, a.e. in $x \in (a, b)$,*

(ii) $\psi_m(y) \Phi_j \left(\left|\frac{f_{1j}(y)}{f_{2j}(y)}\right|\right)$; *and* $\Phi_i \left(\left|\frac{f_{1i}(y)}{f_{2i}(y)}\right|\right)$, *for $i \in \{1, \ldots, m\} - \{j\}$ are all Lebesgue integrable.*

Corollary 21.14 *It holds*

$$\int_a^b u(x) e^{\left(\sum_{i=1}^m \left|\frac{I_{b-}^{\alpha_i} f_{1i}(x)}{I_{b-}^{\alpha_i} f_{2i}(x)}\right|\right)} dx \leq \qquad (21.64)$$

$$\left(\prod_{\substack{i=1 \\ i \neq j}}^m \int_a^b e^{\left|\frac{f_{1i}(y)}{f_{2i}(y)}\right|} dy\right) \left(\int_a^b e^{\left|\frac{f_{1j}(y)}{f_{2j}(y)}\right|} \psi_m(y) \, dy\right),$$

true for all measurable functions, $i = 1, \ldots, m$, $f_{1i}, f_{2i} : (a, b) \to \mathbb{R}$ such that

(i) $\frac{f_{1i}(y)}{f_{2i}(y)}$, $e^{\left|\frac{f_{1i}(y)}{f_{2i}(y)}\right|}$ *are both* $\frac{\chi_{[x,b)}(y)(y-x)^{\alpha_i - 1}}{\Gamma(\alpha_i)} f_{2i}(y) \, dy$ *-integrable, a.e. in $x \in (a, b)$,*

(ii) $\psi_m(y) e^{\left|\frac{f_{1j}(y)}{f_{2j}(y)}\right|}$; and $e^{\left|\frac{f_{1i}(y)}{f_{2i}(y)}\right|}$, for $i \in \{1, \ldots, m\} - \{j\}$ are all Lebesgue integrable.

Corollary 21.15 *Let $p_i \geq 1$, $i = 1, \ldots, m$. It holds*

$$\int_a^b u(x) \left(\prod_{i=1}^m \left|\frac{I_{b-}^{\alpha_i} f_{1i}(x)}{I_{b-}^{\alpha_i} f_{2i}(x)}\right|^{p_i}\right) dx \leq \tag{21.65}$$

$$\left(\prod_{\substack{i=1 \\ i \neq j}}^m \int_a^b \left|\frac{f_{1i}(y)}{f_{2i}(y)}\right|^{p_i} dy\right) \left(\int_a^b \left|\frac{f_{1j}(y)}{f_{2j}(y)}\right|^{p_j} \psi_m(y) dy\right),$$

true for all measurable functions, $i = 1, \ldots, m$, $f_{1i}, f_{2i} : (a, b) \to \mathbb{R}$ such that

(i) $\left|\frac{f_{1i}(y)}{f_{2i}(y)}\right|^{p_i}$ *is* $\frac{\chi_{[x,b)}(y)(y-x)^{\alpha_i-1}}{\Gamma(\alpha_i)} f_{2i}(y) dy$ *-integrable, a.e. in $x \in (a, b)$,*

(ii) $\psi_m(y) \left|\frac{f_{1j}(y)}{f_{2j}(y)}\right|^{p_j}$; *and* $\left|\frac{f_{1i}(y)}{f_{2i}(y)}\right|^{p_i}$, *for $i \in \{1, \ldots, m\} - \{j\}$ are all Lebesgue integrable.*

Let us again assume $0 < f_{2i}(x) < \infty$, a.e. in $x \in (a, b)$, and we choose $u(x) = \prod_{i=1}^m \left(I_{b-}^{\alpha_i} f_{2i}\right)(x)$. Then

$$\psi_m(y) = \left(\prod_{i=1}^m \left(\frac{f_{2i}(y)}{\Gamma(\alpha_i)}\right)\right) \frac{(y-a)^{\left(\sum_{i=1}^m \alpha_i - m + 1\right)}}{\left(\sum_{i=1}^m \alpha_i - m + 1\right)}, \tag{21.66}$$

given that $\sum_{i=1}^m \alpha_i > m - 1$.

Corollary 21.16 *Let $p_i \geq 1$, $i = 1, \ldots, m$, and $\sum_{i=1}^m \alpha_i > m - 1$. Assume $0 < f_{2i}(x) < \infty$, a.e., $i = 1, \ldots, m$. It holds*

$$\int_a^b \prod_{i=1}^m \left(I_{b-}^{\alpha_i} f_{1i}(x)\right)^{p_i} \left(I_{b-}^{\alpha_i} f_{2i}(x)\right)^{1-p_i} dx \leq \tag{21.67}$$

$$\frac{1}{\left(\prod_{i=1}^{m}\Gamma\left(\alpha_i\right)\right)\left(\sum_{i=1}^{m}\alpha_i-m+1\right)}\left(\prod_{\substack{i=1\\i\neq j}}^{m}\int_a^b\left(\frac{f_{1i}\left(y\right)}{f_{2i}\left(y\right)}\right)^{p_i}dy\right)\cdot$$

$$\left(\int_a^b\left(f_{1j}\left(y\right)\right)^{p_j}\left(f_{2j}\left(y\right)\right)^{1-p_j}\left(\prod_{\substack{i=1\\i\neq j}}^{m}f_{2i}\left(y\right)\right)\left(y-a\right)^{\left(\sum_{i=1}^{m}\alpha_i-m+1\right)}dy\right)\leq$$

$$\frac{\left(b-a\right)^{\left(\sum_{i=1}^{m}\alpha_i-m+1\right)}}{\left(\sum_{i=1}^{m}\alpha_i-m+1\right)\left(\prod_{i=1}^{m}\Gamma\left(\alpha_i\right)\right)}\cdot$$

$$\left(\prod_{\substack{i=1\\i\neq j}}^{m}\int_a^b\left(\frac{f_{1i}\left(y\right)}{f_{2i}\left(y\right)}\right)^{p_i}dy\right)\left(\int_a^b\left(f_{1j}\left(y\right)\right)^{p_j}\left(f_{2j}\left(y\right)\right)^{1-p_j}\left(\prod_{\substack{i=1\\i\neq j}}^{m}f_{2i}\left(y\right)\right)dy\right),$$

true for all measurable functions, $i=1,\ldots,m$, f_{1i}, $f_{2i}:(a,b)\to\mathbb{R}$ such that

(i) $\left(\frac{f_{1i}(y)}{f_{2i}(y)}\right)^{p_i}$ *is* $\frac{\chi_{[x,b)}(y)(y-x)^{\alpha_i-1}}{\Gamma(\alpha_i)}$ $f_{2i}(y)\,dy$ *-integrable, a.e. in $x\in(a,b)$, $i=1,\ldots,m$,*

(ii) $\left(f_{1j}(y)\right)^{p_j}\left(f_{2j}(y)\right)^{1-p_j}\left(\prod_{\substack{i=1\\i\neq j}}^{m}f_{2i}(y)\right)$; *and* $\left(\frac{f_{1i}(y)}{f_{2i}(y)}\right)^{p_i}$, *for $i\in\{1,\ldots,$*
$m\}-\{j\}$ *are all Lebesgue integrable.*

We mention

Definition 21.17 ([2], *p. 448*) The left generalized Riemann-Liouville fractional derivative of f of order $\beta>0$ is given by

$$D_a^\beta f(x)=\frac{1}{\Gamma(n-\beta)}\left(\frac{d}{dx}\right)^n\int_a^x(x-y)^{n-\beta-1}f(y)\,dy,\qquad(21.68)$$

where $n=[\beta]+1$, $x\in[a,b]$.

For $a,b\in\mathbb{R}$, we say that $f\in L_1(a,b)$ has an L_∞ fractional derivative $D_a^\beta f$ $(\beta>0)$ in $[a,b]$, if and only if

(1) $D_a^{\beta-k}f\in C([a,b])$, $k=2,\ldots,n=[\beta]+1$,

(2) $D_a^{\beta-1} f \in AC([a, b])$

(3) $D_a^{\beta} f \in L_\infty(a, b)$.

Above we define $D_a^0 f := f$ and $D_a^{-\delta} f := I_{a+}^{\delta} f$, if $0 < \delta \le 1$.

From [2, p. 449] and [12] we mention and use

Lemma 21.18 *Let $\beta > \alpha \ge 0$ and let $f \in L_1(a, b)$ have an L_∞ fractional derivative $D_a^{\beta} f$ in $[a, b]$ and let $D_a^{\beta-k} f(a) = 0, k = 1, \ldots, [\beta] + 1$, then*

$$D_a^{\alpha} f(x) = \frac{1}{\Gamma(\beta - \alpha)} \int_a^x (x - y)^{\beta-\alpha-1} D_a^{\beta} f(y) \, dy, \qquad (21.69)$$

for all $a \le x \le b$.

  Here $D_a^{\alpha} f \in AC([a, b])$ for $\beta - \alpha \ge 1$, and $D_a^{\alpha} f \in C([a, b])$ for $\beta - \alpha \in (0, 1)$. Notice here that

$$D_a^{\alpha} f(x) = \left(I_{a+}^{\beta-\alpha} \left(D_a^{\beta} f \right) \right)(x), \quad a \le x \le b. \qquad (21.70)$$

For more on the last, see [6].

Let $f_{1i}, f_{2i} \in L_1(a, b)$; $\alpha_i, \beta_i : \beta_i > \alpha_i \ge 0, i = 1, \ldots, m$. Here $(j = 1, 2)$ $(f_{ji}, \alpha_i, \beta_i)$ fulfill terminology and assumptions of Definition 21.17 and Lemma 21.18. Indeed we have

$$D_a^{\alpha_i} f_{1i}(x) = \left(I_{a+}^{\beta_i-\alpha_i} \left(D_a^{\beta_i} f_{1i} \right) \right)(x), \qquad (21.71)$$

and

$$D_a^{\alpha_i} f_{2i}(x) = \left(I_{a+}^{\beta_i-\alpha_i} \left(D_a^{\beta_i} f_{2i} \right) \right)(x), \qquad (21.72)$$

$a \le x \le b, i = 1, \ldots, m$.

  Assume $0 < D_a^{\beta_i} f_{2i}(x) < \infty$, a.e., $i = 1, \ldots, m$.

  Let $j \in \{1, \ldots, m\}$ be fixed. Assume that the function

$$x \to \frac{\left(\prod_{i=1}^{m} f_{2i}(y) \right) u(x) \chi_{(a,x]}(y) (x - y)^{\left(\left(\sum_{i=1}^{m}(\beta_i-\alpha_i) \right) - m \right)}}{\left(\prod_{i=1}^{m} (D_a^{\alpha_i} f_{2i})(x) \right) \left(\prod_{i=1}^{m} \Gamma(\beta_i - \alpha_i) \right)} \qquad (21.73)$$

is integrable on (a, b), for each $y \in (a, b)$.

  Here we have

$$\theta_m^*(y) := \left(\prod_{i=1}^{m} \left(\frac{f_{2i}(y)}{\Gamma(\beta_i - \alpha_i)} \right) \right).$$

$$\int_y^b u(x) \left(\frac{(x-y)^{\left(\sum_{i=1}^m (\beta_i - \alpha_i)\right) - m}}{\prod_{i=1}^m (D_a^{\alpha_i} f_{2i})(x)} \right) dx < \infty, \qquad (21.74)$$

for any $y \in (a, b)$.

Here $\Phi_i : \mathbb{R}_+ \to \mathbb{R}_+, i = 1, \ldots, m$, are convex and increasing functions.

Proposition 21.19 *Here all as above. It holds*

$$\int_a^b u(x) \prod_{i=1}^m \Phi_i \left(\left| \frac{D_a^{\alpha_i} f_{1i}(x)}{D_a^{\alpha_i} f_{2i}(x)} \right| \right) dx \leq \qquad (21.75)$$

$$\left(\prod_{\substack{i=1 \\ i \neq j}}^m \int_a^b \Phi_i \left(\left| \frac{D_a^{\beta_i} f_{1i}(y)}{D_a^{\beta_i} f_{2i}(y)} \right| \right) dy \right) \left(\int_a^b \Phi_j \left(\left| \frac{D_a^{\beta_j} f_{1j}(y)}{D_a^{\beta_j} f_{2j}(y)} \right| \right) \theta_m^*(y)\, dy \right),$$

under the properties: $(i = 1, \ldots, m)$

(i) $\Phi_i \left(\left| \frac{D_a^{\beta_i} f_{1i}(y)}{D_a^{\beta_i} f_{2i}(y)} \right| \right)$ *is* $\frac{\chi_{(a,x]}(y)(x-y)^{(\beta_i - \alpha_i - 1)}}{\Gamma(\beta_i - \alpha_i)} \left(D_a^{\beta_i} f_{2i}(y) \right) dy$ *-integrable, a.e. in* $x \in (a, b),$

(ii) $\theta_m^*(y)\, \Phi_j \left(\left| \frac{D_a^{\beta_j} f_{1j}(y)}{D_a^{\beta_j} f_{2j}(y)} \right| \right);$ *and* $\Phi_i \left(\left| \frac{D_a^{\beta_i} f_{1i}(y)}{D_a^{\beta_i} f_{2i}(y)} \right| \right),$ *for* $i \in \{1, \ldots, m\} - \{j\}$ *are all Lebesgue integrable.*

Proof By Proposition 21.9. ∎

Corollary 21.20 *It holds*

$$\int_a^b u(x)\, e^{\left(\sum_{i=1}^m \left| \frac{D_a^{\alpha_i} f_{1i}(x)}{D_a^{\alpha_i} f_{2i}(x)} \right| \right)} dx \leq \qquad (21.76)$$

$$\left(\prod_{\substack{i=1 \\ i \neq j}}^m \int_a^b e^{\left| \frac{D_a^{\beta_i} f_{1i}(y)}{D_a^{\beta_i} f_{2i}(y)} \right|} dy \right) \left(\int_a^b e^{\left| \frac{D_a^{\beta_j} f_{1j}(y)}{D_a^{\beta_j} f_{2j}(y)} \right|} \theta_m^*(y)\, dy \right),$$

under the properties: $(i = 1, \ldots, m)$

(i) $e^{\left| \begin{smallmatrix} D_a^{\beta_i} f_{1i}(y) \\ D_a^{\beta_i} f_{2i}(y) \end{smallmatrix} \right|}$ is $\frac{\chi_{(a,x]}(y)(x-y)^{(\beta_i-\alpha_i-1)}}{\Gamma(\beta_i-\alpha_i)} \left(D_a^{\beta_i} f_{2i}(y) \right) dy$ -integrable, a.e. in $x \in$ (a, b),

(ii) $\theta_m^*(y) e^{\left| \begin{smallmatrix} D_a^{\beta_j} f_{1j}(y) \\ D_a^{\beta_j} f_{2j}(y) \end{smallmatrix} \right|}$; and $e^{\left| \begin{smallmatrix} D_a^{\beta_i} f_{1i}(y) \\ D_a^{\beta_i} f_{2i}(y) \end{smallmatrix} \right|}$, for $i \in \{1, \ldots, m\} - \{j\}$ are all Lebesgue integrable.

We need

Definition 21.21 ([9], p. 50, [2], p. 449) Let $\nu \geq 0$, $n := \lceil \nu \rceil$, $f \in AC^n([a, b])$. Then the left Caputo fractional derivative is given by

$$D_{*a}^\nu f(x) = \frac{1}{\Gamma(n-\nu)} \int_a^x (x-t)^{n-\nu-1} f^{(n)}(t) \, dt$$

$$= \left(I_{a+}^{n-\nu} f^{(n)} \right)(x), \tag{21.77}$$

and it exists almost everywhere for $x \in [a, b]$, in fact $D_{*a}^\nu f \in L_1(a, b)$, ([2], p. 394).

We have $D_{*a}^n f = f^{(n)}$, $n \in \mathbb{Z}_+$.

We also need

Theorem 21.22 ([4, 7]) Let $\nu > \rho > 0$, $\nu, \rho \notin \mathbb{N}$. Call $n := \lceil \nu \rceil$, $m^* := \lceil \rho \rceil$. Assume $f \in AC^n([a, b])$, such that $f^{(k)}(a) = 0$, $k = m^*, m^* + 1, \ldots, n - 1$, and $D_{*a}^\nu f \in L_\infty(a, b)$. Then $D_{*a}^\rho f \in C([a, b])$ if $\nu - \rho \in (0, 1)$, and $D_{*a}^\rho f \in AC([a, b])$, if $\nu - \rho \geq 1$ (where $D_{*a}^\rho f = I_{a+}^{m^*-\rho} f^{(m^*)}(x)$), and

$$D_{*a}^\rho f(x) = \frac{1}{\Gamma(\nu-\rho)} \int_a^x (x-t)^{\nu-\rho-1} D_{*a}^\nu f(t) \, dt$$

$$= \left(I_{a+}^{\nu-\rho} \left(D_{*a}^\nu f \right) \right)(x), \tag{21.78}$$

$\forall \, x \in [a, b]$.

For more on the last, see [7].

Let $\nu_i > \rho_i > 0$, $\nu_i, \rho_i \notin \mathbb{N}$, $n_i := \lceil \nu_i \rceil$, $m_i^* := \lceil \rho_i \rceil$, $i = 1, \ldots, m$. Assume $f_{1i}, f_{2i} \in AC^{n_i}([a, b])$, such that $(j = 1, 2)$ $f_{ji}^{(k_i)}(a) = 0$, $k_i = m_i^*, m_i^* + 1, \ldots, n_i - 1$, and $D_{*a}^{\nu_i} f_{ji} \in L_\infty(a, b)$. Based on Definition 21.21 and Theorem 21.22 we get that

$$D_{*a}^{\rho_i} f_{ji}(x) = \left(I_{a+}^{\nu_i-\rho_i} \left(D_{*a}^{\nu_i} f_{ji} \right) \right)(x) \in \mathbb{R}, \tag{21.79}$$

$\forall \, x \in [a, b]$; $j = 1, 2$; $i = 1, \ldots, m$.

Assume $0 < D_{*a}^{\nu_i} f_{2i}(x) < \infty$, a.e., $i = 1, \ldots, m$.

Let $j \in \{1, \ldots, m\}$ be fixed. Assume that the function

$$x \to \frac{\left(\prod_{i=1}^{m} D_{*a}^{\nu_i} f_{2i}(y)\right) u(x) \chi_{(a,x]}(y) (x-y)^{\left(\left(\sum_{i=1}^{m}(\nu_i - \rho_i)\right) - m\right)}}{\left(\prod_{i=1}^{m} \left(D_{*a}^{\rho_i} f_{2i}\right)(x)\right)\left(\prod_{i=1}^{m} \Gamma(\nu_i - \rho_i)\right)} \tag{21.80}$$

is integrable on (a, b), for each $y \in (a, b)$.

Here we have

$$\psi_m^*(y) := \left(\prod_{i=1}^{m} \left(\frac{D_{*a}^{\nu_i} f_{2i}(y)}{\Gamma(\nu_i - \rho_i)}\right)\right) \int_y^b u(x) \left(\frac{(x-y)^{\left(\left(\sum_{i=1}^{m}(\nu_i - \rho_i)\right) - m\right)}}{\prod_{i=1}^{m} \left(D_{*a}^{\rho_i} f_{2i}\right)(x)}\right) dx < \infty, \tag{21.81}$$

for any $y \in (a, b)$.

Here $\Phi_i : \mathbb{R}_+ \to \mathbb{R}_+, i = 1, \ldots, m$, are convex and increasing functions.

Proposition 21.23 *It holds*

$$\int_a^b u(x) \prod_{i=1}^{m} \Phi_i \left(\left|\frac{D_{*a}^{\rho_i} f_{1i}(x)}{D_{*a}^{\rho_i} f_{2i}(x)}\right|\right) dx \leq \tag{21.82}$$

$$\left(\prod_{\substack{i=1 \\ i \neq j}}^{m} \int_a^b \Phi_i \left(\left|\frac{D_{*a}^{\nu_i} f_{1i}(y)}{D_{*a}^{\nu_i} f_{2i}(y)}\right|\right) dy\right) \left(\int_a^b \Phi_j \left(\left|\frac{D_{*a}^{\nu_j} f_{1j}(y)}{D_{*a}^{\nu_j} f_{2j}(y)}\right|\right) \psi_m^*(y) dy\right),$$

under the properties: $(i = 1, \ldots, m)$

(i) $\Phi_i \left(\left|\frac{D_{*a}^{\nu_i} f_{1i}(y)}{D_{*a}^{\nu_i} f_{2i}(y)}\right|\right)$ *is* $\frac{\chi_{(a,x]}(y)(x-y)^{(\nu_i - \rho_i - 1)}}{\Gamma(\nu_i - \rho_i)} \left(D_{*a}^{\nu_i} f_{2i}(y)\right) dy$ *-integrable, a.e. in $x \in$ (a, b),*

(ii) $\psi_m^*(y) \Phi_j \left(\left|\frac{D_{*a}^{\nu_j} f_{1j}(y)}{D_{*a}^{\nu_j} f_{2j}(y)}\right|\right)$; *and* $\Phi_i \left(\left|\frac{D_{*a}^{\nu_i} f_{1i}(y)}{D_{*a}^{\nu_i} f_{2i}(y)}\right|\right)$, *for $i \in \{1, \ldots, m\} - \{j\}$ are all Lebesgue integrable.*

Proof By Proposition 21.9. ∎

Corollary 21.24 *It holds*

$$
\int_a^b u(x)\, e^{\left(\sum_{i=1}^m \left|\frac{D_{*a}^{\rho_i} f_{1i}(x)}{D_{*a}^{\rho_i} f_{2i}(x)}\right|\right)} dx \le \tag{21.83}
$$

$$
\left(\prod_{\substack{i=1\\i\neq j}}^m \int_a^b e^{\left|\frac{D_{*a}^{\nu_i} f_{1i}(y)}{D_{*a}^{\nu_i} f_{2i}(y)}\right|} dy\right)\left(\int_a^b e^{\left|\frac{D_{*a}^{\nu_j} f_{1j}(y)}{D_{*a}^{\nu_j} f_{2j}(y)}\right|} \psi_m^*(y)\, dy\right),
$$

under the properties: $(i = 1, \ldots, m)$

(i) $e^{\left|\frac{D_{*a}^{\nu_i} f_{1i}(y)}{D_{*a}^{\nu_i} f_{2i}(y)}\right|}$ *is* $\frac{\chi_{(a,x]}(y)(x-y)^{(\nu_i-\rho_i-1)}}{\Gamma(\nu_i-\rho_i)}$ $\left(D_{*a}^{\nu_i} f_{2i}(y)\right)$ *dy -integrable, a.e. in* $x \in (a,b)$,

(ii) $\psi_m^*(y)\, e^{\left|\frac{D_{*a}^{\nu_j} f_{1j}(y)}{D_{*a}^{\nu_j} f_{2j}(y)}\right|}$; *and* $e^{\left|\frac{D_{*a}^{\nu_i} f_{1i}(y)}{D_{*a}^{\nu_i} f_{2i}(y)}\right|}$, *for* $i \in \{1, \ldots, m\} - \{j\}$ *are all Lebesgue integrable.*

We need

Definition 21.25 ([3, 10, 11]) Let $\alpha \ge 0$, $n := \lceil \alpha \rceil$, $f \in AC^n([a,b])$. We define the right Caputo fractional derivative of order $\alpha \ge 0$, by

$$
\overline{D}_{b-}^{\alpha} f(x) := (-1)^n I_{b-}^{n-\alpha} f^{(n)}(x), \tag{21.84}
$$

we set $\overline{D}_{-}^{0} f := f$, i.e.

$$
\overline{D}_{b-}^{\alpha} f(x) = \frac{(-1)^n}{\Gamma(n-\alpha)} \int_x^b (J-x)^{n-\alpha-1} f^{(n)}(J)\, dJ. \tag{21.85}
$$

Notice that $\overline{D}_{b-}^{n} f = (-1)^n f^{(n)}$, $n \in \mathbb{N}$.

We need

Theorem 21.26 ([4]) Let $f \in AC^n([a,b])$, $n \in \mathbb{N}$, $n := \lceil \alpha \rceil$, $\alpha > \rho > 0$, $r = \lceil \rho \rceil$, $\alpha, \rho \notin \mathbb{N}$. Assume $f^{(k)}(b) = 0$, $k = r, r+1, \ldots, n-1$, and $\overline{D}_{b-}^{\alpha} f \in L_\infty([a,b])$. Then

$$
\overline{D}_{b-}^{\rho} f(x) = \left(I_{b-}^{\alpha-\rho}\left(\overline{D}_{b-}^{\alpha} f\right)\right)(x) \in C([a,b]), \tag{21.86}
$$

if $\alpha - \rho \in (0,1)$, *and* $\overline{D}_{b-}^{\rho} f \in AC([a,b])$, *if* $\alpha - \rho \ge 1$, *that is*

$$
\overline{D}_{b-}^{\rho} f(x) = \frac{1}{\Gamma(\alpha-\rho)} \int_x^b (t-x)^{\alpha-\rho-1}\left(\overline{D}_{b-}^{\alpha} f\right)(t)\, dt, \tag{21.87}
$$

$\forall x \in [a,b]$.

Here $i = 1, \ldots, m$. Let $\alpha_i > \rho_i > 0$, $\alpha_i, \rho_i \notin \mathbb{N}$, $n_i = \lceil \alpha_i \rceil$, $r_i = \lceil \rho_i \rceil$. Take $f_{1i}, f_{2i} \in AC^{n_i}([a, b])$, such that $(j = 1, 2)$ $f_{ji}^{(k_i)}(b) = 0$, $k_i = r_i, r_i + 1, \ldots, n_i - 1$. Furthermore assume that $\overline{D}_{b-}^{\alpha_i} f_{1i}, \overline{D}_{b-}^{\alpha_i} f_{2i} \in L_\infty(a, b)$. Then by Theorem 21.26 we get that $(j = 1, 2)$:

$$\overline{D}_{b-}^{\rho_i} f_{ji}(x) = \left(I_{b-}^{\alpha_i - \rho_i} \left(\overline{D}_{b-}^{\alpha_i} f_{ji} \right) \right)(x) \in C([a, b]), \quad \forall x \in [a, b]. \tag{21.88}$$

Assume $0 < \overline{D}_{b-}^{\alpha_i} f_{2i}(x) < \infty$, a.e., $i = 1, \ldots, m$.

Let $j \in \{1, \ldots, m\}$ be fixed. Assume that the function

$$x \to \frac{\left(\prod_{i=1}^{m} \overline{D}_{b-}^{\alpha_i} f_{2i}(y) \right) u(x) \chi_{[x,b)}(y) (y - x)^{\left(\left(\sum_{i=1}^{m} (\alpha_i - \rho_i) \right) - m \right)}}{\left(\prod_{i=1}^{m} \left(\overline{D}_{b-}^{\rho_i} f_{2i} \right)(x) \right) \left(\prod_{i=1}^{m} \Gamma(\alpha_i - \rho_i) \right)} \tag{21.89}$$

is integrable on (a, b), for each $y \in (a, b)$.

Here we have

$$T_m(y) := \left(\prod_{i=1}^{m} \left(\frac{\overline{D}_{b-}^{\alpha_i} f_{2i}(y)}{\Gamma(\alpha_i - \rho_i)} \right) \right) \int_a^y u(x) \left(\frac{(y - x)^{\left(\left(\sum_{i=1}^{m} (\alpha_i - \rho_i) \right) - m \right)}}{\prod_{i=1}^{m} \left(\overline{D}_{b-}^{\rho_i} f_{2i} \right)(x)} \right) dx < \infty, \tag{21.90}$$

for any $y \in (a, b)$.

Here $\Phi_i : \mathbb{R}_+ \to \mathbb{R}_+$, $i = 1, \ldots, m$, are convex and increasing functions.

We get

Proposition 21.27 *Here all as above. It holds*

$$\int_a^b u(x) \prod_{i=1}^{m} \Phi_i \left(\left| \frac{\overline{D}_{b-}^{\rho_i} f_{1i}(x)}{\overline{D}_{b-}^{\rho_i} f_{2i}(x)} \right| \right) dx \leq \tag{21.91}$$

$$\left(\prod_{\substack{i=1 \\ i \neq j}}^{m} \int_a^b \Phi_i \left(\left| \frac{\overline{D}_{b-}^{\alpha_i} f_{1i}(y)}{\overline{D}_{b-}^{\alpha_i} f_{2i}(y)} \right| \right) dy \right) \left(\int_a^b \Phi_j \left(\left| \frac{\overline{D}_{b-}^{\alpha_j} f_{1j}(y)}{\overline{D}_{b-}^{\alpha_j} f_{2j}(y)} \right| \right) T_m(y) \, dy \right),$$

under the properties: $(i = 1, \ldots, m)$

(i) $\Phi_i \left(\left| \dfrac{\overline{D}_{b-}^{\alpha_i} f_{1i}(y)}{\overline{D}_{b-}^{\alpha_i} f_{2i}(y)} \right| \right)$ is $\dfrac{\chi_{[x,b)}(y)(y-x)^{(\alpha_i-\rho_i-1)}}{\Gamma(\alpha_i-\rho_i)} \left(\overline{D}_{b-}^{\alpha_i} f_{2i}(y) \right) dy$ -integrable, a.e. in $x \in (a, b)$,

(ii) $T_m(y) \Phi_j \left(\left| \dfrac{\overline{D}_{b-}^{\alpha_j} f_{1j}(y)}{\overline{D}_{b-}^{\alpha_j} f_{2j}(y)} \right| \right)$; and $\Phi_i \left(\left| \dfrac{\overline{D}_{b-}^{\alpha_i} f_{1i}(y)}{\overline{D}_{b-}^{\alpha_i} f_{2i}(y)} \right| \right)$, for $i \in \{1, \ldots, m\} - \{j\}$ are all Lebesgue integrable.

Proof By Proposition 21.13. ∎

Corollary 21.28 *It holds*

$$\int_a^b u(x) e^{\left(\sum\limits_{i=1}^m \left| \frac{\overline{D}_{b-}^{\rho_i} f_{1i}(x)}{\overline{D}_{b-}^{\rho_i} f_{2i}(x)} \right| \right)} dx \leq \tag{21.92}$$

$$\left(\prod\limits_{\substack{i=1 \\ i\neq j}}^m \int_a^b e^{\left| \frac{\overline{D}_{b-}^{\alpha_i} f_{1i}(y)}{\overline{D}_{b-}^{\alpha_i} f_{2i}(y)} \right|} dy \right) \left(\int_a^b e^{\left| \frac{\overline{D}_{b-}^{\alpha_j} f_{1j}(y)}{\overline{D}_{b-}^{\alpha_j} f_{2j}(y)} \right|} T_m(y) \, dy \right),$$

under the properties: $(i = 1, \ldots, m)$

(i) $e^{\left| \frac{\overline{D}_{b-}^{\alpha_i} f_{1i}(y)}{\overline{D}_{b-}^{\alpha_i} f_{2i}(y)} \right|}$ is $\dfrac{\chi_{[x,b)}(y)(y-x)^{(\alpha_i-\rho_i-1)}}{\Gamma(\alpha_i-\rho_i)} \left(\overline{D}_{b-}^{\alpha_i} f_{2i}(y) \right) dy$ -integrable, a.e. in $x \in (a, b)$,

(ii) $T_m(y) e^{\left| \frac{\overline{D}_{b-}^{\alpha_j} f_{1j}(y)}{\overline{D}_{b-}^{\alpha_j} f_{2j}(y)} \right|}$; and $e^{\left| \frac{\overline{D}_{b-}^{\alpha_i} f_{1i}(y)}{\overline{D}_{b-}^{\alpha_i} f_{2i}(y)} \right|}$, for $i \in \{1, \ldots, m\} - \{j\}$ are all Lebesgue integrable.

Proof By Proposition 21.27. ∎

We give

Definition 21.29 Let $\nu > 0$, $n := [\nu]$, $\alpha := \nu - n$ $(0 \leq \alpha < 1)$. Let $a, b \in \mathbb{R}$, $a \leq x \leq b$, $f \in C([a, b])$. We consider $C_a^\nu([a, b]) := \{f \in C^n([a, b]) : I_{a+}^{1-\alpha} f^{(n)} \in C^1([a, b])\}$. For $f \in C_a^\nu([a, b])$, we define the left generalized ν-fractional derivative of f over $[a, b]$ as

$$\Delta_a^\nu f := \left(I_{a+}^{1-\alpha} f^{(n)} \right)', \tag{21.93}$$

see [2], p. 24, and Canavati derivative in [8].

Notice here $\Delta_a^\nu f \in C([a, b])$.

So that

$$\left(\Delta_a^\nu f \right)(x) = \frac{1}{\Gamma(1-\alpha)} \frac{d}{dx} \int_a^x (x-t)^{-\alpha} f^{(n)}(t) \, dt, \tag{21.94}$$

$\forall x \in [a, b]$.

Notice here that

$$\Delta_a^n f = f^{(n)}, \quad n \in \mathbb{Z}_+. \tag{21.95}$$

We need

Theorem 21.30 ([4]) *Let $f \in C_a^\nu([a, b])$, $n = [\nu]$, such that $f^{(i)}(a) = 0$, $i = r, r+1, \ldots, n-1$, where $r := [\rho]$, with $0 < \rho < \nu$. Then*

$$\left(\Delta_a^\rho f\right)(x) = \frac{1}{\Gamma(\nu - \rho)} \int_a^x (x - t)^{\nu - \rho - 1} \left(\Delta_a^\nu f\right)(t)\, dt, \tag{21.96}$$

i.e.

$$\left(\Delta_a^\rho f\right) = I_{a+}^{\nu - \rho}\left(\Delta_a^\nu f\right) \in C([a, b]). \tag{21.97}$$

Thus $f \in C_a^\rho([a, b])$.

Let $\nu_i > \rho_i > 0$, $n_i := [\nu_i]$, $r_i := [\rho_i]$, $i = 1, \ldots, m$. Let $f_{1i}, f_{2i} \in C_a^{\nu_i}([a, b])$, such that $(j = 1, 2)\, f_{ji}^{(k_i)}(a) = 0$, $k_i = r_i, r_i + 1, \ldots, n_i - 1$. Notice here $\Delta_a^{\nu_i} f_{ji} \in C([a, b])$, and $\Delta_a^{\rho_i} f_{ji} \in C([a, b])$. Based on Definition 21.29 and Theorem 21.30 we get

$$\Delta_a^{\rho_i} f_{ji}(x) = \left(I_{a+}^{\nu_i - \rho_i}\left(\Delta_a^{\nu_i} f_{ji}\right)\right)(x), \tag{21.98}$$

$\forall\, x \in [a, b]$; $j = 1, 2$; $i = 1, \ldots, m$.

Assume $\Delta_a^{\nu_i} f_{2i}(x) > 0$, $\forall\, x \in [a, b]$.

Let $j \in \{1, \ldots, m\}$ be fixed. Assume that the function

$$x \to \frac{\left(\prod_{i=1}^{m} \Delta_a^{\nu_i} f_{2i}(y)\right) u(x)\, \chi_{(a, x]}(y)\, (x - y)^{\left(\left(\sum_{i=1}^{m} (\nu_i - \rho_i)\right) - m\right)}}{\left(\prod_{i=1}^{m} \left(\Delta_a^{\rho_i} f_{2i}\right)(x)\right)\left(\prod_{i=1}^{m} \Gamma(\nu_i - \rho_i)\right)} \tag{21.99}$$

is integrable on (a, b), for each $y \in (a, b)$.

Here we have

$$W_m(y) := \left(\prod_{i=1}^{m} \left(\frac{\Delta_a^{\nu_i} f_{2i}(y)}{\Gamma(\nu_i - \rho_i)}\right)\right) \int_y^b u(x) \left(\frac{(x - y)^{\left(\left(\sum_{i=1}^{m} (\nu_i - \rho_i)\right) - m\right)}}{\prod_{i=1}^{m} \left(\Delta_a^{\rho_i} f_{2i}\right)(x)}\right) dx < \infty, \tag{21.100}$$

for any $y \in (a, b)$.

Here $\Phi_i : \mathbb{R}_+ \to \mathbb{R}_+$, $i = 1, \ldots, m$, are convex and increasing functions.

Proposition 21.31 *It holds*

$$\int_a^b u(x) \prod_{i=1}^m \Phi_i \left(\left| \frac{\Delta_a^{\rho_i} f_{1i}(x)}{\Delta_a^{\rho_i} f_{2i}(x)} \right| \right) dx \le \tag{21.101}$$

$$\left(\prod_{\substack{i=1 \\ i \ne j}}^m \int_a^b \Phi_i \left(\left| \frac{\Delta_a^{\nu_i} f_{1i}(y)}{\Delta_a^{\nu_i} f_{2i}(y)} \right| \right) dy \right) \left(\int_a^b \Phi_j \left(\left| \frac{\Delta_a^{\nu_j} f_{1j}(y)}{\Delta_a^{\nu_j} f_{2j}(y)} \right| \right) W_m(y) \, dy \right),$$

under the properties: ($i = 1, \ldots, m$)

(i) $\Phi_i \left(\left| \frac{\Delta_a^{\nu_i} f_{1i}(y)}{\Delta_a^{\nu_i} f_{2i}(y)} \right| \right)$ *is* $\frac{\chi_{(a,x]}(y)(x-y)^{(\nu_i - \rho_i - 1)}}{\Gamma(\nu_i - \rho_i)} \left(\Delta_a^{\nu_i} f_{2i}(y) \right) dy$ *-integrable, a.e. in* $x \in$
    (a, b),

(ii) $W_m(y) \Phi_j \left(\left| \frac{\Delta_a^{\nu_j} f_{1j}(y)}{\Delta_a^{\nu_j} f_{2j}(y)} \right| \right)$; *and* $\Phi_i \left(\left| \frac{\Delta_a^{\nu_i} f_{1i}(y)}{\Delta_a^{\nu_i} f_{2i}(y)} \right| \right)$, *for* $i \in \{1, \ldots, m\} - \{j\}$ *are*
    all Lebesgue integrable.

Proof By Proposition 21.9. ∎

Corollary 21.32 *Let* τ *be a fixed prime number. It holds*

$$\int_a^b u(x) \tau^{\left(\sum_{i=1}^m \left| \frac{\Delta_a^{\rho_i} f_{1i}(x)}{\Delta_a^{\rho_i} f_{2i}(x)} \right| \right)} dx \le \tag{21.102}$$

$$\left(\prod_{\substack{i=1 \\ i \ne j}}^m \int_a^b \tau^{\left| \frac{\Delta_a^{\nu_i} f_{1i}(y)}{\Delta_a^{\nu_i} f_{2i}(y)} \right|} dy \right) \left(\int_a^b \tau^{\left| \frac{\Delta_a^{\nu_j} f_{1j}(y)}{\Delta_a^{\nu_j} f_{2j}(y)} \right|} W_m(y) \, dy \right),$$

under the properties: ($i = 1, \ldots, m$)

(i) $\tau^{\left| \frac{\Delta_a^{\nu_i} f_{1i}(y)}{\Delta_a^{\nu_i} f_{2i}(y)} \right|}$ *is* $\frac{\chi_{(a,x]}(y)(x-y)^{(\nu_i - \rho_i - 1)}}{\Gamma(\nu_i - \rho_i)} \left(\Delta_a^{\nu_i} f_{2i}(y) \right) dy$ *-integrable, a.e. in* $x \in (a, b)$,

(ii) $W_m(y) \tau^{\left| \frac{\Delta_a^{\nu_j} f_{1j}(y)}{\Delta_a^{\nu_j} f_{2j}(y)} \right|}$; *and* $\tau^{\left| \frac{\Delta_a^{\nu_i} f_{1i}(y)}{\Delta_a^{\nu_i} f_{2i}(y)} \right|}$, *for* $i \in \{1, \ldots, m\} - \{j\}$ *are all Lebesgue*
    integrable.

Proof By Proposition 21.31. ∎

We need

Definition 21.33 ([3]) Let $\nu > 0$, $n := [\nu]$, $\alpha = \nu - n$, $0 < \alpha < 1$, $f \in C([a, b])$.
Consider

$$C_{b-}^{\nu}([a, b]) := \{f \in C^n([a, b]) : I_{b-}^{1-\alpha} f^{(n)} \in C^1([a, b])\}. \qquad (21.103)$$

Define the right generalized ν-fractional derivative of f over $[a, b]$, by

$$\Delta_{b-}^{\nu} f := (-1)^{n-1} \left(I_{b-}^{1-\alpha} f^{(n)}\right)'. \qquad (21.104)$$

We set $\Delta_{b-}^0 f = f$. Notice that

$$\left(\Delta_{b-}^{\nu} f\right)(x) = \frac{(-1)^{n-1}}{\Gamma(1-\alpha)} \frac{d}{dx} \int_x^b (J - x)^{-\alpha} f^{(n)}(J) \, dJ, \qquad (21.105)$$

and $\Delta_{b-}^{\nu} f \in C([a, b])$.

We also need

Theorem 21.34 ([4]) *Let $f \in C_{b-}^{\nu}([a, b])$, $0 < \rho < \nu$. Assume $f^{(i)}(b) = 0$, $i = r, r+1, \ldots, n-1$, where $r := [\rho]$, $n := [\nu]$. Then*

$$\Delta_{b-}^{\rho} f(x) = \frac{1}{\Gamma(\nu - \rho)} \int_x^b (J - x)^{\nu - \rho - 1} \left(\Delta_{b-}^{\nu} f\right)(J) \, dJ, \qquad (21.106)$$

$\forall x \in [a, b]$, *i.e.*

$$\Delta_{b-}^{\rho} f = I_{b-}^{\nu - \rho} \left(\Delta_{b-}^{\nu} f\right) \in C([a, b]), \qquad (21.107)$$

and $f \in C_{b-}^{\rho}([a, b])$.

Let $\nu_i > \rho_i > 0$, $n_i := [\nu_i]$, $r_i := [\rho_i]$, $i = 1, \ldots, m$. Let $f_{1i}, f_{2i} \in C_{b-}^{\nu_i}([a, b])$, such that $(j = 1, 2)$ $f_{ji}^{(k_i)}(b) = 0$, $k_i = r_i, r_i + 1, \ldots, n_i - 1$. Notice here $\Delta_{b-}^{\nu_i} f_{ji} \in C([a, b])$, and $\Delta_{b-}^{\rho_i} f_{ji} \in C([a, b])$. Based on Definition 21.33 and Theorem 21.34 we get

$$\Delta_{b-}^{\rho_i} f_{ji}(x) = \left(I_{b-}^{\nu_i - \rho_i} \left(\Delta_{b-}^{\nu_i} f_{ji}\right)\right)(x), \qquad (21.108)$$

$\forall x \in [a, b]$; $j = 1, 2$; $i = 1, \ldots, m$.

Assume $\Delta_{b-}^{\nu_i} f_{2i}(x) > 0$, $\forall x \in [a, b]$.

Let $j \in \{1, \ldots, m\}$ be fixed. Assume that the function

$$x \to \frac{\left(\prod_{i=1}^m \Delta_{b-}^{\nu_i} f_{2i}(y)\right) u(x) \chi_{[x,b)}(y) (y - x)^{\left(\left(\sum_{i=1}^m (\nu_i - \rho_i)\right) - m\right)}}{\left(\prod_{i=1}^m \left(\Delta_{b-}^{\rho_i} f_{2i}\right)(x)\right) \left(\prod_{i=1}^m \Gamma(\nu_i - \rho_i)\right)} \qquad (21.109)$$

is integrable on (a, b), for each $y \in (a, b)$.

Here we have

$$W_m^*(y) := \left(\prod_{i=1}^{m} \left(\frac{\Delta_{b-}^{\nu_i} f_{2i}(y)}{\Gamma(\nu_i - \rho_i)} \right) \right) \int_a^y u(x) \left(\frac{(y-x)^{\left(\left(\sum_{i=1}^{m} (\nu_i - \rho_i) \right) - m \right)}}{\prod_{i=1}^{m} \left(\Delta_{b-}^{\rho_i} f_{2i} \right)(x)} \right) dx < \infty,$$

(21.110)

for any $y \in (a, b)$.

Here $\Phi_i : \mathbb{R}_+ \to \mathbb{R}_+, i = 1, \ldots, m$, are convex and increasing functions.

Proposition 21.35 *It holds*

$$\int_a^b u(x) \prod_{i=1}^{m} \Phi_i \left(\left| \frac{\Delta_{b-}^{\rho_i} f_{1i}(x)}{\Delta_{b-}^{\rho_i} f_{2i}(x)} \right| \right) dx \leq \qquad (21.111)$$

$$\left(\prod_{\substack{i=1 \\ i \neq j}}^{m} \int_a^b \Phi_i \left(\left| \frac{\Delta_{b-}^{\nu_i} f_{1i}(y)}{\Delta_{b-}^{\nu_i} f_{2i}(y)} \right| \right) dy \right) \left(\int_a^b \Phi_j \left(\left| \frac{\Delta_{b-}^{\nu_j} f_{1j}(y)}{\Delta_{b-}^{\nu_j} f_{2j}(y)} \right| \right) W_m^*(y) \, dy \right),$$

under the properties: $(i = 1, \ldots, m)$

(i) $\Phi_i \left(\left| \frac{\Delta_{b-}^{\nu_i} f_{1i}(y)}{\Delta_{b-}^{\nu_i} f_{2i}(y)} \right| \right)$ *is* $\frac{\chi_{[x,b)}(y)(y-x)^{(\nu_i - \rho_i - 1)}}{\Gamma(\nu_i - \rho_i)} \left(\Delta_{b-}^{\nu_i} f_{2i}(y) \right) dy$ *-integrable, a.e. in* $x \in (a, b)$,

(ii) $W_m^*(y) \Phi_j \left(\left| \frac{\Delta_{b-}^{\nu_j} f_{1j}(y)}{\Delta_{b-}^{\nu_j} f_{2j}(y)} \right| \right)$; *and* $\Phi_i \left(\left| \frac{\Delta_{b-}^{\nu_i} f_{1i}(y)}{\Delta_{b-}^{\nu_i} f_{2i}(y)} \right| \right)$, *for* $i \in \{1, \ldots, m\} - \{j\}$ *are all Lebesgue integrable.*

Proof By Proposition 21.13. ∎

Corollary 21.36 *Let τ be a fixed prime number. It holds*

$$\int_a^b u(x) \tau^{\left(\sum_{i=1}^{m} \left| \frac{\Delta_{b-}^{\rho_i} f_{1i}(x)}{\Delta_{b-}^{\rho_i} f_{2i}(x)} \right| \right)} dx \leq \qquad (21.112)$$

$$\left(\prod_{\substack{i=1 \\ i \neq j}}^{m} \int_a^b \tau^{\left| \frac{\Delta_{b-}^{\nu_i} f_{1i}(y)}{\Delta_{b-}^{\nu_i} f_{2i}(y)} \right|} dy \right) \left(\int_a^b \tau^{\left| \frac{\Delta_{b-}^{\nu_j} f_{1j}(y)}{\Delta_{b-}^{\nu_j} f_{2j}(y)} \right|} W_m^*(y) \, dy \right),$$

under the properties: $(i = 1, \ldots, m)$

(i) $\tau \begin{vmatrix} \Delta_{b-}^{\nu_i} f_{1i}(y) \\ \Delta_{b-}^{\nu_i} f_{2i}(y) \end{vmatrix}$ is $\frac{\chi_{[x,b)}(y)(y-x)^{(\nu_i-\rho_i-1)}}{\Gamma(\nu_i-\rho_i)} \left(\Delta_{b-}^{\nu_i} f_{2i}(y)\right) dy$ -integrable, a.e. in $x \in (a,b)$,

(ii) $W_m^*(y) \tau \begin{vmatrix} \Delta_{b-}^{\nu_j} f_{1j}(y) \\ \Delta_{b-}^{\nu_j} f_{2j}(y) \end{vmatrix}$; and $\tau \begin{vmatrix} \Delta_{b-}^{\nu_i} f_{1i}(y) \\ \Delta_{b-}^{\nu_i} f_{2i}(y) \end{vmatrix}$, for $i \in \{1,\ldots,m\} - \{j\}$ are all Lebesgue integrable.

Proof By Proposition 21.35. ∎

We need

Definition 21.37 ([15], *p.* 99) The fractional integrals of a function f with respect to given function g are defined as follows:

Let $a, b \in \mathbb{R}$, $a < b$, $\alpha > 0$. Here g is an increasing function on $[a, b]$, and $g \in C^1([a, b])$. The left- and right-sided fractional integrals of a function f with respect to another function g in $[a, b]$ are given by

$$\left(I_{a+;g}^{\alpha} f\right)(x) = \frac{1}{\Gamma(\alpha)} \int_a^x \frac{g'(t) f(t) dt}{(g(x) - g(t))^{1-\alpha}}, \quad x > a, \tag{21.113}$$

$$\left(I_{b-;g}^{\alpha} f\right)(x) = \frac{1}{\Gamma(\alpha)} \int_x^b \frac{g'(t) f(t) dt}{(g(t) - g(x))^{1-\alpha}}, \quad x < b, \tag{21.114}$$

respectively.

We make

Remark 21.38 Let f_{1i}, f_{2i} be Lebesgue measurable functions from (a, b) into $\mathbb{R}$, such that $\left(I_{a+;g}^{\alpha_i}(|f_{ji}|)\right)(x) \in \mathbb{R}$, $\forall x \in (a, b)$, $\alpha_i > 0$, $i = 1, \ldots, m$, $j = 1, 2$.

Consider

$$g_{ji}(x) = \left(I_{a+;g}^{\alpha_i}(f_{ji})\right)(x), \tag{21.115}$$

$x \in (a, b)$, $i = 1, \ldots, m$, $j = 1, 2$.

Assume $0 < f_{2i}(y) < \infty$, a.e., $i = 1, \ldots, m$.

Notice that $g_{ji}(x) \in \mathbb{R}$ and it is Lebesgue measurable. We pick again $\Omega_1 = \Omega_2 = (a, b)$, $d\mu_1(x) = dx$, $d\mu_2(y) = dy$, the Lebesgue measure.

Here we have

$$k_i(x, y) = \frac{\chi_{(a,x]}(y) g'(y)}{\Gamma(\alpha_i)(g(x) - g(y))^{1-\alpha_i}}, \quad i = 1, \ldots, m. \tag{21.116}$$

Let $j \in \{1, \ldots, m\}$ be fixed.

Assume that the function

$$
x \to \left(\frac{u(x) \left(\prod_{i=1}^{m} f_{2i}(y) \right) \chi_{(a,x]}(y) \left(g'(y) \right)^{m} \left(g(x) - g(y) \right)^{\left(\sum_{i=1}^{m} \alpha_i \right) - m}}{\left(\prod_{i=1}^{m} \Gamma(\alpha_i) \right) \left(\prod_{i=1}^{m} \left(I_{a+;g}^{\alpha_i} f_{2i} \right)(x) \right)} \right)
$$

(21.117)

is integrable on (a, b), for each $y \in (a, b)$.
   Define ρ_m^g on (a, b) by

$$
\rho_m^g(y) = \frac{\left(\prod_{i=1}^{m} f_{2i}(y) \right) \left(g'(y) \right)^{m}}{\left(\prod_{i=1}^{m} \Gamma(\alpha_i) \right)} \int_{y}^{b} \frac{u(x) \left(g(x) - g(y) \right)^{\left(\sum_{i=1}^{m} \alpha_i \right) - m}}{\prod_{i=1}^{m} \left(I_{a+;g}^{\alpha_i} f_{2i} \right)(x)} dx < \infty,
$$

(21.118)

Here $\Phi_i : \mathbb{R}_+ \to \mathbb{R}_+, i = 1, \ldots, m$, are convex and increasing functions.

Theorem 21.39 *Here all are as in Remark 21.38. It holds*

$$
\int_{a}^{b} u(x) \prod_{i=1}^{m} \Phi_i \left(\left| \frac{\left(I_{a+;g}^{\alpha_i} (f_{1i}) \right)(x)}{\left(I_{a+;g}^{\alpha_i} (f_{2i}) \right)(x)} \right| \right) dx \leq
$$

(21.119)

$$
\left(\prod_{\substack{i=1 \\ i \neq j}}^{m} \int_{a}^{b} \Phi_i \left(\left| \frac{f_{1i}(y)}{f_{2i}(y)} \right| \right) dy \right) \left(\int_{a}^{b} \Phi_j \left(\left| \frac{f_{1j}(y)}{f_{2j}(y)} \right| \right) \rho_m^g(y) dy \right),
$$

true for all measurable functions $f_{1i}, f_{2i} : (a, b) \to \mathbb{R}$ *such that* $\left(I_{a+;g}^{\alpha_i} \left(|f_{ji}| \right) \right)(x)$
$\in \mathbb{R}, \forall x \in (a, b), \alpha_i > 0, i = 1, \ldots, m; j = 1, 2,$ *with:*

(i) $\frac{f_{1i}(y)}{f_{2i}(y)}, \Phi_i \left(\left| \frac{f_{1i}(y)}{f_{2i}(y)} \right| \right)$ *are both* $\frac{\chi_{(a,x]}(y) g'(y)}{\Gamma(\alpha_i)(g(x)-g(y))^{1-\alpha_i}} f_{2i}(y) dy$ *-integrable, a.e. in*
   $x \in (a, b)$,
(ii) $\rho_m^g(y) \Phi_j \left(\left| \frac{f_{1j}(y)}{f_{2j}(y)} \right| \right)$; *and* $\Phi_i \left(\left| \frac{f_{1i}(y)}{f_{2i}(y)} \right| \right),$ *for* $i \in \{1, \ldots, m\} - \{j\}$ *are all inte-*
   grable.

Proof By Theorem 21.3. ∎

   We make

Remark 21.40 Let f_{1i}, f_{2i} be Lebesgue measurable functions from (a, b) into $\mathbb{R}$, such that $\left(I_{b-;g}^{\alpha_i}\left(|f_{ji}|\right)\right)(x) \in \mathbb{R}$, $\forall x \in (a, b)$, $\alpha_i > 0$, $i = 1, \ldots, m$, $j = 1, 2$.

Consider now

$$g_{ji}(x) = \left(I_{b-;g}^{\alpha_i}\left(f_{ji}\right)\right)(x), \tag{21.120}$$

$x \in (a, b)$, $i = 1, \ldots, m$, $j = 1, 2$.

Assume $0 < f_{2i}(y) < \infty$, a.e., $i = 1, \ldots, m$.

Notice that $g_{ji}(x) \in \mathbb{R}$ and it is Lebesgue measurable.

Here we have

$$k_i(x, y) = \frac{\chi_{[x,b)}(y) \, g'(y)}{\Gamma(\alpha_i) \, (g(y) - g(x))^{1-\alpha_i}}, \quad i = 1, \ldots, m. \tag{21.121}$$

Let $j \in \{1, \ldots, m\}$ be fixed.

Assume that the function

$$x \to \left(\frac{u(x) \left(\prod_{i=1}^{m} f_{2i}(y)\right) \chi_{[x,b)}(y) \, (g'(y))^m \, (g(y) - g(x))^{\left(\sum_{i=1}^{m} \alpha_i\right) - m}}{\left(\prod_{i=1}^{m} \Gamma(\alpha_i)\right)\left(\prod_{i=1}^{m} \left(I_{b-;g}^{\alpha_i}(f_{2i})\right)(x)\right)} \right) \tag{21.122}$$

is integrable on (a, b), for each $y \in (a, b)$.

Define $\overline{\rho}_m^g$ on (a, b) by

$$\overline{\rho}_m^g(y) = \frac{\left(\prod_{i=1}^{m} f_{2i}(y)\right)(g'(y))^m}{\left(\prod_{i=1}^{m} \Gamma(\alpha_i)\right)} \int_a^y \frac{u(x)\,(g(y) - g(x))^{\left(\sum_{i=1}^{m}\alpha_i\right)-m}}{\prod_{i=1}^{m}\left(I_{b-;g}^{\alpha_i}(f_{2i})\right)(x)} dx < \infty, \tag{21.123}$$

Here $\Phi_i : \mathbb{R}_+ \to \mathbb{R}_+$, $i = 1, \ldots, m$, are convex and increasing functions.

Theorem 21.41 *Here all are as in Remark 21.40. It holds*

$$\int_a^b u(x) \prod_{i=1}^{m} \Phi_i \left(\left| \frac{\left(I_{b-;g}^{\alpha_i}(f_{1i})\right)(x)}{\left(I_{b-;g}^{\alpha_i}(f_{2i})\right)(x)} \right| \right) dx \leq \tag{21.124}$$

$$\left(\prod_{\substack{i=1 \\ i \neq j}}^{m} \int_a^b \Phi_i \left(\left| \frac{f_{1i}(y)}{f_{2i}(y)} \right| \right) dy \right) \left(\int_a^b \Phi_j \left(\left| \frac{f_{1j}(y)}{f_{2j}(y)} \right| \right) \overline{\rho}_m^g(y)\, dy \right),$$

true for all measurable functions f_{1i}, $f_{2i} : (a, b) \to \mathbb{R}$ *such that* $\left(I_{b-;g}^{\alpha_i} \left(|f_{ji}|\right)\right)(x)$
$\in \mathbb{R}, \forall \, x \in (a, b), \alpha_i > 0, i = 1, \ldots, m; \, j = 1, 2,$ *under the properties:*

(i) $\frac{f_{1i}(y)}{f_{2i}(y)}$, $\Phi_i \left(\left|\frac{f_{1i}(y)}{f_{2i}(y)}\right|\right)$ *are both* $\frac{\chi_{[x,b)}(y)g'(y)}{\Gamma(\alpha_i)(g(y)-g(x))^{1-\alpha_i}} f_{2i}(y) \, dy$ *-integrable, a.e. in*
$x \in (a, b),$

(ii) $\overline{P_m^g}(y) \Phi_j \left(\left|\frac{f_{1j}(y)}{f_{2j}(y)}\right|\right)$; *and* $\Phi_i \left(\left|\frac{f_{1i}(y)}{f_{2i}(y)}\right|\right)$, *for* $i \in \{1, \ldots, m\} - \{j\}$ *are all*
Lebesgue integrable.

Proof By Theorem 21.3. ∎

We need

Definition 21.42 ([13]) Let $0 < a < b < \infty, \alpha > 0$. The left- and right-sided
Hadamard fractional integrals of order α are given by

$$\left(J_{a+}^{\alpha} f\right)(x) = \frac{1}{\Gamma(\alpha)} \int_a^x \left(\ln \frac{x}{y}\right)^{\alpha-1} \frac{f(y)}{y} dy, \quad x > a, \tag{21.125}$$

and

$$\left(J_{b-}^{\alpha} f\right)(x) = \frac{1}{\Gamma(\alpha)} \int_x^b \left(\ln \frac{y}{x}\right)^{\alpha-1} \frac{f(y)}{y} dy, \quad x < b, \tag{21.126}$$

respectively.

Notice that the Hadamard fractional integrals of order α are special cases of left-
and right-sided fractional integrals of a function f with respect to another function,
here $g(x) = \ln x$ on $[a, b], 0 < a < b < \infty$.

Above f is a Lebesgue measurable function from (a, b) into $\mathbb{R}$, such that
$\left(J_{a+}^{\alpha}(|f|)\right)(x), \left(J_{b-}^{\alpha}(|f|)\right)(x) \in \mathbb{R}, \forall \, x \in (a, b)$.

We make

Remark 21.43 Let $(f_{1i}, f_{2i}, \alpha_i), i = 1, \ldots, m,$ and $\left(J_{a+}^{\alpha_i} f_{ji}\right), j = 1, 2,$ all as in
Definition 21.42.

Assume $0 < f_{2i}(y) < \infty$, a.e., $i = 1, \ldots, m$.

Let $j \in \{1, \ldots, m\}$ be fixed.

Assume that the function

$$x \to \left(\frac{u(x) \left(\prod_{i=1}^m f_{2i}(y)\right) \chi_{(a,x]}(y) \ln \left(\frac{x}{y}\right)^{\left(\sum_{i=1}^m \alpha_i\right)-m}}{y^m \left(\prod_{i=1}^m \Gamma(\alpha_i)\right) \left(\prod_{i=1}^m \left(J_{a+}^{\alpha_i} f_{2i}\right)(x)\right)}\right) \tag{21.127}$$

is integrable on (a, b), for each $y \in (a, b)$.

Define γ_m^g on (a, b) by

$$\gamma_m^g(y) = \frac{\left(\prod_{i=1}^{m} f_{2i}(y)\right)}{y^m \left(\prod_{i=1}^{m} \Gamma(\alpha_i)\right)} \int_y^b \frac{u(x) \left(\ln\left(\frac{x}{y}\right)\right)^{\left(\sum_{i=1}^{m} \alpha_i\right) - m}}{\prod_{i=1}^{m} \left(J_{a+}^{\alpha_i} f_{2i}\right)(x)} dx < \infty, \quad (21.128)$$

Here $\Phi_i : \mathbb{R}_+ \to \mathbb{R}_+$, $i = 1, \ldots, m$, are convex and increasing functions.

Theorem 21.44 *Here all are as in Remark 21.43. It holds*

$$\int_a^b u(x) \prod_{i=1}^{m} \Phi_i \left(\left|\frac{\left(J_{a+}^{\alpha_i}(f_{1i})\right)(x)}{\left(J_{a+}^{\alpha_i}(f_{2i})\right)(x)}\right|\right) dx \leq \quad (21.129)$$

$$\left(\prod_{\substack{i=1 \\ i \neq j}}^{m} \int_a^b \Phi_i \left(\left|\frac{f_{1i}(y)}{f_{2i}(y)}\right|\right) dy\right) \left(\int_a^b \Phi_j \left(\left|\frac{f_{1j}(y)}{f_{2j}(y)}\right|\right) \gamma_m^g(y) \, dy\right),$$

under the assumptions:

(i) $\frac{f_{1i}(y)}{f_{2i}(y)}$, $\Phi_i \left(\left|\frac{f_{1i}(y)}{f_{2i}(y)}\right|\right)$ *are both* $\dfrac{\chi_{(a,x]}(y)}{y \Gamma(\alpha_i) \left(\ln\left(\frac{x}{y}\right)\right)^{1-\alpha_i}} f_{2i}(y) \, dy$ *-integrable, a.e. in $x \in (a, b)$,*

(ii) $\gamma_m^g(y) \Phi_j \left(\left|\frac{f_{1j}(y)}{f_{2j}(y)}\right|\right)$; *and* $\Phi_i \left(\left|\frac{f_{1i}(y)}{f_{2i}(y)}\right|\right)$, *for $i \in \{1, \ldots, m\} - \{j\}$ are all integrable.*

Proof By Theorem 21.39. ∎

We make

Remark 21.45 Let $(f_{1i}, f_{2i}, \alpha_i)$, $i = 1, \ldots, m$, and $\left(J_{b-}^{\alpha_i} f_{ji}\right)$, $j = 1, 2$, all as in Definition 21.42.

Assume $0 < f_{2i}(y) < \infty$, a.e., $i = 1, \ldots, m$.

Let $j \in \{1, \ldots, m\}$ be fixed.

Suppose that the function

$$x \to \left(\frac{u(x) \left(\prod_{i=1}^{m} f_{2i}(y)\right) \chi_{[x,b)}(y) \ln\left(\frac{y}{x}\right)^{\left(\sum_{i=1}^{m} \alpha_i\right) - m}}{y^m \left(\prod_{i=1}^{m} \Gamma(\alpha_i)\right) \left(\prod_{i=1}^{m} \left(J_{b-}^{\alpha_i}(f_{2i})\right)(x)\right)}\right) \quad (21.130)$$

is integrable on (a, b), for each $y \in (a, b)$.

Define $\overline{\gamma}_m^g$ on (a, b) by

$$\overline{\gamma}_m^g (y) = \frac{\left(\prod\limits_{i=1}^{m} f_{2i}(y)\right)}{y^m \left(\prod\limits_{i=1}^{m} \Gamma(\alpha_i)\right)} \int_a^y \frac{u(x) \left(\ln\left(\frac{y}{x}\right)\right)^{\left(\sum\limits_{i=1}^{m} \alpha_i\right) - m}}{\prod\limits_{i=1}^{m} \left(J_{b-}^{\alpha_i}(f_{2i})\right)(x)} dx < \infty, \quad (21.131)$$

Here $\Phi_i : \mathbb{R}_+ \to \mathbb{R}_+$, $i = 1, \ldots, m$, are convex and increasing functions.

Theorem 21.46 *Here all as in Remark 21.45. It holds*

$$\int_a^b u(x) \prod_{i=1}^{m} \Phi_i \left(\left|\frac{\left(J_{b-}^{\alpha_i}(f_{1i})\right)(x)}{\left(J_{b-}^{\alpha_i}(f_{2i})\right)(x)}\right|\right) dx \leq \quad (21.132)$$

$$\left(\prod_{\substack{i=1 \\ i \neq j}}^{m} \int_a^b \Phi_i \left(\left|\frac{f_{1i}(y)}{f_{2i}(y)}\right|\right) dy\right) \left(\int_a^b \Phi_j \left(\left|\frac{f_{1j}(y)}{f_{2j}(y)}\right|\right) \overline{\gamma}_m^g(y)\, dy\right),$$

under the assumptions:

(i) $\frac{f_{1i}(y)}{f_{2i}(y)}$, $\Phi_i \left(\left|\frac{f_{1i}(y)}{f_{2i}(y)}\right|\right)$ *are both* $\dfrac{\chi_{[x,b)}(y)}{y \Gamma(\alpha_i)(\ln(\frac{y}{x}))^{1-\alpha_i}} f_{2i}(y)\, dy$ *-integrable, a.e. in $x \in$*
(a, b),

(ii) $\overline{\gamma}_m^g(y) \Phi_j \left(\left|\frac{f_{1j}(y)}{f_{2j}(y)}\right|\right)$; *and* $\Phi_i \left(\left|\frac{f_{1i}(y)}{f_{2i}(y)}\right|\right)$, *for $i \in \{1, \ldots, m\} - \{j\}$ are all integrable.*

Proof By Theorem 21.41. ∎

Corollary 21.47 (to Theorem 21.44) *It holds*

$$\int_a^b u(x)\, e^{\sum\limits_{i=1}^{m} \left|\frac{\left(J_{a+}^{\alpha_i}(f_{1i})\right)(x)}{\left(J_{a+}^{\alpha_i}(f_{2i})\right)(x)}\right|} dx \leq \quad (21.133)$$

$$\left(\prod_{\substack{i=1 \\ i \neq j}}^{m} \int_a^b e^{\left|\frac{f_{1i}(y)}{f_{2i}(y)}\right|} dy\right) \left(\int_a^b e^{\left|\frac{f_{1j}(y)}{f_{2j}(y)}\right|} \gamma_m^g(y)\, dy\right),$$

under the assumptions:

(i) $\frac{f_{1i}(y)}{f_{2i}(y)}$, $e^{\left|\frac{f_{1i}(y)}{f_{2i}(y)}\right|}$ *are both* $\dfrac{\chi_{(a,x]}(y)}{y \Gamma(\alpha_i)\left(\ln\left(\frac{x}{y}\right)\right)^{1-\alpha_i}} f_{2i}(y)\, dy$ *-integrable, a.e. in $x \in (a, b)$,*

(ii) $\gamma_m^g(y)\, e^{\left|\frac{f_{1j}(y)}{f_{2j}(y)}\right|}$; and $e^{\left|\frac{f_{1i}(y)}{f_{2i}(y)}\right|}$, for $i \in \{1, \dots, m\} - \{j\}$ are all integrable.

Corollary 21.48 (to Theorem 21.46) *Let $p_i \geq 1$. It holds*

$$\int_a^b u(x) \left(\prod_{i=1}^m \left| \frac{(J_{b-}^{\alpha_i}(f_{1i}))(x)}{(J_{b-}^{\alpha_i}(f_{2i}))(x)} \right|^{p_i} \right) dx \leq \tag{21.134}$$

$$\left(\prod_{\substack{i=1\\i \neq j}}^m \int_a^b \left| \frac{f_{1i}(y)}{f_{2i}(y)} \right|^{p_i} dy \right) \left(\int_a^b \left| \frac{f_{1j}(y)}{f_{2j}(y)} \right|^{p_j} \overline{\gamma}_m^g(y)\, dy \right),$$

under the assumptions

(i) $\left| \frac{f_{1i}(y)}{f_{2i}(y)} \right|^{p_i}$ *is* $\frac{\chi_{[x,b)}(y)}{y\Gamma(\alpha_i)(\ln(\frac{y}{x}))^{1-\alpha_i}} f_{2i}(y)\, dy$ *-integrable, a.e. in $x \in (a, b)$,*

(ii) $\overline{\gamma}_m^g(y) \left| \frac{f_{1j}(y)}{f_{2j}(y)} \right|^{p_j}$; *and* $\left| \frac{f_{1i}(y)}{f_{2i}(y)} \right|^{p_i}$, *for $i \in \{1, \dots, m\} - \{j\}$ are all integrable.*

Appendix

In this chapter we used a lot the following

Proposition 21.49 *Let $f : [0, \infty) \to \mathbb{R}$ be convex and increasing. Then f is continuous on $[0, \infty)$.*

Proof Fact: f is continuous on $(0, \infty)$, it is known. We want to prove that f is continuous at $x = 0$. Let $\nu > 0$ be fixed. Consider the line (l) through $(0, f(0))$ and $(\nu, f(\nu))$. It has slope $\frac{f(\nu)-f(0)}{\nu} \geq 0$, and equation $y = l(x) = \left(\frac{f(\nu)-f(0)}{\nu}\right) x + f(0)$. We can always pick up $\nu : f(\nu) > f(0)$, otherwise if for all $\nu > 0$ it is $f(\nu) = f(0)$, we have the trivial case of continuity.

By convexity of f we have that for any $0 < x < \nu$, it is $f(x) \leq l(x)$, equivalently,

$$f(x) \leq \left(\frac{f(\nu) - f(0)}{\nu} \right) x + f(0),$$

equivalently,

$$0 \leq f(x) - f(0) \leq \left(\frac{f(\nu) - f(0)}{\nu} \right) x;$$

here $\left(\frac{f(\nu)-f(0)}{\nu} \right) > 0$.

Let $x \to 0$, then $f(x) - f(0) \to 0$. That is $\lim_{x \to 0} f(x) = f(0)$, proving continuity of f at $x = 0$. $\blacksquare$

References

1. G.A. Anastassiou, *Univariate Hardy type fractional inequalities*. To appear, *Advances in Applied Mathematics and Approximation Theory—Contributions from AMAT 2012*, edited volume by G. Anastassiou and O. Duman (Springer, New York, 2013)
2. G.A. Anastassiou, *Fractional Differentiation Inequalities*, Research Monograph (Springer, New York, 2009)
3. G.A. Anastassiou, On right fractional calculus. Chaos, Solitons and Fractals **42**, 365–376 (2009)
4. G.A. Anastassiou, Fractional representation formulae and right fractional inequalities. Math. Comput. Modell. **54**(11–12), 3098–3115 (2011)
5. G.A. Anastassiou, Rational inequalities for integral operators under convexity. Commun. Appl. Anal. **16**(2), 179–210 (2012)
6. M. Andric, J.E. Pecaric, I. Peric, *A multiple Opial type inequality due to Fink for the Riemann-Liouville fractional derivatives*. Submitted (2012)
7. M. Andric, J.E. Pecaric, I. Peric, *Composition identities for the Caputo fractional derivatives and applications to Opial-type inequalities*. Submitted (2012)
8. J.A. Canavati, The Riemann-Liouville integral. Nieuw Archief Voor Wiskunde **5**(1), 53–75 (1987)
9. K. Diethelm, *The Analysis of Fractional Differential Equations*, vol. 2004, 1st edn., Lecture Notes in Mathematics (Springer, New York, 2010)
10. A.M.A. El-Sayed, M. Gaber, On the finite Caputo and finite Riesz derivatives. Electron. J. Theor. Phys. **3**(12), 81–95 (2006)
11. R. Gorenflo, F. Mainardi, *Essentials of Fractional Calculus* (Maphysto Center, Aarhus, 2000). http://www.maphysto.dk/oldpages/events/LevyCAC2000/MainardiNotes/fm2k0a.ps
12. G.D. Handley, J.J. Koliha, J. Pečarić, Hilbert-Pachpatte type integral inequalities for fractional derivatives. Fractional Calculus Appl. Anal. **4**(1), 37–46 (2001)
13. S. Iqbal, K. Krulic, J. Pecaric, On an inequality of H.G. Hardy. J. Inequalities Appl. **2010**(264347), 23
14. S. Iqbal, K. Krulic, J. Pecaric, On an inequality for convex functions with some applications on fractional derivatives and fractional integrals. J. Math. Inequalities **5**(2), 219–230 (2011)
15. A.A. Kilbas, H.M. Srivastava, J.J. Trujillo, *Theory and Applications of Fractional Differential Equations*, North-Holland Mathematics Studies, vol. 204 (Elsevier, New York, 2006)

Chapter 22
Fractional Integral Inequalities with Convexity

Here we present general integral inequalities involving convex and increasing functions applied to products of functions. As specific applications we derive a wide range of fractional inequalities of Hardy type. These involve the left and right: Erdélyi-Kober fractional integrals, mixed Riemann-Liouville fractional multiple integrals. Next we produce multivariate Poincaré type fractional inequalitites involving left fractional radial derivatives of Canavati type, Riemann-Liouville and Caputo types. The exposed inequalities are of L_p type, $p \geq 1$, and exponential type. It follows [6].

22.1 Introduction

We start with some facts about fractional derivatives needed in the sequel, for more details see, for instance [1, 11].

Let $a < b$, $a, b \in \mathbb{R}$. By $C^N ([a, b])$, we denote the space of all functions on $[a, b]$ which have continuous derivatives up to order N, and $AC ([a, b])$ is the space of all absolutely continuous functions on $[a, b]$. By $AC^N ([a, b])$, we denote the space of all functions g with $g^{(N-1)} \in AC ([a, b])$. For any $\alpha \in \mathbb{R}$, we denote by $[\alpha]$ the integral part of α (the integer k satisfying $k \leq \alpha < k + 1$), and $\lceil \alpha \rceil$ is the ceiling of α ($\min\{n \in \mathbb{N}, n \geq \alpha\}$). By $L_1 (a, b)$, we denote the space of all functions integrable on the interval (a, b), and by $L_\infty (a, b)$ the set of all functions measurable and essentially bounded on (a, b). Clearly, $L_\infty (a, b) \subset L_1 (a, b)$.

We start with the definition of the Riemann-Liouville fractional integrals, see [14]. Let $[a, b]$, $(-\infty < a < b < \infty)$ be a finite interval on the real axis $\mathbb{R}$. The Riemann-Liouville fractional integrals $I_{a+}^\alpha f$ and $I_{b-}^\alpha f$ of order $\alpha > 0$ are defined by

$$\left(I_{a+}^\alpha f \right) (x) = \frac{1}{\Gamma (\alpha)} \int_a^x f (t) (x - t)^{\alpha-1} dt, \quad (x > a), \tag{22.1}$$

© Springer International Publishing Switzerland 2016
G.A. Anastassiou, *Intelligent Comparisons: Analytic Inequalities*,
Studies in Computational Intelligence 609,
DOI 10.1007/978-3-319-21121-3_22

$$\left(I_{b-}^{\alpha} f\right)(x) = \frac{1}{\Gamma(\alpha)} \int_x^b f(t)(t-x)^{\alpha-1}\, dt, \quad (x < b), \qquad (22.2)$$

respectively. Here $\Gamma(\alpha)$ is the Gamma function. These integrals are called the left-sided and the right-sided fractional integrals. We mention some properties of the operators $I_{a+}^{\alpha} f$ and $I_{b-}^{\alpha} f$ of order $\alpha > 0$, see also [17]. The first result yields that the fractional integral operators $I_{a+}^{\alpha} f$ and $I_{b-}^{\alpha} f$ are bounded in $L_p(a, b)$, $1 \le p \le \infty$, that is

$$\left\| I_{a+}^{\alpha} f \right\|_p \le K \left\| f \right\|_p, \quad \left\| I_{b-}^{\alpha} f \right\|_p \le K \left\| f \right\|_p, \qquad (22.3)$$

where

$$K = \frac{(b-a)^{\alpha}}{\alpha \Gamma(\alpha)}. \qquad (22.4)$$

Inequality (22.3), that is the result involving the left-sided fractional integral, was proved by Hardy in one of his first papers, see [12]. He did not write down the constant, but the calculation of the constant was hidden inside his proof.

Next we follow [13].

Let $(\Omega_1, \Sigma_1, \mu_1)$ and $(\Omega_2, \Sigma_2, \mu_2)$ be measure spaces with positive σ-finite measures, and let $k : \Omega_1 \times \Omega_2 \to \mathbb{R}$ be a nonnegative measurable function, $k(x, \cdot)$ measurable on Ω_2 and

$$K(x) = \int_{\Omega_2} k(x, y)\, d\mu_2(y), \quad x \in \Omega_1. \qquad (22.5)$$

We suppose that $K(x) > 0$ a.e. on Ω_1, and by a weight function (shortly: a weight), we mean a nonnegative measurable function on the actual set. Let the measurable functions $g : \Omega_1 \to \mathbb{R}$ with the representation

$$g(x) = \int_{\Omega_2} k(x, y) f(y)\, d\mu_2(y), \qquad (22.6)$$

where $f : \Omega_2 \to \mathbb{R}$ is a measurable function.

Theorem 22.1 ([13]) *Let u be a weight function on Ω_1, k a nonnegative measurable function on $\Omega_1 \times \Omega_2$, and K be defined on Ω_1 by (22.5). Assume that the function $x \mapsto u(x) \frac{k(x,y)}{K(x)}$ is integrable on Ω_1 for each fixed $y \in \Omega_2$. Define v on Ω_2 by*

$$v(y) := \int_{\Omega_1} u(x) \frac{k(x, y)}{K(x)} d\mu_1(x) < \infty. \qquad (22.7)$$

If $\Phi : [0, \infty) \to \mathbb{R}$ is convex and increasing function, then the inequality

$$\int_{\Omega_1} u(x) \Phi\left(\left| \frac{g(x)}{K(x)} \right| \right) d\mu_1(x) \le \int_{\Omega_2} v(y) \Phi(|f(y)|)\, d\mu_2(y) \qquad (22.8)$$

holds for all measurable functions $f : \Omega_2 \to \mathbb{R}$ *such that:*

(i) $f, \Phi(|f|)$ *are both* $k(x, y) d\mu_2(y)$ *-integrable,* μ_1 *-a.e. in* $x \in \Omega_1$,
(ii) $\nu\Phi(|f|)$ *is* μ_2 *-integrable,*

and for all corresponding functions g given by (22.6).

Important assumptions (i) and (ii) are missing from Theorem 2.1 of [13].

In this chapter we use and generalize Theorem 22.1 for products of several functions and we give wide applications to Fractional Calculus.

22.2 Main Results

Let $(\Omega_1, \Sigma_1, \mu_1)$ and $(\Omega_2, \Sigma_2, \mu_2)$ be measure spaces with positive σ-finite measures, and let $k_i : \Omega_1 \times \Omega_2 \to \mathbb{R}$ be nonnegative measurable functions, $k_i(x, \cdot)$ measurable on Ω_2, and

$$K_i(x) = \int_{\Omega_2} k_i(x, y) d\mu_2(y), \quad \text{for any } x \in \Omega_1, \tag{22.9}$$

$i = 1, \ldots, m$. We assume that $K_i(x) > 0$ a.e. on Ω_1, and the weight functions are nonnegative measurable functions on the related set.

We consider measurable functions $g_i : \Omega_1 \to \mathbb{R}$ with the representation

$$g_i(x) = \int_{\Omega_2} k_i(x, y) f_i(y) d\mu_2(y), \tag{22.10}$$

where $f_i : \Omega_2 \to \mathbb{R}$ are measurable functions, $i = 1, \ldots, m$.

Here u stands for a weight function on Ω_1.

The first introductory result is proved for $m = 2$.

Theorem 22.2 *Assume that the functions* $(i = 1, 2)$ $x \mapsto \left(u(x) \frac{k_i(x,y)}{K_i(x)} \right)$ *are integrable on* Ω_1, *for each fixed* $y \in \Omega_2$. *Define* u_i *on* Ω_2 *by*

$$u_i(y) := \int_{\Omega_1} u(x) \frac{k_i(x, y)}{K_i(x)} d\mu_1(x) < \infty. \tag{22.11}$$

Let $p, q > 1 : \frac{1}{p} + \frac{1}{q} = 1$. *Let the functions* $\Phi_1, \Phi_2 : \mathbb{R}_+ \to \mathbb{R}_+$, *be convex and increasing.*
Then

$$\int_{\Omega_1} u(x) \Phi_1\left(\left| \frac{g_1(x)}{K_1(x)} \right| \right) \Phi_2\left(\left| \frac{g_2(x)}{K_2(x)} \right| \right) d\mu_1(x) \leq$$

$$\left(\int_{\Omega_2} u_1\left(y\right)\Phi_1\left(\left|f_1\left(y\right)\right|\right)^p d\mu_2\left(y\right)\right)^{\frac{1}{p}}\left(\int_{\Omega_2} u_2\left(y\right)\Phi_2\left(\left|f_2\left(y\right)\right|\right)^q d\mu_2\left(y\right)\right)^{\frac{1}{q}},$$

$$(22.12)$$

for all measurable functions $f_i : \Omega_2 \to \mathbb{R}$ ($i = 1, 2$) such that

   (i) f_1, $\Phi_1\left(\left|f_1\right|\right)^p$ are both $k_1\left(x, y\right) d\mu_2\left(y\right)$ -integrable, μ_1 -a.e. in $x \in \Omega_1$,

   (ii) f_2, $\Phi_2\left(\left|f_2\right|\right)^q$ are both $k_2\left(x, y\right) d\mu_2\left(y\right)$ -integrable, μ_1 -a.e. in $x \in \Omega_1$,

   (iii) $u_1\Phi_1\left(\left|f_1\right|\right)^p$, $u_2\Phi_2\left(\left|f_2\right|\right)^q$, are both μ_2 -integrable,

   and for all corresponding functions g_i ($i = 1, 2$) given by (22.10).

Proof Notice that Φ_1, Φ_2 are continuous functions. Here we use Hölder's inequality. We have

$$\int_{\Omega_1} u\left(x\right)\Phi_1\left(\left|\frac{g_1\left(x\right)}{K_1\left(x\right)}\right|\right)\Phi_2\left(\left|\frac{g_2\left(x\right)}{K_2\left(x\right)}\right|\right)d\mu_1\left(x\right) =$$

$$\int_{\Omega_1} u\left(x\right)^{\frac{1}{p}}\Phi_1\left(\left|\frac{g_1\left(x\right)}{K_1\left(x\right)}\right|\right)u\left(x\right)^{\frac{1}{q}}\Phi_2\left(\left|\frac{g_2\left(x\right)}{K_2\left(x\right)}\right|\right)d\mu_1\left(x\right) \leq \qquad (22.13)$$

$$\left(\int_{\Omega_1} u\left(x\right)\Phi_1\left(\left|\frac{g_1\left(x\right)}{K_1\left(x\right)}\right|\right)^p d\mu_1\left(x\right)\right)^{\frac{1}{p}}\cdot$$

$$\left(\int_{\Omega_1} u\left(x\right)\Phi_2\left(\left|\frac{g_2\left(x\right)}{K_2\left(x\right)}\right|\right)^q d\mu_1\left(x\right)\right)^{\frac{1}{q}} \leq$$

(notice here that Φ_1^p, Φ_2^q are convex, increasing and continuous nonnegative functions, and by Theorem 22.1 we get)

$$\left(\int_{\Omega_2} u_1\left(y\right)\Phi_1\left(\left|f_1\left(y\right)\right|\right)^p d\mu_2\left(y\right)\right)^{\frac{1}{p}}\left(\int_{\Omega_2} u_2\left(y\right)\Phi_2\left(\left|f_2\left(y\right)\right|\right)^q d\mu_2\left(y\right)\right)^{\frac{1}{q}}.$$

$$(22.14)$$

∎

   The general result follows

Theorem 22.3 *Assume that the functions ($i = 1, 2, \ldots, m \in \mathbb{N}$) $x \mapsto \left(u\left(x\right)\frac{k_i\left(x, y\right)}{K_i\left(x\right)}\right)$ are integrable on Ω_1, for each fixed $y \in \Omega_2$. Define u_i on Ω_2 by*

$$u_i\left(y\right) := \int_{\Omega_1} u\left(x\right)\frac{k_i\left(x, y\right)}{K_i\left(x\right)}d\mu_1\left(x\right) < \infty. \qquad (22.15)$$

Let $p_i > 1 : \sum_{i=1}^{m}\frac{1}{p_i} = 1$. Let the functions $\Phi_i : \mathbb{R}_+ \to \mathbb{R}_+$, $i = 1, \ldots, m$, be convex and increasing.

Then

$$\int_{\Omega_1} u(x) \prod_{i=1}^{m} \Phi_i \left(\left| \frac{g_i(x)}{K_i(x)} \right| \right) d\mu_1(x) \le$$

$$\prod_{i=1}^{m} \left(\int_{\Omega_2} u_i(y) \Phi_i(|f_i(y)|)^{p_i} d\mu_2(y) \right)^{\frac{1}{p_i}}, \qquad (22.16)$$

for all measurable functions $f_i : \Omega_2 \to \mathbb{R}$ $(i = 1, \ldots, m)$ *such that*

(i) f_i, $\Phi_i(|f_i|)^{p_i}$ *are both* $k_i(x, y) d\mu_2(y)$ *-integrable,* μ_1 *-a.e. in* $x \in \Omega_1$, $i = 1, \ldots, m$,

(ii) $u_i \Phi_i(|f_i|)^{p_i}$ *is* μ_2 *-integrable,* $i = 1, \ldots, m$,

and for all corresponding functions g_i $(i = 1, \ldots, m)$ *given by* (22.10).

Proof Notice that Φ_i, $i = 1, \ldots, m$, are continuous functions. Here we use the generalized Hölder's inequality. We have

$$\int_{\Omega_1} u(x) \prod_{i=1}^{m} \Phi_i \left(\left| \frac{g_i(x)}{K_i(x)} \right| \right) d\mu_1(x) =$$

$$\int_{\Omega_1} \prod_{i=1}^{m} \left(u(x)^{\frac{1}{p_i}} \Phi_i \left(\left| \frac{g_i(x)}{K_i(x)} \right| \right) \right) d\mu_1(x) \le \qquad (22.17)$$

$$\prod_{i=1}^{m} \left(\int_{\Omega_1} u(x) \Phi_i \left(\left| \frac{g_i(x)}{K_i(x)} \right| \right)^{p_i} d\mu_1(x) \right)^{\frac{1}{p_i}} \le$$

(notice here that $\Phi_i^{p_i}$, $i = 1, \ldots, m$, are convex, increasing and continuous, nonnegative functions, and by Theorem 22.1 we get)

$$\prod_{i=1}^{m} \left(\int_{\Omega_2} u_i(y) \Phi_i(|f_i(y)|)^{p_i} d\mu_2(y) \right)^{\frac{1}{p_i}}. \qquad (22.18)$$

proving the claim. ∎

When $k(x, y) := k_1(x, y) = k_2(x, y) = \cdots = k_m(x, y)$, then $K(x) := K_1(x) = K_2(x) = \cdots = K_m(x)$, we get by Theorems 22.2 and 22.3 the following:

Corollary 22.4 *Assume that the function* $x \mapsto \left(u(x) \frac{k(x,y)}{K(x)} \right)$ *is integrable on* Ω_1, *for each fixed* $y \in \Omega_2$. *Define* U *on* Ω_2 *by*

$$U(y) := \int_{\Omega_1} u(x) \frac{k(x, y)}{K(x)} d\mu_1(x) < \infty. \qquad (22.19)$$

Let $p, q > 1 : \frac{1}{p} + \frac{1}{q} = 1$. Let the functions $\Phi_1, \Phi_2 : \mathbb{R}_+ \to \mathbb{R}_+$, be convex and increasing.
Then

$$\int_{\Omega_1} u(x) \, \Phi_1 \left(\left| \frac{g_1(x)}{K(x)} \right| \right) \Phi_2 \left(\left| \frac{g_2(x)}{K(x)} \right| \right) d\mu_1(x) \leq$$

$$\left(\int_{\Omega_2} U(y) \, \Phi_1 \left(|f_1(y)| \right)^p d\mu_2(y) \right)^{\frac{1}{p}} \left(\int_{\Omega_2} U(y) \, \Phi_2 \left(|f_2(y)| \right)^q d\mu_2(y) \right)^{\frac{1}{q}},$$
(22.20)

for all measurable functions $f_i : \Omega_2 \to \mathbb{R}$ $(i = 1, 2)$ such that

(i) $f_1, f_2, \Phi_1 (|f_1|)^p, \Phi_2 (|f_2|)^q$ are all $k(x, y) \, d\mu_2(y)$ -integrable, μ_1 -a.e. in $x \in \Omega_1$,

(ii) $U \Phi_1 (|f_1|)^p, U \Phi_2 (|f_2|)^q$, are both μ_2 -integrable,

and for all corresponding functions g_i $(i = 1, 2)$ given by (22.10).

Corollary 22.5 Assume that the function $x \mapsto \left(u(x) \frac{k(x,y)}{K(x)} \right)$ is integrable on Ω_1, for each fixed $y \in \Omega_2$. Define U on Ω_2 by

$$U(y) := \int_{\Omega_1} u(x) \frac{k(x, y)}{K(x)} d\mu_1(x) < \infty. \tag{22.21}$$

Let $p_i > 1 : \sum_{i=1}^{m} \frac{1}{p_i} = 1$. Let the functions $\Phi_i : \mathbb{R}_+ \to \mathbb{R}_+, i = 1, \ldots, m$, be convex and increasing.
Then

$$\int_{\Omega_1} u(x) \prod_{i=1}^{m} \Phi_i \left(\left| \frac{g_i(x)}{K(x)} \right| \right) d\mu_1(x) \leq$$

$$\prod_{i=1}^{m} \left(\int_{\Omega_2} U(y) \, \Phi_i \left(|f_i(y)| \right)^{p_i} d\mu_2(y) \right)^{\frac{1}{p_i}}, \tag{22.22}$$

for all measurable functions $f_i : \Omega_2 \to \mathbb{R}, i = 1, \ldots, m$, such that

(i) $f_i, \Phi_i (|f_i|)^{p_i}$ are both $k(x, y) \, d\mu_2(y)$ -integrable, μ_1 -a.e. in $x \in \Omega_1$, for all $i = 1, \ldots, m$,

(ii) $U \Phi_i (|f_i|)^{p_i}$ is μ_2 -integrable, $i = 1, \ldots, m$,

and for all corresponding functions g_i $(i = 1, \ldots, m)$ given by (22.10).

Next we give two applications of Theorem 22.3.

Theorem 22.6 *Assume that the functions* $(i = 1, 2, \ldots, m \in \mathbb{N}) x \mapsto \left(u(x) \frac{k_i(x,y)}{K_i(x)} \right)$
are integrable on Ω_1, *for each fixed* $y \in \Omega_2$. *Define* u_i *on* Ω_2 *by*

$$u_i(y) := \int_{\Omega_1} u(x) \frac{k_i(x,y)}{K_i(x)} d\mu_1(x) < \infty. \tag{22.23}$$

Let $p_i > 1: \sum_{i=1}^{m} \frac{1}{p_i} = 1; \alpha_i \geq 1, i = 1, \ldots, m.$
  Then

$$\int_{\Omega_1} u(x) \left(\prod_{i=1}^{m} \left| \frac{g_i(x)}{K_i(x)} \right|^{\alpha_i} \right) d\mu_1(x) \leq$$

$$\prod_{i=1}^{m} \left(\int_{\Omega_2} u_i(y) |f_i(y)|^{\alpha_i p_i} d\mu_2(y) \right)^{\frac{1}{p_i}}, \tag{22.24}$$

for all measurable functions $f_i : \Omega_2 \to \mathbb{R}, i = 1, \ldots, m,$ *such that*

(i) $f_i, |f_i|^{\alpha_i p_i}$ *are* $k_i(x,y) d\mu_2(y)$ *-integrable,* μ_1 *-a.e. in* $x \in \Omega_1, i = 1, \ldots, m,$
(ii) $u_i |f_i|^{\alpha_i p_i}$ *is* μ_2 *-integrable,* $i = 1, \ldots, m,$

  and for all corresponding functions g_i $(i = 1, \ldots, m)$ *given by* (22.10).

Theorem 22.7 *Assume that the functions* $(i = 1, 2, \ldots, m \in \mathbb{N}) x \mapsto \left(u(x) \frac{k_i(x,y)}{K_i(x)} \right)$
are integrable on Ω_1, *for each fixed* $y \in \Omega_2$. *Define* u_i *on* Ω_2 *by*

$$u_i(y) := \int_{\Omega_1} u(x) \frac{k_i(x,y)}{K_i(x)} d\mu_1(x) < \infty. \tag{22.25}$$

Let $p_i > 1: \sum_{i=1}^{m} \frac{1}{p_i} = 1.$
  Then

$$\int_{\Omega_1} u(x) \left(e^{\sum_{i=1}^{m} \left| \frac{g_i(x)}{K_i(x)} \right|} \right) d\mu_1(x) \leq$$

$$\prod_{i=1}^{m} \left(\int_{\Omega_2} u_i(y) e^{p_i |f_i(y)|} d\mu_2(y) \right)^{\frac{1}{p_i}}, \tag{22.26}$$

for all measurable functions $f_i : \Omega_2 \to \mathbb{R}, i = 1, \dots, m$, *such that*

(i) $f_i, e^{p_i |f_i|}$ *are* $k_i(x, y) d\mu_2(y)$ *-integrable,* μ_1 *-a.e. in* $x \in \Omega_1, i = 1, \dots, m$,
(ii) $u_i e^{p_i |f_i|}$ *is* μ_2 *-integrable,* $i = 1, \dots, m$,

and for all corresponding functions g_i $(i = 1, \dots, m)$ *given by (22.10).*

We need

Definition 22.8 ([17]) Let $(a, b), 0 \le a < b < \infty; \alpha, \sigma > 0$. We consider the left-and right-sided fractional integrals of order α as follows:

(1) for $\eta > -1$, we define

$$\left(I_{a+;\sigma,\eta}^{\alpha} f \right)(x) = \frac{\sigma x^{-\sigma(\alpha+\eta)}}{\Gamma(\alpha)} \int_a^x \frac{t^{\sigma\eta+\sigma-1} f(t) \, dt}{(x^\sigma - t^\sigma)^{1-\alpha}}, \tag{22.27}$$

(2) for $\eta > 0$, we define

$$\left(I_{b-;\sigma,\eta}^{\alpha} f \right)(x) = \frac{\sigma x^{\sigma\eta}}{\Gamma(\alpha)} \int_x^b \frac{t^{\sigma(1-\eta-\alpha)-1} f(t) \, dt}{(t^\sigma - x^\sigma)^{1-\alpha}}. \tag{22.28}$$

These are the Erdélyi-Kober type fractional integrals.

We remind the Beta function

$$B(x, y) := \int_0^1 t^{x-1} (1 - t)^{y-1} \, dt, \tag{22.29}$$

for Re (x), Re $(y) > 0$, and the Incomplete Beta function

$$B(x; \alpha, \beta) = \int_0^x t^{\alpha-1} (1 - t)^{\beta-1} \, dt, \tag{22.30}$$

where $0 < x \le 1; \alpha, \beta > 0$.
    We make

Remark 22.9 Regarding (22.27) we have

$$k(x, y) = \frac{\sigma x^{-\sigma(\alpha+\eta)}}{\Gamma(\alpha)} \chi_{(a,x]}(y) \frac{y^{\sigma\eta+\sigma-1}}{(x^\sigma - y^\sigma)^{1-\alpha}}, \tag{22.31}$$

$x, y \in (a, b), \chi$ stands for the characteristic function.
    Here

$$K(x) = \int_a^b k(x, t) \, dt = \left(I_{a+;\sigma;\eta}^{\alpha} 1 \right)(x)$$

$$= \frac{\sigma x^{-\sigma(\alpha+\eta)}}{\Gamma(\alpha)} \int_a^x \frac{t^{\sigma\eta+\sigma-1}}{(x^\sigma - t^\sigma)^{1-\alpha}} dt \tag{22.32}$$

(setting $z = \frac{t}{x}$)

$$= \frac{\sigma}{\Gamma(\alpha)} \int_{\frac{a}{x}}^{1} z^{\sigma\left((\eta+1)-\frac{1}{\sigma}\right)} \left(1 - z^{\sigma}\right)^{\alpha-1} dz$$

(setting $\lambda = z^{\sigma}$)

$$= \frac{1}{\Gamma(\alpha)} \int_{\left(\frac{a}{x}\right)^{\sigma}}^{1} \lambda^{\eta} \left(1 - \lambda\right)^{\alpha-1} d\lambda. \tag{22.33}$$

Hence

$$K(x) = \frac{1}{\Gamma(\alpha)} \int_{\left(\frac{a}{x}\right)^{\sigma}}^{1} \lambda^{\eta} \left(1 - \lambda\right)^{\alpha-1} d\lambda. \tag{22.34}$$

Indeed it is

$$K(x) = \left(I_{a+;\sigma;\eta}^{\alpha}(1)\right)(x) \tag{22.35}$$

$$= \frac{B(\eta+1, \alpha) - B\left(\left(\frac{a}{x}\right)^{\sigma}; \eta+1, \alpha\right)}{\Gamma(\alpha)}.$$

We also make

Remark 22.10 Regarding (22.28) we have

$$k(x, y) = \frac{\sigma x^{\sigma\eta}}{\Gamma(\alpha)} \chi_{[x,b)}(y) \frac{y^{\sigma(1-\eta-\alpha)-1}}{(y^{\sigma} - x^{\sigma})^{1-\alpha}}, \tag{22.36}$$

$x, y \in (a, b)$.

Here

$$K(x) = \int_{a}^{b} k(x, t) dt = \left(I_{b-;\sigma;\eta}^{\alpha} 1\right)(x)$$

$$= \frac{\sigma x^{\sigma\eta}}{\Gamma(\alpha)} \int_{x}^{b} \frac{t^{\sigma(1-\eta-\alpha)-1}}{(t^{\sigma} - x^{\sigma})^{1-\alpha}} dt \tag{22.37}$$

(setting $z = \frac{t}{x}$)

$$= \frac{\sigma}{\Gamma(\alpha)} \int_{1}^{\left(\frac{b}{x}\right)} \left(z^{\sigma} - 1\right)^{\alpha-1} z^{\sigma(1-\eta-\alpha)-1} dz$$

(setting $\lambda = z^{\sigma}, 1 \leq \lambda < \left(\frac{b}{x}\right)^{\sigma}$)

$$= \frac{1}{\Gamma(\alpha)} \int_{1}^{\left(\frac{b}{x}\right)^{\sigma}} \left(\lambda - 1\right)^{\alpha-1} \lambda^{-\eta-\alpha} d\lambda \tag{22.38}$$

$$= \frac{1}{\Gamma(\alpha)} \int_1^{\left(\frac{b}{x}\right)^{\sigma}} \frac{1}{\lambda^{\eta+1}} \left(1 - \frac{1}{\lambda}\right)^{\alpha-1} d\lambda$$

(setting $w := \frac{1}{\lambda}, 0 < \left(\frac{x}{b}\right)^{\sigma} < w \leq 1$)

$$= \frac{1}{\Gamma(\alpha)} \int_{\left(\frac{x}{b}\right)^{\sigma}}^{1} w^{\eta-1} (1 - w)^{\alpha-1} dw \qquad (22.39)$$

$$= \frac{\left(B(\eta, \alpha) - B\left(\left(\frac{x}{b}\right)^{\sigma}; \eta, \alpha\right)\right)}{\Gamma(\alpha)}.$$

That is

$$K(x) = \left(I_{b-;\sigma;\eta}^{\alpha}(1)\right)(x) \qquad (22.40)$$

$$= \frac{\left(B(\eta, \alpha) - B\left(\left(\frac{x}{b}\right)^{\sigma}; \eta, \alpha\right)\right)}{\Gamma(\alpha)}.$$

We give

Theorem 22.11 *Assume that the function*

$$x \mapsto \left(u(x) \frac{\chi_{(a,x]}(y) \sigma x^{-\sigma(\alpha+\eta)} y^{\sigma\eta+\sigma-1}}{(x^{\sigma} - y^{\sigma})^{1-\alpha} \left[B(\eta+1, \alpha) - B\left(\left(\frac{a}{x}\right)^{\sigma}; \eta+1, \alpha\right)\right]}\right) \qquad (22.41)$$

is integrable on (a, b), for each $y \in (a, b)$. Here $\alpha, \sigma > 0, \eta > -1, 0 \leq a < b < \infty$. Define u_1 on (a, b) by

$$u_1(y) := \sigma y^{\sigma\eta+\sigma-1} \int_y^b \frac{u(x) x^{-\sigma(\alpha+\eta)} (x^{\sigma} - y^{\sigma})^{\alpha-1}}{\left(B(\eta+1, \alpha) - B\left(\left(\frac{a}{x}\right)^{\sigma}; \eta+1, \alpha\right)\right)} dx < \infty. \quad (22.42)$$

Let $p_i > 1 : \sum_{i=1}^{m} \frac{1}{p_i} = 1$. Let the functions $\Phi_i : \mathbb{R}_+ \to \mathbb{R}_+, i = 1, \ldots, m$, be convex and increasing.
Then

$$\int_a^b u(x) \prod_{i=1}^{m} \Phi_i \left(\frac{\left|I_{a+;\sigma;\eta}^{\alpha} f_i(x)\right| \Gamma(\alpha)}{\left(B(\eta+1, \alpha) - B\left(\left(\frac{a}{x}\right)^{\sigma}; \eta+1, \alpha\right)\right)}\right) dx \leq$$

$$\prod_{i=1}^{m} \left(\int_a^b u_1(y) \Phi_i (|f_i(y)|)^{p_i} dy\right)^{\frac{1}{p_i}}, \qquad (22.43)$$

for all measurable functions $f_i : (a, b) \to \mathbb{R}$, $i = 1, \ldots, m$, *such that*

(i) f_i, $\Phi_i (|f_i|)^{p_i}$ *are both* $\frac{\sigma x^{-\sigma(\alpha+\eta)}}{\Gamma(\alpha)} \chi_{(a,x]}(y) \frac{y^{\sigma\eta+\sigma-1}dy}{(x^\sigma-y^\sigma)^{1-\alpha}}$ *-integrable, a.e. in* $x \in$
    (a, b), *for all* $i = 1, \ldots, m$,
(ii) $u_1 \Phi_i (|f_i|)^{p_i}$ *is Lebesgue integrable*, $i = 1, \ldots, m$,

Proof By Corollary 22.5. ∎

Remark 22.12 In (22.42), if we choose

$$u(x) = x^{\sigma(\alpha+\eta+1)-1} \left(B(\eta+1, \alpha) - B\left(\left(\frac{a}{x}\right)^\alpha; \eta+1, \alpha \right) \right), \; x \in (a, b),$$

(22.44)

then

$$u_1(y) = \sigma y^{\sigma\eta+\sigma-1} \int_y^b x^{\sigma-1} (x^\sigma - y^\sigma)^{\alpha-1} dx$$

(setting $w := x^\sigma$, $\frac{dw}{dx} = \sigma x^{\sigma-1}$, $dx = \frac{dw}{\sigma x^{\sigma-1}}$)

$$= y^{\sigma\eta+\sigma-1} \int_{y^\sigma}^{b^\sigma} (w - y^\sigma)^{\alpha-1} dw = y^{\sigma\eta+\sigma-1} \frac{(b^\sigma - y^\sigma)^\alpha}{\alpha}.$$

(22.45)

That is

$$u_1(y) = y^{\sigma\eta+\sigma-1} \frac{(b^\sigma - y^\sigma)^\alpha}{\alpha}, \; y \in (a, b).$$

(22.46)

Based on the above, (22.43) becomes

$$\int_a^b x^{\sigma(\alpha+\eta+1)-1} \left(B(\eta+1, \alpha) - B\left(\left(\frac{a}{x}\right)^\sigma; \eta+1, \alpha \right) \right) \cdot$$

$$\prod_{i=1}^m \Phi_i \left(\frac{\left| I_{a+;\sigma;\eta}^\alpha f_i(x) \right| \Gamma(\alpha)}{\left(B(\eta+1, \alpha) - B\left(\left(\frac{a}{x}\right)^\sigma; \eta+1, \alpha \right) \right)} \right) dx \le$$

$$\frac{1}{\alpha} \prod_{i=1}^m \left(\int_a^b y^{\sigma\eta+\sigma-1} (b^\sigma - y^\sigma)^\alpha \Phi_i (|f_i(y)|)^{p_i} dy \right)^{\frac{1}{p_i}} \le$$

(22.47)

$$\frac{(b^\sigma - a^\sigma)^\alpha}{\alpha} \prod_{i=1}^m \left(\int_a^b y^{\sigma(\eta+1)-1} \Phi_i (|f_i(y)|)^{p_i} dy \right)^{\frac{1}{p_i}},$$

under the assumptions:

(i) following (22.43), and
(ii)* $y^{\sigma(\eta+1)-1} \Phi_i (|f_i(y)|)^{p_i}$ is Lebesgue integrable on (a, b), $i = 1, \ldots, m$.

Corollary 22.13 *Let* $0 \le a < b;\ \alpha, \sigma > 0,\ \eta > -1;\ p_i > 1 : \sum_{i=1}^{m} \dfrac{1}{p_i} = 1;\ \beta_i \ge 1,$
$i = 1, \ldots, m.$
Then

$$\int_a^b x^{\sigma(\alpha+\eta+1)-1} \left(B(\eta+1, \alpha) - B\left(\left(\frac{a}{x}\right)^\sigma ; \eta+1, \alpha \right) \right)^{\left(1- \sum_{i=1}^{m} \beta_i \right)} \cdot$$

$$\left(\prod_{i=1}^{m} \left| I_{a+;\sigma;\eta}^\alpha f_i(x) \right|^{\beta_i} \right) dx \le \frac{1}{\alpha \left(\Gamma(\alpha) \right)^{\sum_{i=1}^{m} \beta_i}} \cdot$$

$$\prod_{i=1}^{m} \left(\int_a^b y^{\sigma\eta+\sigma-1} \left(b^\sigma - y^\sigma \right)^\alpha |f_i(y)|^{\beta_i p_i} dy \right)^{\frac{1}{p_i}} \le \qquad (22.48)$$

$$\left(\frac{(b^\sigma - a^\sigma)^\alpha}{\alpha \left(\Gamma(\alpha) \right)^{\sum_{i=1}^{m} \beta_i}} \right) \prod_{i=1}^{m} \left(\int_a^b y^{\sigma(\eta+1)-1} |f_i(y)|^{\beta_i p_i} dy \right)^{\frac{1}{p_i}},$$

for all measurable functions $f_i : (a, b) \to \mathbb{R},\ i = 1, \ldots, m$ *such that*

(i) $|f_i|^{\beta_i p_i}$ *is* $\left(\frac{\sigma x^{-\sigma(\alpha+\eta)}}{\Gamma(\alpha)} \chi_{(a,x]}(y) \frac{y^{\sigma\eta+\sigma-1} dy}{(x^\sigma-y^\sigma)^{1-\alpha}} \right)$ *-integrable, a.e. in* $x \in (a, b),$
(ii) $y^{\sigma(\eta+1)-1} |f_i(y)|^{\beta_i p_i}$ *is Lebesgue integrable on* $(a, b);\ i = 1, \ldots, m.$

Proof By Theorem 22.11 and (22.47). ∎

Corollary 22.14 *Let* $0 \le a < b;\ \alpha, \sigma > 0,\ \eta > -1;\ p_i > 1 : \sum_{i=1}^{m} \dfrac{1}{p_i} = 1.$
Then

$$\int_a^b x^{\sigma(\alpha+\eta+1)-1} \left(B(\eta+1, \alpha) - B\left(\left(\frac{a}{x}\right)^\sigma ; \eta+1, \alpha \right) \right) \cdot$$

$$e^{\dfrac{\Gamma(\alpha) \left(\sum_{i=1}^{m} \left| I_{a+;\sigma;\eta}^\alpha f_i(x) \right| \right)}{\left(B(\eta+1,\alpha) - B\left(\left(\frac{a}{x}\right)^\sigma ; \eta+1, \alpha \right) \right)}} dx \le$$

$$\frac{1}{\alpha} \prod_{i=1}^{m} \left(\int_a^b y^{\sigma(\eta+1)-1} \left(b^\sigma - y^\sigma \right)^\alpha e^{p_i |f_i(y)|} dy \right)^{\frac{1}{p_i}} \le$$

$$\frac{(b^\sigma - a^\sigma)^\alpha}{\alpha} \prod_{i=1}^{m} \left(\int_a^b y^{\sigma(\eta+1)-1} e^{p_i|f_i(y)|} dy \right)^{\frac{1}{p_i}}, \tag{22.49}$$

for all measurable functions $f_i : (a,b) \to \mathbb{R}$, $i = 1, \ldots, m$ *such that*

(i) f_i, $e^{p_i|f_i|}$ *are both* $\frac{\sigma x^{-\sigma(\alpha+\eta)}}{\Gamma(\alpha)} \chi_{(a,x]}(y) \frac{y^{\sigma\eta+\sigma-1} dy}{(x^\sigma-y^\sigma)^{1-\alpha}}$ *-integrable, a.e. in* $x \in (a,b)$,

(ii) $y^{\sigma(\eta+1)-1} e^{p_i|f_i(y)|}$ *is Lebesgue integrable on* (a,b); $i = 1, \ldots, m$.

Proof By Theorem 22.11 and (22.47). ∎

We present

Theorem 22.15 *Assume that the function*

$$x \mapsto \left(u(x) \frac{\sigma x^{\sigma\eta} \chi_{[x,b)}(y) y^{\sigma(1-\eta-\alpha)-1}}{(y^\sigma - x^\sigma)^{1-\alpha} \left[B(\eta, \alpha) - B\left(\left(\frac{x}{b} \right)^\sigma ; \eta, \alpha \right) \right]} \right)$$

is integrable on (a,b), *for each* $y \in (a,b)$. *Here* $\alpha, \sigma, \eta > 0$, $0 \le a < b < \infty$. *Define* u_2 *on* (a,b) *by*

$$u_2(y) := \sigma y^{\sigma(1-\eta-\alpha)-1} \int_a^y \frac{u(x) x^{\sigma\eta} (y^\sigma - x^\sigma)^{\alpha-1} dx}{\left(B(\eta, \alpha) - B\left(\left(\frac{x}{b} \right)^\sigma ; \eta, \alpha \right) \right)} < \infty. \tag{22.50}$$

Let $p_i > 1 : \sum_{i=1}^{m} \frac{1}{p_i} = 1$. *Let the functions* $\Phi_i : \mathbb{R}_+ \to \mathbb{R}_+$, $i = 1, \ldots, m$, *be convex and increasing.*

Then

$$\int_a^b u(x) \prod_{i=1}^{m} \Phi_i \left(\frac{\left| I_{b-;\sigma;\eta}^\alpha f_i(x) \right| \Gamma(\alpha)}{\left(B(\eta, \alpha) - B\left(\left(\frac{x}{b} \right)^\sigma ; \eta, \alpha \right) \right)} \right) dx \le$$

$$\prod_{i=1}^{m} \left(\int_a^b u_2(y) \Phi_i (|f_i(y)|)^{p_i} dy \right)^{\frac{1}{p_i}}, \tag{22.51}$$

for all measurable functions $f_i : (a,b) \to \mathbb{R}$, $i = 1, \ldots, m$, *such that*

(i) f_i, $\Phi_i (|f_i|)^{p_i}$ *are both* $\left(\frac{\sigma x^{\sigma\eta} \chi_{[x,b)}(y) y^{\sigma(1-\eta-\alpha)-1} dy}{\Gamma(\alpha)(y^\sigma-x^\sigma)^{1-\alpha}} \right)$ *-integrable, a.e. in* $x \in (a,b)$,
    for all $i = 1, \ldots, m$,

(ii) $u_2 \Phi_i (|f_i|)^{p_i}$ *is Lebesgue integrable on* (a,b), $i = 1, \ldots, m$.

Proof By Corollary 22.5. ∎

Remark 22.16 Here $0 < a < b < \infty$; $\alpha, \sigma, \eta > 0$.

In (22.50), if we choose

$$u(x) = x^{\sigma(1-\eta)-1}\left(B(\eta,\alpha) - B\left(\left(\frac{x}{b}\right)^{\alpha};\eta,\alpha\right)\right), \quad x \in (a,b), \quad (22.52)$$

then

$$u_2(y) = \sigma y^{\sigma(1-\eta-\alpha)-1}\int_a^y x^{\sigma-1}\left(y^{\sigma} - x^{\sigma}\right)^{\alpha-1}dx$$

(setting $w := x^{\sigma}, dx = \frac{dw}{\sigma x^{\sigma-1}}$)

$$= y^{\sigma(1-\eta-\alpha)-1}\int_{a^{\sigma}}^{y^{\sigma}}\left(y^{\sigma} - w\right)^{\alpha-1}dw = y^{\sigma(1-\eta-\alpha)-1}\frac{(y^{\sigma} - a^{\sigma})^{\alpha}}{\alpha}. \quad (22.53)$$

That is

$$u_2(y) = y^{\sigma(1-\eta-\alpha)-1}\frac{(y^{\sigma} - a^{\sigma})^{\alpha}}{\alpha}, \quad y \in (a,b). \quad (22.54)$$

Based on the above, (22.51) becomes

$$\int_a^b x^{\sigma(1-\eta)-1}\left(B(\eta,\alpha) - B\left(\left(\frac{x}{b}\right)^{\sigma};\eta,\alpha\right)\right) \cdot$$

$$\prod_{i=1}^m \Phi_i\left(\frac{\left|I_{b-;\sigma;\eta}^{\alpha}f_i(x)\right|\Gamma(\alpha)}{\left(B(\eta,\alpha) - B\left(\left(\frac{x}{b}\right)^{\sigma};\eta,\alpha\right)\right)}\right)dx \leq$$

$$\frac{1}{\alpha}\prod_{i=1}^m\left(\int_a^b y^{\sigma(1-\eta-\alpha)-1}\left(y^{\sigma} - a^{\sigma}\right)^{\alpha}\Phi_i\left(|f_i(y)|\right)^{p_i}dy\right)^{\frac{1}{p_i}} \leq \quad (22.55)$$

$$\frac{(b^{\sigma} - a^{\sigma})^{\alpha}}{\alpha}\prod_{i=1}^m\left(\int_a^b y^{\alpha(1-\eta-\alpha)-1}\Phi_i\left(|f_i(y)|\right)^{p_i}dy\right)^{\frac{1}{p_i}},$$

under the assumptions:

(i) following (22.51), and

(ii)* $y^{\sigma(1-\eta-\alpha)-1}\Phi_i\left(|f_i(y)|\right)^{p_i}$ is Lebesgue integrable on (a,b), $i = 1,\ldots,m$.

Corollary 22.17 *Let* $0 < a < b < \infty$; $\alpha, \sigma, \eta > 0$; $p_i > 1 : \sum_{i=1}^m \frac{1}{p_i} = 1$; $\beta_i \geq 1$, $i = 1,\ldots,m$.

Then

$$\int_a^b x^{\sigma(1-\eta)-1} \left(B(\eta,\alpha) - B\left(\left(\frac{x}{b}\right)^\sigma ; \eta, \alpha\right) \right)^{\left(1-\sum_{i=1}^m \beta_i\right)} \cdot$$

$$\left(\prod_{i=1}^m \left| I_{b-;\sigma;\eta}^\alpha f_i(x) \right|^{\beta_i} \right) dx \leq \frac{1}{\alpha \, (\Gamma(\alpha))^{\sum_{i=1}^m \beta_i}} \cdot$$

$$\prod_{i=1}^m \left(\int_a^b y^{\sigma(1-\eta-\alpha)-1} \left(y^\sigma - a^\sigma \right)^\alpha |f_i(y)|^{\beta_i p_i} \, dy \right)^{\frac{1}{p_i}} \leq \qquad (22.56)$$

$$\frac{(b^\sigma - a^\sigma)^\alpha}{\alpha \, (\Gamma(\alpha))^{\sum_{i=1}^m \beta_i}} \prod_{i=1}^m \left(\int_a^b y^{\sigma(1-\eta-\alpha)-1} |f_i(y)|^{\beta_i p_i} \, dy \right)^{\frac{1}{p_i}},$$

under the assumptions:

(i) $|f_i|^{\beta_i p_i}$ *is* $\left(\frac{\sigma x^{\sigma\eta} \chi_{[x,b)}(y) y^{\sigma(1-\eta-\alpha)-1} dy}{\Gamma(\alpha)(y^\sigma - x^\sigma)^{1-\alpha}} \right)$ *-integrable, a.e. in $x \in (a,b)$, for all $i = 1, \ldots, m$,*

(ii) $y^{\sigma(1-\eta-\alpha)-1} |f_i(y)|^{\beta_i p_i}$ *is Lebesgue integrable on (a,b), $i = 1, \ldots, m$.*

Proof By Theorem 22.15 and (22.55). ∎

Corollary 22.18 *Let* $0 < a < b < \infty$; $\alpha, \sigma, \eta > 0$; $p_i > 1$: $\sum_{i=1}^m \frac{1}{p_i} = 1$.

Then

$$\int_a^b x^{\sigma(1-\eta)-1} \left(B(\eta,\alpha) - B\left(\left(\frac{x}{b}\right)^\sigma ; \eta, \alpha\right) \right) \cdot e^{\frac{\Gamma(\alpha)\left(\sum_{i=1}^m \left| I_{b-;\sigma;\eta}^\alpha f_i(x) \right|\right)}{\left(B(\eta,\alpha) - B\left(\left(\frac{x}{b}\right)^\sigma ; \eta, \alpha\right)\right)}} \, dx \leq$$

$$\frac{1}{\alpha} \prod_{i=1}^m \left(\int_a^b y^{\sigma(1-\eta-\alpha)-1} \left(y^\sigma - a^\sigma \right)^\alpha e^{p_i |f_i(y)|} dy \right)^{\frac{1}{p_i}} \leq$$

$$\frac{(b^\sigma - a^\sigma)^\alpha}{\alpha} \prod_{i=1}^m \left(\int_a^b y^{\sigma(1-\eta-\alpha)-1} e^{p_i |f_i(y)|} dy \right)^{\frac{1}{p_i}}, \qquad (22.57)$$

under the assumptions:

(i) f_i, $e^{p_i|f_i|}$ are both $\left(\frac{\sigma x^{\sigma\eta}\chi_{[x,b)}(y)y^{\sigma(1-\eta-\alpha)-1}dy}{\Gamma(\alpha)(y^\sigma-x^\sigma)^{1-\alpha}}\right)$ -integrable, a.e. in $x \in (a,b)$,
   $i = 1,\dots,m$,

(ii) $y^{\sigma(1-\eta-\alpha)-1}e^{p_i|f_i(y)|}$ is Lebesgue integrable on (a,b); $i = 1,\dots,m$.

Proof By Theorem 22.15 and (22.55). ∎

We make

Remark 22.19 Let $\prod_{i=1}^{N}(a_i,b_i) \subset \mathbb{R}^N$, $N > 1$, $a_i < b_i$, $a_i,b_i \in \mathbb{R}$. Let $\alpha_i > 0$,
$i = 1,\dots,N$; $f \in L_1\left(\prod_{i=1}^{N}(a_i,b_i)\right)$, and set $a = (a_1,\dots,a_N)$, $b = (b_1,\dots,b_N)$,
$\alpha = (\alpha_1,\dots,\alpha_N)$, $x = (x_1,\dots,x_N)$, $t = (t_1,\dots,t_N)$.

We define the left mixed Riemann-Liouville fractional multiple integral of order α (see also [14]):

$$\left(I^\alpha_{a+}f\right)(x) := \frac{1}{\prod_{i=1}^{N}\Gamma(\alpha_i)}\int_{a_1}^{x_1}\dots\int_{a_N}^{x_N}\prod_{i=1}^{N}(x_i-t_i)^{\alpha_i-1}f(t_1,\dots,t_N)\,dt_1\dots dt_N,$$

(22.58)

with $x_i > a_i$, $i = 1,\dots,N$.

We also define the right mixed Riemann-Liouville fractional multiple integral of order α (see also [12]):

$$\left(I^\alpha_{b-}f\right)(x) := \frac{1}{\prod_{i=1}^{N}\Gamma(\alpha_i)}\int_{x_1}^{b_1}\dots\int_{x_N}^{b_N}\prod_{i=1}^{N}(t_i-x_i)^{\alpha_i-1}f(t_1,\dots,t_N)\,dt_1\dots dt_N,$$

(22.59)

with $x_i < b_i$, $i = 1,\dots,N$.

Notice $I^\alpha_{a+}(|f|)$, $I^\alpha_{b-}(|f|)$ are finite if $f \in L_\infty\left(\prod_{i=1}^{N}(a_i,b_i)\right)$.

One can rewrite (22.58) and (22.59) as follows:

$$\left(I^\alpha_{a+}f\right)(x) = \frac{1}{\prod_{i=1}^{N}\Gamma(\alpha_i)}\int_{\prod_{i=1}^{N}(a_i,b_i)}\chi_{\prod_{i=1}^{N}(a_i,x_i]}(t)\prod_{i=1}^{N}(x_i-t_i)^{\alpha_i-1}f(t)\,dt,$$

(22.60)

with $x_i > a_i$, $i = 1,\dots,N$,

and

$$\left(I_{b-}^{\alpha}f\right)(x) = \frac{1}{\prod\limits_{i=1}^{N}\Gamma\left(\alpha_i\right)} \int\limits_{\prod\limits_{i=1}^{N}(a_i,b_i)} \chi_{\prod\limits_{i=1}^{N}[x_i,b_i)}(t) \prod_{i=1}^{N}(t_i-x_i)^{\alpha_i-1} f(t)\,dt,$$

(22.61)

with $x_i < b_i$, $i = 1, \ldots, N$.

The corresponding $k(x,y)$ for I_{a+}^{α}, I_{b-}^{α} are

$$k_{a+}(x,y) = \frac{1}{\prod\limits_{i=1}^{N}\Gamma\left(\alpha_i\right)} \chi_{\prod\limits_{i=1}^{N}(a_i,x_i]}(y) \prod_{i=1}^{N}(x_i-y_i)^{\alpha_i-1},$$

(22.62)

$\forall\, x,y \in \prod\limits_{i=1}^{N}(a_i,b_i)$,

and

$$k_{b-}(x,y) = \frac{1}{\prod\limits_{i=1}^{N}\Gamma\left(\alpha_i\right)} \chi_{\prod\limits_{i=1}^{N}[x_i,b_i)}(y) \prod_{i=1}^{N}(y_i-x_i)^{\alpha_i-1},$$

(22.63)

$\forall\, x,y \in \prod\limits_{i=1}^{N}(a_i,b_i)$.

The corresponding $K(x)$ for I_{a+}^{α} is:

$$K_{a+}(x) = \int_{\prod\limits_{i=1}^{N}(a_i,b_i)} k_{a+}(x,y)\,dy = \left(I_{a+}^{\alpha}1\right)(x) =$$

$$\frac{1}{\prod\limits_{i=1}^{N}\Gamma\left(\alpha_i\right)} \int_{a_1}^{x_1}\cdots\int_{a_N}^{x_N} \prod_{i=1}^{N}(x_i-t_i)^{\alpha_i-1}\,dt_1\ldots dt_N =$$

$$\frac{1}{\prod\limits_{i=1}^{N}\Gamma\left(\alpha_i\right)} \prod_{i=1}^{N}\int_{a_i}^{x_i}(x_i-t_i)^{\alpha_i-1}\,dt_i = \frac{1}{\prod\limits_{i=1}^{N}\Gamma\left(\alpha_i\right)} \prod_{i=1}^{N}\frac{(x_i-a_i)^{\alpha_i}}{\alpha_i}$$

$$= \prod_{i=1}^{N} \left(\frac{(x_i - a_i)^{\alpha_i}}{\Gamma(\alpha_i + 1)} \right),$$

that is

$$K_{a+}(x) = \prod_{i=1}^{N} \frac{(x_i - a_i)^{\alpha_i}}{\Gamma(\alpha_i + 1)}, \qquad (22.64)$$

$\forall\, x \in \prod_{i=1}^{N} (a_i, b_i)$.

Similarly the corresponding $K(x)$ for I_{b-}^{α} is:

$$K_{b-}(x) = \int_{\prod_{i=1}^{N}(a_i, b_i)} k_{b-}(x, y)\, dy = \left(I_{b-}^{\alpha} 1 \right)(x) =$$

$$\frac{1}{\prod_{i=1}^{N} \Gamma(\alpha_i)} \int_{x_1}^{b_1} \cdots \int_{x_N}^{b_N} \prod_{i=1}^{N} (t_i - x_i)^{\alpha_i - 1}\, dt_1 \ldots dt_N =$$

$$\frac{1}{\prod_{i=1}^{N} \Gamma(\alpha_i)} \prod_{i=1}^{N} \int_{x_i}^{b_i} (t_i - x_i)^{\alpha_i - 1}\, dt_i = \frac{1}{\prod_{i=1}^{N} \Gamma(\alpha_i)} \prod_{i=1}^{N} \frac{(b_i - x_i)^{\alpha_i}}{\alpha_i}$$

$$= \prod_{i=1}^{N} \frac{(b_i - x_i)^{\alpha_i}}{\Gamma(\alpha_i + 1)},$$

that is

$$K_{b-}(x) = \prod_{i=1}^{N} \frac{(b_i - x_i)^{\alpha_i}}{\Gamma(\alpha_i + 1)}, \qquad (22.65)$$

$\forall\, x \in \prod_{i=1}^{N} (a_i, b_i)$.

Next we form

$$\frac{k_{a+}(x, y)}{K_{a+}(x)} = \frac{1}{\prod_{i=1}^{N} \Gamma(\alpha_i)} \chi_{\prod_{i=1}^{N}(a_i, x_i]}(y) \prod_{i=1}^{N} (x_i - y_i)^{\alpha_i - 1} \prod_{i=1}^{N} \frac{\Gamma(\alpha_i + 1)}{(x_i - a_i)^{\alpha_i}}$$

$$= \chi_{\displaystyle\prod_{i=1}^{N}(a_i, x_i]}(y) \left(\prod_{i=1}^{N} \alpha_i \right) \left(\prod_{i=1}^{N} \frac{(x_i - y_i)^{\alpha_i - 1}}{(x_i - a_i)^{\alpha_i}} \right),$$

that is

$$\frac{k_{a+}(x, y)}{K_{a+}(x)} = \chi_{\displaystyle\prod_{i=1}^{N}(a_i, x_i]}(y) \left(\prod_{i=1}^{N} \alpha_i \right) \left(\prod_{i=1}^{N} \frac{(x_i - y_i)^{\alpha_i - 1}}{(x_i - a_i)^{\alpha_i}} \right), \qquad (22.66)$$

$\forall\, x, y \in \displaystyle\prod_{i=1}^{N}(a_i, b_i).$

Similarly we form

$$\frac{k_{b-}(x, y)}{K_{b-}(x)} = \frac{1}{\displaystyle\prod_{i=1}^{N}\Gamma(\alpha_i)} \chi_{\displaystyle\prod_{i=1}^{N}[x_i, b_i)}(y) \prod_{i=1}^{N}(y_i - x_i)^{\alpha_i - 1} \prod_{i=1}^{N} \frac{\Gamma(\alpha_i + 1)}{(b_i - x_i)^{\alpha_i}}$$

$$= \chi_{\displaystyle\prod_{i=1}^{N}[x_i, b_i)}(y) \left(\prod_{i=1}^{N} \alpha_i \right) \left(\prod_{i=1}^{N} \frac{(y_i - x_i)^{\alpha_i - 1}}{(b_i - x_i)^{\alpha_i}} \right),$$

that is

$$\frac{k_{b-}(x, y)}{K_{b-}(x)} = \chi_{\displaystyle\prod_{i=1}^{N}[x_i, b_i)}(y) \left(\prod_{i=1}^{N} \alpha_i \right) \left(\prod_{i=1}^{N} \frac{(y_i - x_i)^{\alpha_i - 1}}{(b_i - x_i)^{\alpha_i}} \right), \qquad (22.67)$$

$\forall\, x, y \in \displaystyle\prod_{i=1}^{N}(a_i, b_i).$

We choose the weight function $u_1(x)$ on $\displaystyle\prod_{i=1}^{N}(a_i, b_i)$ such that the function $x \mapsto$

$\left(u_1(x) \frac{k_{a+}(x, y)}{K_{a+}(x)} \right)$ is integrable on $\displaystyle\prod_{i=1}^{N}(a_i, b_i)$, for each fixed $y \in \displaystyle\prod_{i=1}^{N}(a_i, b_i)$. We

define w_1 on $\displaystyle\prod_{i=1}^{N}(a_i, b_i)$ by

$$w_1(y) := \int_{\prod\limits_{i=1}^{N}(a_i,b_i)} u_1(x)\,\frac{k_{a+}(x,y)}{K_{a+}(x)}\,dx < \infty. \qquad (22.68)$$

We have that

$$w_1(y) = \left(\prod_{i=1}^{N}\alpha_i\right)\int_{y_1}^{b_1}\ldots\int_{y_N}^{b_N}u_1(x_1,\ldots,x_N)\left(\prod_{i=1}^{N}\frac{(x_i-y_i)^{\alpha_i-1}}{(x_i-a_i)^{\alpha_i}}\right)dx_1\ldots dx_N,$$

$$(22.69)$$

$\forall\, y \in \prod\limits_{i=1}^{N}(a_i,b_i).$

We also choose the weight function $u_2(x)$ on $\prod\limits_{i=1}^{N}(a_i,b_i)$ such that the function

$x \mapsto \left(u_2(x)\,\frac{k_{b-}(x,y)}{K_{b-}(x)}\right)$ is integrable on $\prod\limits_{i=1}^{N}(a_i,b_i)$, for each fixed $y \in \prod\limits_{i=1}^{N}(a_i,b_i)$.

We define w_2 on $\prod\limits_{i=1}^{N}(a_i,b_i)$ by

$$w_2(y) := \int_{\prod\limits_{i=1}^{N}(a_i,b_i)} u_2(x)\,\frac{k_{b-}(x,y)}{K_{b-}(x)}\,dx < \infty. \qquad (22.70)$$

We have that

$$w_2(y) = \left(\prod_{i=1}^{N}\alpha_i\right)\int_{a_1}^{y_1}\ldots\int_{a_N}^{y_N}u_2(x_1,\ldots,x_N)\left(\prod_{i=1}^{N}\frac{(y_i-x_i)^{\alpha_i-1}}{(b_i-x_i)^{\alpha_i}}\right)dx_1\ldots dx_N,$$

$$(22.71)$$

$\forall\, y \in \prod\limits_{i=1}^{N}(a_i,b_i).$

If we choose as

$$u_1(x) = u_1^*(x) := \prod_{i=1}^{N}(x_i-a_i)^{\alpha_i}, \qquad (22.72)$$

then

$$w_1^*(y) := w_1(y) = \left(\prod_{i=1}^{N} \alpha_i\right) \int_{y_1}^{b_1} \dots \int_{y_N}^{b_N} \left(\prod_{i=1}^{N} (x_i - y_i)^{\alpha_i - 1}\right) dx_1 \dots dx_N$$

$$= \left(\prod_{i=1}^{N} \alpha_i\right) \left(\prod_{i=1}^{N} \int_{y_i}^{b_i} (x_i - y_i)^{\alpha_i - 1} dx_i\right)$$

$$= \left(\prod_{i=1}^{N} \alpha_i\right) \left(\prod_{i=1}^{N} \frac{(b_i - y_i)^{\alpha_i}}{\alpha_i}\right) = \prod_{i=1}^{N} (b_i - y_i)^{\alpha_i}.$$

that is

$$w_1^*(y) = \prod_{i=1}^{N} (b_i - y_i)^{\alpha_i}, \ \forall y \in \prod_{i=1}^{N} (a_i, b_i). \tag{22.73}$$

If we choose as

$$u_2(x) = u_2^*(x) := \prod_{i=1}^{N} (b_i - x_i)^{\alpha_i}, \tag{22.74}$$

then

$$w_2^*(y) := w_2(y) = \left(\prod_{i=1}^{N} \alpha_i\right) \int_{a_1}^{y_1} \dots \int_{a_N}^{y_N} \left(\prod_{i=1}^{N} (y_i - x_i)^{\alpha_i - 1}\right) dx_1 \dots dx_N$$

$$= \left(\prod_{i=1}^{N} \alpha_i\right) \left(\prod_{i=1}^{N} \int_{a_i}^{y_i} (y_i - x_i)^{\alpha_i - 1} dx_i\right)$$

$$= \left(\prod_{i=1}^{N} \alpha_i\right) \left(\prod_{i=1}^{N} \frac{(y_i - a_i)^{\alpha_i}}{\alpha_i}\right) = \prod_{i=1}^{N} (y_i - a_i)^{\alpha_i}.$$

That is

$$w_2^*(y) = \prod_{i=1}^{N} (y_i - a_i)^{\alpha_i}, \ \forall y \in \prod_{i=1}^{N} (a_i, b_i). \tag{22.75}$$

Here we choose $f_j : \prod_{i=1}^{N} (a_i, b_i) \to \mathbb{R}$, $j = 1, \dots, m$, that are Lebesgue measurable and $I_{a+}^\alpha(|f_j|)$, $I_{b-}^\alpha(|f_j|)$ are finite a.e., one or the other, or both.

Let $p_j > 1 : \sum_{j=1}^{m} \frac{1}{p_j} = 1$ and the functions $\Phi_j : \mathbb{R}_+ \to \mathbb{R}_+, j = 1, \ldots, m$, to be convex and increasing.

Then by (22.22) we obtain

$$\int_{\prod\limits_{i=1}^{N}(a_i, b_i)} u_1(x) \prod_{j=1}^{m} \Phi_j \left(\frac{\left| I_{a+}^{\alpha}(f_j)(x) \right| \prod\limits_{i=1}^{N} \Gamma(\alpha_i + 1)}{\prod\limits_{i=1}^{N}(x_i - a_i)^{\alpha_i}} \right) dx \leq$$

$$\prod_{j=1}^{m} \left(\int_{\prod\limits_{i=1}^{N}(a_i, b_i)} w_1(y) \, \Phi_j \left(|f_j(y)| \right)^{p_j} dy \right)^{\frac{1}{p_j}}, \qquad (22.76)$$

under the assumptions:

(i) f_j, $\Phi_j\left(|f_j|\right)^{p_j}$ are both $\dfrac{1}{\prod\limits_{i=1}^{N} \Gamma(\alpha_i)} \chi_{\prod\limits_{i=1}^{N}(a_i, x_i]}(y) \prod\limits_{i=1}^{N}(x_i - y_i)^{\alpha_i - 1} dy$

-integrable, a.e. in $x \in \prod\limits_{i=1}^{N}(a_i, b_i)$, for all $j = 1, \ldots, m$,

(ii) $w_1 \Phi_j \left(|f_j|\right)^{p_j}$ is Lebesgue integrable, $j = 1, \ldots, m$.

Similarly, by (22.22), we obtain

$$\int_{\prod\limits_{i=1}^{N}(a_i, b_i)} u_2(x) \prod_{j=1}^{m} \Phi_j \left(\frac{\left| I_{b-}^{\alpha}(f_j)(x) \right| \prod\limits_{i=1}^{N} \Gamma(\alpha_i + 1)}{\prod\limits_{i=1}^{N}(b_i - x_i)^{\alpha_i}} \right) dx \leq$$

$$\prod_{j=1}^{m} \left(\int_{\prod\limits_{i=1}^{N}(a_i, b_i)} w_2(y) \, \Phi_j \left(|f_j(y)| \right)^{p_j} dy \right)^{\frac{1}{p_j}}, \qquad (22.77)$$

under the assumptions:

(i) f_j, $\Phi_j\left(\left|f_j\right|\right)^{p_j}$ are both $\dfrac{1}{\prod\limits_{i=1}^{N} \Gamma\left(\alpha_i\right)} \chi_{\prod\limits_{i=1}^{N}[x_i, b_i]}(y) \prod\limits_{i=1}^{N}(y_i - x_i)^{\alpha_i - 1} dy$

-integrable, a.e. in $x \in \prod\limits_{i=1}^{N}(a_i, b_i)$, for all $j = 1, \ldots, m$,

(ii) $w_2 \Phi_j\left(\left|f_j\right|\right)^{p_j}$ is Lebesgue integrable, $j = 1, \ldots, m$.

Using (22.72) and (22.73) we rewrite (22.76), as follows

$$\int_{\prod\limits_{i=1}^{N}(a_i, b_i)} \left(\prod_{i=1}^{N}(x_i - a_i)^{\alpha_i}\right) \prod_{j=1}^{m} \Phi_j \left(\frac{\left|I_{a+}^{\alpha}\left(f_j\right)(x)\right| \prod\limits_{i=1}^{N} \Gamma\left(\alpha_i + 1\right)}{\prod\limits_{i=1}^{N}(x_i - a_i)^{\alpha_i}}\right) dx \leq$$

$$\prod_{j=1}^{m} \left(\int_{\prod\limits_{i=1}^{N}(a_i, b_i)} \left(\prod_{i=1}^{N}(b_i - y_i)^{\alpha_i}\right) \Phi_j\left(\left|f_j(y)\right|\right)^{p_j} dy\right)^{\frac{1}{p_j}} \leq \qquad (22.78)$$

$$\left(\prod_{i=1}^{N}(b_i - a_i)^{\alpha_i}\right) \prod_{j=1}^{m} \left(\int_{\prod\limits_{i=1}^{N}(a_i, b_i)} \Phi_j\left(\left|f_j(y)\right|\right)^{p_j} dy\right)^{\frac{1}{p_j}},$$

under the assumptions:

(i)   following (22.76), and
(ii)* $\Phi_j\left(\left|f_j\right|\right)^{p_j}$ is Lebesgue integrable, $j = 1, \ldots, m$.

Similarly, using (22.74) and (22.75) we rewrite (22.77),

$$\int_{\prod\limits_{i=1}^{N}(a_i, b_i)} \left(\prod_{i=1}^{N}(b_i - x_i)^{\alpha_i}\right) \prod_{j=1}^{m} \Phi_j \left(\frac{\left|I_{b-}^{\alpha}\left(f_j\right)(x)\right| \prod\limits_{i=1}^{N} \Gamma\left(\alpha_i + 1\right)}{\prod\limits_{i=1}^{N}(b_i - x_i)^{\alpha_i}}\right) dx \leq$$

$$\prod_{j=1}^{m} \left(\int_{\prod_{i=1}^{N}(a_i,b_i)} \left(\prod_{i=1}^{N}(y_i - a_i)^{\alpha_i} \right) \Phi_j \left(|f_j(y)| \right)^{p_j} dy \right)^{\frac{1}{p_j}} \leq \qquad (22.79)$$

$$\left(\prod_{i=1}^{N}(b_i - a_i)^{\alpha_i} \right) \prod_{j=1}^{m} \left(\int_{\prod_{i=1}^{N}(a_i,b_i)} \Phi_j \left(|f_j(y)| \right)^{p_j} dy \right)^{\frac{1}{p_j}},$$

under the assumptions:

(i)   following (22.77), and
(ii)* $\Phi_j \left(|f_j| \right)^{p_j}$ is Lebesgue integrable, $j = 1, \ldots, m$.

Let now $\beta_j \geq 1$, $j = 1, \ldots, m$.
Then, by (22.78), we obtain

$$\int_{\prod_{i=1}^{N}(a_i,b_i)} \left(\prod_{i=1}^{N}(x_i - a_i)^{\alpha_i} \right)^{\left(1 - \sum_{j=1}^{m} \beta_j \right)} \left(\prod_{j=1}^{m} |I_{a+}^{\alpha}(f_j)(x)|^{\beta_j} \right) dx \leq$$

$$\left(\frac{1}{\left(\prod_{i=1}^{N} \Gamma(\alpha_i + 1) \right)^{\sum_{j=1}^{m} \beta_j}} \right) \cdot$$

$$\prod_{j=1}^{m} \left(\int_{\prod_{i=1}^{N}(a_i,b_i)} \left(\prod_{i=1}^{N}(b_i - y_i)^{\alpha_i} \right) |f_j(y)|^{\beta_j p_j} dy \right)^{\frac{1}{p_j}} \leq \qquad (22.80)$$

$$\left(\frac{\prod_{i=1}^{N}(b_i - a_i)^{\alpha_i}}{\left(\prod_{i=1}^{N}\Gamma(\alpha_i + 1)\right)^{\sum_{j=1}^{m}\beta_j}}\right)\prod_{j=1}^{m}\left(\int_{\prod_{i=1}^{N}(a_i, b_i)}|f_j(y)|^{\beta_j p_j}\,dy\right)^{\frac{1}{p_j}}.$$

But it holds

$$\int_{\prod_{i=1}^{N}(a_i, b_i)}\left(\prod_{i=1}^{N}(x_i - a_i)^{\alpha_i}\right)^{\left(1 - \sum_{j=1}^{m}\beta_j\right)}\left(\prod_{j=1}^{m}|I_{a+}^{\alpha}(f_j)(x)|^{\beta_j}\right)dx \geq$$

$$\left(\prod_{i=1}^{N}(b_i - a_i)^{\alpha_i}\right)^{\left(1 - \sum_{j=1}^{m}\beta_j\right)}\left(\int_{\prod_{i=1}^{N}(a_i, b_i)}\prod_{j=1}^{m}|I_{a+}^{\alpha}(f_j)(x)|^{\beta_j}\,dx\right). \quad (22.81)$$

So by (22.80) and (22.81) we derive

$$\int_{\prod_{i=1}^{N}(a_i, b_i)}\prod_{j=1}^{m}|I_{a+}^{\alpha}(f_j)(x)|^{\beta_j}\,dx \leq \quad (22.82)$$

$$\left(\prod_{i=1}^{N}\frac{(b_i - a_i)^{\alpha_i}}{\Gamma(\alpha_i + 1)}\right)^{\sum_{j=1}^{m}\beta_j}\prod_{j=1}^{m}\left(\int_{\prod_{i=1}^{N}(a_i, b_i)}|f_j(y)|^{\beta_j p_j}\,dy\right)^{\frac{1}{p_j}},$$

under the assumptions:

(i) $|f_j|^{p_j\beta_j}$ is $\dfrac{1}{\prod_{i=1}^{N}\Gamma(\alpha_i)}\chi_{\prod_{i=1}^{N}(a_i, x_i]}(y)\prod_{i=1}^{N}(x_i - y_i)^{\alpha_i - 1}\,dy$ -integrable, a.e. in

$x \in \prod_{i=1}^{N}(a_i, b_i)$, for all $j = 1, \ldots, m$,

(ii) $\left|f_j\right|^{p_j \beta_j}$ is Lebesgue integrable, $j = 1, \ldots, m$.

We also have, by (22.78), that

$$\int_{\prod\limits_{i=1}^{N}(a_i, b_i)} \left(\prod_{i=1}^{N}(x_i - a_i)^{\alpha_i}\right) e^{\left(\sum\limits_{j=1}^{m}\left|I_{a+}^{\alpha}(f_j)(x)\right|\right)} \left(\prod_{i=1}^{N}\frac{\Gamma(\alpha_i + 1)}{(x_i - a_i)^{\alpha_i}}\right) dx \leq$$

$$\prod_{j=1}^{m}\left(\int_{\prod\limits_{i=1}^{N}(a_i, b_i)} \left(\prod_{i=1}^{N}(b_i - y_i)^{\alpha_i}\right) e^{p_j\left|f_j(y)\right|}dy\right)^{\frac{1}{p_j}} \leq \qquad (22.83)$$

$$\left(\prod_{i=1}^{N}(b_i - a_i)^{\alpha_i}\right)\prod_{j=1}^{m}\left(\int_{\prod\limits_{i=1}^{N}(a_i, b_i)} e^{p_j\left|f_j(y)\right|}dy\right)^{\frac{1}{p_j}},$$

under the assumptions:

(i) $f_j, e^{p_j|f_j|}$ are both $\dfrac{1}{\prod\limits_{i=1}^{N}\Gamma(\alpha_i)} \chi_{\prod\limits_{i=1}^{N}(a_i, x_i]}(y)\prod\limits_{i=1}^{N}(x_i - y_i)^{\alpha_i - 1}\, dy$ -integrable,

a.e. in $x \in \prod\limits_{i=1}^{N}(a_i, b_i)$, for all $j = 1, \ldots, m$,

(ii) $e^{p_j|f_j|}$ is Lebesgue integrable, $j = 1, \ldots, m$.

From (22.79) we get

$$\int_{\prod\limits_{i=1}^{N}(a_i, b_i)} \left(\prod_{i=1}^{N}(b_i - x_i)^{\alpha_i}\right)^{\left(1 - \sum\limits_{j=1}^{m}\beta_j\right)} \left(\prod_{j=1}^{m}\left|I_{b-}^{\alpha}(f_j)(x)\right|^{\beta_j}\right) dx \leq$$

$$
\left(\frac{1}{\left(\prod_{i=1}^{N} \Gamma\left(\alpha_i + 1\right) \right)^{\sum_{j=1}^{m} \beta_j}} \right) \cdot
$$

$$
\prod_{j=1}^{m} \left(\int_{\prod_{i=1}^{N} (a_i, b_i)} \left(\prod_{i=1}^{N} (y_i - a_i)^{\alpha_i} \right) \left| f_j(y) \right|^{\beta_j p_j} dy \right)^{\frac{1}{p_j}} \leq \qquad (22.84)
$$

$$
\left(\frac{\prod_{i=1}^{N} (b_i - a_i)^{\alpha_i}}{\left(\prod_{i=1}^{N} \Gamma\left(\alpha_i + 1\right) \right)^{\sum_{j=1}^{m} \beta_j}} \right) \prod_{j=1}^{m} \left(\int_{\prod_{i=1}^{N} (a_i, b_i)} \left| f_j(y) \right|^{\beta_j p_j} dy \right)^{\frac{1}{p_j}} \cdot
$$

But it holds

$$
\int_{\prod_{i=1}^{N} (a_i, b_i)} \left(\prod_{i=1}^{N} (b_i - x_i)^{\alpha_i} \right)^{\left(1 - \sum_{j=1}^{m} \beta_j \right)} \left(\prod_{j=1}^{m} \left| I_{b-}^{\alpha}(f_j)(x) \right|^{\beta_j} \right) dx \geq
$$

$$
\left(\prod_{i=1}^{N} (b_i - a_i)^{\alpha_i} \right)^{\left(1 - \sum_{j=1}^{m} \beta_j \right)} \left(\int_{\prod_{i=1}^{N} (a_i, b_i)} \left(\prod_{j=1}^{m} \left| I_{b-}^{\alpha}(f_j)(x) \right|^{\beta_j} \right) dx \right).
$$

$$
(22.85)
$$

So by (22.84) and (22.85) we obtain

$$
\int_{\prod_{i=1}^{N} (a_i, b_i)} \left(\prod_{j=1}^{m} \left| I_{b-}^{\alpha}(f_j)(x) \right|^{\beta_j} \right) dx \leq \qquad (22.86)
$$

$$\left(\prod_{i=1}^{N} \frac{(b_i - a_i)^{\alpha_i}}{\Gamma(\alpha_i + 1)}\right)^{\sum_{j=1}^{m} \beta_j} \prod_{j=1}^{m} \left(\int_{\prod_{i=1}^{N}(a_i, b_i)} |f_j(y)|^{\beta_j p_j} dy\right)^{\frac{1}{p_j}},$$

under the assumptions:

(i) $|f_j|^{p_j \beta_j}$ is $\dfrac{1}{\prod\limits_{i=1}^{N} \Gamma(\alpha_i) \prod\limits_{i=1}^{N} [x_i, b_i)} \cdot \chi_N$ (y) $\prod\limits_{i=1}^{N} (y_i - x_i)^{\alpha_i - 1} dy$ -integrable, a.e. in

$x \in \prod\limits_{i=1}^{N} (a_i, b_i)$, for all $j = 1, \ldots, m,$

(ii) $|f_j|^{p_j \beta_j}$ is Lebesgue integrable, $j = 1, \ldots, m.$

We also have, by (22.79), that

$$\int_{\prod_{i=1}^{N}(a_i, b_i)} \left(\prod_{i=1}^{N} (b_i - x_i)^{\alpha_i}\right) e^{\left(\sum_{j=1}^{m} |I_{b-}^{\alpha}(f_j)(x)|\right)} \left(\prod_{i=1}^{N} \frac{\Gamma(\alpha_i + 1)}{(b_i - x_i)^{\alpha_i}}\right) dx \le$$

$$\prod_{j=1}^{m} \left(\int_{\prod_{i=1}^{N}(a_i, b_i)} \left(\prod_{i=1}^{N} (y_i - a_i)^{\alpha_i}\right) e^{p_j |f_j(y)|} dy\right)^{\frac{1}{p_j}} \le \qquad (22.87)$$

$$\left(\prod_{i=1}^{N} (b_i - a_i)^{\alpha_i}\right) \prod_{j=1}^{m} \left(\int_{\prod_{i=1}^{N}(a_i, b_i)} e^{p_j |f_j(y)|} dy\right)^{\frac{1}{p_j}},$$

under the assumptions:

(i) $f_j, e^{p_j |f_j|}$ are both $\dfrac{1}{\prod\limits_{i=1}^{N} \Gamma(\alpha_i) \prod\limits_{i=1}^{N} [x_i, b_i)} \cdot \chi_N$ (y) $\prod\limits_{i=1}^{N} (y_i - x_i)^{\alpha_i - 1} dy$ -integrable,

a.e. in $x \in \prod\limits_{i=1}^{N} (a_i, b_i)$, for all $j = 1, \ldots, m,$

(ii) $e^{p_j|f_j|}$ is Lebesgue integrable, $j = 1, \ldots, m$.

Background 22.20 *In order to apply Theorem 22.1 to the case of a spherical shell we need:*

Let $N \geq 2$, $S^{N-1} := \{x \in \mathbb{R}^N : |x| = 1\}$ *the unit sphere on* $\mathbb{R}^N$, *where* $|\cdot|$ *stands for the Euclidean norm in* $\mathbb{R}^N$. *Also denote the ball* $B(0, R) := \{x \in \mathbb{R}^N : |x| < R\} \subseteq \mathbb{R}^N$, $R > 0$, *and the spherical shell*

$$A := B(0, R_2) - \overline{B(0, R_1)}, \ 0 < R_1 < R_2. \tag{22.88}$$

For the following see [15, pp. 149–150], and [17, pp. 87–88].

For $x \in \mathbb{R}^N - \{0\}$ *we can write uniquely* $x = r\omega$, *where* $r = |x| > 0$, *and* $\omega = \frac{x}{r} \in S^{N-1}$, $|\omega| = 1$.

Clearly here

$$\mathbb{R}^N - \{0\} = (0, \infty) \times S^{N-1}, \tag{22.89}$$

and

$$\overline{A} = [R_1, R_2] \times S^{N-1}. \tag{22.90}$$

We will be using

Theorem 22.21 ([1, p. 322]) *Let* $f : A \to \mathbb{R}$ *be a Lebesgue integrable function. Then*

$$\int_A f(x) \, dx = \int_{S^{N-1}} \left(\int_{R_1}^{R_2} f(r\omega) \, r^{N-1} dr \right) d\omega. \tag{22.91}$$

So we are able to write an integral on the shell in polar form using the polar coordinates (r, ω).

We need

Definition 22.22 ([1, p. 458]) Let $\nu > 0$, $n := [\nu]$, $\alpha := \nu - n$, $f \in C^n(\overline{A})$, and A is a spherical shell. Assume that there exists function $\dfrac{\partial_{R_1}^\nu f(x)}{\partial r^\nu} \in C(\overline{A})$, given by

$$\frac{\partial_{R_1}^\nu f(x)}{\partial r^\nu} := \frac{1}{\Gamma(1-\alpha)} \frac{\partial}{\partial r} \left(\int_{R_1}^r (r-t)^{-\alpha} \frac{\partial^n f(t\omega)}{\partial r^n} dt \right), \tag{22.92}$$

where $x \in \overline{A}$; that is $x = r\omega$, $r \in [R_1, R_2]$, $\omega \in S^{N-1}$.

We call $\dfrac{\partial_{R_1}^\nu f}{\partial r^\nu}$ the left radial Canavati-type fractional derivative of f of order ν. If $\nu = 0$, then set $\dfrac{\partial_{R_1}^\nu f(x)}{\partial r^\nu} := f(x)$.

Based on [1, p. 288], and [5] we have

Lemma 22.23 Let $\gamma \geq 0$, $m := [\gamma]$, $\nu > 0$, $n := [\nu]$, with $0 \leq \gamma < \nu$. Let $f \in C^n(\overline{A})$ and there exists $\dfrac{\partial_{R_1}^\nu f(x)}{\partial r^\nu} \in C(\overline{A})$, $x \in \overline{A}$, A a spherical shell. Further *assume that* $\frac{\partial^j f(R_1\omega)}{\partial r^j} = 0$, $j = m, m+1, \ldots, n-1$, $\forall \omega \in S^{N-1}$. *Then there exists* $\frac{\partial^\gamma_{R_1} f(x)}{\partial r^\gamma} \in C\left(\overline{A}\right)$ *such that*

$$\frac{\partial^\gamma_{R_1} f(x)}{\partial r^\gamma} = \frac{\partial^\gamma_{R_1} f(r\omega)}{\partial r^\gamma} = \frac{1}{\Gamma(\nu - \gamma)} \int_{R_1}^r (r-t)^{\nu-\gamma-1} \frac{\partial^\nu_{R_1} f(t\omega)}{\partial r^\nu} dt, \quad (22.93)$$

$\forall \omega \in S^{N-1}$; *all* $R_1 \leq r \leq R_2$, *indeed* $f(r\omega) \in C^\gamma_{R_1}([R_1, R_2])$, $\forall \omega \in S^{N-1}$.

We make

Remark 22.24 In the settings and assumptions of Theorem 22.1 and Lemma 22.23 we have

$$k(r, t) = \frac{1}{\Gamma(\nu - \gamma)} \chi_{[R_1, r]}(t)(r-t)^{\nu-\gamma-1}, \quad (22.94)$$

and

$$K(r) = \frac{(r-R_1)^{\nu-\gamma}}{\Gamma(\nu - \gamma + 1)}, \quad (22.95)$$

$r, t \in [R_1, R_2]$.

Furthermore we get

$$\frac{k(r, t)}{K(r)} = (\nu - \gamma) \chi_{[R_1, r]}(t) \frac{(r-t)^{\nu-\gamma-1}}{(r-R_1)^{\nu-\gamma}}, \quad (22.96)$$

and by choosing

$$u(r) := (r-R_1)^{\nu-\gamma}, \ r \in [R_1, R_2], \quad (22.97)$$

we find

$$U(t) = (\nu - \gamma) \int_t^{R_2} (r-t)^{\nu-\gamma-1} dr = (R_2 - t)^{\nu-\gamma}, \quad (22.98)$$

$t \in [R_1, R_2]$.

Then by (22.8) for $p \geq 1$ we find

$$\int_{R_1}^{R_2} (r-R_1)^{\nu-\gamma} \left| \frac{\partial^\gamma_{R_1} f(r\omega)}{\partial r^\gamma} \right|^p \frac{(\Gamma(\nu - \gamma + 1))^p}{(r-R_1)^{(\nu-\gamma)p}} dr \leq$$

$$\int_{R_1}^{R_2} (R_2 - r)^{\nu-\gamma} \left| \frac{\partial^\nu_{R_1} f(r\omega)}{\partial r^\nu} \right|^p dr, \quad (22.99)$$

and

$$\int_{R_1}^{R_2} (r-R_1)^{(\nu-\gamma)(1-p)} \left| \frac{\partial^\gamma_{R_1} f(r\omega)}{\partial r^\gamma} \right|^p dr \leq$$

$$\frac{1}{(\Gamma(\nu - \gamma + 1))^p} \int_{R_1}^{R_2} (R_2 - r)^{\nu - \gamma} \left| \frac{\partial_{R_1}^{\nu} f(r\omega)}{\partial r^{\nu}} \right|^p dr \leq$$

$$\frac{(R_2 - R_1)^{\nu - \gamma}}{(\Gamma(\nu - \gamma + 1))^p} \int_{R_1}^{R_2} \left| \frac{\partial_{R_1}^{\nu} f(r\omega)}{\partial r^{\nu}} \right|^p dr. \qquad (22.100)$$

But it holds

$$\int_{R_1}^{R_2} (r - R_1)^{(\nu - \gamma)(1-p)} \left| \frac{\partial_{R_1}^{\gamma} f(r\omega)}{\partial r^{\gamma}} \right|^p dr \geq$$

$$(R_2 - R_1)^{(\nu - \gamma)(1-p)} \int_{R_1}^{R_2} \left| \frac{\partial_{R_1}^{\gamma} f(r\omega)}{\partial r^{\gamma}} \right|^p dr. \qquad (22.101)$$

Consequently we derive

$$\int_{R_1}^{R_2} \left| \frac{\partial_{R_1}^{\gamma} f(r\omega)}{\partial r^{\gamma}} \right|^p dr \leq \left(\frac{(R_2 - R_1)^{(\nu - \gamma)}}{\Gamma(\nu - \gamma + 1)} \right)^p \int_{R_1}^{R_2} \left| \frac{\partial_{R_1}^{\nu} f(r\omega)}{\partial r^{\nu}} \right|^p dr, \qquad (22.102)$$

$\forall \omega \in S^{N-1}$.

Here we have $R_1 \leq r \leq R_2$, and $R_1^{N-1} \leq r^{N-1} \leq R_2^{N-1}$, and $R_2^{1-N} \leq r^{1-N} \leq R_1^{1-N}$.

From (22.102) we have

$$R_2^{1-N} \int_{R_1}^{R_2} r^{N-1} \left| \frac{\partial_{R_1}^{\gamma} f(r\omega)}{\partial r^{\gamma}} \right|^p dr \leq$$

$$\int_{R_1}^{R_2} r^{1-N} r^{N-1} \left| \frac{\partial_{R_1}^{\gamma} f(r\omega)}{\partial r^{\gamma}} \right|^p dr \leq$$

$$\left(\frac{(R_2 - R_1)^{(\nu - \gamma)}}{\Gamma(\nu - \gamma + 1)} \right)^p \int_{R_1}^{R_2} r^{1-N} r^{N-1} \left| \frac{\partial_{R_1}^{\nu} f(r\omega)}{\partial r^{\nu}} \right|^p dr \leq$$

$$R_1^{1-N} \left(\frac{(R_2 - R_1)^{(\nu - \gamma)}}{\Gamma(\nu - \gamma + 1)} \right)^p \int_{R_1}^{R_2} r^{N-1} \left| \frac{\partial_{R_1}^{\nu} f(r\omega)}{\partial r^{\nu}} \right|^p dr. \qquad (22.103)$$

So we get

$$\int_{R_1}^{R_2} r^{N-1} \left| \frac{\partial_{R_1}^{\gamma} f(r\omega)}{\partial r^{\gamma}} \right|^p dr \leq$$

$$\left(\frac{R_2}{R_1}\right)^{N-1} \left(\frac{(R_2 - R_1)^{(\nu-\gamma)}}{\Gamma(\nu - \gamma + 1)}\right)^p \int_{R_1}^{R_2} r^{N-1} \left|\frac{\partial_{R_1}^\nu f(r\omega)}{\partial r^\nu}\right|^p dr, \qquad (22.104)$$

$\forall \, \omega \in S^{N-1}$.

Hence

$$\int_{S^{N-1}} \left(\int_{R_1}^{R_2} r^{N-1} \left|\frac{\partial_{R_1}^\gamma f(r\omega)}{\partial r^\gamma}\right|^p dr\right) d\omega \leq$$

$$\left(\frac{R_2}{R_1}\right)^{N-1} \left(\frac{(R_2 - R_1)^{(\nu-\gamma)}}{\Gamma(\nu - \gamma + 1)}\right)^p \int_{S^{N-1}} \left(\int_{R_1}^{R_2} r^{N-1} \left|\frac{\partial_{R_1}^\nu f(r\omega)}{\partial r^\nu}\right|^p dr\right) d\omega.$$

$$(22.105)$$

By Theorem 22.21, equality (22.91), we obtain

$$\int_A \left|\frac{\partial_{R_1}^\gamma f(x)}{\partial r^\gamma}\right|^p dx \leq \left(\frac{R_2}{R_1}\right)^{N-1} \left(\frac{(R_2 - R_1)^{(\nu-\gamma)}}{\Gamma(\nu - \gamma + 1)}\right)^p \int_A \left|\frac{\partial_{R_1}^\nu f(x)}{\partial r^\nu}\right|^p dx.$$

$$(22.106)$$

We have proved the following fractional Poincaré type inequalities on the shell.

Theorem 22.25 *Here all as in Lemma 22.23, $p \geq 1$.*
It holds
(1)

$$\left\|\frac{\partial_{R_1}^\gamma f}{\partial r^\gamma}\right\|_{p,A} \leq \left(\frac{R_2}{R_1}\right)^{\left(\frac{N-1}{p}\right)} \left(\frac{(R_2 - R_1)^{(\nu-\gamma)}}{\Gamma(\nu - \gamma + 1)}\right) \left\|\frac{\partial_{R_1}^\nu f}{\partial r^\nu}\right\|_{p,A}, \qquad (22.107)$$

(2) When $\gamma = 0$, we have

$$\|f\|_{p,A} \leq \left(\frac{R_2}{R_1}\right)^{\left(\frac{N-1}{p}\right)} \left(\frac{(R_2 - R_1)^\nu}{\Gamma(\nu + 1)}\right) \left\|\frac{\partial_{R_1}^\nu f}{\partial r^\nu}\right\|_{p,A}. \qquad (22.108)$$

See the related, and proof, results in [1, pp. 458–459] with different constants and proof in the corresponding inequalities.

Similar results can be produced for the right radial Canavati type fractional derivative.

We choose to omit it.

We make

Remark 22.26 (from [1], p. 460) Here we denote $\lambda_{\mathbb{R}^N}(x) \equiv dx$ the Lebesgue measure on $\mathbb{R}^N$, $N \geq 2$, and by $\lambda_{S^{N-1}}(\omega) = d\omega$ the surface measure on S^{N-1}, where $\mathcal{B}_X$ stands for the Borel class on space X. Define the measure R_N on $\left((0, \infty), \mathcal{B}_{(0,\infty)}\right)$ by

$$R_N(B) = \int_B r^{N-1} dr, \text{ any } B \in \mathcal{B}_{(0,\infty)}.$$

Now let $F \in L_1(A) = L_1\left([R_1, R_2] \times S^{N-1}\right)$.
   Call

$$K(F) := \{\omega \in S^{N-1} : F(\cdot\omega) \notin L_1\left([R_1, R_2], \mathcal{B}_{[R_1, R_2]}, R_N\right)\}. \qquad (22.109)$$

We get, by Fubini's theorem and [18], pp. 87–88, that

$$\lambda_{S^{N-1}}(K(F)) = 0.$$

Of course

$$\theta(F) := [R_1, R_2] \times K(F) \subset A,$$

and

$$\lambda_{\mathbb{R}^N}(\theta(F)) = 0.$$

Above $\lambda_{S^{N-1}}$ is defined as follows: let $A \subset S^{N-1}$ be a Borel set, and let

$$\widetilde{A} := \{ru : 0 < r < 1, u \in A\} \subset \mathbb{R}^N;$$

we define

$$\lambda_{S^{N-1}}(A) := N\lambda_{\mathbb{R}^N}\left(\widetilde{A}\right).$$

We have that

$$\lambda_{S^{N-1}}\left(S^{N-1}\right) = \frac{2\pi^{\frac{N}{2}}}{\Gamma\left(\frac{N}{2}\right)},$$

the surface area of S^{N-1}.
   See also [16, pp. 149–150], [18, pp. 87–88] and [1], p. 320.

Following [1, p. 466] we define the left Riemann-Liouville radial fractional derivative next.

Definition 22.27 Let $\beta > 0$, $m := [\beta] + 1$, $F \in L_1(A)$, and A is the spherical shell. We define

$$\frac{\overline{\partial}_{R_1}^\beta F(x)}{\partial r^\beta} := \begin{cases} \frac{1}{\Gamma(m-\beta)} \left(\frac{\partial}{\partial r}\right)^m \int_{R_1}^r (r-t)^{m-\beta-1} F(t\omega) \, dt, \\ \qquad\qquad \text{for } \omega \in S^{N-1} - K(F), \\ 0, \quad \text{for } \omega \in K(F), \end{cases} \qquad (22.110)$$

where $x = r\omega \in A$, $r \in [R_1, R_2]$, $\omega \in S^{N-1}$; $K(F)$ as in (22.109).

If $\beta = 0$, define

$$\frac{\overline{\partial}_{R_1}^{\beta} F(x)}{\partial r^{\beta}} := F(x).$$

We need the following important representation result for left Riemann-Liouville radial fractional derivatives, by [1, p. 466].

Theorem 22.28 *Let* $\nu \geq \gamma + 1$, $\gamma \geq 0$, $n := [\nu]$, $m := [\gamma]$, $F : \overline{A} \to \mathbb{R}$ *with* $F \in L_1(A)$. *Assume that* $F(\cdot\omega) \in AC^n([R_1, R_2])$, $\forall \omega \in S^{N-1}$, *and that* $\frac{\overline{\partial}_{R_1}^{\nu} F(\cdot\omega)}{\partial r^{\nu}}$ *is measurable on* $[R_1, R_2]$, $\forall \omega \in S^{N-1}$. *Also assume* $\exists \frac{\overline{\partial}_{R_1}^{\nu} F(r\omega)}{\partial r^{\nu}} \in \mathbb{R}$, $\forall r \in [R_1, R_2]$ *and* $\forall \omega \in S^{N-1}$, *and* $\frac{\overline{\partial}_{R_1}^{\nu} F(x)}{\partial r^{\nu}}$ *is measurable on* $\overline{A}$. *Suppose* $\exists M_1 > 0$:

$$\left| \frac{\overline{\partial}_{R_1}^{\nu} F(r\omega)}{\partial r^{\nu}} \right| \leq M_1, \; \forall \; (r, \omega) \in [R_1, R_2] \times S^{N-1}. \tag{22.111}$$

We suppose that $\frac{\overline{\partial}^j F(R_1\omega)}{\partial r^j} = 0$, $j = m, m+1, \ldots, n-1$; $\forall \omega \in S^{N-1}$.
Then

$$\frac{\overline{\partial}_{R_1}^{\gamma} F(x)}{\partial r^{\gamma}} = \overline{D}_{R_1}^{\gamma} F(r\omega) = \frac{1}{\Gamma(\nu - \gamma)} \int_{R_1}^{r} (r - t)^{\nu - \gamma - 1} \left(\overline{D}_{R_1}^{\nu} F \right)(t\omega)\, dt,$$
$$\tag{22.112}$$

valid $\forall x \in \overline{A}$; *that is, true* $\forall r \in [R_1, R_2]$ *and* $\forall \omega \in S^{N-1}$; $\gamma > 0$.
Here

$$\overline{D}_{R_1}^{\gamma} F(\cdot\omega) \in AC([R_1, R_2]), \tag{22.113}$$

$\forall \omega \in S^{N-1}$; $\gamma > 0$.
Furthermore

$$\frac{\overline{\partial}_{R_1}^{\gamma} F(x)}{\partial r^{\gamma}} \in L_{\infty}(A), \; \gamma > 0. \tag{22.114}$$

In particular, it holds

$$F(x) = F(r\omega) = \frac{1}{\Gamma(\nu)} \int_{R_1}^{r} (r - t)^{\nu - 1} \left(\overline{D}_{R_1}^{\nu} F \right)(t\omega)\, dt, \tag{22.115}$$

true $\forall x \in \overline{A}$; *that is, true* $\forall r \in [R_1, R_2]$ *and* $\forall \omega \in S^{N-1}$, *and*

$$F(\cdot\omega) \in AC([R_1, R_2]), \; \forall \omega \in S^{N-1}. \tag{22.116}$$

We give also the following fractional Poincaré type inequalities on the spherical shell.

Theorem 22.29 *Here all as in Theorem 22.28, $p \geq 1$. Then*

(1)

$$\left\| \frac{\overline{\partial}_{R_1}^{\gamma} F}{\partial r^{\gamma}} \right\|_{p,A} \leq \left(\frac{R_2}{R_1} \right)^{\left(\frac{N-1}{p} \right)} \left(\frac{(R_2 - R_1)^{(\nu - \gamma)}}{\Gamma(\nu - \gamma + 1)} \right) \left\| \frac{\overline{\partial}_{R_1}^{\nu} F}{\partial r^{\nu}} \right\|_{p,A}, \qquad (22.117)$$

(2) When $\gamma = 0$, we have

$$\| F \|_{p,A} \leq \left(\frac{R_2}{R_1} \right)^{\left(\frac{N-1}{p} \right)} \left(\frac{(R_2 - R_1)^{\nu}}{\Gamma(\nu + 1)} \right) \left\| \frac{\overline{\partial}_{R_1}^{\nu} F}{\partial r^{\nu}} \right\|_{p,A}. \qquad (22.118)$$

Proof As in Theorem 22.25, based on Theorem 22.28. ∎

See also similar results in [1, p. 468].
We also need (see [1], p. 421).

Definition 22.30 Let $F : \overline{A} \rightarrow \mathbb{R}$, $\nu \geq 0$, $n := \lceil \nu \rceil$ such that $F(\cdot \omega) \in AC^n([R_1, R_2])$, for all $\omega \in S^{N-1}$.

We call the left Caputo radial fractional derivative the following function

$$\frac{\partial_{*R_1}^{\nu} F(x)}{\partial r^{\nu}} := \frac{1}{\Gamma(n - \nu)} \int_{R_1}^{r} (r - t)^{n - \nu - 1} \frac{\partial^n F(t\omega)}{\partial r^n} dt, \qquad (22.119)$$

where $x \in \overline{A}$, i.e. $x = r\omega, r \in [R_1, R_2], \omega \in S^{N-1}$.

Clearly

$$\frac{\partial_{*R_1}^{0} F(x)}{\partial r^0} = F(x), \qquad (22.120)$$

$$\frac{\partial_{*R_1}^{\nu} F(x)}{\partial r^{\nu}} = \frac{\partial^{\nu} F(x)}{\partial r^{\nu}}, \text{ if } \nu \in \mathbb{N}.$$

Above function (22.119) exists almost everywhere for $x \in \overline{A}$, see [1], p. 422.

We mention the following fundamental representation result (see [1], pp. 422–423 and [5]).

Theorem 22.31 *Let $\nu \geq \gamma + 1$, $\gamma \geq 0$, $n := \lceil \nu \rceil$, $m := \lceil \gamma \rceil$, $F : \overline{A} \rightarrow \mathbb{R}$ with $F \in L_1(A)$. Assume that $F(\cdot \omega) \in AC^n([R_1, R_2])$, for all $\omega \in S^{N-1}$, and that $\frac{\partial_{*R_1}^{\nu} F(\cdot \omega)}{\partial r^{\nu}} \in L_{\infty}(R_1, R_2)$ for all $\omega \in S^{N-1}$.*

Further assume that $\frac{\partial_{*R_1}^{\nu} F(x)}{\partial r^{\nu}} \in L_{\infty}(A)$. More precisely, for these $r \in [R_1, R_2]$, for each $\omega \in S^{N-1}$, for which $D_{*R_1}^{\nu} F(r\omega)$ takes real values, there exists $M_1 > 0$ such that $\left| D_{*R_1}^{\nu} F(r\omega) \right| \leq M_1$.

We suppose that $\frac{\partial^j F(R_1 \omega)}{\partial r^j} = 0$, $j = m, m+1, \ldots, n-1$; for every $\omega \in S^{N-1}$.

Then

$$\frac{\partial^\gamma_{*R_1} F(x)}{\partial r^\gamma} = D^\gamma_{*R_1} F(r\omega) = \frac{1}{\Gamma(\nu - \gamma)} \int_{R_1}^r (r-t)^{\nu-\gamma-1} \left(D^\nu_{*R_1} F\right)(t\omega)\, dt,$$

$$(22.121)$$

valid $\forall\, x \in \overline{A}$; *i.e. true* $\forall\, r \in [R_1, R_2]$ *and* $\forall\, \omega \in S^{N-1}$; $\gamma > 0$.
  Here

$$D^\gamma_{*R_1} F(\cdot\omega) \in AC\left([R_1, R_2]\right),$$

$$(22.122)$$

$\forall\, \omega \in S^{N-1}$; $\gamma > 0$.
  Furthermore

$$\frac{\partial^\gamma_{*R_1} F(x)}{\partial r^\gamma} \in L_\infty(A), \gamma > 0.$$

$$(22.123)$$

In particular, it holds

$$F(x) = F(r\omega) = \frac{1}{\Gamma(\nu)} \int_{R_1}^r (r-t)^{\nu-1} \left(D^\nu_{*R_1} F\right)(t\omega)\, dt,$$

$$(22.124)$$

true $\forall\, x \in \overline{A}$; *i.e., true* $\forall\, r \in [R_1, R_2]$ *and* $\forall\, \omega \in S^{N-1}$, *and*

$$F(\cdot\omega) \in AC\left([R_1, R_2]\right), \ \forall\, \omega \in S^{N-1}.$$

$$(22.125)$$

We finish with the following Poincaré type inequalities involving left Caputo radial fractional derivatives.

Theorem 22.32 *Here all as in Theorem 22.31, $p \geq 1$. Then*
  (1)

$$\left\| \frac{\partial^\gamma_{*R_1} F}{\partial r^\gamma} \right\|_{p,A} \leq \left(\frac{R_2}{R_1}\right)^{\left(\frac{N-1}{p}\right)} \left(\frac{(R_2 - R_1)^{(\nu-\gamma)}}{\Gamma(\nu-\gamma+1)}\right) \left\| \frac{\partial^\nu_{*R_1} F}{\partial r^\nu} \right\|_{p,A},$$

$$(22.126)$$

(2) When $\gamma = 0$, *we have*

$$\|F\|_{p,A} \leq \left(\frac{R_2}{R_1}\right)^{\left(\frac{N-1}{p}\right)} \left(\frac{(R_2 - R_1)^\nu}{\Gamma(\nu+1)}\right) \left\| \frac{\partial^\nu_{*R_1} F}{\partial r^\nu} \right\|_{p,A}.$$

$$(22.127)$$

Proof As in Theorem 22.25, based on Theorem 22.31. ∎

See also similar results in [1, p. 464].

References

1. G.A. Anastassiou, *Fractional Differentiation Inequalities*, Research Monograph (Springer, New York, 2009)
2. G.A. Anastassiou, On right fractional calculus. Chaos Solitons Fractals **42**, 365–376 (2009)
3. G.A. Anastassiou, Balanced fractional Opial inequalities. Chaos Solitons Fractals **42**(3), 1523–1528 (2009)
4. G.A. Anastassiou, Fractional Korovkin theory. Chaos Solitons Fractals **42**, 2080–2094 (2009)
5. G.A. Anastassiou, Fractional representation formulae and right fractional inequalities. Math. Comput. Model. **54**(11–12), 3098–3115 (2011)
6. G.A. Anastassiou, Fractional integral inequalities involving convexity. Sarajevo J. Math. **8**(21), 203–233 (2012)
7. J.A. Canavati, The Riemann-Liouville integral. Nieuw Archief Voor Wiskunde **5**(1), 53–75 (1987)
8. K. Diethelm, *The Analysis of Fractional Differential Equations*, vol. 2004, 1st edn., Lecture Notes in Mathematics (Springer, New York, Heidelberg, 2010)
9. A.M.A. El-Sayed, M. Gaber, On the finite Caputo and finite Riesz derivatives. Electron. J. Theor. Phys. **3**(12), 81–95 (2006)
10. R. Gorenflo, F. Mainardi, *Essentials of Fractional Calculus* (Maphysto Center, Aarhus, 2000). http://www.maphysto.dk/oldpages/events/LevyCAC2000/MainardiNotes/fm2k0a.ps
11. G.D. Handley, J.J. Koliha, J. Pečarić, Hilbert-Pachpatte type integral inequalities for fractional derivatives. Fractional Calculus Appl Anal **4**(1), 37–46 (2001)
12. H.G. Hardy, Notes on some points in the integral calculus. Messenger Math. **47**(10), 145–150 (1918)
13. S. Iqbal, K. Krulic, J. Pecaric, On an inequality of H.G. Hardy. J. Inequalities Appl. 2010(264347), 23
14. A.A. Kilbas, H.M. Srivastava, J.J. Trujillo, *Theory and Applications of Fractional Differential Equations*, North-Holland Mathematics Studies, vol. 204 (Elsevier, New York, 2006)
15. T. Mamatov, S. Samko, Mixed fractional integration operators in mixed weighted Hölder spaces. Fractional Calculus Appl. Anal. **13**(3), 245–259 (2010)
16. W. Rudin, *Real and Complex Analysis, International*, Student edn. (Mc Graw Hill, New York, 1970)
17. S.G. Samko, A.A. Kilbas, O.I. Marichev, *Fractional Integral and Derivatives: Theory and Applications* (Gordon and Breach Science Publishers, Yverdon, 1993)
18. D. Stroock, *A Concise Introduction to the Theory of Integration*, 3rd edn. (Birkhäuser, Boston, 1999)

Chapter 23
Vectorial Inequalities for Integral Operators Involving Ratios of Functions Using Convexity

Here we present vectorial integral inequalities for products of multivariate convex and increasing functions applied to vectors of ratios of functions. As applications we derive a wide range of vectorial fractional inequalities of Hardy type. They involve the left and right Riemann-Liouville fractional integrals and their generalizations, in particular the Hadamard fractional integrals. Also inequalities for Riemann-Liouville, Caputo, Canavati and their generalizations fractional derivatives. These application inequalities are of L_p type, $p \geq 1$, and exponential type. It follows [5].

23.1 Introduction

Let $(\Omega_1, \Sigma_1, \mu_1)$ and $(\Omega_2, \Sigma_2, \mu_2)$ be measure spaces with positive σ-finite measures, and let $k : \Omega_1 \times \Omega_2 \to \mathbb{R}$ be a nonnegative measurable function, $k(x, \cdot)$ measurable on Ω_2 and

$$K(x) = \int_{\Omega_2} k(x, y) \, d\mu_2(y), \quad x \in \Omega_1. \tag{23.1}$$

We suppose that $K(x) > 0$ a.e. on Ω_1, and by a weight function (shortly: a weight), we mean a nonnegative measurable function on the actual set. Let the measurable functions $g_i : \Omega_1 \to \mathbb{R}, i = 1, \ldots, n$, with the representation

$$g_i(x) = \int_{\Omega_2} k(x, y) f_i(y) \, d\mu_2(y), \tag{23.2}$$

where $f_i : \Omega_2 \to \mathbb{R}$ are measurable functions, $i = 1, \ldots, n$.

Denote by $\overrightarrow{x} = x := (x_1, \ldots, x_n) \in \mathbb{R}^n$, $\overrightarrow{g} := (g_1, \ldots, g_n)$ and $\overrightarrow{f} := (f_1, \ldots, f_n)$.

© Springer International Publishing Switzerland 2016
G.A. Anastassiou, *Intelligent Comparisons: Analytic Inequalities*,
Studies in Computational Intelligence 609,
DOI 10.1007/978-3-319-21121-3_23

We consider here $\Phi : \mathbb{R}_+^n \to \mathbb{R}$ a convex function, which is increasing per coordinate, i.e. if $x_i \leq y_i, i = 1, \ldots, n$, then

$$\Phi (x_1, \ldots, x_n) \leq \Phi (y_1, \ldots, y_n) .$$

In [6] we proved that

Theorem 23.1 *Let u be a weight function on Ω_1, and k, K, g_i, f_i, $i = 1, \ldots, n \in \mathbb{N}$, and Φ defined as above. Assume that the function $x \to u (x) \frac{k(x,y)}{K(x)}$ is integrable on Ω_1 for each fixed $y \in \Omega_2$. Define v on Ω_2 by*

$$v (y) := \int_{\Omega_1} u (x) \frac{k (x, y)}{K (x)} d\mu_1 (x) < \infty. \tag{23.3}$$

Then

$$\int_{\Omega_1} u (x) \Phi \left(\frac{|g_1 (x)|}{K (x)}, \ldots, \frac{|g_n (x)|}{K (x)} \right) d\mu_1 (x) \leq$$

$$\int_{\Omega_2} v (y) \Phi (| f_1 (y)|, \ldots, | f_n (y)|) d\mu_2 (y) , \tag{23.4}$$

under the assumptions:

(i) f_i, $\Phi (|f_1|, \ldots, |f_n|)$, are $k (x, y) d\mu_2 (y)$ -integrable, μ_1 -a.e. in $x \in \Omega_1$, for all $i = 1, \ldots, n$,
(ii) $v (y) \Phi (| f_1 (y)|, \ldots, | f_n (y)|)$ is μ_2 -integrable.

Notation 23.2 *From now on we may write*

$$\vec{g} (x) = \int_{\Omega_2} k (x, y) \vec{f} (y) d\mu_2 (y) , \tag{23.5}$$

which means

$$(g_1 (x), \ldots, g_n (x)) = \left(\int_{\Omega_2} k (x, y) f_1 (y) d\mu_2 (y) , \ldots, \int_{\Omega_2} k (x, y) f_n (y) d\mu_2 (y) \right) . \tag{23.6}$$

Similarly, we may write

$$|\vec{g} (x)| = \left| \int_{\Omega_2} k (x, y) \vec{f} (y) d\mu_2 (y) \right| , \tag{23.7}$$

and we mean

$$(|g_1 (x)|, \ldots, |g_n (x)|) =$$

$$\left(\left|\int_{\Omega_2} k\,(x,\,y)\,f_1\,(y)\,d\mu_2\,(y)\right|,\,\ldots,\,\left|\int_{\Omega_2} k\,(x,\,y)\,f_n\,(y)\,d\mu_2\,(y)\right|\right). \qquad (23.8)$$

We also can write that

$$|\overrightarrow{g}\,(x)| \le \int_{\Omega_2} k\,(x,\,y)\left|\overrightarrow{f}\,(y)\right|d\mu_2\,(y), \qquad (23.9)$$

and we mean the fact that

$$|g_i\,(x)| \le \int_{\Omega_2} k\,(x,\,y)\,|f_i\,(y)|\,d\mu_2\,(y), \qquad (23.10)$$

for all $i = 1, \ldots, n$, etc.

Notation 23.3 *Next let $(\Omega_1, \Sigma_1, \mu_1)$ and $(\Omega_2, \Sigma_2, \mu_2)$ be measure spaces with positive σ-finite measures, and let $k_j : \Omega_1 \times \Omega_2 \to \mathbb{R}$ be a nonnegative measurable function, $k_j\,(x, \cdot)$ measurable on Ω_2 and*

$$K_j\,(x) = \int_{\Omega_2} k_j\,(x,\,y)\,d\mu_2\,(y),\,x \in \Omega_1,\,j = 1, \ldots, m. \qquad (23.11)$$

We suppose that $K_j\,(x) > 0$ a.e. on Ω_1. Let the measurable functions $g_{ji} : \Omega_1 \to \mathbb{R}$ with the representation

$$g_{ji}\,(x) = \int_{\Omega_2} k_j\,(x,\,y)\,f_{ji}\,(y)\,d\mu_2\,(y), \qquad (23.12)$$

where $f_{ji} : \Omega_2 \to \mathbb{R}$ are measurable functions, $i = 1, \ldots, n$ and $j = 1, \ldots, m$.

Denote the function vectors $\overrightarrow{g_j} := (g_{j1}, g_{j2}, \ldots, g_{jn})$ and $\overrightarrow{f_j} := (f_{j1}, \ldots, f_{jn})$, $j = 1, \ldots, m$.

We say $\overrightarrow{f_j}$ is integrable with respect to measure μ, iff all f_{ji} are integrable with respect to μ.

We also consider here $\Phi_j : \mathbb{R}_+^n \to \mathbb{R}_+$, $j = 1, \ldots, m$, convex functions that are increasing per coordinate. Again u is a weight function on Ω_1.

For $m \in \mathbb{N}$, in [6] we proved

Theorem 23.4 *Here we follow Notation 23.3. Let $\rho \in \{1, \ldots, m\}$ be fixed. Assume that the function*

$$x \mapsto \left(\frac{u\,(x)\displaystyle\prod_{j=1}^{m} k_j\,(x,\,y)}{\displaystyle\prod_{j=1}^{m} K_j\,(x)}\right)$$

is integrable on Ω_1 for each fixed $y \in \Omega_2$. Define λ_m on Ω_2 by

$$\lambda_m (y) := \int_{\Omega_1} \left(\frac{u(x) \prod_{j=1}^{m} k_j (x, y)}{\prod_{j=1}^{m} K_j (x)} \right) d\mu_1 (x) < \infty. \qquad (23.13)$$

Then

$$\int_{\Omega_1} u(x) \prod_{j=1}^{m} \Phi_j \left(\left| \frac{\overrightarrow{g_j} (x)}{K_j (x)} \right| \right) d\mu_1 (x) \leq \qquad (23.14)$$

$$\left(\prod_{\substack{j=1 \\ j \neq \rho}}^{m} \int_{\Omega_2} \Phi_j \left(\left| \overrightarrow{f_j} (y) \right| \right) d\mu_2 (y) \right) \left(\int_{\Omega_2} \Phi_\rho \left(\left| \overrightarrow{f_\rho} (y) \right| \right) \lambda_m (y) \, d\mu_2 (y) \right),$$

under the assumptions:

(i) $\overrightarrow{f_j}, \Phi_j \left(\left| \overrightarrow{f_j} \right| \right)$ *are both $k_j (x, y) \, d\mu_2 (y)$ -integrable, μ_1 -a.e. in $x \in \Omega_1$, $j = 1, \ldots, m$,*

(ii) $\lambda_m \Phi_\rho \left(\left| \overrightarrow{f_\rho} \right| \right)$; $\Phi_1 \left(\left| \overrightarrow{f_1} \right| \right), \Phi_2 \left(\left| \overrightarrow{f_2} \right| \right), \Phi_3 \left(\left| \overrightarrow{f_3} \right| \right), \ldots, \widehat{\Phi_\rho \left(\left| \overrightarrow{f_\rho} \right| \right)}, \ldots, \Phi_m$ $\left(\left| \overrightarrow{f_m} \right| \right)$ *are all μ_2 -integrable, where $\widehat{\Phi_\rho \left(\left| \overrightarrow{f_\rho} \right| \right)}$ means a missing item.*

We make

Remark 23.5 Following Notation 23.3, let $F_j : \Omega_2 \to \mathbb{R} \cup \{\pm\infty\}$ be measurable functions, $j = 1, \ldots, m$, with $0 < F_j (y) < \infty$, a.e. on Ω_2. In (23.11) we replace $k_j (x, y)$ by $k_j (x, y) F_j (y)$, $j = 1, \ldots, m$, and we have the modified $K_j (x)$ as

$$L_j (x) := \int_{\Omega_2} k_j (x, y) F_j (y) \, d\mu_2 (y), x \in \Omega_1. \qquad (23.15)$$

We assume $L_j (x) > 0$ a.e. on Ω_1.

As new $\overrightarrow{f_j}$ we consider now $\gamma_j := \frac{\overrightarrow{f_j}}{F_j}, j = 1, \ldots, m$, where $\overrightarrow{f_j} = (f_{j1}, \ldots, f_{jn})$; $\overrightarrow{\gamma_j} = \left(\frac{f_{j1}}{F_j}, \ldots, \frac{f_{jn}}{F_j} \right).$

Notice that

$$g_{ji} (x) = \int_{\Omega_2} k_j (x, y) f_{ji} (y) \, d\mu_2 (y) \qquad (23.16)$$

$$= \int_{\Omega_2} (k_j (x, y) F_j (y)) \left(\frac{f_{ji} (y)}{F_j (y)} \right) d\mu_2 (y),$$

$x \in \Omega_1$, all $j = 1, \ldots, m; i = 1, \ldots, n$.

So we can write

$$\overrightarrow{g_j}(x) = \int_{\Omega_2} \left(k_j(x, y) F_j(y) \right) \overrightarrow{\gamma_j}(y) \, d\mu_2(y), \, j = 1, \ldots, m. \qquad (23.17)$$

In this chapter we get first general results by applying Theorem 23.4 to $(\overrightarrow{\gamma_j}, \overrightarrow{g_j})$, $j = 1, \ldots, m$, and on other various important settings, then we give wide applications to Fractional Calculus. This chapter is inspired by [15, 16].

23.2 Main Results

We present our first main result

Theorem 23.6 *Here we follow Remark 23.5. Let $\rho \in \{1, \ldots, m\}$ be fixed. Assume that the function*

$$x \mapsto \left(\frac{u(x) \left(\prod_{j=1}^{m} F_j(y) \right) \left(\prod_{j=1}^{m} k_j(x, y) \right)}{\prod_{j=1}^{m} L_j(x)} \right)$$

is integrable on Ω_1, for each $y \in \Omega_2$. Define U_m on Ω_2 by

$$U_m(y) := \left(\prod_{j=1}^{m} F_j(y) \right) \int_{\Omega_1} \frac{u(x) \prod_{j=1}^{m} k_j(x, y)}{\prod_{j=1}^{m} L_j(x)} \, d\mu_1(x) < \infty. \qquad (23.18)$$

Then

$$\int_{\Omega_1} u(x) \prod_{j=1}^{m} \Phi_j \left(\left| \frac{\overrightarrow{g_j}(x)}{L_j(x)} \right| \right) d\mu_1(x) \leq \qquad (23.19)$$

$$\left(\prod_{\substack{j=1 \\ j \neq \rho}}^{m} \int_{\Omega_2} \Phi_j \left(\left| \frac{\overrightarrow{f_j}(y)}{F_j(y)} \right| \right) d\mu_2(y) \right) \cdot \left(\int_{\Omega_2} \Phi_\rho \left(\left| \frac{\overrightarrow{f_\rho}(y)}{F_\rho(y)} \right| \right) U_m(y) \, d\mu_2(y) \right),$$

under the assumptions:

(i) $\frac{\vec{f_j}}{F_j}, \Phi_j\left(\frac{|\vec{f_j}|}{F_j}\right)$ *are both* $k_j(x, y) F_j(y) d\mu_2(y)$ *-integrable,* μ_1 *-a.e. in* $x \in \Omega_1$,

$j = 1, \ldots, m$,

(ii) $U_m \Phi_\rho\left(\frac{|\vec{f_\rho}|}{F_\rho}\right); \Phi_1\left(\frac{|\vec{f_1}|}{F_1}\right), \Phi_2\left(\frac{|\vec{f_2}|}{F_2}\right), \ldots, \widehat{\Phi_\rho\left(\frac{|\vec{f_\rho}|}{F_\rho}\right)}, \ldots, \Phi_m\left(\frac{|\vec{f_m}|}{F_m}\right)$, *are*

μ_2 *-integrable, where* $\widehat{\Phi_\rho\left(\frac{|\vec{f_\rho}|}{F_\rho}\right)}$ *is absent.*

Proof By Theorem 23.4. ∎

We give the general applications.

Theorem 23.7 *All as in Theorem 23.6,* $p \geq 1$. *It holds*

$$\int_{\Omega_1}\left(\frac{u(x)}{\prod_{j=1}^m L_j(x)}\right) \prod_{j=1}^m \left(\sum_{i=1}^n |g_{ji}(x)|^p\right)^{\frac{1}{p}} d\mu_1(x) \leq \qquad (23.20)$$

$$\left(\prod_{\substack{j=1 \\ j \neq \rho}}^m \int_{\Omega_2} \frac{\left(\sum_{i=1}^n |f_{ji}(y)|^p\right)^{\frac{1}{p}}}{F_j(y)} d\mu_2(y)\right) \cdot$$

$$\left(\int_{\Omega_2} \left(\frac{\left(\sum_{i=1}^n |f_{\rho i}(y)|^p\right)^{\frac{1}{p}}}{F_\rho(y)}\right) U_m(y) d\mu_2(y)\right),$$

under the assumptions:

(i) $\left\|\vec{f_j}\right\|_p$ *is* $k_j(x, y) d\mu_2(y)$ *-integrable,* μ_1 *-a.e. in* $x \in \Omega_1$, $j = 1, \ldots, m$,

(ii) $\left(\frac{U_m}{F_\rho}\right)\left\|\vec{f_\rho}\right\|_p; \frac{\left\|\vec{f_j}\right\|_p}{F_j}$, $j \neq \rho$, $j = 1, \ldots, m$, *are all* μ_2 *-integrable.*

Proof By Theorem 23.6 with $\Phi_j(x_1, \ldots, x_n) = \left\|\vec{x}\right\|_p$, $\vec{x} = (x_1, \ldots, x_n)$, $j = 1, \ldots, m$. ∎

We furthermore give

Theorem 23.8 *All as in Theorem 23.6. It holds*

$$\int_{\Omega_1} u(x) \prod_{j=1}^{m} \left(\sum_{i=1}^{n} e^{\left(\frac{|g_{ji}(x)|}{L_j(x)} \right)} \right) d\mu_1(x) \leq \tag{23.21}$$

$$\left(\prod_{\substack{j=1 \\ j \neq \rho}}^{m} \int_{\Omega_2} \left(\sum_{i=1}^{n} e^{\left(\frac{|f_{ji}(y)|}{F_j(y)} \right)} \right) d\mu_2(y) \right) \left(\int_{\Omega_2} \left(\sum_{i=1}^{n} e^{\left(\frac{|f_{\rho i}(y)|}{F_\rho(y)} \right)} \right) U_m(y) d\mu_2(y) \right),$$

under the assumptions:

(i) $\frac{\vec{f_j}}{F_j}$, $\left(\sum_{i=1}^{n} e^{\left(\frac{|f_{ji}(y)|}{F_j(y)} \right)} \right)$ *are $k_j(x, y) F_j(y) d\mu_2(y)$ -integrable, μ_1 -a.e. in $x \in$*
Ω_1, $j = 1, \ldots, m$,

(ii) $U_m(y) \left(\sum_{i=1}^{n} e^{\left(\frac{|f_{\rho i}(y)|}{F_\rho(y)} \right)} \right)$ *and* $\left(\sum_{i=1}^{n} e^{\left(\frac{|f_{ji}(y)|}{F_j(y)} \right)} \right)$ *for $j \neq \rho$, $j = 1, \ldots, m$, are*
all μ_2-integrable.

Proof Apply Theorem 23.6 with $\Phi_j(x_1, \ldots, x_n) = \sum_{i=1}^{n} e^{x_i}$, $j = 1, \ldots, m$. ∎

We make

Remark 23.9 Following Notation 23.3 and Remark 23.5, we choose as

$$F_j(y) = \left\| \vec{f_j}(y) \right\|_\infty := \max\{ |f_{j1}(y)|, \ldots, |f_{jn}(y)| \},$$

or

$$F_j(y) = \left\| \vec{f_j}(y) \right\|_q := \left(\sum_{i=1}^{n} |f_{ji}(y)|^q \right)^{\frac{1}{q}}, q \geq 1, \tag{23.22}$$

$y \in \Omega_2$, which are measurable function, $j = 1, \ldots, m$. We assume that

$$0 < \left\| \vec{f_j}(y) \right\|_q < \infty, \text{ a.e. on } \Omega_2, \tag{23.23}$$

$j = 1, \ldots, m; 1 \leq q \leq \infty$ fixed.

Now in (23.11) we replace $k_j(x, y)$ by $k_j(x, y) \left\| \vec{f_j}(y) \right\|_q$, $j = 1, \ldots, m$, and
the new modified $K_j(x)$ is

$$L_{jq}(x) := \int_{\Omega_2} k_j(x, y) \left\| \vec{f_j}(y) \right\|_q d\mu_2(y), \, x \in \Omega_1, 1 \le q \le \infty. \quad (23.24)$$

We assume $L_{jq}(x) > 0$ a.e. on Ω_1.

Let $\rho \in \{1, \ldots, m\}$ be fixed. Assume that the function

$$x \mapsto \left(\frac{u(x) \left(\prod_{j=1}^{m} \left\| \vec{f_j}(y) \right\|_q \right) \left(\prod_{j=1}^{m} k_j(x, y) \right)}{\prod_{j=1}^{m} L_{jq}(x)} \right) \quad (23.25)$$

is integrable on Ω_1, for each $y \in \Omega_2$.

So here the corresponding U_m is U_{mq} on Ω_2, defined by

$$U_{mq}(y) := \left(\prod_{j=1}^{m} \left\| \vec{f_j}(y) \right\|_q \right) \int_{\Omega_1} \frac{u(x) \prod_{j=1}^{m} k_j(x, y)}{\prod_{j=1}^{m} L_{jq}(x)} d\mu_1(x) < \infty, \quad (23.26)$$

$1 \le q \le \infty$.

We give also two more general applications.

Theorem 23.10 *Here all as in Remark 23.9, $1 \le q \le \infty$, $p \ge 1$. It holds*

$$\int_{\Omega_1} \left(\frac{u(x)}{\prod_{j=1}^{m} L_{jq}(x)} \right) \prod_{j=1}^{m} \left(\sum_{i=1}^{n} |g_{ji}(x)|^p \right)^{\frac{1}{p}} d\mu_1(x) \le \quad (23.27)$$

$$\left(\prod_{\substack{j=1 \\ j \ne \rho}}^{m} \int_{\Omega_2} \frac{\left(\sum_{i=1}^{n} |f_{ji}(y)|^p \right)^{\frac{1}{p}}}{\left\| \vec{f_j}(y) \right\|_q} d\mu_2(y) \right).$$

$$\left(\int_{\Omega_2} \left(\frac{\left(\sum_{i=1}^{n} \left| f_{\rho i}\left(y\right) \right|^{p} \right)^{\frac{1}{p}}}{\left\| \vec{f_\rho}\left(y\right) \right\|_q} \right) U_{mq}\left(y\right) d\mu_2\left(y\right) \right),$$

under the assumptions:

(i) $\left\| \vec{f_j} \right\|_p$ *is* $k_j\left(x, y\right) d\mu_2\left(y\right)$ *-integrable,* μ_1 *-a.e. in* $x \in \Omega_1$, $j = 1, \ldots, m$,

(ii) $\left(\frac{U_{mq}}{\left\| \vec{f_\rho} \right\|_q} \right) \left\| \vec{f_\rho} \right\|_p$; $\frac{\left\| \vec{f_j} \right\|_p}{\left\| \vec{f_j} \right\|_q}$, $j \neq \rho$, $j = 1, \ldots, m$, *are all* μ_2-*integrable.*

Proof By Theorem 23.7. ∎

We continue with

Theorem 23.11 *Here all as in Remark 23.9,* $1 \leq q \leq \infty$. *It holds*

$$\int_{\Omega_1} u\left(x\right) \prod_{j=1}^{m} \left(\sum_{i=1}^{n} e^{\left(\frac{\left| g_{ji}\left(x\right) \right|}{L_{jq}\left(x\right)} \right)} \right) d\mu_1\left(x\right) \leq \qquad (23.28)$$

$$\left(\prod_{\substack{j=1 \\ j \neq \rho}}^{m} \int_{\Omega_2} \left(\sum_{i=1}^{n} e^{\left(\frac{\left| f_{ji}\left(y\right) \right|}{\left\| \vec{f_j}\left(y\right) \right\|_q} \right)} \right) d\mu_2\left(y\right) \right) \cdot$$

$$\left(\int_{\Omega_2} \left(\sum_{i=1}^{n} e^{\left(\frac{\left| f_{\rho i}\left(y\right) \right|}{\left\| \vec{f_\rho}\left(y\right) \right\|_q} \right)} \right) U_{mq}\left(y\right) d\mu_2\left(y\right) \right),$$

under the assumptions:

(i) $\frac{\vec{f_j}}{\left\| \vec{f_j} \right\|_q}$, $\left(\sum_{i=1}^{n} e^{\left(\frac{\left| f_{ji}\left(y\right) \right|}{\left\| \vec{f_j}\left(y\right) \right\|_q} \right)} \right)$ *are* $k_j\left(x, y\right) \left\| \vec{f_j}\left(y\right) \right\|_q d\mu_2\left(y\right)$ *-integrable,* μ_1 *-a.e.*

in $x \in \Omega_1$, $j = 1, \ldots, m$,

(ii) $U_{mq}\left(y\right) \left(\sum_{i=1}^{n} e^{\left(\frac{\left| f_{\rho i}\left(y\right) \right|}{\left\| \vec{f_\rho}\left(y\right) \right\|_q} \right)} \right)$ *and* $\left(\sum_{i=1}^{n} e^{\left(\frac{\left| f_{ji}\left(y\right) \right|}{\left\| \vec{f_j}\left(y\right) \right\|_q} \right)} \right)$ *for* $j \neq \rho$, $j = 1, \ldots, m$,

are all μ_2-*integrable.*

Proof By Theorem 23.8. ∎

Terminology 23.12 *To keep exposition short, in the rest of this chapter we give only applications of Theorems 23.10 and 23.11 to Fractional Calculus. We need the following:*

Let $a < b, a, b \in \mathbb{R}$. By $C^N ([a, b])$, we denote the space of all functions on $[a, b]$ which have continuous derivatives up to order N, and $AC ([a, b])$ is the space of all absolutely continuous functions on $[a, b]$. By $AC^N ([a, b])$, we denote the space of all functions g with $g^{(N-1)} \in AC ([a, b])$. For any $\alpha \in \mathbb{R}$, we denote by $[\alpha]$ the integral part of α (the integer k satisfying $k \leq \alpha < k + 1$), and $\lceil \alpha \rceil$ is the ceiling of α ($\min\{n \in \mathbb{N}, n \geq \alpha\}$). By $L_1 (a, b)$, we denote the space of all functions integrable on the interval (a, b), and by $L_\infty (a, b)$ the set of all functions measurable and essentially bounded on (a, b). Clearly, $L_\infty (a, b) \subset L_1 (a, b)$.

We give the definition of the Riemann-Liouville fractional integrals, see [17]. Let $[a, b], (-\infty < a < b < \infty)$ be a finite interval on the real axis $\mathbb{R}$.

The Riemann-Liouville fractional integrals $I_{a+}^\alpha f$ and $I_{b-}^\alpha f$ of order $\alpha > 0$ are defined by

$$\left(I_{a+}^\alpha f\right) (x) = \frac{1}{\Gamma (\alpha)} \int_a^x f (t) (x - t)^{\alpha-1} dt, (x > a), \qquad (23.29)$$

$$\left(I_{b-}^\alpha f\right) (x) = \frac{1}{\Gamma (\alpha)} \int_x^b f (t) (t - x)^{\alpha-1} dt, (x < b), \qquad (23.30)$$

respectively. Here $\Gamma (\alpha)$ is the Gamma function. These integrals are called the left-sided and the right-sided fractional integrals.

We make

Remark 23.13 Let f_{ji} be Lebesgue measurable functions from (a, b) into $\mathbb{R}$, such that $\left(I_{a+}^{\alpha_j} (|f_{ji}|)\right) (x) \in \mathbb{R}$, $\forall\, x \in (a, b), \alpha_j > 0, j = 1, \ldots, m; i = 1, \ldots, n$, e.g. when $f_{ji} \in L_\infty (a, b)$.

Consider here

$$g_{ji} (x) = \left(I_{a+}^{\alpha_j} f_{ji}\right) (x) = \frac{1}{\Gamma (\alpha_j)} \int_a^x (x - t)^{\alpha_j-1} f_{ji} (t) dt, \qquad (23.31)$$

$x \in (a, b), j = 1, \ldots, m; i = 1, \ldots, n$.

Notice that $g_{ji} (x) \in \mathbb{R}$ and it is Lebesgue measurable.

We pick $\Omega_1 = \Omega_2 = (a, b)$, $d\mu_1 (x) = dx$, $d\mu_2 (y) = dy$, the Lebesgue measure. We see that

$$\left(I_{a+}^{\alpha_j} f_{ji}\right) (x) = \int_a^b \frac{\chi_{(a,x]} (t) (x - t)^{\alpha_j-1}}{\Gamma (\alpha_j)} f_{ji} (t) dt, \qquad (23.32)$$

where χ stands for the characteristic function.

So, we pick as

$$k_j(x, t) := \frac{\chi_{(a,x]}(t)(x-t)^{\alpha_j-1}}{\Gamma(\alpha_j)}, j = 1, \ldots, m. \tag{23.33}$$

We assume here that

$$0 < \left\|\overrightarrow{f_j}(y)\right\|_q < \infty, \text{ a.e. on } (a,b), \tag{23.34}$$

$j = 1, \ldots, m; 1 \le q \le \infty$ fixed.

We call

$$L_{jq}^+(x) := \int_a^b \left(\frac{\chi_{(a,x]}(y)(x-y)^{\alpha_j-1}}{\Gamma(\alpha_j)}\right) \left\|\overrightarrow{f_j}(y)\right\|_q dy = I_{a+}^{\alpha_j}\left(\left\|\overrightarrow{f_j}\right\|_q\right)(x), \tag{23.35}$$

$x \in (a,b), 1 \le q \le \infty, j = 1, \ldots, m.$

We have that $L_{jq}^+(x) > 0$ a.e. on (a,b).

Let $\rho \in \{1, \ldots, m\}$ be fixed. Assume that the function

$$x \to \frac{u(x)\left(\prod_{j=1}^m \left\|\overrightarrow{f_j}(y)\right\|_q\right)\chi_{(a,x]}(y)(x-y)^{\sum_{j=1}^m \alpha_j - m}}{\prod_{j=1}^m \left(L_{jq}^+(x)\Gamma(\alpha_j)\right)} \tag{23.36}$$

is integrable on (a,b), for each $y \in (a,b)$.

So we get

$$U_{mq}^+(y) := \left(\prod_{i=1}^m \left(\frac{\left\|\overrightarrow{f_j}(y)\right\|_q}{\Gamma(\alpha_j)}\right)\right)\int_a^b \frac{u(x)\chi_{(a,x]}(y)(x-y)^{\sum_{j=1}^m \alpha_j - m}}{\prod_{j=1}^m L_{jq}^+(x)}dx < \infty, \tag{23.37}$$

$1 \le q \le \infty, y \in (a,b).$

Here again $p \ge 1$.

We give

Theorem 23.14 *All as in Remark 23.13. It holds*

$$\int_a^b u(x) \prod_{j=1}^m \left(\frac{\left\| \overrightarrow{\left(I_{a+}^{\alpha_j} f_j \right)(x)} \right\|_p}{I_{a+}^{\alpha_j}\left(\left\| \overrightarrow{f_j} \right\|_q \right)(x)} \right) dx \le \qquad (23.38)$$

$$\left(\prod_{\substack{j=1 \\ j \ne \rho}}^m \int_a^b \left(\frac{\left\| \overrightarrow{f_j}(y) \right\|_p}{\left\| \overrightarrow{f_j}(y) \right\|_q} \right) dy \right) \left(\int_a^b U_{mq}^+(y) \frac{\left\| \overrightarrow{f_\rho}(y) \right\|_p}{\left\| \overrightarrow{f_\rho}(y) \right\|_q} dy \right),$$

under the assumptions:

(i) $\left\| \overrightarrow{f_j} \right\|_p$ *is* $\dfrac{\chi_{(a,x]}(y)(x-y)^{\alpha_j-1}}{\Gamma(\alpha_j)} dy$ *-integrable, a.e. in* $x \in (a,b)$, $j = 1, \ldots, m$,

(ii) $\left(\dfrac{U_{mq}^+}{\left\| \overrightarrow{f_\rho} \right\|_q} \right) \left\| \overrightarrow{f_\rho} \right\|_p$; $\left(\dfrac{\left\| \overrightarrow{f_j}(y) \right\|_p}{\left\| \overrightarrow{f_j}(y) \right\|_q} \right)$, $j \ne \rho$, $j = 1, \ldots, m$, *are all Lebesgue integrable.*

Proof By Theorem 23.10. ∎

Furthermore we present

Theorem 23.15 *All as in Remark 23.13. It holds*

$$\int_a^b u(x) \prod_{j=1}^m \left(\sum_{i=1}^n e^{\left(\frac{\left| I_{a+}^{\alpha_j}(f_{ji})(x) \right|}{I_{a+}^{\alpha_j}\left(\left\| \overrightarrow{f_j} \right\|_q \right)(x)} \right)} \right) dx \le \qquad (23.39)$$

$$\left(\prod_{\substack{j=1 \\ j \ne \rho}}^m \int_a^b \left(\sum_{i=1}^n e^{\left| \frac{f_{ji}(y)}{\left\| \overrightarrow{f_j}(y) \right\|_q} \right|} \right) dy \right) \left(\int_a^b \left(\sum_{i=1}^n e^{\left| \frac{f_{\rho i}(y)}{\left\| \overrightarrow{f_\rho}(y) \right\|_q} \right|} \right) U_{mq}^+(y)\, dy \right),$$

under the assumptions:

(i) $\dfrac{\overrightarrow{f_j}}{\left\| \overrightarrow{f_j} \right\|_q}$, $\left(\sum_{i=1}^n e^{\left| \frac{f_{ji}(y)}{\left\| \overrightarrow{f_j}(y) \right\|_q} \right|} \right)$ *are* $\left\| \overrightarrow{f_j}(y) \right\|_q \dfrac{\chi_{(a,x]}(y)(x-y)^{\alpha_j-1}}{\Gamma(\alpha_j)} dy$ *-integrable, a.e. in*

$x \in (a,b)$, $j = 1, \ldots, m$,

(ii) $U_{mq}^+(y) \left(\sum_{i=1}^n e^{\left| \frac{f_{\rho i}(y)}{\left\| \overrightarrow{f_\rho}(y) \right\|_q} \right|} \right)$ *and* $\left(\sum_{i=1}^n e^{\left| \frac{f_{ji}(y)}{\left\| \overrightarrow{f_j}(y) \right\|_q} \right|} \right)$ *for* $j \ne \rho$, $j = 1, \ldots, m$, *are*

all Lebesgue integrable.

Proof By Theorem 23.11. ∎

We make

Remark 23.16 (continuation of Remark 23.13) If we choose

$$u(x) = \prod_{j=1}^{m} L_{jq}^{+}(x) = \prod_{j=1}^{m} I_{a+}^{\alpha_j}\left(\left\|\vec{f_j}\right\|_q\right)(x),$$ (23.40)

then

$$U_{mq}^{+}(y) = \prod_{j=1}^{m}\left(\frac{\left\|\vec{f_j}(y)\right\|_q}{\Gamma(\alpha_j)}\right)\frac{(b-y)^{\sum_{j=1}^{m}\alpha_j - m + 1}}{\left(\sum_{j=1}^{m}\alpha_j - m + 1\right)},$$ (23.41)

given that $\sum_{j=1}^{m}\alpha_j > m - 1, \forall\, x, y \in (a, b)$.

Notice that

$$U_{mq}^{+}(y) \leq \prod_{j=1}^{m}\left(\frac{\left\|\vec{f_j}(y)\right\|_q}{\Gamma(\alpha_j)}\right)\frac{(b-a)^{\sum_{j=1}^{m}\alpha_j - m + 1}}{\left(\sum_{j=1}^{m}\alpha_j - m + 1\right)},$$ (23.42)

if $\sum_{j=1}^{m}\alpha_j > m - 1, \forall\, y \in (a, b)$.

Thus (23.38) becomes

$$\int_a^b\left(\prod_{j=1}^{m}\left\|\overrightarrow{\left(I_{a+}^{\alpha_j}f_j\right)(x)}\right\|_p\right)dx \leq$$ (23.43)

$$\left(\frac{(b-a)^{\sum_{j=1}^{m}\alpha_j - m + 1}}{\left(\sum_{j=1}^{m}\alpha_j - m + 1\right)\prod_{j=1}^{m}\Gamma(\alpha_j)}\right)\left(\prod_{\substack{j=1\\j\neq\rho}}^{m}\int_a^b\frac{\left\|\vec{f_j}(y)\right\|_p}{\left\|\vec{f_j}(y)\right\|_q}dy\right).$$

$$\left(\int_a^b \left(\prod_{\substack{j=1 \\ j \neq \rho}}^{m} \left\| \vec{f}_j (y) \right\|_q \right) \left\| \vec{f}_\rho (y) \right\|_p dy \right),$$

under the assumptions:

(i) following (23.38), and

(ii) $*$ $\left(\prod_{\substack{j=1 \\ j \neq \rho}}^{m} \left\| \vec{f}_j (y) \right\|_q \right) \left\| \vec{f}_\rho (y) \right\|_p ; \left(\dfrac{\left\| \vec{f}_j \right\|_p}{\left\| \vec{f}_j \right\|_q} \right)$, $j = 1, \ldots, m$, with $j \neq \rho$, are all

Lebesgue integrable.

Remark 23.17 Let f_{ji} be Lebesgue measurable functions from (a, b) into $\mathbb{R}$, such that $\left(I_{b-}^{\alpha_j} (|f_{ji}|) \right) (x) \in \mathbb{R}$, $\forall x \in (a, b)$, $\alpha_j > 0$, $j = 1, \ldots, m$; $i = 1, \ldots, n$, e.g. when $f_{ji} \in L_\infty (a, b)$.

Consider here

$$g_{ji} (x) = \left(I_{b-}^{\alpha_j} f_{ji} \right) (x) = \frac{1}{\Gamma (\alpha_j)} \int_x^b (t - x)^{\alpha_j - 1} f_{ji} (t) dt, \qquad (23.44)$$

$x \in (a, b)$, $j = 1, \ldots, m$; $i = 1, \ldots, n$.

Notice that $g_{ji} (x) \in \mathbb{R}$ and it is Lebesgue measurable.

We pick $\Omega_1 = \Omega_2 = (a, b)$, $d\mu_1 (x) = dx$, $d\mu_2 (y) = dy$, the Lebesgue measure. We see that

$$\left(I_{b-}^{\alpha_j} f_{ji} \right) (x) = \int_a^b \frac{\chi_{[x,b)} (t) (t - x)^{\alpha_j - 1}}{\Gamma (\alpha_j)} f_{ji} (t) dt, \qquad (23.45)$$

where χ stands for the characteristic function.

So, we pick as

$$k_j (x, t) := \frac{\chi_{[x,b)} (t) (y - x)^{\alpha_j - 1}}{\Gamma (\alpha_j)}, j = 1, \ldots, m. \qquad (23.46)$$

We assume here that

$$0 < \left\| \vec{f}_j (y) \right\|_q < \infty, \text{ a.e. on } (a, b), \qquad (23.47)$$

$j = 1, \ldots, m$; $1 \leq q \leq \infty$ fixed.

We call

$$L_{jq}^- (x) := \int_a^b \left(\frac{\chi_{[x,b)} (y) (y - x)^{\alpha_j - 1}}{\Gamma (\alpha_j)} \right) \left\| \vec{f}_j (y) \right\|_q dy = I_{b-}^{\alpha_j} \left(\left\| \vec{f}_j \right\|_q \right) (x),$$

$$(23.48)$$

$x \in (a, b)$, $1 \leq q \leq \infty$, $j = 1, \ldots, m$.

We have that $L_{jq}^-(x) > 0$ a.e. on (a, b).

Let $\rho \in \{1, \ldots, m\}$ be fixed. Assume that the function

$$x \to \frac{u(x) \left(\prod_{j=1}^{m} \left\| \vec{f_j}(y) \right\|_q \right) \chi_{[x,b)}(y)(y-x)^{\sum_{j=1}^{m} \alpha_j - m}}{\prod_{j=1}^{m} \left(L_{jq}^-(x) \Gamma(\alpha_j) \right)} \tag{23.49}$$

is integrable on (a, b), for each $y \in (a, b)$.

So we get

$$U_{mq}^-(y) := \left(\prod_{j=1}^{m} \frac{\left\| \vec{f_j}(y) \right\|_q}{\Gamma(\alpha_j)} \right) \int_a^b \frac{u(x) \chi_{[x,b)}(y)(y-x)^{\sum_{j=1}^{m} \alpha_j - m}}{\prod_{j=1}^{m} L_{jq}^-(x)} dx < \infty,$$

$$\tag{23.50}$$

$1 \leq q \leq \infty$, $y \in (a, b)$.

Here again $p \geq 1$.

We give

Theorem 23.18 *All as in Remark 23.17. It holds*

$$\int_a^b u(x) \prod_{j=1}^{m} \left(\frac{\left\| \overrightarrow{\left(I_{b-}^{\alpha_j} f_j \right)(x)} \right\|_p}{I_{b-}^{\alpha_j} \left(\left\| \vec{f_j} \right\|_q \right)(x)} \right) dx \leq \tag{23.51}$$

$$\left(\prod_{\substack{j=1 \\ j \neq \rho}}^{m} \int_a^b \left(\frac{\left\| \vec{f_j}(y) \right\|_p}{\left\| \vec{f_j}(y) \right\|_q} \right) dy \right) \left(\int_a^b U_{mq}^-(y) \frac{\left\| \vec{f_\rho}(y) \right\|_p}{\left\| \vec{f_\rho}(y) \right\|_q} dy \right),$$

under the assumptions:

(i) $\left\| \vec{f_j} \right\|_p$ is $\frac{\chi_{[x,b)}(y)(y-x)^{\alpha_j - 1}}{\Gamma(\alpha_j)} dy$ -integrable, a.e. in $x \in (a, b)$, $j = 1, \ldots, m$,

(ii) $\left(\frac{U_{mq}^-}{\left\| \vec{f_\rho} \right\|_q} \right) \left\| \vec{f_\rho} \right\|_p$; $\left(\frac{\left\| \vec{f_j} \right\|_p}{\left\| \vec{f_j} \right\|_q} \right)$, $j \neq \rho$, $j = 1, \ldots, m$, are all Lebesgue integrable.

Proof By Theorem 23.10. ∎

Furthermore we give

Theorem 23.19 *All as in Remark 23.17. It holds*

$$\int_a^b u(x) \prod_{j=1}^m \left(\sum_{i=1}^n e^{\left(\frac{\left| I_{b-}^{\alpha j} f_{ji}(x) \right|}{I_{b-}^{\alpha j} \left(\left\| \vec{f}_j \right\|_q \right)(x)} \right)} \right) dx \leq \tag{23.52}$$

$$\left(\prod_{\substack{j=1 \\ j \neq \rho}}^m \int_a^b \left(\sum_{i=1}^n e^{\left| \frac{f_{ji}(y)}{\left\| \vec{f}_j(y) \right\|_q} \right|} \right) dy \right) \left(\int_a^b \left(\sum_{i=1}^n e^{\left| \frac{f_{\rho i}(y)}{\left\| \vec{f}_\rho(y) \right\|_q} \right|} \right) U_{mq}^-(y)\, dy \right),$$

under the assumptions:

(i) $\dfrac{\vec{f}_j}{\left\| \vec{f}_j \right\|_q}, \left(\displaystyle\sum_{i=1}^n e^{\left| \frac{f_{ji}(y)}{\left\| \vec{f}_j(y) \right\|_q} \right|} \right)$ *are* $\left\| \vec{f}_j(y) \right\|_q \dfrac{\chi_{[x,b)}(y)(y-x)^{\alpha_j - 1}}{\Gamma(\alpha_j)} dy$ *-integrable, a.e. in*

  $x \in (a, b)$, $j = 1, \ldots, m$,

(ii) $U_{mq}^-(y) \left(\displaystyle\sum_{i=1}^n e^{\left| \frac{f_{\rho i}(y)}{\left\| \vec{f}_\rho(y) \right\|_q} \right|} \right)$ *and* $\left(\displaystyle\sum_{i=1}^n e^{\left| \frac{f_{ji}(y)}{\left\| \vec{f}_j(y) \right\|_q} \right|} \right)$ *for* $j \neq \rho$, $j = 1, \ldots, m$, *are*

  all Lebesgue integrable.

Proof By Theorem 23.11. ∎

We make

Remark 23.20 (continuation of Remark 23.17) If we choose

$$u(x) = \prod_{j=1}^m L_{jq}^-(x) = \prod_{j=1}^m I_{b-}^{\alpha j}\left(\left\| \vec{f}_j \right\|_q \right)(x), \tag{23.53}$$

then

$$U_{mq}^-(y) = \prod_{j=1}^m \left(\frac{\left\| \vec{f}_j(y) \right\|_q}{\Gamma(\alpha_j)} \right) \frac{(y-a)^{\sum_{j=1}^m \alpha_j - m + 1}}{\left(\sum_{j=1}^m \alpha_j - m + 1 \right)}, \tag{23.54}$$

given that $\sum_{j=1}^{m} \alpha_j > m - 1, \forall x, y \in (a, b)$.

Notice that

$$U_{mq}^{-}(y) \leq \prod_{j=1}^{m} \left(\frac{\left\| \overrightarrow{f_j}(y) \right\|_q}{\Gamma(\alpha_j)} \right) \frac{(b-a)^{\sum_{j=1}^{m} \alpha_j - m + 1}}{\left(\sum_{j=1}^{m} \alpha_j - m + 1 \right)}, \tag{23.55}$$

if $\sum_{j=1}^{m} \alpha_j > m - 1, \forall y \in (a, b)$.

Thus (23.51) becomes

$$\int_a^b \left(\prod_{j=1}^{m} \left\| \overrightarrow{\left(I_{b-}^{\alpha_j} f_j \right)(x)} \right\|_p \right) dx \leq \tag{23.56}$$

$$\left(\frac{(b-a)^{\sum_{j=1}^{m} \alpha_j - m + 1}}{\left(\sum_{j=1}^{m} \alpha_j - m + 1 \right) \prod_{j=1}^{m} \Gamma(\alpha_j)} \right) \left(\prod_{\substack{j=1 \\ j \neq \rho}}^{m} \int_a^b \frac{\left\| \overrightarrow{f_j}(y) \right\|_p}{\left\| \overrightarrow{f_j}(y) \right\|_q} dy \right) \cdot$$

$$\left(\int_a^b \left(\prod_{\substack{j=1 \\ j \neq \rho}}^{m} \left\| \overrightarrow{f_j}(y) \right\|_q \right) \left\| \overrightarrow{f_\rho}(y) \right\|_p dy \right),$$

under the assumptions:

(i) following (23.51), and

(ii) $*\left(\prod_{\substack{j=1 \\ j \neq \rho}}^{m} \left\| \overrightarrow{f_j}(y) \right\|_q \right) \left\| \overrightarrow{f_\rho}(y) \right\|_p$; $\left(\frac{\left\| \overrightarrow{f_j} \right\|_p}{\left\| \overrightarrow{f_j} \right\|_q} \right)$, $j = 1, \ldots, m$, with $j \neq \rho$, are all Lebesgue measurable.

We mention

Definition 23.21 ([1], *p. 448*) The left generalized Riemann-Liouville fractional derivative of f of order $\beta > 0$ is given by

$$D_a^\beta f(x) = \frac{1}{\Gamma(n-\beta)} \left(\frac{d}{dx}\right)^n \int_a^x (x-y)^{n-\beta-1} f(y) \, dy, \qquad (23.57)$$

where $n = [\beta] + 1, x \in [a, b]$.

For $a, b \in \mathbb{R}$, we say that $f \in L_1(a, b)$ has an L_∞ fractional derivative $D_a^\beta f$ ($\beta > 0$) in $[a, b]$, if and only if

(1) $D_a^{\beta-k} f \in C([a, b])$, $k = 2, \ldots, n = [\beta] + 1$,
(2) $D_a^{\beta-1} f \in AC([a, b])$
(3) $D_a^\beta f \in L_\infty(a, b)$.

Above we define $D_a^0 f := f$ and $D_a^{-\delta} f := I_{a+}^\delta f$, if $0 < \delta \leq 1$.

From [1, p. 449] and [14] we mention and use

Lemma 23.22 *Let $\beta > \alpha \geq 0$ and let $f \in L_1(a, b)$ have an L_∞ fractional derivative $D_a^\beta f$ in $[a, b]$ and let $D_a^{\beta-k} f(a) = 0$, $k = 1, \ldots, [\beta] + 1$, then*

$$D_a^\alpha f(x) = \frac{1}{\Gamma(\beta-\alpha)} \int_a^x (x-y)^{\beta-\alpha-1} D_a^\beta f(y) \, dy, \qquad (23.58)$$

for all $a \leq x \leq b$.

Here $D_a^\alpha f \in AC([a, b])$ for $\beta - \alpha \geq 1$, and $D_a^\alpha f \in C([a, b])$ for $\beta - \alpha \in (0, 1)$.
Notice here that

$$D_a^\alpha f(x) = \left(I_{a+}^{\beta-\alpha}\left(D_a^\beta f\right)\right)(x), a \leq x \leq b. \qquad (23.59)$$

For more on the last, see [7].
We make

Remark 23.23 Let f_{ji} be Lebesgue measurable functions from (a, b) into $\mathbb{R}$, $j = 1, \ldots, m; i = 1, \ldots, n$, as in Definition 23.21 and Lemma 23.22. More precisely, let $\beta_j > \alpha_j \geq 0$, $j = 1, \ldots, m$ and let $f_{ji} \in L_1(a, b)$ have an L_∞ fractional derivative $D_a^{\beta_j} f_{ji}$ in $[a, b]$ and let $D_a^{\beta_j-k_j} f_{ji}(a) = 0$, $k_j = 1, \ldots, [\beta_j] + 1$, for all $i = 1, \ldots, n$; for $j = 1, \ldots, m$.

By Lemma 23.22 we get

$$D_a^{\alpha_j} f_{ji}(x) = \left(I_{a+}^{\beta_j-\alpha_j}\left(D_a^{\beta_j} f_{ji}\right)\right)(x), \qquad (23.60)$$

$a \leq x \leq b, i = 1, \ldots, n; j = 1, \ldots, m$.
We assume here that

$$0 < \left\| \overrightarrow{D_a^{\beta_j} f_j(y)} \right\|_q < \infty, \text{ a.e. on } (a, b), \qquad (23.61)$$

$j = 1, \ldots, m$; where $1 \leq q \leq \infty, p \geq 1$, both be fixed.

We further assume that

$$\sum_{j=1}^{m} (\beta_j - \alpha_j) > m - 1, \tag{23.62}$$

here $\rho \in \{1, \ldots, m\}$ is fixed.

We give

Proposition 23.24 *Here all as in Remark 23.23. It holds*

$$\int_a^b \left(\prod_{j=1}^{m} \left\| \overrightarrow{\left(D_a^{\alpha_j} f_j\right)(x)} \right\|_p \right) dx \leq \tag{23.63}$$

$$\left(\frac{(b-a)^{\sum_{j=1}^{m}(\beta_j-\alpha_j)-m+1}}{\left(\sum_{j=1}^{m}(\beta_j-\alpha_j) - m + 1 \right) \prod_{j=1}^{m} \Gamma(\beta_j - \alpha_j)} \right) \left(\prod_{\substack{j=1 \\ j \neq \rho}}^{m} \int_a^b \frac{\left\| \overrightarrow{D_a^{\beta_j} f_j(y)} \right\|_p}{\left\| \overrightarrow{D_a^{\beta_j} f_j(y)} \right\|_q} dy \right) \cdot$$

$$\left(\int_a^b \left(\prod_{\substack{j=1 \\ j \neq \rho}}^{m} \left\| \overrightarrow{D_a^{\beta_j} f_j(y)} \right\|_q \right) \left\| \overrightarrow{D_a^{\beta_\rho} f_\rho(y)} \right\|_p dy \right),$$

under the assumption:

$$\left\| \overrightarrow{D_a^{\beta_j} f_j} \right\|_q^{-1}, \text{ for all } j \in \{1, \ldots, m\} - \{\rho\} \text{ are Lebesgue integrable.}$$

Proof See Remark 23.16, in particular we use (23.43). ∎

We need

Definition 23.25 ([11], p. 50, [1], p. 449) Let $\nu \geq 0$, $n := \lceil \nu \rceil$, $f \in AC^n([a, b])$. Then the left Caputo fractional derivative is given by

$$D_{*a}^{\nu} f(x) = \frac{1}{\Gamma(n-\nu)} \int_a^x (x-t)^{n-\nu-1} f^{(n)}(t)\, dt$$

$$= \left(I_{a+}^{n-\nu} f^{(n)} \right)(x), \tag{23.64}$$

and it exists almost everywhere for $x \in [a, b]$, in fact $D_{*a}^{\nu} f \in L_1(a, b)$, ([1], p. 394). We have $D_{*a}^{n} f = f^{(n)}$, $n \in \mathbb{Z}_+$.

We also need

Theorem 23.26 ([4, 8]) *Let $\nu > \rho > 0$, $\nu, \rho \notin \mathbb{N}$. Call $n := \lceil \nu \rceil$, $m^* := \lceil \rho \rceil$. Assume $f \in AC^n([a, b])$, such that $f^{(k)}(a) = 0$, $k = m^*, m^* + 1, \ldots, n - 1$, and $D^{\nu}_{*a} f \in L_{\infty}(a, b)$. Then $D^{\rho}_{*a} f \in C([a, b])$ if $\nu - \rho \in (0, 1)$, and $D^{\rho}_{*a} f \in AC([a, b])$, if $\nu - \rho \geq 1$ (where $D^{\rho}_{*a} f = I^{m^*-\rho}_{a+} f^{(m^*)}(x)$), and*

$$D^{\rho}_{*a} f(x) = \frac{1}{\Gamma(\nu - \rho)} \int_a^x (x - t)^{\nu - \rho - 1} D^{\nu}_{*a} f(t) \, dt$$

$$= \left(I^{\nu-\rho}_{a+} \left(D^{\nu}_{*a} f \right) \right)(x), \qquad (23.65)$$

$\forall x \in [a, b]$.

For more on the last, see [8].
We make

Remark 23.27 Let $\nu_j > \rho_j > 0$, $\nu_j, \rho_j \notin \mathbb{N}$, $j = 1, \ldots, m$. Call $n_j := \lceil \nu_j \rceil$, $m^*_j := \lceil \rho_j \rceil$. Assume $f_{ji} \in AC^{n_j}([a, b])$, $i = 1, \ldots, n$; $j = 1, \ldots, m$; such that $f^{(k_j)}_{ji}(a) = 0$, $k_j = m^*_j, m^*_j + 1, \ldots, n_j - 1$, and $D^{\nu_j}_{*a} f_{ji} \in L_{\infty}(a, b)$, $j = 1, \ldots, m$; $i = 1, \ldots, n$. Then on Theorem 23.26 we get that $D^{\rho_j}_{*a} f_{ji} \in C([a, b])$, if $\nu_j - \rho_j \in (0, 1)$, and $D^{\rho_j}_{*a} f_{ji} \in AC([a, b])$, if $\nu_j - \rho_j \geq 1$, and

$$D^{\rho_j}_{*a} f_{ji}(x) = \left(I^{\nu_j-\rho_j}_{a+} \left(D^{\nu_j}_{*a} f_{ji} \right) \right)(x), \qquad (23.66)$$

$\forall x \in [a, b]$; $j = 1, \ldots, m$; $i = 1, \ldots, n$.

We assume here that

$$0 < \left\| \overrightarrow{D^{\nu_j}_{*a} f_j}(y) \right\|_q < \infty, \text{ a.e. on } (a, b), \qquad (23.67)$$

$j = 1, \ldots, m$; where $1 \leq q \leq \infty$, $p \geq 1$, both be fixed.

We further assume that

$$\sum_{j=1}^m (\nu_j - \rho_j) > m - 1, \qquad (23.68)$$

here $\rho \in \{1, \ldots, m\}$ is fixed.

We give

Proposition 23.28 *Here all as in Remark 23.27. It holds*

$$\int_a^b \left(\prod_{j=1}^m \left\| \overrightarrow{\left(D_{*a}^{\rho_j} f_j \right)(x)} \right\|_p \right) dx \le \qquad (23.69)$$

$$\left(\frac{(b-a)^{\sum_{j=1}^m (\nu_j - \rho_j) - m + 1}}{\left(\sum_{j=1}^m (\nu_j - \rho_j) - m + 1 \right) \prod_{j=1}^m \Gamma(\nu_j - \rho_j)} \right) \left(\prod_{\substack{j=1 \\ j \ne \rho}}^m \int_a^b \frac{\left\| \overrightarrow{D_{*a}^{\nu_j} f_j}(y) \right\|_p}{\left\| \overrightarrow{D_{*a}^{\nu_j} f_j}(y) \right\|_q} dy \right) \cdot$$

$$\left(\int_a^b \left(\prod_{\substack{j=1 \\ j \ne \rho}}^m \left\| \overrightarrow{D_{*a}^{\nu_j} f_j}(y) \right\|_q \right) \left\| \overrightarrow{D_{*a}^{\nu_\rho} f_\rho}(y) \right\|_p dy \right),$$

under the assumption:

$$\left\| \overrightarrow{D_{*a}^{\nu_j} f_j} \right\|_q^{-1}, \textit{ for all } j \in \{1, \ldots, m\} - \{\rho\} \textit{ are Lebesgue integrable.}$$

Proof Use of (23.43). ∎

We need

Definition 23.29 ([2, 12, 13]) Let $\alpha \ge 0$, $n := \lceil \alpha \rceil$, $f \in AC^n([a, b])$. We define the right Caputo fractional derivative of order $\alpha \ge 0$, by

$$\overline{D}_{b-}^\alpha f(x) := (-1)^n I_{b-}^{n-\alpha} f^{(n)}(x), \qquad (23.70)$$

we set $\overline{D}_-^0 f := f$, i.e.

$$\overline{D}_{b-}^\alpha f(x) = \frac{(-1)^n}{\Gamma(n-\alpha)} \int_x^b (J - x)^{n-\alpha-1} f^{(n)}(J) \, dJ. \qquad (23.71)$$

Notice that $\overline{D}_{b-}^n f = (-1)^n f^{(n)}$, $n \in \mathbb{N}$.

We need

Theorem 23.30 ([4]) *Let* $f \in AC^n([a, b])$, $n \in \mathbb{N}$, $n := \lceil \alpha \rceil$, $\alpha > \rho > 0$, $r = \lceil \rho \rceil$, $\alpha, \rho \notin \mathbb{N}$. *Assume* $f^{(k)}(b) = 0$, $k = r, r + 1, \ldots, n - 1$, *and* $\overline{D}_{b-}^\alpha f \in L_\infty([a, b])$. *Then*

$$\overline{D}_{b-}^\rho f(x) = \left(I_{b-}^{\alpha-\rho} \left(\overline{D}_{b-}^\alpha f \right) \right)(x) \in C([a, b]), \qquad (23.72)$$

if $\alpha - \rho \in (0, 1)$, *and* $\overline{D}_{b-}^\rho f \in AC([a, b])$, *if* $\alpha - \rho \ge 1$, *that is*

$$\overline{D}_{b-}^{\rho} f(x) = \frac{1}{\Gamma(\alpha - \rho)} \int_x^b (t - x)^{\alpha - \rho - 1} \left(\overline{D}_{b-}^{\alpha} f\right)(t)\, dt, \qquad (23.73)$$

$\forall\, x \in [a, b]$.

We make

Remark 23.31 Let $\alpha_j > \rho_j > 0$, $n_j = \lceil \alpha_j \rceil$, $r_j = \lceil \rho_j \rceil$, $\alpha_j, \rho_j \notin \mathbb{N}$. Assume $f_{ji} \in AC^{n_j}([a, b])$, $j = 1, \ldots, m$; $i = 1, \ldots, n$. Suppose that $f_{ji}^{(k_j)}(b) = 0$, $k_j = r_j, r_j + 1, \ldots, n_j - 1$, and $\overline{D}_{b-}^{\alpha_j} f_{ji} \in L_\infty(a, b)$. Then by (23.72) we have that

$$\overline{D}_{b-}^{\rho_j} f_{ji}(x) = \left(I_{b-}^{\alpha_j - \rho_j}\left(\overline{D}_{b-}^{\alpha_j} f_{ji}\right)\right)(x), \qquad (23.74)$$

$\forall\, x \in [a, b]$; $j = 1, \ldots, m$; $i = 1, \ldots, n$.

Here $\overline{D}_{b-}^{\rho_j} f_{ji} \in C([a, b])$, if $\alpha_j - \rho_j \in (0, 1)$, and $\overline{D}_{b-}^{\rho_j} f_{ji} \in AC([a, b])$, if $\alpha_j - \rho_j \geq 1$.

We assume here that

$$0 < \left\|\overrightarrow{\overline{D}_{b-}^{\alpha_j} f_j}(y)\right\|_q < \infty, \text{ a.e. on } (a, b), \qquad (23.75)$$

$j = 1, \ldots, m$; where $1 \leq q \leq \infty$, $p \geq 1$, both be fixed.

We further assume that

$$\sum_{j=1}^m (\alpha_j - \rho_j) > m - 1, \qquad (23.76)$$

here $\rho \in \{1, \ldots, m\}$ is fixed.

We give

Proposition 23.32 *Here all as in Remark 23.31. It holds*

$$\int_a^b \left(\prod_{j=1}^m \left\|\overrightarrow{\overline{D}_{b-}^{\rho_j} f_j}(x)\right\|_p\right) dx \leq \qquad (23.77)$$

$$\left(\frac{(b-a)^{\sum\limits_{j=1}^m (\alpha_j - \rho_j) - m + 1}}{\left(\sum\limits_{j=1}^m (\alpha_j - \rho_j) - m + 1\right)\prod\limits_{j=1}^m \Gamma(\alpha_j - \rho_j)}\right)\left(\prod_{\substack{j=1 \\ j \neq \rho}}^m \int_a^b \frac{\left\|\overrightarrow{\overline{D}_{b-}^{\alpha_j} f_j}(y)\right\|_p}{\left\|\overrightarrow{\overline{D}_{b-}^{\alpha_j} f_j}(y)\right\|_q} dy\right).$$

$$\left(\int_a^b \left(\prod_{\substack{j=1\\j\neq\rho}}^m \left\|\overrightarrow{D_{b-}^{\alpha_j}f_j}(y)\right\|_q\right)\left\|\overrightarrow{D_{b-}^{\alpha_\rho}f_\rho}(y)\right\|_p dy\right),$$

under the assumption:

$$\left\|\overrightarrow{D_{b-}^{\alpha_j}f_j}\right\|_q^{-1}, \text{ for all } j \in \{1, \ldots, m\} - \{\rho\} \text{ are Lebesgue integrable.}$$

Proof Use of (23.56). ∎

We need

Definition 23.33 Let $\nu > 0$, $n := [\nu]$, $\alpha := \nu - n$ ($0 \leq \alpha < 1$). Let $a, b \in \mathbb{R}$, $a \leq x \leq b$, $f \in C([a, b])$. We consider $C_a^\nu([a, b]) := \{f \in C^n([a, b]) : I_{a+}^{1-\alpha} f^{(n)} \in C^1([a, b])\}$. For $f \in C_a^\nu([a, b])$, we define the left generalized ν-fractional derivative of f over $[a, b]$ as

$$\Delta_a^\nu f := \left(I_{a+}^{1-\alpha} f^{(n)}\right)', \tag{23.78}$$

see [1], p. 24, and Canavati derivative in [10].
Notice here $\Delta_a^\nu f \in C([a, b])$.
So that

$$\left(\Delta_a^\nu f\right)(x) = \frac{1}{\Gamma(1-\alpha)} \frac{d}{dx} \int_a^x (x-t)^{-\alpha} f^{(n)}(t)\, dt, \tag{23.79}$$

$\forall\, x \in [a, b]$.
Notice here that

$$\Delta_a^n f = f^{(n)}, n \in \mathbb{Z}_+.$$

We need

Theorem 23.34 ([4]) *Let $f \in C_a^\nu([a, b])$, $n = [\nu]$, such that $f^{(i)}(a) = 0$, $i = r, r+1, \ldots, n-1$, where $r := [\rho]$, with $0 < \rho < \nu$. Then*

$$\left(\Delta_a^\rho f\right)(x) = \frac{1}{\Gamma(\nu-\rho)} \int_a^x (x-t)^{\nu-\rho-1} \left(\Delta_a^\nu f\right)(t)\, dt, \tag{23.80}$$

i.e.

$$\left(\Delta_a^\rho f\right) = I_{a+}^{\nu-\rho}\left(\Delta_a^\nu f\right) \in C([a, b]). \tag{23.81}$$

Thus $f \in C_a^\rho([a, b])$.

We make

Remark 23.35 Let $f_{ji} \in C_a^{\nu_j}([a,b])$, $n_j := [\nu_j]$, such that $f_{ji}^{(k_j)}(a) = 0$, $k_j = r_j, r_j + 1, \ldots, n_j - 1$, where $r_j := [\rho_j]$, with $0 < \rho_j < \nu_j$; $j = 1, \ldots, m$, $i = 1, \ldots, n$. Then by (23.81) we get

$$\left(\Delta_a^{\rho_j} f_{ji}\right)(x) = \left(I_{a+}^{\nu_j - \rho_j}\left(\Delta_a^{\nu_j} f_{ji}\right)\right)(x) \in C([a,b]). \qquad (23.82)$$

Thus $f_{ji} \in C_a^{\rho_j}([a,b])$, $j = 1, \ldots, m$; $i = 1, \ldots, n$.

We assume that

$$0 < \left\|\overrightarrow{\Delta_a^{\nu_j} f_j}(y)\right\|_q < \infty, \text{ a.e. on } (a,b), \qquad (23.83)$$

$j = 1, \ldots, m$; where $1 \le q \le \infty$, $p \ge 1$, both be fixed.

We further assume that

$$\sum_{j=1}^{m}(\nu_j - \rho_j) > m - 1, \qquad (23.84)$$

here $\rho \in \{1, \ldots, m\}$ is fixed.

We give

Proposition 23.36 *Let all here as in Remark 23.35. It holds*

$$\int_a^b \left(\prod_{j=1}^{m}\left\|\overrightarrow{\Delta_a^{\rho_j} f_j}(x)\right\|_p\right) dx \le \qquad (23.85)$$

$$\left(\frac{(b-a)^{\sum_{j=1}^{m}(\nu_j - \rho_j) - m + 1}}{\left(\sum_{j=1}^{m}(\nu_j - \rho_j) - m + 1\right)\prod_{j=1}^{m}\Gamma(\nu_j - \rho_j)}\right)\left(\prod_{\substack{j=1 \\ j \ne \rho}}^{m}\int_a^b \frac{\left\|\overrightarrow{\Delta_a^{\nu_j} f_j}(y)\right\|_p}{\left\|\overrightarrow{\Delta_a^{\nu_j} f_j}(y)\right\|_q} dy\right) \cdot$$

$$\left(\int_a^b \left(\prod_{\substack{j=1 \\ j \ne \rho}}^{m}\left\|\overrightarrow{\Delta_a^{\nu_j} f_j}(y)\right\|_q\right)\left\|\overrightarrow{\Delta_a^{\nu_\rho} f_\rho}(y)\right\|_p dy\right),$$

under the assumption:

$$\left\|\overrightarrow{\Delta_a^{\nu_j} f_j}\right\|_q^{-1}, \text{ for all } j \in \{1, \ldots, m\} - \{\rho\} \text{ are Lebesgue integrable.}$$

Proof Use of (23.43). ∎

We need

Definition 23.37 ([2]) Let $\nu > 0$, $n := [\nu]$, $\alpha = \nu - n$, $0 < \alpha < 1$, $f \in C([a, b])$. Consider

$$C_{b-}^{\nu}([a, b]) := \{f \in C^n([a, b]) : I_{b-}^{1-\alpha} f^{(n)} \in C^1([a, b])\}. \tag{23.86}$$

Define the right generalized ν-fractional derivative of f over $[a, b]$, by

$$\Delta_{b-}^{\nu} f := (-1)^{n-1} \left(I_{b-}^{1-\alpha} f^{(n)}\right)'. \tag{23.87}$$

We set $\Delta_{b-}^{0} f = f$. Notice that

$$\left(\Delta_{b-}^{\nu} f\right)(x) = \frac{(-1)^{n-1}}{\Gamma(1-\alpha)} \frac{d}{dx} \int_x^b (J - x)^{-\alpha} f^{(n)}(J) \, dJ, \tag{23.88}$$

and $\Delta_{b-}^{\nu} f \in C([a, b])$.

We also need

Theorem 23.38 ([4]) *Let* $f \in C_{b-}^{\nu}([a, b])$, $0 < \rho < \nu$. *Assume* $f^{(i)}(b) = 0$, $i = r, r+1, \ldots, n-1$, *where* $r := [\rho]$, $n := [\nu]$. *Then*

$$\Delta_{b-}^{\rho} f(x) = \frac{1}{\Gamma(\nu - \rho)} \int_x^b (J - x)^{\nu - \rho - 1} \left(\Delta_{b-}^{\nu} f\right)(J) \, dJ, \tag{23.89}$$

$\forall \, x \in [a, b]$, *i.e.*

$$\Delta_{b-}^{\rho} f = I_{b-}^{\nu - \rho} \left(\Delta_{b-}^{\nu} f\right) \in C([a, b]), \tag{23.90}$$

and $f \in C_{b-}^{\rho}([a, b])$.

We make

Remark 23.39 Let $f_{ji} \in C_{b-}^{\nu_j}([a, b])$, $0 < \rho_j < \nu_j$. Assume $f_{ji}^{(k_j)}(b) = 0$, $k_j = r_j, r_j + 1, \ldots, n_j - 1$, where $r_j := [\rho_j]$, $n_j := [\nu_j]$; $j = 1, \ldots, m$, $i = 1, \ldots, n$. Then by (23.90) we get

$$\Delta_{b-}^{\rho_j} f_{ji} = I_{b-}^{\nu_j - \rho_j} \left(\Delta_{b-}^{\nu_j} f_{ji}\right) \in C([a, b]), \tag{23.91}$$

and $f_{ji} \in C_{b-}^{\rho_j}([a, b])$, $j = 1, \ldots, m$; $i = 1, \ldots, n$.

We assume that

$$0 < \left\| \overrightarrow{\Delta_{b-}^{\nu_j} f_j(y)} \right\|_q < \infty, \text{ a.e. on } (a, b), \tag{23.92}$$

$j = 1, \ldots, m$; where $1 \leq q \leq \infty$, $p \geq 1$, both be fixed.

We further assume that

$$\sum_{j=1}^{m} (\nu_j - \rho_j) > m - 1, \tag{23.93}$$

here $\rho \in \{1, \ldots, m\}$ is fixed.

We present

Proposition 23.40 *Let all here as in Remark 23.39. It holds*

$$\int_a^b \left(\prod_{j=1}^{m} \left\| \overrightarrow{\Delta_{b-}^{\rho_j} f_j (x)} \right\|_p \right) dx \leq \tag{23.94}$$

$$\left(\frac{(b-a)^{\sum\limits_{j=1}^{m}(\nu_j - \rho_j) - m + 1}}{\left(\sum\limits_{j=1}^{m} (\nu_j - \rho_j) - m + 1 \right) \prod\limits_{j=1}^{m} \Gamma (\nu_j - \rho_j)} \right) \left(\prod_{\substack{j=1 \\ j \neq \rho}}^{m} \int_a^b \frac{\left\| \overrightarrow{\Delta_{b-}^{\nu_j} f_j (y)} \right\|_p}{\left\| \overrightarrow{\Delta_{b-}^{\nu_j} f_j (y)} \right\|_q} dy \right) \cdot$$

$$\left(\int_a^b \left(\prod_{\substack{j=1 \\ j \neq \rho}}^{m} \left\| \overrightarrow{\Delta_{b-}^{\nu_j} f_j (y)} \right\|_q \right) \left\| \overrightarrow{\Delta_{b-}^{\nu_\rho} f_\rho (y)} \right\|_p dy \right),$$

under the assumption:

$\left\| \overrightarrow{\Delta_{b-}^{\nu_j} f_j} \right\|_q^{-1}$, *for all $j \in \{1, \ldots, m\} - \{\rho\}$ are Lebesgue integrable.*

Proof Use of (23.56). ∎

We need

Definition 23.41 ([16], p. 99) The fractional integrals of a function f with respect to given function g are defined as follows:

Let $a, b \in \mathbb{R}$, $a < b$, $\alpha > 0$. Here g is an increasing function on $[a, b]$, and $g \in C^1 ([a, b])$. The left- and right-sided fractional integrals of a function f with respect to another function g in $[a, b]$ are given by

$$\left(I_{a+;g}^{\alpha} f \right) (x) = \frac{1}{\Gamma (\alpha)} \int_a^x \frac{g' (t) f (t) \, dt}{(g (x) - g (t))^{1-\alpha}}, x > a, \tag{23.95}$$

$$\left(I_{b-;g}^{\alpha} f\right)(x) = \frac{1}{\Gamma(\alpha)} \int_x^b \frac{g'(t) f(t) \, dt}{(g(t) - g(x))^{1-\alpha}}, \; x < b, \tag{23.96}$$

respectively.

We also need

Definition 23.42 ([16]) Let $0 < a < b < \infty$, $\alpha > 0$. The left- and right-sided Hadamard fractional integrals of order α are given by

$$\left(J_{a+}^{\alpha} f\right)(x) = \frac{1}{\Gamma(\alpha)} \int_a^x \left(\ln \frac{x}{y}\right)^{\alpha-1} \frac{f(y)}{y} \, dy, \; x > a, \tag{23.97}$$

and

$$\left(J_{b-}^{\alpha} f\right)(x) = \frac{1}{\Gamma(\alpha)} \int_x^b \left(\ln \frac{y}{x}\right)^{\alpha-1} \frac{f(y)}{y} \, dy, \; x < b, \tag{23.98}$$

respectively.

Notice that the Hadamard fractional integrals of order α are special cases of left- and right-sided fractional integrals of a function f with respect to another function, here $g(x) = \ln x$ on $[a, b]$, $0 < a < b < \infty$.

Above f is a Lebesgue measurable function from (a, b) into $\mathbb{R}$, such that $\left(J_{a+}^{\alpha}(|f|)\right)(x)$, $\left(J_{b-}^{\alpha}(|f|)\right)(x) \in \mathbb{R}$, $\forall \, x \in (a, b)$.

We make

Remark 23.43 Let f_{ji} be Lebesgue measurable functions from (a, b) into $\mathbb{R}$, such that $\left(I_{a+;g}^{\alpha_j}(|f_{ji}|)\right)(x) \in \mathbb{R}$, $\forall \, x \in (a, b)$, $\alpha_j > 0$, $j = 1, \ldots, m$, $j = 1, \ldots, n$.

Consider here

$$g_{ji}(x) = \left(I_{a+;g}^{\alpha_j}(f_{ji})\right)(x) = \frac{1}{\Gamma(\alpha_j)} \int_a^x (g(x) - g(t))^{\alpha_j-1} g'(t) f_{ji}(t) \, dt, \tag{23.99}$$

$x \in (a, b)$, $j = 1, \ldots, m$, $i = 1, \ldots, n$.

Notice that $g_{ji}(x) \in \mathbb{R}$ and it is Lebesgue measurable. We pick $\Omega_1 = \Omega_2 = (a, b)$, $d\mu_1(x) = dx$, $d\mu_2(y) = dy$, the Lebesgue measure.

We see that

$$\left(I_{a+;g}^{\alpha_j} f_{ji}\right)(x) = \int_a^b \frac{\chi_{(a,x]}(t) (g(x) - g(t))^{\alpha_j-1} g'(t)}{\Gamma(\alpha_j)} f_{ji}(t) \, dt. \tag{23.100}$$

So here it is

$$k_j(x, t) = \frac{\chi_{(a,x]}(t) (g(x) - g(t))^{\alpha_j-1} g'(t)}{\Gamma(\alpha_j)}, \; j = 1, \ldots, m. \tag{23.101}$$

We assume here that

$$0 < \left\| \overrightarrow{f_j\,(y)} \right\|_q < \infty, \text{ a.e. on } (a, b),\qquad (23.102)$$

$j = 1, \ldots, m;\ 1 \le q \le \infty,\ p \ge 1$ fixed.
We call

$$L_{jq}^{g+}(x) := \int_a^b \left(\frac{\chi_{(a,x]}(y)\,(g(x) - g(y))^{\alpha_j - 1}\,g'(y)}{\Gamma(\alpha_j)} \right) \left\| \overrightarrow{f_j}(y) \right\|_q dy \quad (23.103)$$

$$= \left(I_{a+;g}^{\alpha_j} \left(\left\| \overrightarrow{f_j} \right\|_q \right) \right)(x),$$

$x \in (a, b),\ j = 1, \ldots, m.$
We assume that $L_{jq}^{g+}(x) > 0$ a.e. on (a, b).
Let $\rho \in \{1, \ldots, m\}$ be fixed. Assume more that the function

$$x \to \frac{u(x) \left(\prod\limits_{j=1}^m \left\| \overrightarrow{f_j}(y) \right\|_q \right) \chi_{(a,x]}(y)\,(g(x) - g(y))^{\sum\limits_{j=1}^m \alpha_j - m}\,(g'(y))^m}{\prod\limits_{j=1}^m \left(L_{jq}^{g+}(x)\,\Gamma(\alpha_j) \right)}$$

$$(23.104)$$

is integrable on (a, b), for each $y \in (a, b)$.
So we get

$$U_{mq}^{g+}(y) := \left(\prod_{j=1}^m \left(\frac{\left\| \overrightarrow{f_j}(y) \right\|_q}{\Gamma(\alpha_j)} \right) \right) (g'(y))^m \cdot \qquad (23.105)$$

$$\int_a^b \frac{u(x)\,\chi_{(a,x]}(y)\,(g(x) - g(y))^{\sum\limits_{j=1}^m \alpha_j - m}}{\prod\limits_{j=1}^m L_{jq}^{g+}(x)} dx < \infty,$$

$y \in (a, b)$.

We give

Theorem 23.44 *All as in Remark 23.43. It holds*

$$\int_a^b u(x) \prod_{j=1}^m \left(\frac{\left\| \overrightarrow{\left(I_{a+;g}^{\alpha_j}(f_j) \right)(x)} \right\|_p}{I_{a+;g}^{\alpha_j} \left(\left\| \overrightarrow{f_j} \right\|_q \right)(x)} \right) dx \le \qquad (23.106)$$

$$\left(\prod_{\substack{j=1 \\ j \ne \rho}}^m \int_a^b \left(\frac{\left\| \overrightarrow{f_j}(y) \right\|_p}{\left\| \overrightarrow{f_j}(y) \right\|_q} \right) dy \right) \left(\int_a^b U_{mq}^{g+}(y) \frac{\left\| \overrightarrow{f_\rho}(y) \right\|_p}{\left\| \overrightarrow{f_\rho}(y) \right\|_q} dy \right),$$

under the assumptions:

(i) $\left\| \overrightarrow{f_j} \right\|_p$ is $\frac{\chi_{(a,x]}(y)(g(x)-g(y))^{\alpha_j-1}g'(y)}{\Gamma(\alpha_j)} dy$ -integrable, a.e. in $x \in (a,b)$, $j = 1, \dots, m$,

(ii) $\left(\frac{U_{mq}^{g+}}{\left\| \overrightarrow{f_\rho} \right\|_q} \right) \left\| \overrightarrow{f_\rho} \right\|_p$; $\left(\frac{\left\| \overrightarrow{f_j} \right\|_p}{\left\| \overrightarrow{f_j} \right\|_q} \right)$, $j \ne \rho$, $j = 1, \dots, m$, are all Lebesgue integrable.

Proof By Theorem 23.10. ∎

Furthermore we present

Theorem 23.45 *All as in Remark 23.43. It holds*

$$\int_a^b u(x) \prod_{j=1}^m \left(\sum_{i=1}^n e^{\left(\frac{\left| I_{a+;g}^{\alpha_j} f_{ji}(x) \right|}{I_{a+;g}^{\alpha_j} \left(\left\| \overrightarrow{f_j} \right\|_q \right)(x)} \right)} \right) dx \le \qquad (23.107)$$

$$\left(\prod_{\substack{j=1 \\ j \ne \rho}}^m \int_a^b \left(\sum_{i=1}^n e^{\left| \frac{f_{ji}(y)}{\left\| \overrightarrow{f_j}(y) \right\|_q} \right|} \right) dy \right) \left(\int_a^b \left(\sum_{i=1}^n e^{\left| \frac{f_{\rho i}(y)}{\left\| \overrightarrow{f_\rho}(y) \right\|_q} \right|} \right) U_{mq}^{g+}(y) dy \right),$$

under the assumptions:

(i) $\frac{\overrightarrow{f_j}}{\left\| \overrightarrow{f_j} \right\|_q}$, $\left(\sum_{i=1}^n e^{\left| \frac{f_{ji}(y)}{\left\| \overrightarrow{f_j}(y) \right\|_q} \right|} \right)$ are $\left\| \overrightarrow{f_j}(y) \right\|_q \frac{\chi_{(a,x]}(y)(g(x)-g(y))^{\alpha_j-1}g'(y)}{\Gamma(\alpha_j)} dy$

-integrable, a.e. in $x \in (a,b)$, $j = 1, \dots, m$,

(ii) $U_{mq}^{g+}(y)\left(\displaystyle\sum_{i=1}^{n}e^{\left|\frac{f_{\rho i}(y)}{\left\|\overrightarrow{f_{\rho}(y)}\right\|_{q}}\right|}\right)$ and $\left(\displaystyle\sum_{i=1}^{n}e^{\left|\frac{f_{ji}(y)}{\left\|\overrightarrow{f_j(y)}\right\|_{q}}\right|}\right)$ for $j\neq\rho$, $j=1,\dots,m$, are

all Lebesgue integrable.

Proof By Theorem 23.11. ∎

We make

Remark 23.46 (continuation of Remark 23.43) Here we assume that $\displaystyle\sum_{j=1}^{m}\alpha_j>m-1$.

We also choose

$$u(x)=g'(x)\prod_{j=1}^{m}L_{jq}^{g+}(x)=g'(x)\prod_{j=1}^{m}I_{a+;g}^{\alpha_j}\left(\left\|\overrightarrow{f_j}\right\|_q\right)(x),\qquad(23.108)$$

$x\in(a,b)$.

Hence

$$U_{mq}^{g+}(y)=\prod_{j=1}^{m}\left(\frac{\left\|\overrightarrow{f_j}(y)\right\|_q}{\Gamma(\alpha_j)}\right)(g'(y))^m\frac{(g(b)-g(y))^{\sum_{j=1}^{m}\alpha_j-m+1}}{\left(\displaystyle\sum_{j=1}^{m}\alpha_j-m+1\right)},\qquad(23.109)$$

$y\in(a,b)$.

Notice that

$$U_{mq}^{g+}(y)\leq\prod_{j=1}^{m}\left(\frac{\left\|\overrightarrow{f_j}(y)\right\|_q}{\Gamma(\alpha_j)}\right)\|g'\|_\infty^m\frac{(g(b)-g(a))^{\sum_{j=1}^{m}\alpha_j-m+1}}{\left(\displaystyle\sum_{j=1}^{m}\alpha_j-m+1\right)},\qquad(23.110)$$

$\forall\,y\in(a,b)$.

Then (23.106) becomes

$$\int_a^b g'(x)\left(\prod_{j=1}^{m}\left\|\overrightarrow{\left(I_{a+;g}^{\alpha_j}f_j\right)(x)}\right\|_p\right)dx\leq\qquad(23.111)$$

$$
\left(\frac{\|g'\|_\infty^m (g(b) - g(a))^{\sum\limits_{j=1}^m \alpha_j - m + 1}}{\left(\sum\limits_{j=1}^m \alpha_j - m + 1 \right) \prod\limits_{j=1}^m \Gamma(\alpha_j)} \right) \left(\prod\limits_{\substack{j=1 \\ j \neq \rho}}^m \int_a^b \frac{\left\| \overrightarrow{f_j}(y) \right\|_p}{\left\| \overrightarrow{f_j}(y) \right\|_q} dy \right) \cdot
$$

$$
\left(\int_a^b \left(\prod\limits_{\substack{j=1 \\ j \neq \rho}}^m \left\| \overrightarrow{f_j}(y) \right\|_q \right) \left\| \overrightarrow{f_\rho}(y) \right\|_p dy \right),
$$

under the assumptions:

(i) following (23.106), and

(ii) $*\left(\prod\limits_{\substack{j=1 \\ j \neq \rho}}^m \left\| \overrightarrow{f_j}(y) \right\|_q \right) \left\| \overrightarrow{f_\rho}(y) \right\|_p$; $\left(\dfrac{\left\| \overrightarrow{f_j}(y) \right\|_p}{\left\| \overrightarrow{f_j}(y) \right\|_q} \right)$, $j = 1, \ldots, m$, with $j \neq \rho$, are all Lebesgue integrable.

We make

Remark 23.47 Let f_{ji} be Lebesgue measurable functions from (a, b) into $\mathbb{R}$, such that $\left(I_{b-;g}^{\alpha_j} (|f_{ji}|) \right)(x) \in \mathbb{R}$, $\forall x \in (a, b)$, $\alpha_j > 0$, $j = 1, \ldots, m$; $i = 1, \ldots, n$. Consider here

$$
g_{ji}(x) = \left(I_{b-;g}^{\alpha_j} f_{ji} \right)(x) = \frac{1}{\Gamma(\alpha_j)} \int_x^b (g(t) - g(x))^{\alpha_j - 1} g'(t) f_{ji}(t) dt,
$$
(23.112)

$x \in (a, b)$, $j = 1, \ldots, m$; $i = 1, \ldots, n$.

Notice that $g_{ji}(x) \in \mathbb{R}$ and it is Lebesgue measurable.

We pick $\Omega_1 = \Omega_2 = (a, b)$, $d\mu_1(x) = dx$, $d\mu_2(y) = dy$, the Lebesgue measure. We see that

$$
\left(I_{b-;g}^{\alpha_j} f_{ji} \right)(x) = \int_a^b \frac{\chi_{[x,b)}(t)(g(t) - g(x))^{\alpha_j - 1} g'(t)}{\Gamma(\alpha_j)} f_{ji}(t) dt. \quad (23.113)
$$

So here it is

$$
k_j(x, t) := \frac{\chi_{[x,b)}(t)(g(t) - g(x))^{\alpha_j - 1} g'(t)}{\Gamma(\alpha_j)}, \quad j = 1, \ldots, m. \quad (23.114)
$$

We assume here that

$$
0 < \left\| \overrightarrow{f_j}(y) \right\|_q < \infty, \text{ a.e. on } (a, b), \quad (23.115)
$$

$j = 1, \ldots, m$; $1 \le q \le \infty$, $p \ge 1$ fixed.

We call

$$L_{jq}^{g-}(x) := \int_a^b \left(\frac{\chi_{[x,b)}(t)(g(t) - g(x))^{\alpha_j - 1} g'(t)}{\Gamma(\alpha_j)} \right) \left\| \vec{f_j}(t) \right\|_q dt \quad (23.116)$$

$$= I_{b-;g}^{\alpha_j} \left(\left\| \vec{f_j} \right\|_q \right)(x),$$

$x \in (a, b)$, $j = 1, \ldots, m$.

We assume that $L_{jq}^{g-}(x) > 0$, a.e. on (a, b).

Let $\rho \in \{1, \ldots, m\}$ be fixed. Assume more that the function

$$x \to \frac{u(x) \left(\prod_{j=1}^m \left\| \vec{f_j}(y) \right\|_q \right) \chi_{[x,b)}(y)(g(y) - g(x))^{\sum_{j=1}^m \alpha_j - m} (g'(y))^m}{\prod_{j=1}^m \left(L_{jq}^{g-}(x) \Gamma(\alpha_j) \right)}$$

$$(23.117)$$

is integrable on (a, b), for each $y \in (a, b)$.

So we get

$$U_{mq}^{g-}(y) := \left(\prod_{j=1}^m \frac{\left\| \vec{f_j}(y) \right\|_q}{\Gamma(\alpha_j)} \right) (g'(y))^m \cdot \quad (23.118)$$

$$\int_a^b \frac{u(x) \chi_{[x,b)}(y)(g(y) - g(x))^{\sum_{j=1}^m \alpha_j - m}}{\prod_{j=1}^m L_{jq}^{g-}(x)} dx < \infty,$$

$y \in (a, b)$.

We give

Theorem 23.48 *All as in Remark 23.47. It holds*

$$\int_a^b u(x) \prod_{j=1}^m \left(\frac{\left\| \left(I_{b-;g}^{\alpha_j} \vec{f_j} \right)(x) \right\|_p}{I_{b-;g}^{\alpha_j} \left(\left\| \vec{f_j} \right\|_q \right)(x)} \right) dx \le \quad (23.119)$$

$$\left(\prod_{\substack{j=1 \\ j\neq\rho}}^{m}\int_{a}^{b}\left(\frac{\left\|\overrightarrow{f_{j}}\,(y)\right\|_{p}}{\left\|\overrightarrow{f_{j}}\,(y)\right\|_{q}}\right)dy\right)\left(\int_{a}^{b}U_{mq}^{g-}\,(y)\,\frac{\left\|\overrightarrow{f_{\rho}}\,(y)\right\|_{p}}{\left\|\overrightarrow{f_{\rho}}\,(y)\right\|_{q}}dy\right),$$

under the assumptions:

(i) $\left\|\overrightarrow{f_{j}}\right\|_{p}$ is $\dfrac{\chi_{[x,b)}(y)(g(y)-g(x))^{\alpha_{j}-1}g'(y)}{\Gamma(\alpha_{j})}dy$ -integrable, a.e. in $x\in(a,b)$, $j=1,\ldots,m$,

(ii) $\left(\dfrac{U_{mq}^{g-}}{\left\|\overrightarrow{f_{\rho}}\right\|_{q}}\right)\left\|\overrightarrow{f_{\rho}}\right\|_{p};\left(\dfrac{\left\|\overrightarrow{f_{j}}\right\|_{p}}{\left\|\overrightarrow{f_{j}}\right\|_{q}}\right)$, $j\neq\rho$, $j=1,\ldots,m$, are all Lebesgue integrable.

Proof By Theorem 23.10. ∎

Furthermore we present

Theorem 23.49 *All here as in Remark 23.47. It holds*

$$\int_{a}^{b}u(x)\prod_{j=1}^{m}\left(\sum_{i=1}^{n}e^{\left(\frac{\left|I_{b-;g}^{\alpha_{j}}f_{ji}(x)\right|}{I_{b-;g}^{\alpha_{j}}\left(\left\|\overrightarrow{f_{j}}\right\|_{q}\right)(x)}\right)}\right)dx\leq \qquad (23.120)$$

$$\left(\prod_{\substack{j=1 \\ j\neq\rho}}^{m}\int_{a}^{b}\left(\sum_{i=1}^{n}e^{\left|\frac{f_{ji}(y)}{\left\|\overrightarrow{f_{j}}(y)\right\|_{q}}\right|}\right)dy\right)\left(\int_{a}^{b}\left(\sum_{i=1}^{n}e^{\left|\frac{f_{\rho i}(y)}{\left\|\overrightarrow{f_{\rho}}(y)\right\|_{q}}\right|}\right)U_{mq}^{g-}\,(y)\,dy\right),$$

under the assumptions:

(i) $\dfrac{\overrightarrow{f_{j}}}{\left\|\overrightarrow{f_{j}}\right\|_{q}}$, $\left(\sum_{i=1}^{n}e^{\left|\frac{f_{ji}(y)}{\left\|\overrightarrow{f_{j}}(y)\right\|_{q}}\right|}\right)$ are $\left\|\overrightarrow{f_{j}}\,(y)\right\|_{q}\dfrac{\chi_{[x,b)}(y)(g(y)-g(x))^{\alpha_{j}-1}g'(y)}{\Gamma(\alpha_{j})}dy$ -integrable,

a.e. in $x\in(a,b)$, $j=1,\ldots,m$,

(ii) $U_{mq}^{g-}\,(y)\left(\sum_{i=1}^{n}e^{\left|\frac{f_{\rho i}(y)}{\left\|\overrightarrow{f_{\rho}}(y)\right\|_{q}}\right|}\right)$ and $\left(\sum_{i=1}^{n}e^{\left|\frac{f_{ji}(y)}{\left\|\overrightarrow{f_{j}}(y)\right\|_{q}}\right|}\right)$ for $j\neq\rho$, $j=1,\ldots,m$, are

all Lebesgue integrable.

Proof By Theorem 23.11. ∎

We make

Remark 23.50 (continuation of Remark 23.47) Here we assume that $\sum\limits_{j=1}^{m} \alpha_j > m-1$.

We also choose

$$u(x) = g'(x) \prod_{j=1}^{m} L_{jq}^{g-}(x) = g'(x) \prod_{j=1}^{m} I_{b-;g}^{\alpha_j}\left(\left\|\overrightarrow{f_j}\right\|_q\right)(x), \qquad (23.121)$$

$x \in (a,b)$.

Hence

$$U_{mq}^{g-}(y) = \left(\prod_{j=1}^{m}\left(\frac{\left\|\overrightarrow{f_j}(y)\right\|_q}{\Gamma(\alpha_j)}\right)\right)(g'(y))^m \frac{(g(y)-g(a))^{\sum\limits_{j=1}^{m}\alpha_j-m+1}}{\left(\sum\limits_{j=1}^{m}\alpha_j-m+1\right)}, \qquad (23.122)$$

$y \in (a,b)$.

Notice that

$$U_{mq}^{g-}(y) \leq \left(\prod_{j=1}^{m}\left(\frac{\left\|\overrightarrow{f_j}(y)\right\|_q}{\Gamma(\alpha_j)}\right)\right)\|g'\|_\infty^m \frac{(g(b)-g(a))^{\sum\limits_{j=1}^{m}\alpha_j-m+1}}{\left(\sum\limits_{j=1}^{m}\alpha_j-m+1\right)}, \qquad (23.123)$$

$\forall\, y \in (a,b)$.

Then (23.119) becomes

$$\int_a^b g'(x)\left(\prod_{j=1}^{m}\left\|\overrightarrow{\left(I_{b-;g}^{\alpha_j} f_j\right)(x)}\right\|_p\right)dx \leq \qquad (23.124)$$

$$\left(\frac{\|g'\|_\infty^m (g(b)-g(a))^{\sum\limits_{j=1}^{m}\alpha_j-m+1}}{\left(\sum\limits_{j=1}^{m}\alpha_j-m+1\right)\prod\limits_{j=1}^{m}\Gamma(\alpha_j)}\right)\left(\prod_{\substack{j=1\\j\neq\rho}}^{m}\int_a^b \frac{\left\|\overrightarrow{f_j}(y)\right\|_p}{\left\|\overrightarrow{f_j}(y)\right\|_q}dy\right).$$

$$\left(\int_a^b \left(\prod_{\substack{j=1 \\ j \neq \rho}}^m \left\| \vec{f_j}(y) \right\|_q \right) \left\| \vec{f_\rho}(y) \right\|_p dy \right),$$

under the assumptions:

(i) following (23.119),

(ii) $* \left(\prod_{\substack{j=1 \\ j \neq \rho}}^m \left\| \vec{f_j}(y) \right\|_q \right) \left\| \vec{f_\rho}(y) \right\|_p ; \left(\frac{\left\| \vec{f_j}(y) \right\|_p}{\left\| \vec{f_j}(y) \right\|_q} \right), j = 1, \ldots, m,$ with $j \neq \rho$, are all

   Lebesgue integrable.

We finish with

Proposition 23.51 *Let* $0 < a < b < \infty$, $\alpha_j > 0$, $\sum_{j=1}^m \alpha_j > m - 1$, $m \in \mathbb{N}$. *Let* f_{ji}

Lebesgue measurable functions from (a, b) *into* $\mathbb{R}$, *such that* $\left(J_{a+}^{\alpha_j} \left(|f_{ji}| \right) \right)(x) \in \mathbb{R}$,
$\forall x \in (a, b), j = 1, \ldots, m, i = 1, \ldots, n$. *We take fixed* $1 \leq q \leq \infty, p \geq 1$. *We assume that* $0 < \left\| \vec{f_j}(y) \right\|_q < \infty$, *a.e. on* $(a, b), j = 1, \ldots, m, \rho \in \{1, \ldots, m\}$.
Then

$$\int_a^b \left(\prod_{j=1}^m \left\| \overrightarrow{\left(J_{a+}^{\alpha_j}(f_j) \right)(x)} \right\|_p \right) dx \leq \tag{23.125}$$

$$\left(\frac{b \left(\ln \left(\frac{b}{a} \right) \right)^{\sum_{j=1}^m \alpha_j - m + 1}}{a^m \left(\sum_{j=1}^m \alpha_j - m + 1 \right) \prod_{j=1}^m \Gamma(\alpha_j)} \right) \left(\prod_{\substack{j=1 \\ j \neq \rho}}^m \int_a^b \frac{\left\| \vec{f_j}(y) \right\|_p}{\left\| \vec{f_j}(y) \right\|_q} dy \right) \cdot$$

$$\left(\int_a^b \left(\prod_{\substack{j=1 \\ j \neq \rho}}^m \left\| \vec{f_j}(y) \right\|_q \right) \left\| \vec{f_\rho}(y) \right\|_p dy \right),$$

under the assumptions:

(i) $\left\| \vec{f_j} \right\|_p$ *is* $\dfrac{\chi_{(a,x]}(y) \left(\ln \frac{x}{y} \right)^{\alpha_j - 1}}{y \Gamma(\alpha_j)}$ *-dy-integrable, a.e. in* $x \in (a, b), j = 1, \ldots, m,$

(ii) $\left(\prod_{\substack{j=1 \\ j \neq \rho}}^{m} \left\| \vec{f}_j \left(y \right) \right\|_q \right) \left\| \vec{f}_\rho \left(y \right) \right\|_p ; \left(\dfrac{\left\| \vec{f}_j(y) \right\|_p}{\left\| \vec{f}_j(y) \right\|_q} \right)$, $j = 1, \ldots, m$, with $j \neq \rho$, are all
Lebesgue integrable.

Proof By Remark 23.46 and (23.111) in particular. ∎

Proposition 23.52 *Let* $0 < a < b < \infty$, $\alpha_j > 0$, $\sum\limits_{j=1}^{m} \alpha_j > m - 1$, $m \in \mathbb{N}$. *Let* f_{ji}

Lebesgue measurable functions from (a, b) *into* $\mathbb{R}$, *such that* $\left(J_{b-}^{\alpha_j} \left(\left| f_{ji} \right| \right) \right) (x) \in \mathbb{R}$,
$\forall \, x \in (a, b)$, $j = 1, \ldots, m$, $i = 1, \ldots, n$. *We take fixed* $1 \leq q \leq \infty$, $p \geq 1$. *We*
assume that $0 < \left\| \vec{f}_j \left(y \right) \right\|_q < \infty$, *a.e. on* (a, b), $j = 1, \ldots, m$, $\rho \in \{1, \ldots, m\}$.
Then

$$\int_a^b \left(\prod_{j=1}^{m} \left\| \overrightarrow{\left(J_{b-}^{\alpha_j} \left(f_j \right) \right) (x)} \right\|_p \right) dx \leq \qquad (23.126)$$

$$\left(\dfrac{b \left(\ln \left(\frac{b}{a} \right) \right)^{\sum\limits_{j=1}^{m} \alpha_j - m + 1}}{a^m \left(\sum\limits_{j=1}^{m} \alpha_j - m + 1 \right) \prod\limits_{j=1}^{m} \Gamma \left(\alpha_j \right)} \right) \left(\prod_{\substack{j=1 \\ j \neq \rho}}^{m} \int_a^b \dfrac{\left\| \vec{f}_j \left(y \right) \right\|_p}{\left\| \vec{f}_j \left(y \right) \right\|_q} dy \right) \cdot$$

$$\left(\int_a^b \left(\prod_{\substack{j=1 \\ j \neq \rho}}^{m} \left\| \vec{f}_j \left(y \right) \right\|_q \right) \left\| \vec{f}_\rho \left(y \right) \right\|_p dy \right),$$

under the assumptions:

(i) $\left\| \vec{f}_j \right\|_p$ *is* $\dfrac{\chi_{[x,b)}(y) \left(\ln \frac{y}{x} \right)^{\alpha_j - 1}}{y \Gamma (\alpha_j)}$ *dy-integrable, a.e. in* $x \in (a, b)$, $j = 1, \ldots, m$,

(ii) $\left(\prod\limits_{\substack{j=1 \\ j \neq \rho}}^{m} \left\| \vec{f}_j \left(y \right) \right\|_q \right) \left\| \vec{f}_\rho \left(y \right) \right\|_p ; \left(\dfrac{\left\| \vec{f}_j(y) \right\|_p}{\left\| \vec{f}_j(y) \right\|_q} \right)$, $j = 1, \ldots, m$, with $j \neq \rho$, are all
Lebesgue integrable.

Proof By Remark 23.50 and (23.124) in particular. ∎

References

1. G.A. Anastassiou, *Fractional Differentiation Inequalities*, Research Monograph (Springer, New York, 2009)
2. G.A. Anastassiou, On right fractional calculus. Chaos, Solitons Fractals **42**, 365–376 (2009)
3. G.A. Anastassiou, Balanced fractional Opial inequalities. Chaos Solitons Fractals **42**(3), 1523–1528 (2009)
4. G.A. Anastassiou, Fractional representation formulae and right fractional inequalities. Math. Comput. Model. **54**(11–12), 3098–3115 (2011)
5. G.A. Anastassiou, Vectorial inequalities for integral operators involving ratios of functions and convexity. Discontinuity, Nonlinearity Complex. **1**(3), 279–304 (2012)
6. G.A. Anastassiou, Vectorial Hardy type fractional inequalities. Bull. Tbilisi Int. Cent. Math. Inf. **16**(2), 21–57 (2012)
7. M. Andric, J.E. Pecaric, I. Peric, A multiple Opial type inequality due to Fink for the Riemann-Liouville fractional derivatives. Submitted (2012)
8. M. Andric, J.E. Pecaric, I. Peric, Composition identities for the Caputo fractional derivatives and applications to Opial-type inequalities. Submitted (2012)
9. D. Baleanu, K. Diethelm, E. Scalas, J.J. Trujillo, *Fractional Calculus Models and Numerical Methods*, Series on Complexity: Nonlinearity and Chaos (World Scientific, Singapore, 2012)
10. J.A. Canavati, The Riemann-Liouville integral. Nieuw Archief Voor Wiskunde **5**(1), 53–75 (1987)
11. K. Diethelm, *The Analysis of Fractional Differential Equations*, Lecture Notes in Mathematics, vol. 2004, 1st edn. (Springer, New York, 2010)
12. A.M.A. El-Sayed, M. Gaber, On the finite Caputo and finite Riesz derivatives. Electron. J. Theor. Phys. **3**(12), 81–95 (2006)
13. R. Gorenflo, F. Mainardi, Essentials of Fractional Calculus. Maphysto Center (2000). http://www.maphysto.dk/oldpages/events/LevyCAC2000/MainardiNotes/fm2k0a.ps
14. G.D. Handley, J.J. Koliha, J. Pečarić, Hilbert-Pachpatte type integral inequalities for fractional derivatives. Fract. Calc. Appl. Anal. **4**(1), 37–46 (2001)
15. H.G. Hardy, Notes on some points in the integral calculus. Messenger Math. **47**(10), 145–150 (1918)
16. S. Iqbal, K. Krulic, J. Pecaric, On an inequality of H.G. Hardy. J. Inequalities Appl. **2010**, Article ID 264347, 23 p
17. A.A. Kilbas, H.M. Srivastava, J.J. Trujillo, *Theory and Applications of Fractional Differential Equations*, North-Holland Mathematics Studies, vol. 204 (Elsevier, New York, 2006)
18. S.G. Samko, A.A. Kilbas, O.I. Marichev, *Fractional Integral and Derivatives: Theory and Applications* (Gordon and Breach Science Publishers, Yverdon, 1993)

Chapter 24
About Vectorial Splitting Rational L_p Inequalities for Integral Operators

Here we present L_p, $p > 1$, vectorial integral inequalitites for products of multivariate convex and increasing functions applied to vectors of ratios of functions. As applications we derive a wide range of vectorial fractional inequalities of Hardy type. They involve the left and right Erdelyi-Kober fractional integrals and left and right mixed Riemann-Liouville fractional multiple integrals. Also we give vectorial inequalities for Riemann-Liouville, Caputo, Canavati radial fractional derivatives. Some inequalities are of exponential type. It follows [6].

24.1 Introduction

Let $(\Omega_1, \Sigma_1, \mu_1)$ and $(\Omega_2, \Sigma_2, \mu_2)$ be measure spaces with positive σ-finite measures, and let $k : \Omega_1 \times \Omega_2 \to \mathbb{R}$ be a nonnegative measurable function, $k(x, \cdot)$ measurable on Ω_2 and

$$K(x) = \int_{\Omega_2} k(x, y) \, d\mu_2(y), \quad x \in \Omega_1. \tag{24.1}$$

We suppose that $K(x) > 0$ a.e. on Ω_1, and by a weight function (shortly: a weight), we mean a nonnegative measurable function on the actual set. Let the measurable functions $g_i : \Omega_1 \to \mathbb{R}$, $i = 1, \ldots, n$, with the representation

$$g_i(x) = \int_{\Omega_2} k(x, y) f_i(y) \, d\mu_2(y), \tag{24.2}$$

where $f_i : \Omega_2 \to \mathbb{R}$ are measurable functions, $i = 1, \ldots, n$.

Denote by $\overrightarrow{x} = x := (x_1, \ldots, x_n) \in \mathbb{R}^n$, $\overrightarrow{g} := (g_1, \ldots, g_n)$ and $\overrightarrow{f} := (f_1, \ldots, f_n)$.

© Springer International Publishing Switzerland 2016
G.A. Anastassiou, *Intelligent Comparisons: Analytic Inequalities*,
Studies in Computational Intelligence 609,
DOI 10.1007/978-3-319-21121-3_24

We consider here $\Phi : \mathbb{R}^n_+ \to \mathbb{R}$ a convex function, which is increasing per coordinate, i.e. if $x_i \leq y_i, i = 1, \ldots, n$, then

$$\Phi(x_1, \ldots, x_n) \leq \Phi(y_1, \ldots, y_n).$$

In [5] we proved:

Theorem 24.1 *Let u be a weight function on Ω_1, and k, K, g_i, f_i, $i = 1, \ldots, n \in \mathbb{N}$, and Φ defined as above. Assume that the function $x \to u(x) \frac{k(x,y)}{K(x)}$ is integrable on Ω_1 for each fixed $y \in \Omega_2$. Define v on Ω_2 by*

$$v(y) := \int_{\Omega_1} u(x) \frac{k(x,y)}{K(x)} d\mu_1(x) < \infty. \qquad (24.3)$$

Then

$$\int_{\Omega_1} u(x) \Phi\left(\frac{|g_1(x)|}{K(x)}, \ldots, \frac{|g_n(x)|}{K(x)}\right) d\mu_1(x) \leq$$

$$\int_{\Omega_2} v(y) \Phi(|f_1(y)|, \ldots, |f_n(y)|) d\mu_2(y), \qquad (24.4)$$

under the assumptions:

(i) f_i, $\Phi(|f_1|, \ldots, |f_n|)$, are $k(x,y) d\mu_2(y)$ -integrable, μ_1 -a.e. in $x \in \Omega_1$, for all $i = 1, \ldots, n$,

(ii) $v(y) \Phi(|f_1(y)|, \ldots, |f_n(y)|)$ is μ_2 -integrable.

Notation 24.2 From now on we may write

$$\overrightarrow{g}(x) = \int_{\Omega_2} k(x,y) \overrightarrow{f}(y) d\mu_2(y), \qquad (24.5)$$

which means

$$(g_1(x), \ldots, g_n(x)) = \left(\int_{\Omega_2} k(x,y) f_1(y) d\mu_2(y), \ldots, \int_{\Omega_2} k(x,y) f_n(y) d\mu_2(y)\right). \qquad (24.6)$$

Similarly, we may write

$$|\overrightarrow{g}(x)| = \left|\int_{\Omega_2} k(x,y) \overrightarrow{f}(y) d\mu_2(y)\right|, \qquad (24.7)$$

and we mean

$$(|g_1(x)|, \ldots, |g_n(x)|) =$$

$$\left(\left|\int_{\Omega_2} k(x,y) f_1(y) d\mu_2(y)\right|, \ldots, \left|\int_{\Omega_2} k(x,y) f_n(y) d\mu_2(y)\right|\right). \qquad (24.8)$$

We also can write that

$$\left|\overrightarrow{g}(x)\right| \le \int_{\Omega_2} k(x, y) \left|\overrightarrow{f}(y)\right| d\mu_2(y), \qquad (24.9)$$

and we mean the fact that

$$|g_i(x)| \le \int_{\Omega_2} k(x, y) |f_i(y)| d\mu_2(y), \qquad (24.10)$$

for all $i = 1, \ldots, n$, etc.

Notation 24.3 Next let $(\Omega_1, \Sigma_1, \mu_1)$ and $(\Omega_2, \Sigma_2, \mu_2)$ be measure spaces with positive σ-finite measures, and let $k_j : \Omega_1 \times \Omega_2 \to \mathbb{R}$ be a nonnegative measurable function, $k_j(x, \cdot)$ measurable on Ω_2 and

$$K_j(x) = \int_{\Omega_2} k_j(x, y) d\mu_2(y), \quad x \in \Omega_1, j = 1, \ldots, m. \qquad (24.11)$$

We suppose that $K_j(x) > 0$ a.e. on Ω_1.

Let the measurable functions $g_{ji} : \Omega_1 \to \mathbb{R}$ with the representation

$$g_{ji}(x) = \int_{\Omega_2} k_j(x, y) f_{ji}(y) d\mu_2(y), \qquad (24.12)$$

where $f_{ji} : \Omega_2 \to \mathbb{R}$ are measurable functions, $i = 1, \ldots, n$ and $j = 1, \ldots, m$.

Denote the function vectors $\overrightarrow{g_j} := (g_{j1}, g_{j2}, \ldots, g_{jn})$ and $\overrightarrow{f_j} := (f_{j1}, \ldots, f_{jn})$, $j = 1, \ldots, m$.

We say $\overrightarrow{f_j}$ is integrable with respect to measure μ, iff all f_{ji} are integrable with respect to μ.

We also consider here $\Phi_j : \mathbb{R}^n_+ \to \mathbb{R}_+$, $j = 1, \ldots, m$, convex functions that are increasing per coordinate. Again u is a weight function on Ω_1.

In [7], we also proved the following general result:

Theorem 24.4 *All as in Notation 24.3. Assume that the functions ($j = 1, 2, \ldots, m \in \mathbb{N}$)*

$$x \mapsto \left(\frac{u(x) k_j(x, y)}{K_j(x)}\right)$$

are integrable on Ω_1 for each fixed $y \in \Omega_2$. Define u_j on Ω_2 by

$$u_j(y) := \int_{\Omega_1} u(x) \frac{k_j(x, y)}{K_j(x)} d\mu_1(x) < \infty. \qquad (24.13)$$

Let $p_j > 1 : \sum_{j=1}^{m} \dfrac{1}{p_j} = 1$. Let the functions $\Phi_j : \mathbb{R}_+^n \to \mathbb{R}_+$, $j = 1, \ldots, m$, be convex and increasing per coordinate.

Then

$$\int_{\Omega_1} u(x) \prod_{j=1}^{m} \Phi_j \left(\left| \dfrac{\overrightarrow{g_j}(x)}{K_j(x)} \right| \right) d\mu_1(x) \leq \qquad (24.14)$$

$$\prod_{j=1}^{m} \left(\int_{\Omega_2} u_j(y) \, \Phi_j \left(\left| \overrightarrow{f_j}(y) \right| \right)^{p_j} d\mu_2(y) \right)^{\frac{1}{p_j}},$$

under the assumptions:

(i) $\overrightarrow{f_j}, \Phi_j \left(\left| \overrightarrow{f_j} \right| \right)^{p_j}$ are both $k_j(x, y) \, d\mu_2(y)$ -integrable, μ_1 -a.e. in $x \in \Omega_1$, $j = 1, \ldots, m$,

(ii) $u_j \Phi_j \left(\left| \overrightarrow{f_j} \right| \right)^{p_j}$ is μ_2 -integrable, $j = 1, \ldots, m$.

When $k(x, y) := k_1(x, y) = k_2(x, y) = \cdots = k_m(x, y)$, then $K(x) := K_1(x) = K_2(x) = \cdots = K_m(x)$, we get by Theorem 24.4 the following:

Corollary 24.5 ([7]) *Assume that the function*

$$x \mapsto \left(\dfrac{u(x) k(x, y)}{K(x)} \right)$$

is integrable on Ω_1 for each fixed $y \in \Omega_2$. Define U on Ω_2 by

$$U(y) := \int_{\Omega_1} u(x) \dfrac{k(x, y)}{K(x)} d\mu_1(x) < \infty. \qquad (24.15)$$

Let $p_j > 1 : \sum_{j=1}^{m} \dfrac{1}{p_j} = 1$. Let the functions $\Phi_j : \mathbb{R}_+^n \to \mathbb{R}_+$, $j = 1, \ldots, m$, be convex and increasing per coordinate.

Then

$$\int_{\Omega_1} u(x) \prod_{j=1}^{m} \Phi_j \left(\left| \dfrac{\overrightarrow{g_j}(x)}{K(x)} \right| \right) d\mu_1(x) \leq \qquad (24.16)$$

$$\prod_{j=1}^{m} \left(\int_{\Omega_2} U(y) \, \Phi_j \left(\left| \overrightarrow{f_j}(y) \right| \right)^{p_j} d\mu_2(y) \right)^{\frac{1}{p_j}},$$

under the assumptions:

(i) $\overrightarrow{f_j}, \Phi_j \left(\left|\overrightarrow{f_j}\right|\right)^{p_j}$ are both $k(x, y) \, d\mu_2(y)$ -integrable, μ_1 -a.e. in $x \in \Omega_1$, for all $j = 1, \ldots, m$,

(ii) $U\Phi_j \left(\left|\overrightarrow{f_j}\right|\right)^{p_j}$ is μ_2 -integrable, $j = 1, \ldots, m$.

We make

Remark 24.6 Following Notation 24.3, let $F_j : \Omega_2 \to \mathbb{R} \cup \{\pm\infty\}$ be measurable functions, $j = 1, \ldots, m$, with $0 < F_j(y) < \infty$, a.e. on Ω_2. In (24.11) we replace $k_j(x, y)$ by $k_j(x, y) F_j(y)$, $j = 1, \ldots, m$, and we have the modified $K_j(x)$ as

$$L_j(x) := \int_{\Omega_2} k_j(x, y) F_j(y) \, d\mu_2(y), \quad x \in \Omega_1. \tag{24.17}$$

We assume $L_j(x) > 0$ a.e. on Ω_1.

As new $\overrightarrow{f_j}$ we consider now $\gamma_j := \frac{\overrightarrow{f_j}}{F_j}, j = 1, \ldots, m$, where $\overrightarrow{f_j} = (f_{j1}, \ldots, f_{jn})$; $\overrightarrow{\gamma_j} = \left(\frac{f_{j1}}{F_j}, \ldots, \frac{f_{jn}}{F_j}\right)$.

Notice that

$$g_{ji}(x) = \int_{\Omega_2} k_j(x, y) f_{ji}(y) \, d\mu_2(y) \tag{24.18}$$

$$= \int_{\Omega_2} \left(k_j(x, y) F_j(y)\right) \left(\frac{f_{ji}(y)}{F_j(y)}\right) d\mu_2(y),$$

$x \in \Omega_1$, all $j = 1, \ldots, m$; $i = 1, \ldots, n$.

So we can write

$$\overrightarrow{g_j}(x) = \int_{\Omega_2} \left(k_j(x, y) F_j(y)\right) \overrightarrow{\gamma_j}(y) \, d\mu_2(y), \quad j = 1, \ldots, m. \tag{24.19}$$

In this chapter we get first general L_p, $p > 1$, results by applying Theorem 24.4 and Corollary 24.5, for $\left(\overrightarrow{\gamma_j}, \overrightarrow{g_j}\right)$, $j = 1, \ldots, m$, and on other various important settings, then we give wide applications to Fractional Calculus. This chapter is inspired by [7, 15–17].

24.2 Main Results

We present our first main result

Theorem 24.7 *Here all as in Notation 24.3 and Remark 24.6. Assume that the functions ($j = 1, 2, \ldots, m \in \mathbb{N}$)*

$$x \mapsto \left(\frac{u(x) k_j(x, y) F_j(y)}{K_j(x)}\right)$$

are integrable on Ω_1, for almost each fixed $y \in \Omega_2$. Define W_j on Ω_2 by

$$W_j(y) := \left(\int_{\Omega_1} \frac{u(x) k_j(x, y)}{K_j(x)} d\mu_1(x) \right) F_j(y) < \infty, \qquad (24.20)$$

a.e. on Ω_2.

Let $p_j > 1 : \sum_{j=1}^{m} \frac{1}{p_j} = 1$. *Let the functions* $\Phi_j : \mathbb{R}_+^n \to \mathbb{R}_+$, $j = 1, \ldots, m$, *be convex and increasing per coordinate.*

Then

$$\int_{\Omega_1} u(x) \prod_{j=1}^{m} \Phi_j \left(\left| \frac{\overrightarrow{g_j}(x)}{L_j(x)} \right| \right) d\mu_1(x) \leq \qquad (24.21)$$

$$\prod_{j=1}^{m} \left(\int_{\Omega_2} W_j(y) \Phi_j \left(\left| \frac{\overrightarrow{f_j}(y)}{F_j(y)} \right| \right)^{p_j} d\mu_2(y) \right)^{\frac{1}{p_j}},$$

under the assumptions:

(i) $\frac{\overrightarrow{f_j}}{F_j}$, $\Phi_j \left(\frac{|\overrightarrow{f_j}|}{F_j} \right)^{p_j}$ *are both $k_j(x, y) F_j(y) d\mu_2(y)$ -integrable, μ_1 -a.e. in $x \in \Omega_1$, $j = 1, \ldots, m$,*

(ii) $W_j \Phi_j \left(\frac{|\overrightarrow{f_j}|}{F_j} \right)^{p_j}$ *is μ_2 -integrable, $j = 1, \ldots, m$.*

Proof Direct application of Theorem 24.4 for $(\overrightarrow{\gamma_j}, \overrightarrow{g_j})$, $j = 1, \ldots, m$. ∎

When $k(x, y) := k_1(x, y) = \cdots = k_m(x, y)$, then $K(x) := K_1(x) = \cdots = K_m(x)$, and we take $F(y) := F_1(y) = \cdots = F_m(y)$; $L(x) := L_1(x) = \cdots = L_m(x)$, we get by Theorem 24.7 the following:

Corollary 24.8 *Assume that the function*

$$x \mapsto \left(\frac{u(x) k(x, y)}{K(x)} F(y) \right)$$

is integrable on Ω_1, for almost each fixed $y \in \Omega_2$. Define W on Ω_2 by

$$W(y) := \left(\int_{\Omega_1} \frac{u(x) k(x, y)}{K(x)} d\mu_1(x) \right) F(y) < \infty, \qquad (24.22)$$

a.e. on Ω_2.

Let $p_j > 1 : \sum_{j=1}^{m} \frac{1}{p_j} = 1$. *Let the functions* $\Phi_j : \mathbb{R}_+^n \to \mathbb{R}_+$, $j = 1, \ldots, m$, *be convex and increasing per coordinate.*

Then

$$\int_{\Omega_1} u(x) \prod_{j=1}^{m} \Phi_j \left(\left\| \frac{\overrightarrow{g_j}(x)}{L(x)} \right\| \right) d\mu_1(x) \leq \qquad (24.23)$$

$$\prod_{j=1}^{m} \left(\int_{\Omega_2} W(y) \Phi_j \left(\left\| \frac{\overrightarrow{f_j}(y)}{F(y)} \right\| \right)^{p_j} d\mu_2(y) \right)^{\frac{1}{p_j}},$$

under the assumptions:

(i) $\frac{\overrightarrow{f_j}}{F}, \Phi_j \left(\frac{\left\| \overrightarrow{f_j} \right\|}{F} \right)^{p_j}$ *are both* $k(x, y) F(y) d\mu_2(y)$ *-integrable,* μ_1 *-a.e. in* $x \in \Omega_1$,

   for all $j = 1, \ldots, m$,

(ii) $W \Phi_j \left(\frac{\left\| \overrightarrow{f_j} \right\|}{F} \right)^{p_j}$ *is* μ_2 *-integrable,* $j = 1, \ldots, m$.

We give the general applications.

Theorem 24.9 *All as in Theorem 24.7,* $p \geq 1$. *It holds*

$$\int_{\Omega_1} \left(\frac{u(x)}{\prod\limits_{j=1}^{m} L_j(x)} \right) \prod_{j=1}^{m} \left(\sum_{i=1}^{n} |g_{ji}(x)|^p \right)^{\frac{1}{p}} d\mu_1(x) \leq \qquad (24.24)$$

$$\prod_{j=1}^{m} \left(\int_{\Omega_2} \frac{W_j(y)}{(F_j(y))^{p_j}} \left(\sum_{i=1}^{n} |f_{ji}(y)|^p \right)^{\frac{p_j}{p}} d\mu_2(y) \right)^{\frac{1}{p_j}},$$

under the assumptions:

(i) $\left(\frac{\left\| \overrightarrow{f_j} \right\|_p}{F_j} \right)^{p_j}$ *is* $k_j(x, y) F_j(y) d\mu_2(y)$ *-integrable,* μ_1 *-a.e. in* $x \in \Omega_1$, $j = 1, \ldots, m$,

(ii) $W_j \left(\frac{\left\| \overrightarrow{f_j} \right\|_p}{F_j} \right)^{p_j}$ *is* μ_2-*integrable,* $j = 1, \ldots, m$.

Proof By Theorem 24.7 with $\Phi_j(x_1, \ldots, x_n) = \left\| \overrightarrow{x} \right\|_p$, $\overrightarrow{x} = (x_1, \ldots, x_n)$, $j = 1, \ldots, m$. ∎

We furthermore give

Theorem 24.10 *All as in Theorem 24.7. It holds*

$$\int_{\Omega_1} u(x) \prod_{j=1}^{m} \ln \left(\sum_{i=1}^{n} e^{\left(\frac{|g_{ji}(x)|}{L_j(x)}\right)} \right) d\mu_1(x) \leq \qquad (24.25)$$

$$\prod_{j=1}^{m} \left(\int_{\Omega_2} W_j(y) \left(\ln \left(\sum_{i=1}^{n} e^{\left(\frac{|f_{ji}(y)|}{F_j(y)}\right)} \right) \right)^{p_j} d\mu_2(y) \right)^{\frac{1}{p_j}},$$

under the assumptions:

(i) $\dfrac{\vec{f_j}}{F_j}, \left(\ln \left(\sum_{i=1}^{n} e^{\left(\frac{|f_{ji}(y)|}{F_j(y)}\right)} \right) \right)^{p_j}$ *are both* $k_j(x,y) F_j(y) d\mu_2(y)$ *-integrable,* μ_1

-a.e. in $x \in \Omega_1$, $j = 1, \ldots, m$,

(ii) $W_j(y) \left(\ln \left(\sum_{i=1}^{n} e^{\left(\frac{|f_{ji}(y)|}{F_j(y)}\right)} \right) \right)^{p_j}$ *is* μ_2*-integrable,* $j = 1, \ldots, m$.

Proof Apply Theorem 24.7 with $\Phi_j(x_1, \ldots, x_n) = \ln \left(\sum_{i=1}^{n} e^{x_i} \right)$, $x_i \geq 0$, $j = 1, \ldots, m$. ∎

We make

Remark 24.11 Following Notation 24.3 and Remark 24.6, we choose as

$$F_j(y) = \left\| \vec{f_j}(y) \right\|_{\infty} := \max\{|f_{j1}(y)|, \ldots, |f_{jn}(y)|\}, \qquad (24.26)$$

or

$$F_j(y) = \left\| \vec{f_j}(y) \right\|_{q} := \left(\sum_{i=1}^{n} |f_{ji}(y)|^q \right)^{\frac{1}{q}}, \quad q \geq 1, \qquad (24.27)$$

$y \in \Omega_2$, which are measurable function, $j = 1, \ldots, m$. We assume that

$$0 < \left\| \vec{f_j}(y) \right\|_{q} < \infty, \text{ a.e. on } \Omega_2, \qquad (24.28)$$

$j = 1, \ldots, m$; $1 \leq q \leq \infty$ fixed.

Now in (24.11) we replace $k_j(x,y)$ by $k_j(x,y) \left\| \vec{f_j}(y) \right\|_{q}$, $j = 1, \ldots, m$, and the new modified $K_j(x)$ is

$$L_{jq}(x) := \int_{\Omega_2} k_j(x,y) \left\| \vec{f}_j(y) \right\|_q d\mu_2(y), \ x \in \Omega_1, \ 1 \le q \le \infty. \quad (24.29)$$

We assume $L_{jq}(x) > 0$ a.e. on Ω_1.

Here we assume that the functions $(j = 1, \ldots, m)$

$$x \mapsto \left(u(x) \frac{k_j(x,y) \left\| \vec{f}_j(y) \right\|_q}{L_{jq}(x)} \right) \quad (24.30)$$

are integrable on Ω_1, for almost each fixed $y \in \Omega_2$.

Define W_{jq} on Ω_2 by

$$W_{jq}(y) := \left(\int_{\Omega_1} \frac{u(x) k_j(x,y)}{L_{jq}(x)} d\mu_1(x) \right) \left\| \vec{f}_j(y) \right\|_q < \infty, \quad (24.31)$$

a.e. on Ω_2.

Let $p_j > 1 : \sum_{j=1}^{m} \frac{1}{p_j} = 1$. Also Notation 24.3 is in place.

We give

Theorem 24.12 *Here all as in Remark 24.11, $p \ge 1$. It holds*

$$\int_{\Omega_1} u(x) \prod_{j=1}^{m} \left(\frac{\left\| \vec{g}_j(x) \right\|_p}{L_{jq}(x)} \right) d\mu_1(x) \le \quad (24.32)$$

$$\prod_{j=1}^{m} \left(\int_{\Omega_2} W_{jq}(y) \left(\frac{\left\| \vec{f}_j(y) \right\|_p}{\left\| \vec{f}_j(y) \right\|_q} \right)^{p_j} d\mu_2(y) \right)^{\frac{1}{p_j}},$$

under the assumptions:

(i) $\left(\frac{\left\| \vec{f}_j(y) \right\|_p}{\left\| \vec{f}_j(y) \right\|_q} \right)^{p_j}$ *is* $k_j(x,y) \left\| \vec{f}_j(y) \right\|_q d\mu_2(y)$ *-integrable, μ_1 -a.e. in $x \in \Omega_1$, $j = 1, \ldots, m$,*

(ii) $W_{jq} \left(\frac{\left\| \vec{f}_j \right\|_p}{\left\| \vec{f}_j \right\|_q} \right)^{p_j}$ *is* μ_2-integrable, $j = 1, \ldots, m$.

Proof By Theorem 24.9. ■

We also give

Theorem 24.13 *Here all as in Remark 24.11. It holds*

$$\int_{\Omega_1} u(x) \prod_{j=1}^{m} \ln \left(\sum_{i=1}^{n} e^{\left(\frac{|g_{ji}(x)|}{L_{jq}(x)} \right)} \right) d\mu_1(x) \leq \qquad (24.33)$$

$$\prod_{j=1}^{m} \left(\int_{\Omega_2} W_{jq}(y) \left(\ln \left(\sum_{i=1}^{n} e^{\left(\frac{|f_{ji}(y)|}{\|\vec{f}_j(y)\|_q} \right)} \right) \right)^{p_j} d\mu_2(y) \right)^{\frac{1}{p_j}}$$

under the assumptions:

(i) $\dfrac{\vec{f}_j(y)}{\left\|\vec{f}_j(y)\right\|_q}, \left(\ln \left(\sum\limits_{i=1}^{n} e^{\left(\frac{|f_{ji}(y)|}{\|\vec{f}_j(y)\|_q} \right)} \right) \right)^{p_j}$ *are* $k_j(x, y) \left\|\vec{f}_j(y)\right\|_q d\mu_2(y)$-*integrable,*

μ_1-*a.e. in* $x \in \Omega_1$, $j = 1, \ldots, m$,

(ii) $W_{jq}(y) \left(\ln \left(\sum\limits_{i=1}^{n} e^{\left(\frac{|f_{ji}(y)|}{\|\vec{f}_j(y)\|_q} \right)} \right) \right)^{p_j}$ *is* μ_2-*integrable,* $j = 1, \ldots, m$.

Proof By Theorem 24.10. ∎

We need the following:

Let $a < b$, $a, b \in \mathbb{R}$. By $C^N([a, b])$, we denote the space of all functions on $[a, b]$ which have continuous derivatives up to order N, and $AC([a, b])$ is the space of all absolutely continuous functions on $[a, b]$. By $AC^N([a, b])$, we denote the space of all functions g with $g^{(N-1)} \in AC([a, b])$. For any $\alpha \in \mathbb{R}$, we denote by $[\alpha]$ the integral part of α (the integer k satisfying $k \leq \alpha < k + 1$), and $\lceil \alpha \rceil$ is the ceiling of α ($\min\{n \in \mathbb{N}, n \geq \alpha\}$). By $L_1(a, b)$, we denote the space of all functions integrable on the interval (a, b), and by $L_\infty(a, b)$ the set of all functions measurable and essentially bounded on (a, b). Clearly, $L_\infty(a, b) \subset L_1(a, b)$.

We need

Definition 24.14 ([21]) Let (a, b), $0 \leq a < b < \infty$; $\alpha, \sigma > 0$. We consider the left and right-sided fractional integrals of order α as follows:

(1) for $\eta > -1$, we define

$$\left(I_{a+;\sigma,\eta}^{\alpha} f \right)(x) = \frac{\sigma x^{-\sigma(\alpha+\eta)}}{\Gamma(\alpha)} \int_{a}^{x} \frac{t^{\sigma\eta+\sigma-1} f(t) \, dt}{(x^\sigma - t^\sigma)^{1-\alpha}}, \qquad (24.34)$$

(2) for $\eta > 0$, we define

$$\left(I_{b-;\sigma,\eta}^{\alpha} f\right)(x) = \frac{\sigma x^{\sigma\eta}}{\Gamma(\alpha)} \int_{x}^{b} \frac{t^{\sigma(1-\eta-\alpha)-1} f(t)\, dt}{(t^{\sigma} - x^{\sigma})^{1-\alpha}}. \tag{24.35}$$

These are the Erdélyi-Kober type fractional integrals.

We make

Remark 24.15 Regarding (24.34) we have all

$$k_j(x, y) = k_1(x, y) := \frac{\sigma x^{-\sigma(\alpha+\eta)}}{\Gamma(\alpha)} \chi_{(a,x]}(y) \frac{y^{\sigma\eta+\sigma-1}}{(x^{\sigma} - y^{\sigma})^{1-\alpha}}, \tag{24.36}$$

$x, y \in (a, b)$, χ stands for the characteristic function, $j = 1, \ldots, m$.
In this case

$$g_{ji}(x) = \left(I_{a+;\sigma,\eta}^{\alpha} f_{ji}\right)(x) = \int_{a}^{b} k_1(x, y)\, f_{ji}(y)\, dy, \tag{24.37}$$

$i = 1, \ldots, n$ and $j = 1, \ldots, m$.
We assume that

$$0 < \left\| \overrightarrow{f_j}(y) \right\|_q < \infty, \text{ a.e. on } (a, b), \tag{24.38}$$

$j = 1, \ldots, m$; $1 \leq q \leq \infty$ fixed.
We further assume

$$\left(I_{a+;\sigma,\eta}^{\alpha} \left(\left\| \overrightarrow{f_j} \right\|_q\right)\right)(x) > 0, \text{ a.e. on } (a, b), \tag{24.39}$$

$j = 1, \ldots, m$.

We also make

Remark 24.16 Regarding (24.35) we have all

$$k_j(x, y) = k_2(x, y) := \frac{\sigma x^{\sigma\eta}}{\Gamma(\alpha)} \chi_{[x,b)}(y) \frac{y^{\sigma(1-\eta-\sigma)-1}}{(y^{\sigma} - x^{\sigma})^{1-\alpha}}, \tag{24.40}$$

$x, y \in (a, b)$, $j = 1, \ldots, m$.
In this case

$$g_{ji}(x) = \left(I_{b-;\sigma,\eta}^{\alpha} f_{ji}\right)(x) = \int_{a}^{b} k_2(x, y)\, f_{ji}(y)\, dy, \tag{24.41}$$

$j = 1, \ldots, m$, $i = 1, \ldots, n$.

We assume again (24.38). And we further assume that

$$\left(I_{b-;\sigma,\eta}^{\alpha} \left(\left\| \vec{f}_j \right\|_q \right) \right)(x) > 0, \text{ a.e. on } (a, b), \tag{24.42}$$

$j = 1, \dots, m$.

We give

Theorem 24.17 *Here all as in Remark 24.15. Assume that the functions ($j = 1, \dots, m \in \mathbb{N}$)*

$$x \mapsto \left(\frac{u(x) k_1(x, y) \left\| \vec{f}_j(y) \right\|_q}{I_{a+;\sigma,\eta}^{\alpha} \left(\left\| \vec{f}_j \right\|_q \right)(x)} \right) \tag{24.43}$$

are integrable on (a, b), for almost each fixed $y \in (a, b)$.
Define ψ_{jq}^{+} on (a, b) by

$$\psi_{jq}^{+}(y) := \left(\int_a^b \frac{u(x) k_1(x, y)\, dx}{I_{a+;\sigma,\eta}^{\alpha} \left(\left\| \vec{f}_j \right\|_q \right)(x)} \right) \left\| \vec{f}_j(y) \right\|_q < \infty, \tag{24.44}$$

a.e. on (a, b).
Let $p_j > 1 : \sum\limits_{j=1}^m \frac{1}{p_j} = 1; \ p \geq 1, 1 \leq q \leq \infty$, fixed.
Then

$$\int_a^b u(x) \prod_{j=1}^m \left(\frac{\left\| I_{a+;\sigma,\eta}^{\alpha} \left(\vec{f}_j \right)(x) \right\|_p}{\left(I_{a+;\sigma,\eta}^{\alpha} \left(\left\| \vec{f}_j \right\|_q \right) \right)(x)} \right) dx \leq \tag{24.45}$$

$$\prod_{i=1}^m \left(\int_a^b \psi_{jq}^{+}(y) \left(\frac{\left\| \vec{f}_j(y) \right\|_p}{\left\| \vec{f}_j(y) \right\|_q} \right)^{p_j} dy \right)^{\frac{1}{p_j}},$$

under the assumptions:

(i) $\left(\frac{\left\| \vec{f}_j(y) \right\|_p}{\left\| \vec{f}_j(y) \right\|_q} \right)^{p_j}$ *is $k_1(x, y) \left\| \vec{f}_j(y) \right\|_q dy$ -integrable, a.e. in $x \in (a, b)$, $j = 1, \dots, m$,*

(ii) $\psi_{jq}^{+} \left(\frac{\left\| \vec{f}_j \right\|_p}{\left\| \vec{f}_j \right\|_q} \right)^{p_j}$ *is integrable, $j = 1, \dots, m$.*

Proof By Theorem 24.12. ∎

We also give

Theorem 24.18 *All here as in Theorem 24.17. It holds*

$$\int_a^b u(x) \prod_{j=1}^m \ln \left(\sum_{i=1}^n e^{\frac{\left|\left(I_{a+;\sigma,\eta}^\alpha f_{ji}\right)(x)\right|}{\left(I_{a+;\sigma,\eta}^\alpha \left(\|\vec{f_j}\|_q\right)\right)(x)}} \right) dx \le \qquad (24.46)$$

$$\prod_{j=1}^m \left(\int_a^b \psi_{jq}^+(y) \left(\ln \left(\sum_{i=1}^n e^{\frac{|f_{ji}(y)|}{\|\vec{f_j}(y)\|_q}} \right) \right)^{p_j} dy \right)^{\frac{1}{p_j}},$$

under the assumptions:

(i) $\dfrac{\vec{f_j}(y)}{\|\vec{f_j}(y)\|_q}$, $\left(\ln \left(\displaystyle\sum_{i=1}^n e^{\frac{|f_{ji}(y)|}{\|\vec{f_j}(y)\|_q}} \right) \right)^{p_j}$ *are both* $k_1(x,y) \|\vec{f_j}(y)\|_q dy$ *-integrable,*

  a.e. in $x \in (a,b)$, $j = 1, \ldots, m$,

(ii) $\psi_{jq}^+(y) \left(\ln \left(\displaystyle\sum_{i=1}^n e^{\frac{|f_{ji}(y)|}{\|\vec{f_j}(y)\|_q}} \right) \right)^{p_j}$ *is integrable,* $j = 1, \ldots, m$.

Proof By Theorem 24.13. ∎

We continue with

Theorem 24.19 *Here all as in Remark 24.16. Assume that the functions* $(j = 1, \ldots, m \in \mathbb{N})$

$$x \mapsto \left(\frac{u(x) k_2(x,y) \|\vec{f_j}(y)\|_q}{\left(I_{b-;\sigma,\eta}^\alpha \left(\|\vec{f_j}\|_q \right) \right)(x)} \right) \qquad (24.47)$$

are integrable on (a,b), *for almost each fixed* $y \in (a,b)$.
  Define ψ_{jq}^- *on* (a,b) *by*

$$\psi_{jq}^-(y) := \left(\int_a^b \frac{u(x) k_2(x,y) dx}{\left(I_{b-;\sigma,\eta}^\alpha \left(\|\vec{f_j}\|_q \right) \right)(x)} \right) \|\vec{f_j}(y)\|_q < \infty, \qquad (24.48)$$

a.e. on (a,b).

Let $p_j > 1 : \sum_{j=1}^{m} \dfrac{1}{p_j} = 1; \ p \ge 1, \ 1 \le q \le \infty, \ fixed.$

Then

$$\int_a^b u(x) \prod_{j=1}^{m} \left(\frac{\left\| I_{b-;\sigma,\eta}^{\alpha} \left(\vec{f}_j \right)(x) \right\|_p}{\left(I_{b-;\sigma,\eta}^{\alpha} \left(\left\| \vec{f}_j \right\|_q \right) \right)(x)} \right) dx \le \qquad (24.49)$$

$$\prod_{j=1}^{m} \left(\int_a^b \psi_{jq}^-(y) \left(\frac{\left\| \vec{f}_j(y) \right\|_p}{\left\| \vec{f}_j(y) \right\|_q} \right)^{p_j} dy \right)^{\frac{1}{p_j}},$$

under the assumptions:

(i) $\left(\dfrac{\left\| \vec{f}_j(y) \right\|_p}{\left\| \vec{f}_j(y) \right\|_q} \right)^{p_j}$ *is $k_2(x,y) \left\| \vec{f}_j(y) \right\|_q dy$ -integrable, a.e. in $x \in (a,b)$, $j = 1, \ldots, m$,*

(ii) $\psi_{jq}^- \left(\dfrac{\left\| \vec{f}_j(y) \right\|_p}{\left\| \vec{f}_j(y) \right\|_q} \right)^{p_j}$ *is integrable, $j = 1, \ldots, m$.*

Proof By Theorem 24.12. ■

We also give

Theorem 24.20 *All here as in Theorem 24.19. It holds*

$$\int_a^b u(x) \prod_{j=1}^{m} \ln \left(\sum_{i=1}^{n} e^{\frac{\left| \left(I_{b-;\sigma,\eta}^{\alpha} f_{ji} \right)(x) \right|}{\left(I_{b-;\sigma,\eta}^{\alpha} \left\| \vec{f}_j \right\|_q \right)(x)}} \right) dx \le \qquad (24.50)$$

$$\prod_{j=1}^{m} \left(\int_a^b \psi_{jq}^-(y) \left(\ln \left(\sum_{i=1}^{n} e^{\frac{\left| f_{ji}(y) \right|}{\left\| \vec{f}_j(y) \right\|_q}} \right) \right)^{p_j} dy \right)^{\frac{1}{p_j}},$$

under the assumptions:

(i) $\dfrac{\vec{f}_j(y)}{\left\| \vec{f}_j(y) \right\|_q}, \ \left(\ln \left(\sum_{i=1}^{n} e^{\frac{\left| f_{ji}(y) \right|}{\left\| \vec{f}_j(y) \right\|_q}} \right) \right)^{p_j}$ *are both $k_2(x,y) \left\| \vec{f}_j(y) \right\|_q dy$ -integrable,*

a.e. in $x \in (a,b)$, $j = 1, \ldots, m$,

(ii) $\psi_{jq}^{-}(y)\left(\ln\left(\sum_{i=1}^{n}e^{\frac{|f_{ji}(y)|}{\|\vec{f}_{j}(y)\|_{q}}}\right)\right)^{p_j}$ is integrable, $j = 1, \ldots, m$.

Proof By Theorem 24.13. ∎

We make

Remark 24.21 Let $\prod_{i=1}^{N}(a_i, b_i) \subset \mathbb{R}^N$, $N > 1$, $a_i < b_i$, $a_i, b_i \in \mathbb{R}$. Let $\alpha_i > 0$,

$i = 1, \ldots, N$; $f \in L_1\left(\prod_{i=1}^{N}(a_i, b_i)\right)$, and set $a = (a_1, \ldots, a_N)$, $b = (b_1, \ldots, b_N)$,

$\alpha = (\alpha_1, \ldots, \alpha_N)$, $x = (x_1, \ldots, x_N)$, $t = (t_1, \ldots, t_N)$.

We define the left mixed Riemann-Liouville fractional multiple integral of order α (see also [19]):

$$\left(I_{a+}^{\alpha}f\right)(x) := \frac{1}{\prod_{i=1}^{N}\Gamma(\alpha_i)} \int_{a_1}^{x_1} \cdots \int_{a_N}^{x_N} \prod_{i=1}^{N}(x_i - t_i)^{\alpha_i - 1} f(t_1, \ldots, t_N) \, dt_1, \ldots, dt_N,$$

(24.51)

with $x_i > a_i$, $i = 1, \ldots, N$.

We also define the right mixed Riemann-Liouville fractional multiple integral of order α (see also [16]):

$$\left(I_{b-}^{\alpha}f\right)(x) := \frac{1}{\prod_{i=1}^{N}\Gamma(\alpha_i)} \int_{x_1}^{b_1} \cdots \int_{x_N}^{b_N} \prod_{i=1}^{N}(t_i - x_i)^{\alpha_i - 1} f(t_1, \ldots, t_N) \, dt_1 \ldots dt_N,$$

(24.52)

with $x_i < b_i$, $i = 1, \ldots, N$.

Notice $I_{a+}^{\alpha}(|f|)$, $I_{b-}^{\alpha}(|f|)$ are finite if $f \in L_\infty\left(\prod_{i=1}^{N}(a_i, b_i)\right)$.

One can rewrite (24.51) and (24.52) as follows:

$$\left(I_{a+}^{\alpha}f\right)(x) = \frac{1}{\prod_{i=1}^{N}\Gamma(\alpha_i)} \int_{\prod_{i=1}^{N}(a_i, b_i)} \chi_{\prod_{i=1}^{N}(a_i, x_i]}(t) \prod_{i=1}^{N}(x_i - t_i)^{\alpha_i - 1} f(t) \, dt,$$

(24.53)

with $x_i > a_i$, $i = 1, \ldots, N$,

and

$$
\left(I_{b-}^{\alpha} f\right)(x) = \frac{1}{\prod_{i=1}^{N} \Gamma(\alpha_i)} \int_{\prod_{i=1}^{N} (a_i, b_i)} \chi_{\prod_{i=1}^{N} [x_i, b_i)} (t) \prod_{i=1}^{N} (t_i - x_i)^{\alpha_i - 1} f(t)\, dt,
$$

(24.54)

with $x_i < b_i$, $i = 1, \dots, N$.

The corresponding $k(x, y)$ for I_{a+}^{α}, I_{b-}^{α} are

$$
k_{a+}(x, y) = \frac{1}{\prod_{i=1}^{N} \Gamma(\alpha_i)} \chi_{\prod_{i=1}^{N} (a_i, x_i]}(y) \prod_{i=1}^{N} (x_i - y_i)^{\alpha_i - 1},
$$

(24.55)

$\forall\, x, y \in \prod_{i=1}^{N} (a_i, b_i)$,

and

$$
k_{b-}(x, y) = \frac{1}{\prod_{i=1}^{N} \Gamma(\alpha_i)} \chi_{\prod_{i=1}^{N} [x_i, b_i)}(y) \prod_{i=1}^{N} (y_i - x_i)^{\alpha_i - 1},
$$

(24.56)

$\forall\, x, y \in \prod_{i=1}^{N} (a_i, b_i)$.

We make

Remark 24.22 In the case of (24.51) we choose

$$
g_{jr}(x) = \left(I_{a+}^{\alpha} f_{jr}\right)(x)
$$

(24.57)

$j = 1, \dots, m$, $r = 1, \dots, n$, $\forall\, x \in \prod_{i=1}^{N} (a_i, b_i)$.

Here $f_{jr} : \prod_{i=1}^{N} (a_i, b_i) \to \mathbb{R}$, $j = 1, \dots, m$, $r = 1, \dots, n$, are Lebesgue measurable functions and $I_{a+}^{\alpha}\left(\left|\overrightarrow{f_j}\right|\right)$ is finite a.e.; $\overrightarrow{f_j} = (f_{j1}, f_{j2}, \dots, f_{jn})$.

We assume that $0 < \left\| \vec{f_j} (y) \right\|_q < \infty$, a.e. on $\prod_{i=1}^{N} (a_i, b_i)$, $1 \leq q \leq \infty$ and that

$\left(I_{a+}^{\alpha} \left(\left\| \vec{f_j} \right\|_q \right) \right) (x) > 0$, a.e. on $\prod_{i=1}^{N} (a_i, b_i)$, $j = 1, \ldots, m$.

Let also $p \geq 1$.

We also make

Remark 24.23 In the case of (24.52) we choose

$$g_{jr} (x) = \left(I_{b-}^{\alpha} f_{jr} \right) (x),\tag{24.58}$$

$j = 1, \ldots, m, r = 1, \ldots, n, \forall\, x \in \prod_{i=1}^{N} (a_i, b_i)$.

Here $f_{jr} : \prod_{i=1}^{N} (a_i, b_i) \to \mathbb{R}$, $j = 1, \ldots, m, r = 1, \ldots, n$, are Lebesgue measurable functions and $I_{b-}^{\alpha} \left(\left| \vec{f_j} \right| \right)$ is finite a.e.; $\vec{f_j} = (f_{j1}, f_{j2}, \ldots, f_{jn})$.

We assume that $0 < \left\| \vec{f_j} (y) \right\|_q < \infty$, a.e. on $\prod_{i=1}^{N} (a_i, b_i)$, $1 \leq q \leq \infty$ and that

$\left(I_{b-}^{\alpha} \left(\left\| \vec{f_j} \right\|_q \right) \right) (x) > 0$, a.e. on $\prod_{i=1}^{N} (a_i, b_i)$, $j = 1, \ldots, m$.

Let also $p \geq 1$.

We present

Theorem 24.24 *Here all as in Remark 24.22. Assume that the functions* $(j = 1, \ldots, m \in \mathbb{N})$

$$x \mapsto \left(\frac{u (x) \, k_{a+} (x, y) \left\| \vec{f_j} (y) \right\|_q}{\left(I_{a+}^{\alpha} \left(\left\| \vec{f_j} \right\|_q \right) \right) (x)} \right) \tag{24.59}$$

are integrable on $\prod_{i=1}^{N} (a_i, b_i)$, *for almost each fixed* $y \in \prod_{i=1}^{N} (a_i, b_i)$.

Define T_{jq}^{+} on $\prod_{i=1}^{N} (a_i, b_i)$ by

$$T_{jq}^{+}(y) := \left(\int_{\prod_{i=1}^{N}(a_i, b_i)} \frac{u(x)\, k_{a+}(x, y)\, dx}{\left(I_{a+}^{\alpha} \left(\left\| \vec{f_j} \right\|_q \right) \right)(x)} \right) \left\| \vec{f_j}(y) \right\|_q < \infty, \qquad (24.60)$$

a.e. on $\prod_{i=1}^{N} (a_i, b_i)$.

Let $p_j > 1 : \sum_{j=1}^{m} \dfrac{1}{p_j} = 1$. Then

$$\int_{\prod_{i=1}^{N}(a_i, b_i)} u(x) \prod_{j=1}^{m} \left(\frac{\left\| I_{a+}^{\alpha} \left(\vec{f_j} \right)(x) \right\|_p}{\left(I_{a+}^{\alpha} \left(\left\| \vec{f_j} \right\|_q \right) \right)(x)} \right) dx \leq \qquad (24.61)$$

$$\prod_{j=1}^{m} \left(\int_{\prod_{i=1}^{N}(a_i, b_i)} T_{jq}^{+}(y) \left(\frac{\left\| \vec{f_j}(y) \right\|_p}{\left\| \vec{f_j}(y) \right\|_q} \right)^{p_j} dy \right)^{\frac{1}{p_j}},$$

under the assumptions:

(i) $\left(\dfrac{\left\| \vec{f_j}(y) \right\|_p}{\left\| \vec{f_j}(y) \right\|_q} \right)^{p_j}$ *is $k_{a+}(x, y) \left\| \vec{f_j}(y) \right\|_q dy$ -integrable, a.e. in $x \in \prod_{i=1}^{N} (a_i, b_i)$, $j = 1, \ldots, m$,*

(ii) $T_{jq}^{+} \left(\dfrac{\left\| \vec{f_j} \right\|_p}{\left\| \vec{f_j} \right\|_q} \right)^{p_j}$ *is integrable, $j = 1, \ldots, m$.*

Proof By Theorem 24.12. ∎

We also give

Theorem 24.25 *Here all as in Theorem 24.24. It holds*

$$\int_{\prod_{i=1}^{N}(a_i,b_i)} u(x) \prod_{j=1}^{m} \ln\left(\sum_{r=1}^{n} e^{\frac{|(I_{a+}^{\alpha} f_{jr})(x)|}{(I_{a+}^{\alpha}(\|\vec{f_j}\|_q))(x)}}\right) dx \leq \qquad (24.62)$$

$$\prod_{j=1}^{m}\left(\int_{\prod_{i=1}^{N}(a_i,b_i)} T_{jq}^{+}(y)\left(\ln\left(\sum_{r=1}^{n} e^{\frac{|f_{jr}(y)|}{\|\vec{f_j}(y)\|_q}}\right)\right)^{p_j} dy\right)^{\frac{1}{p_j}},$$

under the assumptions:

(i) $\dfrac{\vec{f_j}(y)}{\|\vec{f_j}(y)\|_q}$, $\left(\ln\left(\sum_{r=1}^{n} e^{\frac{|f_{jr}(y)|}{\|\vec{f_j}(y)\|_q}}\right)\right)^{p_j}$ *are both* $k_{a+}(x,y)\left\|\vec{f_j}(y)\right\|_q dy$ *-integr-*

able, a.e. in $x \in \prod_{i=1}^{N}(a_i,b_i)$, $j=1,\ldots,m$,

(ii) $T_{jq}^{+}(y)\left(\ln\left(\sum_{r=1}^{n} e^{\frac{|f_{jr}(y)|}{\|\vec{f_j}(y)\|_q}}\right)\right)^{p_j}$ *is integrable,* $j=1,\ldots,m$.

Proof By Theorem 24.13. ∎

We continue with

Theorem 24.26 *Here all as in Remark 24.23. Assume that the functions* ($j = 1,\ldots,m \in \mathbb{N}$)

$$x \mapsto \left(\frac{u(x)k_{b-}(x,y)\left\|\vec{f_j}(y)\right\|_q}{\left(I_{b-}^{\alpha}\left(\|\vec{f_j}\|_q\right)\right)(x)}\right) \qquad (24.63)$$

are integrable on $\prod_{i=1}^{N}(a_i,b_i)$, *for almost each fixed* $y \in \prod_{i=1}^{N}(a_i,b_i)$.

Define T_{jq}^{-} *on* $\prod_{i=1}^{N}(a_i,b_i)$ *by*

$$T_{jq}^{-}(y) := \left(\int_{\prod_{i=1}^{N}(a_i,b_i)} \frac{u(x)k_{b-}(x,y)\,dx}{\left(I_{b-}^{\alpha}\left(\|\vec{f_j}\|_q\right)\right)(x)}\right)\left\|\vec{f_j}(y)\right\|_q < \infty, \qquad (24.64)$$

a.e. on $\prod_{i=1}^{N} (a_i, b_i)$.

Let $p_j > 1 : \sum_{j=1}^{m} \frac{1}{p_j} = 1$. *Then*

$$
\int_{\prod_{i=1}^{N} (a_i, b_i)} u(x) \prod_{j=1}^{m} \left(\frac{\left\| I_{b-}^{\alpha} \left(\vec{f}_j \right) (x) \right\|_p}{\left(I_{b-}^{\alpha} \left(\left\| \vec{f}_j \right\|_q \right) \right) (x)} \right) dx \leq \tag{24.65}
$$

$$
\prod_{j=1}^{m} \left(\int_{\prod_{i=1}^{N} (a_i, b_i)} T_{jq}^{-} (y) \left(\frac{\left\| \vec{f}_j (y) \right\|_p}{\left\| \vec{f}_j (y) \right\|_q} \right)^{p_j} dy \right)^{\frac{1}{p_j}},
$$

under the assumptions:

(i) $\left(\frac{\left\| \vec{f}_j(y) \right\|_p}{\left\| \vec{f}_j(y) \right\|_q} \right)^{p_j}$ *is* $k_{b-}(x, y) \left\| \vec{f}_j (y) \right\|_q dy$ *-integrable, a.e. in* $x \in \prod_{i=1}^{N} (a_i, b_i)$, $j = 1, \ldots, m$,

(ii) $T_{jq}^{-} \left(\frac{\left\| \vec{f}_j \right\|_p}{\left\| \vec{f}_j \right\|_q} \right)^{p_j}$ *is integrable,* $j = 1, \ldots, m$.

Proof By Theorem 24.12. ∎

We also give

Theorem 24.27 *Here all as in Theorem 24.26. It holds*

$$
\int_{\prod_{i=1}^{N} (a_i, b_i)} u(x) \prod_{j=1}^{m} \ln \left(\sum_{r=1}^{n} e^{\frac{\left| \left(I_{b-}^{\alpha} f_{jr} \right)(x) \right|}{\left(I_{b-}^{\alpha} \left(\left\| \vec{f}_j \right\|_q \right) \right)(x)}} \right) dx \leq \tag{24.66}
$$

$$
\prod_{j=1}^{m} \left(\int_{\prod_{i=1}^{N} (a_i, b_i)} T_{jq}^{-} (y) \left(\ln \left(\sum_{r=1}^{n} e^{\frac{\left| f_{jr}(y) \right|}{\left\| \vec{f}_j(y) \right\|_q}} \right) \right)^{p_j} dy \right)^{\frac{1}{p_j}},
$$

under the assumptions:

(i) $\dfrac{\overrightarrow{f_j}(y)}{\left\|\overrightarrow{f_j}(y)\right\|_q}$, $\left(\ln \left(\displaystyle\sum_{r=1}^{n} e^{\frac{|f_{jr}(y)|}{\left\|\overrightarrow{f_j}(y)\right\|_q}} \right) \right)^{p_j}$ are both $k_{b-}(x,y)\left\|\overrightarrow{f_j}(y)\right\|_q dy$ -integra-

ble, a.e. in $x \in \displaystyle\prod_{i=1}^{N} (a_i, b_i)$, $j = 1, \ldots, m$,

(ii) $T_{jq}^{-}(y) \left(\ln \left(\displaystyle\sum_{r=1}^{n} e^{\frac{|f_{jr}(y)|}{\left\|\overrightarrow{f_j}(y)\right\|_q}} \right) \right)^{p_j}$ is integrable, $j = 1, \ldots, m$.

Proof By Theorem 24.13. ∎

We make

Remark 24.28 Next we follow [17] and our introduction.

Let $(\Omega_1, \Sigma_1, \mu_1)$ and $(\Omega_2, \Sigma_2, \mu_2)$ be measure spaces with positive σ-finite measures, and let $k : \Omega_1 \times \Omega_2 \to \mathbb{R}$ be a nonnegative measurable function, $k(x, \cdot)$ measurable on Ω_2 and

$$K(x) = \int_{\Omega_2} k(x, y) \, d\mu_2(y), \text{ for any } x \in \Omega_1. \qquad (24.67)$$

We assume $K(x) > 0$, a.e. on Ω_1, and the weight functions are nonnegative functions on the related set. We consider measurable functions $g_i : \Omega_1 \to \mathbb{R}$ with the representation

$$g_i(x) = \int_{\Omega_2} k(x, y) f_i(y) \, d\mu_2(y),$$

where $f_i : \Omega_2 \to \mathbb{R}$ are measurable, $i = 1, \ldots, n$. Here u stands for a weight function on Ω_1. So we follow Notation 24.3 for $j = m = 1$. We write here $\overrightarrow{g} = (g_1, \ldots, g_n)$, $\overrightarrow{f} = (f_1, \ldots, f_n)$.

We assume that

$$0 < \left\|\overrightarrow{f}(y)\right\|_q < \infty, \text{ a.e. on } (a, b), \qquad (24.68)$$

$1 \le q \le \infty$ fixed.
Let

$$L_q(x) := \int_{\Omega_2} k(x, y) \left\|\overrightarrow{f}(y)\right\|_q d\mu_2(y), \ x \in \Omega_1, \qquad (24.69)$$

$1 \le q \le \infty$ fixed.
We assume $L_q(x) > 0$ a.e. on Ω_1.

We further assume that the function

$$
x \mapsto \left(\frac{u(x) k(x, y) \left\| \vec{f}(y) \right\|_q}{L_q(x)} \right) \tag{24.70}
$$

is integrable on Ω_1, for almost each fixed $y \in \Omega_2$.

Define W_q on Ω_2 by

$$
W_q(y) := \left(\int_{\Omega_1} \frac{u(x) k(x, y)}{L_q(x)} d\mu_1(x) \right) \left\| \vec{f}(y) \right\|_q < \infty, \tag{24.71}
$$

a.e. on Ω_2.

Let

$$
\vec{\gamma} := \left(\frac{f_1}{\left\| \vec{f}(y) \right\|_q}, \frac{f_2}{\left\| \vec{f}(y) \right\|_q}, \dots, \frac{f_n}{\left\| \vec{f}(y) \right\|_q} \right), \tag{24.72}
$$

i.e. $\vec{\gamma} = \dfrac{\vec{f}}{\left\| \vec{f}(y) \right\|_q}$.

Here $\Phi : \mathbb{R}_+^n \to \mathbb{R}$ is a convex and increasing per coordinate function.
We apply Theorem 24.1 for $(\vec{\gamma}, \vec{g})$ to obtain:

Theorem 24.29 *Let all here as in Remark 24.28. It holds*

$$
\int_{\Omega_1} u(x) \Phi \left(\frac{\left| \vec{g}(x) \right|}{L_q(x)} \right) d\mu_1(x) \leq \tag{24.73}
$$

$$
\int_{\Omega_2} W_q(y) \Phi \left(\frac{\left| \vec{f}(y) \right|}{\left\| \vec{f}(y) \right\|_q} \right) d\mu_2(y),
$$

under the assumptions:

(i) $\dfrac{\vec{f}(y)}{\left\| \vec{f}(y) \right\|_q}$, $\Phi \left(\dfrac{\left| \vec{f}(y) \right|}{\left\| \vec{f}(y) \right\|_q} \right)$ *are* $k(x, y) \left\| \vec{f}(y) \right\|_q d\mu_2(y)$ *-integrable,* μ_1 *-a.e. in* $x \in \Omega_1$,

(ii) $W_q(y) \Phi \left(\dfrac{\left| \vec{f}(y) \right|}{\left\| \vec{f}(y) \right\|_q} \right)$ *is* μ_2*-integrable.*

Proof By Theorem 24.1. ∎

Next we deal with the spherical shell

Background 24.30 *We need:*

Let $N \geq 2$, $S^{N-1} := \{x \in \mathbb{R}^N : |x| = 1\}$ *the unit sphere on* $\mathbb{R}^N$, *where* $|\cdot|$ *stands for the Euclidean norm in* $\mathbb{R}^N$. *Also denote the ball* $B(0, R) := \{x \in \mathbb{R}^N : |x| < R\} \subseteq \mathbb{R}^N$, $R > 0$, *and the spherical shell*

$$A := B(0, R_2) - \overline{B(0, R_1)}, \quad 0 < R_1 < R_2. \tag{24.74}$$

For the following see [20, pp. 149–150], and [22, pp. 87–88].

For $x \in \mathbb{R}^N - \{0\}$ *we can write uniquely* $x = r\omega$, *where* $r = |x| > 0$, *and* $\omega = \frac{x}{r} \in S^{N-1}$, $|\omega| = 1$.

Clearly here

$$\mathbb{R}^N - \{0\} = (0, \infty) \times S^{N-1}, \tag{24.75}$$

and

$$\overline{A} = [R_1, R_2] \times S^{N-1}. \tag{24.76}$$

We will be using

Theorem 24.31 [1, p. 322] *Let* $f : A \to \mathbb{R}$ *be a Lebesgue integrable function. Then*

$$\int_A f(x)\, dx = \int_{S^{N-1}} \left(\int_{R_1}^{R_2} f(r\omega)\, r^{N-1} dr \right) d\omega. \tag{24.77}$$

So we are able to write an integral on the shell in polar form using the polar coordinates (r, ω).

We need

Definition 24.32 [1, p. 458] *Let* $\nu > 0$, $n := [\nu]$, $\alpha := \nu - n$, $f \in C^n(\overline{A})$, *and* A *is a spherical shell. Assume that there exists function* $\frac{\partial_{R_1}^{\nu} f(x)}{\partial r^{\nu}} \in C(\overline{A})$, *given by*

$$\frac{\partial_{R_1}^{\nu} f(x)}{\partial r^{\nu}} := \frac{1}{\Gamma(1-\alpha)} \frac{\partial}{\partial r} \left(\int_{R_1}^{r} (r-t)^{-\alpha} \frac{\partial^n f(t\omega)}{\partial r^n}\, dt \right), \tag{24.78}$$

where $x \in \overline{A}$; *that is* $x = r\omega$, $r \in [R_1, R_2]$, $\omega \in S^{N-1}$.

We call $\frac{\partial_{R_1}^{\nu} f}{\partial r^{\nu}}$ *the left radial Canavati-type fractional derivative of* f *of order* ν. *If* $\nu = 0$, *then set* $\frac{\partial_{R_1}^{\nu} f(x)}{\partial r^{\nu}} := f(x)$.

Based on [1, p. 288], and [4] we have

Lemma 24.33 *Let* $\gamma \geq 0$, $m := [\gamma]$, $\nu > 0$, $n := [\nu]$, *with* $0 \leq \gamma < \nu$. *Let* $f \in C^n(\overline{A})$ *and there exists* $\frac{\partial_{R_1}^{\nu} f(x)}{\partial r^{\nu}} \in C(\overline{A})$, $x \in \overline{A}$, A *a spherical shell. Further* *assume that* $\dfrac{\partial^j f(R_1 \omega)}{\partial r^j} = 0$, $j = m, m+1, \ldots, n-1$, $\forall \omega \in S^{N-1}$. *Then there exists* $\dfrac{\partial^\gamma_{R_1} f(x)}{\partial r^\gamma} \in C\left(\overline{A}\right)$ *such that*

$$\frac{\partial^\gamma_{R_1} f(x)}{\partial r^\gamma} = \frac{\partial^\gamma_{R_1} f(r\omega)}{\partial r^\gamma} = \frac{1}{\Gamma(\nu - \gamma)} \int_{R_1}^r (r-t)^{\nu-\gamma-1} \frac{\partial^\nu_{R_1} f(t\omega)}{\partial r^\nu} dt, \quad (24.79)$$

$\forall \omega \in S^{N-1}$; *all* $R_1 \le r \le R_2$, *indeed* $f(r\omega) \in C^\gamma_{R_1}([R_1, R_2])$, $\forall \omega \in S^{N-1}$.

We make

Remark 24.34 In the settings and assumptions of Theorem 24.29 and Lemma 24.33 we have

$$k(r, t) = k_*(r, t) := \frac{1}{\Gamma(\nu - \gamma)} \chi_{[R_1, r]}(t) (r-t)^{\nu-\gamma-1}, \quad (24.80)$$

$r, t \in [R_1, R_2]$.

Let f_i, $i = 1, \ldots, n$, as in Lemma 24.33. Denote $\overrightarrow{f} = (f_1, \ldots, f_n)$. Assume that $\left\| \dfrac{\partial^\nu_{R_1} \overrightarrow{f}(y)}{\partial r^\nu} \right\|_q > 0$ on $\overline{A}$, $1 \le q \le \infty$ fixed. We take

$$g_i(r, \omega) := \frac{\partial^\gamma_{R_1} f_i(r\omega)}{\partial r^\gamma}, \quad i = 1, \ldots, n, \quad (24.81)$$

for every $(r, \omega) \in [R_1, R_2] \times S^{N-1}$.

Let

$$C_q(x) = C_q(r\omega) := \int_{R_1}^{R_2} k_*(r, t) \left\| \frac{\partial^\nu_{R_1} \overrightarrow{f}(t\omega)}{\partial r^\nu} \right\|_q dt, \quad (24.82)$$

$r \in [R_1, R_2]$, $1 \le q \le \infty$ fixed, $\omega \in S^{N-1}$.

We have that $C_q(r\omega) > 0$ on $(R_1, R_2]$, for every $\omega \in S^{N-1}$.
Here we choose $u(r, \omega) = C_q(r\omega)$.
So that the function

$$\lambda_q(t, \omega) := \left(\int_{R_1}^{R_2} k_*(r, t) \, dr \right) \left\| \frac{\partial^\nu_{R_1} \overrightarrow{f}(t\omega)}{\partial r^\nu} \right\|_q = \quad (24.83)$$

$$\left(\int_t^{R_2} \frac{1}{\Gamma(\nu - \gamma)} (r-t)^{\nu-\gamma-1} \, dr \right) \left\| \frac{\partial^\nu_{R_1} \overrightarrow{f}(t\omega)}{\partial r^\nu} \right\|_q =$$

$$\frac{(R_2 - t)^{\nu - \gamma}}{\Gamma(\nu - \gamma + 1)} \left\| \frac{\partial_{R_1}^{\nu} \overrightarrow{f(t\omega)}}{\partial r^{\nu}} \right\|_q < \infty, \tag{24.84}$$

for every $t \in [R_1, R_2]$.

Let $\Gamma : \mathbb{R}_+^n \to \mathbb{R}_+$ convex and increasing per coordinate. By Theorem 24.29 we get that

$$\int_{R_1}^{R_2} C_q(r\omega) \Phi \left(\frac{\left| \frac{\partial_{R_1}^{\gamma} \overrightarrow{f(r\omega)}}{\partial r^{\gamma}} \right|}{C_q(r\omega)} \right) dr \leq \tag{24.85}$$

$$\int_{R_1}^{R_2} \frac{(R_2 - t)^{\nu - \gamma}}{\Gamma(\nu - \gamma + 1)} \left\| \frac{\partial_{R_1}^{\nu} \overrightarrow{f(t\omega)}}{\partial r^{\nu}} \right\|_q \Phi \left(\frac{\left| \frac{\partial_{R_1}^{\nu} \overrightarrow{f(t\omega)}}{\partial r^{\nu}} \right|}{\left\| \frac{\partial_{R_1}^{\nu} \overrightarrow{f(t\omega)}}{\partial r^{\nu}} \right\|_q} \right) dt \leq$$

$$\frac{(R_2 - R_1)^{\nu - \gamma}}{\Gamma(\nu - \gamma + 1)} \int_{R_1}^{R_2} \left\| \frac{\partial_{R_1}^{\nu} \overrightarrow{f(t\omega)}}{\partial r^{\nu}} \right\|_q \Phi \left(\frac{\left| \frac{\partial_{R_1}^{\nu} \overrightarrow{f(t\omega)}}{\partial r^{\nu}} \right|}{\left\| \frac{\partial_{R_1}^{\nu} \overrightarrow{f(t\omega)}}{\partial r^{\nu}} \right\|_q} \right) dt, \tag{24.86}$$

true for every $\omega \in S^{N-1}$.

Here we have $R_1 \leq r \leq R_2$, and $R_1^{N-1} \leq r^{N-1} \leq R_2^{N-1}$, and $R_2^{1-N} \leq r^{1-N} \leq R_1^{1-N}$. So by (24.85)–(24.86) and $r^{N-1}r^{1-N} = 1$, we have

$$\int_{R_1}^{R_2} C_q(r\omega) \Phi \left(\frac{\left| \frac{\partial_{R_1}^{\gamma} \overrightarrow{f(r\omega)}}{\partial r^{\gamma}} \right|}{C_q(r\omega)} \right) r^{N-1} dr \leq$$

$$\frac{(R_2 - R_1)^{\nu - \gamma}}{\Gamma(\nu - \gamma + 1)} \left(\frac{R_2}{R_1} \right)^{N-1} \int_{R_1}^{R_2} \left\| \frac{\partial_{R_1}^{\nu} \overrightarrow{f(r\omega)}}{\partial r^{\nu}} \right\|_q \Phi \left(\frac{\left| \frac{\partial_{R_1}^{\nu} \overrightarrow{f(r\omega)}}{\partial r^{\nu}} \right|}{\left\| \frac{\partial_{R_1}^{\nu} \overrightarrow{f(r\omega)}}{\partial r^{\nu}} \right\|_q} \right) r^{N-1} dr, \tag{24.87}$$

$\forall\, \omega \in S^{N-1}$. Therefore it holds

$$
\int_{S^{N-1}} \left(\int_{R_1}^{R_2} C_q\left(r\omega\right) \Phi \left(\frac{\left| \frac{\partial_{R_1}^{\gamma} \overrightarrow{f(r\omega)}}{\partial r^{\gamma}} \right|}{C_q\left(r\omega\right)} \right) r^{N-1} dr \right) d\omega \leq \tag{24.88}
$$

$$
\frac{(R_2 - R_1)^{\nu - \gamma}}{\Gamma\left(\nu - \gamma + 1\right)} \left(\frac{R_2}{R_1} \right)^{N-1} \cdot
$$

$$
\int_{S^{N-1}} \left(\int_{R_1}^{R_2} \left\| \frac{\partial_{R_1}^{\nu} \overrightarrow{f}\,(r\omega)}{\partial r^{\nu}} \right\|_q \Phi \left(\frac{\left| \frac{\partial_{R_1}^{\nu} \overrightarrow{f(r\omega)}}{\partial r^{\nu}} \right|}{\left\| \frac{\partial_{R_1}^{\nu} \overrightarrow{f(r\omega)}}{\partial r^{\nu}} \right\|_q} \right) r^{N-1} dr \right) d\omega.
$$

Using Theorem 24.31 we derive

$$
\int_A C_q\left(x\right) \Phi \left(\frac{\left| \frac{\partial_{R_1}^{\gamma} \overrightarrow{f(x)}}{\partial r^{\gamma}} \right|}{C_q\left(x\right)} \right) dx \leq \tag{24.89}
$$

$$
\frac{(R_2 - R_1)^{\nu - \gamma}}{\Gamma\left(\nu - \gamma + 1\right)} \left(\frac{R_2}{R_1} \right)^{N-1} \int_A \left\| \frac{\partial_{R_1}^{\nu} \overrightarrow{f}\,(x)}{\partial r^{\nu}} \right\|_q \Phi \left(\frac{\left| \frac{\partial_{R_1}^{\nu} \overrightarrow{f(x)}}{\partial r^{\nu}} \right|}{\left\| \frac{\partial_{R_1}^{\nu} \overrightarrow{f(x)}}{\partial r^{\nu}} \right\|_q} \right) dx.
$$

We have proved the following result.

Theorem 24.35 Let $\overrightarrow{f} = (f_1, \ldots, f_n)$, with f_i, $i = 1, \ldots, n$, as in Lemma 24.33, $\nu > \gamma \geq 0$. Assume that $\left\| \frac{\partial_{R_1}^{\nu} \overrightarrow{f(y)}}{\partial r^{\nu}} \right\|_q > 0$, $\forall\, y \in \overline{A}$, $1 \leq q \leq \infty$ fixed.
Let $\Phi : \mathbb{R}_+^n \to \mathbb{R}_+$ be convex and increasing per coordinate.
Define

$$
C_q\left(x\right) = C_q\left(r\omega\right) := \frac{1}{\Gamma\left(\nu - \gamma\right)} \int_{R_1}^{r} (r - t)^{\nu - \gamma - 1} \left\| \frac{\partial_{R_1}^{\nu} \overrightarrow{f}\,(t\omega)}{\partial r^{\nu}} \right\|_q dt, \tag{24.90}
$$

$\forall x \in \overline{A}$. *Then*

$$\int_A C_q(x) \, \Phi\left(\frac{\left|\frac{\partial_{R_1}^\gamma \overrightarrow{f(x)}}{\partial r^\gamma}\right|}{C_q(x)}\right) dx \leq$$

$$\frac{(R_2 - R_1)^{\nu-\gamma}}{\Gamma(\nu - \gamma + 1)}\left(\frac{R_2}{R_1}\right)^{N-1}\int_A \left\|\frac{\partial_{R_1}^\nu \overrightarrow{f(x)}}{\partial r^\nu}\right\|_q \Phi\left(\frac{\left|\frac{\partial_{R_1}^\nu \overrightarrow{f(x)}}{\partial r^\nu}\right|}{\left\|\frac{\partial_{R_1}^\nu \overrightarrow{f(x)}}{\partial r^\nu}\right\|_q}\right) dx. \qquad (24.91)$$

Corollary 24.36 *All as in Theorem 24.35. It holds*

$$\int_A C_q(x) \ln\left(\sum_{i=1}^n e^{\frac{\left|\frac{\partial_{R_1}^\gamma f_i(x)}{\partial r^\gamma}\right|}{C_q(x)}}\right) dx \leq \qquad (24.92)$$

$$\frac{(R_2 - R_1)^{\nu-\gamma}}{\Gamma(\nu - \gamma + 1)}\left(\frac{R_2}{R_1}\right)^{N-1}\int_A \left\|\frac{\partial_{R_1}^\nu \overrightarrow{f(x)}}{\partial r^\nu}\right\|_q \ln\left(\sum_{i=1}^n e^{\frac{\left|\frac{\partial_{R_1}^\nu f_i(x)}{\partial r^\nu}\right|}{\left\|\frac{\partial_{R_1}^\nu \overrightarrow{f(x)}}{\partial r^\nu}\right\|_q}}\right) dx.$$

Proof By Theorem 24.35, when $\Phi(x) = \ln\left(\sum_{i=1}^n e^{x_i}\right)$, $x_i \geq 0$. ∎

Similar results can be produced for the right radial Canavati type fractional derivative. We omit this treatment.

We make

Remark 24.37 (from [1], p. 460) Here we denote $\lambda_{\mathbb{R}^N}(x) \equiv dx$ the Lebesgue measure on $\mathbb{R}^N$, $N \geq 2$, and by $\lambda_{S^{N-1}}(\omega) = d\omega$ the surface measure on S^{N-1}, where $\mathcal{B}_X$ stands for the Borel class on space X. Define the measure R_N on $((0, \infty), \mathcal{B}_{(0,\infty)})$ by

$$R_N(B) = \int_B r^{N-1} dr, \text{ any } B \in \mathcal{B}_{(0,\infty)}.$$

Now let $F \in L_1(A) = L_1([R_1, R_2] \times S^{N-1})$.

Call

$$K(F) := \{\omega \in S^{N-1} : F(\cdot \omega) \notin L_1 ([R_1, R_2], \mathcal{B}_{[R_1, R_2]}, R_N)\}. \tag{24.93}$$

We get, by Fubini's theorem and [22], pp. 87–88, that

$$\lambda_{S^{N-1}} (K(F)) = 0.$$

Of course

$$\theta(F) := [R_1, R_2] \times K(F) \subset A,$$

and

$$\lambda_{\mathbb{R}^N} (\theta(F)) = 0.$$

Above $\lambda_{S^{N-1}}$ is defined as follows: let $A \subset S^{N-1}$ be a Borel set, and let

$$\widetilde{A} := \{ru : 0 < r < 1, u \in A\} \subset \mathbb{R}^N;$$

we define

$$\lambda_{S^{N-1}} (A) := N \lambda_{\mathbb{R}^N} (\widetilde{A}).$$

We have that

$$\lambda_{S^{N-1}} \left(S^{N-1}\right) = \frac{2\pi^{\frac{N}{2}}}{\Gamma\left(\frac{N}{2}\right)},$$

the surface area of S^{N-1}.

See also [20, pp. 149–150], [22, pp. 87–88] and [1], p. 320.

Following [1, p. 466] we define the left Riemann-Liouville radial fractional derivative next.

Definition 24.38 Let $\beta > 0$, $m := [\beta] + 1$, $F \in L_1 (A)$, and A is the spherical shell. We define

$$\frac{\overline{\partial}^{\beta}_{R_1} F(x)}{\partial r^{\beta}} := \begin{cases} \frac{1}{\Gamma(m-\beta)} \left(\frac{\partial}{\partial r}\right)^m \int_{R_1}^r (r-t)^{m-\beta-1} F(t\omega) \, dt, \\ \qquad \text{for } \omega \in S^{N-1} - K(F), \\ 0, \quad \text{for } \omega \in K(F), \end{cases} \tag{24.94}$$

where $x = r\omega \in A$, $r \in [R_1, R_2]$, $\omega \in S^{N-1}$; $K(F)$ as in (24.93).
  If $\beta = 0$, define

$$\frac{\overline{\partial}^{\beta}_{R_1} F(x)}{\partial r^{\beta}} := F(x).$$

Definition 24.39 ([1], p. 327) We say that $f \in L_1 (a, w)$, $a < w$; $a, w \in \mathbb{R}$ has an L_∞ left Riemann-Liouville fractional derivative $\overline{D}^{\beta}_a f$ ($\beta > 0$) in $[a, w]$, iff

(1) $\overline{D}_a^{\beta-k} f \in C([a, w])$, $k = 1, \ldots, m := [\beta] + 1$;
(2) $\overline{D}_a^{\beta-1} f \in AC([a, w])$; and
(3) $\overline{D}_a^{\beta} f \in L_\infty(a, w)$.

Define $\overline{D}_a^0 f := f$ and $\overline{D}_a^{-\delta} f := I_{a+}^{\delta} f$, if $0 < \delta \leq 1$; here $I_{a+}^{\delta} f$ is the usual left univariate Riemann-Liouville fractional integral of f.

We need the following representation result.

Theorem 24.40 ([1, p. 331]) *Let* $\beta > \alpha > 0$ *and* $F \in L_1(A)$. *Assume that* $\dfrac{\overline{\partial}_{R_1}^{\beta} F(x)}{\partial r^{\beta}} \in L_\infty(A)$. *Further assume that* $\overline{D}_{R_1}^{\beta} F(r\omega)$ *takes real values for almost all* $r \in [R_1, R_2]$, *for each* $\omega \in S^{N-1}$, *and for these* $\left|\overline{D}_{R_1}^{\beta} F(r\omega)\right| \leq M_1$ *for some* $M_1 > 0$. *For each* $\omega \in S^{N-1} - K(F)$, *we assume that* $F(\cdot\omega)$ *have an* L_∞ *fractional derivative* $\overline{D}_{R_1}^{\beta} F(\cdot\omega)$ *in* $[R_1, R_2]$, *and that*

$$\overline{D}_{R_1}^{\beta-k} F(R_1\omega) = 0, \quad k = 1, \ldots, [\beta] + 1.$$

Then

$$\frac{\overline{\partial}_{R_1}^{\alpha} F(x)}{\partial r^{\alpha}} = \left(\overline{D}_{R_1}^{\alpha} F\right)(r\omega) = \frac{1}{\Gamma(\beta - \alpha)} \int_{R_1}^{r} (r - t)^{\beta - \alpha - 1} \left(\overline{D}_{R_1}^{\beta} F\right)(t\omega) \, dt,$$

$$(24.95)$$

is true for all $x \in A$; *i.e. true for all* $r \in [R_1, R_2]$ *and for all* $\omega \in S^{N-1}$.
  Here

$$\left(\overline{D}_{R_1}^{\alpha} F\right)(\cdot\omega) \in AC([R_1, R_2]), \quad \text{for } \beta - \alpha \geq 1$$

and

$$\left(\overline{D}_{R_1}^{\alpha} F\right)(\cdot\omega) \in C([R_1, R_2]), \quad \text{for } \beta - \alpha \in (0, 1),$$

for all $\omega \in S^{N-1}$. *Furthermore*

$$\frac{\overline{\partial}_{R_1}^{\alpha} F(x)}{\partial r^{\alpha}} \in L_\infty(A).$$

In particular, it holds

$$F(x) = F(r\omega) = \frac{1}{\Gamma(\beta)} \int_{R_1}^{r} (r - t)^{\beta - 1} \left(\overline{D}_{R_1}^{\beta} F\right)(t\omega) \, dt, \qquad (24.96)$$

for all $r \in [R_1, R_2]$ *and* $\omega \in S^{N-1} - K(F)$; $x = r\omega$, *and*

$$F(\cdot\omega) \in AC([R_1, R_2]), \quad \text{for } \beta \geq 1$$

and

$$F(\cdot\omega) \in C([R_1, R_2]), \ \text{for } \beta \in (0, 1),$$

for all $\omega \in S^{N-1} - K(F)$.

Similarly to Theorem 24.35 we obtain

Theorem 24.41 *Let $\overrightarrow{F} = (F_1, \ldots, F_n)$, with F_i, $i = 1, \ldots, n$, as in Theorem 24.40, $\beta > \alpha > 0$. Assume that $0 < \left\| \dfrac{\overline{\partial}^{\beta}_{R_1} \overrightarrow{F(y)}}{\partial r^{\beta}} \right\|_q < \infty$, a.e. in $y \in \overline{A}$, $1 \leq q \leq \infty$ fixed.*

Let $\Phi : \mathbb{R}^n_+ \to \mathbb{R}_+$ be convex and increasing per coordinate. Also assume that

$$M_q(x) = M_q(r\omega) := \frac{1}{\Gamma(\beta - \alpha)} \int_{R_1}^{r} (r - t)^{\beta - \alpha - 1} \left\| \frac{\overline{\partial}^{\beta}_{R_1} \overrightarrow{F(t\omega)}}{\partial r^{\beta}} \right\|_q dt > 0, \tag{24.97}$$

a.e. in $x \in \overline{A}$. Then

$$\int_A M_q(x) \, \Phi\left(\frac{\left| \frac{\overline{\partial}^{\alpha}_{R_1} \overrightarrow{F(x)}}{\partial r^{\alpha}} \right|}{M_q(x)} \right) dx \leq$$

$$\frac{(R_2 - R_1)^{\beta - \alpha}}{\Gamma(\beta - \alpha + 1)} \left(\frac{R_2}{R_1} \right)^{N-1} \int_A \left\| \frac{\overline{\partial}^{\beta}_{R_1} \overrightarrow{F(x)}}{\partial r^{\beta}} \right\|_q \Phi\left(\frac{\left| \frac{\overline{\partial}^{\beta}_{R_1} \overrightarrow{F(x)}}{\partial r^{\beta}} \right|}{\left\| \frac{\overline{\partial}^{\beta}_{R_1} \overrightarrow{F(x)}}{\partial r^{\beta}} \right\|_q} \right) dx. \tag{24.98}$$

under the assumptions:

for every $\omega \in S^{N-1}$ we have

(i) $\dfrac{\left| \frac{\overline{\partial}^{\beta}_{R_1} \overrightarrow{F(t\omega)}}{\partial r^{\beta}} \right|}{\left\| \frac{\overline{\partial}^{\beta}_{R_1} \overrightarrow{F(t\omega)}}{\partial r^{\beta}} \right\|_q}$, $\Phi\left(\dfrac{\left| \frac{\overline{\partial}^{\beta}_{R_1} \overrightarrow{F(t\omega)}}{\partial r^{\beta}} \right|}{\left\| \frac{\overline{\partial}^{\beta}_{R_1} \overrightarrow{F(t\omega)}}{\partial r^{\beta}} \right\|_q} \right)$ *are both* $\frac{1}{\Gamma(\beta - \alpha)} \chi_{[R_1, r]}(t)(r - t)^{\beta - \alpha - 1}$.

$\left\| \dfrac{\overline{\partial}^{\beta}_{R_1} \overrightarrow{F(t\omega)}}{\partial r^{\beta}} \right\|_q$ *dt -integrable in $t \in [R_1, R_2]$, a.e. in $r [R_1, R_2]$,*

(ii) $\left\| \dfrac{\overline{\partial^{\beta}_{R_1} F(t\omega)}}{\partial r^{\beta}} \right\|_q \Phi \left(\dfrac{\left| \dfrac{\overline{\partial^{\beta}_{R_1} F(t\omega)}}{\partial r^{\beta}} \right|}{\left\| \dfrac{\overline{\partial^{\beta}_{R_1} F(t\omega)}}{\partial r^{\beta}} \right\|_q} \right)$ is integrable in t on $[R_1, R_2]$.

One can give varous applications of Theorem 24.41 for different specific Φ's, we omit this task.

We also need (see [1], p. 421).

Definition 24.42 Let $F : \overline{A} \to \mathbb{R}$, $\nu \geq 0$, $n := \lceil \nu \rceil$ such that $F(\cdot\omega) \in AC^n([R_1, R_2])$, for all $\omega \in S^{N-1}$.

We call the left Caputo radial fractional derivative the following function

$$\frac{\partial^{\nu}_{*R_1} F(x)}{\partial r^{\nu}} := \frac{1}{\Gamma(n-\nu)} \int_{R_1}^{r} (r-t)^{n-\nu-1} \frac{\partial^n F(t\omega)}{\partial r^n} dt, \qquad (24.99)$$

where $x \in \overline{A}$, i.e. $x = r\omega$, $r \in [R_1, R_2]$, $\omega \in S^{N-1}$.

Clearly

$$\frac{\partial^0_{*R_1} F(x)}{\partial r^0} = F(x), \qquad (24.100)$$

$$\frac{\partial^{\nu}_{*R_1} F(x)}{\partial r^{\nu}} = \frac{\partial^{\nu} F(x)}{\partial r^{\nu}}, \quad \text{if } \nu \in \mathbb{N}. \qquad (24.101)$$

Above function (24.99) exists almost everywhere for $x \in \overline{A}$, see [1], p. 422.

We mention the following fundamental representation result (see [1], pp. 422–423, [4, 8]).

Theorem 24.43 *Let $\beta > \alpha \geq 0$, $n := \lceil \beta \rceil$, $m := \lceil \alpha \rceil$, $F : \overline{A} \to \mathbb{R}$ with $F \in L_1(A)$. Assume that $F(\cdot\omega) \in AC^n([R_1, R_2])$, for all $\omega \in S^{N-1}$, and that $\dfrac{\partial^{\beta}_{*R_1} F(\cdot\omega)}{\partial r^{\beta}} \in L_{\infty}(R_1, R_2)$ for all $\omega \in S^{N-1}$.*

Further assume that $\dfrac{\partial^{\beta}_{*R_1} F(x)}{\partial r^{\beta}} \in L_{\infty}(A)$. More precisely, for these $r \in [R_1, R_2]$, for each $\omega \in S^{N-1}$, for which $D^{\beta}_{*R_1} F(r\omega)$ takes real values, there exists $M_1 > 0$ such that $\left| D^{\beta}_{*R_1} F(r\omega) \right| \leq M_1$.

We suppose that $\dfrac{\partial^j F(R_1\omega)}{\partial r^j} = 0$, $j = m, m+1, \ldots, n-1$; for every $\omega \in S^{N-1}$. Then

$$\frac{\partial^{\alpha}_{*R_1} F(x)}{\partial r^{\alpha}} = D^{\alpha}_{*R_1} F(r\omega) = \frac{1}{\Gamma(\beta-\alpha)} \int_{R_1}^{r} (r-t)^{\beta-\alpha-1} \left(D^{\beta}_{*R_1} F \right)(t\omega) dt,$$

$$(24.102)$$

valid $\forall x \in \overline{A}$; i.e. true $\forall r \in [R_1, R_2]$ and $\forall \omega \in S^{N-1}$; $\alpha > 0$.

Here

$$D_{*R_1}^\alpha F \left(\cdot \omega \right) \in C \left([R_1, R_2] \right),$$

$\forall\, \omega \in S^{N-1}; \alpha > 0.$

Furthermore

$$\frac{\partial_{*R_1}^\alpha F(x)}{\partial r^\alpha} \in L_\infty(A), \alpha > 0. \tag{24.103}$$

In particular, it holds

$$F(x) = F(r\omega) = \frac{1}{\Gamma(\beta)} \int_{R_1}^r (r-t)^{\beta-1} \left(D_{*R_1}^\beta F \right)(t\omega)\, dt, \tag{24.104}$$

true $\forall\, x \in \overline{A}$; *i.e. true* $\forall\, r \in [R_1, R_2]$ *and* $\forall\, \omega \in S^{N-1}$, *and*

$$F(\cdot\omega) \in C\left([R_1, R_2]\right), \ \forall\, \omega \in S^{N-1}. \tag{24.105}$$

Similarly to Theorem 24.35 we derive

Theorem 24.44 *Let* $\overrightarrow{F} = (F_1, \ldots, F_n)$, *with* F_i, $i = 1, \ldots, n$, *as in Theorem 24.43,* $\beta > \alpha \geq 0$. *Assume that* $0 < \left\| \dfrac{\partial_{*R_1}^\beta \overrightarrow{F(y)}}{\partial r^\beta} \right\|_q < \infty$, *a.e. in* $y \in \overline{A}$, $1 \leq q \leq \infty$ *fixed.*

Let $\Phi : \mathbb{R}_+^n \to \mathbb{R}_+$ *be convex and increasing per coordinate.*
Also assume

$$\theta_q(x) = \theta_q(r\omega) := \frac{1}{\Gamma(\beta-\alpha)} \int_{R_1}^r (r-t)^{\beta-\alpha-1} \left\| \frac{\partial_{*R_1}^\beta \overrightarrow{F(t\omega)}}{\partial r^\beta} \right\|_q dt > 0, \tag{24.106}$$

a.e. in $x \in \overline{A}$. *Then*

$$\int_A \theta_q(x)\, \Phi \left(\frac{\left\| \dfrac{\partial_{*R_1}^\alpha \overrightarrow{F(x)}}{\partial r^\alpha} \right\|}{\theta_q(x)} \right) dx \leq$$

$$\frac{(R_2 - R_1)^{\beta-\alpha}}{\Gamma(\beta-\alpha+1)} \left(\frac{R_2}{R_1} \right)^{N-1} \int_A \left\| \frac{\partial_{*R_1}^\beta \overrightarrow{F(x)}}{\partial r^\beta} \right\|_q \Phi \left(\frac{\left\| \dfrac{\partial_{*R_1}^\beta \overrightarrow{F(x)}}{\partial r^\beta} \right\|}{\left\| \dfrac{\partial_{*R_1}^\beta \overrightarrow{F(x)}}{\partial r^\beta} \right\|_q} \right) dx, \tag{24.107}$$

under the assumptions:
 for every $\omega \in S^{N-1}$ *we have*

(i) $\dfrac{\left\|\dfrac{\partial^{\beta}_{*R_1} \overrightarrow{F(t\omega)}}{\partial r^{\beta}}\right\|}{\left\|\dfrac{\partial^{\beta}_{*R_1} \overrightarrow{F(t\omega)}}{\partial r^{\beta}}\right\|_q}$, $\Phi\left(\dfrac{\left\|\dfrac{\partial^{\beta}_{*R_1} \overrightarrow{F(t\omega)}}{\partial r^{\beta}}\right\|}{\left\|\dfrac{\partial^{\beta}_{*R_1} \overrightarrow{F(t\omega)}}{\partial r^{\beta}}\right\|_q}\right)$ *are both* $\dfrac{1}{\Gamma(\beta-\alpha)}\chi_{[R_1,r]}(t)(r-t)^{\beta-\alpha-1}$.

$\left\|\dfrac{\partial^{\beta}_{*R_1} \overrightarrow{F(t\omega)}}{\partial r^{\beta}}\right\|_q$ dt *-integrable in* $t \in [R_1, R_2]$, *a.e. in* $r \in [R_1, R_2]$,

(ii) $\left\|\dfrac{\partial^{\beta}_{*R_1} \overrightarrow{F(t\omega)}}{\partial r^{\beta}}\right\|_q \Phi\left(\dfrac{\left\|\dfrac{\partial^{\beta}_{*R_1} \overrightarrow{F(t\omega)}}{\partial r^{\beta}}\right\|}{\left\|\dfrac{\partial^{\beta}_{*R_1} \overrightarrow{F(t\omega)}}{\partial r^{\beta}}\right\|_q}\right)$ *is integrable in* t *on* $[R_1, R_2]$.

References

1. G.A. Anastassiou, *Fractional Differentiation Inequalities*, Research Monograph (Springer, New York, 2009)
2. G.A. Anastassiou, On right fractional calculus. Chaos Solitons Fractals **42**, 365–376 (2009)
3. G.A. Anastassiou, Balanced fractional opial inequalities. Chaos Solitons Fractals **42**(3), 1523–1528 (2009)
4. G.A. Anastassiou, Fractional representation formulae and right fractional inequalities. Math. Comput. Model. **54**(11–12), 3098–3115 (2011)
5. G.A. Anastassiou, Vectorial Hardy type fractional inequalities. Bull. Tbilisi Int. Cent. Math. Inf. **16**(2), 21–57 (2012)
6. G.A. Anastassiou, Vectorial splitting rational L_p inequalities for integral operators. J. Appl. Nonlinear Dyn. **2**(1), 59–81 (2013)
7. G.A. Anastassiou, Vectorial Fractional Integral Inequalities with convexity. Central Eur. J. Phys. **11**(10), 1194–1211 (2013)
8. M. Andric, J.E. Pecaric, I. Peric, *Composition identities for the Caputo fractional derivatives and applications to Opial-type inequalities*. Submitted (2012)
9. D. Baleanu, K. Diethelm, E. Scalas, J.J. Trujillo, *Fractional Calculus Models and Numerical Methods*, Series on Complexity: Nonlinearity and Chaos (World Scientific, Singapore, 2012)
10. J.A. Canavati, The Riemann-Liouville Integral. Nieuw Archief Voor Wiskunde **5**(1), 53–75 (1987)
11. K. Diethelm, in *The Analysis of Fractional Differential Equations*, 1st edn. Lecture Notes in Mathematics, vol. 2004 (Springer, New York, 2010)
12. A.M.A. El-Sayed, M. Gaber, On the finite Caputo and finite Riesz derivatives. Electron. J. Theor. Phys. **3**(12), 81–95 (2006)
13. R. Gorenflo, F. Mainardi, *Essentials of Fractional Calculus*, Maphysto Center (2000), http://www.maphysto.dk/oldpages/events/LevyCAC2000/MainardiNotes/fm2k0a.ps
14. G.D. Handley, J.J. Koliha, J. Pečarić, Hilbert-Pachpatte type integral inequalities for fractional derivatives. Fractional Calc. Appl. Anal. **4**(1), 37–46 (2001)
15. H.G. Hardy, Notes on some points in the integral calculus. Messenger Math. **47**(10), 145–150 (1918)

16. S. Iqbal, K. Krulic, J. Pecaric, On an inequality of H.G. Hardy. J. Inequalities Appl. **2010**, Article ID 264347, 23 p
17. S. Iqbal, K. Krulic, J. Pecaric, On an inequality for convex functions with some applications on fractional derivatives and fractional integrals. J. Math. Inequalities **5**(2), 219–230 (2011)
18. A.A. Kilbas, H.M. Srivastava, J.J. Trujillo, in *Theory and applications of fractional differential equations*. North-Holland Mathematics Studies, vol. 204 (Elsevier, New York, 2006)
19. T. Mamatov, S. Samko, Mixed fractional integration operators in mixed weighted Hölder spaces. Fractional Calc. Appl. Anal. **13**(3), 245–259 (2010)
20. W. Rudin, *Real and Complex Analysis*, International Student Edition (Mc Graw Hill, London, 1970)
21. S.G. Samko, A.A. Kilbas, O.I. Marichev, *Fractional Integral and Derivatives: Theory and Applications* (Gordon and Breach Science Publishers, Yverdon, 1993)
22. D. Stroock, *A Concise Introduction to the Theory of Integration*, 3rd edn. (Birkhäuser, Boston, 1999)

Chapter 25
About Separating Rational L_p Inequalities for Integral Operators

Here we present L_p, $p > 1$, integral inequalities for convex and increasing functions applied to products of ratios of functions and other important mixtures. As applications we derive a wide range of fractional inequalities of Hardy type. They involve the left and right Erdé lyi-Kober fractional integrals and left and right mixed Riemann-Liouville fractional multiple integrals. Also we give inequalities for Riemann-Liouville, Caputo, Canavati radial fractional derivatives. Some inequalities are of exponential type. It follows [3].

25.1 Introduction

Let $(\Omega_1, \Sigma_1, \mu_1)$ and $(\Omega_2, \Sigma_2, \mu_2)$ be measure spaces with positive σ-finite measures, and let $k_i : \Omega_1 \times \Omega_2 \to \mathbb{R}$ be nonnegative measurable functions, $k_i(x, \cdot)$ measurable on Ω_2 and

$$K_i(x) = \int_{\Omega_2} k_i(x, y)\, d\mu_2(y), \text{ for any } x \in \Omega_1, \qquad (25.1)$$

$i = 1, \ldots, m$. We assume that $K_i(x) > 0$ a.e. on Ω_1, and the weight functions are nonnegative measurable functions on the related set.

We consider measurable functions $g_i : \Omega_1 \to \mathbb{R}$ with the representation

$$g_i(x) = \int_{\Omega_2} k_i(x, y)\, f_i(y)\, d\mu_2(y), \qquad (25.2)$$

where $f_i : \Omega_2 \to \mathbb{R}$ are measurable functions, $i = 1, \ldots, m$.

Here u stands for a weight function on Ω_1.

In [4] we proved the following general result.

© Springer International Publishing Switzerland 2016
G.A. Anastassiou, *Intelligent Comparisons: Analytic Inequalities*,
Studies in Computational Intelligence 609,
DOI 10.1007/978-3-319-21121-3_25

Theorem 25.1 *Assume that the functions* $(1 = 1, 2, \ldots, m \in \mathbb{N}) x \mapsto \left(u(x) \frac{k_i(x,y)}{K_i(x)} \right)$
are integrable on Ω_1, *for each fixed* $y \in \Omega_2$. *Define* u_i *on* Ω_2 *by*

$$u_i(y) := \int_{\Omega_1} u(x) \frac{k_i(x, y)}{K_i(x)} d\mu_1(x) < \infty. \tag{25.3}$$

Let $p_i > 1 : \sum_{i=1}^{m} \frac{1}{p_i} = 1$. *Let the functions* $\Phi_i : \mathbb{R}_+ \to \mathbb{R}_+, i = 1, \ldots, m$, *be convex and increasing. Then*

$$\int_{\Omega_1} u(x) \prod_{i=1}^{m} \Phi_i \left(\left| \frac{g_i(x)}{K_i(x)} \right| \right) d\mu_1(x) \leq \tag{25.4}$$

$$\prod_{i=1}^{m} \left(\int_{\Omega_2} u_i(y) \, \Phi_i \left(|f_i(y)| \right)^{p_i} d\mu_2(y) \right)^{\frac{1}{p_i}},$$

for all measurable functions, $f_i : \Omega_2 \to \mathbb{R} \ (i = 1, \ldots, m)$ *such that*

(i) $f_i, \Phi_i (|f_i|)^{p_i}$ *are both* $k_i(x, y) d\mu_2(y)$ *-integrable,* μ_1 *-a.e. in* $x \in \Omega_1, i = 1, \ldots, m$,

(ii) $u_i \Phi_i (|f_i|)^{p_i}$ *is* μ_2 *-integrable,* $i = 1, \ldots, m$, *and for all corresponding functions* $g_i \ (i = 1, \ldots, m)$ *given by (25.2).*

Here $\mathbb{R}^* := \mathbb{R} \cup \{\pm\infty\}$. Let $\varphi : \mathbb{R}^{*2} \to \mathbb{R}^*$ be a Borel measurable function. Let $f_{1i}, f_{2i} : \Omega_2 \to \mathbb{R}$ be measurable functions, $i = 1, \ldots, m$.

The function $\varphi(f_{1i}(y), f_{2i}(y)), y \in \Omega_2, i = 1, \ldots, m$, is Σ_2-measurable. In this chapter we assume that $0 < \varphi(f_{1i}(y), f_{2i}(y)) < \infty$, a.e., $i = 1, \ldots, m$.

We consider

$$f_{3i}(y) := \frac{f_{1i}(y)}{\varphi(f_{1i}(y), f_{2i}(y))}, \tag{25.5}$$

$i = 1, \ldots, m, y \in \Omega_2$, which is a measurable function.

We also consider here

$$k_i^*(x, y) := k_i(x, y) \varphi(f_{1i}(y), f_{2i}(y)), \tag{25.6}$$

$y \in \Omega_2, i = 1, \ldots, m$, which is a nonnegative a.e. measurable function on $\Omega_1 \times \Omega_2$. We have that $k_i^*(x, \cdot)$ is measurable on $\Omega_2, i = 1, \ldots, m$.

Denote by

$$K_i^*(x) := \int_{\Omega_2} k_i^*(x, y) d\mu_2(y) \tag{25.7}$$

$$= \int_{\Omega_2} k_i(x, y) \varphi(f_{1i}(y), f_{2i}(y)) d\mu_2(y), \ i = 1, \ldots, m.$$

We assume that $K_i^*(x) > 0$, a.e. on Ω_1.
So here the function

$$g_{1i}(x) = \int_{\Omega_2} k_i(x, y) f_{1i}(y) d\mu_2(y)$$

$$= \int_{\Omega_2} k_i(x, y) \varphi(f_{1i}(y), f_{2i}(y)) \left(\frac{f_{1i}(y)}{\varphi(f_{1i}(y), f_{2i}(y))}\right) d\mu_2(y)$$

$$= \int_{\Omega_2} k_i^*(x, y) f_{3i}(y) d\mu_2(y), \quad i = 1, \ldots, m. \tag{25.8}$$

A typical example is when

$$\varphi(f_{1i}(y), f_{2i}(y)) = f_{2i}(y), \quad i = 1, \ldots, m, \ y \in \Omega_2. \tag{25.9}$$

In that case we have that

$$f_{3i}(y) = \frac{f_{1i}(y)}{f_{2i}(y)}, \quad i = 1, \ldots, m, \ y \in \Omega_2. \tag{25.10}$$

The latter case was studied in [7], for $m = 1$, which is an article with interesting ideas however containing several mistakes.
In the special case (25.10) we get that

$$K_i^*(x) = g_{2i}(x) := \int_{\Omega_2} k_i(x, y) f_{2i}(y) d\mu_2(y), \quad i = 1, \ldots, m. \tag{25.11}$$

In this chapter we get first general L_p, $p > 1$, results by applying Theorem 25.1 for (f_{3i}, g_{1i}), $i = 1, \ldots, m$, and on other various important settings, then we give wide related application to Fractional Calculus.

25.2 Main Results

We present

Theorem 25.2 *Assume that the functions* $(i = 1, \ldots, m \in \mathbb{N})$

$$x \mapsto \left(u(x) \frac{k_i(x, y) \varphi(f_{1i}(y), f_{2i}(y))}{K_i^*(x)}\right)$$

are integrable on Ω_1, *for each fixed* $y \in \Omega_2$. *Define* u_i^* *on* Ω_2 *by*

$$u_i^*(y) := \varphi(f_{1i}(y), f_{2i}(y)) \int_{\Omega_1} u(x) \frac{k_i(x, y)}{K_i^*(x)} d\mu_1(x) < \infty, \qquad (25.12)$$

a.e. on Ω_2.

Let $p_i > 1 : \sum_{i=1}^{m} \frac{1}{p_i} = 1$. Let the functions $\Phi_i : \mathbb{R}_+ \to \mathbb{R}_+, i = 1, \ldots, m$, be convex and increasing. Then

$$\int_{\Omega_1} u(x) \prod_{i=1}^{m} \Phi_i \left(\left| \frac{g_{1i}(x)}{K_i^*(x)} \right| \right) d\mu_1(x) \leq \qquad (25.13)$$

$$\prod_{i=1}^{m} \left(\int_{\Omega_2} u_i^*(y) \left(\Phi_i \left(\frac{|f_{1i}(y)|}{\varphi(f_{1i}(y), f_{2i}(y))} \right) \right)^{p_i} d\mu_2(y) \right)^{\frac{1}{p_i}},$$

under the assumptions:

(i) $\left(\frac{f_{1i}(y)}{\varphi(f_{1i}(y), f_{2i}(y))} \right)$, $\Phi_i \left(\frac{|f_{1i}(y)|}{\varphi(f_{1i}(y), f_{2i}(y))} \right)^{p_i}$ are both $k_i(x, y) \varphi(f_{1i}(y), f_{2i}(y))$ $d\mu_2(y)$ -integrable, μ_1 -a.e. in $x \in \Omega_1$,

(ii) $u_i^*(y) \Phi_i \left(\frac{|f_{1i}(y)|}{\varphi(f_{1i}(y), f_{2i}(y))} \right)^{p_i}$ is μ_2 -integrable, $i = 1, \ldots, m$.

In the special case of (25.9)–(25.11) we derive

Theorem 25.3 *Here* $0 < f_{2i}(y) < \infty$, *a.e.*, $i = 1, \ldots, m$. *Assume that the functions* $(i = 1, \ldots, m \in \mathbb{N})$

$$x \mapsto \left(\frac{u(x) k_i(x, y) f_{2i}(y)}{g_{2i}(x)} \right)$$

are integrable on Ω_1, *for each fixed* $y \in \Omega_2$; *with* $g_{2i}(x) > 0$, *a.e. on* Ω_1.
Define ψ_i *on* Ω_2 *by*

$$\psi_i(y) := f_{2i}(y) \int_{\Omega_1} u(x) \frac{k_i(x, y)}{g_{2i}(x)} d\mu_1(x) < \infty, \qquad (25.14)$$

a.e. on Ω_2.

Let $p_i > 1 : \sum_{i=1}^{m} \frac{1}{p_i} = 1$. Let the functions $\Phi_i : \mathbb{R}_+ \to \mathbb{R}_+, i = 1, \ldots, m$, be convex and increasing. Then

$$\int_{\Omega_1} u(x) \prod_{i=1}^{m} \Phi_i \left(\left| \frac{g_{1i}(x)}{g_{2i}(x)} \right| \right) d\mu_1(x) \leq \qquad (25.15)$$

$$\prod_{i=1}^{m} \left(\int_{\Omega_2} \psi_i\,(y)\,\Phi_i \left(\left| \frac{f_{1i}\,(y)}{f_{2i}\,(y)} \right| \right)^{p_i} d\mu_2\,(y) \right)^{\frac{1}{p_i}},$$

under the assumptions:

(i) $\frac{f_{1i}(y)}{f_{2i}(y)}$, $\Phi_i \left(\left| \frac{f_{1i}(y)}{f_{2i}(y)} \right| \right)^{p_i}$ *are both* $k_i\,(x,y)\,f_{2i}\,(y)\,d\mu_2\,(y)$ *-integrable,* μ_1 *-a.e. in* $x \in \Omega_1$,

(ii) $\psi_i\,(y)\,\Phi_i \left(\left| \frac{f_{1i}(y)}{f_{2i}(y)} \right| \right)^{p_i}$ *is* μ_2 *-integrable,* $i = 1, \dots, m$.

Next we consider the case of $\varphi\,(s_i, t_i) = |a_{1i}s_i + a_{2i}t_i|^r$, where $r \in \mathbb{R}$; $s_i, t_i \in \mathbb{R}^*$; $a_{1i}, a_{2i} \in \mathbb{R}$, $i = 1, \dots, m$. We assume here that

$$0 < |a_{1i}\,f_{1i}\,(y) + a_{2i}\,f_{2i}\,(y)|^r < \infty, \tag{25.16}$$

a.e., $i = 1, \dots, m$.

We further assume that

$$\overline{K_i^*}\,(x) := \int_{\Omega_2} k_i\,(x,y)\,|a_{1i}\,f_{1i}\,(y) + a_{2i}\,f_{2i}\,(y)|^r\,d\mu_2\,(y) > 0, \tag{25.17}$$

a.e. on Ω_1, $i = 1, \dots, m$.

Here we have

$$f_{3i}\,(y) = \frac{f_{1i}\,(y)}{|a_{1i}\,f_{1i}\,(y) + a_{2i}\,f_{2i}\,(y)|^r}, \tag{25.18}$$

$i = 1, \dots, m$, $y \in \Omega_2$.

Denote by

$$\overline{k_i^*}\,(x,y) := k_i\,(x,y)\,|a_{1i}\,f_{1i}\,(y) + a_{2i}\,f_{2i}\,(y)|^r, \tag{25.19}$$

$i = 1, \dots, m$.

By Theorem 25.2 we obtain

Theorem 25.4 *Assume that the functions* $(i = 1, \dots, m \in \mathbb{N})$

$$x \mapsto \left(u\,(x)\,\frac{k_i\,(x,y)\,|a_{1i}\,f_{1i}\,(y) + a_{2i}\,f_{2i}\,(y)|^r}{\overline{K_i^*}\,(x)} \right)$$

are integrable on Ω_1, *for each fixed* $y \in \Omega_2$. *Define* $\overline{u_i^*}$ *on* Ω_2 *by*

$$\overline{u_i^*}\,(y) := |a_{1i}\,f_{1i}\,(y) + a_{2i}\,f_{2i}\,(y)|^r \int_{\Omega_1} u\,(x)\,\frac{k_i\,(x,y)}{\overline{K_i^*}\,(x)}\,d\mu_1\,(x) < \infty. \tag{25.20}$$

a.e. on Ω_2.

Let $p_i > 1 : \sum_{i=1}^{m} \dfrac{1}{p_i} = 1$. Let the functions $\Phi_i : \mathbb{R}_+ \to \mathbb{R}_+$, $i = 1, \ldots, m$, be convex and increasing. Then

$$\int_{\Omega_1} u(x) \prod_{i=1}^{m} \Phi_i \left(\left| \frac{g_{1i}(x)}{K_i^*(x)} \right| \right) d\mu_1(x) \leq \tag{25.21}$$

$$\prod_{i=1}^{m} \left(\int_{\Omega_2} \overline{u_i^*}(y) \left(\Phi_i \left(\frac{|f_{1i}(y)|}{(a_{1i} f_{1i}(y) + a_{2i} f_{2i}(y))^r} \right) \right)^{p_i} d\mu_2(y) \right)^{\frac{1}{p_i}},$$

under the assumptions:

(i) $\left(\dfrac{f_{1i}(y)}{|a_{1i} f_{1i}(y) + a_{2i} f_{2i}(y)|^r} \right)$, $\Phi_i \left(\left| \dfrac{f_{1i}(y)}{(a_{1i} f_{1i}(y) + a_{2i} f_{2i}(y))^r} \right| \right)^{p_i}$ *are both*
$k_i(x, y) |a_{1i} f_{1i}(y) + a_{2i} f_{2i}(y)|^r d\mu_2(y)$ *-integrable, μ_1 -a.e. in $x \in \Omega_1$,*

(ii) $\overline{u_i^*}(y) \left(\Phi_i \left(\left| \dfrac{f_{1i}(y)}{(a_{1i} f_{1i}(y) + a_{2i} f_{2i}(y))^r} \right| \right) \right)^{p_i}$ *is μ_2 -integrable, $i = 1, \ldots, m$.*

In Theorem 25.4 of great interest is the case of $r \in \mathbb{Z} - \{0\}$ and $a_{1i} = a_{2i} = 1$, all $i = 1, \ldots, m$; or $a_{1i} = 1$, $a_{2i} = -1$, all $i = 1, \ldots, m$.

Another interesting case arises when

$$\varphi(f_{1i}(y), f_{2i}(y)) := |f_{1i}(y)|^{r_1} |f_{2i}(y)|^{r_2}, \tag{25.22}$$

$i = 1, \ldots, m$, where $r_1, r_2 \in \mathbb{R}$. We assume that

$$0 < |f_{1i}(y)|^{r_1} |f_{2i}(y)|^{r_2} < \infty, \text{ a.e., } i = 1, \ldots, m. \tag{25.23}$$

In this case

$$f_{3i}(y) = \frac{f_{1i}(y)}{|f_{1i}(y)|^{r_1} |f_{2i}(y)|^{r_2}}, \tag{25.24}$$

$i = 1, \ldots, m$, $y \in \Omega_2$, also

$$k_i^*(x, y) = k_{pi}^*(x, y) := k_i(x, y) |f_{1i}(y)|^{r_1} |f_{2i}(y)|^{r_2}, \tag{25.25}$$

$y \in \Omega_2$, $i = 1, \ldots, m$.

We have

$$K_i^*(x) = K_{pi}^*(x) := \int_{\Omega_2} k_i(x, y) |f_{1i}(y)|^{r_1} |f_{2i}(y)|^{r_2} d\mu_2(y), \tag{25.26}$$

$i = 1, \ldots, m$.

We assume that $K_{pi}^* > 0$, a.e. on Ω_1.

By Theorem 25.2 we derive

Theorem 25.5 *Assume that the functions* $(i = 1, \ldots, m \in \mathbb{N})$

$$x \mapsto \left(u\,(x)\, \frac{k_i\,(x, y)\,|f_{1i}\,(y)|^{r_1}\,|f_{2i}\,(y)|^{r_2}}{K^*_{p_i}\,(x)} \right)$$

are integrable on Ω_1, *for each fixed* $y \in \Omega_2$. *Define* $u^*_{p_i}$ *on* Ω_2 *by*

$$u^*_{p_i}\,(y) := |f_{1i}\,(y)|^{r_1}\,|f_{2i}\,(y)|^{r_2} \int_{\Omega_1} u\,(x)\, \frac{k_i\,(x, y)}{K^*_{p_i}\,(x)}\,d\mu_1\,(x) < \infty, \qquad (25.27)$$

a.e. on Ω_2.

Let $p_i > 1 : \displaystyle\sum_{i=1}^{m} \frac{1}{p_i} = 1$. *Let the functions* $\Phi_i : \mathbb{R}_+ \to \mathbb{R}_+$, $i = 1, \ldots, m$, *be convex and increasing. Then*

$$\int_{\Omega_1} u\,(x) \prod_{i=1}^{m} \Phi_i\left(\left| \frac{g_{1i}\,(x)}{K^*_{p_i}\,(x)} \right| \right) d\mu_1\,(x) \leq \qquad (25.28)$$

$$\prod_{i=1}^{m} \left(\int_{\Omega_2} u^*_{p_i}\,(y) \left(\Phi_i\left(\frac{|f_{1i}\,(y)|^{1-r_1}}{|f_{2i}\,(y)|^{r_2}} \right) \right)^{p_i} d\mu_2\,(y) \right)^{\frac{1}{p_i}},$$

under the assumptions:

(i) $\left(\frac{f_{1i}(y)}{|f_{1i}(y)|^{r_1}|f_{2i}(y)|^{r_2}} \right)$, $\left(\Phi_i\left(\frac{|f_{1i}(y)|^{1-r_1}}{|f_{2i}(y)|^{r_2}} \right) \right)^{p_i}$ *are both*
$k_i\,(x, y)\,|f_{1i}\,(y)|^{r_1}\,|f_{2i}\,(y)|^{r_2}\,d\mu_2\,(y)$ *-integrable,* μ_1 *-a.e. in* $x \in \Omega_1$,

(ii) $u^*_{p_i}\,(y) \left(\Phi_i\left(\frac{|f_{1i}(y)|^{1-r_1}}{|f_{2i}(y)|^{r_2}} \right) \right)^{p_i}$ *is* μ_2 *-integrable,* $i = 1, \ldots, m$.

In Theorem 25.5 of interest will be the case of $r_1 = 1 - n$, $r_2 = -n$, $n \in \mathbb{N}$. In that case $|f_{1i}\,(y)|^{1-r_1}\,|f_{2i}\,(y)|^{-r_2} = |f_{1i}\,(y)\,f_{2i}\,(y)|^n$, etc.

Next we apply Theorem 25.2 for specific convex functions.

Theorem 25.6 *Assume that the functions* $(i = 1, \ldots, m \in \mathbb{N})$

$$x \mapsto \left(u\,(x)\, \frac{k_i\,(x, y)\,\varphi\,(f_{1i}\,(y),\,f_{2i}\,(y))}{K^*_i\,(x)} \right)$$

are integrable on Ω_1, *for each fixed* $y \in \Omega_2$. *Define* u^*_i *on* Ω_2 *by*

$$u^*_i\,(y) := \varphi\,(f_{1i}\,(y),\,f_{2i}\,(y)) \int_{\Omega_1} u\,(x)\, \frac{k_i\,(x, y)}{K^*_i\,(x)}\,d\mu_1\,(x) < \infty, \qquad (25.29)$$

a.e. on Ω_2.

Let $p_i > 1 : \sum_{i=1}^{m} \dfrac{1}{p_i} = 1.$ *Then*

$$\int_{\Omega_1} u(x)\, e^{\sum_{i=1}^{m} \left| \frac{g_{1i}(x)}{K_i^*(x)} \right|}\, d\mu_1(x) \leq \tag{25.30}$$

$$\prod_{i=1}^{m} \left(\int_{\Omega_2} u_i^*(y) \left(e^{\frac{p_i |f_{1i}(y)|}{\varphi(f_{1i}(y), f_{2i}(y))}} \right) d\mu_2(y) \right)^{\frac{1}{p_i}},$$

under the assumptions:

(i) $\dfrac{f_{1i}(y)}{\varphi(f_{1i}(y), f_{2i}(y))},\ e^{\frac{p_i |f_{1i}(y)|}{\varphi(f_{1i}(y), f_{2i}(y))}}$ *are both* $k_i(x, y)\, \varphi(f_{1i}(y), f_{2i}(y))\, d\mu_2(y)$ - *integrable,* μ_1 -*a.e. in* $x \in \Omega_1$,

(ii) $u_i^*(y)\, e^{\frac{p_i |f_{1i}(y)|}{\varphi(f_{1i}(y), f_{2i}(y))}}$ *is* μ_2-*integrable,* $i = 1, \ldots, m$.

Theorem 25.7 *Assume that the functions* $(i = 1, \ldots, m \in \mathbb{N})$

$$x \mapsto \left(u(x)\, \frac{k_i(x, y)\, \varphi(f_{1i}(y), f_{2i}(y))}{K_i^*(x)} \right)$$

are integrable on Ω_1, *for each fixed* $y \in \Omega_2$. *Define* u_i^* *on* Ω_2 *by*

$$u_i^*(y) := \varphi(f_{1i}(y), f_{2i}(y)) \int_{\Omega_1} u(x)\, \frac{k_i(x, y)}{K_i^*(x)}\, d\mu_1(x) < \infty, \tag{25.31}$$

a.e. on Ω_2. *Let* $p_i > 1 : \sum_{i=1}^{m} \dfrac{1}{p_i} = 1;\ \beta_i \geq 1,\ i = 1, \ldots, m$.

   Then

$$\int_{\Omega_1} u(x) \prod_{i=1}^{m} \left| \frac{g_{1i}(x)}{K_i^*(x)} \right|^{\beta_i}\, d\mu_1(x) \leq \tag{25.32}$$

$$\prod_{i=1}^{m} \left(\int_{\Omega_2} u_i^*(y) \left(\frac{|f_{1i}(y)|}{\varphi(f_{1i}(y), f_{2i}(y))} \right)^{p_i \beta_i}\, d\mu_2(y) \right)^{\frac{1}{p_i}},$$

under the assumptions:

(i) $\left(\frac{f_{1i}(y)}{\varphi(f_{1i}(y),f_{2i}(y))}\right)$, $\left(\frac{|f_{1i}(y)|}{\varphi(f_{1i}(y),f_{2i}(y))}\right)^{p_i\beta_i}$ *are both* $k_i(x,y)\varphi(f_{1i}(y),f_{2i}(y))$
$d\mu_2(y)$ *-integrable,* μ_1 *-a.e. in* $x \in \Omega_1$,

(ii) $u_i^*(y)\left(\frac{|f_{1i}(y)|}{\varphi(f_{1i}(y),f_{2i}(y))}\right)^{p_i\beta_i}$ *is* μ_2 *-integrable,* $i = 1,\ldots,m$.

We continue as follows:
Choosing $r_1 = 0, r_2 = -1, i = 1,\ldots,m$, on (25.22) we have that

$$\varphi(f_{1i}(y),f_{2i}(y)) = |f_{2i}(y)|^{-1}. \tag{25.33}$$

We assume that

$$0 < |f_{2i}(y)|^{-1} < \infty, \quad \text{a.e., } i = 1,\ldots,m, \tag{25.34}$$

equivalently,

$$0 < |f_{2i}(y)| < \infty, \quad \text{a.e., } i = 1,\ldots,m. \tag{25.35}$$

In this case

$$f_{3i} = f_{1i}(y)\,|f_{2i}(y)|, \tag{25.36}$$

$i = 1,\ldots,m, y \in \Omega_2$, also it is

$$k_{pi}^*(x,y) = \overline{k_{pi}^*}(x,y) := \frac{k_i(x,y)}{|f_{2i}(y)|}, \tag{25.37}$$

$y \in \Omega_2, i = 1,\ldots,m$.
We have that

$$K_{pi}^*(x) = \overline{K_{pi}^*}(x) := \int_{\Omega_2}\frac{k_i(x,y)}{|f_{2i}(y)|}d\mu_2(y), \tag{25.38}$$

$i = 1,\ldots,m$.
We assume that $\overline{K_{pi}^*}(x) > 0$, a.e. on Ω_1.
By Theorem 25.5 we obtain

Corollary 25.8 *Assume that the functions* $(i = 1,\ldots,m \in \mathbb{N})$

$$x \mapsto \left(\frac{u(x)k_i(x,y)|f_{2i}(y)|^{-1}}{\overline{K_{pi}^*}(x)}\right)$$

are integrable on Ω_1, for each fixed $y \in \Omega_2$. Define ψ^*_{pi} on Ω_2 by

$$\psi^*_{pi}(y) := |f_{2i}(y)|^{-1} \int_{\Omega_1} u(x) \frac{k_i(x, y)}{K^*_{pi}(x)} d\mu_1(x) < \infty, \tag{25.39}$$

a.e. on Ω_2.

Let $p_i > 1 : \sum_{i=1}^{m} \frac{1}{p_i} = 1$. Let the functions $\Phi_i : \mathbb{R}_+ \to \mathbb{R}_+$, $i = 1, \ldots, m$, be convex and increasing. Then

$$\int_{\Omega_1} u(x) \prod_{i=1}^{m} \Phi_i \left(\left| \frac{g_{1i}(x)}{K^*_{pi}(x)} \right| \right) d\mu_1(x) \leq \tag{25.40}$$

$$\prod_{i=1}^{m} \left(\int_{\Omega_2} \psi^*_{pi}(y) \left(\Phi_i \left(|f_{1i}(y) f_{2i}(y)| \right) \right)^{p_i} d\mu_2(y) \right)^{\frac{1}{p_i}},$$

under the assumptions:

(i) $f_{1i}(y) f_{2i}(y)$, $(\Phi_i(|f_{1i}(y) f_{2i}(y)|))^{p_i}$ *are both* $k_i(x, y) |f_{2i}(y)|^{-1} d\mu_2(y)$ - *integrable,* μ_1 -a.e. *in* $x \in \Omega_1$,
(ii) $\psi^*_{pi}(y) (\Phi_i(|f_{1i}(y) f_{2i}(y)|))^{p_i}$ *is* μ_2 -integrable, $i = 1, \ldots, m$.

To keep exposition short, in the next big part of this chapter we give only applications of Theorem 25.3 to Fractional Calculus. We need the following:

Let $a < b$, $a, b \in \mathbb{R}$. By $C^N([a, b])$, we denote the space of all functions on $[a, b]$ which have continuous derivatives up to order N, and $AC([a, b])$ is the space of all absolutely continuous functions on $[a, b]$. By $AC^N([a, b])$, we denote the space of all functions g with $g^{(N-1)} \in AC([a, b])$. For any $\alpha \in \mathbb{R}$, we denote by $[\alpha]$ the integral part of α (the integer k satisfying $k \leq \alpha < k + 1$), and $\lceil \alpha \rceil$ is the ceiling of α ($\min\{n \in \mathbb{N}, n \geq \alpha\}$). By $L_1(a, b)$, we denote the space of all functions integrable on the interval (a, b), and by $L_\infty(a, b)$ the set of all functions measurable and essentially bounded on (a, b). Clearly, $L_\infty(a, b) \subset L_1(a, b)$.

We need

Definition 25.9 ([10]) Let (a, b), $0 \leq a < b < \infty$; $\alpha, \sigma > 0$. We consider the left and right-sided fractional integrals of order α as follows:

(1) for $\eta > -1$, we define

$$\left(I^\alpha_{a+;\sigma,\eta} f \right)(x) = \frac{\sigma x^{-\sigma(\alpha+\eta)}}{\Gamma(\alpha)} \int_a^x \frac{t^{\sigma\eta+\sigma-1} f(t) dt}{(x^\sigma - t^\sigma)^{1-\alpha}}, \tag{25.41}$$

(2) for $\eta > 0$, we define

$$\left(I^\alpha_{b-;\sigma,\eta} f\right)(x) = \frac{\sigma x^{\sigma\eta}}{\Gamma(\alpha)} \int_x^b \frac{t^{\sigma(1-\eta-\alpha)-1} f(t)\, dt}{(t^\sigma - x^\sigma)^{1-\alpha}}. \qquad (25.42)$$

These are the Erdélyi-Kober type fractional integrals.

We make

Remark 25.10 Regarding (25.41) we have all

$$k_i(x, y) = k_1(x, y) := \frac{\sigma x^{-\sigma(\alpha+\eta)}}{\Gamma(\alpha)} \chi_{(a,x]}(y) \frac{y^{\sigma\eta+\sigma-1}}{(x^\sigma - y^\sigma)^{1-\alpha}}, \qquad (25.43)$$

$x, y \in (a, b)$, χ stands for the characteristic function.
In this case

$$g_{1i}(x) = \left(I^\alpha_{a+;\alpha,\eta} f_{1i}\right)(x) = \int_a^b k_1(x, y)\, f_{1i}(y)\, dy, \qquad (25.44)$$

and

$$g_{2i}(x) = \left(I^\alpha_{a+;\sigma,\eta} f_{2i}\right)(x) = \int_a^b k_1(x, y)\, f_{2i}(y)\, dy, \qquad (25.45)$$

$i = 1, \ldots, m$.
We assume $\left(I^\alpha_{a+;\sigma,\eta} f_{2i}\right)(x) > 0$, a.e. on (a, b), and $0 < f_{2i}(y) < \infty$, a.e., $i = 1, \ldots, m$.

We also make

Remark 25.11 Regarding (25.42) we have all

$$k_i(x, y) = k_2(x, y) := \frac{\sigma x^{\sigma\eta}}{\Gamma(\alpha)} \chi_{[x,b)}(y) \frac{y^{\sigma(1-\eta-\alpha)-1}}{(y^\sigma - x^\sigma)^{1-\alpha}}, \qquad (25.46)$$

$x, y \in (a, b)$.
In this case

$$g_{1i}(x) = \left(I^\alpha_{b-;\alpha,\eta} f_{1i}\right)(x) = \int_a^b k_2(x, y)\, f_{1i}(y)\, dy, \qquad (25.47)$$

and

$$g_{2i}(x) = \left(I^\alpha_{b-;\sigma,\eta} f_{2i}\right)(x) = \int_a^b k_2(x, y)\, f_{2i}(y)\, dy, \qquad (25.48)$$

$i = 1, \ldots, m$.

We assume $\left(I^\alpha_{b-;\sigma,\eta} f_{2i}\right)(x) > 0$, a.e. on (a, b), and $0 < f_{2i}(y) < \infty$, a.e., $i = 1, \ldots, m$.

Next we apply Theorem 25.3.

Theorem 25.12 *Here all as in Remark 25.10. Assume that the functions ($i = 1, \ldots, m \in \mathbb{N}$)*

$$x \mapsto \left(\frac{u(x)\, \sigma x^{-\sigma(\alpha+\eta)}}{\left(I^{\alpha_i}_{a+;\sigma,\eta} f_{2i}\right)(x)\, \Gamma(\alpha)} \chi_{(a,x]}(y) \frac{y^{\sigma\eta+\sigma-1} f_{2i}(y)}{(x^\sigma - y^\sigma)^{1-\alpha}} \right)$$

are integrable on (a, b), for each fixed $y \in (a, b)$.
Define ψ_i^+ on (a, b) by

$$\psi_i^+(y) := \frac{\sigma f_{2i}(y)\, y^{\sigma\eta+\sigma-1}}{\Gamma(\alpha)} \int_y^b u(x) \frac{x^{-\sigma(\alpha+\eta)} (x^\sigma - y^\sigma)^{\alpha-1}}{\left(I^\alpha_{a+;\sigma,\eta} f_{2i}\right)(x)} dx < \infty,$$

(25.49)

a.e. on (a, b). Let $p_i > 1 : \sum_{i=1}^m \dfrac{1}{p_i} = 1$. Let the functions $\Phi_i : \mathbb{R}_+ \to \mathbb{R}_+$, $i = 1, \ldots, m$, be convex and increasing. Then

$$\int_a^b u(x) \prod_{i=1}^m \Phi_i \left(\frac{\left| \left(I^\alpha_{a+;\sigma,\eta} f_{1i}\right)(x) \right|}{\left(I^\alpha_{a+;\sigma,\eta} f_{2i}\right)(x)} \right) dx \leq$$

(25.50)

$$\prod_{i=1}^m \left(\int_a^b \psi_i^+(y)\, \Phi_i \left(\frac{|f_{1i}(y)|}{f_{2i}(y)} \right)^{p_i} dy \right)^{\frac{1}{p_i}},$$

under the assumptions:

(i) $\dfrac{f_{1i}(y)}{f_{2i}(y)}$, $\Phi_i\left(\left| \dfrac{f_{1i}(y)}{f_{2i}(y)} \right| \right)^{p_i}$ *are both* $\dfrac{\sigma x^{-\sigma(\alpha+\eta)}}{\Gamma(\alpha)} \chi_{(a,x]}(y) \dfrac{y^{\sigma\eta+\sigma-1}}{(x^\sigma - y^\sigma)^{1-\alpha}} f_{2i}(y)\, dy$
-*integrable, a.e. in $x \in (a, b)$,*

(ii) $\psi_i^+(y)\, \Phi_i\left(\left| \dfrac{f_{1i}(y)}{f_{2i}(y)} \right| \right)^{p_i}$ *is integrable, $i = 1, \ldots, m$.*

Corollary 25.13 (to Theorem 25.12) *It holds*

$$\int_a^b u(x)\, e^{\sum_{i=1}^m \frac{\left| \left(I^\alpha_{a+;\sigma,\eta} f_{1i}\right)(x) \right|}{\left(I^\alpha_{a+;\sigma,\eta} f_{2i}\right)(x)}} dx \leq$$

(25.51)

$$\prod_{i=1}^{m} \left(\int_a^b \psi_i^+ (y) \, e^{\frac{p_i |f_{1i}(y)|}{f_{2i}(y)}} dy \right)^{\frac{1}{p_i}} ,$$

under the assumptions:

(i) $\frac{f_{1i}(y)}{f_{2i}(y)}$, $e^{p_i \left| \frac{f_{1j}(y)}{f_{2j}(y)} \right|}$ are both $\frac{\sigma x^{-\sigma(\alpha+\eta)}}{\Gamma(\alpha)} \chi_{(a,x]}(y) \, y^{\sigma\eta+\sigma-1} (x^\sigma - y^\sigma)^{\alpha-1} f_{2i}(y) \, dy$ -integrable, a.e. in $x \in (a, b)$,

(ii) $\psi_i^+ (y) \, e^{p_i \left| \frac{f_{1j}(y)}{f_{2j}(y)} \right|}$ is integrable, $i = 1, \ldots, m$.

Corollary 25.14 (to Theorem 25.12) *Let $\beta_i \geq 1$, $i = 1, \ldots, m$. It holds*

$$\int_a^b u(x) \prod_{i=1}^{m} \left(\frac{\left| I_{a+;\sigma,\eta}^\alpha f_{1i}(x) \right|}{I_{a+;\sigma,\eta}^\alpha f_{2i}(x)} \right)^{\beta_i} dx \leq \qquad (25.52)$$

$$\prod_{i=1}^{m} \left(\int_a^b \psi_i^+ (y) \left(\frac{|f_{1i}(y)|}{f_{2i}(y)} \right)^{\beta_i p_i} dy \right)^{\frac{1}{p_i}} ,$$

under the assumptions:

(i) $\left| \frac{f_{1i}(y)}{f_{2i}(y)} \right|^{\beta_i p_i}$ is $\frac{\sigma x^{-\sigma(\alpha+\eta)}}{\Gamma(\alpha)} \chi_{(a,x]}(y) \, \frac{y^{\sigma\eta+\sigma-1}}{(x^\sigma-y^\sigma)^{1-\alpha}} f_{2i}(y) \, dy$ -integrable, a.e. in $x \in (a, b)$,

(ii) $\psi_i^+ (y) \left| \frac{f_{1i}(y)}{f_{2i}(y)} \right|^{\beta_i p_i}$ is integrable, $i = 1, \ldots, m$.

We also give

Theorem 25.15 *Here all as in Remark 25.11. Assume that the functions ($i = 1, \ldots, m \in \mathbb{N}$)*

$$x \mapsto \left(\frac{u(x) \sigma x^{\sigma\eta}}{\left(I_{b-;\sigma,\eta}^\alpha f_{2i} \right)(x)} \chi_{[x,b)}(y) \, \frac{y^{\sigma(1-\eta-\alpha)-1} f_{2i}(y)}{(y^\sigma - x^\sigma)^{1-\alpha} \, \Gamma(\alpha)} \right)$$

are integrable on (a, b), for each fixed $y \in (a, b)$.
   Define ψ_i^- on (a, b) by

$$\psi_i^- (y) := \frac{\sigma f_{2i}(y) \, y^{\sigma(1-\eta-\alpha)-1}}{\Gamma(\alpha)} \int_a^y u(x) \, \frac{x^{\sigma\eta} (y^\sigma - x^\sigma)^{\alpha-1}}{\left(I_{b-;\sigma,\eta}^\alpha f_{2i} \right)(x)} dx < \infty, \quad (25.53)$$

a.e. on (a, b). Let $p_i > 1$: $\sum_{i=1}^{m} \frac{1}{p_i} = 1$. Let the functions $\Phi_i : \mathbb{R}_+ \to \mathbb{R}_+$, $i = 1, \ldots, m$, be convex and increasing. Then

$$\int_a^b u(x) \prod_{i=1}^m \Phi_i \left(\frac{\left| \left(I_{b-;\sigma,\eta}^\alpha f_{1i} \right)(x) \right|}{\left(I_{b-;\sigma,\eta}^\alpha f_{2i} \right)(x)} \right) dx \le \qquad (25.54)$$

$$\prod_{i=1}^m \left(\int_a^b \psi_i^-(y) \, \Phi_i \left(\frac{|f_{1i}(y)|}{f_{2i}(y)} \right)^{p_i} dy \right)^{\frac{1}{p_i}},$$

under the assumptions:

(i) $\frac{f_{1i}(y)}{f_{2i}(y)}$, $\Phi_i \left(\left| \frac{f_{1i}(y)}{f_{2i}(y)} \right| \right)^{p_i}$ *are both* $\frac{\sigma x^{\sigma \eta}}{\Gamma(\alpha)} \chi_{[x,b)}(y) \frac{y^{\sigma(1-\eta-\alpha)-1}}{(y^\sigma - x^\sigma)^{1-\alpha}} f_{2i}(y) dy$ *-integrable,*
   a.e. in $x \in (a, b)$,

(ii) $\psi_i^-(y) \, \Phi_i \left(\left| \frac{f_{1i}(y)}{f_{2i}(y)} \right| \right)^{p_i}$ *is integrable,* $i = 1, \dots, m$.

Proof By Theorem 25.3. ∎

Corollary 25.16 (to Theorem 25.15) *It holds*

$$\int_a^b u(x) \, e^{\sum_{i=1}^m \frac{\left| \left(I_{b-;\sigma,\eta}^\alpha f_{1i} \right)(x) \right|}{\left(I_{b-;\sigma,\eta}^\alpha f_{2i} \right)(x)}} dx \le \qquad (25.55)$$

$$\prod_{i=1}^m \left(\int_a^b \psi_i^-(y) \, e^{\frac{p_i |f_{1i}(y)|}{f_{2i}(y)}} dy \right)^{\frac{1}{p_i}},$$

under the assumptions:

(i) $\frac{f_{1i}(y)}{f_{2i}(y)}$, $e^{p_i \left| \frac{f_{1i}(y)}{f_{2i}(y)} \right|}$ *are both* $\frac{\sigma x^{\sigma \eta}}{\Gamma(\alpha)} \chi_{[x,b)}(y) \frac{y^{\sigma(1-\eta-\alpha)-1}}{(y^\sigma - x^\sigma)^{1-\alpha}} f_{2i}(y) dy$ *-integrable, a.e. in*
   $x \in (a, b)$,

(ii) $\psi_i^-(y) \, e^{p_i \left| \frac{f_{1i}(y)}{f_{2i}(y)} \right|}$ *is integrable,* $i = 1, \dots, m$.

Corollary 25.17 (to Theorem 25.15) *Let* $\beta_i \ge 1$, $i = 1, \dots, m$. *It holds*

$$\int_a^b u(x) \prod_{i=1}^m \left(\frac{\left| I_{b-;\sigma,\eta}^\alpha f_{1i}(x) \right|}{I_{b-;\sigma,\eta}^\alpha f_{2i}(x)} \right)^{\beta_i} dx \le \qquad (25.56)$$

$$\prod_{i=1}^m \left(\int_a^b \psi_i^-(y) \left(\frac{|f_{1i}(y)|}{f_{2i}(y)} \right)^{\beta_i p_i} dy \right)^{\frac{1}{p_i}},$$

under the assumptions:

(i) $\left|\frac{f_{1i}(y)}{f_{2i}(y)}\right|^{\beta_i p_i}$ is $\frac{\sigma x^{\sigma\eta}}{\Gamma(\alpha)}\chi_{[x,b)}(y)\frac{y^{\sigma(1-\eta-\alpha)-1}}{(y^\sigma-x^\sigma)^{1-\alpha}}f_{2i}(y)dy$ *-integrable, a.e. in* $x\in(a,b)$,

(ii) $\psi_i^-(y)\left|\frac{f_{1i}(y)}{f_{2i}(y)}\right|^{\beta_i p_i}$ *is integrable,* $i=1,\ldots,m$.

We make

Remark 25.18 Let $\prod_{i=1}^{N}(a_i,b_i)\subset\mathbb{R}^N$, $N>1$, $a_i<b_i$, $a_i,b_i\in\mathbb{R}$. Let $\alpha_i>0$,

$i=1,\ldots,N$; $f\in L_1\left(\prod_{i=1}^{N}(a_i,b_i)\right)$, and set $a=(a_1,\ldots,a_N)$, $b=(b_1,\ldots,b_N)$,

$\alpha=(\alpha_1,\ldots,\alpha_N)$, $x=(x_1,\ldots,x_N)$, $t=(t_1,\ldots,t_N)$.

We define the left mixed Riemann-Liouville fractional multiple integral of order α (see also [8]):

$$\left(I_{a+}^\alpha f\right)(x):=\frac{1}{\prod\limits_{i=1}^{N}\Gamma(\alpha_i)}\int_{a_1}^{x_1}\cdots\int_{a_N}^{x_N}\prod_{i=1}^{N}(x_i-t_i)^{\alpha_i-1}f(t_1,\ldots,t_N)\,dt_1\ldots dt_N,$$

(25.57)

with $x_i>a_i$, $i=1,\ldots,N$.

We also define the right mixed Riemann-Liouville fractional multiple integral of order α (see also [6]):

$$\left(I_{b-}^\alpha f\right)(x):=\frac{1}{\prod\limits_{i=1}^{N}\Gamma(\alpha_i)}\int_{x_1}^{b_1}\cdots\int_{x_N}^{b_N}\prod_{i=1}^{N}(t_i-x_i)^{\alpha_i-1}f(t_1,\ldots,t_N)\,dt_1\ldots dt_N,$$

(25.58)

with $x_i<b_i$, $i=1,\ldots,N$.

Notice $I_{a+}^\alpha(|f|)$, $I_{b-}^\alpha(|f|)$ are finite if $f\in L_\infty\left(\prod_{i=1}^{N}(a_i,b_i)\right)$.

One can rewrite (25.57) and (25.58) as follows:

$$\left(I_{a+}^\alpha f\right)(x)=\frac{1}{\prod\limits_{i=1}^{N}\Gamma(\alpha_i)}\int_{\prod\limits_{i=1}^{N}(a_i,b_i)}\chi_{\prod\limits_{i=1}^{N}(a_i,x_i]}(t)\prod_{i=1}^{N}(x_i-t_i)^{\alpha_i-1}f(t)\,dt,$$

(25.59)

with $x_i>a_i$, $i=1,\ldots,N$,
and

$$\left(I_{b-}^\alpha f\right)(x) = \frac{1}{\displaystyle\prod_{i=1}^N \Gamma(\alpha_i)} \int_{\prod_{i=1}^N (a_i, b_i)} \chi_{\prod_{i=1}^N [x_i, b_i)}(t) \prod_{i=1}^N (t_i - x_i)^{\alpha_i - 1} f(t)\, dt,$$

(25.60)

with $x_i < b_i$, $i = 1, \ldots, N$.

The corresponding $k(x, y)$ for I_{a+}^α, I_{b-}^α are

$$k_{a+}(x, y) = \frac{1}{\displaystyle\prod_{i=1}^N \Gamma(\alpha_i)} \chi_{\prod_{i=1}^N (a_i, x_i]}(y) \prod_{i=1}^N (x_i - y_i)^{\alpha_i - 1},$$

(25.61)

$$\forall\, x, y \in \prod_{i=1}^N (a_i, b_i),$$

and

$$k_{b-}(x, y) = \frac{1}{\displaystyle\prod_{i=1}^N \Gamma(\alpha_i)} \chi_{\prod_{i=1}^N [x_i, b_i)}(y) \prod_{i=1}^N (y_i - x_i)^{\alpha_i - 1},$$

(25.62)

$$\forall\, x, y \in \prod_{i=1}^N (a_i, b_i).$$

We make

Remark 25.19 In the case of (25.57) we choose

$$g_{1j}(x) = \left(I_{a+}^\alpha f_{1j}\right)(x)$$

(25.63)

and

$$g_{2j}(x) = \left(I_{a+}^\alpha f_{2j}\right)(x),$$

(25.64)

$$\forall\, x \in \prod_{i=1}^N (a_i, b_i).$$

We assume $\left(I_{a+}^\alpha f_{2j}\right)(x) > 0$, a.e. on $\prod_{i=1}^N (a_i, b_i)$, and $0 < f_{2j}(y) < \infty$, a.e.,
$j = 1, \ldots, m$.

Above the functions $f_{1j}, f_{2j} \in L_1\left(\prod_{i=1}^N (a_i, b_i)\right)$, $j = 1, \ldots, m$.

We make

Remark 25.20 In the case of (25.58) we choose

$$g_{1j}(x) = \left(I_{b-}^{\alpha} f_{1j}\right)(x) \tag{25.65}$$

and

$$g_{2j}(x) = \left(I_{b-}^{\alpha} f_{2j}\right)(x), \tag{25.66}$$

$\forall\, x \in \prod_{i=1}^{N}(a_i, b_i).$

We assume $\left(I_{b-}^{\alpha} f_{2j}\right)(x) > 0$, a.e. on $\prod_{i=1}^{N}(a_i, b_i)$, and $0 < f_{2j}(y) < \infty$, a.e., $j = 1, \ldots, m.$

Above the functions $f_{1j}, f_{2j} \in L_1\left(\prod_{i=1}^{N}(a_i, b_i)\right)$, $j = 1, \ldots, m.$

We present

Theorem 25.21 *Here all as in Remark 25.19. Assume that the functions* $(j = 1, \ldots, m \in \mathbb{N})$

$$x \mapsto \left(\frac{u(x)\, f_{2j}(x)\, \chi_{\prod_{i=1}^{N}(a_i, x_i]}(y) \prod_{i=1}^{N}(x_i - y_i)^{\alpha_i - 1}}{\left(I_{a+}^{\alpha} f_{2j}\right)(x) \prod_{i=1}^{N} \Gamma(\alpha_i)} \right)$$

are integrable on $\prod_{i=1}^{N}(a_i, b_i)$, *for each fixed* $y \in \prod_{i=1}^{N}(a_i, b_i)$. *We define* W_j *on* $\prod_{i=1}^{N}(a_i, b_i)$ *by*

$$W_j(y) := \frac{f_{2j}(y)}{\prod_{i=1}^{N} \Gamma(\alpha_i)} \int_{y_1}^{b_1} \cdots \int_{y_N}^{b_N} \frac{u(x_1, \ldots, x_N) \prod_{i=1}^{N}(x_i - y_i)^{\alpha_i - 1}}{\left(I_{a+}^{\alpha} f_{2j}\right)(x_1, \ldots, x_N)} dx_1 \ldots dx_N < \infty,$$

$$\tag{25.67}$$

a.e.

Let $p_j > 1 : \sum_{j=1}^{m} \dfrac{1}{p_j} = 1$. *Let the functions* $\Phi_j : \mathbb{R}_+ \to \mathbb{R}_+$, $j = 1, \ldots, m$, *be convex and increasing. Then*

$$\int_{\prod_{i=1}^{N} (a_i, b_i)} u(x) \prod_{j=1}^{m} \Phi_j \left(\frac{\left| \left(I_{a+}^{\alpha} f_{1j} \right)(x) \right|}{\left(I_{a+}^{\alpha} f_{2j} \right)(x)} \right) dx \leq \tag{25.68}$$

$$\prod_{j=1}^{m} \left(\int_{\prod_{i=1}^{N} (a_i, b_i)} W_j(y) \, \Phi_j \left(\frac{\left| f_{1j}(y) \right|}{f_{2j}(y)} \right)^{p_j} dy \right)^{\frac{1}{p_j}},$$

under the assumptions:

(i) $\dfrac{f_{1j}(y)}{f_{2j}(y)}$, $\Phi_j \left(\left| \dfrac{f_{1j}(y)}{f_{2j}(y)} \right| \right)^{p_j}$ *are both* $\dfrac{1}{\prod_{i=1}^{N} \Gamma(\alpha_i)} \chi_{\prod_{i=1}^{N}(a_i, x_i]} (y) \prod_{i=1}^{N} (x_i - y_i)^{\alpha_i - 1}$

$f_{2j}(y) \, dy$ *-integrable, a.e. in* $x \in \prod_{i=1}^{N} (a_i, b_i)$,

(ii) $W_j(y) \, \Phi_j \left(\left| \dfrac{f_{1j}(y)}{f_{2j}(y)} \right| \right)^{p_j}$ *is integrable,* $j = 1, \ldots, m$.

Corollary 25.22 (to Theorem 25.21) *It holds*

$$\int_{\prod_{i=1}^{N} (a_i, b_i)} u(x) \, e^{\sum_{j=1}^{m} \frac{\left| \left(I_{a+}^{\alpha} f_{1j} \right)(x) \right|}{\left(I_{a+}^{\alpha} f_{2j} \right)(x)}} dx \leq \tag{25.69}$$

$$\prod_{j=1}^{m} \left(\int_{\prod_{i=1}^{N} (a_i, b_i)} W_j(y) \, e^{\frac{p_j \left| f_{1j}(y) \right|}{f_{2j}(y)}} dy \right)^{\frac{1}{p_j}},$$

under the assumptions:

(i) $\frac{f_{1j}(y)}{f_{2j}(y)}, e^{P_j \left| \frac{f_{1j}(y)}{f_{2j}(y)} \right|}$ *are both* $\dfrac{1}{\prod\limits_{i=1}^{N} \Gamma(\alpha_i) \prod\limits_{i=1}^{N}(a_i, x_i]} \chi_N$ $(y) \prod\limits_{i=1}^{N}(x_i - y_i)^{\alpha_i - 1} f_{2j}(y) \, dy$

$-$*integrable, a.e. in* $x \in \prod\limits_{i=1}^{N}(a_i, b_i),$

(ii) $W_j(y) e^{P_j \left| \frac{f_{1j}(y)}{f_{2j}(y)} \right|}$ *is integrable,* $j = 1, \ldots, m.$

Corollary 25.23 (to Theorem 25.21) *Let* $\beta_j \geq 1,\ j = 1, \ldots, m.$ *It holds*

$$\int_{\prod\limits_{i=1}^{N}(a_i, b_i)} u(x) \prod_{j=1}^{m} \left(\frac{|I_{a+}^{\alpha} f_{1j}(x)|}{I_{a+}^{\alpha} f_{2j}(x)} \right)^{\beta_j} dx \leq \qquad (25.70)$$

$$\prod_{j=1}^{m} \left(\int_{\prod\limits_{i=1}^{N}(a_i, b_i)} W_j(y) \left(\frac{|f_{1j}(y)|}{f_{2j}(y)} \right)^{\beta_j P_j} dy \right)^{\frac{1}{P_j}},$$

under the assumptions:

(i) $\left| \frac{f_{1j}(y)}{f_{2j}(y)} \right|^{\beta_j P_j}$ *is* $\dfrac{1}{\prod\limits_{i=1}^{N} \Gamma(\alpha_i) \prod\limits_{i=1}^{N}(a_i, x_i]} \chi_N$ $(y) \prod\limits_{i=1}^{N}(x_i - y_i)^{\alpha_i - 1} f_{2j}(y) \, dy$-*integrable,*

a.e. in $x \in \prod\limits_{i=1}^{N}(a_i, b_i),$

(ii) $W_j(y) \left| \frac{f_{1j}(y)}{f_{2j}(y)} \right|^{\beta_j P_j}$ *is integrable,* $j = 1, \ldots, m.$

We also give

Theorem 25.24 *Here all as in Remark 25.20. Assume that the functions* ($j = 1, \ldots, m \in \mathbb{N}$)

$$x \mapsto \left(\frac{u(x) f_{2j}(y) \chi_{\prod_{i=1}^{N}[x_i, b_i]}(y) \prod_{i=1}^{N}(y_i - x_i)^{\alpha_i - 1}}{\left(I_{b-}^{\alpha} f_{2j}\right)(x) \prod_{i=1}^{N}\Gamma(\alpha_i)} \right)$$

are integrable on $\prod_{i=1}^{N}(a_i, b_i)$, for each fixed $y \in \prod_{i=1}^{N}(a_i, b_i)$.

Define M_j on $\prod_{i=1}^{N}(a_i, b_i)$ by

$$M_j(y) := \frac{f_{2j}(y)}{\prod_{i=1}^{N}\Gamma(\alpha_i)} \int_{a_1}^{y_1} \cdots \int_{a_N}^{y_N} \frac{u(x_1, \ldots, x_N) \prod_{i=1}^{N}(y_i - x_i)^{\alpha_i - 1}}{\left(I_{b-}^{\alpha} f_{2j}\right)(x_1, \ldots, x_N)} dx_1 \ldots dx_N < \infty,$$

(25.71)

a.e.

Let $p_j > 1 : \sum_{j=1}^{m} \frac{1}{p_j} = 1$. Let the functions $\Phi_j : \mathbb{R}_+ \to \mathbb{R}_+$, $j = 1, \ldots, m$, be convex and increasing. Then

$$\int_{\prod_{i=1}^{N}(a_i, b_i)} u(x) \prod_{j=1}^{m} \Phi_j\left(\frac{\left|\left(I_{b-}^{\alpha} f_{1j}\right)(x)\right|}{\left(I_{b-}^{\alpha} f_{2j}\right)(x)}\right) dx \leq \qquad (25.72)$$

$$\prod_{j=1}^{m} \left(\int_{\prod_{i=1}^{N}(a_i, b_i)} M_j(y) \Phi_j\left(\frac{|f_{1j}(y)|}{f_{2j}(y)}\right)^{p_j} dy \right)^{\frac{1}{p_j}},$$

under the assumptions:

(i) $\frac{f_{1j}(y)}{f_{2j}(y)}$, $\Phi_j\left(\left|\frac{f_{1j}(y)}{f_{2j}(y)}\right|\right)^{p_j}$ are both $\frac{1}{\prod_{i=1}^{N}\Gamma(\alpha_i)} \chi_{\prod_{i=1}^{N}[x_i, b_i]}(y) \prod_{i=1}^{N}(y_i - x_i)^{\alpha_i - 1}$

$$f_{2j}(y)\, dy \text{ -integrable, a.e. in } x \in \prod_{i=1}^{N}(a_i, b_i),$$

(ii) $M_j(y) \, \Phi_j \left(\left| \frac{f_{1j}(y)}{f_{2j}(y)} \right| \right)^{p_j}$ is integrable, $j = 1, \ldots, m.$

Corollary 25.25 (to Theorem 25.24) *It holds*

$$\int_{\prod_{i=1}^{N}(a_i, b_i)} u(x)\, e^{\displaystyle \sum_{j=1}^{m} \frac{\left|(I^{\alpha}_{b-}f_{1j})(x)\right|}{(I^{\alpha}_{b-}f_{2j})(x)}}\, dx \le \tag{25.73}$$

$$\prod_{j=1}^{m}\left(\int_{\prod_{i=1}^{N}(a_i, b_i)} M_j(y)\, e^{\displaystyle \frac{p_j\left|f_{1j}(y)\right|}{f_{2j}(y)}}\, dy \right)^{\frac{1}{p_j}},$$

under the assumptions:

(i) $\dfrac{f_{1j}(y)}{f_{2j}(y)}, e^{\, p_j \left| \frac{f_{1j}(y)}{f_{2j}(y)} \right|}$ *are both* $\dfrac{1}{\prod_{i=1}^{N}\Gamma(\alpha_i)} \chi_{\prod_{i=1}^{N}[x_i, b_i)}(y)\, \prod_{i=1}^{N}(y_i - x_i)^{\alpha_i - 1} f_{2j}(y)\, dy$

-integrable, a.e. in $x \in \prod_{i=1}^{N}(a_i, b_i),$

(ii) $M_j(y)\, e^{\, p_j \left| \frac{f_{1j}(y)}{f_{2j}(y)} \right|}$ *is integrable,* $j = 1, \ldots, m.$

Corollary 25.26 (to Theorem 25.24) *Let* $\beta_j \ge 1$, $j = 1, \ldots, m.$ *It holds*

$$\int_{\prod_{i=1}^{N}(a_i, b_i)} u(x) \prod_{j=1}^{m} \left(\frac{\left|I^{\alpha}_{b-}f_{1j}(x)\right|}{I^{\alpha}_{b-}f_{2j}(x)} \right)^{\beta_j}\, dx \le \tag{25.74}$$

$$\prod_{j=1}^{m}\left(\int_{\prod_{i=1}^{N}(a_i, b_i)} M_j(y) \left(\frac{\left|f_{1j}(y)\right|}{f_{2j}(y)} \right)^{\beta_j p_j}\, dy \right)^{\frac{1}{p_j}},$$

under the assumptions:

(i) $\left|\frac{f_{1j}(y)}{f_{2j}(y)}\right|^{\beta_j p_j}$ is $\dfrac{1}{\displaystyle\prod_{i=1}^{N}\Gamma(\alpha_i)\ \prod_{i=1}^{N}[x_i,b_i]} \chi_N$ $(y)\displaystyle\prod_{i=1}^{N}(y_i-x_i)^{\alpha_i-1}\, f_{2j}(y)\,dy$ -integrable,

$$a.e.\ in\ x \in \prod_{i=1}^{N}(a_i,b_i),$$

(ii) $M_j(y)\left|\frac{f_{1j}(y)}{f_{2j}(y)}\right|^{\beta_j p_j}$ is integrable, $j=1,\dots,m.$

We make

Remark 25.27 Next we follow [7] and our introduction.

Let $(\Omega_1,\Sigma_1,\mu_1)$ and $(\Omega_2,\Sigma_2,\mu_2)$ be measure spaces with positive σ-finite measures, and let $k:\Omega_1\times\Omega_2\to\mathbb{R}$ be a nonnegative measurable function, $k(x,\cdot)$ measurable on Ω_2 and

$$K(x)=\int_{\Omega_2}k(x,y)\,d\mu_2(y),\ \text{for any }x\in\Omega_1.\qquad(25.75)$$

We assume $K(x)>0$, a.e. on Ω_1, and the weight functions are nonnegative functions on the related set. We consider measurable functions $g_1,g_2:\Omega_1\to\mathbb{R}$ with the representation

$$g_i(x)=\int_{\Omega_2}k(x,y)\,f_i(y)\,d\mu_2(y),\qquad(25.76)$$

where $f_1,f_2:\Omega_2\to\mathbb{R}$ are measurable, $i=1,2$. Here u stands for a weight function on Ω_1.

The next theorem comes from [7], but it is fixed by us, the terms come from Remark 25.27.

Theorem 25.28 *Here $0<f_2(y)<\infty$, a.e. on Ω_2, and $g_2(x)>0$, a.e. on Ω_1. Assume that the function*

$$x\mapsto u(x)\,\frac{f_2(y)\,k(x,y)}{g_2(x)}$$

is integrable on Ω_1, for each fixed $y\in\Omega_2$.
    Define v on Ω_2 by

$$v(y):=f_2(y)\int_{\Omega_1}\frac{u(x)\,k(x,y)}{g_2(x)}d\mu_1(x)<\infty,\qquad(25.77)$$

a.e. on Ω_2.

Let $\Phi : \mathbb{R}_+ \to \mathbb{R}$ be convex and increasing. Then

$$\int_{\Omega_1} u(x) \Phi \left(\frac{|g_1(x)|}{g_2(x)} \right) d\mu_1(x) \leq \qquad (25.78)$$

$$\int_{\Omega_2} v(y) \Phi \left(\frac{|f_1(y)|}{f_2(y)} \right) d\mu_2(y),$$

under the further assumptions:

(i) $\frac{f_1(y)}{f_2(y)}$, $\Phi \left(\left| \frac{f_1(y)}{f_2(y)} \right| \right)$ are both $k(x, y) f_2(y) d\mu_2(y)$ -integrable, μ_1 -a.e. in $x \in \Omega_1$,

(ii) $v(y) \Phi \left(\left| \frac{f_1(y)}{f_2(y)} \right| \right)$ is μ_2 -integrable.

Next we deal with the spherical shell

Background 25.29 *We need:*
Let $N \geq 2$, $S^{N-1} := \{x \in \mathbb{R}^N : |x| = 1\}$ *the unit sphere on* $\mathbb{R}^N$, *where* $|\cdot|$ *stands for the Euclidean norm in* $\mathbb{R}^N$. *Also denote the ball* $B(0, R) := \{x \in \mathbb{R}^N : |x| < R\} \subseteq \mathbb{R}^N$, $R > 0$, *and the spherical shell*

$$A := B(0, R_2) - \overline{B(0, R_1)}, 0 < R_1 < R_2. \qquad (25.79)$$

For the following see [9, pp. 149–150], and [11, pp. 87–88].
For $x \in \mathbb{R}^N - \{0\}$ *we can write uniquely* $x = r\omega$, *where* $r = |x| > 0$, *and* $\omega = \frac{x}{r} \in S^{N-1}$, $|\omega| = 1$.
Clearly here

$$\mathbb{R}^N - \{0\} = (0, \infty) \times S^{N-1}, \qquad (25.80)$$

and

$$\overline{A} = [R_1, R_2] \times S^{N-1}. \qquad (25.81)$$

We will be using

Theorem 25.30 [1, p. 322] *Let* $f : A \to \mathbb{R}$ *be a Lebesgue integrable function. Then*

$$\int_A f(x) \, dx = \int_{S^{N-1}} \left(\int_{R_1}^{R_2} f(r\omega) r^{N-1} dr \right) d\omega. \qquad (25.82)$$

So we are able to write an integral on the shell in polar form using the polar coordinates (r, ω).

We need

Definition 25.31 [1, p. 458] Let $v > 0$, $n := [v]$, $\alpha := v - n$, $f \in C^n(\overline{A})$, and A is a spherical shell. Assume that there exists function $\frac{\partial^v_{R_1} f(x)}{\partial r^v} \in C(\overline{A})$, given by

$$\frac{\partial^v_{R_1} f(x)}{\partial r^v} := \frac{1}{\Gamma(1 - \alpha)} \frac{\partial}{\partial r} \left(\int_{R_1}^r (r - t)^{-\alpha} \frac{\partial^n f(t\omega)}{\partial r^n} dt \right), \qquad (25.83)$$

where $x \in \overline{A}$; that is $x = r\omega$, $r \in [R_1, R_2]$, $\omega \in S^{N-1}$.

We call $\frac{\partial^v_{R_1} f}{\partial r^v}$ the left radial Canavati-type fractional derivative of f of order v. If $v = 0$, then set $\frac{\partial^v_{R_1} f(x)}{\partial r^v} := f(x)$.

Based on [1, p. 288], and [2] we have

Lemma 25.32 Let $\gamma \geq 0$, $m := [\gamma]$, $v > 0$, $n := [v]$, with $0 \leq \gamma < v$. Let $f \in C^n(\overline{A})$ and there exists $\frac{\partial^v_{R_1} f(x)}{\partial r^v} \in C(\overline{A})$, $x \in \overline{A}$, A a spherical shell. Further assume that $\frac{\partial^j f(R_1\omega)}{\partial r^j} = 0$, $j = m, m+1, \ldots, n-1$, $\forall \omega \in S^{N-1}$. Then there exists $\frac{\partial^\gamma_{R_1} f(x)}{\partial r^\gamma} \in C(\overline{A})$ such that

$$\frac{\partial^\gamma_{R_1} f(x)}{\partial r^\gamma} = \frac{\partial^\gamma_{R_1} f(r\omega)}{\partial r^\gamma} = \frac{1}{\Gamma(v - \gamma)} \int_{R_1}^r (r - t)^{v-\gamma-1} \frac{\partial^v_{R_1} f(t\omega)}{\partial r^v} dt, \quad (25.84)$$

$\forall \omega \in S^{N-1}$; all $R_1 \leq r \leq R_2$, indeed $f(r\omega) \in C^\gamma_{R_1}([R_1, R_2])$, $\forall \omega \in S^{N-1}$.

We make

Remark 25.33 In the settings and assumptions of Theorem 25.28 and Lemma 25.32 we have

$$k(r, t) = k_*(r, t) := \frac{1}{\Gamma(v - \gamma)} \chi_{[R_1, r]}(t) (r - t)^{v-\gamma-1}, \qquad (25.85)$$

$r, t \in [R_1, R_2]$.

Let f_1, f_2 as in Lemma 25.32. Assume $\frac{\partial^v_{R_1} f_2}{\partial r^v} > 0$ on $\overline{A}$. We take

$$g_1(r, \omega) := \frac{\partial^\gamma_{R_1} f_1(r\omega)}{\partial r^\gamma} \qquad (25.86)$$

and

$$g_2(r, \omega) := \frac{\partial^\gamma_{R_1} f_2(r\omega)}{\partial r^\gamma}, \qquad (25.87)$$

for every $(r, \omega) \in [R_1, R_2] \times S^{N-1}$.

We further assume that $g_2(r, \omega) > 0$ for every $(r, \omega) \in [R_1, R_2] \times S^{N-1}$.

We choose here $u(r, \omega) = g_2(r, \omega)$. Then for a fixed $\omega \in S^{N-1}$, the function $r \mapsto \frac{\partial^{\nu}_{R_1}}{\partial r^{\nu}} f_2(t, \omega) k_*(r, t)$ is continuous, hence integrable on $[R_1, R_2]$, for each fixed $t \in [R_1, R_2]$. So we have

$$v(t) = \frac{\partial^{\nu}_{R_1}}{\partial r^{\nu}} f_2(t\omega) \int_t^{R_2} \frac{1}{\Gamma(\nu - \gamma)} (r - t)^{\nu - \gamma - 1} dr \qquad (25.88)$$

$$= \frac{\partial^{\nu}_{R_1}}{\partial r^{\nu}} f_2(t\omega) \frac{(R_2 - t)^{\nu - \gamma}}{\Gamma(\nu - \gamma + 1)} < \infty,$$

for every $t \in [R_1, R_2]$.

Let $\Phi : \mathbb{R}_+ \to \mathbb{R}$ be convex and increasing. Then by (25.78) and (25.88), we get

$$\int_{R_1}^{R_2} \left(\frac{\partial^{\gamma}_{R_1} f_2(r\omega)}{\partial r^{\gamma}} \right) \Phi \left(\frac{\left| \frac{\partial^{\gamma}_{R_1} f_1(r\omega)}{\partial r^{\gamma}} \right|}{\frac{\partial^{\gamma}_{R_1} f_2(r\omega)}{\partial r^{\gamma}}} \right) dr \leq \qquad (25.89)$$

$$\frac{(R_2 - R_1)^{\nu - \gamma}}{\Gamma(\nu - \gamma + 1)} \int_{R_1}^{R_2} \frac{\partial^{\nu}_{R_1}}{\partial r^{\nu}} f_2(t\omega) \Phi \left(\frac{\left| \frac{\partial^{\nu}_{R_1} f_1(t\omega)}{\partial r^{\nu}} \right|}{\frac{\partial^{\nu}_{R_1} f_2(t\omega)}{\partial r^{\nu}}} \right) dt =$$

$$\frac{(R_2 - R_1)^{\nu - \gamma}}{\Gamma(\nu - \gamma + 1)} \int_{R_1}^{R_2} \frac{\partial^{\nu}_{R_1}}{\partial r^{\nu}} f_2(r\omega) \Phi \left(\frac{\left| \frac{\partial^{\nu}_{R_1} f_1(r\omega)}{\partial r^{\nu}} \right|}{\frac{\partial^{\nu}_{R_1} f_2(r\omega)}{\partial r^{\nu}}} \right) dr,$$

true $\forall \omega \in S^{N-1}$.

Here we have $R_1 \leq r \leq R_2$, and $R_1^{N-1} \leq r^{N-1} \leq R_2^{N-1}$, and $R_2^{1-N} \leq r^{1-N} \leq R_1^{1-N}$. So by (25.89) and $r^{N-1} r^{1-N} = 1$, we have

$$\int_{R_1}^{R_2} \left(\frac{\partial^{\gamma}_{R_1} f_2(r\omega)}{\partial r^{\gamma}} \right) \Phi \left(\frac{\left| \frac{\partial^{\gamma}_{R_1} f_1(r\omega)}{\partial r^{\gamma}} \right|}{\frac{\partial^{\gamma}_{R_1} f_2(r\omega)}{\partial r^{\gamma}}} \right) r^{N-1} dr \leq \qquad (25.90)$$

$$\frac{(R_2 - R_1)^{\nu - \gamma}}{\Gamma(\nu - \gamma + 1)} \left(\frac{R_2}{R_1} \right)^{N-1} \int_{R_1}^{R_2} \frac{\partial^{\nu}_{R_1}}{\partial r^{\nu}} f_2(r\omega) \Phi \left(\frac{\left| \frac{\partial^{\nu}_{R_1} f_1(r\omega)}{\partial r^{\nu}} \right|}{\frac{\partial^{\nu}_{R_1} f_2(r\omega)}{\partial r^{\nu}}} \right) r^{N-1} dr.$$

Hence we get

$$\int_{S^{N-1}} \left(\int_{R_1}^{R_2} \left(\frac{\partial_{R_1}^{\gamma} f_2(r\omega)}{\partial r^{\gamma}} \right) \Phi \left(\frac{\left| \frac{\partial_{R_1}^{\gamma} f_1(r\omega)}{\partial r^{\gamma}} \right|}{\frac{\partial_{R_1}^{\gamma} f_2(r\omega)}{\partial r^{\gamma}}} \right) r^{N-1} dr \right) d\omega \leq \qquad (25.91)$$

$$\frac{(R_2 - R_1)^{\nu-\gamma}}{\Gamma(\nu - \gamma + 1)} \left(\frac{R_2}{R_1} \right)^{N-1} \int_{S^{N-1}} \left(\int_{R_1}^{R_2} \frac{\partial_{R_1}^{\nu}}{\partial r^{\nu}} f_2(r\omega) \Phi \left(\frac{\left| \frac{\partial_{R_1}^{\nu} f_1(r\omega)}{\partial r^{\nu}} \right|}{\frac{\partial_{R_1}^{\nu} f_2(r\omega)}{\partial r^{\nu}}} \right) r^{N-1} dr \right) d\omega.$$

Using Theorem 25.30 we obtain

$$\int_A \frac{\partial_{R_1}^{\gamma} f_2(x)}{\partial r^{\gamma}} \Phi \left(\frac{\left| \frac{\partial_{R_1}^{\gamma} f_1(x)}{\partial r^{\gamma}} \right|}{\frac{\partial_{R_1}^{\gamma} f_2(x)}{\partial r^{\gamma}}} \right) dx \leq \qquad (25.92)$$

$$\frac{(R_2 - R_1)^{\nu-\gamma}}{\Gamma(\nu - \gamma + 1)} \left(\frac{R_2}{R_1} \right)^{N-1} \int_A \frac{\partial_{R_1}^{\nu}}{\partial r^{\nu}} f_2(x) \Phi \left(\frac{\left| \frac{\partial_{R_1}^{\nu} f_1(x)}{\partial r^{\nu}} \right|}{\frac{\partial_{R_1}^{\nu} f_2(x)}{\partial r^{\nu}}} \right) dx.$$

We have proved the following result.

Theorem 25.34 Let all and f_1, f_2 as in Lemma 25.32, $\nu > \gamma \geq 0$. Assume $\frac{\partial_{R_1}^{\nu} f_2}{\partial r^{\nu}} > 0$ and $\frac{\partial_{R_1}^{\gamma} f_2}{\partial r^{\gamma}} > 0$ on $\overline{A}$. Let $\Phi : \mathbb{R}_+ \to \mathbb{R}$ be convex and increasing. Then

$$\int_A \frac{\partial_{R_1}^{\gamma} f_2(x)}{\partial r^{\gamma}} \Phi \left(\frac{\left| \frac{\partial_{R_1}^{\gamma} f_1(x)}{\partial r^{\gamma}} \right|}{\frac{\partial_{R_1}^{\gamma} f_2(x)}{\partial r^{\gamma}}} \right) dx \leq \qquad (25.93)$$

$$\frac{(R_2 - R_1)^{\nu-\gamma}}{\Gamma(\nu - \gamma + 1)} \left(\frac{R_2}{R_1} \right)^{N-1} \int_A \frac{\partial_{R_1}^{\nu}}{\partial r^{\nu}} f_2(x) \Phi \left(\frac{\left| \frac{\partial_{R_1}^{\nu} f_1(x)}{\partial r^{\nu}} \right|}{\frac{\partial_{R_1}^{\nu} f_2(x)}{\partial r^{\nu}}} \right) dx.$$

Corollary 25.35 (to Theorem 25.34) *It holds*

$$\int_A \frac{\partial_{R_1}^\gamma f_2(x)}{\partial r^\gamma} e^{\frac{\left|\frac{\partial_{R_1}^\gamma f_1(x)}{\partial r^\gamma}\right|}{\frac{\partial_{R_1}^\gamma f_2(x)}{\partial r^\gamma}}} dx \leq \tag{25.94}$$

$$\frac{(R_2 - R_1)^{\nu-\gamma}}{\Gamma(\nu - \gamma + 1)} \left(\frac{R_2}{R_1}\right)^{N-1} \int_A \frac{\partial_{R_1}^\nu}{\partial r^\nu} f_2(x) e^{\frac{\left|\frac{\partial_{R_1}^\nu f_1(x)}{\partial r^\nu}\right|}{\frac{\partial_{R_1}^\nu f_2(x)}{\partial r^\nu}}} dx.$$

Corollary 25.36 (to Theorem 25.34) *Let $p \geq 1$. It holds*

$$\int_A \left(\frac{\partial_{R_1}^\gamma f_2(x)}{\partial r^\gamma}\right)^{1-p} \left|\frac{\partial_{R_1}^\gamma f_1(x)}{\partial r^\gamma}\right|^p dx \leq \tag{25.95}$$

$$\frac{(R_2 - R_1)^{\nu-\gamma}}{\Gamma(\nu - \gamma + 1)} \left(\frac{R_2}{R_1}\right)^{N-1} \left(\int_A \frac{\partial_{R_1}^\nu}{\partial r^\nu} f_2(x)\right)^{1-p} \left|\frac{\partial_{R_1}^\nu f_1(x)}{\partial r^\nu}\right|^p dx.$$

Similar results can be produced for the right radial Canavati type fractional derivative. We omit this treatment.

We make

Remark 25.37 (from[1], p. 460) Here we denote $\lambda_{\mathbb{R}^N}(x) \equiv dx$ the Lebesgue measure on $\mathbb{R}^N$, $N \geq 2$, and by $\lambda_{S^{N-1}}(\omega) = d\omega$ the surface measure on S^{N-1}, where $\mathcal{B}_X$ stands for the Borel class on space X. Define the measure R_N on $\left((0, \infty), \mathcal{B}_{(0,\infty)}\right)$ by

$$R_N(B) = \int_B r^{N-1} dr, \quad \text{any } B \in \mathcal{B}_{(0,\infty)}.$$

Now let $F \in L_1(A) = L_1\left([R_1, R_2] \times S^{N-1}\right)$.

Call

$$K(F) := \{\omega \in S^{N-1} : F(\cdot\omega) \notin L_1\left([R_1, R_2], \mathcal{B}_{[R_1,R_2]}, R_N\right)\}. \tag{25.96}$$

We get, by Fubini's theorem and [11], pp. 87–88, that

$$\lambda_{S^{N-1}}(K(F)) = 0.$$

Of course

$$\theta(F) := [R_1, R_2] \times K(F) \subset A,$$

and

$$\lambda_{\mathbb{R}^N} (\theta (F)) = 0.$$

Above $\lambda_{S^{N-1}}$ is defined as follows: let $A \subset S^{N-1}$ be a Borel set, and let

$$\widetilde{A} := \{ru : 0 < r < 1, u \in A\} \subset \mathbb{R}^N;$$

we define

$$\lambda_{S^{N-1}} (A) := N \lambda_{\mathbb{R}^N} \left(\widetilde{A}\right).$$

We have that

$$\lambda_{S^{N-1}} \left(S^{N-1}\right) = \frac{2\pi^{\frac{N}{2}}}{\Gamma\left(\frac{N}{2}\right)},$$

the surface area of S^{N-1}.

See also [9, pp. 149–150], [11, pp. 87–88] and [1], p. 320.

Following [1, p. 466] we define the left Riemann-Liouville radial fractional derivative next.

Definition 25.38 Let $\beta > 0$, $m := [\beta] + 1$, $F \in L_1 (A)$, and A is the spherical shell. We define

$$\frac{\overline{\partial}_{R_1}^\beta F (x)}{\partial r^\beta} := \begin{cases} \frac{1}{\Gamma(m-\beta)} \left(\frac{\partial}{\partial r}\right)^m \int_{R_1}^r (r - t)^{m-\beta-1} F (t\omega)\, dt, \\ \qquad\qquad \text{for } \omega \in S^{N-1} - K (F), \\ 0, \quad \text{for } \omega \in K (F), \end{cases} \qquad (25.97)$$

where $x = r\omega \in A$, $r \in [R_1, R_2]$, $\omega \in S^{N-1}$; $K (F)$ as in (25.96).

If $\beta = 0$, define

$$\frac{\overline{\partial}_{R_1}^\beta F (x)}{\partial r^\beta} := F (x).$$

Definition 25.39 ([1], p. 327) We say that $f \in L_1 (a, w)$, $a < w$; $a, w \in \mathbb{R}$ has an L_∞ left Riemann-Liouville fractional derivative $\overline{D}_a^\beta f$ ($\beta > 0$) in $[a, w]$, iff

(1) $\overline{D}_a^{\beta-k} f \in C ([a, w])$, $k = 1, \ldots, m := [\beta] + 1$;
(2) $\overline{D}_a^{\beta-1} f \in AC ([a, w])$; and
(3) $\overline{D}_a^\beta f \in L_\infty (a, w)$.

Define $\overline{D}_a^0 f := f$ and $\overline{D}_a^{-\delta} f := I_{a+}^\delta f$, if $0 < \delta \le 1$; here $I_{a+}^\delta f$ is the left univariate Riemann-Liouville fractional integral of f, see (25.57).

We need the following representation result.

Theorem 25.40 ([1, p. 331]) *Let $\beta > \alpha > 0$ and $F \in L_1(A)$. Assume that $\dfrac{\partial_{R_1}^{\beta} F(x)}{\partial r^{\beta}} \in L_{\infty}(A)$. Further assume that $\overline{D}_{R_1}^{\beta} F(r\omega)$ takes real values for almost all $r \in [R_1, R_2]$, for each $\omega \in S^{N-1}$, and for these $\left| \overline{D}_{R_1}^{\beta} F(r\omega) \right| \le M_1$ for some $M_1 > 0$. For each $\omega \in S^{N-1} - K(F)$, we assume that $F(\cdot\omega)$ have an L_{∞} fractional derivative $\overline{D}_{R_1}^{\beta} F(\cdot\omega)$ in $[R_1, R_2]$, and that*

$$\overline{D}_{R_1}^{\beta-k} F(R_1\omega) = 0, \quad k = 1, \dots, [\beta] + 1.$$

Then

$$\frac{\partial_{R_1}^{\alpha} F(x)}{\partial r^{\alpha}} = \left(\overline{D}_{R_1}^{\alpha} F \right)(r\omega) = \frac{1}{\Gamma(\beta - \alpha)} \int_{R_1}^{r} (r - t)^{\beta-\alpha-1} \left(\overline{D}_{R_1}^{\beta} F \right)(t\omega)\, dt,$$

$$(25.98)$$

is true for all $x \in A$; i.e., true for all $r \in [R_1, R_2]$ and for all $\omega \in S^{N-1}$.
  Here

$$\left(\overline{D}_{R_1}^{\alpha} F \right)(\cdot\omega) \in AC\left([R_1, R_2]\right), \quad \text{for } \beta - \alpha \ge 1$$

and

$$\left(\overline{D}_{R_1}^{\alpha} F \right)(\cdot\omega) \in C\left([R_1, R_2]\right), \quad \text{for } \beta - \alpha \in (0, 1),$$

for all $\omega \in S^{N-1}$. Furthermore

$$\frac{\partial_{R_1}^{\alpha} F(x)}{\partial r^{\alpha}} \in L_{\infty}(A).$$

In particular, it holds

$$F(x) = F(r\omega) = \frac{1}{\Gamma(\beta)} \int_{R_1}^{r} (r - t)^{\beta-1} \left(\overline{D}_{R_1}^{\beta} F \right)(t\omega)\, dt, \qquad (25.99)$$

for all $r \in [R_1, R_2]$ and $\omega \in S^{N-1} - K(F)$; $x = r\omega$, and

$$F(\cdot\omega) \in AC\left([R_1, R_2]\right), \quad \text{for } \beta \ge 1$$

and

$$F(\cdot\omega) \in C\left([R_1, R_2]\right), \quad \text{for } \beta \in (0, 1),$$

for all $\omega \in S^{N-1} - K(F)$.

  Similarly to Theorem 25.34 we obtain

Theorem 25.41 *Let all here and F_1, F_2 as in Theorem 25.40, $\beta > \alpha \geq 0$. Assume that $0 < \dfrac{\overline{\partial}^{\beta}_{R_1} F_2(x)}{\partial r^{\beta}} < \infty$, a.e. on A and $\dfrac{\overline{\partial}^{\alpha}_{R_1} F_2(x)}{\partial r^{\beta}} > 0$, a.e. on A. Let $\Phi : \mathbb{R}_+ \to \mathbb{R}$ be convex and increasing. Then*

$$\int_A \frac{\overline{\partial}^{\alpha}_{R_1} F_2(x)}{\partial r^{\alpha}} \Phi\left(\frac{\left| \frac{\overline{\partial}^{\alpha}_{R_1} F_1(x)}{\partial r^{\alpha}} \right|}{\frac{\overline{\partial}^{\alpha}_{R_1} F_2(x)}{\partial r^{\alpha}}} \right) dx \leq \tag{25.100}$$

$$\frac{(R_2 - R_1)^{\beta - \alpha}}{\Gamma(\beta - \alpha + 1)} \left(\frac{R_2}{R_1} \right)^{N-1} \int_A \frac{\overline{\partial}^{\beta}_{R_1} F_2(x)}{\partial r^{\beta}} \Phi\left(\frac{\left| \frac{\overline{\partial}^{\beta}_{R_1} F_1(x)}{\partial r^{\beta}} \right|}{\frac{\overline{\partial}^{\beta}_{R_1} F_2(x)}{\partial r^{\beta}}} \right) dx,$$

under the assumptions: for every $\omega \in S^{N-1}$ we have

(i) $\dfrac{\overline{\partial}^{\beta}_{R_1} F_1(t\omega)}{\partial r^{\beta}} , \Phi\left(\dfrac{\left| \frac{\overline{\partial}^{\beta}_{R_1} F_1(t\omega)}{\partial r^{\beta}} \right|}{\frac{\overline{\partial}^{\beta}_{R_1} F_2(t\omega)}{\partial r^{\beta}}} \right)$ *are both* $\dfrac{1}{\Gamma(\beta - \alpha)} \chi_{[R_1, r]}(t)(r - t)^{\beta - \alpha - 1}$

$\dfrac{\overline{\partial}^{\beta}_{R_1} F_2(t\omega)}{\partial r^{\beta}}$-*dt-integrable in* $t \in [R_1, R_2]$, *a.e. in* $r \in [R_1, R_2]$,

(ii) $\dfrac{\overline{\partial}^{\beta}_{R_1} F_2(t\omega)}{\partial r^{\beta}} \Phi\left(\dfrac{\left| \frac{\overline{\partial}^{\beta}_{R_1} F_1(t\omega)}{\partial r^{\beta}} \right|}{\frac{\overline{\partial}^{\beta}_{R_1} F_2(t\omega)}{\partial r^{\beta}}} \right)$ *is integrable in* t *on* $[R_1, R_2]$.

Corollary 25.42 (to Theorem 25.41) *It holds*

$$\int_A \frac{\overline{\partial}^{\alpha}_{R_1} F_2(x)}{\partial r^{\alpha}} e^{\frac{\left| \frac{\overline{\partial}^{\alpha}_{R_1} F_1(x)}{\partial r^{\alpha}} \right|}{\frac{\overline{\partial}^{\alpha}_{R_1} F_2(x)}{\partial r^{\alpha}}}} dx \leq \tag{25.101}$$

$$\frac{(R_2 - R_1)^{\beta - \alpha}}{\Gamma(\beta - \alpha + 1)} \left(\frac{R_2}{R_1} \right)^{N-1} \int_A \frac{\overline{\partial}^{\beta}_{R_1} F_2(x)}{\partial r^{\beta}} e^{\frac{\left| \frac{\overline{\partial}^{\beta}_{R_1} F_1(x)}{\partial r^{\beta}} \right|}{\frac{\overline{\partial}^{\beta}_{R_1} F_2(x)}{\partial r^{\beta}}}} dx,$$

under the assumptions: for every $\omega \in S^{N-1}$ we have

(i) $\dfrac{\dfrac{\overline{\partial}^{\beta}_{R_1} F_1 (t\omega)}{\partial r^{\beta}}}{\dfrac{\overline{\partial}^{\beta}_{R_1} F_2 (t\omega)}{\partial r^{\beta}}}$, $e^{\left| \dfrac{\dfrac{\overline{\partial}^{\beta}_{R_1} F_1 (t\omega)}{\partial r^{\beta}}}{\dfrac{\overline{\partial}^{\beta}_{R_1} F_2 (t\omega)}{\partial r^{\beta}}} \right|}$ are both $\dfrac{1}{\Gamma(\beta-\alpha)} \chi_{[R_1,r]} (t) (r - t)^{\beta-\alpha-1} \dfrac{\overline{\partial}^{\beta}_{R_1} F_2 (t\omega)}{\partial r^{\beta}} dt$-

integrable in $t \in [R_1, R_2]$, a.e. in $r \in [R_1, R_2]$,

(ii) $\dfrac{\overline{\partial}^{\beta}_{R_1} F_2 (t\omega)}{\partial r^{\beta}} e^{\left| \dfrac{\dfrac{\overline{\partial}^{\beta}_{R_1} F_1 (t\omega)}{\partial r^{\beta}}}{\dfrac{\overline{\partial}^{\beta}_{R_1} F_2 (t\omega)}{\partial r^{\beta}}} \right|}$ is integrable in t on $[R_1, R_2]$.

Corollary 25.43 (to Theorem 25.41) *Let $p \geq 1$. It holds*

$$\int_A \left(\frac{\overline{\partial}^{\alpha}_{R_1} F_2 (x)}{\partial r^{\alpha}} \right)^{1-p} \left| \frac{\overline{\partial}^{\alpha}_{R_1} F_1 (x)}{\partial r^{\alpha}} \right|^p dx \leq \qquad (25.102)$$

$$\frac{(R_2 - R_1)^{\beta-\alpha}}{\Gamma (\beta - \alpha + 1)} \left(\frac{R_2}{R_1} \right)^{N-1} \int_A \left(\frac{\overline{\partial}^{\beta}_{R_1} F_2 (x)}{\partial r^{\beta}} \right)^{1-p} \left| \frac{\overline{\partial}^{\beta}_{R_1} F_1 (x)}{\partial r^{\beta}} \right|^p dx,$$

under the assumptions: for every $\omega \in S^{N-1}$ we have

(i) $\left(\left| \dfrac{\dfrac{\overline{\partial}^{\beta}_{R_1} F_1 (t\omega)}{\partial r^{\beta}}}{\dfrac{\overline{\partial}^{\beta}_{R_1} F_2 (t\omega)}{\partial r^{\beta}}} \right| \right)^p$ *is* $\dfrac{1}{\Gamma(\beta-\alpha)} \chi_{[R_1,r]} (t) (r - t)^{\beta-\alpha-1} \dfrac{\overline{\partial}^{\beta}_{R_1} F_2 (t\omega)}{\partial r^{\beta}} dt$-*integrable in $t \in$*

$[R_1, R_2]$, a.e. in $r \in [R_1, R_2]$,

(ii) $\left(\left| \dfrac{\overline{\partial}^{\beta}_{R_1} F_2 (t\omega)}{\partial r^{\beta}} \right| \right)^{1-p} \left| \dfrac{\overline{\partial}^{\beta}_{R_1} F_1 (t\omega)}{\partial r^{\beta}} \right|^p$ *is integrable in t on $[R_1, R_2]$.*

We also need (see [1], p. 421).

Definition 25.44 Let $F : \overline{A} \to \mathbb{R}$, $\nu \geq 0$, $n := \lceil \nu \rceil$ such that $F (\cdot \omega) \in AC^n ([R_1, R_2])$, for all $\omega \in S^{N-1}$.

We call the left Caputo radial fractional derivative the following function

$$\frac{\partial^{\nu}_{*R_1} F (x)}{\partial r^{\nu}} := \frac{1}{\Gamma (n - \nu)} \int_{R_1}^r (r - t)^{n-\nu-1} \frac{\partial^n F (t\omega)}{\partial r^n} dt, \qquad (25.103)$$

where $x \in \overline{A}$, i.e., $x = r\omega$, $r \in [R_1, R_2]$, $\omega \in S^{N-1}$.

Clearly

$$\frac{\partial^0_{*R_1} F(x)}{\partial r^0} = F(x), \tag{25.104}$$

$$\frac{\partial^v_{*R_1} F(x)}{\partial r^v} = \frac{\partial^v F(x)}{\partial r^v}, \quad \text{if } v \in \mathbb{N}. \tag{25.105}$$

Above function (25.103) exists almost everywhere for $x \in \overline{A}$, see [1], p. 422.

We mention the following fundamental representation result (see [1], pp. 422–423, [2] and [5]).

Theorem 25.45 *Let $\beta > \alpha \geq 0, n := \lceil \beta \rceil, m := \lceil \alpha \rceil, F : \overline{A} \to \mathbb{R}$ with $F \in L_1(A)$. Assume that $F(\cdot\omega) \in AC^n([R_1, R_2])$, for all $\omega \in S^{N-1}$, and that $\frac{\partial^\beta_{*R_1} F(\cdot\omega)}{\partial r^\beta} \in L_\infty(R_1, R_2)$ for all $\omega \in S^{N-1}$.*

Further assume that $\frac{\partial^\beta_{*R_1} F(x)}{\partial r^\beta} \in L_\infty(A)$. More precisely, for these $r \in [R_1, R_2]$, for each $\omega \in S^{N-1}$, for which $D^\beta_{*R_1} F(r\omega)$ takes real values, there exists $M_1 > 0$ such that $\left| D^\beta_{*R_1} F(r\omega) \right| \leq M_1$.

We suppose that $\frac{\partial^j F(R_1\omega)}{\partial r^j} = 0, j = m, m+1, \ldots, n-1$; for every $\omega \in S^{N-1}$. Then

$$\frac{\partial^\alpha_{*R_1} F(x)}{\partial r^\alpha} = D^\alpha_{*R_1} F(r\omega) = \frac{1}{\Gamma(\beta - \alpha)} \int_{R_1}^r (r-t)^{\beta-\alpha-1} \left(D^\beta_{*R_1} F \right)(t\omega) \, dt, \tag{25.106}$$

valid $\forall x \in \overline{A}$; i.e., true $\forall r \in [R_1, R_2]$ and $\forall \omega \in S^{N-1}$; $\alpha > 0$.

Here

$$D^\alpha_{*R_1} F(\cdot\omega) \in C([R_1, R_2]),$$

$\forall \omega \in S^{N-1}$; $\alpha > 0$.

Furthermore

$$\frac{\partial^\alpha_{*R_1} F(x)}{\partial r^\alpha} \in L_\infty(A), \alpha > 0. \tag{25.107}$$

In particular, it holds

$$F(x) = F(r\omega) = \frac{1}{\Gamma(\beta)} \int_{R_1}^r (r-t)^{\beta-1} \left(D^\beta_{*R_1} F \right)(t\omega) \, dt, \tag{25.108}$$

true $\forall x \in \overline{A}$; i.e., true $\forall r \in [R_1, R_2]$ and $\forall \omega \in S^{N-1}$, and

$$F(\cdot\omega) \in C([R_1, R_2]), \quad \forall \omega \in S^{N-1}. \tag{25.109}$$

Similarly to Theorem 25.34 we obtain

Theorem 25.46 *Let all here and F_1, F_2 as in Theorem 25.45, $\beta > \alpha \geq 0$. Assume that $0 < \dfrac{\partial^{\beta}_{*R_1} F_2(x)}{\partial r^{\beta}} < \infty$, a.e. on A and $\dfrac{\partial^{\alpha}_{*R_1} F_2(x)}{\partial r^{\beta}} > 0$, a.e. on A. Let $\Phi : \mathbb{R}_+ \to \mathbb{R}$ be convex and increasing. Then*

$$\int_A \frac{\partial^{\alpha}_{*R_1} F_2(x)}{\partial r^{\alpha}} \Phi\left(\frac{\left|\frac{\partial^{\alpha}_{*R_1} F_1(x)}{\partial r^{\alpha}}\right|}{\frac{\partial^{\alpha}_{*R_1} F_2(x)}{\partial r^{\alpha}}}\right) dx \leq \tag{25.110}$$

$$\frac{(R_2 - R_1)^{\beta - \alpha}}{\Gamma(\beta - \alpha + 1)} \left(\frac{R_2}{R_1}\right)^{N-1} \int_A \frac{\partial^{\beta}_{*R_1} F_2(x)}{\partial r^{\beta}} \Phi\left(\frac{\left|\frac{\partial^{\beta}_{*R_1} F_1(x)}{\partial r^{\beta}}\right|}{\frac{\partial^{\beta}_{*R_1} F_2(x)}{\partial r^{\beta}}}\right) dx,$$

under the assumptions: for every $\omega \in S^{N-1}$ we have

(i) $\dfrac{\partial^{\beta}_{*R_1} F_1(t\omega)}{\partial r^{\beta}}$, $\Phi\left(\dfrac{\left|\frac{\partial^{\beta}_{*R_1} F_1(t\omega)}{\partial r^{\beta}}\right|}{\frac{\partial^{\beta}_{*R_1} F_2(t\omega)}{\partial r^{\beta}}}\right)$ *are both* $\dfrac{1}{\Gamma(\beta - \alpha)} \chi_{[R_1, r]}(t)(r - t)^{\beta - \alpha - 1}$

$\dfrac{\partial^{\beta}_{*R_1} F_2(t\omega)}{\partial r^{\beta}}$ *-dt-integrable in $t \in [R_1, R_2]$, a.e. in $r \in [R_1, R_2]$,*

(ii) $\dfrac{\partial^{\beta}_{*R_1} F_2(t\omega)}{\partial r^{\beta}} \Phi\left(\dfrac{\left|\frac{\partial^{\beta}_{*R_1} F_1(t\omega)}{\partial r^{\beta}}\right|}{\frac{\partial^{\beta}_{*R_1} F_2(t\omega)}{\partial r^{\beta}}}\right)$ *is integrable in t on $[R_1, R_2]$.*

Corollary 25.47 (to Theorem 25.46) *It holds*

$$\int_A \frac{\partial^{\alpha}_{*R_1} F_2(x)}{\partial r^{\alpha}} e^{\frac{\left|\frac{\partial^{\alpha}_{*R_1} F_1(x)}{\partial r^{\alpha}}\right|}{\frac{\partial^{\alpha}_{*R_1} F_2(x)}{\partial r^{\alpha}}}} dx \leq \tag{25.111}$$

$$\frac{(R_2 - R_1)^{\beta - \alpha}}{\Gamma(\beta - \alpha + 1)} \left(\frac{R_2}{R_1}\right)^{N-1} \int_A \frac{\partial^{\beta}_{*R_1} F_2(x)}{\partial r^{\beta}} e^{\frac{\left|\frac{\partial^{\beta}_{*R_1} F_1(x)}{\partial r^{\beta}}\right|}{\frac{\partial^{\beta}_{*R_1} F_2(x)}{\partial r^{\beta}}}} dx,$$

under the assumptions: for every $\omega \in S^{N-1}$ we have

(i) $\dfrac{\dfrac{\partial^{\beta}_{*R_1} F_1(t\omega)}{\partial r^{\beta}}}{\dfrac{\partial^{\beta}_{*R_1} F_2(t\omega)}{\partial r^{\beta}}}, \ e^{\left|\dfrac{\dfrac{\partial^{\beta}_{*R_1} F_1(t\omega)}{\partial r^{\beta}}}{\dfrac{\partial^{\beta}_{*R_1} F_2(t\omega)}{\partial r^{\beta}}}\right|}$ *are both* $\dfrac{1}{\Gamma(\beta-\alpha)}\chi_{[R_1,r]}(t)(r-t)^{\beta-\alpha-1}\dfrac{\partial^{\beta}_{*R_1} F_2(t\omega)}{\partial r^{\beta}}\,dt$ -

integrable in $t \in [R_1, R_2]$, *a.e. in* $r \in [R_1, R_2]$,

(ii) $\dfrac{\partial^{\beta}_{*R_1} F_2(t\omega)}{\partial r^{\beta}}\, e^{\left|\dfrac{\dfrac{\partial^{\beta}_{*R_1} F_1(t\omega)}{\partial r^{\beta}}}{\dfrac{\partial^{\beta}_{*R_1} F_2(t\omega)}{\partial r^{\beta}}}\right|}$ *is integrable in* t *on* $[R_1, R_2]$.

We finish with

Corollary 25.48 (to Theorem 25.46) *Let* $p \geq 1$. *It holds*

$$\int_A \left(\frac{\partial^{\alpha}_{*R_1} F_2(x)}{\partial r^{\alpha}}\right)^{1-p} \left|\frac{\partial^{\alpha}_{*R_1} F_1(x)}{\partial r^{\alpha}}\right|^p dx \leq \qquad (25.112)$$

$$\frac{(R_2 - R_1)^{\beta-\alpha}}{\Gamma(\beta-\alpha+1)} \left(\frac{R_2}{R_1}\right)^{N-1} \int_A \left(\frac{\partial^{\beta}_{*R_1} F_2(x)}{\partial r^{\beta}}\right)^{1-p} \left|\frac{\partial^{\beta}_{*R_1} F_1(x)}{\partial r^{\beta}}\right|^p dx,$$

under the assumptions: for every $\omega \in S^{N-1}$ *we have*

(i) $\left(\left|\dfrac{\dfrac{\partial^{\beta}_{*R_1} F_1(t\omega)}{\partial r^{\beta}}}{\dfrac{\partial^{\beta}_{*R_1} F_2(t\omega)}{\partial r^{\beta}}}\right|\right)^p$ *is* $\dfrac{1}{\Gamma(\beta-\alpha)}\chi_{[R_1,r]}(t)(r-t)^{\beta-\alpha-1}\dfrac{\partial^{\beta}_{*R_1} F_2(t\omega)}{\partial r^{\beta}}dt$-*integrable in* $t \in$

$[R_1, R_2]$, *a.e. in* $r \in [R_1, R_2]$,

(ii) $\left(\left|\dfrac{\partial^{\beta}_{*R_1} F_2(t\omega)}{\partial r^{\beta}}\right|\right)^{1-p} \left|\dfrac{\partial^{\beta}_{*R_1} F_1(t\omega)}{\partial r^{\beta}}\right|^p$ *is integrable in* t *on* $[R_1, R_2]$.

References

1. G.A. Anastassiou, *Fractional Differentiation Inequalities*, Research Monograph (Springer, New York, 2009)
2. G.A. Anastassiou, Fractional representation formulae and right fractional inequalities. Math. Comput. Modell. **54**(11–12), 3098–3115 (2011)
3. G.A. Anastassiou, Separating rational L_p inequalities for integral operators. Panam. Math. J. **22**(3), 117–145 (2012)
4. G.A. Anastassiou, Fractional integral inequalities involving convexity. Sarajevo J. Math. M. Kulenovic 60th birthday **8**(21), 203–233 (2012)

5. M. Andric, J.E. Pecaric, I. Peric, Composition identities for the Caputo fractional derivatives and applications to opial-type inequalities (2012) (Submitted)
6. S. Iqbal, K. Krulic, J. Pecaric, On an inequality of H.G. Hardy. J. Inequalities Appl. **2010** Article ID 264347, p. 23 (2010)
7. S. Iqbal, K. Krulic, J. Pecaric, On an inequality for convex functions with some applications on fractional derivatives and fractional integrals. J. Math. Inequalities **5**(2), 219–230 (2011)
8. T. Mamatov, S. Samko, Mixed fractional integration operators in mixed weighted Hölder spaces. Fractional Calc. Appl. Anal. **13**(3), 245–259 (2010)
9. W. Rudin, *Real and Complex Analysis*, International Student Edition (Mc Graw Hill, London, 1970)
10. S.G. Samko, A.A. Kilbas, O.I. Marichev, *Fractional Integral and Derivatives: Theory and Applications* (Gordon and Breach Science Publishers, Yverdon, 1993)
11. D. Stroock, *A Concise Introduction to the Theory of Integration*, 3rd edn. (Birkhäuser, Boston, 1999)

Chapter 26
About Vectorial Hardy Type Fractional Inequalities

Here we present vectorial integral inequalities for products of multivariate convex and increasing functions applied to vectors of functions. As applications we derive a wide range of vectorial fractional inequalities of Hardy type. They involve the left and right Riemann-Liouville fractional integrals and their generalizations, in particular the Hadamard fractional integrals. Also inequalities for left and right Riemann-Liouville, Caputo, Canavati and their generalizations fractional derivatives. These application inequalities are of L_p type, $p \geq 1$, and exponential type. It follows [5].

26.1 Introduction

We start with some facts about fractional derivatives needed in the sequel, for more details see, for instance [1, 11].

Let $a < b$, $a, b \in \mathbb{R}$. By $C^N([a, b])$, we denote the space of all functions on $[a, b]$ which have continuous derivatives up to order N, and $AC([a, b])$ is the space of all absolutely continuous functions on $[a, b]$. By $AC^N([a, b])$, we denote the space of all functions g with $g^{(N-1)} \in AC([a, b])$. For any $\alpha \in \mathbb{R}$, we denote by $[\alpha]$ the integral part of α (the integer k satisfying $k \leq \alpha < k + 1$), and $\lceil \alpha \rceil$ is the ceiling of α ($\min\{n \in \mathbb{N}, n \geq \alpha\}$). By $L_1(a, b)$, we denote the space of all functions integrable on the interval (a, b), and by $L_\infty(a, b)$ the set of all functions measurable and essentially bounded on (a, b). Clearly, $L_\infty(a, b) \subset L_1(a, b)$.

We start with the definition of the Riemann-Liouville fractional integrals, see [14]. Let $[a, b]$, $(-\infty < a < b < \infty)$ be a finite interval on the real axis $\mathbb{R}$. The Riemann-Liouville fractional integrals $I_{a+}^\alpha f$ and $I_{b-}^\alpha f$ of order $\alpha > 0$ are defined by

$$\left(I_{a+}^\alpha f\right)(x) = \frac{1}{\Gamma(\alpha)} \int_a^x f(t)(x - t)^{\alpha-1} dt, \quad (x > a), \tag{26.1}$$

$$\left(I_{b-}^\alpha f\right)(x) = \frac{1}{\Gamma(\alpha)} \int_x^b f(t)(t - x)^{\alpha-1} dt, \quad (x < b), \tag{26.2}$$

© Springer International Publishing Switzerland 2016
G.A. Anastassiou, *Intelligent Comparisons: Analytic Inequalities*,
Studies in Computational Intelligence 609,
DOI 10.1007/978-3-319-21121-3_26

respectively. Here $\Gamma(\alpha)$ is the Gamma function. These integrals are called the left-sided and the right-sided fractional integrals. We mention some properties of the operators $I_{a+}^{\alpha} f$ and $I_{b-}^{\alpha} f$ of order $\alpha > 0$, see also [16]. The first result yields that the fractional integral operators $I_{a+}^{\alpha} f$ and $I_{b-}^{\alpha} f$ are bounded in $L_p(a, b)$, $1 \leq p \leq \infty$, that is

$$\left\| I_{a+}^{\alpha} f \right\|_p \leq K \left\| f \right\|_p \quad , \quad \left\| I_{b-}^{\alpha} f \right\|_p \leq K \left\| f \right\|_p \tag{26.3}$$

where

$$K = \frac{(b-a)^{\alpha}}{\alpha \Gamma(\alpha)}. \tag{26.4}$$

Inequality (26.3), that is the result involving the left-sided fractional integral, was proved by H.G. Hardy in one of his first papers, see [12]. He did not write down the constant, but the calculation of the constant was hidden inside his proof.

Next we are motivated by [13]. We produce a wide range of vectorial integral inequalities related to integral operators, with applications to vectorial Hardy type fractional inequalities.

26.2 Main Results

Let $(\Omega_1, \Sigma_1, \mu_1)$ and $(\Omega_2, \Sigma_2, \mu_2)$ be measure spaces with positive σ-finite measures, and let $k : \Omega_1 \times \Omega_2 \to \mathbb{R}$ be a nonnegative measurable function, $k(x, \cdot)$ measurable on Ω_2 and

$$K(x) = \int_{\Omega_2} k(x, y) \, d\mu_2(y), \quad x \in \Omega_1. \tag{26.5}$$

We suppose that $K(x) > 0$ a.e. on Ω_1, and by a weight function (shortly: a weight), we mean a nonnegative measurable function on the actual set. Let the measurable functions $g_i : \Omega_1 \to \mathbb{R}$, $i = 1, \ldots, n$, with the representation

$$g_i(x) = \int_{\Omega_2} k(x, y) f_i(y) \, d\mu_2(y), \tag{26.6}$$

where $f_i : \Omega_2 \to \mathbb{R}$ are measurable functions, $i = 1, \ldots, n$.

Denote by $\overrightarrow{x} = x := (x_1, \ldots, x_n) \in \mathbb{R}^n$, $\overrightarrow{g} := (g_1, \ldots, g_n)$ and $\overrightarrow{f} := (f_1, \ldots, f_n)$.

We consider here $\Phi : \mathbb{R}_+^n \to \mathbb{R}$ a convex function, which is increasing per coordinate, i.e. if $x_i \leq y_i$, $i = 1, \ldots, n$, then $\Phi(x_1, \ldots, x_n) \leq \Phi(y_1, \ldots, y_n)$.

Examples for Φ:

(1) Given g_i is convex and increasing on $\mathbb{R}_+$, then $\Phi(x_1, \ldots, x_n) := \sum_{i=1}^{n} g_i(x_i)$ is convex on $\mathbb{R}_+^n$, and increasing per coordinate; the same properties hold for:

(2) $\|x\|_p = \left(\sum_{i=1}^{n} x_i^p\right)^{\frac{1}{p}}$, $p \geq 1$,

(3) $\|x\|_\infty = \max_{i \in \{1, \ldots, n\}} x_i$,

(4) $\sum_{i=1}^{n} x_i^2$,

(5) $\sum_{i=1}^{n} \left(i \cdot x_i^2\right)$,

(6) $\sum_{i=1}^{n} \sum_{j=1}^{i} x_j^2$,

(7) $\ln\left(\sum_{i=1}^{n} e^{x_i}\right)$,

(8) let g_j are convex and increasing per coordinate on $\mathbb{R}_+^n$, then so is $\sum_{j=1}^{m} e^{g_j(x)}$, and so is $\ln\left(\sum_{j=1}^{m} e^{g_j(x)}\right)$, $x \in \mathbb{R}_+^n$.

It is a well known fact that, if $C \subseteq \mathbb{R}^n$ is an open and convex set, and $f : C \to \mathbb{R}$ is a convex function, then f is continuous on C.

Proposition 26.1 *Let* $\Phi : \mathbb{R}_+^n \to \mathbb{R}$ *be a convex function which is increasing per coordinate. Then* Φ *is continuous.*

Proof The set $(0, \infty)^n$ is an open and convex subset of $\mathbb{R}^n$. Thus Φ is continuous there. So we need to prove only that Φ is continuous at the origin $0 = (0, \ldots, 0)$. By $B(0, r)$ we denote the open ball in $\mathbb{R}^n$, $r > 0$. Let $x \in B(0, r) \cap \mathbb{R}_+^n$, $x \neq 0$; that is $0 < \|x\| < r$.

Define $g : [0, r] \to \mathbb{R}$ by $g(t) := \Phi\left(t \cdot \frac{x}{\|x\|}\right)$, $t \in [0, r]$. For $t_1, t_2 \in [0, r]$, $\lambda \in (0, 1)$, we observe that $g(\lambda t_1 + (1 - \lambda) t_2) = \Phi\left((\lambda t_1 + (1 - \lambda) t_2) \frac{x}{\|x\|}\right) = \Phi\left(\lambda \left(t_1 \frac{x}{\|x\|}\right) + (1 - \lambda) \left(t_2 \frac{x}{\|x\|}\right)\right) \leq \lambda \Phi\left(t_1 \frac{x}{\|x\|}\right) + (1 - \lambda) \Phi\left(t_2 \frac{x}{\|x\|}\right) = \lambda g(t_1) + (1 - \lambda) g(t_2)$, that is g is a convex function on $[0, r]$.

Next let $t_1 \leq t_2$, $t_1, t_2 \in [0, r]$, then $g(t_1) = \Phi\left(t_1 \frac{x}{\|x\|}\right) = \Phi\left(t_1 \frac{x_1}{\|x\|}, t_1 \frac{x_2}{\|x\|}, \ldots, t_1 \frac{x_n}{\|x\|}\right) \leq \Phi\left(t_2 \frac{x_1}{\|x\|}, t_2 \frac{x_2}{\|x\|}, \ldots, t_2 \frac{x_n}{\|x\|}\right) = \Phi\left(t_2 \frac{x}{\|x\|}\right) = g(t_2)$, hence $g(t_1) \leq g(t_2)$, that is g is increasing on $[0, r]$. Of course g is continuous on $(0, r)$.

We first prove that g is continuous at zero. Consider the line (l) through $(0, g(0))$ and $(r, g(r))$. It has slope $\frac{g(r) - g(0)}{r} \geq 0$, and equation $y = l(z) = \left(\frac{g(r) - g(0)}{r}\right) z + g(0)$. If $g(r) = g(0)$, then $g(t) = g(0)$, for all $t \in [0, r]$, so trivially g is continuous at zero and r.

We treat the other case of $g(r) > g(0)$. By convexity of g we have that for any $0 < z < r$, it is $g(z) \leq l(z)$, equivalently, $g(z) \leq \left(\frac{g(r) - g(0)}{r}\right) z + g(0)$, equivalently $0 \leq g(z) - g(0) \leq \left(\frac{g(r) - g(0)}{r}\right) z$; here $\frac{g(r) - g(0)}{r} > 0$. Letting $z \to 0$, then $g(z) - g(0) \to 0$. That is $\lim_{z \to 0} g(z) = g(0)$, proving continuity of g at zero. So that g is continuous on $[0, r)$.

Clearly Φ is continuous at $r \cdot \frac{x}{\|x\|} \in (0, \infty)^n$. So we choose $(r_n)_{n \in \mathbb{N}}$ such that $0 < \|x\| < r_n \le r$, with $r_n \to r$, then $\Phi\left(r_n \frac{x}{\|x\|}\right) \to \Phi\left(r \frac{x}{\|x\|}\right)$, proving continuity of g at r.

Therefore g is continuous on $[0, r]$.

Hence there exists $M > 0$ such that $|g(t)| < M$, $\forall\, t \in [0, r]$. Since g is convex on $[0, r]$ it has an increasing slope, therefore $\frac{g(\|x\|) - g(0)}{\|x\|} \le \frac{g(r) - g(0)}{r} < \frac{M - g(0)}{r}$, that is $g(\|x\|) - \Phi(0) < \left(\frac{M - \Phi(0)}{r}\right)\|x\|$. Equivalently, we have $0 \le \Phi(x) - \Phi(0) < \left(\frac{M - \Phi(0)}{r}\right)\|x\|$. Clearly $\lim_{x \to 0} \Phi(x) = \Phi(0)$, proving continuity of Φ at $x = 0$. ∎

We need also

Theorem 26.2 (multivariate Jensen inequality, see also [9, p. 26],[15]) *Let f be a convex function defined on a convex subset $C \subseteq \mathbb{R}^n$, and let $X = (X_1, \ldots, X_n)$ be a random vector such that $P(X \in C) = 1$. Assume also $E(|X|)$, $E(|f(X)|) < \infty$. Then $EX \in C$, and*

$$f(EX) \le Ef(X). \tag{26.7}$$

We give our first main result.

Theorem 26.3 *Let u be a weight function on Ω_1, and k, K, g_i, f_i, $i = 1, \ldots, n \in \mathbb{N}$, and Φ defined as above. Assume that the function $x \to u(x) \frac{k(x,y)}{K(x)}$ is integrable on Ω_1 for each fixed $y \in \Omega_2$. Define v on Ω_2 by*

$$v(y) := \int_{\Omega_1} u(x) \frac{k(x,y)}{K(x)} d\mu_1(x) < \infty. \tag{26.8}$$

Then

$$\int_{\Omega_1} u(x)\, \Phi\left(\frac{|g_1(x)|}{K(x)}, \ldots, \frac{|g_n(x)|}{K(x)}\right) d\mu_1(x) \le$$

$$\int_{\Omega_2} v(y)\, \Phi(|f_1(y)|, \ldots, |f_n(y)|)\, d\mu_2(y), \tag{26.9}$$

under the assumptions:

(i) *f_i, $\Phi(|f_1|, \ldots, |f_n|)$, are $k(x,y)\, d\mu_2(y)$-integrable, μ_1-a.e. in $x \in \Omega_1$, for all $i = 1, \ldots, n$,*

(ii) *$v(y)\, \Phi(|f_1(y)|, \ldots, |f_n(y)|)$ is μ_2-integrable.*

Proof Here we use Proposition 26.1, Jensen's inequality, Tonelli's theorem, Fubini's theorem, and that Φ is increasing per coordinate. We have

$$\int_{\Omega_1} u(x)\, \Phi\left(\left|\frac{\overrightarrow{g}(x)}{K(x)}\right|\right) d\mu_1(x) =$$

$$\int_{\Omega_1} u(x) \, \Phi \left(\frac{1}{K(x)} \left| \int_{\Omega_2} k(x,y) \, f_1(y) \, d\mu_2(y) \right|, \right.$$

$$\left. \frac{1}{K(x)} \left| \int_{\Omega_2} k(x,y) \, f_2(y) \, d\mu_2(y) \right|, \ldots, \frac{1}{K(x)} \left| \int_{\Omega_2} k(x,y) \, f_n(y) \, d\mu_2(y) \right| \right) d\mu_1(x)$$

$$\le \int_{\Omega_1} u(x) \, \Phi \left(\frac{1}{K(x)} \int_{\Omega_2} k(x,y) \, |f_1(y)| \, d\mu_2(y), \ldots, \right.$$

$$\left. \frac{1}{K(x)} \int_{\Omega_2} k(x,y) \, |f_n(y)| \, d\mu_2(y) \right) d\mu_1(x) \le$$

(by Jensen's inequality)

$$\int_{\Omega_1} \frac{u(x)}{K(x)} \left(\int_{\Omega_2} k(x,y) \, \Phi(|f_1(y)|, \ldots, |f_n(y)|) \, d\mu_2(y) \right) d\mu_1(x) =$$

$$\int_{\Omega_1} \frac{u(x)}{K(x)} \left(\int_{\Omega_2} k(x,y) \, \Phi\left(\left| \overrightarrow{f}(y) \right| \right) d\mu_2(y) \right) d\mu_1(x) =$$

$$\int_{\Omega_1} \left(\int_{\Omega_2} \frac{u(x)}{K(x)} k(x,y) \, \Phi\left(\left| \overrightarrow{f}(y) \right| \right) d\mu_2(y) \right) d\mu_1(x) =$$

$$\int_{\Omega_2} \left(\int_{\Omega_1} \frac{u(x)}{K(x)} k(x,y) \, \Phi\left(\left| \overrightarrow{f}(y) \right| \right) d\mu_1(x) \right) d\mu_2(y) =$$

$$\int_{\Omega_2} \Phi\left(\left| \overrightarrow{f}(y) \right| \right) \left(\int_{\Omega_1} \frac{u(x)}{K(x)} k(x,y) \, d\mu_1(x) \right) d\mu_2(y) =$$

$$\int_{\Omega_2} \Phi\left(\left| \overrightarrow{f}(y) \right| \right) v(y) \, d\mu_2(y) =$$

$$\int_{\Omega_2} \Phi(|f_1(y)|, \ldots, |f_n(y)|) \, v(y) \, d\mu_2(y),$$

proving the claim. ■

Notation 26.4 From now on we may write

$$\overrightarrow{g}(x) = \int_{\Omega_2} k(x,y) \, \overrightarrow{f}(y) \, d\mu_2(y), \tag{26.10}$$

which means

$$(g_1(x), \ldots, g_n(x)) = \left(\int_{\Omega_2} k(x, y) f_1(y) d\mu_2(y), \ldots, \int_{\Omega_2} k(x, y) f_n(y) d\mu_2(y) \right).$$
(26.11)

Similarly, we may write

$$|\vec{g}(x)| = \left| \int_{\Omega_2} k(x, y) \vec{f}(y) d\mu_2(y) \right|,$$
(26.12)

and we mean

$$(|g_1(x)|, \ldots, |g_n(x)|) =$$

$$\left(\left| \int_{\Omega_2} k(x, y) f_1(y) d\mu_2(y) \right|, \ldots, \left| \int_{\Omega_2} k(x, y) f_n(y) d\mu_2(y) \right| \right).$$
(26.13)

We also can write that

$$|\vec{g}(x)| \leq \int_{\Omega_2} k(x, y) \left| \vec{f}(y) \right| d\mu_2(y),$$
(26.14)

and we mean the fact that

$$|g_i(x)| \leq \int_{\Omega_2} k(x, y) |f_i(y)| d\mu_2(y),$$
(26.15)

for all $i = 1, \ldots, n$, etc.

Notation 26.5 Next let $(\Omega_1, \Sigma_1, \mu_1)$ and $(\Omega_2, \Sigma_2, \mu_2)$ be measure spaces with positive σ-finite measures, and let $k_j : \Omega_1 \times \Omega_2 \to \mathbb{R}$ be a nonnegative measurable function, $k_j(x, \cdot)$ measurable on Ω_2 and

$$K_j(x) = \int_{\Omega_2} k_j(x, y) d\mu_2(y), \quad x \in \Omega_1, j = 1, \ldots, m.$$
(26.16)

We suppose that $K_j(x) > 0$ a.e. on Ω_1. Let the measurable functions $g_{ji} : \Omega_1 \to \mathbb{R}$ with the representation

$$g_{ji}(x) = \int_{\Omega_2} k_j(x, y) f_{ji}(y) d\mu_2(y),$$
(26.17)

where $f_{ji} : \Omega_2 \to \mathbb{R}$ are measurable functions, $i = 1, \ldots, n$ and $j = 1, \ldots, m$.

Denote the function vectors $\vec{g_j} := (g_{j1}, g_{j2}, \ldots, g_{jn})$ and $\vec{f_j} := (f_{j1}, \ldots, f_{jn})$, $j = 1, \ldots, m$.

We say $\vec{f_j}$ is integrable with respect to measure μ, iff all f_{ji} are integrable with respect to μ.

We also consider here $\Phi_j : \mathbb{R}_+^n \to \mathbb{R}_+$, $j = 1, \ldots, m$, convex functions that are increasing per coordinate. Again u is a weight function on Ω_1.

Our second main result is when $m = 2$.

Theorem 26.6 *Here all as in Notation 26.5. Assume that the function*
$x \mapsto \left(\frac{u(x) k_1(x,y) k_2(x,y)}{K_1(x) K_2(x)} \right)$ *is integrable on Ω_1, for each $y \in \Omega_2$. Define λ_2 on Ω_2 by*

$$\lambda_2(y) := \int_{\Omega_1} \frac{u(x) k_1(x,y) k_2(x,y)}{K_1(x) K_2(x)} d\mu_1(x) < \infty. \qquad (26.18)$$

Then

$$\int_{\Omega_1} u(x) \Phi_1 \left(\left| \frac{\overrightarrow{g_1}(x)}{K_1(x)} \right| \right) \Phi_2 \left(\left| \frac{\overrightarrow{g_2}(x)}{K_2(x)} \right| \right) d\mu_1(x) \leq$$

$$\left(\int_{\Omega_2} \Phi_2 \left(\left| \overrightarrow{f_2}(y) \right| \right) d\mu_2(y) \right) \left(\int_{\Omega_2} \Phi_1 \left(\left| \overrightarrow{f_1}(y) \right| \right) \lambda_2(y) d\mu_2(y) \right), \qquad (26.19)$$

under the assumptions:

(i) $\{f_{1i}, \Phi_1(|f_{11}|, \ldots, |f_{1n}|)\}$, $\{f_{2i}, \Phi_2(|f_{21}|, \ldots, |f_{2n}|)\}$ *are $k_j(x,y) d\mu_2(y)$ - integrable, μ_1 -a.e. in $x \in \Omega_1$, $j = 1, 2$ (respectively), for all $i = 1, \ldots, n$.*
(ii) $\lambda_2 \Phi_1 \left(\left| \overrightarrow{f_1} \right| \right)$, $\Phi_2 \left(\left| \overrightarrow{f_2} \right| \right)$, *are both μ_2 -integrable.*

Proof Acting, similarly as in the proof of Theorem 26.3 we have

$$\int_{\Omega_1} u(x) \Phi_1 \left(\left| \frac{\overrightarrow{g_1}(x)}{K_1(x)} \right| \right) \Phi_2 \left(\left| \frac{\overrightarrow{g_2}(x)}{K_2(x)} \right| \right) d\mu_1(x) =$$

$$\int_{\Omega_1} u(x) \Phi_1 \left(\left| \frac{1}{K_1(x)} \int_{\Omega_2} k_1(x,y) \overrightarrow{f_1}(y) d\mu_2(y) \right| \right) \cdot \qquad (26.20)$$

$$\Phi_2 \left(\left| \frac{1}{K_2(x)} \int_{\Omega_2} k_2(x,y) \overrightarrow{f_2}(y) d\mu_2(y) \right| \right) d\mu_1(x) \leq$$

$$\int_{\Omega_1} u(x) \Phi_1 \left(\frac{1}{K_1(x)} \int_{\Omega_2} k_1(x,y) \left| \overrightarrow{f_1}(y) \right| d\mu_2(y) \right) \cdot$$

$$\Phi_2 \left(\frac{1}{K_2(x)} \int_{\Omega_2} k_2(x,y) \left| \overrightarrow{f_2}(y) \right| d\mu_2(y) \right) d\mu_1(x) \leq$$

$$\int_{\Omega_1} u(x) \frac{1}{K_1(x)} \left(\int_{\Omega_2} k_1(x,y) \Phi_1 \left(\left| \overrightarrow{f_1}(y) \right| \right) d\mu_2(y) \right) \cdot$$

$$\frac{1}{K_2(x)} \left(\int_{\Omega_2} k_2(x, y) \Phi_2\left(\left| \overrightarrow{f_2}(y) \right| \right) d\mu_2(y) \right) d\mu_1(x) = \qquad (26.21)$$

(calling $\gamma_1(x) := \int_{\Omega_2} k_1(x, y) \Phi_1\left(\left| \overrightarrow{f_1}(y) \right| \right) d\mu_2(y)$)

$$\int_{\Omega_1} \int_{\Omega_2} \frac{u(x) \gamma_1(x)}{K_1(x) K_2(x)} k_2(x, y) \Phi_2\left(\left| \overrightarrow{f_2}(y) \right| \right) d\mu_2(y) d\mu_1(x) =$$

$$\int_{\Omega_2} \int_{\Omega_1} \frac{u(x) \gamma_1(x)}{K_1(x) K_2(x)} k_2(x, y) \Phi_2\left(\left| \overrightarrow{f_2}(y) \right| \right) d\mu_1(x) d\mu_2(y) =$$

$$\int_{\Omega_2} \Phi_2\left(\left| \overrightarrow{f_2}(y) \right| \right) \left(\int_{\Omega_1} \frac{u(x) \gamma_1(x)}{K_1(x) K_2(x)} k_2(x, y) d\mu_1(x) \right) d\mu_2(y) =$$

$$\int_{\Omega_2} \Phi_2\left(\left| \overrightarrow{f_2}(y) \right| \right) \cdot \qquad (26.22)$$

$$\left(\int_{\Omega_1} \frac{u(x) k_2(x, y)}{K_1(x) K_2(x)} \left(\int_{\Omega_2} k_1(x, y) \Phi_1\left(\left| \overrightarrow{f_1}(y) \right| \right) d\mu_2(y) \right) d\mu_1(x) \right) d\mu_2(y) =$$

$$\int_{\Omega_2} \Phi_2\left(\left| \overrightarrow{f_2}(y) \right| \right) \cdot$$

$$\left[\int_{\Omega_1} \left(\int_{\Omega_2} \frac{u(x) k_1(x, y) k_2(x, y)}{K_1(x) K_2(x)} \Phi_1\left(\left| \overrightarrow{f_1}(y) \right| \right) d\mu_2(y) \right) d\mu_1(x) \right] d\mu_2(y) =$$

$$\left(\int_{\Omega_2} \Phi_2\left(\left| \overrightarrow{f_2}(y) \right| \right) d\mu_2(y) \right) \cdot$$

$$\left[\int_{\Omega_1} \left(\int_{\Omega_2} \frac{u(x) k_1(x, y) k_2(x, y)}{K_1(x) K_2(x)} \Phi_1\left(\left| \overrightarrow{f_1}(y) \right| \right) d\mu_2(y) \right) d\mu_1(x) \right] =$$

$$\left(\int_{\Omega_2} \Phi_2\left(\left| \overrightarrow{f_2}(y) \right| \right) d\mu_2(y) \right) \cdot$$

$$\left[\int_{\Omega_2} \left(\int_{\Omega_1} \frac{u(x) k_1(x, y) k_2(x, y)}{K_1(x) K_2(x)} \Phi_1\left(\left| \overrightarrow{f_1}(y) \right| \right) d\mu_1(x) \right) d\mu_2(y) \right] =$$

$$\left(\int_{\Omega_2} \Phi_2\left(\left| \overrightarrow{f_2}(y) \right| \right) d\mu_2(y) \right) \cdot \qquad (26.23)$$

$$\left[\int_{\Omega_2} \Phi_1\left(\left| \overrightarrow{f_1}(y) \right| \right) \left(\int_{\Omega_1} \frac{u(x) k_1(x, y) k_2(x, y)}{K_1(x) K_2(x)} d\mu_1(x) \right) d\mu_2(y) \right] =$$

$$\left(\int_{\Omega_2} \Phi_2 \left(\left| \overrightarrow{f_2}(y) \right| \right) d\mu_2(y) \right) \left[\int_{\Omega_2} \Phi_1 \left(\left| \overrightarrow{f_1}(y) \right| \right) \lambda_2(y) d\mu_2(y) \right],$$

proving the claim. ∎

When $m = 3$, the corresponding result follows.

Theorem 26.7 *Here all as in Notation 26.5. Assume that the function*
$x \mapsto \left(\frac{u(x)k_1(x,y)k_2(x,y)k_3(x,y)}{K_1(x)K_2(x)K_3(x)} \right)$ *is integrable on Ω_1, for each $y \in \Omega_2$. Define λ_3 on Ω_2 by*

$$\lambda_3(y) := \int_{\Omega_1} \frac{u(x)\,k_1(x,y)\,k_2(x,y)\,k_3(x,y)}{K_1(x)\,K_2(x)\,K_3(x)} d\mu_1(x) < \infty. \tag{26.24}$$

Then

$$\int_{\Omega_1} u(x) \prod_{j=1}^{3} \Phi_j \left(\left| \frac{\overrightarrow{g_j}(x)}{K_j(x)} \right| \right) d\mu_1(x) \le \tag{26.25}$$

$$\left(\prod_{j=2}^{3} \int_{\Omega_2} \Phi_j \left(\left| \overrightarrow{f_j}(y) \right| \right) d\mu_2(y) \right) \left(\int_{\Omega_2} \Phi_1 \left(\left| \overrightarrow{f_1}(y) \right| \right) \lambda_3(y) d\mu_2(y) \right),$$

under the assumptions:

(i) $\overrightarrow{f_j}, \Phi_j \left(\left| \overrightarrow{f_j} \right| \right)$, *are $k_j(x,y)\,d\mu_2(y)$-integrable, μ_1-a.e. in $x \in \Omega_1$, $j = 1, 2, 3$,*
(ii) $\lambda_3 \Phi_1 \left(\left| \overrightarrow{f_1} \right| \right)$, $\Phi_2 \left(\left| \overrightarrow{f_2} \right| \right)$, $\Phi_3 \left(\left| \overrightarrow{f_3} \right| \right)$, *are all μ_2-integrable.*

Proof We also have

$$\int_{\Omega_1} u(x) \prod_{j=1}^{3} \Phi_j \left(\left| \frac{\overrightarrow{g_j}(x)}{K_j(x)} \right| \right) d\mu_1(x) =$$

$$\int_{\Omega_1} u(x) \prod_{j=1}^{3} \Phi_j \left(\left| \frac{1}{K_j(x)} \int_{\Omega_2} k_j(x,y)\,\overrightarrow{f_j}(y)\,d\mu_2(y) \right| \right) d\mu_1(x) \le \tag{26.26}$$

$$\int_{\Omega_1} u(x) \prod_{j=1}^{3} \Phi_j \left(\frac{1}{K_j(x)} \int_{\Omega_2} k_j(x,y) \left| \overrightarrow{f_j}(y) \right| d\mu_2(y) \right) d\mu_1(x) \le$$

$$\int_{\Omega_1} u(x) \prod_{j=1}^{3} \left(\frac{1}{K_j(x)} \int_{\Omega_2} k_j(x,y)\,\Phi_j \left(\left| \overrightarrow{f_j}(y) \right| \right) d\mu_2(y) \right) d\mu_1(x) =$$

$$\int_{\Omega_1} \left(\frac{u(x)}{\prod\limits_{j=1}^{3} K_j(x)} \right) \left(\prod_{j=1}^{3} \int_{\Omega_2} k_j(x,y) \, \Phi_j \left(\left| \overrightarrow{f_j}(y) \right| \right) d\mu_2(y) \right) d\mu_1(x) =$$

(calling $\theta(x) := \dfrac{u(x)}{\prod\limits_{j=1}^{3} K_j(x)}$)

$$\int_{\Omega_1} \theta(x) \left(\prod_{j=1}^{3} \int_{\Omega_2} k_j(x,y) \, \Phi_j \left(\left| \overrightarrow{f_j}(y) \right| \right) d\mu_2(y) \right) d\mu_1(x) = \qquad (26.27)$$

$$\int_{\Omega_1} \theta(x) \left[\int_{\Omega_2} \left(\prod_{j=1}^{2} \int_{\Omega_2} k_j(x,y) \, \Phi_j \left(\left| \overrightarrow{f_j}(y) \right| \right) d\mu_2(y) \right) \right.$$

$$\left. k_3(x,y) \, \Phi_3 \left(\left| \overrightarrow{f_3}(y) \right| \right) d\mu_2(y) \right] d\mu_1(x) =$$

$$\int_{\Omega_1} \left(\int_{\Omega_2} \theta(x) \left(\prod_{j=1}^{2} \int_{\Omega_2} k_j(x,y) \, \Phi_j \left(\left| \overrightarrow{f_j}(y) \right| \right) d\mu_2(y) \right) \right.$$

$$\left. k_3(x,y) \, \Phi_3 \left(\left| \overrightarrow{f_3}(y) \right| \right) d\mu_2(y) \right) d\mu_1(x) =$$

$$\int_{\Omega_2} \left(\int_{\Omega_1} \theta(x) \left(\prod_{j=1}^{2} \int_{\Omega_2} k_j(x,y) \, \Phi_j \left(\left| \overrightarrow{f_j}(y) \right| \right) d\mu_2(y) \right) \right.$$

$$\left. k_3(x,y) \, \Phi_3 \left(\left| \overrightarrow{f_3}(y) \right| \right) d\mu_1(x) \right) d\mu_2(y) =$$

$$\int_{\Omega_2} \Phi_3 \left(\left| \overrightarrow{f_3}(y) \right| \right) \left(\int_{\Omega_1} \theta(x) \, k_3(x,y) \left(\prod_{j=1}^{2} \int_{\Omega_2} k_j(x,y) \, \Phi_j \left(\left| \overrightarrow{f_j}(y) \right| \right) d\mu_2(y) \right) \right.$$

(26.28)

$$d\mu_1(x) \right) d\mu_2(y) =$$

$$\int_{\Omega_2} \Phi_3 \left(\left| \overrightarrow{f_3}(y) \right| \right) \left[\int_{\Omega_1} \theta(x) \, k_3(x,y) \left(\int_{\Omega_2} \left\{ \int_{\Omega_2} k_1(x,y) \, \Phi_1 \left(\left| \overrightarrow{f_1}(y) \right| \right) d\mu_2(y) \right\} \right. \right. \cdot$$

$$k_2\,(x,\,y)\,\Phi_2\left(\left|\overrightarrow{f_2}\,(y)\right|\right)d\mu_2\,(y)\right)d\mu_1\,(x)\Big]d\mu_2\,(y)=$$

$$\int_{\Omega_2}\Phi_3\left(\left|\overrightarrow{f_3}\,(y)\right|\right)\Big[\int_{\Omega_1}\left(\int_{\Omega_2}\theta\,(x)\,k_2\,(x,\,y)\,k_3\,(x,\,y)\,\Phi_2\left(\left|\overrightarrow{f_2}\,(y)\right|\right)\cdot\right.\quad(26.29)$$

$$\left\{\int_{\Omega_2}k_1\,(x,\,y)\,\Phi_1\left(\left|\overrightarrow{f_1}\,(y)\right|\right)d\mu_2\,(y)\right\}d\mu_2\,(y)\right)d\mu_1\,(x)\Big]d\mu_2\,(y)=$$

$$\left(\int_{\Omega_2}\Phi_3\left(\left|\overrightarrow{f_3}\,(y)\right|\right)d\mu_2\,(y)\right)\Big[\int_{\Omega_1}\left(\int_{\Omega_2}\theta\,(x)\,k_2\,(x,\,y)\,k_3\,(x,\,y)\,\Phi_2\left(\left|\overrightarrow{f_2}\,(y)\right|\right)\cdot\right.$$

$$\left\{\int_{\Omega_2}k_1\,(x,\,y)\,\Phi_1\left(\left|\overrightarrow{f_1}\,(y)\right|\right)d\mu_2\,(y)\right\}d\mu_2\,(y)\right)d\mu_1\,(x)\Big]=$$

$$\left(\int_{\Omega_2}\Phi_3\left(\left|\overrightarrow{f_3}\,(y)\right|\right)d\mu_2\,(y)\right)\Big[\int_{\Omega_2}\left(\int_{\Omega_1}\theta\,(x)\,k_2\,(x,\,y)\,k_3\,(x,\,y)\,\Phi_2\left(\left|\overrightarrow{f_2}\,(y)\right|\right)\cdot\right.$$

$$\left\{\int_{\Omega_2}k_1\,(x,\,y)\,\Phi_1\left(\left|\overrightarrow{f_1}\,(y)\right|\right)d\mu_2\,(y)\right\}d\mu_1\,(x)\right)d\mu_2\,(y)\Big]=\quad(26.30)$$

$$\left(\int_{\Omega_2}\Phi_3\left(\left|\overrightarrow{f_3}\,(y)\right|\right)d\mu_2\,(y)\right)\Big[\int_{\Omega_2}\Phi_2\left(\left|\overrightarrow{f_2}\,(y)\right|\right)\left(\int_{\Omega_1}\theta\,(x)\,k_2\,(x,\,y)\,k_3\,(x,\,y)\cdot\right.$$

$$\left(\int_{\Omega_2}k_1\,(x,\,y)\,\Phi_1\left(\left|\overrightarrow{f_1}\,(y)\right|\right)d\mu_2\,(y)\right)d\mu_1\,(x)\right)d\mu_2\,(y)\Big]=$$

$$\left(\int_{\Omega_2}\Phi_3\left(\left|\overrightarrow{f_3}\,(y)\right|\right)d\mu_2\,(y)\right)\Big[\int_{\Omega_2}\Phi_2\left(\left|\overrightarrow{f_2}\,(y)\right|\right)\left\{\int_{\Omega_1}\left(\int_{\Omega_2}\theta\,(x)\prod_{j=1}^{3}k_j\,(x,\,y)\cdot\right.\right.$$

$$\Phi_1\left(\left|\overrightarrow{f_1}\,(y)\right|\right)d\mu_2\,(y)\right)d\mu_1\,(x)\Big\}d\mu_2\,(y)\Big]=$$

$$\left(\int_{\Omega_2}\Phi_3\left(\left|\overrightarrow{f_3}\,(y)\right|\right)d\mu_2\,(y)\right)\left(\int_{\Omega_2}\Phi_2\left(\left|\overrightarrow{f_2}\,(y)\right|\right)d\mu_2\,(y)\right)\cdot$$

$$\left(\int_{\Omega_1}\left(\int_{\Omega_2}\theta\,(x)\prod_{j=1}^{3}k_j\,(x,\,y)\,\Phi_1\left(\left|\overrightarrow{f_1}\,(y)\right|\right)d\mu_2\,(y)\right)d\mu_1\,(x)\right)=\quad(26.31)$$

$$\left(\prod_{j=2}^{3}\int_{\Omega_2}\Phi_j\left(\left|\overrightarrow{f_j}(y)\right|\right)d\mu_2(y)\right)\cdot$$

$$\left(\int_{\Omega_2}\left(\int_{\Omega_1}\theta(x)\prod_{j=1}^{3}k_j(x,y)\,\Phi_1\left(\left|\overrightarrow{f_1}(y)\right|\right)d\mu_1(x)\right)d\mu_2(y)\right)=$$

$$\left(\prod_{j=2}^{3}\int_{\Omega_2}\Phi_j\left(\left|\overrightarrow{f_j}(y)\right|\right)d\mu_2(y)\right)\cdot$$

$$\left(\int_{\Omega_2}\Phi_1\left(\left|\overrightarrow{f_1}(y)\right|\right)\left(\int_{\Omega_1}\theta(x)\prod_{j=1}^{3}k_j(x,y)\,d\mu_1(x)\right)d\mu_2(y)\right)=$$

$$\left(\prod_{j=2}^{3}\int_{\Omega_2}\Phi_j\left(\left|\overrightarrow{f_j}(y)\right|\right)d\mu_2(y)\right)\left(\int_{\Omega_2}\Phi_1\left(\left|\overrightarrow{f_1}(y)\right|\right)\lambda_3(y)\,d\mu_2(y)\right),\quad(26.32)$$

proving the claim. ∎

For general $m\in\mathbb{N}$, the following result is valid.

Theorem 26.8 *Again here we follow Notation 26.5. Assume that the function $x\mapsto$*

$$\left(\frac{u(x)\prod_{j=1}^{m}k_j(x,y)}{\prod_{j=1}^{m}K_j(x)}\right)$$ *is integrable on Ω_1, for each $y\in\Omega_2$. Define λ_m on Ω_2 by*

$$\lambda_m(y):=\int_{\Omega_1}\left(\frac{u(x)\prod_{j=1}^{m}k_j(x,y)}{\prod_{j=1}^{m}K_j(x)}\right)d\mu_1(x)<\infty.\qquad(26.33)$$

Then

$$\int_{\Omega_1}u(x)\prod_{j=1}^{m}\Phi_j\left(\left|\frac{\overrightarrow{g_j}(x)}{K_j(x)}\right|\right)d\mu_1(x)\le\qquad(26.34)$$

$$\left(\prod_{j=2}^{m}\int_{\Omega_2}\Phi_j\left(\left|\overrightarrow{f_j}(y)\right|\right)d\mu_2(y)\right)\left(\int_{\Omega_2}\Phi_1\left(\left|\overrightarrow{f_1}(y)\right|\right)\lambda_m(y)\,d\mu_2(y)\right),$$

under the assumptions:

(i) $\overrightarrow{f_j}$, $\Phi_j\left(\left|\overrightarrow{f_j}\right|\right)$, *are* $k_j(x,y)\,d\mu_2(y)$ *-integrable,* μ_1 *-a.e. in* $x \in \Omega_1$, $j = 1,\ldots,m$,

(ii) $\lambda_m \Phi_1\left(\left|\overrightarrow{f_1}\right|\right)$, $\Phi_2\left(\left|\overrightarrow{f_2}\right|\right)$, $\Phi_3\left(\left|\overrightarrow{f_3}\right|\right),\ldots,\Phi_m\left(\left|\overrightarrow{f_m}\right|\right)$, *are all* μ_2 *-integrable.*

When $k(x,y) = k_1(x,y) = k_2(x,y) = \cdots = k_m(x,y)$, then $K(x) := K_1(x) = K_2(x) = \cdots = K_m(x)$. Then from Theorem 26.8 we get:

Corollary 26.9 *Assume that the function* $x \mapsto \left(\frac{u(x)k^m(x,y)}{K^m(x)}\right)$ *is integrable on* Ω_1, *for each* $y \in \Omega_2$. *Define* U_m *on* Ω_2 *by*

$$U_m(y) := \int_{\Omega_1}\left(\frac{u(x)\,k^m(x,y)}{K^m(x)}\right)d\mu_1(x) < \infty. \qquad (26.35)$$

Then

$$\int_{\Omega_1} u(x) \prod_{j=1}^{m} \Phi_j\left(\left|\frac{\overrightarrow{g_j}(x)}{K(x)}\right|\right)d\mu_1(x) \leq \qquad (26.36)$$

$$\left(\prod_{j=2}^{m}\int_{\Omega_2}\Phi_j\left(\left|\overrightarrow{f_j}(y)\right|\right)d\mu_2(y)\right)\left(\int_{\Omega_2}\Phi_1\left(\left|\overrightarrow{f_1}(y)\right|\right)U_m(y)\,d\mu_2(y)\right),$$

under the assumptions:

(i) $\overrightarrow{f_j}$, $\Phi_j\left(\left|\overrightarrow{f_j}\right|\right)$, *are* $k(x,y)\,d\mu_2(y)$ *-integrable,* μ_1 *-a.e. in* $x \in \Omega_1$, $j = 1,\ldots,m$,

(ii) $U_m\Phi_1\left(\left|\overrightarrow{f_1}\right|\right)$, $\Phi_2\left(\left|\overrightarrow{f_2}\right|\right)$, $\Phi_3\left(\left|\overrightarrow{f_3}\right|\right),\ldots,\Phi_m\left(\left|\overrightarrow{f_m}\right|\right)$, *are all* μ_2 *-integrable.*

When $m = 2$ from Corollary 26.9 we obtain

Corollary 26.10 *Assume that the function* $x \mapsto \left(\frac{u(x)k^2(x,y)}{K^2(x)}\right)$ *is integrable on* Ω_1, *for each* $y \in \Omega_2$. *Define* U_2 *on* Ω_2 *by*

$$U_2(y) := \int_{\Omega_1}\left(\frac{u(x)\,k^2(x,y)}{K^2(x)}\right)d\mu_1(x) < \infty. \qquad (26.37)$$

Then

$$\int_{\Omega_1} u(x)\,\Phi_1\left(\left|\frac{\overrightarrow{g_1}(x)}{K(x)}\right|\right)\Phi_2\left(\left|\frac{\overrightarrow{g_2}(x)}{K(x)}\right|\right)d\mu_1(x) \leq \qquad (26.38)$$

$$\left(\int_{\Omega_2} \Phi_2 \left(\left| \overrightarrow{f_2}(y) \right| \right) d\mu_2(y) \right) \left(\int_{\Omega_2} \Phi_1 \left(\left| \overrightarrow{f_1}(y) \right| \right) U_2(y) d\mu_2(y) \right),$$

under the assumptions:

(i) $\overrightarrow{f_1}$, $\overrightarrow{f_2}$, $\Phi_1 \left(\left| \overrightarrow{f_1} \right| \right)$, $\Phi_2 \left(\left| \overrightarrow{f_2} \right| \right)$ *are all* $k(x, y) d\mu_2(y)$ *-integrable,* μ_1 *-a.e. in* $x \in \Omega_1$,

(ii) $U_2 \Phi_1 \left(\left| \overrightarrow{f_1} \right| \right)$, $\Phi_2 \left(\left| \overrightarrow{f_2} \right| \right)$, *are both* μ_2 *-integrable.*

For $m \in \mathbb{N}$, the following more general result is also valid.

Theorem 26.11 *Let* $\rho \in \{1, \dots, m\}$ *be fixed. Assume that the function* $x \mapsto$

$$\left(\frac{u(x) \prod_{j=1}^{m} k_j(x,y)}{\prod_{j=1}^{m} K_j(x)} \right)$$ *is integrable on* Ω_1, *for each* $y \in \Omega_2$. *Define* λ_m *on* Ω_2 *by*

$$\lambda_m(y) := \int_{\Omega_1} \left(\frac{u(x) \prod_{j=1}^{m} k_j(x, y)}{\prod_{j=1}^{m} K_j(x)} \right) d\mu_1(x) < \infty. \qquad (26.39)$$

Then

$$I := \int_{\Omega_1} u(x) \prod_{j=1}^{m} \Phi_j \left(\left| \frac{\overrightarrow{g_j}(x)}{K_j(x)} \right| \right) d\mu_1(x) \leq \qquad (26.40)$$

$$\left(\prod_{\substack{j=1 \\ j \neq \rho}}^{m} \int_{\Omega_2} \Phi_j \left(\left| \overrightarrow{f_j}(y) \right| \right) d\mu_2(y) \right) \left(\int_{\Omega_2} \Phi_\rho \left(\left| \overrightarrow{f_\rho}(y) \right| \right) \lambda_m(y) d\mu_2(y) \right) := I_\rho,$$

under the assumptions:

(i) $\overrightarrow{f_j}$, $\Phi_j \left(\left| \overrightarrow{f_j} \right| \right)$, *are* $k_j(x, y) d\mu_2(y)$ *-integrable,* μ_1 *-a.e. in* $x \in \Omega_1$, $j = 1, \dots, m$,

(ii) $\lambda_m \Phi_\rho \left(\left| \overrightarrow{f_\rho} \right| \right)$; $\Phi_1 \left(\left| \overrightarrow{f_1} \right| \right)$, $\Phi_2 \left(\left| \overrightarrow{f_2} \right| \right)$, $\Phi_3 \left(\left| \overrightarrow{f_3} \right| \right)$, $\dots$, $\widehat{\Phi_\rho \left(\left| \overrightarrow{f_\rho} \right| \right)}$, $\dots$, $\Phi_m \left(\left| \overrightarrow{f_m} \right| \right)$, *are all* μ_2 *-integrable, where* $\widehat{\Phi_\rho \left(\left| \overrightarrow{f_\rho} \right| \right)}$ *means a missing item.*

We make

Remark 26.12 In the notations and assumptions of Theorem 26.11, replace assumption (ii) by the assumption,

(iii) $\Phi_1 \left(\left| \overrightarrow{f_1} \right| \right), \ldots, \Phi_m \left(\left| \overrightarrow{f_m} \right| \right)$; $\lambda_m \Phi_1 \left(\left| \overrightarrow{f_1} \right| \right), \ldots, \lambda_m \Phi_m \left(\left| \overrightarrow{f_m} \right| \right)$, are all μ_2 -integrable functions.

Then, clearly it holds,

$$I \leq \frac{\sum\limits_{\rho=1}^{m} I_\rho}{m}. \tag{26.41}$$

Two general applications of Theorem 26.11 follow for specific Φ_j.

Theorem 26.13 *Here all as in Theorem 26.11. It holds*

$$\int_{\Omega_1} u(x) \prod_{j=1}^{m} \left(\sum_{i=1}^{n} e^{\left| \frac{g_{ji}(x)}{K_j(x)} \right|} \right) d\mu_1(x) \leq \tag{26.42}$$

$$\left(\prod_{\substack{j=1 \\ j \neq \rho}}^{m} \int_{\Omega_2} \left(\sum_{i=1}^{n} e^{|f_{ji}(y)|} \right) d\mu_2(y) \right) \left(\int_{\Omega_2} \left(\sum_{i=1}^{n} e^{|f_{\rho i}(y)|} \right) \lambda_m(y) \, d\mu_2(y) \right),$$

under the assumptions:

(i) $\overrightarrow{f_j}, \left(\sum\limits_{i=1}^{n} e^{|f_{ji}(y)|} \right)$, *are* $k_j(x, y) \, d\mu_2(y)$ *-integrable,* μ_1 *-a.e. in* $x \in \Omega_1$, $j = 1, \ldots, m$,

(ii) $\lambda_m(y) \left(\sum\limits_{i=1}^{n} e^{|f_{\rho i}(y)|} \right)$ *and* $\left(\sum\limits_{i=1}^{n} e^{|f_{ji}(y)|} \right)$ *for* $j \neq \rho$, $j = 1, \ldots, m$, *are all* μ_2 *-integrable.*

Proof Apply Theorem 26.11 with $\Phi_j(x_1, \ldots, x_n) = \sum\limits_{i=1}^{m} e^{x_i}$, for all $j = 1, \ldots, m$. ∎

We continue with

Theorem 26.14 *Here all as in Theorem 26.11 and $p \geq 1$. It holds*

$$\int_{\Omega_1} u(x) \prod_{j=1}^{m} \left\| \frac{\overrightarrow{g_j}(x)}{K_j(x)} \right\|_p d\mu_1(x) \leq \tag{26.43}$$

$$\left(\prod_{\substack{j=1\\j\neq\rho}}^{m}\int_{\Omega_2}\left\|\overrightarrow{f_j}(y)\right\|_p d\mu_2(y)\right)\left(\int_{\Omega_2}\left\|\overrightarrow{f_\rho}(y)\right\|_p \lambda_m(y)\,d\mu_2(y)\right),$$

under the assumptions:

(i) $\left\|\overrightarrow{f_j}\right\|_p$ is $k_j(x,y)\,d\mu_2(y)$ -integrable, μ_1 -a.e. in $x\in\Omega_1$, $j=1,\dots,m$,

(ii) $\lambda_m\left\|\overrightarrow{f_\rho}\right\|_p;\left\|\overrightarrow{f_j}\right\|_p$ $j\neq\rho$, $j=1,\dots,m$, *are all* μ_2 *-integrable.*

Proof Apply Theorem 26.11 with $\Phi_j(x_1,\dots,x_n)=\left\|\overrightarrow{x}\right\|_p$, $\overrightarrow{x}=(x_1,\dots,x_n)$, for all $j=1,\dots,m$. ∎

We make

Remark 26.15 Let f_{ji} be Lebesgue measurable functions from (a,b) into $\mathbb{R}$, such that $\left(I_{a+}^{\alpha_j}\left(|f_{ji}|\right)\right)(x)\in\mathbb{R}$, $\forall\,x\in(a,b)$, $\alpha_j>0$, $j=1,\dots,m$, $i=1,\dots,n$, e.g. when $f_{ji}\in L_\infty(a,b)$.

Consider here

$$g_{ji}(x)=\left(I_{a+}^{\alpha_j}f_{ji}\right)(x),\quad x\in(a,b),\,j=1,\dots,m,\,i=1,\dots,n,$$

we remind

$$\left(I_{a+}^{\alpha_j}f_{ji}\right)(x)=\frac{1}{\Gamma(\alpha_j)}\int_a^x(x-t)^{\alpha_j-1}f_{ji}(t)\,dt.\qquad(26.44)$$

Notice that $g_{ji}(x)\in\mathbb{R}$ and it is Lebesgue measurable.

We pick $\Omega_1=\Omega_2=(a,b), d\mu_1(x)=dx, d\mu_2(y)=dy$, the Lebesgue measure. We see that

$$\left(I_{a+}^{\alpha_j}f_{ji}\right)(x)=\int_a^b\frac{\chi_{(a,x]}(t)(x-t)^{\alpha_j-1}}{\Gamma(\alpha_j)}f_{ji}(t)\,dt,\qquad(26.45)$$

where χ stands for the characteristic function.

So, we pick here

$$k_j(x,t):=\frac{\chi_{(a,x]}(t)(x-t)^{\alpha_j-1}}{\Gamma(\alpha_j)},\quad j=1,\dots,m.\qquad(26.46)$$

In fact

$$k_j(x,y)=\begin{cases}\dfrac{(x-y)^{\alpha_j-1}}{\Gamma(\alpha_j)}, & a<y\leq x,\\[2mm]0, & x<y<b.\end{cases}\qquad(26.47)$$

Clearly it holds

$$K_j(x) = \int_{(a,b)} \frac{\chi_{(a,x]}(y)(x-y)^{\alpha_j-1}}{\Gamma(\alpha_j)} dy = \frac{(x-a)^{\alpha_j}}{\Gamma(\alpha_j+1)}, \qquad (26.48)$$

$a < x < b$, $j = 1, \ldots, m$.

Notice that

$$\prod_{j=1}^{m} \frac{k_j(x,y)}{K_j(x)} = \prod_{j=1}^{m} \left(\frac{\chi_{(a,x]}(y)(x-y)^{\alpha_j-1}}{\Gamma(\alpha_j)} \cdot \frac{\Gamma(\alpha_j+1)}{(x-a)^{\alpha_j}} \right) =$$

$$\prod_{j=1}^{m} \left(\frac{\chi_{(a,x]}(y)(x-y)^{\alpha_j-1}}{(x-a)^{\alpha_j}} \alpha_j \right) = \frac{\chi_{(a,x]}(y)(x-y)^{\left(\sum\limits_{j=1}^{m}\alpha_j-m\right)} \left(\prod\limits_{j=1}^{m}\alpha_j\right)}{(x-a)^{\left(\sum\limits_{j=1}^{m}\alpha_j\right)}}. $$

$$(26.49)$$

Calling

$$\alpha := \sum_{j=1}^{m} \alpha_j > 0, \ \gamma := \prod_{j=1}^{m} \alpha_j > 0, \qquad (26.50)$$

we have that

$$\prod_{j=1}^{m} \frac{k_j(x,y)}{K_j(x)} = \frac{\chi_{(a,x]}(y)(x-y)^{\alpha-m}\gamma}{(x-a)^{\alpha}}. \qquad (26.51)$$

Therefore, for (26.33), we get for appropiate weight u that

$$\lambda_m(y) = \gamma \int_y^b u(x) \frac{(x-y)^{\alpha-m}}{(x-a)^{\alpha}} dx < \infty, \qquad (26.52)$$

for all $a < y < b$.

Let now

$$u(x) = (x-a)^{\alpha}, \ x \in (a,b). \qquad (26.53)$$

Then

$$\lambda_m(y) = \gamma \int_y^b (x-y)^{\alpha-m} dx = \frac{\gamma(b-y)^{\alpha-m+1}}{\alpha-m+1}, \qquad (26.54)$$

$y \in (a,b)$, where $\alpha > m-1$.

By Theorem 26.13 we get

$$
\int_a^b (x-a)^\alpha \prod_{j=1}^m \left(\sum_{i=1}^n e^{\left(\frac{\left| \left(I_{a+}^{\alpha_j} f_{ji} \right)(x) \right| \Gamma(\alpha_j+1)}{(x-a)^{\alpha_j}} \right)} \right) dx \le \tag{26.55}
$$

$$
\left(\frac{\gamma}{\alpha-m+1} \right) \left(\prod_{\substack{j=1 \\ j \neq \rho}}^m \int_a^b \left(\sum_{i=1}^n e^{|f_{ji}(y)|} \right) dy \right) \cdot
$$

$$
\left(\int_a^b (b-y)^{\alpha-m+1} \left(\sum_{i=1}^n e^{|f_{\rho i}(y)|} \right) dy \right) \le
$$

$$
\left(\frac{\gamma(b-a)^{\alpha-m+1}}{\alpha-m+1} \right) \prod_{j=1}^m \left(\int_a^b \left(\sum_{i=1}^n e^{|f_{ji}(y)|} \right) dy \right), \tag{26.56}
$$

under the assumptions:

(i) $\alpha > m-1$,

(ii) $\left(\sum_{i=1}^n e^{|f_{ji}(y)|} \right)$ is $\frac{\chi_{(a,x]}(y)(x-y)^{\alpha_j-1}}{\Gamma(\alpha_j)} dy$ -integrable, a.e. in $x \in (a,b)$, $j = 1, \ldots, m$,

(iii) $\left(\sum_{i=1}^n e^{|f_{ji}(y)|} \right)$, $j = 1, \ldots, m$, are all Lebesgue integrable on (a,b).

Let $p \ge 1$, by Theorem 26.14 we get

$$
\int_a^b (x-a)^\alpha \left(\prod_{j=1}^m \left\| \frac{\left(\overrightarrow{I_{a+}^{\alpha_j} f_j} \right)(x)}{(x-a)^{\alpha_j}} \right\|_p \right) \left(\prod_{j=1}^m \Gamma(\alpha_j+1) \right) dx \le
$$

$$
\left(\frac{\gamma(b-a)^{\alpha-m+1}}{\alpha-m+1} \right) \prod_{j=1}^m \left(\int_a^b \left\| \overrightarrow{f_j}(y) \right\|_p dy \right). \tag{26.57}
$$

Above $\overrightarrow{I_{a+}^{\alpha_j} f_j} := \left(I_{a+}^{\alpha_j} f_{j1}, \ldots, I_{a+}^{\alpha_j} f_{jn} \right)$, $j = 1, \ldots, m$, etc.

But we see that

$$
\prod_{j=1}^m \left\| \frac{\left(\overrightarrow{I_{a+}^{\alpha_j} f_j} \right)(x)}{(x-a)^{\alpha_j}} \right\|_p = \left(\frac{1}{(x-a)^\alpha} \right) \prod_{j=1}^m \left\| \overrightarrow{I_{a+}^{\alpha_j} f_j}(x) \right\|_p . \tag{26.58}
$$

We have proved that

$$
\int_a^b \left(\prod_{j=1}^m \left\| \left(\overrightarrow{I_{a+}^{\alpha_j} f_j} \right)(x) \right\|_p \right) dx \le
$$

$$
\left(\frac{\gamma (b-a)^{\alpha-m+1}}{(\alpha-m+1)\left(\prod_{j=1}^m \Gamma(\alpha_j+1)\right)} \right) \prod_{j=1}^m \left(\int_a^b \left\| \overrightarrow{f_j}(y) \right\|_p dy \right). \tag{26.59}
$$

Thus we derive that

$$
\left\| \prod_{j=1}^m \left\| \left(\overrightarrow{I_{a+}^{\alpha_j} f_j} \right) \right\|_p \right\|_{1,(a,b)} \le
$$

$$
\left(\frac{\gamma (b-a)^{\alpha-m+1}}{(\alpha-m+1)\left(\prod_{j=1}^m \Gamma(\alpha_j+1)\right)} \right) \prod_{j=1}^m \left\| \left\| \overrightarrow{f_j} \right\|_p \right\|_{1,(a,b)}, \tag{26.60}
$$

under the assumptions:

(i) $\alpha > m - 1$, $p \ge 1$,

(ii) $\left\| \overrightarrow{f_j} \right\|_p$ is $\frac{\chi_{(a,x]}(y)(x-y)^{\alpha_j-1}}{\Gamma(\alpha_j)}$-$dy$-integrable, a.e. in $x \in (a,b)$, $j = 1, \ldots, m$,

(iii) $\left\| \overrightarrow{f_j} \right\|_p$, $j = 1, \ldots, m$, are all Lebesgue integrable on (a,b).

Using the last condition (iii), we derive that $f_{ji} \in L_1(a,b)$, for all $j = 1, \ldots, m$; $i = 1, \ldots, n$ and by assuming $\alpha_j \ge 1$, we obtain that $I_{a+}^{\alpha_j}(|f_{ji}|)$ is finite on (a,b).

We continue with

Remark 26.16 Let f_{ji} be Lebesgue measurable functions : $(a,b) \to \mathbb{R}$, such that $I_{b-}^{\alpha_j}(|f_{ji}|)(x) < \infty$, $\forall x \in (a,b)$, $\alpha_j > 0$, $j = 1, \ldots, m$, $i = 1, \ldots, n$, e.g. when $f_{ji} \in L_\infty(a,b)$.

Consider here

$$
g_{ji}(x) = \left(I_{b-}^{\alpha_j} f_{ji} \right)(x), \quad x \in (a,b), \, j = 1, \ldots, m, \, i = 1, \ldots, n, \tag{26.61}
$$

we remind

$$
\left(I_{b-}^{\alpha_j} f_{ji} \right)(x) = \frac{1}{\Gamma(\alpha_j)} \int_x^b f_{ji}(t)(t-x)^{\alpha_j-1} dt, \tag{26.62}
$$

$(x < b)$.

Notice that $g_{ji}(x) \in \mathbb{R}$ and it is Lebesgue measurable.

We pick $\Omega_1 = \Omega_2 = (a, b), d\mu_1(x) = dx, d\mu_2(y) = dy$, the Lebesgue measure.

We see that

$$\left(I_{b-}^{\alpha_j} f_{ji}\right)(x) = \int_a^b \chi_{[x,b)}(t) \frac{(t-x)^{\alpha_j-1}}{\Gamma(\alpha_j)} f_{ji}(t) dt. \qquad (26.63)$$

So, we pick here

$$k_j(x, t) := \chi_{[x,b)}(t) \frac{(t-x)^{\alpha_j-1}}{\Gamma(\alpha_j)}, \quad j = 1, \dots, m. \qquad (26.64)$$

In fact

$$k_j(x, y) = \begin{cases} \frac{(y-x)^{\alpha_j-1}}{\Gamma(\alpha_j)}, & x \le y < b, \\ 0, & a < y < x. \end{cases} \qquad (26.65)$$

Clearly it holds

$$K_j(x) = \int_{(a,b)} \chi_{[x,b)}(y) \frac{(y-x)^{\alpha_j-1}}{\Gamma(\alpha_j)} dy = \frac{(b-x)^{\alpha_j}}{\Gamma(\alpha_j+1)}, \qquad (26.66)$$

$a < x < b, j = 1, \dots, m$.

Notice that

$$\prod_{j=1}^m \frac{k_j(x, y)}{K_j(x)} = \prod_{j=1}^m \left(\chi_{[x,b)}(y) \frac{(y-x)^{\alpha_j-1}}{\Gamma(\alpha_j)} \cdot \frac{\Gamma(\alpha_j+1)}{(b-x)^{\alpha_j}}\right) =$$

$$\prod_{j=1}^m \left(\chi_{[x,b)}(y) \frac{(y-x)^{\alpha_j-1} \alpha_j}{(b-x)^{\alpha_j}}\right) = \chi_{[x,b)}(y) \frac{(y-x)^{\left(\sum_{j=1}^m \alpha_j - m\right)} \left(\prod_{j=1}^m \alpha_j\right)}{(b-x)^{\left(\sum_{j=1}^m \alpha_j\right)}}. \qquad (26.67)$$

Calling

$$\alpha := \sum_{j=1}^m \alpha_j > 0, \quad \gamma := \prod_{j=1}^m \alpha_j > 0, \qquad (26.68)$$

we have that

$$\prod_{j=1}^{m} \frac{k_j(x, y)}{K_j(x)} = \frac{\chi_{[x,b)}(y)(y-x)^{\alpha-m}\gamma}{(b-x)^{\alpha}}. \tag{26.69}$$

Therefore, for (26.33), we get for appropiate weight u that

$$\lambda_m(y) = \gamma \int_a^y u(x) \frac{(y-x)^{\alpha-m}}{(b-x)^{\alpha}} dx < \infty, \tag{26.70}$$

for all $a < y < b$.
   Let now

$$u(x) = (b-x)^{\alpha}, \quad x \in (a, b). \tag{26.71}$$

Then

$$\lambda_m(y) = \gamma \int_a^y (y-x)^{\alpha-m} dx = \frac{\gamma(y-a)^{\alpha-m+1}}{\alpha-m+1}, \tag{26.72}$$

$y \in (a, b)$, where $\alpha > m - 1$.
   By Theorem 26.13 we get

$$\int_a^b (b-x)^{\alpha} \prod_{j=1}^{m} \left(\sum_{i=1}^{n} e^{\left(\frac{\left| \left(I_{b-}^{\alpha j} f_{ji} \right)(x) \right| \Gamma(\alpha_j+1)}{(b-x)^{\alpha j}} \right)} \right) dx \leq \tag{26.73}$$

$$\left(\frac{\gamma}{\alpha-m+1} \right) \left(\prod_{\substack{j=1 \\ j \neq \rho}}^{m} \int_a^b \left(\sum_{i=1}^{n} e^{|f_{ji}(y)|} \right) dy \right) \cdot$$

$$\left(\int_a^b (y-a)^{\alpha-m+1} \left(\sum_{i=1}^{n} e^{|f_{\rho i}(y)|} \right) dy \right) \leq$$

$$\left(\frac{\gamma(b-a)^{\alpha-m+1}}{\alpha-m+1} \right) \prod_{j=1}^{m} \left(\int_a^b \left(\sum_{i=1}^{n} e^{|f_{ji}(y)|} \right) dy \right), \tag{26.74}$$

under the assumptions:

(i) $\alpha > m - 1$,

(ii) $\left(\sum_{i=1}^{n} e^{|f_{ji}(y)|}\right)$ is $\dfrac{\chi_{[x,b)}(y)(y-x)^{\alpha_j-1}}{\Gamma(\alpha_j)} dy$ -integrable, a.e. in $x \in (a,b)$, $j = 1, \ldots, m$,

(iii) $\left(\sum_{i=1}^{n} e^{|f_{ji}(y)|}\right)$, $j = 1, \ldots, m$, are all Lebesgue integrable on (a,b).

Let $p \geq 1$, by Theorem 26.14 we get

$$\int_a^b (b-x)^\alpha \left(\prod_{j=1}^m \left\|\frac{\left(\overrightarrow{I_{b-}^{\alpha_j} f_j}\right)(x)}{(b-x)^{\alpha_j}}\right\|_p\right) \left(\prod_{j=1}^m \Gamma(\alpha_j+1)\right) dx \leq$$

$$\left(\frac{\gamma (b-a)^{\alpha-m+1}}{\alpha - m + 1}\right) \prod_{j=1}^m \left(\int_a^b \left\|\overrightarrow{f_j}(y)\right\|_p dy\right). \tag{26.75}$$

But we see that

$$\prod_{j=1}^m \left\|\frac{\left(I_{b-}^{\alpha_j} f_{ji}\right)(x)}{(b-x)^{\alpha_j}}\right\|_p = \left(\frac{1}{(b-x)^\alpha}\right) \prod_{j=1}^m \left\|I_{b-}^{\alpha_j} f_{ji}(x)\right\|_p. \tag{26.76}$$

We have proved that

$$\int_a^b \left(\prod_{j=1}^m \left\|\left(I_{b-}^{\alpha_j} f_{ji}\right)(x)\right\|_p\right) dx \leq$$

$$\left(\frac{\gamma (b-a)^{\alpha-m+1}}{(\alpha - m + 1)\left(\prod_{j=1}^m \Gamma(\alpha_j+1)\right)}\right) \prod_{j=1}^m \left(\int_a^b \left\|\overrightarrow{f_j}(y)\right\|_p dy\right). \tag{26.77}$$

Thus we derive that

$$\left\|\prod_{j=1}^m \left\|\left(I_{b-}^{\alpha_j} f_{ji}\right)\right\|_p\right\|_{1,(a,b)} \leq$$

$$\left(\frac{\gamma (b-a)^{\alpha-m+1}}{(\alpha - m + 1)\left(\prod_{j=1}^m \Gamma(\alpha_j+1)\right)}\right) \prod_{j=1}^m \left\|\left\|\overrightarrow{f_j}\right\|_p\right\|_{1,(a,b)}, \tag{26.78}$$

under the assumptions:

(i) $\alpha > m - 1$, $p \geq 1$,

(ii) $\left\| \vec{f_j} \right\|_p$ is $\dfrac{\chi_{[x,b)}(y)(y-x)^{\alpha_j - 1}}{\Gamma(\alpha_j)}$-$dy$-integrable, a.e. in $x \in (a, b)$, $j = 1, \ldots, m$,

(iii) $\left\| \vec{f_j} \right\|_p$, $j = 1, \ldots, m$, are all Lebesgue integrable on (a, b).

Using the last assumption (iii), we derive again that $f_{ji} \in L_1(a, b)$, for all $j = 1, \ldots, m$; $i = 1, \ldots, n$, and by assuming $\alpha_j \geq 1$, we obtain that $I_{b-}^{\alpha_j}\left(|f_{ji}|\right)$ is finite on (a, b).

We mention

Definition 26.17 ([1, p. 448],) The left generalized Riemann-Liouville fractional derivative of f of order $\beta > 0$ is given by

$$D_a^\beta f(x) = \frac{1}{\Gamma(n-\beta)}\left(\frac{d}{dx}\right)^n \int_a^x (x-y)^{n-\beta-1} f(y)\, dy, \qquad (26.79)$$

where $n = [\beta] + 1$, $x \in [a, b]$.

For $a, b \in \mathbb{R}$, we say that $f \in L_1(a, b)$ has an L_∞ fractional derivative $D_a^\beta f$ ($\beta > 0$) in $[a, b]$, if and only if

(1) $D_a^{\beta-k} f \in C([a, b])$, $k = 2, \ldots, n = [\beta] + 1$,
(2) $D_a^{\beta-1} f \in AC([a, b])$
(3) $D_a^\beta f \in L_\infty(a, b)$.

Above we define $D_a^0 f := f$ and $D_a^{-\delta} f := I_{a+}^\delta f$, if $0 < \delta \leq 1$.

From [1, p. 449] and [11] we mention and use

Lemma 26.18 Let $\beta > \alpha \geq 0$ and let $f \in L_1(a, b)$ have an L_∞ fractional derivative $D_a^\beta f$ in $[a, b]$ and let $D_a^{\beta-k} f(a) = 0$, $k = 1, \ldots, [\beta] + 1$, then

$$D_a^\alpha f(x) = \frac{1}{\Gamma(\beta-\alpha)} \int_a^x (x-y)^{\beta-\alpha-1} D_a^\beta f(y)\, dy, \qquad (26.80)$$

for all $a \leq x \leq b$.

Here $D_a^\alpha f \in AC([a, b])$ for $\beta - \alpha \geq 1$, and $D_a^\alpha f \in C([a, b])$ for $\beta - \alpha \in (0, 1)$. Notice here that

$$D_a^\alpha f(x) = \left(I_{a+}^{\beta-\alpha}\left(D_a^\beta f\right)\right)(x), \quad a \leq x \leq b. \qquad (26.81)$$

We give

Theorem 26.19 Let $f_{ji} \in L_1(a, b)$, $\alpha_j, \beta_j : \beta_j > \alpha_j \geq 0$, $j = 1, \ldots, m$; $i = 1, \ldots, n$. Here $\left(f_{ji}, \alpha_j, \beta_j\right)$ fulfill terminology and assumptions of Definition

26.17 and Lemma 26.18. Let $\overline{\alpha} := \sum_{j=1}^{m} (\beta_j - \alpha_j)$, $\overline{\gamma} := \prod_{j=1}^{m} (\beta_j - \alpha_j)$, *assume*

$\overline{\alpha} > m - 1$. *Then*

$$\int_a^b (x-a)^{\overline{\alpha}} \prod_{j=1}^{m} \left(\sum_{i=1}^{n} e^{\left(\frac{\left| \left(D_a^{\alpha_j} f_{ji} \right)(x) \right| \Gamma(\beta_j - \alpha_j + 1)}{(x-a)^{(\beta_j - \alpha_j)}} \right)} \right) dx \leq$$

$$\left(\frac{\overline{\gamma}(b-a)^{\overline{\alpha}-m+1}}{\overline{\alpha}-m+1} \right) \left(\prod_{j=1}^{m} \left(\int_a^b \left(\sum_{i=1}^{n} e^{\left| \left(D_a^{\beta_j} f_{ji} \right)(y) \right|} \right) dy \right) \right). \tag{26.82}$$

Proof Use of (26.55)–(26.56). ∎

We also give

Theorem 26.20 *All here as in Theorem 26.19, plus* $p \geq 1$. *Then*

$$\left\| \prod_{j=1}^{m} \left\| \overrightarrow{D_a^{\alpha_j} f_j} \right\|_p \right\|_{1,(a,b)} \leq \tag{26.83}$$

$$\left(\frac{\overline{\gamma}(b-a)^{(\overline{\alpha}-m+1)}}{(\overline{\alpha}-m+1) \prod_{j=1}^{m} (\Gamma(\beta_j - \alpha_j + 1))} \right) \left(\prod_{j=1}^{m} \left\| \left\| \overrightarrow{D_a^{\beta_j} f_j} \right\|_p \right\|_{1,(a,b)} \right).$$

Above $\overrightarrow{D_a^{\beta_j} f_j} := \left(D_a^{\beta_j} f_{j1}, \ldots, D_a^{\beta_j} f_{jn} \right)$, $j = 1, \ldots, m$, *etc.*

Proof By (26.60). ∎

We need

Definition 26.21 ([1, p. 449], [7, p. 50]) Let $\nu \geq 0$, $n := \lceil \nu \rceil$, $f \in AC^n([a,b])$. Then the left Caputo fractional derivative is given by

$$D_{*a}^{\nu} f(x) = \frac{1}{\Gamma(n-\nu)} \int_a^x (x-t)^{n-\nu-1} f^{(n)}(t) \, dt$$

$$= \left(I_{a+}^{n-\nu} f^{(n)} \right)(x), \tag{26.84}$$

and it exists almost everywhere for $x \in [a,b]$, in fact $D_{*a}^{\nu} f \in L_1(a,b)$, ([1, p. 394]).

We have $D_{*a}^n f = f^{(n)}, n \in \mathbb{Z}_+.$

We also need

Theorem 26.22 *([4]) Let* $\nu \geq \rho + 1, \rho > 0, \nu, \rho \notin \mathbb{N}.$ *Call* $n := \lceil \nu \rceil, m^* := \lceil \rho \rceil.$ *Assume* $f \in AC^n ([a, b])$, *such that* $f^{(k)} (a) = 0, k = m^*, m^* + 1, \ldots, n - 1,$ *and* $D_{*a}^\nu f \in L_\infty (a, b).$ *Then* $D_{*a}^\rho f \in AC ([a, b])$ *(where* $D_{*a}^\rho f = \left(I_{a+}^{m^*-\rho} f^{(m^*)} \right) (x)$), *and*

$$D_{*a}^\rho f (x) = \frac{1}{\Gamma (\nu - \rho)} \int_a^x (x - t)^{\nu-\rho-1} D_{*a}^\nu f (t) \, dt$$

$$= \left(I_{a+}^{\nu-\rho} \left(D_{*a}^\nu f \right) \right) (x), \tag{26.85}$$

$\forall \, x \in [a, b].$

We give

Theorem 26.23 *Let* $\left(f_{ji}, \nu_j, \rho_j \right), j = 1, \ldots, m, m \geq 2, i = 1, \ldots, n,$ *as in the assumptions of Theorem 26.22. Set* $\alpha^* := \sum_{j=1}^m \left(\nu_j - \rho_j \right), \gamma^* := \prod_{j=1}^m \left(\nu_j - \rho_j \right).$ *Here* $a, b \in \mathbb{R}, a < b.$ *Then*

$$\int_a^b (x - a)^{\alpha^*} \prod_{j=1}^m \left(\sum_{i=1}^n e^{\left(\left| D_{*a}^{\rho_j} f_{ji}(x) \right| \left(\frac{\Gamma(\nu_j - \rho_j + 1)}{(x-a)^{(\nu_j - \rho_j)}} \right) \right)} \right) dx \leq$$

$$\left(\frac{\gamma^* (b - a)^{\alpha^* - m + 1}}{(\alpha^* - m + 1)} \right) \left(\prod_{j=1}^m \left(\int_a^b \left(e^{\left| D_{*a}^{\nu_j} f_{ji}(y) \right|} \right) dy \right) \right). \tag{26.86}$$

Proof Use of (26.55), (26.56). See here that $\alpha^* \geq m > m - 1.$ ∎

We continue with

Theorem 26.24 *All as in Theorem 26.23, plus* $p \geq 1.$ *Then*

$$\left\| \prod_{j=1}^m \left\| \overrightarrow{D_{*a}^{\rho_j} f_j} \right\| \right\|_{p \, \Big|_{1,(a,b)}} \leq \tag{26.87}$$

$$\left(\frac{\gamma^* (b-a)^{(\alpha^*-m+1)}}{(\alpha^*-m+1)\left(\prod_{j=1}^{m} \left(\Gamma\left(\nu_j - \rho_j + 1\right)\right)\right)} \right) \prod_{j=1}^{m} \left\| \left\| \overrightarrow{D_{*a}^{\nu_j} f_j} \right\|_p \right\|_{1,(a,b)}.$$

Proof By (26.60). ∎

We need

Definition 26.25 ([2, 8, 10]) Let $\alpha \geq 0$, $n := \lceil \alpha \rceil$, $f \in AC^n ([a, b])$. We define the right Caputo fractional derivative of order $\alpha \geq 0$, by

$$\overline{D}_{b-}^{\alpha} f (x) := (-1)^n I_{b-}^{n-\alpha} f^{(n)} (x), \tag{26.88}$$

we set $\overline{D}_{-}^{0} f := f$, i.e.

$$\overline{D}_{b-}^{\alpha} f (x) = \frac{(-1)^m}{\Gamma (n-\alpha)} \int_{x}^{b} (J-x)^{n-\alpha-1} f^{(n)} (J)\, dJ. \tag{26.89}$$

Notice that $\overline{D}_{b-}^{n} f = (-1)^n f^{(n)}$, $n \in \mathbb{N}$.

We need

Theorem 26.26 ([4]) *Let $f \in AC^n ([a, b])$, $\alpha > 0$, $n \in \mathbb{N}$, $n := \lceil \alpha \rceil$, $\alpha \geq \rho + 1$, $\rho > 0$, $r = \lceil \rho \rceil$, $\alpha, \rho \notin \mathbb{N}$. Assume $f^{(k)} (b) = 0$, $k = r, r+1, \ldots, n-1$, and $\overline{D}_{b-}^{\alpha} f \in L_\infty ([a, b])$. Then*

$$\overline{D}_{b-}^{\rho} f (x) = \left(I_{b-}^{\alpha-\rho} \left(\overline{D}_{b-}^{\alpha} f \right)\right) (x) \in AC ([a, b]), \tag{26.90}$$

that is

$$\overline{D}_{b-}^{\rho} f (x) = \frac{1}{\Gamma (\alpha-\rho)} \int_{x}^{b} (t-x)^{\alpha-\rho-1} \left(\overline{D}_{b-}^{\alpha} f \right) (t)\, dt, \tag{26.91}$$

$\forall\, x \in [a, b]$.

We give

Theorem 26.27 *Let $\left(f_{ji}, \alpha_j, \rho_j \right)$, $j = 1, \ldots, m$, $m \geq 2$, $i = 1, \ldots, n$, as in the assumptions of Theorem 26.26. Set $A := \sum_{j=1}^{m} (\alpha_j - \rho_j)$, $B := \prod_{j=1}^{m} (\alpha_j - \rho_j)$. Here $a, b \in \mathbb{R}$, $a < b$, $p \geq 1$. Then*

$$\left\|\prod_{j=1}^{m}\left\|\overrightarrow{\overline{D}_{b-}^{\rho_j}f_j}\right\|_p\right\|_{1,(a,b)} \leq \tag{26.92}$$

$$\left(\frac{B\,(b-a)^{(A-m+1)}}{(A-m+1)\left(\prod_{j=1}^{m}\left(\Gamma\left(\alpha_j-\rho_j+1\right)\right)\right)}\right)\prod_{j=1}^{m}\left\|\left\|\overrightarrow{\overline{D}_{b-}^{\nu_j}f_j}\right\|_p\right\|_{1,(a,b)}.$$

Proof By (26.60), plus $A \geq m > m - 1$. ∎

We continue with

Theorem 26.28 *All here as in Theorem 26.27. Then*

$$\int_a^b (b-x)^A \left(\prod_{j=1}^{m}\ln\left(\sum_{i=1}^{n}e^{\left(\left|\overline{D}_{b-}^{\rho_j}f_{ji}(x)\right|\left(\frac{\Gamma(\alpha_j-\rho_j+1)}{(b-x)^{(\alpha_j-\rho_j)}}\right)\right)}\right)\right)dx \leq$$

$$\left(\frac{B\,(b-a)^{A-m+1}}{(A-m+1)}\right)\left(\prod_{j=1}^{m}\left(\int_a^b\ln\left(\sum_{i=1}^{n}e^{\left|\overline{D}_{b-}^{\alpha_j}f_{ji}(y)\right|}\right)dy\right)\right). \tag{26.93}$$

Proof Using Theorem 26.11. ∎

We give

Definition 26.29 Let $\nu > 0$, $n := [\nu]$, $\alpha := \nu - n$ ($0 \leq \alpha < 1$). Let $a, b \in \mathbb{R}$, $a \leq x \leq b$, $f \in C([a, b])$. We consider $C_a^{\nu}([a, b]) := \{f \in C^n([a, b]) : I_{a+}^{1-\alpha}f^{(n)} \in C^1([a, b])\}$. For $f \in C_a^{\nu}([a, b])$, we define the left generalized ν-fractional derivative of f over $[a, b]$ as

$$\Delta_a^{\nu}f := \left(I_{a+}^{1-\alpha}f^{(n)}\right)', \tag{26.94}$$

see [1, p. 24], and Canavati derivative in [6].
Notice here $\Delta_a^{\nu}f \in C([a, b])$.
So that

$$\left(\Delta_a^{\nu}f\right)(x) = \frac{1}{\Gamma(1-\alpha)}\frac{d}{dx}\int_a^x (x-t)^{-\alpha}f^{(n)}(t)\,dt, \tag{26.95}$$

$\forall\, x \in [a, b]$.

Notice here that

$$\Delta_a^n f = f^{(n)}, \quad n \in \mathbb{Z}_+. \tag{26.96}$$

We need

Theorem 26.30 ([4]) *Let* $f \in C_a^\nu([a, b])$, $n = [\nu]$, *such that* $f^{(i)}(a) = 0$, $i = r, r+1, \ldots, n-1$, *where* $r := [\rho]$, *with* $0 < \rho < \nu$. *Then*

$$\left(\Delta_a^\rho f\right)(x) = \frac{1}{\Gamma(\nu - \rho)} \int_a^x (x - t)^{\nu - \rho - 1} \left(\Delta_a^\nu f\right)(t)\, dt, \tag{26.97}$$

i.e.

$$\left(\Delta_a^\rho f\right) = I_{a+}^{\nu - \rho} \left(\Delta_a^\nu f\right) \in C([a, b]). \tag{26.98}$$

Thus $f \in C_a^\rho([a, b])$.

We present

Theorem 26.31 *Let* $\left(f_{ji}, \nu_j, \rho_j\right)$, $j = 1, \ldots, m$, $m \geq 2$; $i = 1, \ldots, n$, *as in the assumptions of Theorem 26.30. Set* $A := \sum_{j=1}^m (\nu_j - \rho_j)$, $B := \prod_{j=1}^m (\nu_j - \rho_j)$. *Here* $a, b \in \mathbb{R}$, $a < b$, $p \geq 1$, *and* $A > m - 1$. *Then*

$$\left\| \prod_{j=1}^m \left\| \overrightarrow{\Delta_a^{\rho_j} f_j} \right\|_p \right\|_{1,(a,b)} \leq \tag{26.99}$$

$$\left(\frac{B(b-a)^{(A-m+1)}}{(A-m+1)\left(\prod_{j=1}^m \left(\Gamma(\nu_j - \rho_j + 1) \right) \right)} \right) \prod_{j=1}^m \left\| \left\| \overrightarrow{\Delta_a^{\nu_j} f_j} \right\|_p \right\|_{1,(a,b)}.$$

Proof By (26.60). ∎

We continue with

Theorem 26.32 *All here as in Theorem 26.31. Then*

$$\int_a^b (x - a)^A \left(\prod_{j=1}^m \ln \left(\sum_{i=1}^n e^{\left(\left| \Delta_a^{\rho_j} f_{ji}(x) \right| \frac{\Gamma(\nu_j - \rho_j + 1)}{(x-a)^{(\nu_j - \rho_j)}} \right)} \right) \right) dx \leq$$

$$\left(\frac{B\,(b-a)^{A-m+1}}{A-m+1}\right)\left(\prod_{j=1}^{m}\left(\int_{a}^{b}\ln\left(\sum_{i=1}^{n}e^{\left|\Delta_{a}^{\nu_{j}}f_{ji}(y)\right|}\right)\right)dy\right). \qquad (26.100)$$

Proof Using Theorem 26.11. ■

We need

Definition 26.33 ([2]) Let $\nu > 0$, $n := [\nu]$, $\alpha = \nu - n$, $0 < \alpha < 1$, $f \in C([a, b])$. Consider

$$C_{b-}^{\nu}([a, b]) := \{f \in C^{n}([a, b]) : I_{b-}^{1-\alpha} f^{(n)} \in C^{1}([a, b])\}. \qquad (26.101)$$

Define the right generalized ν-fractional derivative of f over $[a, b]$, by

$$\Delta_{b-}^{\nu} f := (-1)^{n-1}\left(I_{b-}^{1-\alpha} f^{(n)}\right)'. \qquad (26.102)$$

We set $\Delta_{b-}^{0} f = f$. Notice that

$$\left(\Delta_{b-}^{\nu} f\right)(x) = \frac{(-1)^{n-1}}{\Gamma(1-\alpha)}\frac{d}{dx}\int_{x}^{b}(J-x)^{-\alpha} f^{(n)}(J)\,dJ, \qquad (26.103)$$

and $\Delta_{b-}^{\nu} f \in C([a, b])$.

We also need

Theorem 26.34 ([4]) *Let* $f \in C_{b-}^{\nu}([a, b])$, $0 < \rho < \nu$. *Assume* $f^{(i)}(b) = 0$, $i = r, r+1, \ldots, n-1$, *where* $r := [\rho]$, $n := [\nu]$. *Then*

$$\Delta_{b-}^{\rho} f(x) = \frac{1}{\Gamma(\nu-\rho)}\int_{x}^{b}(J-x)^{\nu-\rho-1}\left(\Delta_{b-}^{\nu} f\right)(J)\,dJ, \qquad (26.104)$$

$\forall\, x \in [a, b]$, *i.e.*

$$\Delta_{b-}^{\rho} f = I_{b-}^{\nu-\rho}\left(\Delta_{b-}^{\nu} f\right) \in C([a, b]), \qquad (26.105)$$

and $f \in C_{b-}^{\rho}([a, b])$.

We give

Theorem 26.35 *Let* $\left(f_{ji}, \nu_{j}, \rho_{j}\right)$, $j = 1, \ldots, m$, $m \geq 2$; $i = 1, \ldots, n$, *as in the assumptions of Theorem 26.34. Set* $A := \sum_{j=1}^{m}\left(\nu_{j} - \rho_{j}\right)$, $B := \prod_{j=1}^{m}\left(\nu_{j} - \rho_{j}\right)$. *Here* $a, b \in \mathbb{R}$, $a < b$, $p \geq 1$, *and* $A > m - 1$. *Then*

$$\left\|\prod_{j=1}^{m}\left\|\overrightarrow{\Delta_{b-}^{\rho_j} f_j}\right\|_p\right\|_{1,(a,b)} \leq \qquad (26.106)$$

$$\left(\frac{B\,(b-a)^{(A-m+1)}}{(A-m+1)\left(\prod_{j=1}^{m}\left(\Gamma\left(\nu_j-\rho_j+1\right)\right)\right)}\right)\prod_{j=1}^{m}\left\|\left\|\overrightarrow{\Delta_{b-}^{\nu_j} f_j}\right\|_p\right\|_{1,(a,b)}.$$

Proof By (26.60). ■

Theorem 26.36 *All here as in Theorem 26.35. Then*

$$\int_a^b (b-x)^A \left(\prod_{j=1}^{m}\ln\left(\sum_{i=1}^{n} e^{\left(\left|\Delta_{b-}^{\rho_j} f_{ji}(x)\right|\frac{\Gamma(\nu_j-\rho_j+1)}{(b-x)^{(\nu_j-\rho_j)}}\right)}\right)\right) dx \leq$$

$$\left(\frac{B\,(b-a)^{A-m+1}}{A-m+1}\right)\left(\prod_{j=1}^{m}\left(\int_a^b \ln\left(\sum_{i=1}^{n} e^{\left|\Delta_{b-}^{\nu_j} f_{ji}(y)\right|}\right) dy\right)\right). \qquad (26.107)$$

Proof Using Theorem 26.11. ■

We make

Definition 26.37 [14, p. 99] The fractional integrals of a function f with respect to given function g are defined as follows:

Let $a, b \in \mathbb{R}$, $a < b$, $\alpha > 0$. Here g is a strictly increasing function on $[a, b]$ and $g \in C^1([a, b])$. The left- and right-sided fractional integrals of a function f with respect to another function g in $[a, b]$ are given by

$$\left(I_{a+;g}^{\alpha} f\right)(x) = \frac{1}{\Gamma(\alpha)}\int_a^x \frac{g'(t)\,f(t)\,dt}{(g(x)-g(t))^{1-\alpha}}, \quad x > a, \qquad (26.108)$$

$$\left(I_{b-;g}^{\alpha} f\right)(x) = \frac{1}{\Gamma(\alpha)}\int_x^b \frac{g'(t)\,f(t)\,dt}{(g(t)-g(x))^{1-\alpha}}, \quad x < b, \qquad (26.109)$$

respectively.

We make

Remark 26.38 Let f_{ji} be Lebesgue measurable functions from (a, b) into $\mathbb{R}$, such that $\left(I_{a+;g}^{\alpha_j}\left(|f_{ji}|\right)\right)(x) \in \mathbb{R}$, $\forall\, x \in (a, b)$, $\alpha_j > 0$, $j = 1, \ldots, m$, $i = 1, \ldots, n$.

Consider

$$g_{ji}(x) := \left(I_{a+;g}^{\alpha_j} f_{ji}\right)(x), \quad x \in (a,b), \ j = 1, \dots, m, \ i = 1, \dots, n. \quad (26.110)$$

where

$$\left(I_{a+;g}^{\alpha_j} f_{ji}\right)(x) = \frac{1}{\Gamma(\alpha_j)} \int_a^x \frac{g'(t) f_{ji}(t) \, dt}{(g(x) - g(t))^{1-\alpha_j}}, \quad x > a. \quad (26.111)$$

Notice that $g_{ji}(x) \in \mathbb{R}$ and it is Lebesgue measurable.

We pick $\Omega_1 = \Omega_2 = (a,b), d\mu_1(x) = dx, d\mu_2(y) = dy$, the Lebesgue measure. We see that

$$\left(I_{a+;g}^{\alpha_j} f_{ji}\right)(x) = \int_a^b \frac{\chi_{(a,x]}(t) g'(t) f_{ji}(t)}{\Gamma(\alpha_j)(g(x) - g(t))^{1-\alpha_j}} \, dt, \quad (26.112)$$

where χ is the characteristic function.

So, we pick here

$$k_j(x,t) := \frac{\chi_{(a,x]}(t) g'(t)}{\Gamma(\alpha_j)(g(x) - g(t))^{1-\alpha_j}}, \quad j = 1, \dots, m. \quad (26.113)$$

In fact

$$k_j(x,y) = \begin{cases} \frac{g'(y)}{\Gamma(\alpha_j)(g(x)-g(y))^{1-\alpha_j}}, & a < y \le x, \\ 0, & x < y < b. \end{cases} \quad (26.114)$$

Clearly it holds

$$K_j(x) = \int_a^b \frac{\chi_{(a,x]}(y) g'(y)}{\Gamma(\alpha_j)(g(x) - g(y))^{1-\alpha_j}} \, dy =$$

$$\int_a^x \frac{g'(y)}{\Gamma(\alpha_j)(g(x) - g(y))^{1-\alpha_j}} \, dy = \frac{1}{\Gamma(\alpha_j)} \int_a^x (g(x) - g(y))^{\alpha_j - 1} \, dg(y) =$$

$$(26.115)$$

$$\frac{1}{\Gamma(\alpha_j)} \int_{g(a)}^{g(x)} (g(x) - z)^{\alpha_j - 1} \, dz = \frac{(g(x) - g(a))^{\alpha_j}}{\Gamma(\alpha_j + 1)}.$$

So for $a < x < b, j = 1, \dots, m$, we get

$$K_j(x) = \frac{(g(x) - g(a))^{\alpha_j}}{\Gamma(\alpha_j + 1)}. \quad (26.116)$$

Notice that

$$\prod_{j=1}^{m} \frac{k_j\,(x,y)}{K_j\,(x)} = \prod_{j=1}^{m} \left(\frac{\chi_{(a,x]}\,(y)\,g'\,(y)}{\Gamma\,(\alpha_j)\,(g\,(x) - g\,(y))^{1-\alpha_j}} \cdot \frac{\Gamma\,(\alpha_j + 1)}{(g\,(x) - g\,(a))^{\alpha_j}} \right) =$$

$$\frac{\chi_{(a,x]}\,(y)\,(g\,(x) - g\,(y))^{\left(\sum\limits_{j=1}^{m} \alpha_j - m \right)} (g'\,(y))^{m} \left(\prod\limits_{j=1}^{m} \alpha_j \right)}{(g\,(x) - g\,(a))^{\left(\sum\limits_{j=1}^{m} \alpha_j \right)}}. \qquad (26.117)$$

Calling

$$\alpha := \sum_{j=1}^{m} \alpha_j > 0, \ \ \gamma := \prod_{j=1}^{m} \alpha_j > 0, \qquad (26.118)$$

we have that

$$\prod_{j=1}^{m} \frac{k_j\,(x,y)}{K_j\,(x)} = \frac{\chi_{(a,x]}\,(y)\,(g\,(x) - g\,(y))^{\alpha-m} (g'\,(y))^{m}\,\gamma}{(g\,(x) - g\,(a))^{\alpha}}. \qquad (26.119)$$

Therefore, for (26.39), we get for appropiate weight u that (denote λ_m by λ_m^g)

$$\lambda_m^g\,(y) = \gamma\,(g'\,(y))^{m} \int_y^b u\,(x)\,\frac{(g\,(x) - g\,(y))^{\alpha-m}}{(g\,(x) - g\,(a))^{\alpha}} dx < \infty, \qquad (26.120)$$

for all $a < y < b$.
  Let now

$$u\,(x) = (g\,(x) - g\,(a))^{\alpha}\,g'\,(x), \ \ x \in (a,b). \qquad (26.121)$$

Then

$$\lambda_m^g\,(y) = \gamma\,(g'\,(y))^{m} \int_y^b (g\,(x) - g\,(y))^{\alpha-m}\,g'\,(x)\,dx =$$

$$\gamma\,(g'\,(y))^{m} \int_{g(y)}^{g(b)} (z - g\,(y))^{\alpha-m}\,dz = \qquad (26.122)$$

$$\gamma\,(g'\,(y))^{m}\,\frac{(g\,(b) - g\,(y))^{\alpha-m+1}}{\alpha - m + 1},$$

with $\alpha > m - 1$. That is

$$\lambda_m^g(y) = \gamma \left(g'(y)\right)^m \frac{(g(b) - g(y))^{\alpha-m+1}}{\alpha - m + 1}, \tag{26.123}$$

$\alpha > m - 1, y \in (a, b)$.

By Theorem 26.11 we get, for $p \geq 1$, that

$$\int_a^b (g(x) - g(a))^\alpha g'(x) \prod_{j=1}^m \left\| \frac{\left(\overrightarrow{I_{a+;g}^{\alpha_j} f_j(x)}\right) \Gamma(\alpha_j + 1)}{(g(x) - g(a))^{\alpha_j}} \right\|_p dx \leq$$

$$\left(\frac{\gamma \|g'\|_\infty^m (g(b) - g(x))^{\alpha-m+1}}{\alpha - m + 1} \right) \left(\prod_{j=1}^m \int_a^b \left\| \overrightarrow{f_j}(y) \right\|_p dy \right). \tag{26.124}$$

So we have proved that

$$\int_a^b g'(x) \prod_{j=1}^m \left\| \left(\overrightarrow{I_{a+;g}^{\alpha_j} f_j(x)} \right) \right\|_p dx \leq$$

$$\left(\frac{\gamma \|g'\|_\infty^m (g(b) - g(a))^{\alpha-m+1}}{(\alpha - m + 1) \prod_{j=1}^m \left(\Gamma(\alpha_j + 1) \right)} \right) \left(\prod_{j=1}^m \int_a^b \left\| \overrightarrow{f_j}(y) \right\|_p dy \right), \tag{26.125}$$

under the asumptions:

(i) $p \geq 1$, $\alpha > m - 1$, f_{ji} with $I_{a+;g}^{\alpha_j} \left(|f_{ji}| \right)$ finite, $j = 1, \ldots, m$; $i = 1, \ldots, n$,

(ii) $\left\| \overrightarrow{f_j} \right\|_p$ are $\frac{\chi_{(a,x]}(y)g'(y)dy}{\Gamma(\alpha_j)(g(x)-g(y))^{1-\alpha_j}}$ -integrable, a.e. in $x \in (a, b)$, $j = 1, \ldots, m$,

(iii) $\left\| \overrightarrow{f_j} \right\|_p$ are Lebesgue integrable, $j = 1, \ldots, m$.

We need

Definition 26.39 ([13]) Let $0 < a < b < \infty$, $\alpha > 0$. The left- and right-sided Hadamard fractional integrals of order α are given by

$$\left(J_{a+}^\alpha f\right)(x) = \frac{1}{\Gamma(\alpha)} \int_a^x \left(\ln \frac{x}{y}\right)^{\alpha-1} \frac{f(y)}{y} dy, \quad x > a, \tag{26.126}$$

and

$$\left(J_{b-}^{\alpha} f\right)(x) = \frac{1}{\Gamma(\alpha)} \int_x^b \left(\ln \frac{y}{x}\right)^{\alpha-1} \frac{f(y)}{y} dy, \quad x < b, \tag{26.127}$$

respectively.

Notice that the Hadamard fractional integrals of order α are special cases of left- and right-sided fractional integrals of a function f with respect to another function, here $g(x) = \ln x$ on $[a, b]$, $0 < a < b < \infty$.

Above f is a Lebesgue measurable function from (a, b) into $\mathbb{R}$, such that $\left(J_{a+}^{\alpha}(|f|)\right)(x)$ and/or $\left(J_{b-}^{\alpha}(|f|)\right)(x) \in \mathbb{R}, \forall x \in (a, b)$.

We give

Theorem 26.40 *Let* $\left(f_{ji}, \alpha_j\right)$, $j = 1, \dots, m$, $i = 1, \dots, n$, *and* $J_{a+}^{\alpha_j} f_{ji}$ *as in Definition 26.39. Set* $\alpha := \sum_{j=1}^m \alpha_j$, $\gamma := \prod_{j=1}^m \alpha_j$; $p \geq 1$, $\alpha > m - 1$. *Then*

$$\int_a^b \prod_{j=1}^m \left\| \overrightarrow{J_{a+}^{\alpha_j} f_j}(x) \right\|_p dx \leq \tag{26.128}$$

$$\left(\frac{b\gamma \left(\ln\left(\frac{b}{a}\right)\right)^{\alpha-m+1}}{a^m (\alpha - m + 1) \prod_{j=1}^m \left(\Gamma(\alpha_j + 1)\right)} \right) \left(\prod_{j=1}^m \int_a^b \left\| \overrightarrow{f_j}(y) \right\|_p dy \right),$$

under the assumptions:

(i) $\left(J_{a+}^{\alpha_j} |f_{ji}|\right)$ *finite,* $j = 1, \dots, m$; $i = 1, \dots, n$,

(ii) $\left\| \overrightarrow{f_j} \right\|_p$ *are* $\left(\frac{\chi_{(a,x]}(y)dy}{\Gamma(\alpha_j) y \left(\ln\left(\frac{x}{y}\right)\right)^{1-\alpha_j}} \right)$ *-integrable, a.e. in* $x \in (a, b)$, $j = 1, \dots, m$,

(iii) $\left\| \overrightarrow{f_j} \right\|_p$ *are Lebesgue integrable,* $j = 1, \dots, m$.

Proof By (26.125). ∎

We make

Remark 26.41 Let f_{ji} be Lebesgue measurable functions from (a, b) into $\mathbb{R}$, such that $\left(I_{b-;g}^{\alpha_j}(|f_{ji}|)\right)(x) \in \mathbb{R}, \forall x \in (a, b), \alpha_j > 0, j = 1, \dots, m, i = 1, \dots, n$.

Consider

$$g_{ji}(x) := \left(I_{b-;g}^{\alpha_j} f_{ji}\right)(x), \quad x \in (a, b), j = 1, \dots, m, i = 1, \dots, n, \tag{26.129}$$

where

$$\left(I_{b-;g}^{\alpha_j} f_{ji}\right)(x) = \frac{1}{\Gamma(\alpha_j)} \int_x^b \frac{g'(t) f_{ji}(t) dt}{(g(t) - g(x))^{1-\alpha_j}}, \quad x < b. \qquad (26.130)$$

Notice that $g_{ji}(x) \in \mathbb{R}$ and it is Lebesgue measurable.

We pick $\Omega_1 = \Omega_2 = (a, b), d\mu_1(x) = dx, d\mu_2(y) = dy$, the Lebesgue measure. We see that

$$\left(I_{b-;g}^{\alpha_j} f_{ji}\right)(x) = \int_a^b \frac{\chi_{[x,b)}(t) g'(t) f_{ji}(t) dt}{\Gamma(\alpha_j) (g(t) - g(x))^{1-\alpha_j}}, \qquad (26.131)$$

where χ is the characteristic function.

So, we pick here

$$k_j(x, y) := \frac{\chi_{[x,b)}(y) g'(y)}{\Gamma(\alpha_j)(g(y) - g(x))^{1-\alpha_j}}, \quad j = 1, \ldots, m. \qquad (26.132)$$

In fact

$$k_j(x, y) = \begin{cases} \frac{g'(y)}{\Gamma(\alpha_j)(g(y)-g(x))^{1-\alpha_j}}, & x \leq y < b, \\ 0, & a < y < x. \end{cases} \qquad (26.133)$$

Clearly it holds

$$K_j(x) = \int_a^b \frac{\chi_{[x,b)}(y) g'(y) dy}{\Gamma(\alpha_j)(g(y) - g(x))^{1-\alpha_j}} =$$

$$\frac{1}{\Gamma(\alpha_j)} \int_x^b g'(y)(g(y) - g(x))^{\alpha_j-1} dy = \qquad (26.134)$$

$$\frac{1}{\Gamma(\alpha_j)} \int_{g(x)}^{g(b)} (z - g(x))^{\alpha_j-1} dg(y) = \frac{(g(b) - g(x))^{\alpha_j}}{\Gamma(\alpha_j+1)}.$$

So for $a < x < b, j = 1, \ldots, m$, we get

$$K_j(x) = \frac{(g(b) - g(x))^{\alpha_j}}{\Gamma(\alpha_j + 1)}. \qquad (26.135)$$

Notice that

$$\prod_{j=1}^m \frac{k_j(x, y)}{K_j(x)} = \prod_{j=1}^m \left(\frac{\chi_{[x,b)}(y) g'(y)}{\Gamma(\alpha_j)(g(y) - g(x))^{1-\alpha_j}} \cdot \frac{\Gamma(\alpha_j + 1)}{(g(b) - g(x))^{\alpha_j}} \right) =$$

$$\frac{\chi_{[x,b)}\left(y\right)\left(g'\left(y\right)\right)^{m}\left(g\left(y\right)-g\left(x\right)\right)^{\left(\sum_{j=1}^{m}\alpha_j-m\right)}\prod_{j=1}^{m}\alpha_j}{\left(g\left(b\right)-g\left(x\right)\right)^{\sum_{j=1}^{m}\alpha_j}}. \tag{26.136}$$

Calling

$$\alpha := \sum_{j=1}^{m}\alpha_j > 0, \ \gamma := \prod_{j=1}^{m}\alpha_j > 0, \tag{26.137}$$

we have that

$$\prod_{j=1}^{m}\frac{k_j\left(x,y\right)}{K_j\left(x\right)} = \frac{\chi_{[x,b)}\left(y\right)\left(g'\left(y\right)\right)^{m}\left(g\left(y\right)-g\left(x\right)\right)^{\alpha-m}\gamma}{\left(g\left(b\right)-g\left(x\right)\right)^{\alpha}}. \tag{26.138}$$

Therefore, for (26.39), we get for appropiate weight u that (denote λ_m by λ_m^g)

$$\lambda_m^g\left(y\right) = \gamma\left(g'\left(y\right)\right)^{m}\int_a^y u\left(x\right)\frac{\left(g\left(y\right)-g\left(x\right)\right)^{\alpha-m}}{\left(g\left(b\right)-g\left(x\right)\right)^{\alpha}}dx < \infty, \tag{26.139}$$

for all $a < y < b$.

Let now

$$u\left(x\right) = \left(g\left(b\right)-g\left(x\right)\right)^{\alpha}g'\left(x\right), \ x \in \left(a,b\right). \tag{26.140}$$

Then

$$\lambda_m^g\left(y\right) = \gamma\left(g'\left(y\right)\right)^{m}\int_a^y g'\left(x\right)\left(g\left(y\right)-g\left(x\right)\right)^{\alpha-m}dx =$$

$$\gamma\left(g'\left(y\right)\right)^{m}\int_a^y\left(g\left(y\right)-g\left(x\right)\right)^{\alpha-m}dg\left(x\right) = \gamma\left(g'\left(y\right)\right)^{m}\int_{g(a)}^{g(y)}\left(g\left(y\right)-z\right)^{\alpha-m}dz =$$
$$\tag{26.141}$$

$$\gamma\left(g'\left(y\right)\right)^{m}\frac{\left(g\left(y\right)-g\left(a\right)\right)^{\alpha-m+1}}{\alpha-m+1},$$

with $\alpha > m - 1$. That is

$$\lambda_m^g\left(y\right) = \gamma\left(g'\left(y\right)\right)^{m}\frac{\left(g\left(y\right)-g\left(a\right)\right)^{\alpha-m+1}}{\alpha-m+1}, \tag{26.142}$$

$\alpha > m - 1, y \in \left(a,b\right)$.

By Theorem 26.11 we get, for $p \geq 1$, that

$$\int_a^b (g(b) - g(x))^\alpha g'(x) \prod_{j=1}^m \left\| \frac{\left(\overrightarrow{I_{b-;g}^{\alpha_j} f_j}(x) \right) \Gamma(\alpha_j + 1)}{(g(b) - g(x))^{\alpha_j}} \right\|_p dx \leq$$

$$\left(\frac{\gamma \|g'\|_\infty^m (g(b) - g(a))^{\alpha - m + 1}}{\alpha - m + 1} \right) \left(\prod_{j=1}^m \int_a^b \left\| \overrightarrow{f_j}(y) \right\|_p dy \right). \qquad (26.143)$$

So we have proved that

$$\int_a^b g'(x) \prod_{j=1}^m \left\| \left(\overrightarrow{I_{b-;g}^{\alpha_j} f_j}(x) \right) \right\|_p dx \leq$$

$$\left(\frac{\gamma \|g'\|_\infty^m (g(b) - g(a))^{\alpha - m + 1}}{(\alpha - m + 1) \prod_{j=1}^m \Gamma(\alpha_j + 1)} \right) \left(\prod_{j=1}^m \int_a^b \left\| \overrightarrow{f_j}(y) \right\|_p dy \right). \qquad (26.144)$$

under the assumptions:

(i) $p \geq 1, \alpha > m - 1$, f_{ij} with $I_{b-;g}^{\alpha_j} |f_{ji}|$ finite, $j = 1, \ldots, m;\ i = 1, \ldots, n$,

(ii) $\left\| \overrightarrow{f_j} \right\|_p$ are $\left(\frac{\chi_{[x,b)}(y) g'(y) dy}{\Gamma(\alpha_j)(g(y) - g(x))^{1 - \alpha_j}} \right)$-integrable, a.e. in $x \in (a, b)$, $j = 1, \ldots, m$,

(iii) $\left\| \overrightarrow{f_j} \right\|_p$ are Lebesgue integrable, $j = 1, \ldots, m$.

Theorem 26.42 Let (f_{ji}, α_j), $j = 1, \ldots, m$, $i = 1, \ldots, n$, and $J_{b-}^{\alpha_j} f_{ji}$ as in Definition 26.39. Set $\alpha := \sum_{j=1}^m \alpha_j$, $\gamma := \prod_{j=1}^m \alpha_j$; $p \geq 1$, $\alpha > m - 1$. Then

$$\int_a^b \prod_{j=1}^m \left\| \overrightarrow{J_{b-}^{\alpha_j} f_j}(x) \right\|_p dx \leq \qquad (26.145)$$

$$\left(\frac{b\gamma \left(\ln \left(\frac{b}{a} \right) \right)^{\alpha - m + 1}}{a^m (\alpha - m + 1) \prod_{j=1}^m \Gamma(\alpha_j + 1)} \right) \left(\prod_{j=1}^m \int_a^b \left\| \overrightarrow{f_j}(y) \right\|_p dy \right),$$

under the assumptions:

(i) $\left(J_{b-}^{\alpha_j}|f_{ji}|\right)$ *finite,* $j = 1, \ldots, m$; $i = 1, \ldots, n$,

(ii) $\left\|\overrightarrow{f_j}\right\|_p$ *are* $\left(\dfrac{\chi_{[x,b)}(y)dy}{\Gamma(\alpha_j)y(\ln(\frac{y}{x}))^{1-\alpha_j}}\right)$ *-integrable, a.e. in* $x \in (a, b)$, $j = 1, \ldots, m$,

(iii) $\left\|\overrightarrow{f_j}\right\|_p$ *are Lebesgue integrable,* $j = 1, \ldots, m$.

Proof By (26.144). ■

References

1. G.A. Anastassiou, *Fractional Differentiation Inequalities*, Research Monograph (Springer, New York, 2009)
2. G.A. Anastassiou, On right fractional calculus. Chaos, Solitons Fractals **42**, 365–376 (2009)
3. G.A. Anastassiou, Balanced fractional Opial inequalities. Chaos, Solitons Fractals **42**(3), 1523–1528 (2009)
4. G.A. Anastassiou, Fractional Representation formulae and right fractional inequalities. Math. Comput. Model. **54**(11–12), 3098–3115 (2011)
5. G.A. Anastassiou, Vectorial hardy type fractional inequalities. Bull. Tbilisi Int. Cent Math. Inf **16**(2) 21–57 (2012)
6. J.A. Canavati, The Riemann-Liouville integral. Nieuw Archief Voor Wiskunde **5**(1), 53–75 (1987)
7. K. Diethelm, *The Analysis of Fractional Differential Equations*, vol. 2004, 1st edn., Lecture Notes in Mathematics (Springer, New York, 2010)
8. A.M.A. El-Sayed, M. Gaber, On the finite Caputo and finite Riesz derivatives. Electron. J. Theor. Phys. **3**(12), 81–95 (2006)
9. T.S. Ferguson, *Mathematical Statistics* (Academic Press, New York, 1967)
10. R. Gorenflo, F. Mainardi, *Essentials of Fractional Calculus* (Maphysto Center, 2000), http://www.maphysto.dk/oldpages/events/LevyCAC2000/MainardiNotes/fm2k0a.ps
11. G.D. Handley, J.J. Koliha, J. Pečarić, Hilbert-Pachpatte type integral inequalities for fractional derivatives. Fract. Calc. Appl. Anal. **4**(1), 37–46 (2001)
12. H.G. Hardy, Notes on some points in the integral calculus. Messenger Math. **47**(10), 145–150 (1918)
13. S. Iqbal, K. Krulic, J. Pecaric, On an inequality of H.G. Hardy. J. Inequal. Appl. **2010**, Article ID 264347, 23 (2010)
14. A.A. Kilbas, H.M. Srivastava, J.J. Trujillo, *Theory and Applications of Fractional Differential Equations*, vol. 204, North-Holland Mathematics Studies (Elsevier, New York, 2006)
15. M. Perlman, Jensen's Inequality for a Convex Vector-valued function on an infinite-dimensional space. J. of Multivar. Anal. **4**, 52–65 (1974)
16. S.G. Samko, A.A. Kilbas, O.I. Marichev, *Fractional Integral and Derivatives: Theory and Applications* (Gordon and Breach Science Publishers, Yverdon, 1993)

Chapter 27
About Vectorial Fractional Integral Inequalities Using Convexity

Here we present vectorial general integral inequalities involving products of multi-variate convex and increasing functions applied to vectors of functions. As specific applications we derive a wide range of vectorial fractional inequalities of Hardy type. These involve the left and right: Erdélyi-Kober fractional integrals, mixed Riemann-Liouville fractional multiple integrals. Next we produce multivariate Poincaré type vectorial fractional inequalities involving left fractional radial derivatives of Cana-vati type, Riemann-Liouville and Caputo types. The exposed inequalities are of L_p type, $p \geq 1$, and exponential type. It follows [6].

27.1 Introduction

We start with some facts about fractional derivatives needed in the sequel, for more details see, for instance [1, 12].

Let $a < b$, $a, b \in \mathbb{R}$. By $C^N ([a, b])$, we denote the space of all functions on $[a, b]$ which have continuous derivatives up to order N, and $AC ([a, b])$ is the space of all absolutely continuous functions on $[a, b]$. By $AC^N ([a, b])$, we denote the space of all functions g with $g^{(N-1)} \in AC ([a, b])$. For any $\alpha \in \mathbb{R}$, we denote by $[\alpha]$ the integral part of α (the integer k satisfying $k \leq \alpha < k + 1$), and $\lceil \alpha \rceil$ is the ceiling of α ($\min\{n \in \mathbb{N}, n \geq \alpha\}$). By $L_1 (a, b)$, we denote the space of all functions integrable on the interval (a, b), and by $L_\infty (a, b)$ the set of all functions measurable and essentially bounded on (a, b). Clearly, $L_\infty (a, b) \subset L_1 (a, b)$.

We start with the definition of the Riemann-Liouville fractional integrals, see [15]. Let $[a, b]$, $(-\infty < a < b < \infty)$ be a finite interval on the real axis $\mathbb{R}$. The Riemann-Liouville fractional integrals $I_{a+}^\alpha f$ and $I_{b-}^\alpha f$ of order $\alpha > 0$ are defined by

$$\left(I_{a+}^\alpha f \right) (x) = \frac{1}{\Gamma (\alpha)} \int_a^x f (t) (x - t)^{\alpha-1} \, dt, \quad (x > a), \qquad (27.1)$$

© Springer International Publishing Switzerland 2016
G.A. Anastassiou, *Intelligent Comparisons: Analytic Inequalities*,
Studies in Computational Intelligence 609,
DOI 10.1007/978-3-319-21121-3_27

$$\left(I_{b-}^{\alpha} f\right)(x) = \frac{1}{\Gamma(\alpha)} \int_x^b f(t)(t-x)^{\alpha-1} dt, \quad (x < b), \tag{27.2}$$

respectively. Here $\Gamma(\alpha)$ is the Gamma function. These integrals are called the left-sided and the right-sided fractional integrals. We mention some properties of the operators $I_{a+}^{\alpha} f$ and $I_{b-}^{\alpha} f$ of order $\alpha > 0$, see also [18]. The first result yields that the fractional integral operators $I_{a+}^{\alpha} f$ and $I_{b-}^{\alpha} f$ are bounded in $L_p(a, b)$, $1 \le p \le \infty$, that is

$$\left\| I_{a+}^{\alpha} f \right\|_p \le K \left\| f \right\|_p, \left\| I_{b-}^{\alpha} f \right\|_p \le K \left\| f \right\|_p \tag{27.3}$$

where

$$K = \frac{(b-a)^{\alpha}}{\alpha \Gamma(\alpha)}. \tag{27.4}$$

Inequality (27.3), that is the result involving the left-sided fractional integral, was proved by Hardy in one of his first papers, see [13]. He did not write down the constant, but the calculation of the constant was hidden inside his proof.

Next we are motivated by [14].

We produce a wide range of vectorial integral inequalities related to integral operators, with applications to vectorial Hardy type fractional inequalities.

27.2 Main Results

Let $(\Omega_1, \Sigma_1, \mu_1)$ and $(\Omega_2, \Sigma_2, \mu_2)$ be measure spaces with positive σ-finite measures, and let $k : \Omega_1 \times \Omega_2 \to \mathbb{R}$ be a nonnegative measurable function, $k(x, \cdot)$ measurable on Ω_2 and

$$K(x) = \int_{\Omega_2} k(x, y) d\mu_2(y), \quad x \in \Omega_1. \tag{27.5}$$

We suppose that $K(x) > 0$ a.e. on Ω_1, and by a weight function (shortly: a weight), we mean a nonnegative measurable function on the actual set. Let the measurable functions $g_i : \Omega_1 \to \mathbb{R}$, $i = 1, \ldots, n$, with the representation

$$g_i(x) = \int_{\Omega_2} k(x, y) f_i(y) d\mu_2(y), \tag{27.6}$$

where $f_i : \Omega_2 \to \mathbb{R}$ is a measurable functions, $i = 1, \ldots, n$.

Denote by $\overrightarrow{x} = x := (x_1, \ldots, x_n) \in \mathbb{R}^n$, $\overrightarrow{g} := (g_1, \ldots, g_n)$ and $\overrightarrow{f} := (f_1, \ldots, f_n)$.

We consider here $\Phi : \mathbb{R}_+^n \to \mathbb{R}$ a convex function, which is increasing per coordinate, i.e. if $x_i \le y_i$, $i = 1, \ldots, n$, then

$$\Phi(x_1, \ldots, x_n) \le \Phi(y_1, \ldots, y_n). \tag{27.7}$$

Examples for Φ:

(1) Given g_i is convex and increasing on $\mathbb{R}_+$, then $\Phi(x_1, \ldots, x_n) := \sum_{i=1}^n g_i(x_i)$ is convex on $\mathbb{R}_+^n$, and increasing per coordinate; the same properties hold for:

(2) $\|x\|_p = \left(\sum_{i=1}^n x_i^p\right)^{\frac{1}{p}}$, $p \geq 1$,

(3) $\|x\|_\infty = \max\limits_{i \in \{1,\ldots,n\}} x_i$,

(4) $\sum_{i=1}^n x_i^2$,

(5) $\sum_{i=1}^n \left(i \cdot x_i^2\right)$,

(6) $\sum_{i=1}^n \sum_{j=1}^i x_j^2$,

(7) $\ln\left(\sum_{i=1}^n e^{x_i}\right)$,

(8) let g_j are convex and increasing per coordinate on $\mathbb{R}_+^n$, then so is $\sum_{j=1}^m e^{g_j(x)}$, and so is $\ln\left(\sum_{j=1}^m e^{g_j(x)}\right)$, $x \in \mathbb{R}_+^n$.

It is a well known fact that, if $C \subseteq \mathbb{R}^n$ is an open and convex set, and $f : C \to \mathbb{R}$ is a convex function, then f is continuous on C.

We need

Proposition 27.1 ([7]) *Let* $\Phi : \mathbb{R}_+^n \to \mathbb{R}$ *be a convex function which is increasing per coordinate. Then* Φ *is continuous.*

We also need

Theorem 27.2 ([7]) *Let* u *be a weight function on* Ω_1, *and* $k, K, g_i, f_i, i = 1, \ldots, n \in \mathbb{N}$, *and* Φ *defined as above. Assume that the function* $x \to u(x) \frac{k(x,y)}{K(x)}$ *is integrable on* Ω_1 *for each fixed* $y \in \Omega_2$. *Define* v *on* Ω_2 *by*

$$v(y) := \int_{\Omega_1} u(x) \frac{k(x,y)}{K(x)} d\mu_1(x) < \infty. \qquad (27.8)$$

Then

$$\int_{\Omega_1} u(x) \, \Phi\left(\frac{|g_1(x)|}{K(x)}, \ldots, \frac{|g_n(x)|}{K(x)}\right) d\mu_1(x) \leq$$

$$\int_{\Omega_2} v(y) \, \Phi\left(|f_1(y)|, \ldots, |f_n(y)|\right) d\mu_2(y), \qquad (27.9)$$

under the assumptions:

(i) f_i, $\Phi(|f_1|, \ldots, |f_n|)$, *are* $k(x,y) d\mu_2(y)$ *-integrable,* μ_1 *-a.e. in* $x \in \Omega_1$, *for all* $i = 1, \ldots, n$,

(ii) $v(y) \, \Phi(|f_1(y)|, \ldots, |f_n(y)|)$ *is* μ_2 *-integrable.*

Notation 27.3 *From now on we may write*

$$\vec{g}(x) = \int_{\Omega_2} k(x,y) \, \vec{f}(y) d\mu_2(y), \qquad (27.10)$$

which means

$$(g_1(x), \ldots, g_n(x)) = \left(\int_{\Omega_2} k(x, y) f_1(y) d\mu_2(y), \ldots, \int_{\Omega_2} k(x, y) f_n(y) d\mu_2(y) \right).$$
(27.11)

Similarly, we may write

$$\left| \overrightarrow{g}(x) \right| = \left| \int_{\Omega_2} k(x, y) \overrightarrow{f}(y) d\mu_2(y) \right|,$$
(27.12)

and we mean

$$(|g_1(x)|, \ldots, |g_n(x)|) =$$

$$\left(\left| \int_{\Omega_2} k(x, y) f_1(y) d\mu_2(y) \right|, \ldots, \left| \int_{\Omega_2} k(x, y) f_n(y) d\mu_2(y) \right| \right).$$
(27.13)

We also can write that

$$\left| \overrightarrow{g}(x) \right| \leq \int_{\Omega_2} k(x, y) \left| \overrightarrow{f}(y) \right| d\mu_2(y),$$
(27.14)

and we mean the fact that

$$|g_i(x)| \leq \int_{\Omega_2} k(x, y) |f_i(y)| d\mu_2(y),$$
(27.15)

for all $i = 1, \ldots, n$, etc.

Notation 27.4 *Next let $(\Omega_1, \Sigma_1, \mu_1)$ and $(\Omega_2, \Sigma_2, \mu_2)$ be measure spaces with positive σ-finite measures, and let $k_j : \Omega_1 \times \Omega_2 \to \mathbb{R}$ be a nonnegative measurable function, $k_j(x, \cdot)$ measurable on Ω_2 and*

$$K_j(x) = \int_{\Omega_2} k_j(x, y) d\mu_2(y), \quad x \in \Omega_1, j = 1, \ldots, m.$$
(27.16)

We suppose that $K_j(x) > 0$ a.e. on Ω_1. Let the measurable functions $g_{ji} : \Omega_1 \to \mathbb{R}$ with the representation

$$g_{ji}(x) = \int_{\Omega_2} k_j(x, y) f_{ji}(y) d\mu_2(y),$$
(27.17)

where $f_{ji} : \Omega_2 \to \mathbb{R}$ are measurable functions, $i = 1, \ldots, n$ and $j = 1, \ldots, m$.
   Denote the function vectors $\overrightarrow{g_j} := (g_{j1}, g_{j2}, \ldots, g_{jn})$ and $\overrightarrow{f_j} := (f_{j1}, \ldots, f_{jn})$, $j = 1, \ldots, m$.

We say $\vec{f}_j$ is integrable with respect to measure μ, iff all f_{ji} are integrable with respect to μ.

We also consider here $\Phi_j : \mathbb{R}^n_+ \to \mathbb{R}_+$, $j = 1, \ldots, m$, convex functions that are increasing per coordinate. Again u is a weight function on Ω_1.

Our first main result is when $m = 2$.

Theorem 27.5 *All as in Notation 27.4. Assume that the functions* $(j = 1, 2)$ $x \mapsto$ $\left(u(x) \frac{k_j(x,y)}{K_j(x)} \right)$ *are integrable on* Ω_1, *for each fixed* $y \in \Omega_2$. *Define* u_j *on* Ω_2 *by*

$$u_j(y) := \int_{\Omega_1} u(x) \frac{k_j(x, y)}{K_j(x)} d\mu_1(x) < \infty. \tag{27.18}$$

Let $p_1, q_1 > 1 : \frac{1}{p_1} + \frac{1}{q_1} = 1$. *Let the functions* $\Phi_1, \Phi_2 : \mathbb{R}^n_+ \to \mathbb{R}_+$, *be convex and increasing per coordinate.*
Then

$$\int_{\Omega_1} u(x) \Phi_1 \left(\left| \frac{\vec{g_1}(x)}{K_1(x)} \right| \right) \Phi_2 \left(\left| \frac{\vec{g_2}(x)}{K_2(x)} \right| \right) d\mu_1(x) \le$$

$$\left(\int_{\Omega_2} u_1(y) \Phi_1 \left(\left| \vec{f_1}(y) \right| \right)^{p_1} d\mu_2(y) \right)^{\frac{1}{p_1}} \left(\int_{\Omega_2} u_2(y) \Phi_2 \left(\left| \vec{f_2}(y) \right| \right)^{q_1} d\mu_2(y) \right)^{\frac{1}{q_1}}, \tag{27.19}$$

under the assumptions:

(i) $\vec{f_1}$, $\Phi_1 \left(\left| \vec{f_1} \right| \right)^{p_1}$ *are both* $k_1(x, y) d\mu_2(y)$ *-integrable,* μ_1 *-a.e. in* $x \in \Omega_1$,

(ii) $\vec{f_2}$, $\Phi_2 \left(\left| \vec{f_2} \right| \right)^{q_1}$ *are both* $k_2(x, y) d\mu_2(y)$ *-integrable,* μ_1 *-a.e. in* $x \in \Omega_1$,

(iii) $u_1 \Phi_1 \left(\left| \vec{f_1} \right| \right)^{p_1}$, $u_2 \Phi_2 \left(\left| \vec{f_2} \right| \right)^{q_1}$, *are both* μ_2 *-integrable.*

Proof Notice that Φ_1, Φ_2 are continuous functions by Proposition 27.1. Here we use Hölder's inequality. We have

$$\int_{\Omega_1} u(x) \Phi_1 \left(\left| \frac{\vec{g_1}(x)}{K_1(x)} \right| \right) \Phi_2 \left(\left| \frac{\vec{g_2}(x)}{K_2(x)} \right| \right) d\mu_1(x) =$$

$$\int_{\Omega_1} u(x)^{\frac{1}{p_1}} \Phi_1 \left(\left| \frac{\vec{g_1}(x)}{K_1(x)} \right| \right) u(x)^{\frac{1}{q_1}} \Phi_2 \left(\left| \frac{\vec{g_2}(x)}{K_2(x)} \right| \right) d\mu_1(x) \le \tag{27.20}$$

$$\left(\int_{\Omega_1} u(x) \Phi_1 \left(\left| \frac{\vec{g_1}(x)}{K_1(x)} \right| \right)^{p_1} d\mu_1(x) \right)^{\frac{1}{p_1}}.$$

$$\left(\int_{\Omega_1} u\,(x)\,\Phi_2 \left(\left| \frac{\overrightarrow{g_2}\,(x)}{K_2\,(x)} \right| \right)^{q_1} d\mu_1\,(x) \right)^{\frac{1}{q_1}} \leq$$

(notice here that Φ_1^p, Φ_2^q are convex, increasing per coordinate and continuous non-negative functions, and by Theorem 27.2 we get)

$$\left(\int_{\Omega_2} u_1\,(y)\,\Phi_1 \left(\left| \overrightarrow{f_1}\,(y) \right| \right)^{p_1} d\mu_2\,(y) \right)^{\frac{1}{p_1}} \left(\int_{\Omega_2} u_2\,(y)\,\Phi_2 \left(\left| \overrightarrow{f_2}\,(y) \right| \right)^{q_1} d\mu_2\,(y) \right)^{\frac{1}{q_1}}.$$
(27.21)

∎

The general result follows

Theorem 27.6 *All as in Notation 27.4. Assume that the functions ($j = 1, 2, \ldots, m \in$* $\mathbb{N}$*) $x \mapsto \left(u\,(x)\,\frac{k_j(x,y)}{K_j(x)} \right)$ are integrable on Ω_1, for each fixed $y \in \Omega_2$. Define u_j on* Ω_2 *by*

$$u_j\,(y) := \int_{\Omega_1} u\,(x)\,\frac{k_j\,(x, y)}{K_j\,(x)} d\mu_1\,(x) < \infty. \tag{27.22}$$

Let $p_j > 1 : \displaystyle\sum_{j=1}^{m} \frac{1}{p_j} = 1$. Let the functions $\Phi_j : \mathbb{R}_+^n \to \mathbb{R}_+$, $j = 1, \ldots, m$, be convex and increasing per coordinate.
Then

$$\int_{\Omega_1} u\,(x)\,\prod_{j=1}^{m} \Phi_j \left(\left| \frac{\overrightarrow{g_j}\,(x)}{K_j\,(x)} \right| \right) d\mu_1\,(x) \leq$$

$$\prod_{j=1}^{m} \left(\int_{\Omega_2} u_j\,(y)\,\Phi_j \left(\left| \overrightarrow{f_j}\,(y) \right| \right)^{p_j} d\mu_2\,(y) \right)^{\frac{1}{p_j}}, \tag{27.23}$$

under the assumptions:

(i) *$\overrightarrow{f_j}$, $\Phi_j \left(\left| \overrightarrow{f_j} \right| \right)^{p_j}$ are both $k_j\,(x, y)\,d\mu_2\,(y)$ -integrable, μ_1 -a.e. in $x \in \Omega_1$,* $j = 1, \ldots, m,$
(ii) *$u_j\Phi_j \left(\left| \overrightarrow{f_j} \right| \right)^{p_j}$ is μ_2 -integrable, $j = 1, \ldots, m$.*

Proof Notice that Φ_j, $j = 1, \ldots, m$, are continuous functions. Here we use the generalized Hölder's inequality. We have

$$\int_{\Omega_1} u\,(x)\,\prod_{j=1}^{m} \Phi_j \left(\left| \frac{\overrightarrow{g_j}\,(x)}{K_j\,(x)} \right| \right) d\mu_1\,(x) =$$

$$\int_{\Omega_1} \prod_{j=1}^{m} \left(u(x)^{\frac{1}{p_j}} \, \Phi_j \left(\left| \frac{\overrightarrow{g_j}(x)}{K_j(x)} \right| \right) \right) d\mu_1(x) \leq \qquad (27.24)$$

$$\prod_{j=1}^{m} \left(\int_{\Omega_1} u(x) \, \Phi_j \left(\left| \frac{\overrightarrow{g_j}(x)}{K_j(x)} \right| \right)^{p_j} d\mu_1(x) \right)^{\frac{1}{p_j}} \leq$$

(notice here that $\Phi_j^{p_j}$, $j = 1, \ldots, m$, are convex, increasing per coordinate and continuous, nonnegative functions, and by Theorem 27.2 we get)

$$\prod_{j=1}^{m} \left(\int_{\Omega_2} u_j(y) \, \Phi_j \left(\left| \overrightarrow{f_j}(y) \right| \right)^{p_j} d\mu_2(y) \right)^{\frac{1}{p_j}}. \qquad (27.25)$$

proving the claim. ∎

When $k(x, y) := k_1(x, y) = k_2(x, y) = \ldots = k_m(x, y)$, then $K(x) := K_1(x) = K_2(x) = \ldots = K_m(x)$, we get by Theorems 27.5 and 27.6 the following:

Corollary 27.7 *Assume that the function* $x \mapsto \left(u(x) \frac{k(x,y)}{K(x)} \right)$ *is integrable on* Ω_1, *for each fixed* $y \in \Omega_2$. *Define U on* Ω_2 *by*

$$U(y) := \int_{\Omega_1} u(x) \frac{k(x, y)}{K(x)} d\mu_1(x) < \infty. \qquad (27.26)$$

Let $p_1, q_1 > 1 : \frac{1}{p_1} + \frac{1}{q_1} = 1$. *Let the functions* $\Phi_1, \Phi_2 : \mathbb{R}_+^n \to \mathbb{R}_+$, *be convex and increasing per coordinate.*
Then

$$\int_{\Omega_1} u(x) \, \Phi_1 \left(\left| \frac{\overrightarrow{g_1}(x)}{K(x)} \right| \right) \Phi_2 \left(\left| \frac{\overrightarrow{g_2}(x)}{K(x)} \right| \right) d\mu_1(x) \leq$$

$$\left(\int_{\Omega_2} U(y) \, \Phi_1 \left(\left| \overrightarrow{f_1}(y) \right| \right)^{p_1} d\mu_2(y) \right)^{\frac{1}{p_1}} \left(\int_{\Omega_2} U(y) \, \Phi_2 \left(\left| \overrightarrow{f_2}(y) \right| \right)^{q_1} d\mu_2(y) \right)^{\frac{1}{q_1}},$$

$$(27.27)$$

under the assumptions:

(i) $\overrightarrow{f_1}, \overrightarrow{f_2}, \Phi_1 \left(\left| \overrightarrow{f_1} \right| \right)^{p_1}, \Phi_2 \left(\left| \overrightarrow{f_2} \right| \right)^{q_1}$ *are all* $k(x, y) \, d\mu_2(y)$ *-integrable,* μ_1 *-a.e. in* $x \in \Omega_1$,

(ii) $U \Phi_1 \left(\left| \overrightarrow{f_1} \right| \right)^{p_1}, U \Phi_2 \left(\left| \overrightarrow{f_2} \right| \right)^{q_1}$, *are both* μ_2 *-integrable.*

Corollary 27.8 *Assume that the function* $x \mapsto \left(u(x) \frac{k(x,y)}{K(x)} \right)$ *is integrable on* Ω_1, *for each fixed* $y \in \Omega_2$. *Define U on* Ω_2 *by*

$$U (y) := \int_{\Omega_1} u (x) \frac{k (x, y)}{K (x)} d\mu_1 (x) < \infty. \tag{27.28}$$

Let $p_j > 1 : \sum_{j=1}^{m} \frac{1}{p_j} = 1$. Let the functions $\Phi_j : \mathbb{R}_+^n \to \mathbb{R}_+$, $j = 1, \ldots, m$, be convex and increasing per coordinate.

Then

$$\int_{\Omega_1} u (x) \prod_{j=1}^{m} \Phi_j \left(\left| \frac{\overrightarrow{g_j} (x)}{K (x)} \right| \right) d\mu_1 (x) \le$$

$$\prod_{j=1}^{m} \left(\int_{\Omega_2} U (y) \Phi_j \left(\left| \overrightarrow{f_j} (y) \right| \right)^{p_j} d\mu_2 (y) \right)^{\frac{1}{p_j}}, \tag{27.29}$$

under the assumptions:

(i) $\overrightarrow{f_j}$, $\Phi_j \left(\left| \overrightarrow{f_j} \right| \right)^{p_j}$ *are both $k (x, y) d\mu_2 (y)$ -integrable, μ_1 -a.e. in $x \in \Omega_1$, for all $j = 1, \ldots, m$,*

(ii) $U\Phi_j \left(\left| \overrightarrow{f_j} \right| \right)^{p_j}$ *is μ_2 -integrable, $j = 1, \ldots, m$.*

We give the general application

Theorem 27.9 *Here all as in Theorem 27.6. It holds*

$$\int_{\Omega_1} \left(\frac{u (x)}{\prod_{j=1}^{m} K_j (x)} \right) \prod_{j=1}^{m} \left(\sum_{i=1}^{n} |g_{ji} (x)|^{p_j} \right)^{\frac{1}{p_j}} d\mu_1 (x) \le$$

$$\prod_{j=1}^{m} \left(\int_{\Omega_2} u_j (y) \left(\sum_{i=1}^{n} |f_{ji} (y)|^{p_j} \right) d\mu_2 (y) \right)^{\frac{1}{p_j}}, \tag{27.30}$$

under the assumptions:

(i) $p_j > 1 : \sum_{j=1}^{m} \frac{1}{p_j} = 1$,

(ii) $|f_{ji} (y)|^{p_j}$ *are $k_j (x, y) d\mu_2 (y)$ -integrable, μ_1 -a.e. in $x \in \Omega_1$, all $i = 1, \ldots, n$; $j = 1, \ldots, m$,*

(iii) $u_j |f_{ji}|^{p_j}$ *is μ_2 -integrable, for all $i = 1, \ldots, n$; $j = 1, \ldots, m$.*

Proof Apply Theorem 27.6 with $\Phi_j (x_1, \ldots, x_n) := \left\| \overrightarrow{x} \right\|_{p_j}$, $\overrightarrow{x} = (x_1, \ldots, x_n)$, for all $j = 1, \ldots, m$. ■

Another general application follows.

Theorem 27.10 *Here all as in Theorem 27.6. It holds*

$$\int_{\Omega_1} u(x) \prod_{j=1}^{m} \ln \left(\sum_{i=1}^{n} e^{\left| \frac{g_{ji}(x)}{K_j(x)} \right|} \right) d\mu_1(x) \leq$$

$$\prod_{j=1}^{m} \left(\int_{\Omega_2} u_j(y) \left(\ln \left(\sum_{i=1}^{n} e^{|f_{ji}(y)|} \right) \right)^{p_j} d\mu_2(y) \right)^{\frac{1}{p_j}}, \qquad (27.31)$$

under the assumptions:

(i) $p_j > 1 : \sum_{j=1}^{m} \frac{1}{p_j} = 1$,

(ii) $\overrightarrow{f_j}$, $\left(\ln \left(\sum_{i=1}^{n} e^{|f_{ji}(y)|} \right) \right)^{p_j}$ *are both* $k_j(x, y) d\mu_2(y)$ *-integrable,* μ_1 *-a.e. in* $x \in \Omega_1$, $j = 1, \ldots, m$,

(iii) $u_j(y) \left(\ln \left(\sum_{i=1}^{n} e^{|f_{ji}(y)|} \right) \right)^{p_j}$ *is* μ_2 *-integrable,* $j = 1, \ldots, m$.

Proof Apply Theorem 27.6 with $\Phi_j(x_1, \ldots, x_n) := \ln \left(\sum_{i=1}^{n} e^{x_i} \right)$, for all $j = 1, \ldots, m$. ∎

We need

Definition 27.11 ([18]) Let (a, b), $0 \leq a < b < \infty$; $\alpha, \sigma > 0$. We consider the left- and right-sided fractional integrals of order α as follows:

(1) for $\eta > -1$, we define

$$\left(I_{a+;\sigma,\eta}^{\alpha} f \right)(x) = \frac{\sigma x^{-\sigma(\alpha+\eta)}}{\Gamma(\alpha)} \int_{a}^{x} \frac{t^{\sigma\eta+\sigma-1} f(t) \, dt}{(x^{\sigma} - t^{\sigma})^{1-\alpha}}, \qquad (27.32)$$

(2) for $\eta > 0$, we define

$$\left(I_{b-;\sigma,\eta}^{\alpha} f \right)(x) = \frac{\sigma x^{\sigma\eta}}{\Gamma(\alpha)} \int_{x}^{b} \frac{t^{\sigma(1-\eta-\alpha)-1} f(t) \, dt}{(t^{\sigma} - x^{\sigma})^{1-\alpha}}. \qquad (27.33)$$

These are the Erdélyi-Kober type fractional integrals.

We remind the Beta function

$$B(x, y) := \int_{0}^{1} t^{x-1} (1-t)^{y-1} \, dt, \qquad (27.34)$$

for Re (x), Re $(y) > 0$, and the Incomplete Beta function

$$B(x; \alpha, \beta) = \int_{0}^{x} t^{\alpha-1} (1-t)^{\beta-1} \, dt, \qquad (27.35)$$

where $0 < x \leq 1$; $\alpha, \beta > 0$.

We make

Remark 27.12 Regarding (27.32) we have

$$k(x, y) = \frac{\sigma x^{-\sigma(\alpha+\eta)}}{\Gamma(\alpha)} \chi_{(a,x]}(y) \frac{y^{\sigma\eta+\sigma-1}}{(x^\sigma - y^\sigma)^{1-\alpha}}, \tag{27.36}$$

$x, y \in (a, b)$, χ stands for the characteristic function.
Here

$$K(x) = \int_a^b k(x, t) \, dt = \left(I_{a+;\sigma;\eta}^\alpha 1\right)(x)$$

$$= \frac{\sigma x^{-\sigma(\alpha+\eta)}}{\Gamma(\alpha)} \int_a^x \frac{t^{\sigma\eta+\sigma-1}}{(x^\sigma - t^\sigma)^{1-\alpha}} \, dt \tag{27.37}$$

(setting $z = \frac{t}{x}$)

$$= \frac{\sigma}{\Gamma(\alpha)} \int_{\frac{a}{x}}^1 z^{\sigma\left((\eta+1)-\frac{1}{\sigma}\right)} \left(1 - z^\sigma\right)^{\alpha-1} dz$$

(setting $\lambda = z^\sigma$)

$$= \frac{1}{\Gamma(\alpha)} \int_{\left(\frac{a}{x}\right)^\sigma}^1 \lambda^\eta (1 - \lambda)^{\alpha-1} \, d\lambda. \tag{27.38}$$

Hence

$$K(x) = \frac{1}{\Gamma(\alpha)} \int_{\left(\frac{a}{x}\right)^\sigma}^1 \lambda^\eta (1 - \lambda)^{\alpha-1} \, d\lambda. \tag{27.39}$$

Indeed it is

$$K(x) = \left(I_{a+;\sigma;\eta}^\alpha (1)\right)(x) \tag{27.40}$$

$$= \frac{B(\eta+1, \alpha) - B\left(\left(\frac{a}{x}\right)^\sigma ; \eta+1, \alpha\right)}{\Gamma(\alpha)}.$$

We also make

Remark 27.13 Regarding (27.33) we have

$$k(x, y) = \frac{\sigma x^{\sigma\eta}}{\Gamma(\alpha)} \chi_{[x,b)}(y) \frac{y^{\sigma(1-\eta-\alpha)-1}}{(y^\sigma - x^\sigma)^{1-\alpha}}, \tag{27.41}$$

$x, y \in (a, b)$.
Here

$$K(x) = \int_a^b k(x, t) \, dt = \left(I_{b-;\sigma;\eta}^\alpha 1\right)(x)$$

$$= \frac{\sigma x^{\sigma \eta}}{\Gamma (\alpha)} \int_x^b \frac{t^{\sigma(1-\eta-\alpha)-1}}{(t^\sigma - x^\sigma)^{1-\alpha}} dt \qquad (27.42)$$

(setting $z = \frac{t}{x}$)

$$= \frac{\sigma}{\Gamma (\alpha)} \int_1^{\left(\frac{b}{x}\right)} (z^\sigma - 1)^{\alpha-1} z^{\sigma(1-\eta-\alpha)-1} dz$$

(setting $\lambda = z^\sigma, 1 \le \lambda < \left(\frac{b}{x}\right)^\sigma$)

$$= \frac{1}{\Gamma (\alpha)} \int_1^{\left(\frac{b}{x}\right)^\sigma} (\lambda - 1)^{\alpha-1} \lambda^{-\eta-\alpha} d\lambda \qquad (27.43)$$

$$= \frac{1}{\Gamma (\alpha)} \int_1^{\left(\frac{b}{x}\right)^\sigma} \frac{1}{\lambda^{\eta+1}} \left(1 - \frac{1}{\lambda}\right)^{\alpha-1} d\lambda$$

(setting $w := \frac{1}{\lambda}, 0 < \left(\frac{x}{b}\right)^\sigma < w \le 1$)

$$= \frac{1}{\Gamma (\alpha)} \int_{\left(\frac{x}{b}\right)^\sigma}^1 w^{\eta-1} (1 - w)^{\alpha-1} dw \qquad (27.44)$$

$$= \frac{\left(B (\eta, \alpha) - B \left(\left(\frac{x}{b}\right)^\sigma ; \eta, \alpha\right)\right)}{\Gamma (\alpha)}.$$

That is

$$K (x) = \left(I^\alpha_{b-;\sigma;\eta} (1)\right) (x) \qquad (27.45)$$

$$= \frac{\left(B (\eta, \alpha) - B \left(\left(\frac{x}{b}\right)^\sigma ; \eta, \alpha\right)\right)}{\Gamma (\alpha)}.$$

We give

Theorem 27.14 *Assume that the function*

$$x \mapsto \left(u (x) \frac{\chi_{(a,x]} (y) \sigma x^{-\sigma(\alpha+\eta)} y^{\sigma\eta+\sigma-1}}{(x^\sigma - y^\sigma)^{1-\alpha} \left[B (\eta + 1, \alpha) - B \left(\left(\frac{a}{x}\right)^\sigma ; \eta + 1, \alpha\right)\right]} \right) \qquad (27.46)$$

is integrable on (a, b), *for each* $y \in (a, b)$. *Here* $\alpha, \sigma > 0, \eta > -1, 0 \le a < b < \infty$. *Define* u_1 *on* (a, b) *by*

$$u_1 (y) := \sigma y^{\sigma\eta+\sigma-1} \int_y^b \frac{u (x) x^{-\sigma(\alpha+\eta)} (x^\sigma - y^\sigma)^{\alpha-1}}{\left(B (\eta + 1, \alpha) - B \left(\left(\frac{a}{x}\right)^\sigma ; \eta + 1, \alpha\right)\right)} dx < \infty. \qquad (27.47)$$

Let $p_j > 1 : \sum\limits_{j=1}^{m} \frac{1}{p_j} = 1$. Let the functions $\Phi_j : \mathbb{R}_+^n \to \mathbb{R}_+$, $j = 1, \ldots, m$, be convex and increasing per coordinate.

Then

$$\int_a^b u(x) \prod_{j=1}^{m} \Phi_j \left(\frac{\left| \overrightarrow{I_{a+;\sigma;\eta}^{\alpha} f_j}(x) \right| \Gamma(\alpha)}{(B(\eta+1,\alpha) - B((\frac{a}{x})^{\sigma}; \eta+1,\alpha))} \right) dx \le$$

$$\prod_{j=1}^{m} \left(\int_a^b u_1(y) \Phi_j \left(\left| \overrightarrow{f_j}(y) \right| \right)^{p_j} dy \right)^{\frac{1}{p_j}}, \tag{27.48}$$

under the assumptions:

(i) $\overrightarrow{f_j}, \Phi_j \left(\left| \overrightarrow{f_j} \right| \right)^{p_j}$ *are both* $\frac{\sigma x^{-\sigma(\alpha+\eta)}}{\Gamma(\alpha)} \chi_{(a,x]}(y) \frac{y^{\sigma\eta+\sigma-1}dy}{(x^{\sigma}-y^{\sigma})^{1-\alpha}}$ *-integrable, a.e. in $x \in (a,b)$, for all $j = 1, \ldots, m$,*

(ii) $u_1 \Phi_j \left(\left| \overrightarrow{f_j} \right| \right)^{p_j}$ *is Lebesgue integrable, $j = 1, \ldots, m$.*

Proof By Corollary 27.8. ∎

Remark 27.15 In (27.47), if we choose

$$u(x) = x^{\sigma(\alpha+\eta+1)-1} \left(B(\eta+1,\alpha) - B\left(\left(\frac{a}{x}\right)^{\alpha}; \eta+1,\alpha \right) \right), \quad x \in (a,b), \tag{27.49}$$

then

$$u_1(y) = \sigma y^{\sigma\eta+\sigma-1} \int_y^b x^{\sigma-1}(x^{\sigma}-y^{\sigma})^{\alpha-1} dx$$

(setting $w := x^{\sigma}, \frac{dw}{dx} = \sigma x^{\sigma-1}, dx = \frac{dw}{\sigma x^{\sigma-1}}$)

$$= y^{\sigma\eta+\sigma-1} \int_{y^{\sigma}}^{b^{\sigma}} (w - y^{\sigma})^{\alpha-1} dw = y^{\sigma\eta+\sigma-1} \frac{(b^{\sigma} - y^{\sigma})^{\alpha}}{\alpha}. \tag{27.50}$$

That is

$$u_1(y) = y^{\sigma\eta+\sigma-1} \frac{(b^{\sigma} - y^{\sigma})^{\alpha}}{\alpha}, \quad y \in (a,b). \tag{27.51}$$

Based on the above, (27.48) becomes

$$\int_a^b x^{\sigma(\alpha+\eta+1)-1} \left(B(\eta+1,\alpha) - B\left(\left(\frac{a}{x}\right)^{\sigma}; \eta+1,\alpha \right) \right) \cdot$$

$$\prod_{j=1}^{m} \Phi_j \left(\frac{\left| I^{\alpha}_{a+;\sigma;\eta} \vec{f_j}(x) \right| \Gamma(\alpha)}{\left(B(\eta+1,\alpha) - B\left(\left(\frac{a}{x}\right)^{\sigma}; \eta+1, \alpha\right) \right)} \right) dx \le$$

$$\frac{1}{\alpha} \prod_{j=1}^{m} \left(\int_a^b y^{\sigma\eta+\sigma-1} \left(b^{\sigma} - y^{\sigma}\right)^{\alpha} \Phi_j \left(\left| \vec{f_j}(y) \right| \right)^{p_j} dy \right)^{\frac{1}{p_j}} \le \qquad (27.52)$$

$$\frac{(b^{\sigma} - a^{\sigma})^{\alpha}}{\alpha} \prod_{j=1}^{m} \left(\int_a^b y^{\sigma(\eta+1)-1} \Phi_j \left(\left| \vec{f_j}(y) \right| \right)^{p_j} dy \right)^{\frac{1}{p_j}},$$

under the assumptions:

(i)   following (27.48), and

(ii)* $y^{\sigma(\eta+1)-1} \Phi_j \left(\left| \vec{f_j}(y) \right| \right)^{p_j}$ is Lebesgue integrable on (a, b), $j = 1, \ldots, m$.

We further present

Theorem 27.16 *Let* $0 \le a < b$; $\alpha, \sigma > 0$, $\eta > -1$; $p_j > 1 : \sum_{j=1}^{m} \frac{1}{p_j} = 1$.

Then

$$\int_a^b x^{\sigma(\alpha+\eta+1)-1} \left(B(\eta+1,\alpha) - B\left(\left(\frac{a}{x}\right)^{\sigma}; \eta+1, \alpha\right) \right)^{(1-m)} \cdot$$

$$\prod_{j=1}^{m} \left(\sum_{i=1}^{n} \left| I^{\alpha}_{a+;\sigma;\eta} f_{ji}(x) \right|^{p_j} \right)^{\frac{1}{p_j}} dx \le \qquad (27.53)$$

$$\left(\frac{(b^{\sigma} - a^{\sigma})^{\alpha}}{\alpha\,(\Gamma(\alpha))^m} \right) \prod_{j=1}^{m} \left(\int_a^b y^{\sigma(\eta+1)-1} \left(\sum_{i=1}^{n} \left| f_{ji}(y) \right|^{p_j} \right) dy \right)^{\frac{1}{p_j}},$$

under the assumptions:

(i) $\left| f_{ji}(y) \right|^{p_j}$ *is* $\left(\frac{\sigma x^{-\sigma(\alpha+\eta)}}{\Gamma(\alpha)} \chi_{(a,x]}(y) \frac{y^{\sigma(\eta+1)-1} dy}{(x^{\sigma} - y^{\sigma})^{1-\alpha}} \right)$ *-integrable, a.e. in* $x \in (a, b)$, *for all* $i = 1, \ldots, n$; $j = 1, \ldots, m$,

(ii) $y^{\sigma(\eta+1)-1} \left| f_{ji} \right|^{p_j}$ *is Lebesgue integrable on* (a, b), *for all* $i = 1, \ldots, n$; $j = 1, \ldots, m$.

Proof By Theorem 27.9. ∎

We also give

Theorem 27.17 Let $0 \leq a < b$; $\alpha, \sigma > 0$, $\eta > -1$; $p_j > 1 : \sum_{j=1}^{m} \frac{1}{p_j} = 1$.

Then

$$\int_a^b x^{\sigma(\alpha+\eta+1)-1} \left(B(\eta+1, \alpha) - B\left(\left(\frac{a}{x}\right)^{\sigma} ; \eta+1, \alpha \right) \right) \cdot$$

$$\prod_{j=1}^{m} \ln \left(\sum_{i=1}^{n} e^{\frac{\Gamma(\alpha)\left(\left| I_{a+;\sigma;\eta}^{\alpha} f_{ji}(x) \right| \right)}{(B(\eta+1,\alpha) - B\left(\left(\frac{a}{x}\right)^{\sigma} ; \eta+1,\alpha \right))}} \right) dx \leq$$

$$\frac{(b^{\sigma} - a^{\sigma})^{\alpha}}{\alpha} \prod_{j=1}^{m} \left(\int_a^b y^{\sigma(\eta+1)-1} \left(\ln \left(\sum_{i=1}^{n} e^{|f_{ji}(y)|} \right) \right)^{p_j} dy \right)^{\frac{1}{p_j}}, \qquad (27.54)$$

under the assumptions:

(i) $\overrightarrow{f_j}$, $\left(\ln \left(\sum_{i=1}^{n} e^{|f_{ji}(y)|} \right) \right)^{p_j}$ *are both* $\frac{\sigma x^{-\sigma(\alpha+\eta)}}{\Gamma(\alpha)} \chi_{(a,x]}(y) \frac{y^{\sigma\eta+\sigma-1}dy}{(x^{\sigma}-y^{\sigma})^{1-\alpha}}$ *-integrable, a.e. in $x \in (a, b)$, for all $i = 1, \ldots, n$; $j = 1, \ldots, m$,*

(ii) $y^{\sigma(\eta+1)-1} \left(\ln \left(\sum_{i=1}^{n} e^{|f_{ji}(y)|} \right) \right)^{p_j}$ *is Lebesgue integrable, $j = 1, \ldots, m$.*

Proof By Theorem 27.10. ■

We present

Theorem 27.18 *Assume that the function*

$$x \mapsto \left(u(x) \frac{\sigma x^{\sigma\eta} \chi_{[x,b)}(y) y^{\sigma(1-\eta-\alpha)-1}}{(y^{\sigma} - x^{\sigma})^{1-\alpha} \left[B(\eta, \alpha) - B\left(\left(\frac{x}{b}\right)^{\sigma} ; \eta, \alpha \right) \right]} \right)$$

is integrable on (a, b), for each $y \in (a, b)$. Here $\alpha, \sigma, \eta > 0$, $0 \leq a < b < \infty$. Define u_2 on (a, b) by

$$u_2(y) := \sigma y^{\sigma(1-\eta-\alpha)-1} \int_a^y \frac{u(x) x^{\sigma\eta} (y^{\sigma} - x^{\sigma})^{\alpha-1} dx}{\left(B(\eta, \alpha) - B\left(\left(\frac{x}{b}\right)^{\sigma} ; \eta, \alpha \right) \right)} < \infty. \qquad (27.55)$$

Let $p_j > 1 : \sum_{j=1}^{m} \frac{1}{p_j} = 1$. Let the functions $\Phi_j : \mathbb{R}_+^n \to \mathbb{R}_+$, $j = 1, \ldots, m$, be convex and increasing per coordinate.

Then

$$\int_a^b u(x) \prod_{j=1}^{m} \Phi_j \left(\frac{\left| I_{b-;\sigma;\eta}^{\alpha} \overrightarrow{f_j}(x) \right| \Gamma(\alpha)}{\left(B(\eta, \alpha) - B\left(\left(\frac{x}{b}\right)^{\sigma} ; \eta, \alpha \right) \right)} \right) dx \leq$$

$$\prod_{j=1}^{m} \left(\int_a^b u_2(y) \, \Phi_j \left(\left| \overrightarrow{f_j}(y) \right| \right)^{p_j} dy \right)^{\frac{1}{p_j}},$$ (27.56)

under the assumptions:

(i) $\overrightarrow{f_j}, \Phi_j \left(\left| \overrightarrow{f_j} \right| \right)^{p_j}$ are both $\left(\frac{\sigma x^{\sigma \eta} \chi_{[x,b)}(y) y^{\sigma(1-\eta-\alpha)-1} dy}{\Gamma(\alpha)(y^\sigma - x^\sigma)^{1-\alpha}} \right)$-integrable, a.e. in $x \in (a, b)$, for all $j = 1, \ldots, m$,

(ii) $u_2 \Phi_j \left(\left| \overrightarrow{f_j} \right| \right)^{p_j}$ is Lebesgue integrable on (a, b), $j = 1, \ldots, m$.

Proof By Corollary 27.8. ∎

Remark 27.19 Here $0 < a < b < \infty$; $\alpha, \sigma, \eta > 0$.
  In (27.55), if we choose

$$u(x) = x^{\sigma(1-\eta)-1} \left(B(\eta, \alpha) - B \left(\left(\frac{x}{b} \right)^\alpha ; \eta, \alpha \right) \right), \, x \in (a, b),$$ (27.57)

then

$$u_2(y) = \sigma y^{\sigma(1-\eta-\alpha)-1} \int_a^y x^{\sigma-1} \left(y^\sigma - x^\sigma \right)^{\alpha-1} dx$$

(setting $w := x^\sigma$, $dx = \frac{dw}{\sigma x^{\sigma-1}}$)

$$= y^{\sigma(1-\eta-\alpha)-1} \int_{a^\sigma}^{y^\sigma} \left(y^\sigma - w \right)^{\alpha-1} dw = y^{\sigma(1-\eta-\alpha)-1} \frac{(y^\sigma - a^\sigma)^\alpha}{\alpha}.$$ (27.58)

That is

$$u_2(y) = y^{\sigma(1-\eta-\alpha)-1} \frac{(y^\sigma - a^\sigma)^\alpha}{\alpha}, \, y \in (a, b).$$ (27.59)

Based on the above, (27.56) becomes

$$\int_a^b x^{\sigma(1-\eta)-1} \left(B(\eta, \alpha) - B \left(\left(\frac{x}{b} \right)^\sigma ; \eta, \alpha \right) \right) \cdot$$

$$\prod_{j=1}^{m} \Phi_j \left(\frac{\left| I_{b-;\sigma;\eta}^{\alpha} \overrightarrow{f_j}(x) \right| \Gamma(\alpha)}{\left(B(\eta, \alpha) - B \left(\left(\frac{x}{b} \right)^\sigma ; \eta, \alpha \right) \right)} \right) dx \le$$

$$\frac{1}{\alpha} \prod_{j=1}^{m} \left(\int_a^b y^{\sigma(1-\eta-\alpha)-1} \left(y^\sigma - a^\sigma \right)^\alpha \Phi_j \left(\left| \overrightarrow{f_j}(y) \right| \right)^{p_j} dy \right)^{\frac{1}{p_j}} \le$$ (27.60)

$$\frac{(b^\sigma - a^\sigma)^\alpha}{\alpha} \prod_{j=1}^{m} \left(\int_a^b y^{\sigma(1-\eta-\alpha)-1} \Phi_j \left(\left| \overrightarrow{f_j}(y) \right| \right)^{p_j} dy \right)^{\frac{1}{p_j}},$$

under the assumptions:

(i)   following (27.56), and

(ii)* $y^{\sigma(1-\eta-\alpha)-1} \Phi_j \left(\left| \overrightarrow{f_j}(y) \right| \right)^{p_j}$ is Lebesgue integrable on (a, b), $j = 1, \dots, m$.

We further present

Theorem 27.20 *Let* $0 < a < b < \infty$; $\alpha, \sigma, \eta > 0$; $p_j > 1 : \sum_{j=1}^{m} \frac{1}{p_j} = 1$. *Then*

$$\int_a^b x^{\sigma(1-\eta)-1} \left(B(\eta, \alpha) - B \left(\left(\frac{x}{b} \right)^\sigma ; \eta, \alpha \right) \right)^{(1-m)} \cdot$$

$$\prod_{j=1}^{m} \left(\sum_{i=1}^{n} \left| I_{b-;\sigma;\eta}^\alpha f_{ji}(x) \right|^{p_j} \right)^{\frac{1}{p_j}} dx \leq$$

$$\frac{(b^\sigma - a^\sigma)^\alpha}{\alpha (\Gamma(\alpha))^m} \prod_{j=1}^{m} \left(\int_a^b y^{\sigma(1-\eta-\alpha)-1} \left(\sum_{i=1}^{n} |f_{ji}(y)|^{p_j} \right) dy \right)^{\frac{1}{p_j}}, \qquad (27.61)$$

under the assumptions:

(i) $|f_{ji}(y)|^{p_j}$ is $\left(\frac{\sigma x^{\sigma \eta} \chi_{[x,b)}(y) y^{\sigma(1-\eta-\alpha)-1} dy}{\Gamma(\alpha)(y^\sigma - x^\sigma)^{1-\alpha}} \right)$ -integrable, a.e. in $x \in (a, b)$, for all
     $i = 1, \dots, n$; $j = 1, \dots, m$,
(ii) $y^{\sigma(1-\eta-\alpha)-1} |f_{ji}|^{p_j}$ is Lebesgue integrable on (a, b), for all $i = 1, \dots, n$;
     $j = 1, \dots, m$.

Proof By Theorem 27.9. ∎

We also give

Theorem 27.21 *Let* $0 < a < b < \infty$; $\alpha, \sigma, \eta > 0$; $p_j > 1 : \sum_{j=1}^{m} \frac{1}{p_j} = 1$. *Then*

$$\int_a^b x^{\sigma(1-\eta)-1} \left(B(\eta, \alpha) - B \left(\left(\frac{x}{b} \right)^\sigma ; \eta, \alpha \right) \right) \cdot$$

$$\prod_{j=1}^{m} \ln \left(\sum_{i=1}^{n} e^{\frac{\Gamma(\alpha) \left(\left| I_{b-;\sigma;\eta}^\alpha f_{ji}(x) \right| \right)}{(B(\eta, \alpha) - B((\frac{x}{b})^\sigma ; \eta, \alpha))}} \right) dx \leq$$

$$\frac{(b^\sigma - a^\sigma)^\alpha}{\alpha} \prod_{j=1}^{m} \left(\int_a^b y^{\sigma(1-\eta-\alpha)-1} \left(\ln \left(\sum_{i=1}^{n} e^{|f_{ji}(y)|} \right) \right)^{p_j} dy \right)^{\frac{1}{p_j}}, \qquad (27.62)$$

under the assumptions:

(i) $\overrightarrow{f_j}, \left(\ln \left(\sum_{i=1}^{n} e^{|f_{ji}(y)|} \right) \right)^{p_j}$ *are both* $\left(\frac{\sigma x^{\sigma\eta} \chi_{[x,b)}(y) y^{\sigma(1-\eta-\alpha)-1} dy}{\Gamma(\alpha)(y^\sigma - x^\sigma)^{1-\alpha}} \right)$ *-integrable, a.e.*
     in $x \in (a, b),$ *for all* $i = 1, \ldots, n;\ j = 1, \ldots, m,$

(ii) $y^{\sigma(1-\eta-\alpha)-1} \left(\ln \left(\sum_{i=1}^{n} e^{|f_{ji}(y)|} \right) \right)^{p_j}$ *is Lebesgue integrable,* $j = 1, \ldots, m.$

Proof By Theorem 27.10. ∎

We make

Remark 27.22 Let $\prod_{i=1}^{N} (a_i, b_i) \subset \mathbb{R}^N,\ N > 1,\ a_i < b_i,\ a_i, b_i \in \mathbb{R}.$ Let $\alpha_i > 0,$
$i = 1, \ldots, N;\ f \in L_1 \left(\prod_{i=1}^{N} (a_i, b_i) \right),$ and set $a = (a_1, \ldots, a_N),\ b = (b_1, \ldots, b_N),$
$\alpha = (\alpha_1, \ldots, \alpha_N),\ x = (x_1, \ldots, x_N),\ t = (t_1, \ldots, t_N).$

We define the left mixed Riemann-Liouville fractional multiple integral of order α (see also [16]):

$$\left(I_{a+}^\alpha f \right)(x) := \frac{1}{\prod_{i=1}^{N} \Gamma(\alpha_i)} \int_{a_1}^{x_1} \cdots \int_{a_N}^{x_N} \prod_{i=1}^{N} (x_i - t_i)^{\alpha_i - 1} f(t_1, \ldots, t_N) \, dt_1 \ldots dt_N,$$

$$(27.63)$$

with $x_i > a_i,\ i = 1, \ldots, N.$

We also define the right mixed Riemann-Liouville fractional multiple integral of order α (see also [14]):

$$\left(I_{b-}^\alpha f \right)(x) := \frac{1}{\prod_{i=1}^{N} \Gamma(\alpha_i)} \int_{x_1}^{b_1} \cdots \int_{x_N}^{b_N} \prod_{i=1}^{N} (t_i - x_i)^{\alpha_i - 1} f(t_1, \ldots, t_N) \, dt_1 \ldots dt_N,$$

$$(27.64)$$

with $x_i < b_i,\ i = 1, \ldots, N.$

Notice $I_{a+}^\alpha (|f|),\ I_{b-}^\alpha (|f|)$ are finite if $f \in L_\infty \left(\prod_{i=1}^{N} (a_i, b_i) \right).$

One can rewrite (27.63) and (27.64) as follows:

$$\left(I_{a+}^{\alpha} f\right)(x) = \frac{1}{\prod\limits_{i=1}^{N} \Gamma\left(\alpha_i\right)} \int_{\prod\limits_{i=1}^{N}(a_i,b_i)} \chi_{\prod\limits_{i=1}^{N}(a_i,x_i]}(t) \prod_{i=1}^{N}(x_i - t_i)^{\alpha_i - 1} f(t)\, dt,$$

(27.65)

with $x_i > a_i,\ i = 1, \ldots, N$,
and

$$\left(I_{b-}^{\alpha} f\right)(x) = \frac{1}{\prod\limits_{i=1}^{N} \Gamma\left(\alpha_i\right)} \int_{\prod\limits_{i=1}^{N}(a_i,b_i)} \chi_{\prod\limits_{i=1}^{N}[x_i,b_i)}(t) \prod_{i=1}^{N}(t_i - x_i)^{\alpha_i - 1} f(t)\, dt,$$

(27.66)

with $x_i < b_i,\ i = 1, \ldots, N$.
The corresponding $k(x, y)$ for I_{a+}^{α}, I_{b-}^{α} are

$$k_{a+}(x, y) = \frac{1}{\prod\limits_{i=1}^{N} \Gamma\left(\alpha_i\right)} \chi_{\prod\limits_{i=1}^{N}(a_i,x_i]}(y) \prod_{i=1}^{N}(x_i - y_i)^{\alpha_i - 1},$$

(27.67)

$$\forall\, x, y \in \prod_{i=1}^{N}(a_i, b_i),$$

and

$$k_{b-}(x, y) = \frac{1}{\prod\limits_{i=1}^{N} \Gamma\left(\alpha_i\right)} \chi_{\prod\limits_{i=1}^{N}[x_i,b_i)}(y) \prod_{i=1}^{N}(y_i - x_i)^{\alpha_i - 1},$$

(27.68)

$$\forall\, x, y \in \prod_{i=1}^{N}(a_i, b_i).$$

The corresponding $K(x)$ for I_{a+}^{α} is:

$$K_{a+}(x) = \int_{\prod\limits_{i=1}^{N}(a_i,b_i)} k_{a+}(x, y)\, dy = \left(I_{a+}^{\alpha} 1\right)(x) =$$

$$\frac{1}{\prod\limits_{i=1}^{N} \Gamma\left(\alpha_i\right)} \int_{a_1}^{x_1} \cdots \int_{a_N}^{x_N} \prod_{i=1}^{N}(x_i - t_i)^{\alpha_i - 1}\, dt_1 \ldots dt_N =$$

$$\frac{1}{\prod\limits_{i=1}^{N} \Gamma(\alpha_i)} \prod_{i=1}^{N} \int_{a_i}^{x_i} (x_i - t_i)^{\alpha_i - 1} \, dt_i = \frac{1}{\prod\limits_{i=1}^{N} \Gamma(\alpha_i)} \prod_{i=1}^{N} \frac{(x_i - a_i)^{\alpha_i}}{\alpha_i}$$

$$= \prod_{i=1}^{N} \left(\frac{(x_i - a_i)^{\alpha_i}}{\Gamma(\alpha_i + 1)} \right),$$

that is

$$K_{a+}(x) = \prod_{i=1}^{N} \frac{(x_i - a_i)^{\alpha_i}}{\Gamma(\alpha_i + 1)}, \tag{27.69}$$

$\forall \, x \in \prod\limits_{i=1}^{N} (a_i, b_i)$.

Similarly the corresponding $K(x)$ for I_{b-}^{α} is:

$$K_{b-}(x) = \int_{\prod\limits_{i=1}^{N} (a_i, b_i)} k_{b-}(x, y) \, dy = \left(I_{b-}^{\alpha} 1 \right)(x) =$$

$$\frac{1}{\prod\limits_{i=1}^{N} \Gamma(\alpha_i)} \int_{x_1}^{b_1} \cdots \int_{x_N}^{b_N} \prod_{i=1}^{N} (t_i - x_i)^{\alpha_i - 1} \, dt_1 ... dt_N =$$

$$\frac{1}{\prod\limits_{i=1}^{N} \Gamma(\alpha_i)} \prod_{i=1}^{N} \int_{x_i}^{b_i} (t_i - x_i)^{\alpha_i - 1} \, dt_i = \frac{1}{\prod\limits_{i=1}^{N} \Gamma(\alpha_i)} \prod_{i=1}^{N} \frac{(b_i - x_i)^{\alpha_i}}{\alpha_i}$$

$$= \prod_{i=1}^{N} \frac{(b_i - x_i)^{\alpha_i}}{\Gamma(\alpha_i + 1)},$$

that is

$$K_{b-}(x) = \prod_{i=1}^{N} \frac{(b_i - x_i)^{\alpha_i}}{\Gamma(\alpha_i + 1)}, \tag{27.70}$$

$\forall \, x \in \prod\limits_{i=1}^{N} (a_i, b_i)$.

Next we form

$$\frac{k_{a+}(x,y)}{K_{a+}(x)} = \frac{1}{\displaystyle\prod_{i=1}^{N}\Gamma(\alpha_i)} \cdot \chi_{\prod_{i=1}^{N}(a_i,x_i]}(y) \prod_{i=1}^{N}(x_i - y_i)^{\alpha_i-1} \prod_{i=1}^{N}\frac{\Gamma(\alpha_i+1)}{(x_i - a_i)^{\alpha_i}}$$

$$= \chi_{\prod_{i=1}^{N}(a_i,x_i]}(y)\left(\prod_{i=1}^{N}\alpha_i\right)\left(\prod_{i=1}^{N}\frac{(x_i - y_i)^{\alpha_i-1}}{(x_i - a_i)^{\alpha_i}}\right),$$

that is

$$\frac{k_{a+}(x,y)}{K_{a+}(x)} = \chi_{\prod_{i=1}^{N}(a_i,x_i]}(y)\left(\prod_{i=1}^{N}\alpha_i\right)\left(\prod_{i=1}^{N}\frac{(x_i - y_i)^{\alpha_i-1}}{(x_i - a_i)^{\alpha_i}}\right), \qquad (27.71)$$

$$\forall\, x, y \in \prod_{i=1}^{N}(a_i, b_i).$$

Similarly we form

$$\frac{k_{b-}(x,y)}{K_{b-}(x)} = \frac{1}{\displaystyle\prod_{i=1}^{N}\Gamma(\alpha_i)} \cdot \chi_{\prod_{i=1}^{N}[x_i,b_i)}(y) \prod_{i=1}^{N}(y_i - x_i)^{\alpha_i-1} \prod_{i=1}^{N}\frac{\Gamma(\alpha_i+1)}{(b_i - x_i)^{\alpha_i}}$$

$$= \chi_{\prod_{i=1}^{N}[x_i,b_i)}(y)\left(\prod_{i=1}^{N}\alpha_i\right)\left(\prod_{i=1}^{N}\frac{(y_i - x_i)^{\alpha_i-1}}{(b_i - x_i)^{\alpha_i}}\right),$$

that is

$$\frac{k_{b-}(x,y)}{K_{b-}(x)} = \chi_{\prod_{i=1}^{N}[x_i,b_i)}(y)\left(\prod_{i=1}^{N}\alpha_i\right)\left(\prod_{i=1}^{N}\frac{(y_i - x_i)^{\alpha_i-1}}{(b_i - x_i)^{\alpha_i}}\right), \qquad (27.72)$$

$$\forall\, x, y \in \prod_{i=1}^{N}(a_i, b_i).$$

We choose the weight function $u_1(x)$ on $\prod\limits_{i=1}^{N}(a_i, b_i)$ such that the function $x \mapsto$

$\left(u_1(x) \frac{k_{a+}(x,y)}{K_{a+}(x)}\right)$ is integrable on $\prod\limits_{i=1}^{N}(a_i, b_i)$, for each fixed $y \in \prod\limits_{i=1}^{N}(a_i, b_i)$. We

define w_1 on $\prod\limits_{i=1}^{N}(a_i, b_i)$ by

$$w_1(y) := \int_{\prod\limits_{i=1}^{N}(a_i, b_i)} u_1(x) \frac{k_{a+}(x, y)}{K_{a+}(x)} dx < \infty. \qquad (27.73)$$

We have that

$$w_1(y) = \left(\prod_{i=1}^{N} \alpha_i\right) \int_{y_1}^{b_1} \cdots \int_{y_N}^{b_N} u_1(x_1, \ldots, x_N) \left(\prod_{i=1}^{N} \frac{(x_i - y_i)^{\alpha_i - 1}}{(x_i - a_i)^{\alpha_i}}\right) dx_1 \ldots dx_N, \qquad (27.74)$$

$\forall\, y \in \prod\limits_{i=1}^{N}(a_i, b_i)$.

We also choose the weight function $u_2(x)$ on $\prod\limits_{i=1}^{N}(a_i, b_i)$ such that the function

$x \mapsto \left(u_2(x) \frac{k_{b-}(x,y)}{K_{b-}(x)}\right)$ is integrable on $\prod\limits_{i=1}^{N}(a_i, b_i)$, for each fixed $y \in \prod\limits_{i=1}^{N}(a_i, b_i)$.

We define w_2 on $\prod\limits_{i=1}^{N}(a_i, b_i)$ by

$$w_2(y) := \int_{\prod\limits_{i=1}^{N}(a_i, b_i)} u_2(x) \frac{k_{b-}(x, y)}{K_{b-}(x)} dx < \infty. \qquad (27.75)$$

We have that

$$w_2(y) = \left(\prod_{i=1}^{N} \alpha_i\right) \int_{a_1}^{y_1} \cdots \int_{a_N}^{y_N} u_2(x_1, \ldots, x_N) \left(\prod_{i=1}^{N} \frac{(y_i - x_i)^{\alpha_i - 1}}{(b_i - x_i)^{\alpha_i}}\right) dx_1 \ldots dx_N, \qquad (27.76)$$

$\forall\, y \in \prod\limits_{i=1}^{N}(a_i, b_i)$.

If we choose as

$$u_1(x) = u_1^*(x) := \prod_{i=1}^{N}(x_i - a_i)^{\alpha_i}, \tag{27.77}$$

then

$$w_1^*(y) := w_1(y) = \left(\prod_{i=1}^{N}\alpha_i\right)\int_{y_1}^{b_1}\cdots\int_{y_N}^{b_N}\left(\prod_{i=1}^{N}(x_i - y_i)^{\alpha_i-1}\right)dx_1...dx_N$$

$$= \left(\prod_{i=1}^{N}\alpha_i\right)\left(\prod_{i=1}^{N}\int_{y_i}^{b_i}(x_i - y_i)^{\alpha_i-1}\,dx_i\right)$$

$$= \left(\prod_{i=1}^{N}\alpha_i\right)\left(\prod_{i=1}^{N}\frac{(b_i - y_i)^{\alpha_i}}{\alpha_i}\right) = \prod_{i=1}^{N}(b_i - y_i)^{\alpha_i}.$$

that is

$$w_1^*(y) = \prod_{i=1}^{N}(b_i - y_i)^{\alpha_i},\ \forall y \in \prod_{i=1}^{N}(a_i, b_i). \tag{27.78}$$

If we choose as

$$u_2(x) = u_2^*(x) := \prod_{i=1}^{N}(b_i - x_i)^{\alpha_i}, \tag{27.79}$$

then

$$w_2^*(y) := w_2(y) = \left(\prod_{i=1}^{N}\alpha_i\right)\int_{a_1}^{y_1}\cdots\int_{a_N}^{y_N}\left(\prod_{i=1}^{N}(y_i - x_i)^{\alpha_i-1}\right)dx_1...dx_N$$

$$= \left(\prod_{i=1}^{N}\alpha_i\right)\left(\prod_{i=1}^{N}\int_{a_i}^{y_i}(y_i - x_i)^{\alpha_i-1}\,dx_i\right)$$

$$= \left(\prod_{i=1}^{N}\alpha_i\right)\left(\prod_{i=1}^{N}\frac{(y_i - a_i)^{\alpha_i}}{\alpha_i}\right) = \prod_{i=1}^{N}(y_i - a_i)^{\alpha_i}.$$

That is

$$w_2^*(y) = \prod_{i=1}^{N}(y_i - a_i)^{\alpha_i},\ \forall y \in \prod_{i=1}^{N}(a_i, b_i). \tag{27.80}$$

Here we choose $f_{jr} : \prod_{i=1}^{N} (a_i, b_i) \rightarrow \mathbb{R}$, $j = 1, \ldots, m$, $r = 1, \ldots, n$, that are Lebesgue measurable and $I_{a+}^{\alpha} \left(\left| \overrightarrow{f_j} \right| \right)$, $I_{b-}^{\alpha} \left(\left| \overrightarrow{f_j} \right| \right)$ are finite a.e., one or the other, or both; $\overrightarrow{f_j} = (f_{j1}, f_{j2}, \ldots, f_{jn})$.

Let $p_j > 1 : \sum_{j=1}^{m} \frac{1}{p_j} = 1$ and the functions $\Phi_j : \mathbb{R}_+^n \rightarrow \mathbb{R}_+$, $j = 1, \ldots, m$, to be convex and increasing per coordinate.

Then by (27.29) we obtain

$$\int_{\prod_{i=1}^{N}(a_i,b_i)} u_1(x) \prod_{j=1}^{m} \Phi_j \left(\frac{\left| I_{a+}^{\alpha} \left(\overrightarrow{f_j} \right)(x) \right| \prod_{i=1}^{N} \Gamma(\alpha_i + 1)}{\prod_{i=1}^{N} (x_i - a_i)^{\alpha_i}} \right) dx \le$$

$$\prod_{j=1}^{m} \left(\int_{\prod_{i=1}^{N}(a_i,b_i)} w_1(y) \, \Phi_j \left(\left| \overrightarrow{f_j}(y) \right| \right)^{p_j} dy \right)^{\frac{1}{p_j}}, \qquad (27.81)$$

under the assumptions:

(i) $\overrightarrow{f_j}$, $\Phi_j \left(\left| \overrightarrow{f_j} \right| \right)^{p_j}$ are both $\frac{1}{\prod_{i=1}^{N} \Gamma(\alpha_i)} \chi_{\prod_{i=1}^{N}(a_i, x_i]}$ (y) $\prod_{i=1}^{N} (x_i - y_i)^{\alpha_i - 1} dy$ -integrable,

a.e. in $x \in \prod_{i=1}^{N} (a_i, b_i)$, for all $j = 1, \ldots, m$,

(ii) $w_1 \Phi_j \left(\left| \overrightarrow{f_j} \right| \right)^{p_j}$ is Lebesgue integrable, $j = 1, \ldots, m$.

Similarly, by (27.29), we obtain

$$\int_{\prod_{i=1}^{N}(a_i,b_i)} u_2(x) \prod_{j=1}^{m} \Phi_j \left(\frac{\left| I_{b-}^{\alpha} \left(\overrightarrow{f_j} \right)(x) \right| \prod_{i=1}^{N} \Gamma(\alpha_i + 1)}{\prod_{i=1}^{N} (b_i - x_i)^{\alpha_i}} \right) dx \le$$

$$\prod_{j=1}^{m}\left(\int_{\prod\limits_{i=1}^{N}(a_i,b_i)} w_2(y)\,\Phi_j\left(\left|\overrightarrow{f_j}(y)\right|\right)^{p_j} dy\right)^{\frac{1}{p_j}}, \qquad (27.82)$$

under the assumptions:

(i) $\overrightarrow{f_j},\ \Phi_j\left(\left|\overrightarrow{f_j}\right|\right)^{p_j}$ are both $\dfrac{1}{\prod\limits_{i=1}^{N}\Gamma(\alpha_i)}\chi_{\prod\limits_{i=1}^{N}[x_i,b_i)}(y)\prod\limits_{i=1}^{N}(y_i-x_i)^{\alpha_i-1}\,dy$ -integrable,

a.e. in $x \in \prod\limits_{i=1}^{N}(a_i,b_i)$, for all $j = 1, \ldots, m$,

(ii) $w_2\Phi_j\left(\left|\overrightarrow{f_j}\right|\right)^{p_j}$ is Lebesgue integrable, $j = 1, \ldots, m$.

Using (27.77) and (27.78) we rewrite (27.81), as follows

$$\int_{\prod\limits_{i=1}^{N}(a_i,b_i)}\left(\prod_{i=1}^{N}(x_i-a_i)^{\alpha_i}\right)\prod_{j=1}^{m}\Phi_j\left(\frac{\left|I_{a+}^{\alpha}\left(\overrightarrow{f_j}\right)(x)\right|\prod\limits_{i=1}^{N}\Gamma(\alpha_i+1)}{\prod\limits_{i=1}^{N}(x_i-a_i)^{\alpha_i}}\right)dx \le$$

$$\prod_{j=1}^{m}\left(\int_{\prod\limits_{i=1}^{N}(a_i,b_i)}\left(\prod_{i=1}^{N}(b_i-y_i)^{\alpha_i}\right)\Phi_j\left(\left|\overrightarrow{f_j}(y)\right|\right)^{p_j} dy\right)^{\frac{1}{p_j}} \le \qquad (27.83)$$

$$\left(\prod_{i=1}^{N}(b_i-a_i)^{\alpha_i}\right)\prod_{j=1}^{m}\left(\int_{\prod\limits_{i=1}^{N}(a_i,b_i)}\Phi_j\left(\left|\overrightarrow{f_j}(y)\right|\right)^{p_j} dy\right)^{\frac{1}{p_j}},$$

under the assumptions:

(i)   following (27.81) and
(ii)* $\Phi_j\left(\left|\overrightarrow{f_j}\right|\right)^{p_j}$ is Lebesgue integrable, $j = 1, \ldots, m$.

Similarly, using (27.79) and (27.80) we rewrite (27.82),

$$\int_{\prod\limits_{i=1}^{N}(a_i,b_i)} \left(\prod_{i=1}^{N}(b_i - x_i)^{\alpha_i}\right)\prod_{j=1}^{m}\Phi_j\left(\frac{\left|I_{b-}^{\alpha}\left(\overrightarrow{f_j}\right)(x)\right|\prod\limits_{i=1}^{N}\Gamma(\alpha_i+1)}{\prod\limits_{i=1}^{N}(b_i - x_i)^{\alpha_i}}\right)dx \leq$$

$$\prod_{j=1}^{m}\left(\int_{\prod\limits_{i=1}^{N}(a_i,b_i)}\left(\prod_{i=1}^{N}(y_i - a_i)^{\alpha_i}\right)\Phi_j\left(\left|\overrightarrow{f_j}(y)\right|\right)^{p_j}dy\right)^{\frac{1}{p_j}} \leq \qquad (27.84)$$

$$\left(\prod_{i=1}^{N}(b_i - a_i)^{\alpha_i}\right)\prod_{j=1}^{m}\left(\int_{\prod\limits_{i=1}^{N}(a_i,b_i)}\Phi_j\left(\left|\overrightarrow{f_j}(y)\right|\right)^{p_j}dy\right)^{\frac{1}{p_j}},$$

under the assumptions:

(i)   following (27.82), and

(ii)* $\Phi_j\left(\left|\overrightarrow{f_j}\right|\right)^{p_j}$ is Lebesgue integrable, $j = 1, \ldots, m$.

We give

Theorem 27.23 *All as in Remark 27.22. It holds*

$$\int_{\prod\limits_{i=1}^{N}(a_i,b_i)}\prod_{j=1}^{m}\left(\sum_{r=1}^{n}\left|I_{a+}^{\alpha}(f_{jr})(x)\right|^{p_j}\right)^{\frac{1}{p_j}}dx \leq \qquad (27.85)$$

$$\left(\prod_{i=1}^{N}\frac{(b_i - a_i)^{\alpha_i}}{\Gamma(\alpha_i+1)}\right)^{m}\prod_{j=1}^{m}\left(\int_{\prod\limits_{i=1}^{N}(a_i,b_i)}\left(\sum_{r=1}^{n}|f_{jr}(y)|^{p_j}\right)dy\right)^{\frac{1}{p_j}},$$

under the assumptions:

(i) $p_j > 1 : \sum_{j=1}^{m} \frac{1}{p_j} = 1$,

(ii) $\left| f_{jr}(y) \right|^{p_j}$ is $\dfrac{1}{\prod_{i=1}^{N} \Gamma(\alpha_i) \prod_{i=1}^{N} (a_i, x_i]} \chi_{\prod_{i=1}^{N} (a_i, x_i]}(y) \prod_{i=1}^{N} (x_i - y_i)^{\alpha_i - 1} \, dy$ -integrable, a.e. in

$$x \in \prod_{i=1}^{N} (a_i, b_i), \text{ for all } r = 1, \ldots, n; \ j = 1, \ldots, m,$$

(iii) $\left| f_{jr}(y) \right|^{p_j}$ is Lebesgue integrable, for all $r = 1, \ldots, n; \ j = 1, \ldots, m$.

Proof By Theorem 27.9 and $\left(\prod_{i=1}^{N} (x_i - a_i)^{\alpha_i} \right)^{1-m} \geq \left(\prod_{i=1}^{N} (b_i - a_i)^{\alpha_i} \right)^{1-m}$. ∎

We further present

Theorem 27.24 *All as in Remark 27.22. It holds*

$$\int_{\prod_{i=1}^{N}(a_i,b_i)} \left(\prod_{i=1}^{N} (x_i - a_i)^{\alpha_i} \right) \prod_{j=1}^{m} \ln \left(\sum_{r=1}^{n} e^{\left(|I_{a+}^{\alpha}(f_{jr})(x)| \right) \left(\prod_{i=1}^{N} \frac{\Gamma(\alpha_i + 1)}{(x_i - a_i)^{\alpha_i}} \right)} \right) dx \leq$$

$$\left(\prod_{i=1}^{N} (b_i - a_i)^{\alpha_i} \right) \prod_{j=1}^{m} \left(\int_{\prod_{i=1}^{N}(a_i,b_i)} \left(\ln \left(\sum_{r=1}^{n} e^{|f_{jr}(y)|} \right) \right)^{p_j} dy \right)^{\frac{1}{p_j}}, \qquad (27.86)$$

under the assumptions:

(i) $p_j > 1 : \sum_{j=1}^{m} \frac{1}{p_j} = 1$,

(ii) $\overrightarrow{f_j}, \left(\ln \left(\sum_{r=1}^{n} e^{|f_{jr}(y)|} \right) \right)^{p_j}$ are both $\dfrac{1}{\prod_{i=1}^{N} \Gamma(\alpha_i) \prod_{i=1}^{N} (a_i, x_i]} \chi_{\prod_{i=1}^{N} (a_i, x_i]}(y) \prod_{i=1}^{N} (x_i - y_i)^{\alpha_i - 1}$

dy -integrable, a.e. in $x \in \prod_{i=1}^{N} (a_i, b_i)$, for all $j = 1, \ldots, m$,

(iii) $\left(\ln \left(\sum_{r=1}^{n} e^{|f_{jr}(y)|} \right) \right)^{p_j}$ is Lebesgue integrable, $j = 1, \ldots, m$.

Proof By Theorem 27.10. ∎

Similarly we obtain

Theorem 27.25 *All as in Remark 27.22. It holds*

$$\int_{\prod_{i=1}^{N}(a_i,b_i)} \prod_{j=1}^{m}\left(\sum_{r=1}^{n}\left|I_{b-}^{\alpha}\left(f_{jr}\right)(x)\right|^{p_j}\right)^{\frac{1}{p_j}} dx \leq \qquad (27.87)$$

$$\left(\prod_{i=1}^{N}\frac{(b_i-a_i)^{\alpha_i}}{\Gamma(\alpha_i+1)}\right)^{m}\prod_{j=1}^{m}\left(\int_{\prod_{i=1}^{N}(a_i,b_i)}\left(\sum_{r=1}^{n}\left|f_{jr}(y)\right|^{p_j}\right)dy\right)^{\frac{1}{p_j}},$$

under the assumptions:

(i) $p_j > 1 : \sum_{j=1}^{m}\frac{1}{p_j}=1$,

(ii) $\left|f_{jr}(y)\right|^{p_j}$ *is* $\dfrac{1}{\prod_{i=1}^{N}\Gamma(\alpha_i)}\chi_{\prod_{i=1}^{N}[x_i,b_i)}(y)\prod_{i=1}^{N}(y_i-x_i)^{\alpha_i-1}dy$ *-integrable, a.e. in*

$x \in \prod_{i=1}^{N}(a_i,b_i)$, *for all* $r = 1,\ldots,n$; $j = 1,\ldots,m$,

(iii) $\left|f_{jr}(y)\right|^{p_j}$ *is Lebesgue integrable, for all* $r = 1,\ldots,n$; $j = 1,\ldots,m$.

Proof By Theorem 27.9 and $\left(\prod_{i=1}^{N}(b_i-x_i)^{\alpha_i}\right)^{1-m} \geq \left(\prod_{i=1}^{N}(b_i-a_i)^{\alpha_i}\right)^{1-m}$. ∎

Furthermore we derive

Theorem 27.26 *All as in Remark 27.22. It holds*

$$\int_{\prod_{i=1}^{N}(a_i,b_i)}\left(\prod_{i=1}^{N}(b_i-x_i)^{\alpha_i}\right)\prod_{j=1}^{m}\ln\left(\sum_{r=1}^{n}e^{\left(\left|I_{b-}^{\alpha}(f_{jr})(x)\right|\right)\left(\prod_{i=1}^{N}\frac{\Gamma(\alpha_i+1)}{(b_i-x_i)^{\alpha_i}}\right)}\right) dx \leq$$

$$\left(\prod_{i=1}^{N}(b_i-a_i)^{\alpha_i}\right)\prod_{j=1}^{m}\left(\int_{\prod_{i=1}^{N}(a_i,b_i)}\left(\ln\left(\sum_{r=1}^{n}e^{\left|f_{jr}(y)\right|}\right)\right)^{p_j}dy\right)^{\frac{1}{p_j}}, \qquad (27.88)$$

under the assumptions:

(i) $p_j > 1 : \sum_{j=1}^{m} \frac{1}{p_j} = 1$,

(ii) $\overrightarrow{f_j}, \left(\ln \left(\sum_{r=1}^{n} e^{|f_{jr}(y)|} \right) \right)^{p_j}$ *are both* $\dfrac{1}{\prod\limits_{i=1}^{N} \Gamma(\alpha_i) \prod\limits_{i=1}^{N} [x_i, b_i)} \chi_N \quad (y) \prod\limits_{i=1}^{N} (y_i - x_i)^{\alpha_i - 1}$

dy -integrable, a.e. in $x \in \prod\limits_{i=1}^{N} (a_i, b_i)$, *for all* $j = 1, \ldots, m$,

(iii) $\left(\ln \left(\sum_{r=1}^{n} e^{|f_{jr}(y)|} \right) \right)^{p_j}$ *is Lebesgue integrable,* $j = 1, \ldots, m$.

Proof By Theorem 27.10. ∎

Background 27.27 *In order to apply Theorem 27.2 to the case of a spherical shell we need:*

Let $N \geq 2$, $S^{N-1} := \{x \in \mathbb{R}^N : |x| = 1\}$ *the unit sphere on* $\mathbb{R}^N$, *where* $|\cdot|$ *stands for the Euclidean norm in* $\mathbb{R}^N$. *Also denote the ball* $B(0, R) := \{x \in \mathbb{R}^N : |x| < R\} \subseteq \mathbb{R}^N$, $R > 0$, *and the spherical shell*

$$A := B(0, R_2) - \overline{B(0, R_1)}, \, 0 < R_1 < R_2. \tag{27.89}$$

For the following see [17, pp. 149–150], *and* [19, pp. 87–88].

For $x \in \mathbb{R}^N - \{0\}$ *we can write uniquely* $x = r\omega$, *where* $r = |x| > 0$, *and* $\omega = \frac{x}{r} \in S^{N-1}$, $|\omega| = 1$.

Clearly here

$$\mathbb{R}^N - \{0\} = (0, \infty) \times S^{N-1}, \tag{27.90}$$

and

$$\overline{A} = [R_1, R_2] \times S^{N-1}. \tag{27.91}$$

We will be using

Theorem 27.28 [1, p. 322] *Let* $f : A \to \mathbb{R}$ *be a Lebesgue integrable function. Then*

$$\int_A f(x) \, dx = \int_{S^{N-1}} \left(\int_{R_1}^{R_2} f(r\omega) r^{N-1} dr \right) d\omega. \tag{27.92}$$

So we are able to write an integral on the shell in polar form using the polar coordinates (r, ω).

We need

Definition 27.29 [1, p. 458] *Let* $\nu > 0$, $n := [\nu]$, $\alpha := \nu - n$, $f \in C^n(\overline{A})$, *and* A *is a spherical shell. Assume that there exists function* $\dfrac{\partial_{R_1}^{\nu} f(x)}{\partial r^{\nu}} \in C(\overline{A})$, *given by*

$$\frac{\partial_{R_1}^{\nu} f(x)}{\partial r^{\nu}} := \frac{1}{\Gamma(1 - \alpha)} \frac{\partial}{\partial r} \left(\int_{R_1}^{r} (r - t)^{-\alpha} \frac{\partial^n f(t\omega)}{\partial r^n} dt \right), \tag{27.93}$$

where $x \in \overline{A}$; that is $x = r\omega$, $r \in [R_1, R_2]$, $\omega \in S^{N-1}$.

We call $\frac{\partial_{R_1}^{\nu} f}{\partial r^{\nu}}$ the left radial Canavati-type fractional derivative of f of order ν. If $\nu = 0$, then set $\frac{\partial_{R_1}^{\nu} f(x)}{\partial r^{\nu}} := f(x)$.

Based on [1, p. 288], and [5] we have

Lemma 27.30 Let $\gamma \geq 0$, $m := [\gamma]$, $\nu > 0$, $n := [\nu]$, with $0 \leq \gamma < \nu$. Let $f \in C^n(\overline{A})$ and there exists $\frac{\partial_{R_1}^{\nu} f(x)}{\partial r^{\nu}} \in C(\overline{A})$, $x \in \overline{A}$, A a spherical shell. Further assume that $\frac{\partial^j f(R_1\omega)}{\partial r^j} = 0$, $j = m, m+1, \ldots, n-1$, $\forall \, \omega \in S^{N-1}$. Then there exists $\frac{\partial_{R_1}^{\gamma} f(x)}{\partial r^{\gamma}} \in C(\overline{A})$ such that

$$\frac{\partial_{R_1}^{\gamma} f(x)}{\partial r^{\gamma}} = \frac{\partial_{R_1}^{\gamma} f(r\omega)}{\partial r^{\gamma}} = \frac{1}{\Gamma(\nu - \gamma)} \int_{R_1}^{r} (r-t)^{\nu-\gamma-1} \frac{\partial_{R_1}^{\nu} f(t\omega)}{\partial r^{\nu}} dt, \quad (27.94)$$

$\forall \, \omega \in S^{N-1}$; all $R_1 \leq r \leq R_2$, indeed $f(r\omega) \in C_{R_1}^{\gamma}([R_1, R_2])$, $\forall \, \omega \in S^{N-1}$.

We make

Remark 27.31 In the settings and assumptions of Theorem 27.2 and Lemma 27.30 we have

$$k(r, t) = \frac{1}{\Gamma(\nu - \gamma)} \chi_{[R_1, r]}(t) (r-t)^{\nu-\gamma-1}, \quad (27.95)$$

and

$$K(r) = \frac{(r - R_1)^{\nu-\gamma}}{\Gamma(\nu - \gamma + 1)}, \quad (27.96)$$

$r, t \in [R_1, R_2]$.

Furthermore we get

$$\frac{k(r, t)}{K(r)} = (\nu - \gamma) \chi_{[R_1, r]}(t) \frac{(r-t)^{\nu-\gamma-1}}{(r - R_1)^{\nu-\gamma}}, \quad (27.97)$$

and by choosing

$$u(r) := (r - R_1)^{\nu-\gamma}, \quad r \in [R_1, R_2], \quad (27.98)$$

we find

$$u(t) = (\nu - \gamma) \int_{t}^{R_2} (r-t)^{\nu-\gamma-1} dr = (R_2 - t)^{\nu-\gamma}, \quad (27.99)$$

$t \in [R_1, R_2]$.

Let here f_i $i = 1, \ldots, n$, as in Definition 27.29 and Lemma 27.30. Let also $p \geq 1$. We define the functions

$$G(x) = G(r\omega) := \left(\sum_{i=1}^{n} \left| \frac{\partial_{R_1}^{\gamma} f_i(r\omega)}{\partial r^{\gamma}} \right|^p \right)^{\frac{1}{p}}, \tag{27.100}$$

and

$$F(x) = F(r\omega) := \left(\sum_{i=1}^{n} \left| \frac{\partial_{R_1}^{\nu} f_i(r\omega)}{\partial r^{\nu}} \right|^p \right)^{\frac{1}{p}} \tag{27.101}$$

$\forall \, \omega \in S^{N-1}$; all $r \in [R_1, R_2]$.

Then by (27.9) we find

$$\int_{R_1}^{R_2} (r - R_1)^{\nu-\gamma} G(r\omega) \frac{(\Gamma(\nu-\gamma+1))^p}{(r-R_1)^{(\nu-\gamma)p}} dr \le$$

$$\int_{R_1}^{R_2} (R_2 - r)^{\nu-\gamma} F(r\omega) dr, \tag{27.102}$$

and

$$\int_{R_1}^{R_2} (r - R_1)^{(\nu-\gamma)(1-p)} G(r\omega) dr \le$$

$$\frac{1}{(\Gamma(\nu-\gamma+1))^p} \int_{R_1}^{R_2} (R_2 - r)^{\nu-\gamma} F(r\omega) dr \le$$

$$\frac{(R_2 - R_1)^{\nu-\gamma}}{(\Gamma(\nu-\gamma+1))^p} \int_{R_1}^{R_2} F(r\omega) dr. \tag{27.103}$$

But it holds

$$\int_{R_1}^{R_2} (r - R_1)^{(\nu-\gamma)(1-p)} G(r\omega) dr \ge$$

$$(R_2 - R_1)^{(\nu-\gamma)(1-p)} \int_{R_1}^{R_2} G(r\omega) dr. \tag{27.104}$$

Consequently we derive

$$\int_{R_1}^{R_2} G(r\omega) dr \le \left(\frac{(R_2 - R_1)^{(\nu-\gamma)}}{\Gamma(\nu-\gamma+1)} \right)^p \int_{R_1}^{R_2} F(r\omega) dr, \tag{27.105}$$

$\forall \, \omega \in S^{N-1}$.

Here we have $R_1 \le r \le R_2$, and $R_1^{N-1} \le r^{N-1} \le R_2^{N-1}$, and $R_2^{1-N} \le r^{1-N} \le R_1^{1-N}$.

From (27.105) we have

$$R_2^{1-N} \int_{R_1}^{R_2} r^{N-1} G(r\omega)\, dr \leq$$

$$\int_{R_1}^{R_2} r^{1-N} r^{N-1} G(r\omega)\, dr \leq$$

$$\left(\frac{(R_2 - R_1)^{(\nu - \gamma)}}{\Gamma(\nu - \gamma + 1)} \right)^p \int_{R_1}^{R_2} r^{1-N} r^{N-1} F(r\omega)\, dr \leq$$

$$R_1^{1-N} \left(\frac{(R_2 - R_1)^{(\nu - \gamma)}}{\Gamma(\nu - \gamma + 1)} \right)^p \int_{R_1}^{R_2} r^{N-1} F(r\omega)\, dr. \qquad (27.106)$$

So we get

$$\int_{R_1}^{R_2} r^{N-1} G(r\omega)\, dr \leq$$

$$\left(\frac{R_2}{R_1} \right)^{N-1} \left(\frac{(R_2 - R_1)^{(\nu - \gamma)}}{\Gamma(\nu - \gamma + 1)} \right)^p \int_{R_1}^{R_2} r^{N-1} F(r\omega)\, dr, \qquad (27.107)$$

$\forall \, \omega \in S^{N-1}$.

Hence

$$\int_{S^{N-1}} \left(\int_{R_1}^{R_2} r^{N-1} G(r\omega)\, dr \right) d\omega \leq$$

$$\left(\frac{R_2}{R_1} \right)^{N-1} \left(\frac{(R_2 - R_1)^{(\nu - \gamma)}}{\Gamma(\nu - \gamma + 1)} \right)^p \int_{S^{N-1}} \left(\int_{R_1}^{R_2} r^{N-1} F(r\omega)\, dr \right) d\omega. \qquad (27.108)$$

By Theorem 27.28, equality (27.92), we obtain

$$\int_A G(x)\, dx \leq \left(\frac{R_2}{R_1} \right)^{N-1} \left(\frac{(R_2 - R_1)^{(\nu - \gamma)}}{\Gamma(\nu - \gamma + 1)} \right)^p \int_A F(x)\, dx. \qquad (27.109)$$

We have proved the following vectorial fractional Poincaré type inequalities on the shell.

Theorem 27.32 *Here all as in Lemma 27.30 and Remark 27.31.*

It holds

(1)

$$\left\| \sum_{i=1}^{n} \left(\left| \frac{\partial_{R_1}^{\gamma} f_i}{\partial r^{\gamma}} \right|^p \right)^{\frac{1}{p}} \right\|_{1,A} \leq$$

$$\left(\frac{R_2}{R_1} \right)^{(N-1)} \left(\frac{(R_2 - R_1)^{(\nu-\gamma)}}{\Gamma(\nu - \gamma + 1)} \right)^p \left\| \left(\sum_{i=1}^{n} \left| \frac{\partial_{R_1}^{\nu} f_i}{\partial r^{\nu}} \right|^p \right)^{\frac{1}{p}} \right\|_{1,A}, \qquad (27.110)$$

(2) *When $\gamma = 0$, we have*

$$\left\| \left(\sum_{i=1}^{n} |f_i|^p \right)^{\frac{1}{p}} \right\|_{1,A} \leq \left(\frac{R_2}{R_1} \right)^{(N-1)} \left(\frac{(R_2 - R_1)^{\nu}}{\Gamma(\nu + 1)} \right)^p \left\| \left(\sum_{i=1}^{n} \left| \frac{\partial_{R_1}^{\nu} f_i}{\partial r^{\nu}} \right|^p \right)^{\frac{1}{p}} \right\|_{1,A}.$$

$$(27.111)$$

Similar results can be produced for the right radial Canavati type fractional derivative.

We choose to omit it.

We make

Remark 27.33 (from [1], p. 460) Here we denote $\lambda_{\mathbb{R}^N}(x) \equiv dx$ the Lebesgue measure on $\mathbb{R}^N$, $N \geq 2$, and by $\lambda_{S^{N-1}}(\omega) = d\omega$ the surface measure on S^{N-1}, where $\mathcal{B}_X$ stands for the Borel class on space X. Define the measure R_N on $((0, \infty), \mathcal{B}_{(0,\infty)})$ by

$$R_N(B) = \int_B r^{N-1} dr, \text{ any } B \in \mathcal{B}_{(0,\infty)}.$$

Now let $F \in L_1(A) = L_1([R_1, R_2] \times S^{N-1})$.

Call

$$K(F) := \{\omega \in S^{N-1} : F(\cdot\omega) \notin L_1([R_1, R_2], \mathcal{B}_{[R_1, R_2]}, R_N)\}. \qquad (27.112)$$

We get, by Fubini's theorem and [19], pp. 87–88, that

$$\lambda_{S^{N-1}}(K(F)) = 0.$$

Of course

$$\theta(F) := [R_1, R_2] \times K(F) \subset A,$$

and

$$\lambda_{\mathbb{R}^N}(\theta(F)) = 0.$$

Above $\lambda_{S^{N-1}}$ is defined as follows: let $A \subset S^{N-1}$ be a Borel set, and let

$$\widetilde{A} := \{ru : 0 < r < 1, u \in A\} \subset \mathbb{R}^N;$$

we define

$$\lambda_{S^{N-1}}(A) := N \lambda_{\mathbb{R}^N}(\widetilde{A}).$$

We have that

$$\lambda_{S^{N-1}}\left(S^{N-1}\right) = \frac{2\pi^{\frac{N}{2}}}{\Gamma\left(\frac{N}{2}\right)},$$

the surface area of S^{N-1}.

See also [17, pp. 149–150], [19, pp. 87–88] and [1], p. 320.

Following [1, p. 466] we define the left Riemann-Liouville radial fractional derivative next.

Definition 27.34 Let $\beta > 0$, $m := [\beta] + 1$, $F \in L_1(A)$, and A is the spherical shell. We define

$$\frac{\overline{\partial}_{R_1}^{\beta} F(x)}{\partial r^{\beta}} := \begin{cases} \frac{1}{\Gamma(m-\beta)} \left(\frac{\partial}{\partial r}\right)^m \int_{R_1}^r (r-t)^{m-\beta-1} F(t\omega) \, dt, \\ \qquad for \, \omega \in S^{N-1} - K(F), \\ 0, \; for \, \omega \in K(F), \end{cases} \tag{27.113}$$

where $x = r\omega \in A$, $r \in [R_1, R_2]$, $\omega \in S^{N-1}$; $K(F)$ as in (27.112).
If $\beta = 0$, define

$$\frac{\overline{\partial}_{R_1}^{\beta} F(x)}{\partial r^{\beta}} := F(x).$$

We need the following important representation result for left Riemann-Liouville radial fractional derivatives, by [1, p. 466].

Theorem 27.35 Let $\nu \geq \gamma + 1$, $\gamma \geq 0$, $n := [\nu]$, $m := [\gamma]$, $F : \overline{A} \to \mathbb{R}$ with $F \in L_1(A)$. Assume that $F(\cdot\omega) \in AC^n([R_1, R_2])$, $\forall \, \omega \in S^{N-1}$, and that $\frac{\overline{\partial}_{R_1}^{\nu} F(\cdot\omega)}{\partial r^{\nu}}$ is measurable on $[R_1, R_2]$, $\forall \, \omega \in S^{N-1}$. Also assume $\exists \, \frac{\overline{\partial}_{R_1}^{\nu} F(r\omega)}{\partial r^{\nu}} \in \mathbb{R}$, $\forall \, r \in [R_1, R_2]$ and $\forall \, \omega \in S^{N-1}$, and $\frac{\overline{\partial}_{R_1}^{\nu} F(x)}{\partial r^{\nu}}$ is measurable on $\overline{A}$. Suppose $\exists \, M_1 > 0$:

$$\left| \frac{\overline{\partial}_{R_1}^{\nu} F(r\omega)}{\partial r^{\nu}} \right| \leq M_1, \quad \forall \, (r, \omega) \in [R_1, R_2] \times S^{N-1}. \tag{27.114}$$

We suppose that $\frac{\overline{\partial}^j F(R_1\omega)}{\partial r^j} = 0$, $j = m, m+1, \ldots, n-1$; $\forall \, \omega \in S^{N-1}$.

Then

$$\frac{\overline{\partial}_{R_1}^{\gamma} F(x)}{\partial r^{\gamma}} = \overline{D}_{R_1}^{\gamma} F(r\omega) = \frac{1}{\Gamma(\nu - \gamma)} \int_{R_1}^{r} (r-t)^{\nu-\gamma-1} \left(\overline{D}_{R_1}^{\nu} F\right)(t\omega) \, dt,$$

$$(27.115)$$

valid $\forall x \in \overline{A}$; *that is, true* $\forall r \in [R_1, R_2]$ *and* $\forall \omega \in S^{N-1}$; $\gamma > 0$.

Here

$$\overline{D}_{R_1}^{\gamma} F(\cdot\omega) \in AC([R_1, R_2]),$$

$$(27.116)$$

$\forall \omega \in S^{N-1}$; $\gamma > 0$.

Furthermore

$$\frac{\overline{\partial}_{R_1}^{\gamma} F(x)}{\partial r^{\gamma}} \in L_{\infty}(A), \gamma > 0.$$

$$(27.117)$$

In particular, it holds

$$F(x) = F(r\omega) = \frac{1}{\Gamma(\nu)} \int_{R_1}^{r} (r-t)^{\nu-1} \left(\overline{D}_{R_1}^{\nu} F\right)(t\omega) \, dt,$$

$$(27.118)$$

true $\forall x \in \overline{A}$; *that is, true* $\forall r \in [R_1, R_2]$ *and* $\forall \omega \in S^{N-1}$, *and*

$$F(\cdot\omega) \in AC([R_1, R_2]), \quad \forall \omega \in S^{N-1}.$$

$$(27.119)$$

We give also the following vectorial fractional Poincaré type inequalities on the spherical shell.

Theorem 27.36 *Here all and* F_i, $i = 1, \ldots, n$, *as in Theorem 27.35,* $p \geq 1$. *Then*

(1)

$$\left\| \left(\sum_{i=1}^{n} \left| \frac{\overline{\partial}_{R_1}^{\gamma} F_i}{\partial r^{\gamma}} \right|^p \right)^{\frac{1}{p}} \right\|_{1,A} \leq$$

$$\left(\frac{R_2}{R_1} \right)^{(N-1)} \left(\frac{(R_2 - R_1)^{(\nu-\gamma)}}{\Gamma(\nu - \gamma + 1)} \right)^p \left\| \left(\sum_{i=1}^{n} \left| \frac{\overline{\partial}_{R_1}^{\nu} F_i}{\partial r^{\nu}} \right|^p \right)^{\frac{1}{p}} \right\|_{1,A},$$

$$(27.120)$$

(2) When $\gamma = 0$, *we have*

$$\left\| \left(\sum_{i=1}^{n} |F_i|^p \right)^{\frac{1}{p}} \right\|_{1,A} \leq \left(\frac{R_2}{R_1} \right)^{(N-1)} \left(\frac{(R_2 - R_1)^{\nu}}{\Gamma(\nu + 1)} \right)^p \left\| \left(\sum_{i=1}^{n} \left| \frac{\overline{\partial}_{R_1}^{\nu} F_i}{\partial r^{\nu}} \right|^p \right)^{\frac{1}{p}} \right\|_{1,A}.$$

$$(27.121)$$

Proof As in Theorem 27.32, based on Theorem 27.35. ■

We also need (see [1], p. 421).

Definition 27.37 Let $F : \overline{A} \to \mathbb{R}$, $\nu \geq 0$, $n := [\nu]$ such that $F(\cdot\omega) \in AC^n([R_1, R_2])$, for all $\omega \in S^{N-1}$.

We call the left Caputo radial fractional derivative the following function

$$\frac{\partial^{\nu}_{*R_1} F(x)}{\partial r^{\nu}} := \frac{1}{\Gamma(n-\nu)} \int_{R_1}^{r} (r-t)^{n-\nu-1} \frac{\partial^n F(t\omega)}{\partial r^n} dt, \tag{27.122}$$

where $x \in \overline{A}$, i.e. $x = r\omega$, $r \in [R_1, R_2]$, $\omega \in S^{N-1}$.

Clearly

$$\frac{\partial^0_{*R_1} F(x)}{\partial r^0} = F(x), \tag{27.123}$$

$$\frac{\partial^{\nu}_{*R_1} F(x)}{\partial r^{\nu}} = \frac{\partial^{\nu} F(x)}{\partial r^{\nu}}, \text{ if } \nu \in \mathbb{N}.$$

Above function (27.122) exists almost everywhere for $x \in \overline{A}$, see [1], p. 422.

We mention the following fundamental representation result (see [1], p. 422–423 and [5]).

Theorem 27.38 Let $\nu \geq \gamma + 1$, $\gamma \geq 0$, $n := [\nu]$, $m := [\gamma]$, $F : \overline{A} \to \mathbb{R}$ with $F \in L_1(A)$. Assume that $F(\cdot\omega) \in AC^n([R_1, R_2])$, for all $\omega \in S^{N-1}$, and that $\frac{\partial^{\nu}_{*R_1} F(\cdot\omega)}{\partial r^{\nu}} \in L_{\infty}(R_1, R_2)$ for all $\omega \in S^{N-1}$.

Further assume that $\frac{\partial^{\nu}_{*R_1} F(x)}{\partial r^{\nu}} \in L_{\infty}(A)$. More precisely, for these $r \in [R_1, R_2]$, for each $\omega \in S^{N-1}$, for which $D^{\nu}_{*R_1} F(r\omega)$ takes real values, there exists $M_1 > 0$ such that $\left| D^{\nu}_{*R_1} F(r\omega) \right| \leq M_1$.

We suppose that $\frac{\partial^j F(R_1\omega)}{\partial r^j} = 0$, $j = m, m+1, \ldots, n-1$; for every $\omega \in S^{N-1}$. Then

$$\frac{\partial^{\gamma}_{*R_1} F(x)}{\partial r^{\gamma}} = D^{\gamma}_{*R_1} F(r\omega) = \frac{1}{\Gamma(\nu-\gamma)} \int_{R_1}^{r} (r-t)^{\nu-\gamma-1} \left(D^{\nu}_{*R_1} F \right)(t\omega) dt, \tag{27.124}$$

valid $\forall\, x \in \overline{A}$; i.e. true $\forall\, r \in [R_1, R_2]$ and $\forall\, \omega \in S^{N-1}$; $\gamma > 0$.

Here

$$D^{\gamma}_{*R_1} F(\cdot\omega) \in AC([R_1, R_2]), \tag{27.125}$$

$\forall\, \omega \in S^{N-1}$; $\gamma > 0$.

Furthermore

$$\frac{\partial^{\gamma}_{*R_1} F(x)}{\partial r^{\gamma}} \in L_{\infty}(A), \gamma > 0. \tag{27.126}$$

In particular, it holds

$$F(x) = F(r\omega) = \frac{1}{\Gamma(\nu)} \int_{R_1}^r (r - t)^{\nu - 1} \left(D_{*R_1}^\nu F \right)(t\omega)\, dt, \qquad (27.127)$$

true $\forall\, x \in \overline{A}$; i.e. true $\forall\, r \in [R_1, R_2]$ and $\forall\, \omega \in S^{N-1}$, and

$$F(\cdot\omega) \in AC\left([R_1, R_2]\right), \ \forall\, \omega \in S^{N-1}. \qquad (27.128)$$

We finish with the following vectorial Poincaré type inequalities involving left Caputo radial fractional derivatives.

Theorem 27.39 *Here all and F_i, $i = 1, \ldots, n$, as in Theorem 27.38, $p \geq 1$. Then*

(1)

$$\left\| \left(\sum_{i=1}^n \left| \frac{\partial_{*R_1}^\gamma F_i}{\partial r^\gamma} \right|^p \right)^{\frac{1}{p}} \right\|_{1,A} \leq$$

$$\left(\frac{R_2}{R_1} \right)^{(N-1)} \left(\frac{(R_2 - R_1)^{(\nu - \gamma)}}{\Gamma(\nu - \gamma + 1)} \right)^p \left\| \left(\sum_{i=1}^n \left| \frac{\partial_{*R_1}^\nu F_i}{\partial r^\nu} \right|^p \right)^{\frac{1}{p}} \right\|_{1,A}, \qquad (27.129)$$

(2) When $\gamma = 0$, we have

$$\left\| \left(\sum_{i=1}^n |F_i|^p \right)^{\frac{1}{p}} \right\|_{1,A} \leq \left(\frac{R_2}{R_1} \right)^{(N-1)} \left(\frac{(R_2 - R_1)^\nu}{\Gamma(\nu + 1)} \right)^p \left\| \left(\sum_{i=1}^n \left| \frac{\partial_{*R_1}^\nu F_i}{\partial r^\nu} \right|^p \right)^{\frac{1}{p}} \right\|_{1,A}.$$

$$(27.130)$$

Proof As in Theorem 27.32, based on Theorem 27.38. ∎

References

1. G.A. Anastassiou, *Fractional Differentiation Inequalities*, Research Monograph (Springer, New York, 2009)
2. G.A. Anastassiou, On right fractional calculus. Chaos, Solitons Fractals **42**, 365–376 (2009)
3. G.A. Anastassiou, Balanced fractional Opial inequalities. Chaos, Solitons Fractals **42**(3), 1523–1528 (2009)
4. G.A. Anastassiou, Fractional Korovkin theory. Chaos, Solitons Fractals **42**, 2080–2094 (2009)
5. G.A. Anastassiou, Fractional representation formulae and right fractional inequalities. Math. Comput.Model. **54**(11–12), 3098–3115 (2011)
6. G.A. Anastassiou, Vectorial fractional integral inequalities with convexity. Cent. Eur. J. Phys. **11**(10), 1194–1211 (2013)

7. G.A. Anastassiou, Vectorial Hardy type fractional inequalities. Bull. Tbilisi Int. Cent. Math. Inf. **16**(2), 21–57 (2012)
8. J.A. Canavati, The Riemann-Liouville integral. Nieuw Archief Voor Wiskunde **5**(1), 53–75 (1987)
9. K. Diethelm, *The Analysis of Fractional Differential Equations*, Lecture Notes in Mathematics, vol. 2004, 1st edn. (Springer, New York, 2010)
10. A.M.A. El-Sayed, M. Gaber, On the finite Caputo and finite Riesz derivatives. Electron. J. Theor. Phys. **3**(12), 81–95 (2006)
11. R. Gorenflo, F. Mainardi, Essentials of Fractional Calculus. Maphysto Center (2000). http://www.maphysto.dk/oldpages/events/LevyCAC2000/MainardiNotes/fm2k0a.ps
12. G.D. Handley, J.J. Koliha, J. Pečarić, Hilbert-Pachpatte type integral inequalities for fractional derivatives. Fract. Calc. Appl. Anal. **4**(1), 37–46 (2001)
13. H.G. Hardy, Notes on some points in the integral calculus. Messenger Math. **47**(10), 145–150 (1918)
14. S. Iqbal, K. Krulic, J. Pecaric, On an inequality of H.G. Hardy. J. Inequalities Appl. **2010**, Article ID 264347, 23 p
15. A.A. Kilbas, H.M. Srivastava, J.J. Trujillo, *Theory and Applications of Fractional Differential Equations*, North-Holland Mathematics Studies, vol. 204 (Elsevier, New York, 2006)
16. T. Mamatov, S. Samko, Mixed fractional integration operators in mixed weighted Hölder spaces. Fract. Calc. Appl. Anal. **13**(3), 245–259 (2010)
17. W. Rudin, *Real and Complex Analysis*, International Student Edition (Mc Graw Hill, New York, 1970)
18. S.G. Samko, A.A. Kilbas, O.I. Marichev, *Fractional Integral and Derivatives: Theory and Applications* (Gordon and Breach Science Publishers, Yverdon, 1993)
19. D. Stroock, *A Concise Introduction to the Theory of Integration*, 3rd edn. (Birkhäuser, Boston, 1999)

Index

B

Balanced Canavati type fractional derivative, 11

C

Cosine function, 241
Cosine operator function's Taylor formula, 232

E

Erdelyi-Kober type fractional integrals, 110

F

Fractional integrals, 95

G

Gaussian hypergeometric function, 396
Gauss-Weierstrass singular integral, 228
Generalized Fink type representation formula, 133

H

Hadamard fractional integrals, 108

I

Infinitesimal generator, 218

L

Left Caputo fractional derivative, 123
Left Caputo radial fractional derivative, 473
Left generalized α-fractional derivative, 2
Left generalized Riemann-Liouville fractional derivative, 122
Left mixed Riemann-Liouville fractional multiple integral, 113
Left radial Canavati-type fractional derivative, 467
Left radial generalised fractional derivative, 50
Left Riemann-Liouville integral, 2
Left Riemann-Liouville radial fractional derivatives, 471

O

One-parameter semi-group of operators, 218

P

Poincare type right Caputo fractional inequality, 342

R

Riemann-Liouville fractional integrals, 95
Riemann-Liouville integral operator, 67
Right Caputo fractional derivative, 125
Right generalized α-fractional derivative, 3
Right mixed Riemann-Liouville fractional multiple integral, 113

G.A. Anastassiou, *Intelligent Comparisons: Analytic Inequalities*, Studies in Computational Intelligence 609, DOI 10.1007/978-3-319-21121-3

Printed in the United States
By Bookmasters